Atomic Numbers and Atomic Masses
of the Elements

Based on $^{12}_{6}C$. Numbers in parentheses are the mass numbers of the most stable isotopes of radioactive elements.

Element	Symbol	Atomic Number	Atomic Mass	Element	Symbol	Atomic Number	Atomic Mass
Actinium	Ac	89	(227)	Mendelevium	Md	101	(260)
Aluminum	Al	13	26.98	Mercury	Hg	80	200.59
Americium	Am	95	(243)	Molybdenum	Mo	42	95.94
Antimony	Sb	51	121.76	Neodymium	Nd	60	144.24
Argon	Ar	18	39.95	Neon	Ne	10	20.18
Arsenic	As	33	74.92	Neptunium	Np	93	(237)
Astatine	At	85	(210)	Nickel	Ni	28	58.69
Barium	Ba	56	137.33	Niobium	Nb	41	92.91
Berkelium	Bk	97	(247)	Nitrogen	N	7	14.01
Beryllium	Be	4	9.01	Nobelium	No	102	(259)
Bismuth	Bi	83	208.98	Osmium	Os	76	190.23
Bohrium	Bh	107	(267)	Oxygen	O	8	16.00
Boron	B	5	10.81	Palladium	Pd	46	106.42
Bromine	Br	35	79.90	Phosphorus	P	15	30.97
Cadmium	Cd	48	112.41	Platinum	Pt	78	195.08
Calcium	Ca	20	40.08	Plutonium	Pu	94	(242)
Californium	Cf	98	(251)	Polonium	Po	84	(209)
Carbon	C	6	12.01	Potassium	K	19	39.10
Cerium	Ce	58	140.12	Praseodymium	Pr	59	140.91
Cesium	Cs	55	132.91	Promethium	Pm	61	(145)
Chlorine	Cl	17	35.45	Protactinium	Pa	91	(231)
Chromium	Cr	24	52.00	Radium	Ra	88	(226)
Cobalt	Co	27	58.93	Radon	Rn	86	(222)
Copper	Cu	29	63.55	Rhenium	Re	75	186.21
Curium	Cm	96	(248)	Rhodium	Rh	45	102.91
Darmstadtium	Ds	110	(271)	Roentgenium	Rg	111	(272)
Dubnium	Db	105	(262)	Rubidium	Rb	37	85.47
Dysprosium	Dy	66	162.50	Ruthenium	Ru	44	101.07
Einsteinium	Es	99	(252)	Rutherfordium	Rf	104	(263)
Erbium	Er	68	167.26	Samarium	Sm	62	150.36
Europium	Eu	63	151.96	Scandium	Sc	21	44.96
Fermium	Fm	100	(257)	Seaborgium	Sg	106	(266)
Fluorine	F	9	19.00	Selenium	Se	34	78.96
Francium	Fr	87	(223)	Silicon	Si	14	28.09
Gadolinium	Gd	64	157.25	Silver	Ag	47	107.87
Gallium	Ga	31	69.72	Sodium	Na	11	22.99
Germanium	Ge	32	72.59	Strontium	Sr	38	87.62
Gold	Au	79	196.97	Sulfur	S	16	32.07
Hafnium	Hf	72	178.49	Tantalum	Ta	73	180.95
Hassium	Hs	108	(269)	Technetium	Tc	43	(98)
Helium	He	2	4.00	Tellurium	Te	52	127.60
Holmium	Ho	67	164.93	Terbium	Tb	65	158.93
Hydrogen	H	1	1.01	Thallium	Tl	81	204.38
Indium	In	49	114.82	Thorium	Th	90	(232)
Iodine	I	53	126.90	Thulium	Tm	69	168.93
Iridium	Ir	77	192.22	Tin	Sn	50	118.71
Iron	Fe	26	55.85	Titanium	Ti	22	47.87
Krypton	Kr	36	83.80	Tungsten	W	74	183.84
Lanthanum	La	57	138.91	Uranium	U	92	(238)
Lawrencium	Lr	103	(262)	Vanadium	V	23	50.94
Lead	Pb	82	207.19	Xenon	Xe	54	131.29
Lithium	Li	3	6.94	Ytterbium	Yb	70	173.04
Lutetium	Lu	71	174.97	Yttrium	Y	39	88.91
Magnesium	Mg	12	24.30	Zinc	Zn	30	65.38
Manganese	Mn	25	54.94	Zirconium	Zr	40	91.22
Meitnerium	Mt	109	(276)				

Common Functional Groups

Name of Class	Structural Feature
Alkane	$-\overset{\displaystyle |}{\underset{\displaystyle |}{C}}-$
Alkene	$\overset{\diagdown}{\underset{\diagup}{C}}=\overset{\diagup}{\underset{\diagdown}{C}}$
Alkyne	$-C\equiv C-$
Aromatic hydrocarbon	(benzene ring) or (benzene ring)
Alcohol	$-\overset{\displaystyle |}{\underset{\displaystyle |}{C}}-OH$
Phenol	(benzene ring) $-OH$
Ether	$-\overset{\displaystyle |}{\underset{\displaystyle |}{C}}-O-\overset{\displaystyle |}{\underset{\displaystyle |}{C}}-$
Thiol	$-\overset{\displaystyle |}{\underset{\displaystyle |}{C}}-SH$
Aldehyde	$-\overset{\displaystyle O}{\overset{\displaystyle \|}{C}}-H \ (-CHO)$
Ketone	$-\overset{\displaystyle |}{\underset{\displaystyle |}{C}}-\overset{\displaystyle O}{\overset{\displaystyle \|}{C}}-\overset{\displaystyle |}{\underset{\displaystyle |}{C}}-$
Carboxylic acid	$-\overset{\displaystyle O}{\overset{\displaystyle \|}{C}}-OH \ (-COOH \text{ or } -CO_2H)$
Ester	$-\overset{\displaystyle O}{\overset{\displaystyle \|}{C}}-O-\overset{\displaystyle |}{\underset{\displaystyle |}{C}}- \ (-COOR \text{ or } -CO_2R)$
Amine	$-\overset{\displaystyle |}{\underset{\displaystyle |}{C}}-NH_2$
Amide	$-\overset{\displaystyle O}{\overset{\displaystyle \|}{C}}-NH_2$

Organic and Biological Chemistry

Organic and Biological Chemistry

FOURTH EDITION

H. STEPHEN STOKER

Weber State University

Houghton Mifflin Company   Boston   New York

Publisher: Charles Hartford

Executive Editor: Richard Stratton

Development Editor: Rebecca Berardy Schwartz

Assistant Editor: Liz Hogan

Project Editor: Andrea Cava

Art and Design Manager: Gary Crespo

Senior Art and Design Coordinator: Jill Haber

Senior Photo Editor: Jennifer Meyer Dare

Composition Buyer: Chuck Dutton

Manufacturing Manager: Karen B. Fawcett

Senior Marketing Manager: Katherine Greig

Marketing Assistant: Naveen Hariprasad

Cover image: © David Madison/Getty Images

Photo Credits appear on page A-23, which is considered an extension of the copyright page.

Printed in the U.S.A.

Library of Congress Control Number: 2005933135

Instructor's exam copy:
ISBN 13: 978-0-618-73241-8
ISBN 10: 0-618-73241-1

For orders, use student text ISBNs:
ISBN 13: 978-0-618-60607-8
ISBN 10: 0-618-60607-6

23456789-WC-09 08 07 06

Brief Contents

Contents

Preface

Writing an introductory college text, particularly one that encompasses as wide a range of topics as *Organic and Biological Chemistry* does, is a huge undertaking. When the first edition of the text was published nine years ago, my hopes were high. Thus, the positive responses of instructors and students who used the three previous editions of this text have been gratifying—and have led to the new edition you hold in your hands. This fourth edition represents a renewed commitment to the goals I initially set out to meet when writing the first edition. These goals have not changed with the passage of time. My initial and ongoing goals are to write a text in which

- The needs are simultaneously met for the many students in the fields of nursing, allied health, biological sciences, agricultural sciences, food sciences, and public health who are required to take such a course.
- The development of chemical topics always starts out at ground level. The students who will use this text often have little or no background in chemistry and hence approach the course with a good deal of trepidation. This "ground level" approach addresses this situation.
- The amount and level of mathematics is purposefully restricted. Clearly, some chemical principles cannot be divorced entirely from mathematics and, when this is the case, appropriate mathematical coverage is included.
- The early chapters focus on fundamental chemical principles and the later chapters, built on these principles, develop the concepts and applications central to the fields of organic chemistry and biochemistry.

Focus on Biochemistry   Most students taking this course have a greater interest in the biochemistry portion of the course than the preceding two parts. But biochemistry, of course, cannot be understood without a knowledge of the fundamentals of organic chemistry, and understanding organic chemistry in turn depends on knowing the key concepts of general chemistry. Thus, in writing this text, I essentially started from the back and worked forward. I began by determining what topics would be considered in the biochemistry chapters and then tailored the organic and then general sections to support that presentation. Users of the previous editions confirm that this approach ensures an efficient but thorough coverage of the principles needed to understand biochemistry.

Emphasis on Visual Support   I believe strongly in visual reinforcement of key concepts in a textbook; thus, this book uses art and photos wherever possible to teach key concepts. Artwork is used to make connections and highlight what is important for the student to know. Reaction equations use color to emphasize the portions of a molecule that undergo change. Colors are likewise assigned to things like valence shells and classes of compounds to help students follow trends. Computer-generated, three-dimensional molecular models accompany many discussions in the organic and biochemistry sections of the text. Color photographs show applications of chemistry to help make concepts real and more readily remembered.

Visual summary features, called *Chemistry at a Glance,* pull together material from several sections of a chapter to help students see the larger picture. For example, Chapter 2 includes IUPAC nomenclature for alkanes, alkenes, and alkynes and Chapter 11 summarizes DNA replication. The *Chemistry at a Glance* feature serves both as an overview for the student reading the material for the first time and as a review tool for the student preparing for exams. Given the popularity of the *Chemistry at a Glance* summaries in the previous editions, many of these features have been greatly expanded and new ones added.

Commitment to Student Learning   In addition to the study help *Chemistry at a Glance* offers, the text is built on a strong foundation of learning aids designed to help students master the course material.

- **Problem-solving pedagogy.** Because problem solving is often difficult for students in this course to master, I have taken special care to provide support to help students build their skills. Within the chapters, worked *Examples* follow the explanation of many concepts. These examples walk students through the thought processes involved in problem solving, carefully outlining all the steps involved. Each is immediately followed by a *Practice Exercise,* to reinforce the information just presented. A dozen new *Examples* have been added to this edition.

- **Chemical Connections.** In every chapter *Chemical Connections* show chemistry as it appears in everyday life. These boxes focus on topics that are relevant to students' future careers in the health and environmental fields and on those that are important for informed citizens to understand. Many of the health-related *Chemical Connections* have been updated to include the latest research findings, and include a number of new boxes on health benefits of garlic and onions, antioxidants present in chocolate, *H. pylori* and stomach ulcers, and anti-inflammatory COX-inhibitor drugs.

- **Margin notes.** Liberally distributed throughout the text, *margin notes* provide tips for remembering and distinguishing between concepts, highlight links across chapters, and describe interesting historical background information.

- **Defined terms.** All definitions are highlighted in the text when they are first presented, using boldface and italic type. Each defined term appears as a complete sentence; students are never forced to deduce a definition from context. In addition, the definitions of all terms appear in the combined *Index/Glossary* found at the end of the text. A major emphasis in this new edition has been "refinements" in the defined terms arena. All defined terms were reexamined to see if they could be stated with greater clarity. The result was a "rewording" of many defined terms. In addition, the number of defined terms has been expanded with 75 new definitions having been added to the text.

- **Review aids.** Several review aids appear at the ends of the chapters. *Concepts to Remember* and *Key Reactions and Equations* provide concise review of the material presented in the chapter. A *Key Terms Review* lists all the key terms in the chapter alphabetically and cross-references the section of the chapter in which they appear. These aids help students prepare for exams.

- **End-of-chapter problems.** An extensive set of end-of-chapter problems complements the worked examples within the chapters. Each end-of-chapter problem set is divided into two sections: *Exercises and Problems* and *Additional Problems*. The *Exercises and Problems* are organized by topic and paired, with each pair testing similar material and the answer to the odd-numbered member of the pair at the back of the book. These problems always involve only a single concept. The *Additional Problems* involve more than one concept and are more difficult than the Exercises and Problems.

- **Multiple-choice practice tests.** New practice tests have been added to the end of each chapter as a cumulative overview and as a preparation aid for exams.

Content Changes   Coverage of a number of topics has been expanded in this edition. The two driving forces in expanded coverage considerations were (1) the requests of users and reviewers of the previous editions and (2) my desire to incorporate new research findings, particularly in the area of biochemistry, into the text. Topics with expanded coverage include

- Constitutional isomerism for hydrocarbons and hydrocarbon derivatives
- Line-angle formula use for hydrocarbons and hydrocarbon derivatives
- Thioethers
- Acid chlorides and acid anhydrides
- Polysaccharides

- Cell membranes
- Polypeptides
- Factors affecting enzyme activity
- DNA
- Electron transport chain reactions

The Package

Study Help for Students

Online Study Center (accessible through college.hmco.com/pic/stokerGOB4e).   Available at no additional cost, this dedicated website offers a wealth of resources to help students succeed, including

- Self-quizzing using Houghton Mifflin's ACE system
- Glossary of key terms
- Flashcards of key terms
- Career preparation information to help students learn about opportunities for study and work in medical-related fields as well as other chemistry-related areas

Interactive Math Tutorials.   Accessed through the Online Study Center, these brief tutorials cover basic mathematical concepts that appear in the text, including solving simple algebraic equations, scientific notation, conversions, reading a graph, and ratio and proportion.

Eduspace® (powered by Blackboard®).   Houghton Mifflin's complete course management solution features algorithmic, end-of-chapter homework questions. Also included are Visualization tutorials, which include videos and animations with practice exercises to help students visualize key chemistry concepts.

SMARTHINKING®.   Live, online tutoring from experienced chemistry instructors is available with SMARTHINKING during peak study hours (upon instructor request with new books). Limits apply; terms and hours of SMARTHINKING service are subject to change.

Course Support for Instructors

Online Teaching Center (accessible through college.hmco.com/pic/stokerGOB4e).   This website allows access to all the student resources listed above, as well as additional instructor classroom resources such as PowerPoint slides and virtually all the art, tables, and photos in JPEG format.

Instructor's Resource Manual with Test Bank.   Prepared by H. Stephen Stoker, the *Instructor's Resource Manual* includes answers to all end-of-chapter exercises and a printed test bank of over 1,500 multiple-choice and matching problems.

HM Class Prep with HM Testing (by Diploma®) (ISBN 10: 0-618-606106; ISBN 13: 978-0-618-60610-8).   HM Class Prep and HM Testing allows an instructor to access both lecture aids and testing software in one place. These components include

- HM Class Prep includes everything an instructor will need to develop lectures—PowerPoint slides of virtually all the text, figures, and tables from the text; JPEGs of virtually all of the art from the text; and an Instructor's Resource Guide with solutions and test bank questions.
- HM Testing provides instructors with all the tools they will need to create, author/edit, customize, and deliver multiple types of tests. Instructors can import questions directly from the test bank, create their own questions, or edit existing algorithmic questions, all within Diploma's powerful electronic platform. Tests can be delivered in print or electronic formats, online and saved to multiple locations. The HM Testing test bank for this title comes preloaded with algorithmic content.

Acknowledgments

I would like to gratefully acknowledge reviewers of earlier editions, whose influence continues to be felt.

◼ Reviewers of the First Edition

Hugh Akers, *Lamar University*; Steven Albrecht, *Oregon State University*; Margaret Asirvatham, *University of Colorado*; George Bandik, *University of Pittsburgh*; Gerald Berkowitz, *Erie Community College*; Robert Bogess, *Radford University*; Christine Brzezowski, *University of Utah*; Harry Conley, *Murray State*; Karen Eichstadt, *Ohio University*; William Euler, *University of Rhode Island*; Arthur Glasfeld, *Reed College*; Fabian Fang, *California State University—Long Beach*; John Fulkrod, *University of Minnesota—Duluth*; Marvin Hackert, *University of Texas at Austin*; Henry Harris, *Armstrong State College*; Leland Harris, *University of Arizona*; Larry Jackson, *Montana State University*; James Jacob, *University of Rhode Island*; James Johnson, *Sinclair Community College;* Eugene Klein, *Tennessee Technological University*; Norman Kulevsky, *University of North Dakota*; James W. Long, *University of Oregon*; Ralph Martinez, *Humboldt State University*; Scott Mohr, *Boston University*; Melvyn Mosher, *Missouri Southern University*; Elva Mae Nicholson, *Eastern Michigan University*; Frasier Nyasulu, *University of Washington*; John Ohlsson, *University of Colorado*; Roger Penn, *Sinclair Community College*; Helen Place, *Washington State University*; John Reasoner, *Western Kentucky University;* Norman Rose, *Portland State University*; Michael Ryan, *Marquette University*; John Searle, *College of San Mateo;* Dan Sullivan, *University of Nebraska at Omaha*; Emanuel Terezakis, *Community College of Rhode Island;* Ruiess Van Fossen Bravo, *Indiana University of Pennsylvania*; Donald Williams, *University of Louisville;* Les Wynston, *California State University—Long Beach.*

◼ Reviewers of the Second Edition

Vicky L.H. Bevilacqua, *Kennesaw State University*; David R. Bjorkman, *East Carolina University*; Frank D. Bay, *North Shore Community College*; Tim Champion, *Johnson C. Smith University*; Alison J. Dobson, *Georgia Southern University*; Naomi Eliezer, *Oakland University*; Wes Fritz, *College of DuPage*; Caroline Gil, *Lexington Community College*; Robert Gooden, *Southern University—Baton Rouge*; Ellen Kime-Hunt, *Riverside Community College*; Peter Krieger, *Palm Beach Community College*; Cathy MacGowan, *Armstrong Atlantic State University*; Lawrence L. Mack, *Bloomsburg University*; Charmaine B. Mamantov, *University of Tennessee, Knoxville*; Joannn S. Monko, *Kutztown University*; Elva Mae Nicholson, *Eastern Michigan University*; Michael Shanklin, *Palo Alto College*; Hugh Akers, *Lamar University*; Eric Holmberg, *University of Alaska*; Marvin Jaffe, *Borough of Manhattan Community College*; David Johnson, *Biola University*; Fred Johnson, *Brevard Community College*; Daniel Jones, *University of North Carolina—Charlotte*; Peter Krieger, *Palm Beach Community College*; Da-hong Lu, *Fitchburg State College*; Cynthia Martin, *Des Moines Area Community College*; Elva Mae Nicholson, *Eastern Michigan University*; Mary Palaszek, *Grand Valley State University*; Diane Payne, *Villa Julie College*; Janet Rogers, *Edinboro University of Pennsylvania*; Jackie Scholars, *Bellevue University*; Michelle Sulikowski, *Texas A&M University*; Joanne Tscherne, *Bergen Community College*; James Yuan, *Old Dominion University.*

◼ Reviewers of the Third Edition

Diane Payne, *Villa Julie College*; Kristan Lenning, *Lexington Community College*; Sidney Alozie, *Bronx Community College*; Barbara Keller, *Lake Superior State University*; Naomi Eliezer, *Oakland University*; Josh Smith, *Humboldt State University*; Garon Smith, *University of Montana*; Mundiyath Venugopalan, *Western Illinois University*; Renee Rosentreter, *Idaho State University*; Laura Kibler-Herzog, *Georgia State University*;

Peter Krieger, *Palm Beach Community College*; Peter Olds, *Laney College*; Marcia Miller, *University of Wisconsin—Eau Claire*; Sara Hein, *Winona State University*.

I also want to thank the following reviewers for their valuable comments and suggestions which helped to guide my revision efforts for this edition: Jennifer Adamski, *Old Dominion University*; M. Reza Asdjodi, *University of Wisconsin—Eau Claire*; Irene Gerow, *East Carolina University*; Ernest Kho, *University of Hawaii at Hilo*; Larry L. Land, *University of Florida*; Michael Myers, *California State University—Long Beach*; H. A. Peoples, *Las Positas College*; Shashi Rishi, *Greenville Technical College*; Steven M. Socol, *McHenry County College*.

Special thanks go to Stephen Z. Goldberg, *Adelphi University*, for his help in ensuring this book's accuracy by reviewing manuscript, proofs, and artwork.

I also give special thanks to the people at Houghton Mifflin who guided the revision through various stages of development and production: Richard Stratton, Executive Editor, Chemistry; Development Editors Kellie Cardone and Rebecca Berardy Schwartz; Katherine Greig, Senior Marketing Manager; Naveen Hariprasad, Marketing Assistant; Charlotte Miller, Art Editor, for her thoughtful and creative contributions to the illustration program; Naomi Kornhauser, Photo Researcher; Jean Hammond, Designer, for the complementary design; Andrea Cava, Project Editor, for making the production process for this text a smooth one; and Peggy Flanagan, Copyeditor.

H. Stephen Stoker
Weber State University

Throughout the text, **an exciting photo program** helps students see the everyday applications of the chemistry they are learning.

Chapter Outlines give students a road map for where they are going.

10 Enzymes and Vitamins

Yellow- and orange-colored vegetables such as pumpkins and squash have significant vitamin A activity due to the presence of the molecule beta-carotene.

In this chapter we consider two topics: enzymes and vitamins. Enzymes govern all chemical reactions in living organisms. They are specialized proteins that, with fascinating specificity and selectivity, catalyze biochemical reactions that store and release energy, build components in our hair and eyes, digest the food we eat, synthesize cellular building blocks, and protect us by repairing cellular damage and clotting our blood. Enzymes respond to their environment, responding quickly to changes in the cell. The deficiency of particular enzymes can cause certain diseases or signal problems such as heart or organ damage. Our knowledge of protein structure (Chapter 9) can help us better understand how enzymes function in living cells.

Vitamins, which are necessary components of a healthful diet, play important roles in metabolism. In most cases, they function as enzyme cofactors or carriers of molecules during biosynthesis.

al Characteristics of Enzymes

an organic compound that acts as a catalyst for a biochemical reaction. The human body contains thousands of different enzymes because almost each reaction in a cell requires its own specific enzyme. Enzymes cause cellular reactions to occur millions of times faster than corresponding uncatalyzed reactions. Enzymes are not consumed during the reaction but merely help the reaction occur more rapidly.

349

---

356   Chapter 10   Enzymes and Vitamins

CHEMICAL CONNECTIONS — *H. pylori* and Stomach Ulcers

Helicobacter pylori, commonly called *H. pylori,* is a bacterium that can function in the highly acidic environment of the stomach. The discovery in 1982 of the existence of this bacterium in the stomach was startling to the medical profession because conventional thought at the time was that bacteria could not survive at the stomach's pH of about 1.4.

It is now known that *H. pylori* causes more than 90% of duodenal ulcers and up to 80% of gastric ulcers. Before this discovery, it was thought that most ulcers were caused by excess stomach acid eating the stomach lining. Contributory causes were thought to be spicy food and stress. Conventional treatment involved acid-suppression or acid-neutralization medications. Now, treatment regimens involve antibiotics. The medical profession was slow to accept the concept of a bacterial cause for most ulcers, and it was not until the mid-1990s that antibiotic treatment became common.

How the enzymes present in the *H. pylori* bacterium can function in the acidic environment of the stomach (where they should be denatured) is now known. Present on the surface of the bacterium is the enzyme *urease,* an enzyme that converts urea to the basic substance ammonia. The ammonia then neutralizes acid present in its immediate vicinity; a protective barrier is thus created. The *urease* itself is protected from denaturation by its complex quaternary structure.

H. pylori causes ulcers by weakening the protective mucous coating of the stomach and duodenum, which allows acid to get through to the sensitive lining beneath. Both the acid and the bacteria irritate the lining and cause a sore—the ulcer. Ultimately the *H. pylori* themselves burrow into the lining to an acid-safe area within the lining.

Approximately two-thirds of the world's population is infected with *H. pylori.* In the United States 30% of the adult population is infected, with the infection most prevalent among older adults. About 20% of people under the age of 40 and half of those over 60 have it. Only one out of every six people infected with *H. pylori* ever suffer symptoms related to ulcers. Why *H. pylori* does not cause ulcers in every infected person is not known.

H. pylori bacteria are most likely spread from person to person through fecal–oral or oral–oral routes. Possible environmental sources include contaminated water sources. The infection is more common in crowded living conditions with poor sanitation. In countries with poor sanitation, 90% of the adult population can be infected.

H. pylori bacteria.

substrate must occupy an enzyme active site for a finite amount of time, and the products must leave the site before the cycle can be repeated. When each enzyme molecule is working at full capacity, the incoming substrate molecules must "wait their turn" for an empty active site. At this point, the enzyme is said to be under saturating conditions.

The rate at which an enzyme accepts and releases substrate molecules at substrate saturation is given by its turnover number. An enzyme's **turnover number** *is the number of substrate molecules transformed per minute by one molecule of enzyme under optimum conditions of temperature, pH, and saturation.* Table 10.2 gives

TABLE 10.2
Turnover Numbers for Selected Enzymes

Enzyme	Turnover Number (per minute)	Reaction Catalyzed
carbonic anhydrase	36,000,000	$CO_2 + H_2O \rightleftharpoons H_2CO_3$
catalase	5,600,000	$2H_2O_2 \rightleftharpoons 2H_2O + O_2$
cholinesterase	1,500,000	hydrolysis of acetylcholine
penicillinase	120,000	hydrolysis of penicillin
lactate dehydrogenase	60,000	conversion of pyruvate to lactate
DNA polymerase I	900	addition of nucleotides to DNA chains

Chemical Connections boxes show chemistry as it appears in everyday life. Topics are relevant to students' future careers in the health and environmental fields and are important for informed citizens to understand.

Chemistry at a Glance pulls together material from a group of sections or a whole chapter to help students see the larger picture through a visual summary. Many Chemistry at a Glance features have been revised and several new ones have been added.

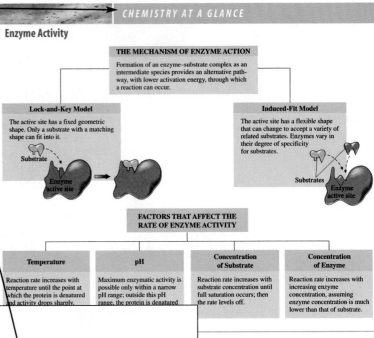

CHEMISTRY AT A GLANCE

Enzyme Activity

THE MECHANISM OF ENZYME ACTION

Formation of an enzyme–substrate complex as an intermediate species provides an alternative pathway, with lower activation energy, through which a reaction can occur.

Lock-and-Key Model

The active site has a fixed geometric shape. Only a substrate with a matching shape can fit into it.

Substrate

Enzyme active site

Induced-Fit Model

The active site has a flexible shape that can change to accept a variety of related substrates. Enzymes vary in their degree of specificity for substrates.

Substrates

Enzyme active site

FACTORS THAT AFFECT THE RATE OF ENZYME ACTIVITY

Temperature	pH	Concentration of Substrate	Concentration of Enzyme
Reaction rate increases with temperature until the point at which the protein is denatured and activity drops sharply.	Maximum enzymatic activity is possible only within a narrow pH range; outside this pH range, the protein is denatured	Reaction rate increases with substrate concentration until full saturation occurs; then the rate levels off.	Reaction rate increases with increasing enzyme concentration, assuming enzyme concentration is much lower than that of substrate.

CHEMISTRY AT A GLANCE

Enzyme Inhibition

ENZYME INHIBITORS

Substances that bind to an enzyme and stop or slow its normal catalytic activity

Competitive Enzyme Inhibitor

A molecule closely resembling the substrate. Binds to the active site and temporarily prevents substrates from occupying it, thus blocking the reaction.

Substrate   Competitive inhibitor

Enzyme

Noncompetitive Enzyme Inhibitor

A molecule that binds to a site on an enzyme that is not the active site. The normal substrate still occupies the active site but the enzyme cannot catalyze the reaction due to the presence of the inhibitor.

Substrate

Noncompetitive inhibitor

Irreversible Enzyme Inhibitor

A molecule that forms a covalent bond to a part of the active site, permanently preventing substrates from occupying it.

Substrate   Irreversible inhibitor   Enzyme active site

---

ed reactions can be *decreased* by a group of substances called
bitor *is a substance that slows or stops the normal catalytic*
ding to it. In this section, we consider three modes by which
ible competitive inhibition, reversible noncompetitive inhibi-
ion.

ve Inhibition

at enzymes are quite specific about the molecules they accept
lar shape and charge distribution are key determining factors
ts a molecule. A **competitive enzyme inhibitor** *is a molecule*
n enzyme substrate in shape and charge distribution that it
ate for occupancy of the enzyme's active site.

Examples of noncompetitive inhibitors include the heavy metal ions Pb^{2+}, Ag^+, and Hg^{10}. The binding sites for these ions are sulfhydryl (—SH) groups located away from the active site. Metal disulfide linkages are formed, an effect that disrupts secondary and tertiary structure.

■ Irreversible Inhibition

An **irreversible enzyme inhibitor** *is a molecule that inactivates enzymes by forming a strong covalent bond to an amino acid side-chain group at the enzyme's active site.* In general, such inhibitors do *not* have structures similar to that of the enzyme's normal substrate. The inhibitor–active site bond is sufficiently strong that addition of excess substrate does not reverse the inhibition process. Thus the enzyme is permanently deactivated. The actions of chemical warfare agents (nerve gases) and organophosphate insecticides are based on irreversible inhibition.

The Chemistry at a Glance feature summarizes what we have considered concerning enzyme inhibition.

10.8 Regulation of Enzyme Activity: Allosteric Enzymes

In the previous section, we looked at the decrease in enzyme activity caused by inhibiting agents that were "foreign" to normal cells. In this section, we consider the regulation of enzyme activity by substances produced within a cell—that is, regulation by "normal" cell components. The concept of noncompetitive inhibition that was developed in the previous section will be part of our discussion.

Margin Notes summarize key information, give tips for remembering or distinguishing between similar ideas, and provide additional details and links between concepts.

10.9 Regulation of Enzyme Activity: Zymogens **361**

■ Allosteric Enzymes

Many, but not all, of the molecules responsible for regulating cellular processes are a special group of enzymes called *allosteric enzymes*. Characteristics of such enzymes are as follows:

1. All allosteric enzymes have quaternary structure; that is, they are composed of two or more protein chains.
2. All allosteric enzymes have two kinds of binding sites: those for substrate and those for regulators.
3. Active and regulatory binding sites are distinct from each other in both location and shape. Often the regulatory site is on one protein chain and the active site is on another.
4. Binding of a molecule at the regulatory site causes changes in the overall three-dimensional structure of the enzyme, including structural changes at the active site.

The term *allosteric* comes from the Greek *allo*, which means "other," and *stereos*, which means "site or space."

Thus an **allosteric enzyme** *is an enzyme with two or more protein chains (quaternary structure) and two kinds of binding sites (substrate and regulator).*

Substances that bind at regulatory sites of allosteric enzymes are called *regulators*. The binding of a *positive regulator* increases enzyme activity; the shape of the active site is changed such that it can more readily accept substrate. The binding of a *negative regulator* (a noncompetitive inhibitor) decreases enzyme activity; changes to the active site are such that substrate is less readily accepted.

Some regulators of allosteric enzyme function are inhibitors (negative regulators), and some increase enzyme activity (positive regulators).

■ Feedback Control

One of the mechanisms by which allosteric enzyme activity is regulated is feedback control. **Feedback control** *is a process in which activation or inhibition of the first reaction in a reaction sequence is controlled by a product of the reaction sequence.*

To illustrate the feedback control mechanism, let us consider a biochemical process within a cell that occurs in several steps, each step catalyzed by a different enzyme.

Most biochemical processes within cells take place in several steps rather than in a single step. A different enzyme is required for each step of the process.

$$A \xrightarrow{\text{Enzyme 1}} B \xrightarrow{\text{Enzyme 2}} C \xrightarrow{\text{Enzyme 3}} D$$

The product of each s...

What will happe...
of the first enzyme (...
ceeds rapidly. At hig...
(by feedback), and ev...
present in the cell to...
use in other cell react...

The general term *allosteric control* is often used to describe a process in which a regulatory molecule that binds at one site in an enzyme influences substrate binding at the active site in the enzyme.

Feedback contro...
regulated; it is just o...
enzyme may be prod...
may even be compou...

10.9 Regulation

A **proteolytic enzyme**...
maintain the primar...
tissues that produce...

Within the chapters, worked-out **Examples** follow the explanation of many concepts. These examples walk students through the thought process involved in problem solving, carefully outlining all the steps involved. They are immediately followed by a **Practice Exercise** to reinforce the information just presented.

---

230 Chapter 7 Carbohydrates

EXAMPLE 7.4

Recognizing Enantiomers and Diastereomers

■ Characterize each of the following pairs of structures as enantiomers, diastereomers, or neither enantiomers nor diastereomers.

a.
```
    CHO              CHO
H——OH            H——OH
HO——H     and    H——OH
H——OH            HO——H
  CH2OH            CH2OH
```

b.
```
    CHO              CHO
H——OH            HO——H
HO——H     and    H——OH
H——OH            HO——H
  CH2OH            CH2OH
```

c.
```
    CHO              CHO
H——OH            H—C—H
HO——H     and    HO——H
H——OH            HO——H
  CH2OH            CH2OH
```

Solution

a. These two structures represent *diastereomers*—the arrangement of —H and —OH substituents is identical for at least one chiral center, whereas the arrangement of —H and —OH substituents at remaining chiral centers is that of mirror images. The —H and —OH substituent arrangement is the same at the first chiral center and is that of mirror images at the second and third chiral centers.

b. These two structures represent *enantiomers*—a mirror-image substituent relationship exists between the two isomers at *every* chiral center.

c. These two structures are *neither enantiomers nor diastereomers*. The connectivity of atoms differs in the two structures at carbon 2. Stereoisomers (enantiomers and diastereomers) must have the same connectivity throughout both structures. (The two structures are not even constitutional isomers because the first structure contains one more oxygen atom than the second.)

Practice Exercise 7.4

Characterize the following pairs of structures as enantiomers, diastereomers, or neither enantiomers nor diastereomers.

a.
```
    CHO              CHO
H——OH            HO——H
H——OH     and    HO——H
H——OH            HO——H
  CH2OH            CH2OH
```

b.
```
    CHO              CHO
H——OH            HO——H
H——OH     and    HO——H
H——OH            HO——H
  CH2OH            CH2OH
```

c.
```
    CHO              CHO
H——OH            HO——H
HO——H     and    H——OH
HO——H            H——OH
  CH2OH            CH2OH
```

We calculate 2^n to predict the maximum possible number of stereoisomers for a molecule with n chiral atoms. In a few cases, the actual number of stereoisomers is less than the maximum because of symmetry considerations that make some mirror images superimposable.

In general, a compound that has n chiral centers may exist in a *maximum* of 2^n stereoisomeric forms. For example, when three chiral centers are present, at most eight stereoisomers ($2^3 = 8$) are possible (four pairs of enantiomers).

The Chemistry at a Glance feature on page 231 summarizes information about the various types of isomers we have encountered so far in the text—the various subtypes of constitutional isomers and the various subtypes of stereoisomers.

Biochemistry. Biochemistry is the study of the chemical substances found in living systems and the chemical interactions of these substances with each other (Section 7.1).

Carbohydrates. Carbohydrates are polyhydroxy aldehydes, polyhydroxy ketones, or compounds that yield such substances upon hydrolysis. Plants contain large quantities of carbohydrates produced via photosynthesis (Section 7.2).

Carbohydrate classification. Carbohydrates are classified into three groups: monosaccharides, oligosaccharides, and polysaccharides (Section 7.3).

Chirality and achirality. A chiral object is not identical to its mirror image. An achiral object is identical to its mirror image (Section 7.4).

Chiral center. A chiral center is an atom in a molecule that has four different groups tetrahedrally bonded to it. Molecules that contain a single chiral center exist in a left-handed and a right-handed form (Section 7.4).

Stereoisomerism. The atoms of stereoisomers are connected in the same way but are arranged differently in space. The major causes of stereoisomerism in molecules are structural rigidity and the presence of a chiral center (Section 7.5).

Enantiomers and diastereomers. Two types of stereoisomers exist: enantiomers and diastereomers. Enantiomers have structures that are nonsuperimposable mirror images of each other. Enantiomers have identical achiral properties but different chiral properties. Diastereomers have structures that are not mirror images of each other (Section 7.5).

Fischer projections. Fischer projections are two-dimensional structural formulas used to depict the three-dimensional shapes of molecules with chiral centers (Section 7.6).

Chirality of monosaccharides. Monosaccharides are classified as D or L stereoisomers on the basis of the configuration of the chiral center farthest from the carbonyl group (Section 7.6).

Optical activity. Chiral compounds are optically active—that is, they rotate the plane of polarized light. Enantiomers rotate the plane of polarized light in opposite directions. The prefix (+) indicates that

the compound rotates the plane of polarized light in a clockwise direction, whereas compounds that rotate the plane of polarized light in a counterclockwise direction have the prefix (−) (Section 7.7).

Classification of monosaccharides. Monosaccharides are classified as aldoses or ketoses on the basis of the type of carbonyl group present. They are further classified as trioses, tetroses, pentoses, etc. on the basis of the number of carbon atoms present (Section 7.8).

Important monosaccharides. Important monosaccharides include glucose, galactose, fructose, and ribose. Glucose and galactose are aldohexoses, fructose is a ketohexose, and ribose is an aldopentose (Section 7.9).

Cyclic monosaccharides. Cyclic monosaccharides form through an intramolecular reaction between the carbonyl group and an alcohol group of an open-chain monosaccharide. These cyclic forms predominate in solution (Section 7.10).

Reactions of monosaccharides. Five important reactions of monosaccharides are (1) oxidation to an acidic sugar, (2) reduction to a sugar alcohol, (3) glycoside formation, (4) phosphate ester formation, and (5) amino sugar formation (Section 7.12).

Disaccharides. Disaccharides are glycosides formed from the linkage of two monosaccharides. The most important disaccharides are maltose, cellobiose, lactose, and sucrose. Each of these has at least one glucose unit in its structure (Section 7.13).

Polysaccharides. Polysaccharides are polymers in which monosaccharides are the monomers. In homopolysaccharides only one type of monomer is present. Two or more monosaccharide monomers are present in heteropolysaccharides. Storage polysaccharides (starch, glycogen) are storage molecules for monosaccharides. Structural polysaccharides (cellulose, chitin) serve as structural elements in plant cell walls and animal exoskeletons (Sections 7.14 to 7.17).

Glycolipids and glycoproteins. Glycolipids and glycoproteins are molecules in which oligosaccharides are attached through glycosidic linkages to lipids and proteins, respectively. Such molecules often govern how cells of differing function interact with each other (Section 7.18).

Concepts to Remember and **Key Reactions and Equations** provide concise review of the material presented in the chapter, helping students prepare for exams.

Exercises and Problems **261**

1. Monosaccharide oxidation (Section 7.12)

   Aldose or ketose + weak oxidizing agent $\longrightarrow$ acidic sugar

2. Monosaccharide reduction (Section 7.12)

   Aldose or ketose + H_2 $\xrightarrow{\text{Catalyst}}$ sugar alcohol

3. Glycoside (acetal) formation (Section 7.12)

   Cyclic monosaccharide + alcohol $\longrightarrow$ glycoside (acetal) + H_2O

4. Monosaccharide ester formation (Section 7.12)

   Monosaccharide + oxyacid $\longrightarrow$ ester + H_2O

5. Hydrolysis of disaccharide (Section 7.13)

   Disaccharide + H_2O $\xrightarrow{\text{Catalyst}}$ two monosaccharides

6. Hydrolysis of maltose (Section 7.13)

   D-Maltose + H_2O $\xrightarrow{\text{H}^+ \text{ or maltase}}$ 2 D-glucose

7. Hydrolysis of cellobiose (Section 7.13)

   D-Cellobiose + H_2O $\xrightarrow{\text{H}^+ \text{ or cellobiase}}$ 2 D-glucose

8. Hydrolysis of lactose (Section 7.13)

   D-Lactose + H_2O $\xrightarrow{\text{H}^+ \text{ or lactase}}$ D-galactose + D-glucose

9. Hydrolysis of sucrose (Section 7.13)

   D-Sucrose + H_2O $\xrightarrow{\text{H}^+ \text{ or sucrase}}$ D-fructose + D-glucose

10. Complete hydrolysis of starch (Section 7.15)

    Starch + H_2O $\xrightarrow{\text{H}^+ \text{ or enzymes}}$ many D-glucose

11. Complete hydrolysis of glycogen (Section 7.15)

    Glycogen + H_2O $\xrightarrow{\text{H}^+ \text{ or enzymes}}$ many D-glucose

Extensive and varied **Exercises and Problems** at the end of each chapter are organized by topic and paired. These problems always involve only a single concept. Answers to the odd-numbered problems can be found at the back of the book.

The members of each pair of problems in this section test similar material.

■ **Biochemical Substances (Section 7.1)**

7.1   Define the term *biochemistry*.

7.2   What are the two general groups of biochemical substances?

7.3   What are the four major types of bioorganic substances?

a. Nail, han
b. Your han
c. The wor

7.16 In each of th
       that are chir

7.117 Indicate whether each of the following compounds is chiral or achiral.
       a. 1-Chloro-2-methylpentane   b. 2-Chloro-2-methylpentane
       c. 2-Chloro-3-methylpentane   d. 3-Chloro-2-methylpentane

7.118 Indicate whether each of the following compounds is optically active or optically inactive.

a. Chitin   b. Amylopectin
c. Hyaluronic acid   d. Glycogen

7.124 List the reactant(s) necessary to effect the following chemical changes.

268   Chapter 7   Carbohydrates

Additional Problems involve more than one concept and are more difficult than the Exercises and Problems.

New **Multiple-Choice Practice Tests** have been added to the end of each chapter as a cumulative review.

7.125 Which of the following statements is *incorrect*?
       a. A chiral center is an atom in a molecule that has four different groups tetrahedrally bonded to it.
       b. A chiral molecule is a molecule whose mirror images are superimposable.
       c. Naturally occurring monosaccharides are almost always "right-handed."
       d. The simplest example of a chiral monosaccharide is glyceraldehyde.

7.126 Which of the following statements concerning the D and L forms of a monosaccharide is *incorrect*?
       a. Structurally they are nonsuperimposable mirror images of each other.
       b. They must contain the same number of chiral centers.
       c. They are enantiomers.
       d. They are diastereomers.

7.127 Which of the following is a correct characterization for the monosaccharide glucose?

a. One   b. Two
c. Three   d. Four

7.130 Which of the following disaccharides produces both D-glucose and D-fructose upon hydrolysis?
       a. Sucrose   b. Lactose
       c. Maltose   d. Cellobiose

7.131 In which of the following pairs of disaccharides do both members of the pair have the same type of glycosidic linkage?
       a. Sucrose and lactose   b. Cellobiose and maltose
       c. Lactose and cellobiose   d. Sucrose and maltose

7.132 In which of the following pairs of carbohydrates are both members of the pair heteropolysaccharides?
       a. Cellulose and amylose
       b. Starch and chitin
       c. Hyaluronic acid and heparin
       d. Glycogen and amylopectin

7.133 In which of the following pairs of polysaccharides are both members of the pair structural polysaccharides?

xviii

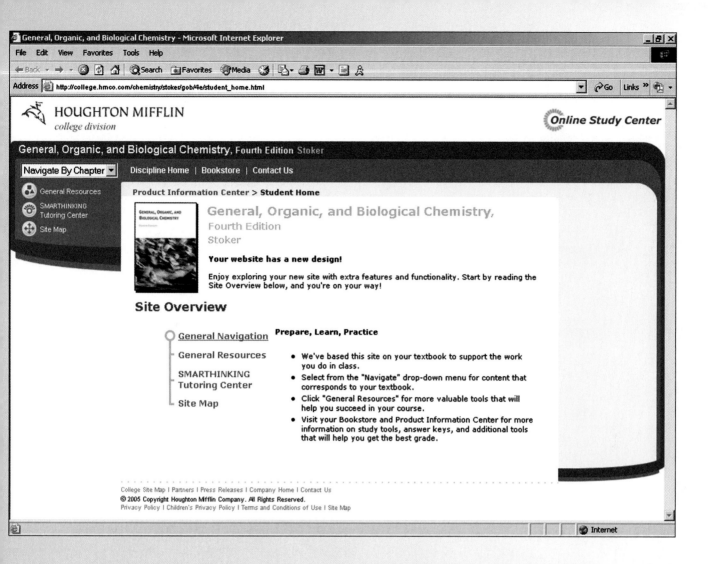

An Online Study Center is accessible through college.hmco.com/pic/stokerGOB4e. It includes a wealth of resources to help students in the course, including:

* Self-quizzing using Houghton Mifflin's ACE system
* Electronic flashcards of key terms, reactions, and concepts
* Career-related information

The **Instructor Website**, accessible through the address above, allows access to all student resources, plus instructor/classroom resources such as downloadable PowerPoint slides and virtually all the art, tables, and photos in JPEG format.

Organic and Biological Chemistry

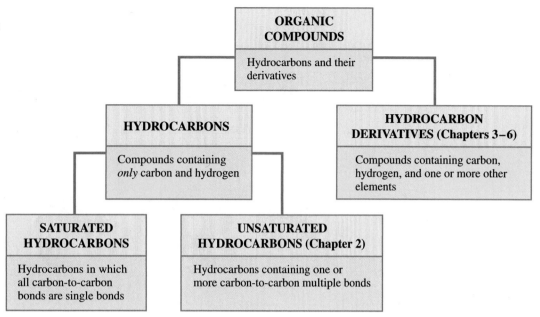

The term *saturated* has the general meaning that there is no more room for something. Its use with hydrocarbons comes from early studies in which chemists tried to add hydrogen atoms to various hydrocarbon molecules. Compounds to which no more hydrogen atoms could be added (because they already contained the maximum number) were called saturated, and those to which hydrogen could be added were called unsaturated.

elements. Additional elements commonly found in hydrocarbon derivatives include O, N, S, P, F, Cl, and Br. Millions of hydrocarbon derivatives are known.

Hydrocarbons may be divided into two large classes: saturated and unsaturated. A **saturated hydrocarbon** *is a hydrocarbon in which all carbon–carbon bonds are single bonds.* Saturated hydrocarbons are the simplest type of organic compound. An **unsaturated hydrocarbon** *is a hydrocarbon in which one or more carbon–carbon multiple bonds (double bonds, triple bonds, or both) are present.* In general, saturated and unsaturated hydrocarbons undergo distinctly different chemical reactions.

Saturated hydrocarbons are the subject of this chapter. Unsaturated hydrocarbons are considered in the next chapter. Figure 1.2 summarizes the terminology presented in this section.

1.4 Alkanes: Acyclic Saturated Hydrocarbons

In a saturated hydrocarbon, the carbon atom arrangement may be *acyclic* or *cyclic.* The term *acyclic* means "not cyclic." Examples of acyclic and cyclic carbon atom arrangements are

$$C{-}C{-}C{-}C{-}C{-}C \qquad \begin{array}{c} C \\ C \quad\quad C \\ C \quad\quad C \\ C \end{array}$$

Acyclic   Cyclic

In this section we consider acyclic saturated hydrocarbons. Cyclic saturated hydrocarbons are considered in Sections 1.12 and 1.13.

An **alkane** *is a saturated hydrocarbon in which the carbon atom arrangement is acyclic.* Thus an alkane is a hydrocarbon that contains only carbon–carbon single bonds (saturated) and has no rings of carbon atoms (acyclic).

The molecular formulas of all alkanes fit the general formula C_nH_{2n+2}, where n is the number of carbon atoms present. The number of hydrogen atoms present in an alkane is always twice the number of carbon atoms plus two more, as in C_4H_{10}, C_5H_{12}, and C_8H_{18}.

FIGURE 1.3   Ball-and-stick and space-filling models showing the molecular structures of (a) methane, (b) ethane, and (c) propane, the three simplest alkanes.

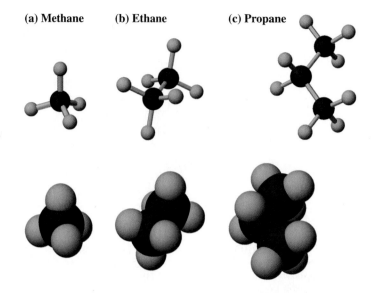

(a) Methane   **(b) Ethane**   **(c) Propane**

The three simplest alkanes are methane (CH_4), ethane (C_2H_6), and propane (C_3H_8). Ball-and-stick and space-filling models showing the molecular structures of these three alkanes are given in Figure 1.3. Note how each carbon atom in each of the models participates in four bonds (Section 1.2). Note also that the geometrical arrangement of atoms about each carbon atom is tetrahedral, an arrangement consistent with the principles of VSEPR theory. The tetrahedral arrangement of the atoms bonded to alkane carbon atoms is fundamental to understanding the structural aspects of organic chemistry.

1.5 Structural Formulas

The structures of alkanes, as well as other types of organic compounds, are generally represented in two dimensions rather than three (Figure 1.3) because of the difficulty in drawing the latter. These two-dimensional structural representations make no attempt to portray accurately the bond angles or molecular geometry of molecules. Their purpose is to convey information about which atoms in a molecule are bonded to which other atoms.

Two-dimensional structural representations for organic molecules are called structural formulas. A **structural formula** *is a two-dimensional structural representation that shows how the various atoms in a molecule are bonded to each other.* Structural formulas are of two types: expanded structural formulas and condensed structural formulas. An **expanded structural formula** *is a structural formula that shows all atoms in a molecule and all bonds connecting the atoms.* When written out, expanded structural formulas generally occupy a lot of space, and condensed structural formulas represent a shorthand method for conveying the same information. A **condensed structural formula** *is a structural formula that uses groupings of atoms, in which central atoms and the atoms connected to them are written as a group, to convey molecular structural information.* The expanded and condensed structural formulas for methane, ethane, and propane follow.

Structural formulas, whether expanded or condensed, do not show the geometry (shape) of the molecule. That information can be conveyed only by 3-D drawings or models such as those in Figure 1.3.

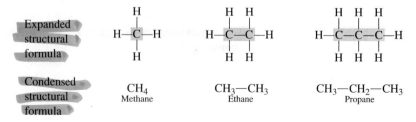

The Occurrence of Methane

Methane (CH_4), the simplest of all hydrocarbons, is a major component of the atmospheres of Jupiter, Saturn, Uranus, and Neptune but only a minor component of Earth's atmosphere (see the accompanying table). Earth's gravitational field, being weaker than that of the large outer planets, cannot retain enough hydrogen (H_2) in its atmosphere to permit the formation of large amounts of methane; H_2 molecules (the smallest and fastest-moving of all molecules) escape from it into outer space.

The small amount of methane present in Earth's atmosphere comes from terrestrial sources. The decomposition of animal and plant matter in an oxygen-deficient environment—swamps, marshes, bogs, and the sediments of lakes—produces methane. A common name for methane, marsh gas, refers to the production of methane in this manner.

Composition of Earth's Atmosphere (in parts per million by volume)

Major components		Minor components	
nitrogen	780,800	argon	9340
oxygen	209,500	carbon dioxide	314
		neon	18
		helium	5
		methane	2
		krypton	1

Bacteria that live in termites and in the digestive tracts of plant-eating animals have the ability to produce methane from plant materials (cellulose). The methane output of a large cow (belching and flatulence) can reach 20 liters per day.

Methane entering the atmosphere from terrestrial sources presents an environmental problem. Methane is a "greenhouse gas" that contributes to global warming. Methane is 15 to 30 times more efficient than carbon dioxide (the primary greenhouse gas) in trapping heat radiated from Earth. Fortunately, its atmospheric level of 2.0 ppm by volume is much lower than that of carbon dioxide (over 300 ppm).

Methane gas is also found associated with coal and petroleum deposits. Methane associated with coal mines is considered a hazard. If left to accumulate, it can form pockets where air is not present, and asphyxiation of miners can occur. When mixed with air in certain ratios, it can also present an explosion hazard. Methane associated with petroleum deposits is most often recovered, processed, and marketed as *natural gas*. The processed natural gas used in the heating of homes is 85% to 95% methane by volume. Because methane is odorless, an odorant (smelly compound) must be added to the processed natural gas used in home heating. Otherwise, natural gas leaks could not be detected.

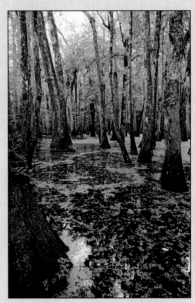

Decomposition of plant and animal matter in marshes is a source of methane gas.

The condensed structural formula for propane, $CH_3—CH_2—CH_3$, is interpreted in the following manner: The first carbon atom is bonded to three hydrogen atoms, and its fourth bond is to the middle carbon atom. The middle carbon atom, besides its bond to the first carbon atom, is also bonded to two hydrogen atoms and to the last carbon atom. The last carbon atom has bonds to three hydrogen atoms in addition to its bond to the middle carbon atom. As is always the case, each carbon atom has four bonds (Section 1.2).

The condensed structural formulas of hydrocarbons in which a long chain of carbon atoms is present are often condensed even more. The formula

$$CH_3—CH_2—CH_2—CH_2—CH_2—CH_2—CH_2—CH_3$$

can be further abbreviated as

$$CH_3—(CH_2)_6—CH_3$$

where parentheses and a subscript are used to denote the number of —CH_2— groups in the chain.

It is important to note that expanded structural formulas show all bonds within a molecule and that condensed structural formulas show only certain bonds—the bonds between carbon atoms. Specifically, the bond line in the condensed structural formula

$$CH_3—CH_3$$

denotes the bond between the first carbon atom and the second carbon atom; it is not a bond between hydrogen atoms and the second carbon atom.

In situations where the focus is solely on the arrangement of carbon atoms in an alkane, *skeletal structural formulas* that omit the hydrogen atoms are often used. A **skeletal structural formula** *is a structural formula that shows the arrangement and bonding of carbon atoms present in an organic molecule but does not show the hydrogen atoms attached to the carbon atoms.*

$$C—C—C—C—C \quad \text{means the same as} \quad CH_3—CH_2—CH_2—CH_2—CH_3$$
Skeletal structural formula   Condensed structural formula

The skeletal structural formula still represents a unique alkane because we know that each carbon atom shown must have enough hydrogen atoms attached to it to give the carbon four bonds.

1.6 Alkane Isomerism

The molecular formulas CH_4, C_2H_6, and C_3H_8 represent the alkanes methane, ethane, and propane, respectively. Next in the alkane molecular formula sequence (C_nH_{2n+2}) is C_4H_{10}, which would be expected to be the molecular formula of the four-carbon alkane. A new phenomenon arises, however, when an alkane has four or more carbon atoms. There is more than one structural formula that is consistent with the molecular formula. Consequently, more than one compound exists with that molecular formula. This situation brings us to the topic of *isomerism.*

> The word *isomer* comes from the Greek *isos,* which means "the same," and *meros,* which means "parts." Isomers have the same parts put together in different ways.

Isomers *are compounds that have the same molecular formula* (that is, the same numbers and kinds of atoms) *but that differ in the way the atoms are arranged.* Isomers, even though they have the same molecular formula, are always different compounds with different properties.

There are two four-carbon alkane isomers, the compounds *butane* and *isobutane.* Both have the molecular formula C_4H_{10}.

$$CH_3—CH_2—CH_2—CH_3 \qquad \begin{array}{c} CH_3—CH—CH_3 \\ | \\ CH_3 \end{array}$$
Butane   Isobutane

Butane and isobutane are different compounds with different properties. Butane has a boiling point of $-1°C$ and a melting point of $-138°C$, whereas the corresponding values for isobutane are $-12°C$ and $-159°C$.

> The existence of isomers necessitates the use of structural formulas in organic chemistry. Isomers always have the same molecular formula and different structural formulas.

Contrasting the two C_4H_{10} isomers structurally, note that butane has a chain of four carbon atoms. It is an example of a continuous-chain alkane. A **continuous-chain alkane** *is an alkane in which all carbon atoms are connected in a continuous nonbranching chain.* The other C_4H_{10} isomer, isobutane, has a chain of three carbon atoms with the fourth carbon attached as a branch on the middle carbon of the three-carbon chain. It is an example of a branched-chain alkane. A **branched-chain alkane** *is an alkane in which one or more branches (of carbon atoms) are attached to a continuous chain of carbon atoms.*

There are three isomers for alkanes with five carbon atoms (C_5H_{12}):

$$CH_3—CH_2—CH_2—CH_2—CH_3 \qquad \begin{array}{c} CH_3—CH—CH_2—CH_3 \\ | \\ CH_3 \end{array} \qquad \begin{array}{c} CH_3 \\ | \\ CH_3—C—CH_3 \\ | \\ CH_3 \end{array}$$
Pentane   Isopentane   Neopentane

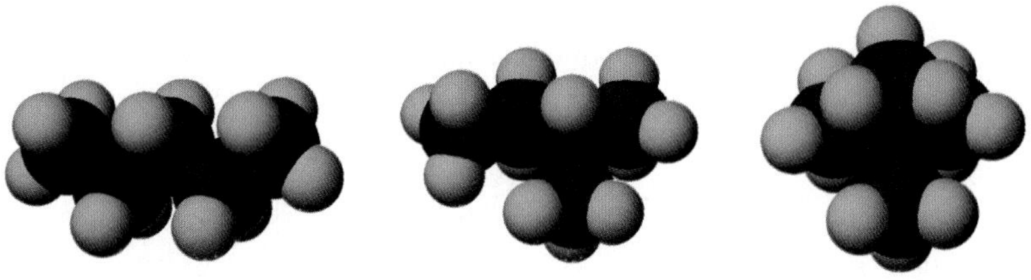

FIGURE 1.4 Space-filling models for the three isomeric C_5H_{12} alkanes: (a) pentane, (b) isopentane, and (c) neopentane.

(a) Pentane
Boiling point = 36.1°C
Density = 0.626 g/mL

(b) Isopentane
Boiling point = 27.8°C
Density = 0.620 g/mL

(c) Neopentane
Boiling point = 9.5°C
Density = 0.614 g/mL

Figure 1.4 shows space-filling models for the three isometric C_5 alkanes. Note how neopentane, the most branched isomer, has the most compact, most spherical three-dimensional shape.

The number of possible alkane isomers increases dramatically with increasing numbers of carbon atoms in the alkane, as shown in Table 1.1. Such isomerism is one of the major reasons for the existence of so many organic compounds.

Several different types of isomerism exist. The alkane isomerism examples discussed in this section are examples of *constitutional isomerism*. **Constitutional isomers** *are isomers that differ in the connectivity of atoms, that is, in the order in which atoms are attached to each other within molecules.* We will see shortly (Section 1.14) and in later chapters that other types of isomers are also possible, even among compounds whose atoms are connected in the same order. In the biochemistry portion of the text, where carbohydrates, lipids, and proteins are considered, we will find that different isomers elucidate different responses within the human body. Often, when many isomers are possible with the same molecular formula, only one isomer will be physiologically active.

Constitutional isomers are also frequently called *structural isomers*. The general characteristics of such isomers, independent of which name is used, are the same molecular formula and different structural formulas.

TABLE 1.1
Number of Isomers Possible for Alkanes of Various Carbon-Chain Lengths

Molecular Formula	Possible Number of Isomers
CH_4	1
C_2H_6	1
C_3H_8	1
C_4H_{10}	2
C_5H_{12}	3
C_6H_{14}	5
C_7H_{16}	9
C_8H_{18}	18
C_9H_{20}	35
$C_{10}H_{22}$	75
$C_{15}H_{32}$	4,347
$C_{20}H_{42}$	336,319
$C_{25}H_{52}$	36,797,588
$C_{30}H_{62}$	4,111,846,763

1.7 Conformations of Alkanes

Rotation about carbon–carbon single bonds is an important property of alkane molecules. Two groups of atoms in an alkane connected by a carbon–carbon single bond can rotate with respect to one another around that bond, much as a wheel rotates around an axle.

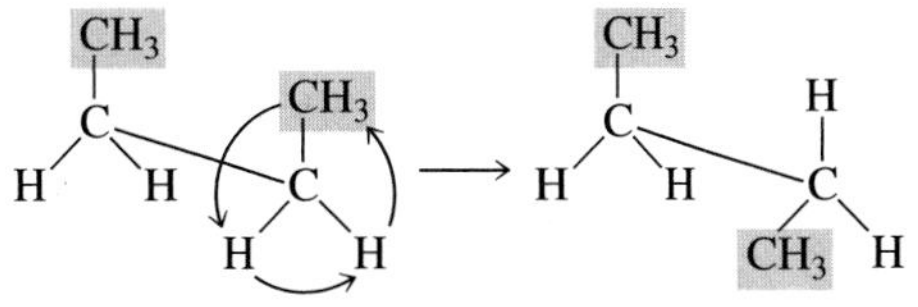

As a result of rotation around single bonds, alkane molecules (except for methane) can exist in infinite numbers of orientations, or conformations. A **conformation** *is the specific three-dimensional arrangement of atoms in an organic molecule at a given instant that results from rotations about carbon–carbon single bonds.*

The following skeletal formulas represent four different conformations for a continuous-chain six-carbon alkane molecule.

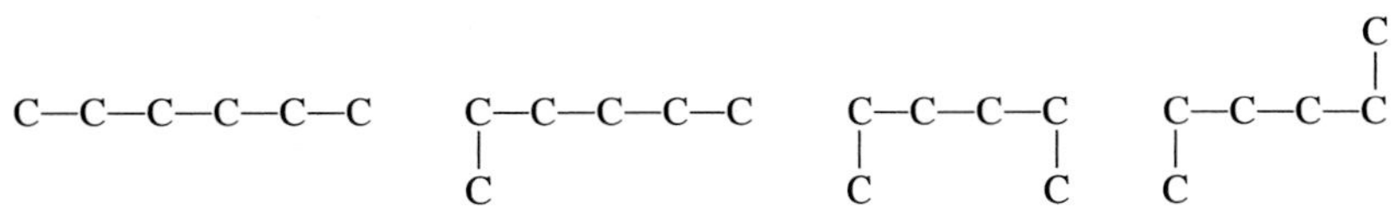

All four skeletal formulas represent the same molecule; that is, they are different conformations of the same molecule. In all four cases, a continuous chain of six carbon atoms is present. In all except the first case, the chain is "bent," but bends do not disrupt the continuity of the chain.

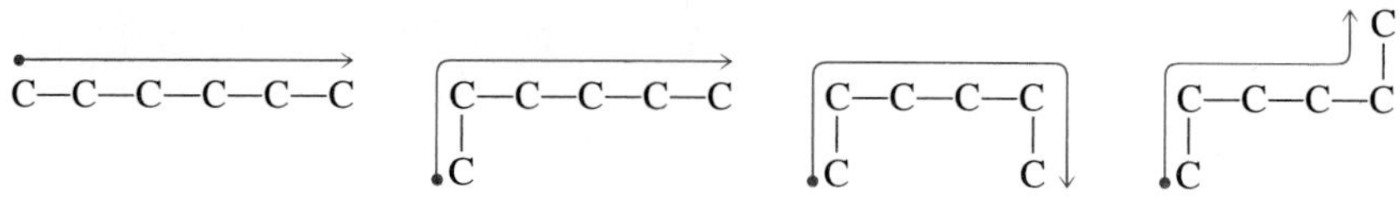

Note that the structures

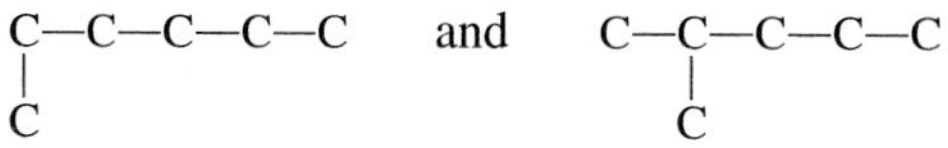

$$C-C-C-C-C \quad \text{and} \quad C-C-C-C-C$$
$$\qquad | \qquad\qquad\qquad\qquad\qquad |$$
$$\qquad C \qquad\qquad\qquad\qquad\qquad C$$

> You should learn to recognize molecules drawn in several different ways (conformations). Like friends, they can be recognized whether they are sitting, reclining, or standing.

are not two conformations of the same alkane but, rather, represent two different alkanes. The first structure involves a continuous chain of six carbon atoms, and the second structure involves a continuous chain of five carbon atoms to which a branch is attached. There is no way that you can get a continuous chain of six carbon atoms out of the second structure without "back-tracking," and "back-tracking" is not allowed.

EXAMPLE 1.1

Recognizing Different Conformations of a Molecule and Constitutional Isomers

■ Determine whether the members of each of the following pairs of structural formulas represent (1) different conformations of the same molecule, (2) different compounds that are constitutional isomers, or (3) different compounds that are not constitutional isomers.

a. 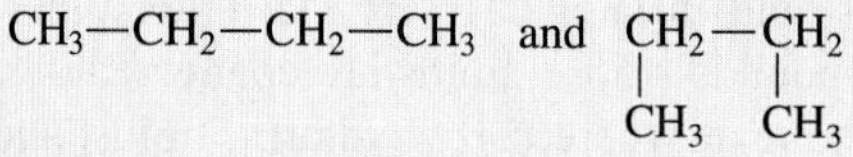 $CH_3-CH_2-CH_2-CH_3$ and CH_2-CH_2
$\qquad\qquad\qquad\qquad\qquad\qquad\qquad\quad | \qquad |$
$\qquad\qquad\qquad\qquad\qquad\qquad\qquad\; CH_3 \quad CH_3$

b. 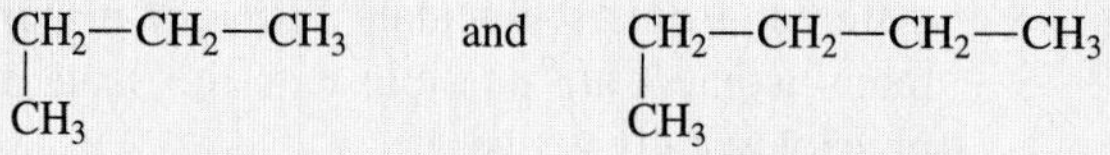 $CH_2-CH_2-CH_3$ and $CH_2-CH_2-CH_2-CH_3$
$\qquad |$ $\qquad\qquad\qquad\qquad\qquad\qquad |$
$\quad CH_3$ $\qquad\qquad\qquad\qquad\quad CH_3$

c. 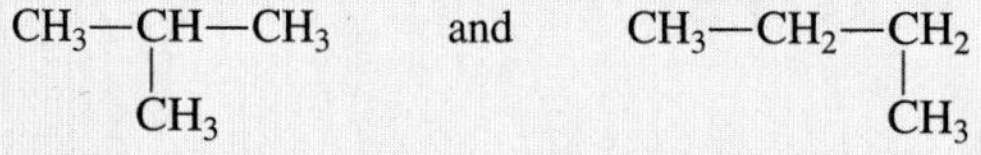 $CH_3-CH-CH_3$ and $CH_3-CH_2-CH_2$
$\qquad\qquad\quad |$ $\qquad\qquad\qquad\qquad\qquad\qquad |$
$\qquad\qquad CH_3$ $\qquad\qquad\qquad\qquad\qquad CH_3$

Solution

a. Both molecules have the molecular formula C_4H_{10}. The connectivity of carbon atoms is the same for both molecules: a continuous chain of four carbon atoms. For the second structural formula, we need to go around two corners to get a four-carbon-atom chain, which is fine because of the free rotation associated with single bonds in alkanes.

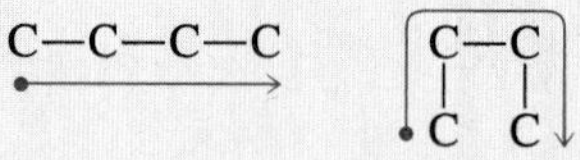

With the same molecular formula and the same connectivity of atoms, these two structural formulas are conformations of the same molecule.

b. The molecular formula of the first compound is C_4H_{10}, and that of the second compound is C_5H_{1}. Thus the two structural formulas represent different compounds that are not constitutional isomers. Constitutional isomers must have the same molecular formula.

c. Both molecules have the same molecular formula, C_4H_{10}. The connectivity of atoms is different. In the first case, we have a chain of three carbon atoms with a branch off the chain. In the second case, a continuous chain of four carbon atoms is present.

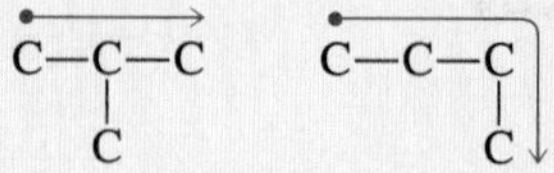

These two structural formulas are those of constitutional isomers.

Practice Exercise 1.1

Determine whether the members of each of the following pairs of structural formulas represent (1) different conformations of the same molecule, (2) different compounds that are constitutional isomers, or (3) different compounds that are not constitutional isomers.

a. $CH_3-CH_2-CH_2-CH_2-CH_3$ and CH_3-CH_2
CH_2-CH_2
CH_3

b. $CH_3-CH-CH_2-CH_3$ and $CH_3-CH-CH_2$
CH_3 CH_3 CH_3

c. $CH_3-CH-CH_2-CH_3$ and $CH_2-CH_2-CH_2$
CH_3 CH_3 CH_3

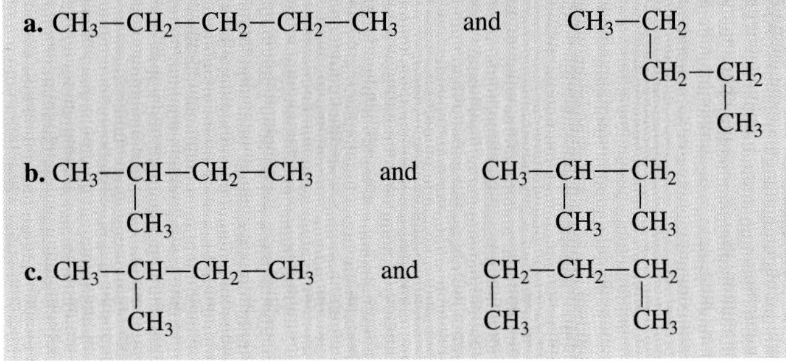

1.8 IUPAC Nomenclature for Alkanes

When relatively few organic compounds were known, chemists arbitrarily named them using what today are called *common names*. These common names gave no information about the structures of the compounds they described. However, as more organic compounds became known, this nonsystematic approach to naming compounds became unwieldy.

Today, formal systematic rules exist for generating names for organic compounds. These rules, which were formulated and are updated periodically by the International Union of Pure and Applied Chemistry (IUPAC), are known as *IUPAC rules*. The advantage of the IUPAC naming system is that it assigns each compound a name that not only identifies it but also enables one to draw its structural formula.

IUPAC names for the first ten *continuous-chain* alkanes are given in Table 1.2. Note that all of these names end in *-ane,* the characteristic ending for all alkane names. Note also that beginning with the five-carbon alkane, Greek numerical prefixes are used to denote the actual number of carbon atoms in the continuous chain.

To name *branched-chain* alkanes, we must be able to name the branch or branches that are attached to the main carbon chain. These branches are formally called substituents. A **substituent** *is an atom or group of atoms attached to a chain (or ring) of carbon atoms.* Note that *substituent* is a general term that applies to carbon-chain attachments in all organic molecules, not just alkanes.

IUPAC is pronounced "eye-you-pack."

Continuous-chain alkanes are also frequently called *straight-chain alkanes* and *normal-chain alkanes.*

You need to memorize the prefixes in column two of Table 1.2. This is the way to count from 1 to 10 in "organic chemistry language."

TABLE 1.2
IUPAC Names for the First Ten Continuous-Chain Alkanes[a]

Molecular Formula	IUPAC Prefix	IUPAC Name	Structural Formula
CH_4	meth-	methane	CH_4
C_2H_6	eth-	ethane	CH_3-CH_3
C_3H_8	prop-	propane	$CH_3-CH_2-CH_3$
C_4H_{10}	but-	butane	$CH_3-CH_2-CH_2-CH_3$
C_5H_{12}	pent-	pentane	$CH_3-CH_2-CH_2-CH_2-CH_3$
C_6H_{14}	hex-	hexane	$CH_3-CH_2-CH_2-CH_2-CH_2-CH_3$
C_7H_{16}	hept-	heptane	$CH_3-CH_2-CH_2-CH_2-CH_2-CH_2-CH_3$
C_8H_{18}	oct-	octane	$CH_3-CH_2-CH_2-CH_2-CH_2-CH_2-CH_2-CH_3$
C_9H_{20}	non-	nonane	$CH_3-CH_2-CH_2-CH_2-CH_2-CH_2-CH_2-CH_2-CH_3$
$C_{10}H_{22}$	dec-	decane	$CH_3-CH_2-CH_2-CH_2-CH_2-CH_2-CH_2-CH_2-CH_2-CH_3$

[a]The IUPAC naming system also includes prefixes for naming continuous-chain alkanes that have more than 10 carbon atoms, but we will not consider them in this text.

TABLE 1.3
Names for the First Six Continuous-Chain Alkyl Groups

Number of Carbons	Structural Formula	Stem of Alkane Name	Suffix	Alkyl Group Name
1	—CH_3	meth-	—yl	methyl
2	—CH_2—CH_3	eth-	—yl	ethyl
3	—CH_2—CH_2—CH_3	prop-	—yl	propyl
4	—CH_2—CH_2—CH_2—CH_3	but-	—yl	butyl
5	—CH_2—CH_2—CH_2—CH_2—CH_3	pent-	—yl	pentyl
6	—CH_2—CH_2—CH_2—CH_2—CH_2—CH_3	hex-	—yl	hexyl

The ending *-yl*, as in meth*yl*, eth*yl*, prop*yl*, and but*yl*, appears in the names of all alkyl groups.

For branched-chain alkanes, the substituents are specifically called *alkyl groups*. An **alkyl group** *is the group of atoms that would be obtained by removing a hydrogen atom from an alkane.*

The two most commonly encountered alkyl groups are the two simplest: the one-carbon and two-carbon alkyl groups. Their formulas and names are

$$——CH_3 \qquad ——CH_2—CH_3$$

Methyl group   Ethyl group

The extra long bond in these formulas (on the left) denotes the point of attachment to the carbon chain. Note that alkyl groups do not lead a stable, independent existence; that is, they are not molecules. They are always found attached to another entity (usually a carbon chain).

Alkyl groups are named by taking the stem of the name of the alkane that contains the same number of carbon atoms and adding the ending *-yl*. Table 1.3 gives the names for small continuous-chain alkyl groups.

We are now ready for the IUPAC rules for naming branched-chain alkanes.

An additional guideline for identifying the longest continuous carbon chain: If two different carbon chains in a molecule have the same largest number of carbon atoms, select as the parent chain the one with the larger number of substituents (alkyl groups) attached to the chain.

Rule 1:   *Identify the longest continuous carbon chain (the parent chain), which may or may not be shown in a straight line, and name the chain.*

$$CH_3—CH_2—CH_2—CH—CH_3$$
$$|$$
$$CH_3$$

The parent chain name is *pentane*, because it has five carbon atoms.

$$CH_3—CH—CH_2—CH_2—CH_3$$
$$|$$
$$CH_2$$
$$|$$
$$CH_3$$

The parent chain name is *hexane*, because it has six carbon atoms.

Additional guidelines for numbering carbon atom chains:

1. If both ends of the chain have a substituent the same distance in, number from the end closest to the second-encountered substituent.
2. If there are substituents equidistant from each end of the chain and there is no third substituent to use as the "tie-breaker," begin numbering at the end nearest the substituent that has alphabetical priority—that is, the substituent whose name occurs first in the alphabet.

Rule 2:   *Number the carbon atoms in the parent chain from the end of the chain nearest a substituent (alkyl group).*

There always are two ways to number the chain (either from left to right or from right to left). This rule gives the first-encountered alkyl group the lowest possible number.

$$\overset{5}{C}H_3—\overset{4}{C}H_2—\overset{3}{C}H_2—\overset{2}{C}H—\overset{1}{C}H_3$$
$$|$$
$$CH_3$$

Right-to-left numbering system

$$CH_3—\overset{3}{C}H—\overset{4}{C}H_2—\overset{5}{C}H_2—\overset{6}{C}H_3$$
$$|$$
$$\overset{2}{C}H_2$$
$$|$$
$$\overset{1}{C}H_3$$

Left-to-right numbering system

Rule 3:   *If only one alkyl group is present, name and locate it (by number), and prefix the number and name to that of the parent carbon chain.*

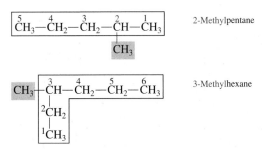

Note that the name is written as one word, with a hyphen between the number (location) and the name of the alkyl group.

Rule 4: *If two or more of the* same kind *of alkyl group are present in a molecule, indicate the number with a Greek numerical prefix (di-, tri-, tetra-, penta-, and so forth). In addition, a number specifying the location of each identical group must be included. These position numbers, separated by commas, precede the numerical prefix. Numbers are separated from words by hyphens.*

$\overset{1}{C}H_3-\overset{2}{C}H-\overset{3}{C}H_2-\overset{4}{C}H-\overset{5}{C}H_3$ 2,4-Dimethylpentane
with CH₃ below positions 2 and 4

$\overset{1}{C}H_3-\overset{2}{C}H_2-\overset{3}{C}-\overset{4}{C}H_2-\overset{5}{C}H_3$ 3,3-Dimethylpentane
with CH₃ above and below position 3

Note that the numerical prefix *di-* must always be accompanied by two numbers, *tri-* by three, and so on, even if the same number is written twice, as in 3,3-dimethylpentane.

> There must be as many numbers as there are alkyl groups in the IUPAC name of a branched-chain alkane.

Rule 5: *When two kinds of alkyl groups are present on the same carbon chain, number each group separately, and list the names of the alkyl groups in alphabetical order.*

3-Ethyl-2-methylpentane structure

Note that ethyl is named first in accordance with the alphabetical rule.

3-Ethyl-4,5-dipropyloctane structure

> Numerical prefixes that designate numbers of alkyl groups, such as *di-*, *tri-*, and *tetra-*, are not considered when determining alphabetical priority for alkyl groups.

Note that the prefix *di-* does not affect the alphabetical order; *ethyl* precedes *propyl*.

Rule 6: *Follow IUPAC punctuation rules, which include the following: (1) Separate numbers from each other by commas. (2) Separate numbers from letters by hyphens. (3) Do not add a hyphen or a space between the last-named substituent and the name of the parent alkane that follows.*

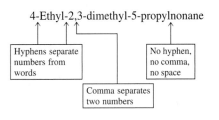

4-Ethyl-2,3-dimethyl-5-propylnonane

EXAMPLE 1.2

Determining IUPAC Names for Branched-Chain Alkanes

■ Give the IUPAC name for each of the following branched-chain alkanes.

a.
$$CH_3-CH-CH-CH_3$$
$$\quad\quad\; |\quad\;\; |$$
$$\quad\quad CH_2\;\; CH_3$$
$$\quad\quad\; |$$
$$\quad\quad CH_3$$

b.
$$CH_3-CH-CH_2-CH_2-CH-CH_2-CH-CH_3$$
$$\quad\quad\; |\quad\quad\quad\quad\quad\quad\; |\quad\quad\;\; |$$
$$\quad\quad CH_3\quad\quad\quad\quad\quad CH_2\quad\; CH_3$$
$$\quad\quad\quad\quad\quad\quad\quad\quad\quad\; |$$
$$\quad\quad\quad\quad\quad\quad\quad\quad\; CH_3$$

Solution

a. The longest carbon chain possesses five carbon atoms. Thus the parent-chain name is pentane.

$$CH_3-\boxed{CH-CH-CH_3}$$
$$\quad\quad\; |\quad\;\; |$$
$$\quad\quad CH_2\;\; CH_3$$
$$\quad\quad\; |$$
$$\quad\quad CH_3$$

This parent chain is numbered from right to left because an alkyl substituent is closer to the right end of the chain than to the left end.

$$\boxed{CH_3}-\overset{3}{CH}-\overset{2}{CH}-\overset{1}{CH_3}$$
$$\quad\quad\quad\; \overset{4}{|}\quad\;\; |$$
$$\quad\quad\quad CH_2\;\; \boxed{CH_3}$$
$$\quad\quad\quad\; \overset{5}{|}$$
$$\quad\quad\quad CH_3$$

There are two methyl group substituents (circled). One methyl group is located on carbon 2 and the other on carbon 3. The IUPAC name for the compound is 2,3-dimethylpentane.

b. There are eight carbon atoms in the longest carbon chain, so the parent name is octane. There are three alkyl groups present (circled).

$$\boxed{CH_3-CH-CH_2-CH_2-CH-CH_2-CH-CH_3}$$
$$\quad\quad\quad (CH_3)\quad\quad\quad\quad\; (CH_2)\quad\; (CH_3)$$
$$\quad\quad\quad\quad\quad\quad\quad\quad\quad\quad\; |$$
$$\quad\quad\quad\quad\quad\quad\quad\quad\quad\; CH_3$$

Selection of the numbering system to be used cannot be made based on the "first-encountered-alkyl-group rule" because an alkyl group is equidistant from each end of the chain. Thus the second-encountered alkyl group is used as the "tie-breaker." It is closer to the right end of the parent chain (carbon 4) than to the left end (carbon 5). Thus we use the right-to-left numbering system.

$$\boxed{\overset{8}{CH_3}-\overset{7}{CH}-\overset{6}{CH_2}-\overset{5}{CH_2}-\overset{4}{CH}-\overset{3}{CH_2}-\overset{2}{CH}-\overset{1}{CH_3}}$$
$$\quad\quad\quad CH_3\quad\quad\quad\quad\quad\; CH_2\quad\; CH_3$$
$$\quad\quad\quad\quad\quad\quad\quad\quad\quad\quad\;\; |$$
$$\quad\quad\quad\quad\quad\quad\quad\quad\quad\;\; CH_3$$

Two different kinds of alkyl groups are present: ethyl and methyl. Ethyl has alphabetical priority over methyl and precedes methyl in the IUPAC name. The IUPAC name is 4-ethyl-2,7-dimethyloctane.

> Always compare the total number of carbon atoms in the name with the number of carbon atoms in the structure to make sure they match. The name 4-ethyl-2,7-dimethyloctane indicates the presence of $2 + 2(1) + 8 = 12$ carbon atoms. The structure does have 12 carbon atoms.

Practice Exercise 1.2

Give the IUPAC name for each of the following alkanes.

a.
$$CH_3-CH-CH_2-CH_2-CH-CH_3$$
$$\quad\quad\; |\quad\quad\quad\quad\quad\;\; |$$
$$\quad\quad CH_2\quad\quad\quad\quad CH_2$$
$$\quad\quad\; |\quad\quad\quad\quad\quad\quad |$$
$$\quad\quad CH_3\quad\quad\quad\quad CH_3$$

b.
$$\quad\quad\quad\quad\quad\quad\quad\quad\quad\; CH_3$$
$$\quad\quad\quad\quad\quad\quad\quad\quad\quad\;\; |$$
$$CH_3-CH_2-CH-C-CH-CH_2-CH_2-CH_3$$
$$\quad\quad\quad\quad\quad |\quad\; |\quad\; |$$
$$\quad\quad\quad\quad CH_3\; CH_3\; CH_3$$

After you learn the rules for naming alkanes, it is relatively easy to reverse the procedure and translate the name of an alkane into a structural formula. Example 1.3 shows how this is done.

EXAMPLE 1.3

Generating the Structural Formula of an Alkane from Its IUPAC Name

A few smaller branched alkanes have common names—that is, non-IUPAC names—that still have widespread use. They make use of the prefixes *iso* and *neo,* as in isobutane, isopentane, and neohexane. These prefixes denote particular end-of-chain carbon atom arrangements.

$$CH_3-CH-(CH_2)_n-CH_3$$
$$\quad\quad |$$
$$\quad\quad CH_3$$

An isoalkane
(e.g., *n* = 1, Isopentane)

$$CH_3-C-(CH_2)_n-CH_3$$
$$\quad\quad |$$
$$\quad\quad CH_3$$

A neoalkane
(e.g., *n* = 1, Neohexane)

■ Draw the condensed structural formula for 3-ethyl-2,3-dimethylpentane.

Solution

Step 1:   The name of this compound ends in *pentane,* so the longest continuous chain has five carbon atoms. Draw this chain of five carbon atoms and number it.

$$\overset{1}{C}-\overset{2}{C}-\overset{3}{C}-\overset{4}{C}-\overset{5}{C}$$

Step 2:   Complete the carbon skeleton by attaching alkyl groups as they are specified in the name. An ethyl group goes on carbon 3, and methyl groups are attached to carbons 2 and 3.

$$\overset{1}{C}-\overset{2}{C}-\overset{3}{C}-\overset{4}{C}-\overset{5}{C}$$

Step 3:   Add hydrogen atoms to the carbon skeleton so that each carbon atom has four bonds.

$$\quad\quad\quad CH_3$$
$$\overset{1}{CH_3}-\overset{2}{CH}-\overset{3}{C}-\overset{4}{CH_2}-\overset{5}{CH_3}$$
$$\quad\quad CH_3 \;\; CH_2$$
$$\quad\quad\quad\quad CH_3$$

Practice Exercise 1.3

Draw the condensed structural formula for 4,5-diethyl-3,4,5-trimethyloctane.

1.9 Line-Angle Formulas for Alkanes

Three two-dimensional methods for denoting alkane structures have been used in previous sections of this chapter. They are expanded structural formulas, condensed structural formulas, and skeletal structural formulas. An even more concise method for denoting molecular structure of alkanes (and other hydrocarbons and their derivatives) exists. This method, *line-angle formulas,* is particularly useful for molecules in which several carbon atoms are present.

A **line-angle formula** *is a structural representation in which a line represents a carbon–carbon bond and a carbon atom is understood to be present at every point where two lines meet and at the ends of lines.* Ball-and-stick-models and line-angle formulas for the alkanes propane, butane, and pentane are as follows:

Ball-and-stick model

Line-angle formula

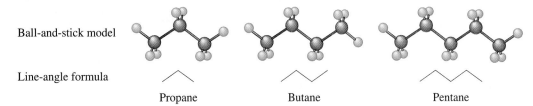

Propane   Butane   Pentane

Note that the zigzag (sawtooth) pattern used in line-angle formulas has a relationship to the three-dimensional shape of the molecules that are represented. The line-angle formula for an unbranched chain of eight carbon atoms would be

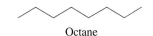

Octane

The structures of branched-chain alkanes can also be designated using line-angle formulas. The five constitutional alkane isomers in which six carbon atoms are present (C_6H_{14}) have the following line-angle formulas:

Six carbons in an
unbranched chain

Five carbons in a chain;
one carbon as a branch

Four carbons in a chain;
two carbons as branches

Hexane

2-Methylpentane

2,2-Dimethylbutane

3-Methylpentane

2,3-Dimethylbutane

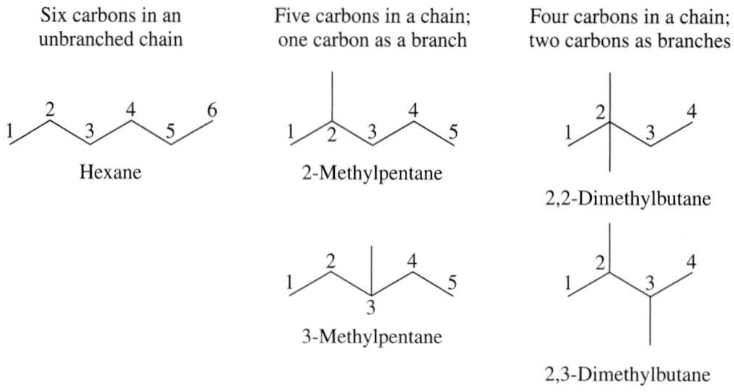

Example 1.4 gives further insights concerning the use and interpretation of line-angle formulas.

EXAMPLE 1.4

Generating Condensed Structural Formulas from Line-Angle Formulas for Alkanes

■ For each of the following alkanes, determine the number of hydrogen atoms present on each carbon atom and then write the condensed structural formula for the alkane.

a. 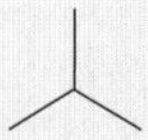   **b.**

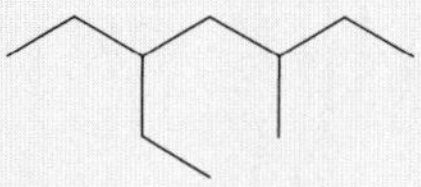

Solution

a. Each carbon atom in an alkane must be bonded to four atoms. Thus, carbon atoms bonded to only one carbon atom have three hydrogen atoms attached; those bonded to two other carbon atoms have two hydrogen atoms attached; those bonded to three other carbon atoms have only one atom attached; and those bonded to four other carbon atoms bear no hydrogen atoms. For this alkane, each carbon atom's hydrogen content is indicated by circled numbers as follows.

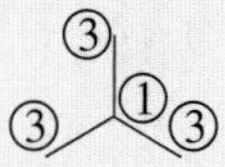

With this information on hydrogen content, the condensed structural formula is written as

$$CH_3-\overset{\displaystyle CH_3}{\underset{\displaystyle |}{CH}}-CH_3$$

b. Using the methods of part **a,** the hydrogen content of this alkane is

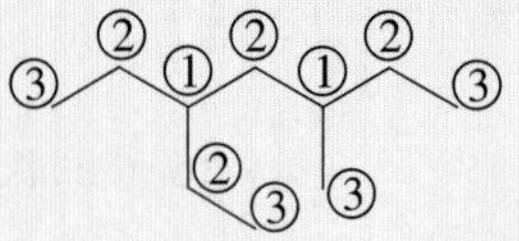

and the condensed structural formula becomes

$$CH_3-CH_2-\underset{\underset{\displaystyle CH_3}{\displaystyle |}}{\underset{\displaystyle |}{CH}}-CH_2-\underset{\underset{\displaystyle}{\displaystyle |}}{\underset{\displaystyle CH_2}{CH}}-CH_2-CH_3$$

$$CH_2$$
$$|$$
$$CH_3$$

Practice Exercise 1.4

For each of the following alkanes, determine the number of hydrogen atoms present on each carbon atom and then write the condensed structural formula for the alkane.

a. 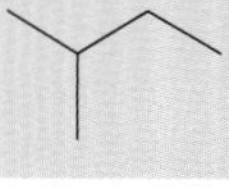   **b.**

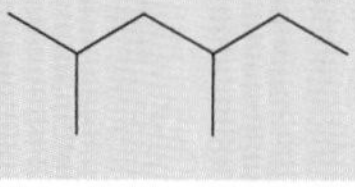

1.10 Classification of Carbon Atoms

Each of the carbon atoms within a hydrocarbon structure can be classified as a primary, secondary, tertiary, or quaternary carbon atom. A **primary carbon atom** *is a carbon atom in an organic molecule that is bonded to only one other carbon atom.* Each of the "end" carbon atoms in the three-carbon propane structure is a primary carbon atom, whereas the middle carbon atom of propane is a secondary carbon atom. A **secondary carbon atom** *is a carbon atom in an organic molecule that is bonded to two other carbon atoms.*

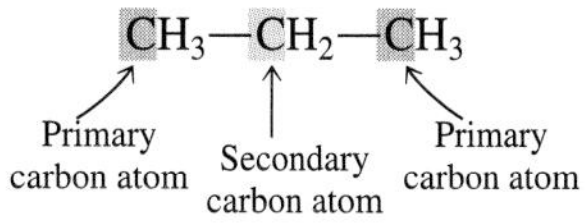

$$CH_3 - CH_2 - CH_3$$

Primary carbon atom   Secondary carbon atom   Primary carbon atom

A **tertiary carbon atom** *is a carbon atom in an organic molecule that is bonded to three other carbon atoms.* The molecule 2-methylpropane contains a tertiary carbon atom.

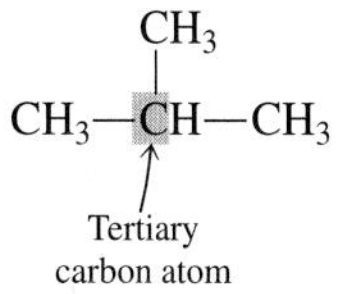

$$CH_3 - \underset{\underset{CH_3}{|}}{CH} - CH_3$$

Tertiary carbon atom

A **quaternary carbon atom** *is a carbon atom in an organic molecule that is bonded to four other carbon atoms.* The molecule 2,2-dimethylpropane contains a quaternary carbon atom.

$$CH_3 - \underset{\underset{CH_3}{|}}{\overset{\overset{CH_3}{|}}{C}} - CH_3$$

Quaternary carbon atom

> The notations 1°, 2°, 3°, and 4° are often used as designations for the terms *primary, secondary, tertiary,* and *quaternary.* Thus we can write
>
> 1° carbon atom
> 2° carbon atom
> 3° carbon atom
> 4° carbon atom

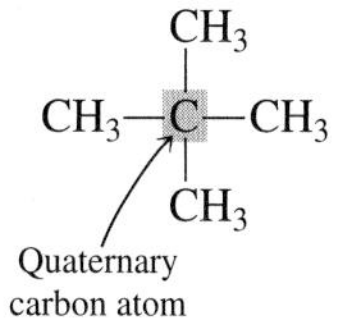

1.11 Branched-Chain Alkyl Groups

To this point in the chapter, all alkyl groups encountered in structures have been continuous-chain alkyl groups (Table 1.3), the simplest type of alkyl group. Just as there are continuous-chain and branched-chain alkanes, there are continuous-chain and branched-chain alkyl groups. Four branched-chain alkyl groups, shown in Figure 1.5, are so common that you should know their names and structures.

For the *sec*-butyl group, the point of attachment of the group to the main carbon chain involves a *secondary* carbon atom. For the *tert*-butyl group, the point of attachment of the group to the main carbon chain involves a *tertiary* carbon atom.

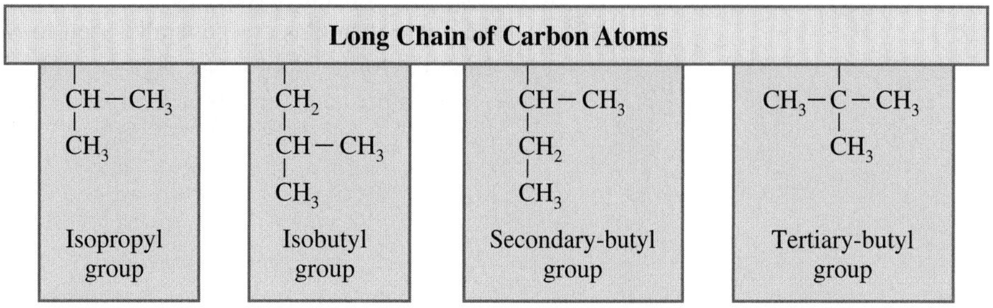

FIGURE 1.5 The four most common branched-chain alkyl groups and their IUPAC names.

Long Chain of Carbon Atoms			
$CH-CH_3$ $\|$ CH_3	CH_2 $\|$ $CH-CH_3$ $\|$ CH_3	$CH-CH_3$ $\|$ CH_2 $\|$ CH_3	CH_3-C-CH_3 $\|$ CH_3
Isopropyl group	Isobutyl group	Secondary-butyl group	Tertiary-butyl group

You need to be able to recognize various conformations of branched-chain alkyl groups. For example, these structures all represent an isopropyl group:

$$CH_3—CH—CH_3$$

$$CH_3—CH \quad\quad CH$$
$$\quad\quad CH_3 \quad\quad CH_3 \; CH_3$$

In each case, you have a chain of three carbon atoms with an attachment point (the long bond) involving the middle carbon atom of the chain.

Two examples of alkanes containing branched-chain alkyl groups follow.

$$\overset{1}{CH_3}—\overset{2}{CH_2}—\overset{3}{CH}—\overset{4}{CH_2}—\overset{5}{CH_2}—\overset{6}{CH}—\overset{7}{CH_2}—\overset{8}{CH_2}—\overset{9}{CH_3}$$

3-Isopropyl-6-propylnonane

$$\overset{1}{CH_3}—\overset{2}{CH_2}—\overset{3}{CH_2}—\overset{4}{CH}—\overset{5}{CH_2}—\overset{6}{CH_2}—\overset{7}{CH_2}—\overset{8}{CH_3}$$

4-*tert*-Butyloctane

In IUPAC naming, hyphenated prefixes, such as *sec-* and *tert-*, are not considered when alphabetizing. The prefixes *iso* and *neo* are not hyphenated prefixes and are included when alphabetizing. The following IUPAC name is thus correct:

5-*sec*-Butyl-4-isopropyl-3-methyldecane

■ Complex Branched Alkyl Groups

Complex branched alkyl groups, for which no "simple" name is available (Figure 1.5), are occasionally encountered. The IUPAC system provision for such groups involves naming them as though they were themselves compounds. Select the *longest alkyl chain* in the complex substituent as the base alkyl group. The base alkyl group is then numbered beginning with the carbon atom attached to the main carbon chain. The substituents on the base alkyl group are listed with appropriate numbers, and parentheses are used to set off the name of the complex alkyl group. Two examples of such nomenclature follow.

$$—\overset{1}{C}—\overset{2}{CH_2}—\overset{3}{CH_3}$$

(1,1-Dimethylpropyl) group

$$—\overset{1}{C}—\overset{2}{CH_2}—\overset{3}{CH}—\overset{4}{CH_3}$$

(1,1,3-Trimethylbutyl) group

1.12 Cycloalkanes

A **cycloalkane** *is a saturated hydrocarbon in which carbon atoms connected to one another in a cyclic (ring) arrangement are present.* The simplest cycloalkane is cyclopropane, which contains a cyclic arrangement of three carbon atoms. Figure 1.6 shows a three-dimensional model of cyclopropane's structure and those of the four-, five-, and six-carbon cycloalkanes.

It takes a minimum of three carbon atoms to form a cyclic arrangement of carbon atoms.

Cyclopropane's three carbon atoms lie in a flat ring. In all other cycloalkane molecules, some puckering of the ring occurs; that is, the ring systems are nonplanar, as shown in Figure 1.6.

The general formula for cycloalkanes is C_nH_{2n}. Thus a given cycloalkane contains two fewer hydrogen atoms than an alkane with the same number of hydrogen atoms (C_nH_{2n+2}). Butane (C_4H_{10}) and cyclobutane (C_4H_8) are not isomers; isomers must have the same molecular formula (Section 1.6).

Line-angle formulas are generally used to represent cycloalkane structures. The line angle formula for cyclopropane is a triangle, that for cyclobutane a square, that for cyclopentane a pentagon, and that for cyclohexane a hexagon.

Cyclopropane   Cyclobutane   Cyclopentane   Cyclohexane

(a) Cyclopropane, C_3H_6

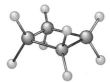
(b) Cyclobutane, C_4H_8

(c) Cyclopentane, C_5H_{10}

(d) Cyclohexane, C_6H_{12}

FIGURE 1.6 Three-dimensional representations of the structures of simple cycloalkanes.

EXAMPLE 1.5

Generating Condensed Structural Formulas from Line-Angle Formulas for Cycloalkanes

■ Generate the condensed structural formula for each of the following cycloalkanes.

a.

b.

Solution

a. First replace each angle and line terminus with a carbon atom, and then add hydrogens as necessary to give each carbon four bonds. The molecular formula of this compound is C_8H_{16}.

b. Similarly, we have

Practice Exercise 1.5

Generate the condensed structural formula for each of the following cycloalkanes.

a.

b.

The observed C—C—C bond angles in cyclopropane are 60°, and those in cyclobutane are 90°, values that are considerably smaller than the 109° angle associated with a tetrahedral arrangement of bonds about a carbon atom. Consequently, cyclopropane and cyclobutane are relatively unstable compounds. Five- and six-membered cycloalkane structures are much more stable, and these structural entities are encountered in many organic molecules.

1.13 IUPAC Nomenclature for Cycloalkanes

IUPAC naming procedures for cycloalkanes are similar to those for alkanes. The ring portion of a cycloalkane molecule serves as the name base, and the prefix *cyclo-* is used to indicate the presence of the ring. Alkyl substituents are named in the same

Cycloalkanes of ring sizes ranging from 3 to over 30 are found in nature, and in principle, there is no limit to ring size. Five-membered rings (cyclopentanes) and six-membered rings (cyclohexanes) are especially abundant in nature.

manner as in alkanes. Numbering conventions used in locating substituents on the ring include the following:

1. If there is just one ring substituent, it is not necessary to locate it by number.
2. When two ring substituents are present, the carbon atoms in the ring are numbered beginning with the substituent of higher alphabetical priority and proceeding in the direction (clockwise or counterclockwise) that gives the other substituent the lower number.
3. When three or more ring substituents are present, ring numbering begins at the substituent that leads to the lowest set of location numbers. When two or more equivalent numbering sets exist, alphabetical priority among substituents determines the set used.

Example 1.6 illustrates the use of the ring-numbering guidelines.

EXAMPLE 1.6

Determining IUPAC Names for Cycloalkanes

■ Assign IUPAC names to each of the following cycloalkanes.

a. 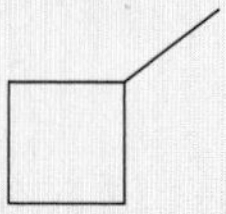   b. 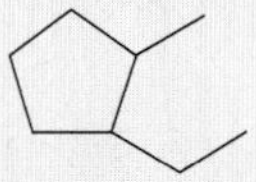   c.

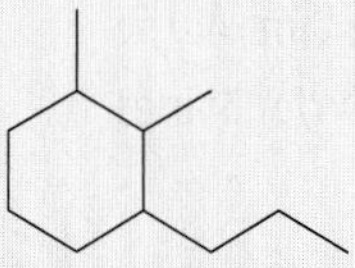

Solution

a. This molecule is a cyclobutane (four-carbon ring) with a methyl substituent. The IUPAC name is simply methylcyclobutane. No number is needed to locate the methyl group, because all four ring positions are equivalent.
b. This molecule is a cyclopentane with ethyl and methyl substituents. The numbers for the carbon atoms that bear the substituents are 1 and 2. On the basis of alphabetical priority, the number 1 is assigned to the carbon atom that bears the ethyl group. The IUPAC name for the compound is 1-ethyl-2-methylcyclopentane.
c. This molecule is a dimethylpropylcyclohexane. Two different 1,2,3 numbering systems exist for locating the substituents. On the basis of alphabetical priority, we use the numbering system that has carbon 1 bearing a methyl group; methyl has alphabetical priority over propyl. Thus the compound name is 1,2-dimethyl-3-propylcyclohexane.

Practice Exercise 1.6

Assign IUPAC names to each of the following cycloalkanes.

a. 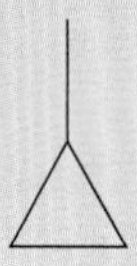   b. 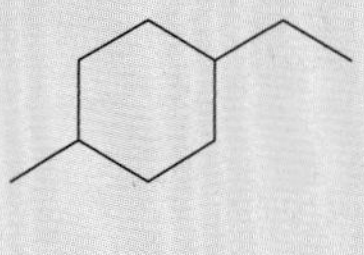   c.

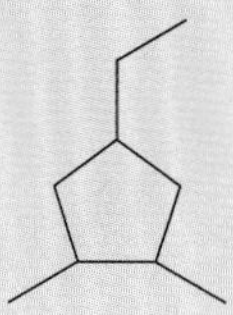

1.14 Isomerism in Cycloalkanes

Constitutional isomers are possible for cycloalkanes that contain four or more carbon atoms. For example, there are five cycloalkane constitutional isomers that have the formula C_5H_{10}: one based on a five-membered ring, one based on a four-membered ring, and three based on a three-membered ring. These isomers are

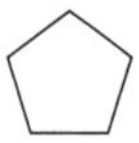

Cyclopentane 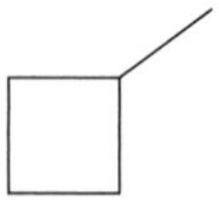   Methylcyclobutane 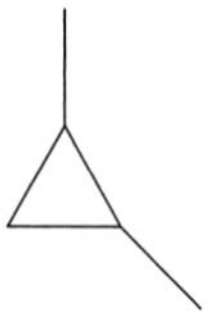   1,2-Dimethyl-cyclopropane   1,1-Dimethyl-cyclopropane   Ethylcyclopropane

A second type of isomerism, called *stereoisomerism,* is possible for some *substituted* cycloalkanes. Whereas constitutional isomerism results from differences in *connectivity,* stereoisomerism results from differences in *configuration.* **Stereoisomers** *are isomers that have the same molecular and structural formulas but different orientations of atoms in space.* Several forms of stereoisomerism exist. The form associated with cycloalkanes is called *cis–trans isomerism.* **Cis–trans isomers** *are isomers that have the same molecular and structural formulas but different orientations of atoms in space because of restricted rotation about bonds.*

In alkanes, there is free rotation about all carbon–carbon bonds (Section 1.7). In cycloalkanes, the ring structure restricts rotation for the carbon atoms in the ring. The consequence of this lack of rotation in a cycloalkane is the creation of "top" and "bottom" positions for the two attachments on each of the ring carbon atoms. This "top–bottom" situation leads to *cis–trans* isomerism in cycloalkanes in which each of two ring carbon atoms bears two different attachments.

Consider the following two structures for the molecule 1,2-dimethylcyclopentane.

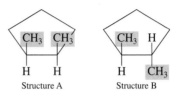

In structure A, both methyl groups are above the plane of the ring (the "top" side). In structure B, one methyl group is above the plane of the ring (the "top" side) and the other below it (the "bottom" side). Structure A cannot be converted into structure B without breaking bonds. Hence structures A and B are isomers; there are two 1,2-dimethylcyclopentanes. The first isomer is called *cis*-1,2-dimethylcyclopentane and the second *trans*-1,2-dimethyl-cyclopentane.

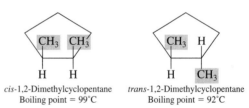

***Cis-** is a prefix that means "on the same side."* In *cis*-1,2-dimethylcyclopentane, the two methyl groups are on the same side of the ring. ***Trans-** is a prefix that means "across from."* In *trans*-1,2-dimethylcyclopentane, the two methyl groups are on opposite sides of the ring.

Cis–trans isomerism can occur in rings of all sizes. The presence of a substituent on each of two carbon atoms in the ring is the requirement for its occurrence. In biochemistry, we will find that the human body often selectively distinguishes between the *cis* and *trans* isomers of a compound. One isomer will be active in the body and the other inactive.

Cis–trans isomers have the same molecular formula and the same structural formula. The only difference between them is the orientation of atoms in space. *Constitutional isomers* have the same molecular formula but different structural formulas.

The Latin *cis* means "on the same side," and the Latin *trans* means "across from." Consider the use of the prefix *trans-* in the phrase "transatlantic voyage."

Cis–trans isomerism will also be encountered in the next chapter (Section 2.4), where the required restricted rotation barrier will be a carbon–carbon double bond rather than a ring of carbon atoms. Another type of stereoisomerism called enantiomerism (left- and right-handed forms of a molecule) will be considered in the discussion of carbohydrates in Chapter 7.

EXAMPLE 1.7

Identifying and Naming Cycloalkane *Cis–Trans* Isomers

■ Determine whether *cis–trans* isomerism is possible for each of the following cycloalkanes. If so, then draw structural formulas for the *cis* and *trans* isomers.

a. Methylcyclohexane   **b.** 1,1-Dimethylcyclohexane
c. 1,3-Dimethylcyclobutane   **d.** 1-Ethyl-2-methylcyclobutane

Solution

a. *Cis–trans* isomerism is not possible because we do not have two substituents on the ring.
b. *Cis–trans* isomerism is not possible. We have two substituents on the ring, but they are on the same carbon atom. Each of two different carbons must bear substituents.

(continued)

FIGURE 1.7 A rock formation such as this is necessary for the accumulation of petroleum and natural gas.

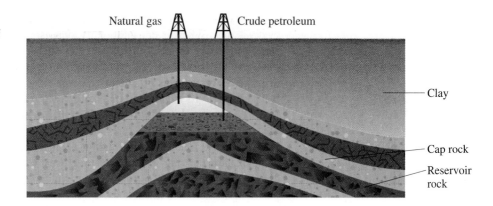

In cycloalkanes, *cis–trans* isomerism can also be denoted by using wedges and dotted lines. A heavy wedge-shaped bond to a ring structure indicates a bond *above* the plane of the ring, and a broken dotted line indicates a bond *below* the plane of the ring.

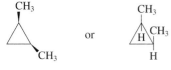

cis-1,2-Dimethylcyclopropane

trans-1,2-Dimethylcyclopropane

c. *Cis–trans* isomerism does exist.

d. *Cis–trans* isomerism does exist.

cis-1-Ethyl-2-methylcyclobutane   *trans*-1-Ethyl-2-methylcyclobutane

Practice Exercise 1.7

Determine whether *cis–trans* isomerism is possible for each of the following cycloalkanes. If so, then draw structural formulas for the *cis* and *trans* isomers.

a. 1-Ethyl-1-methylcyclopentane   **b.** Ethylcyclohexane
c. 1,4-Dimethylcyclohexane   **d.** 1,1-Dimethylcyclooctane

The word *petroleum* comes from the Latin *petra,* which means "rock," and *oleum,* which means "oil."

FIGURE 1.8 An oil rig pumping oil from an underground rock formation.

1.15 Sources of Alkanes and Cycloalkanes

Alkanes and cycloalkanes are not "laboratory curiosities" but rather two families of extremely important naturally occurring compounds. Natural gas and petroleum (crude oil) constitute their largest and most important natural source. Deposits of these resources are usually associated with underground dome-shaped rock formations (Figure 1.7). When a hole is drilled into such a rock formation, it is possible to recover some of the trapped hydrocarbons—that is, the natural gas and/or petroleum (Figure 1.8). Note that petroleum and natural gas do not occur in the earth in the form of "liquid pools" but rather are dispersed throughout a porous rock formation.

Unprocessed natural gas contains 50%–90% methane, 1%–10% ethane, and up to 8% higher-molecular-mass alkanes (predominantly propane and butanes). The higher alkanes found in crude natural gas are removed prior to release of the gas into the pipeline distribution systems. Because the removed alkanes can be liquefied by the use of moderate pressure, they are stored as liquids under pressure in steel cylinders and are marketed as bottled gas.

FIGURE 1.9 The complex hydrocarbon mixture present in petroleum is separated into simpler mixtures by means of a fractionating column.

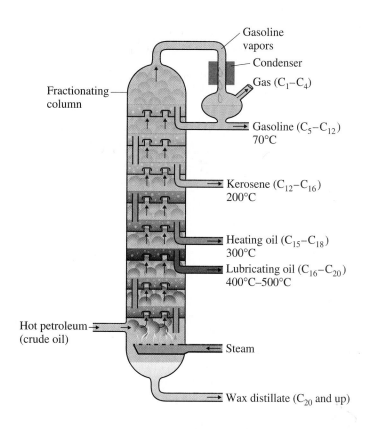

Gasoline vapors

Condenser

Gas (C_1–C_4)

Fractionating column

Gasoline (C_5–C_{12}) 70°C

Kerosene (C_{12}–C_{16}) 200°C

Heating oil (C_{15}–C_{18}) 300°C

Lubricating oil (C_{16}–C_{20}) 400°C–500°C

Hot petroleum (crude oil)

Steam

Wax distillate (C_{20} and up)

FIGURE 1.10 The insolubility of alkanes in water is used to advantage by many plants, which produce unbranched long-chain alkanes that serve as protective coatings on leaves and fruits. Such protective coatings minimize water loss for plants. Apples can be "polished" because of the long-chain alkane coating on their skin, which involves the unbranched alkanes $C_{27}H_{56}$ and $C_{29}H_{60}$. The leaf wax of cabbage and broccoli is mainly unbranched $C_{29}H_{60}$.

Crude petroleum is a complex mixture of hydrocarbons (both cyclic and acyclic) that can be separated into useful fractions through refining. During refining, the physical separation of the crude into component fractions is accomplished by fractional distillation, a process that takes advantage of boiling-point differences between the components of the crude petroleum. Each fraction contains hydrocarbons within a specific boiling-point range. The fractions obtained from a typical fractionation process are shown in Figure 1.9.

1.16 Physical Properties of Alkanes and Cycloalkanes

In this section, we consider a number of generalizations about the physical properties of alkanes and cycloalkanes.

1. *Alkanes and cycloalkanes are insoluble in water.* Water molecules are polar, and alkane and cycloalkane molecules are nonpolar. Molecules of unlike polarity have limited solubility in one another. The water insolubility of alkanes makes them good preservatives for metals. They prevent water from reaching the metal surface and causing corrosion. They also have biological functions as protective coatings (see Figure 1.10).
2. *Alkanes and cycloalkanes have densities lower than that of water.* Alkane and cycloalkane densities fall in the range 0.6 g/mL to 0.8 g/mL, compared with water's density of 1.0 g/mL. When alkanes and cycloalkanes are mixed with water, two layers form (because of insolubility), with the hydrocarbon layer on top (because of its lower density). This density difference between alkanes/cycloalkanes and water explains why oil spills in aqueous environments spread so quickly. The *floating* oil follows the movement of the water.
3. *The boiling points of continuous-chain alkanes and cycloalkanes increase with an increase in carbon chain length or ring size.* For continuous-chain alkanes, the boiling point increases roughly 30°C for every carbon atom added to the chain.

FIGURE 1.11 Trends in normal boiling points for continuous-chain alkanes, 2-methyl branched alkanes, and unsubstituted cycloalkanes as a function of the number of carbon atoms present. For a series of alkanes or cycloalkanes, melting point increases as carbon chain length increases.

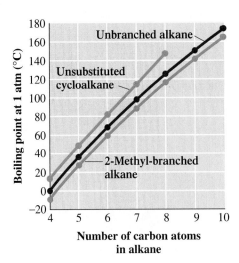

Unbranched Alkanes			
C_1	C_3	C_5	C_7
C_2	C_4	C_6	C_8

Unsubstituted Cycloalkanes			
✕	C_3	C_5	C_7
✕	C_4	C_6	C_8

☐ Gas   ☐ Liquid

FIGURE 1.12 A physical-state summary for unbranched alkanes and unsubstituted cycloalkanes at room temperature and pressure.

The term *paraffins* is an older name for the alkane family of compounds. This name comes from the Latin *parum affinis*, which means "little activity." That is a good summary of the general chemical properties of alkanes.

This trend, shown in Figure 1.11, is the result of increasing London force strength. London forces become stronger as molecular surface area increases. Short, continuous-chain alkanes (1 to 4 carbon atoms) are gases at room temperature. Continuous-chain alkanes containing 5 to 17 carbon atoms are liquids, and alkanes that have carbon chains longer than this are solids at room temperature.

Branching on a carbon chain lowers the boiling point of an alkane. A comparison of the boiling points of unbranched alkanes and their 2-methyl-branched isomers is given in Figure 1.11. Branched alkanes are more compact, with smaller surface areas than their straight-chain isomers.

Cycloalkanes have higher boiling points than their noncyclic counterparts with the same number of carbon atoms (Figure 1.11). These differences are due in large part to cyclic systems having more rigid and more symmetrical structures.

Cyclopropane and cyclobutane are gases at room temperature, and cyclopentane through cyclooctane are liquids at room temperature. Figure 1.12 is a physical-state summary for unbranched alkanes or unsubstituted cycloalkanes with 8 or fewer carbon atoms.

The alkanes and cycloalkanes whose boiling points are compared in Figure 1.11 constitute *homologous series* of organic compounds. In a homologous series, the members differ structurally only in the number of —CH₂— groups present. Members exhibit gradually changing physical properties and usually have very similar chemical properties.

The existence of homologous series of organic compounds gives organization to organic chemistry in the same way that the periodic table gives organization to the chemistry of the elements. Knowing something about a few members of a homologous series usually enables us to deduce the properties of other members in the series.

1.17 Chemical Properties of Alkanes and Cycloalkanes

Alkanes are the least reactive type of organic compound. They can be heated for long periods of time in strong acids and bases with no appreciable reaction. Strong oxidizing agents and reducing agents have little effect on alkanes.

Alkanes are not absolutely unreactive. Two important reactions that they undergo are combustion, which is reaction with oxygen, and halogenation, which is reaction with halogens.

■ Combustion

A **combustion reaction** *is a chemical reaction between a substance and oxygen (usually from air) that proceeds with the evolution of heat and light (usually as a flame).* Alkanes

CHEMICAL CONNECTIONS   The Physiological Effects of Alkanes

The simplest alkanes (methane, ethane, propane, and butane) are gases at room temperature and pressure. Methane and ethane are difficult to liquefy, so they are usually handled as compressed gases. Propane and butane are easily liquefied at room temperature under a moderate pressure. They are stored in low-pressure cylinders in a liquefied form. These four gases are colorless, odorless, and nontoxic, and they have limited physiological effects. The danger in inhaling them lies in potential suffocation due to lack of oxygen. The major immediate danger associated with a natural gas leak is the potential formation of an explosive air–alkane mixture rather than the formation of a toxic air–alkane mixture.

The C_5 to C_8 alkanes, of which there are many isomeric forms, are free-flowing, nonpolar, volatile liquids. They are the primary constituents of gasoline. These compounds are not particularly toxic, but gasoline should not be swallowed because (1) some of the additives present are harmful and (2) liquid alkanes can damage lung tissue because of physical rather than chemical effects. Physical effects include the dissolving of lipid molecules of cell membranes (see Chapter 8), causing pneumonia-like symptoms. Liquid alkanes can also affect the skin for related reasons. These alkanes dissolve natural body oils, causing the skin to dry out. (This "drying out" effect is easily noticed when paint thinner, a mixture of hydrocarbons, is used to remove paint from the hands.)

In direct contrast to liquid alkanes, solid alkanes are used to protect the skin. Pharmaceutical-grade *petrolatum* and *mineral oil* (also called liquid petrolatum), obtained as products from petroleum distillation, have such a function. Petrolatum is a mixture of C_{25} to C_{30} alkanes, and mineral oil involves alkanes in the C_{18} to C_{24} range.

Petrolatum (Vaseline is a well-known brand name) is a semisolid hydrocarbon mixture that is useful both as a skin softener and as a skin protector. Many moisturizing hand

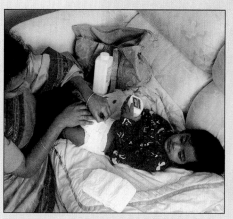

A semi-solid alkane mixture, such as Vaseline, is useful as a skin protector because neither water nor water solutions will penetrate a coating of it. Here, Vaseline is applied to a baby's bottom as a protection against diaper rash.

lotions and some medicated salves contain petrolatum. Neither water nor water solutions (for example, urine) will penetrate protective petrolatum coatings. This explains why petrolatum products protect a baby's bottom from diaper rash.

Mineral oil is often used to replace natural skin oils washed away by frequent bathing and swimming. Too much mineral oil, however, can be detrimental; it will dissolve nonpolar skin materials. Mineral oil has some use as a laxative; it effectively softens and lubricates hard stools. When taken by mouth, it passes through the gastrointestinal tract unchanged and is excreted chemically intact. Loss of fat-soluble vitamins (A, D, E, and K) can occur if mineral oil is consumed while these vitamins are in the digestive tract. Using a mineral oil enema instead avoids this drawback.

readily undergo combustion when ignited. When sufficient oxygen is present to support total combustion, carbon dioxide and water are the products.

$$CH_4 + 2O_2 \longrightarrow CO_2 + 2H_2O + energy$$

$$2C_6H_{14} + 19O_2 \longrightarrow 12CO_2 + 14H_2O + energy$$

The exothermic nature of alkane combustion reactions explains the extensive use of alkanes as fuels. Natural gas, used in home heating, is predominantly methane. Propane is used in home heating in rural areas and in gas barbecue units (see Figure 1.13). Butane fuels portable camping stoves. Gasoline is a complex mixture of many alkanes and other types of hydrocarbons.

Incomplete combustion can occur if insufficient oxygen is present during the combustion process. When this is the case, some carbon monoxide (CO) and/or elemental carbon are reaction products along with carbon dioxide (CO_2). In a chemical laboratory setting, incomplete combustion is often observed. The appearance of deposits of carbon black (soot) on the bottom of glassware is physical evidence that incomplete combustion

FIGURE 1.13 Propane fuel tank on a home barbecue unit.

is occurring. The problem is that the air-to-fuel ratio for the Bunsen burner is not correct. It is too rich; it contains too much fuel and not enough oxygen (air).

Halogenation

The halogens are the elements in Group VIIA of the periodic table: fluorine (F_2), chlorine (Cl_2), bromine (Br_2), and iodine (I_2). A **halogenation reaction** *is a chemical reaction between a substance and a halogen in which one or more halogen atoms are incorporated into molecules of the substance.*

Halogenation of an alkane produces a hydrocarbon derivative in which one or more halogen atoms have been substituted for hydrogen atoms. An example of an alkane halogenation reaction is

$$H-\underset{\underset{H}{|}}{\overset{\overset{H}{|}}{C}}-\underset{\underset{H}{|}}{\overset{\overset{H}{|}}{C}}-H + Br_2 \xrightarrow[\text{light}]{\text{Heat or}} H-\underset{\underset{H}{|}}{\overset{\overset{H}{|}}{C}}-\underset{\underset{H}{|}}{\overset{\overset{H}{|}}{C}}-Br + HBr$$

Alkane halogenation is an example of a substitution reaction, a type of reaction that occurs often in organic chemistry. A **substitution reaction** *is a chemical reaction in which part of a small reacting molecule replaces an atom or a group of atoms on a hydrocarbon or hydrocarbon derivative.* A diagrammatic representation of a substitution reaction is shown in Figure 1.14.

A *general* equation for the substitution of a single halogen atom for one of the hydrogen atoms of an alkane is

$$\underset{\text{Alkane}}{R-H} + \underset{\text{Halogen}}{X_2} \xrightarrow[\text{light}]{\text{Heat or}} \underset{\substack{\text{Halogenated}\\\text{alkane}}}{R-X} + \underset{\substack{\text{Hydrogen}\\\text{halide}}}{H-X}$$

Occasionally, it is useful to represent alkyl groups in a nonspecific way. The symbol R is used for this purpose. Just as *city* is a generic term for Chicago, New York, or San Francisco, the symbol R is a generic designation for any alkyl group. The symbol R comes from the German word *radikal,* which means, in a chemical context, "molecular fragment."

Note the following features of this general equation:

1. The notation R—H is a general formula for an alkane. R— in this case represents an alkyl group. Addition of a hydrogen atom to an alkyl group produces the parent hydrocarbon of the alkyl group.
2. The notation R—X on the product side is the general formula for a halogenated alkane. X is the general symbol for a halogen atom.
3. Reaction conditions are noted by placing these conditions on the equation arrow that separates reactants from products. Halogenation of an alkane requires the presence of heat or light.

(The symbol R is used frequently in organic chemistry and will be encountered in numerous generalized formulas in subsequent chapters; it always represents a generalized organic group in a structural formula. An R group can be an alkyl group—methyl, ethyl, propyl, etc.—or any number of other organic groups. Consider the symbol R to represent the Rest of an organic molecule, which is not specifically specified because it is not the focal point of the discussion occurring at that time.)

FIGURE 1.14 In an alkane substitution reaction, an incoming atom or group of atoms (represented by the orange sphere) replaces a hydrogen atom in the alkane molecule.

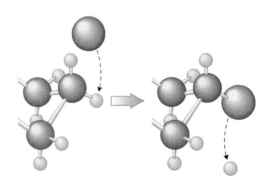

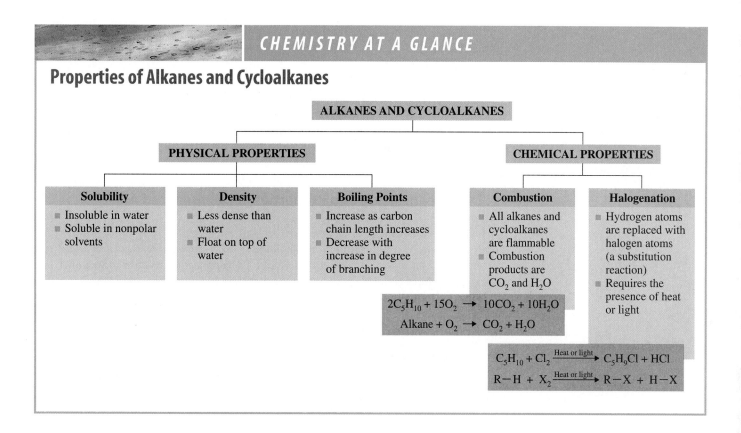

CHEMISTRY AT A GLANCE

Properties of Alkanes and Cycloalkanes

ALKANES AND CYCLOALKANES

PHYSICAL PROPERTIES

Solubility
- Insoluble in water
- Soluble in nonpolar solvents

Density
- Less dense than water
- Float on top of water

Boiling Points
- Increase as carbon chain length increases
- Decrease with increase in degree of branching

CHEMICAL PROPERTIES

Combustion
- All alkanes and cycloalkanes are flammable
- Combustion products are CO_2 and H_2O

$$2C_5H_{10} + 15O_2 \longrightarrow 10CO_2 + 10H_2O$$
$$\text{Alkane} + O_2 \longrightarrow CO_2 + H_2O$$

Halogenation
- Hydrogen atoms are replaced with halogen atoms (a substitution reaction)
- Requires the presence of heat or light

$$C_5H_{10} + Cl_2 \xrightarrow{\text{Heat or light}} C_5H_9Cl + HCl$$
$$R-H + X_2 \xrightarrow{\text{Heat or light}} R-X + H-X$$

In halogenation of an alkane, the alkane is said to undergo *fluorination, chlorination, bromination,* or *iodination,* depending on the identity of the halogen reactant. Chlorination and bromination are the two widely used alkane halogenation reactions. Fluorination reactions generally proceed too quickly to be useful, and iodination reactions go too slowly.

Halogenation usually results in the formation of a mixture of products rather than a single product. More than one product results because more than one hydrogen atom on an alkane can be replaced with halogen atoms. To illustrate this concept, let us consider the chlorination of methane, the simplest alkane.

Methane and chlorine, when heated to a high temperature or in the presence of light, react as follows:

$$CH_4 + Cl_2 \xrightarrow[\text{light}]{\text{Heat or}} CH_3Cl + HCl$$

The reaction does not stop at this stage, however, because the chlorinated methane product can react with additional chlorine to produce polychlorinated products.

$$CH_3Cl + Cl_2 \xrightarrow[\text{light}]{\text{Heat or}} CH_2Cl_2 + HCl$$

$$CH_2Cl_2 + Cl_2 \xrightarrow[\text{light}]{\text{Heat or}} CHCl_3 + HCl$$

$$CHCl_3 + Cl_2 \xrightarrow[\text{light}]{\text{Heat or}} CCl_4 + HCl$$

By controlling the reaction conditions and the ratio of chlorine to methane, it is possible to *favor* formation of one or another of the possible chlorinated methane products.

The chemical properties of cycloalkanes are similar to those of alkanes. Cycloalkanes readily undergo combustion as well as chlorination and bromination. With unsubstituted cycloalkanes, monohalogenation produces a single product because all hydrogen atoms present in the cycloalkane are equivalent to one another.

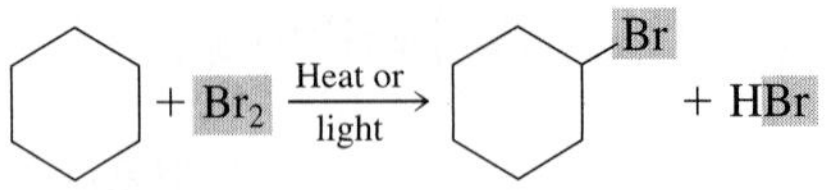

The Chemistry at a Glance feature on page 25 summarizes the physical properties and chemical reactions of alkanes and cycloalkanes.

1.18 Nomenclature and Properties of Halogenated Alkanes

A **halogenated alkane** *is an alkane derivative in which one or more halogen atoms are present.* Similarly, a **halogenated cycloalkane** *is a cycloalkane derivative in which one or more halogen atoms are present.* Produced by halogenation reactions (Section 1.17), these two types of compounds represent the first class of hydrocarbon derivatives (Section 1.3) that we formally consider in this text.

Nomenclature of Halogenated Alkanes

The IUPAC rules for naming halogenated alkanes are similar to those for naming branched alkanes, with the following modifications:

1. Halogen atoms, treated as substituents on a carbon chain, are called *fluoro-*, *chloro-*, *bromo-*, and *iodo-*.
2. When a carbon chain bears both a halogen and an alkyl substituent, the two substituents are considered of equal rank in determining the numbering system for the chain. The chain is numbered from the end closer to a substituent, whether it be a *halo-* or an alkyl group.
3. Alphabetical priority determines the order in which all substituents present are listed.

The following names are derived using these rule adjustments.

$CH_3—CH(Cl)—CH(CH_3)—CH_3$
2-Chloro-3-methylbutane

$CH_3—CH(Br)—CH_2—CH_2(Cl)$
3-Bromo-1-chlorobutane

1-Ethyl-2-fluorocyclohexane

> The contrast between IUPAC and common names for halogenated hydrocarbons is as follows:
>
> IUPAC (one word)
> haloalkane
> chloromethane
> Common (two words)
> alkyl halide
> methyl chloride

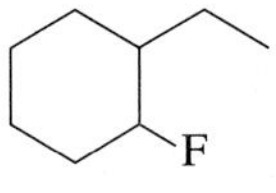

Simple halogenated alkanes can also be named as *alkyl halides.* These non-IUPAC names have two parts. The first part is the name of the hydrocarbon portion of the molecule (the alkyl group). The second part (as a separate word) identifies the halogen portion, which is named as if it were an ion (chloride, bromide, and so on), even though no ions are present (all bonds are covalent bonds). The following examples contrast the IUPAC names and the common names (in parentheses) of selected halogenated alkanes.

$CH_3—CH_2—Cl$
Chloroethane
(ethyl chloride)

$CH_3—CH_2—CH_2—Br$
1-Bromopropane
(propyl bromide)

$CH_3—CH(Cl)—CH_3$
2-Chloropropane
(isopropyl chloride)

> An alternative designation for a halogenated alkane is *alkyl halide.*

Several polyhalogenated methanes have acquired common names that are not clearly related to their structures. Five important examples of this additional nomenclature are CH_2Cl_2 (methylene chloride), $CHCl_3$ (chloroform), CCl_4 (carbon tetrachloride), CCl_3F (Freon-11), and CCl_2F_2 (Freon-12). The compounds Freon-11 and Freon-12 are examples of chlorofluorocarbons (CFCs). CFCs are synthetic compounds that have been heavily used as refrigerants and as air conditioning chemicals. We now know that CFCs are factors in the destruction of stratospheric (high-altitude) ozone, as is discussed in the Chemical Connections feature, "Chlorofluorocarbons and the Ozone Layer" found on the next page.

CHEMICAL CONNECTIONS
Chlorofluorocarbons and the Ozone Layer

Chlorofluorocarbons (CFCs) are compounds composed of the elements chlorine, fluorine, and carbon. CFCs are synthetic compounds that have been developed primarily for use as refrigerants. The two most widely used of the CFCs are trichlorofluoromethane and dichlorodifluoromethane. Both of these compounds are marketed under the trade name Freon.

Trichlorofluoromethane
(Freon-11)

Dichlorodifluoromethane
(Freon-12)

Freon-11 and Freon-12 possess ideal properties for use as a refrigerant gas. Both are inert, nontoxic, and easily compressible. Prior to their development, ammonia was used in refrigeration. Ammonia is toxic, and leaking ammonia-based refrigeration units have been fatal.

We now know that CFCs contribute to a serious environmental problem: destruction of the stratospheric (high-altitude) ozone that we commonly call the ozone layer. Once released into the atmosphere, CFCs persist for long periods without reaction. Consequently, they slowly drift upward in the atmosphere, finally reaching the stratosphere.

It is in the stratosphere, the location of the "ozone layer," that environmental problems occur. At these high altitudes, the CFCs are exposed to ultraviolet light (from the sun), which activates them. The ultraviolet light breaks carbon–chlorine bonds within the CFCs, releasing chlorine atoms.

$$CCl_2F_2 + \text{ultraviolet light} \longrightarrow CClF_2 + Cl$$

The Cl atoms so produced (called atomic chlorine) are extremely reactive species. One of the molecules with which they react is ozone (O_3).

$$Cl + O_3 \longrightarrow ClO + O_2$$

A reaction such as this upsets the O_3–O_2 equilibrium in the stratosphere.

Recent international treaties (this is a worldwide problem) limit, and in some cases ban, future production and use of CFCs. Replacements for the phased-out CFCs are HFCs (hydrogen-fluorocarbons) such as

1,1,1,2-Tetrafluoroethane

Haloalkanes with some carbon–hydrogen bonds are more reactive than CFCs and are generally destroyed at lower altitudes before they reach the stratosphere. Unfortunately, however, their refrigeration properties are not as good as those of the CFCs.

■ Physical Properties of Halogenated Alkanes

Halogenated alkane boiling points are generally higher than those of the corresponding alkane. An important factor contributing to this effect is the polarity of carbon–halogen bonds, which results in increased dipole–dipole interactions.

Some halogenated alkanes have densities that are greater than that of water, a situation not common for organic compounds. Chloroalkanes containing two or more chlorine atoms, bromoalkanes, and iodoalkanes are all more dense than water.

CONCEPTS TO REMEMBER

Carbon atom bonding characteristics. Carbon atoms in organic compounds must have four bonds (Section 1.2).

Types of hydrocarbons. Hydrocarbons, binary compounds of carbon and hydrogen, are of two types: saturated and unsaturated. In saturated hydrocarbons, all carbon–carbon bonds are single bonds. Unsaturated hydrocarbons have one or more carbon–carbon multiple bonds — double bonds, triple bonds, or both (Section 1.3).

Alkanes. Alkanes are saturated hydrocarbons in which the carbon atom arrangement is that of an unbranched or branched chain. The formulas of all alkanes can be represented by the general formula C_nH_{2n+2}, where n is the number of carbon atoms present (Section 1.4).

Structural formulas. Structural formulas are two-dimensional representations of the arrangement of the atoms in molecules. These formulas give complete information about the arrangement of the atoms in a molecule but not the spatial orientation of the atoms. Two types of structural formulas are commonly encountered: expanded and condensed (Section 1.5).

Isomers. Isomers are compounds that have the same molecular formula, (that is, the same numbers and kinds of atoms) but that differ in the way the atoms are arranged (Section 1.5).

Constitutional isomers. Constitutional isomers are isomers that differ in the connectivity of atoms, that is, in the order in which atoms are attached to each other within molecules (Section 1.6).

Conformations. Conformations are differing orientations of the same molecule made possible by free rotation about single bonds in the molecule (Section 1.7).

Alkane nomenclature. The IUPAC name for an alkane is based on the longest continuous chain of carbon atoms in the molecule. A group of carbon atoms attached to the chain is an alkyl group. Both the position and the identity of the alkyl group are prefixed to the name of the longest carbon chain (Section 1.8).

Line-angle formulas. A line-angle formula is a structural representation in which a line represents a carbon–carbon bond and a carbon atom is understood to be present at every point where two lines meet and at the ends of the line. Line-angle formulas are the most concise method for representing the structure of a hydrocarbon or hydrocarbon derivative (Section 1.9).

Cycloalkanes. Cycloalkanes are saturated hydrocarbons in which at least one cyclic arrangement of carbon atoms is present. The formulas of all cycloalkanes can be represented by the general formula C_nH_{2n}, where n is the number of carbon atoms present (Section 1.12).

Cycloalkane nomenclature. The IUPAC name for a cycloalkane is obtained by placing the prefix *cyclo-* before the alkane name that corresponds to the number of carbon atoms in the ring. Alkyl groups attached to the ring are located by using a ring-numbering system (Section 1.13).

Cis–trans isomerism. For certain disubstituted cycloalkanes, *cis–trans* isomers exist. *Cis–trans* isomers are compounds that have the same molecular and structural formulas but different arrangements of atoms in space because of restricted rotation about bonds (Section 1.14).

Natural sources of saturated hydrocarbons. Natural gas and petroleum are the largest and most important natural sources of both alkanes and cycloalkanes (Section 1.15).

Physical properties of saturated hydrocarbons. Saturated hydrocarbons are not soluble in water and have lower densities than water. Melting and boiling points increase with increasing carbon chain length or ring size (Section 1.16).

Chemical properties of saturated hydrocarbons. Two important reactions that saturated hydrocarbons undergo are combustion and halogenation. In combustion, saturated hydrocarbons burn in air to produce CO_2 and H_2O. Halogenation is a substitution reaction in which one or more hydrogen atoms of the hydrocarbon are replaced by halogen atoms (Section 1.17).

Halogenated alkanes. Halogenated alkanes are hydrocarbon derivatives in which one or more halogen atoms have replaced hydrogen atoms of the alkane (Section 1.18).

Halogenated alkane nomenclature. Halogenated alkanes are named by using the rules that apply to branched-chain alkanes, with halogen substituents being treated the same as alkyl groups (Section 1.18).

KEY REACTIONS AND EQUATIONS

1. Combustion (rapid reaction with O_2) of alkanes (Section 1.17)

$$\text{Alkane} + O_2 \longrightarrow CO_2 + H_2O$$

2. Halogenation of alkanes (Section 1.17)

$$R-H + X_2 \xrightarrow[\text{light}]{\text{Heat or}} R-X + H-X$$

EXERCISES AND PROBLEMS

The members of each pair of problems in this section test similar material.

■ Organic and Inorganic Compounds (Section 1.1)

1.1 Indicate whether each of the following statements is true or false.
   a. The number of organic compounds exceeds the number of inorganic compounds by a factor of about 2.
   b. Chemists now believe that a special "vital force" is needed to form an organic compound.
   c. Historically, the *org-* of the term *organic* was conceptually paired with the *org-* in the term *living organism.*
   d. Most but not all compounds found in living organisms are organic compounds.

1.2 Indicate whether each of the following statements is true or false.
   a. Over 7 million organic compounds have been characterized.
   b. The number of known organic compounds and the number of known inorganic compounds are approximately the same.
   c. In essence, organic chemistry is the study of the compounds of one element.
   d. Numerous organic compounds are known that do not occur in living organisms.

■ Bonding Characteristics of the Carbon Atom (Section 1.2)

1.3 Indicate whether each of the following situations meet or do not meet the "bonding requirement" for carbon atoms.
   a. Two single bonds and a double bond
   b. A single bond and two double bonds
   c. Three single bonds and a triple bond
   d. A double bond and a triple bond

1.4 Indicate whether each of the following situations meet or do not meet the "bonding requirement" for carbon atoms.
   a. Four single bonds
   b. Three single bonds and a double bond
   c. Two double bonds and two single bonds
   d. Two double bonds

■ Hydrocarbons and Hydrocarbon Derivatives (Section 1.3)

1.5 What is the difference between a *hydrocarbon* and a *hydrocarbon derivative*?

1.6 Contrast hydrocarbons and hydrocarbon derivatives in terms of number of compounds that are known.

1.7 What is the difference between a *saturated hydrocarbon* and an *unsaturated hydrocarbon*?

1.8   What structural feature is present in an unsaturated hydrocarbon that is not present in a saturated hydrocarbon?

1.9   Classify each of the following hydrocarbons as saturated or unsaturated.

a.
$$H-\underset{\underset{H}{|}}{\overset{\overset{H}{|}}{C}}-\underset{\underset{H}{|}}{\overset{\overset{H}{|}}{C}}-H$$

b.
$$H-\overset{\overset{H}{|}}{C}=\overset{\overset{H}{|}}{C}-\underset{\underset{H}{|}}{\overset{\overset{H}{|}}{C}}-H$$

c.
$$H-\overset{\overset{H}{|}}{C}=\overset{\overset{H}{|}}{C}-H$$

d.
$$H-C\equiv C-\underset{\underset{H}{|}}{\overset{\overset{H}{|}}{C}}-\underset{\underset{H}{|}}{\overset{\overset{H}{|}}{C}}-H$$

1.10   Classify each of the following hydrocarbons as saturated or unsaturated.

a. $H-C\equiv C-H$

b.
$$H-\underset{\underset{H}{|}}{\overset{\overset{H}{|}}{C}}-\underset{\underset{H}{|}}{\overset{\overset{H}{|}}{C}}-\underset{\underset{H}{|}}{\overset{\overset{H}{|}}{C}}-H$$

c.
$$H-\overset{\overset{H}{|}}{C}=\overset{\overset{H}{|}}{C}-\overset{\overset{H}{|}}{C}=\overset{\overset{H}{|}}{C}-H$$

d.
$$H-\underset{\underset{H}{|}}{\overset{\overset{H}{|}}{C}}-\underset{\underset{H}{|}}{\overset{\overset{H}{|}}{C}}-\underset{\underset{H}{|}}{\overset{\overset{H}{|}}{C}}-\underset{\underset{H}{|}}{\overset{\overset{H}{|}}{C}}-H$$

■ **Formulas for Alkanes (Section 1.4)**

1.11   Using the general formula for an alkane, derive the following for specific alkanes.
  a. Number of hydrogen atoms present when 8 carbon atoms are present
  b. Number of carbon atoms present when 10 hydrogen atoms are present
  c. Number of carbon atoms present when 41 total atoms are present
  d. Total number of covalent bonds present in the molecule when 7 carbon atoms are present

1.12   Using the general formula for an alkane, derive the following for specific alkanes.
  a. Number of carbon atoms present when 14 hydrogen atoms are present
  b. Number of hydrogen atoms present when 6 carbon atoms are present
  c. Number of hydrogen atoms present when 32 total atoms are present
  d. Total number of covalent bonds present in the molecule when 16 hydrogen atoms are present

■ **Structural Formulas (Section 1.5)**

1.13   Convert each of the following expanded structural formulas into a condensed structural formula.

a.
$$H-\underset{\underset{H}{|}}{\overset{\overset{H}{|}}{C}}-\underset{\underset{H}{|}}{\overset{\overset{H}{|}}{C}}-\underset{\underset{H}{|}}{\overset{\overset{H}{|}}{C}}-\underset{\underset{H}{|}}{\overset{\overset{H}{|}}{C}}-H$$

b.
$$H-\underset{\underset{H}{|}}{\overset{\overset{H}{|}}{C}}-\underset{\underset{H}{|}}{\overset{\overset{H}{|}}{C}}-\underset{\underset{H-\overset{|}{\underset{H}{C}}-H}{|}}{\overset{\overset{H}{|}}{C}}-\underset{\underset{H}{|}}{\overset{\overset{H}{|}}{C}}-\underset{\underset{H}{|}}{\overset{\overset{H}{|}}{C}}-H$$

c.
$$H-\underset{\underset{H}{|}}{\overset{\overset{H}{|}}{C}}-\underset{\underset{H}{|}}{\overset{\overset{H}{|}}{C}}-\underset{\underset{H-\overset{|}{\underset{H}{C}}-H}{|}}{\overset{\overset{H-\overset{|}{\underset{H}{C}}-H}{|}}{C}}-\underset{\underset{H}{|}}{\overset{\overset{H}{|}}{C}}-\underset{\underset{H}{|}}{\overset{\overset{H}{|}}{C}}-\underset{\underset{H}{|}}{\overset{\overset{H}{|}}{C}}-H$$

d.
$$H-\underset{\underset{H}{|}}{\overset{\overset{H}{|}}{C}}-\underset{\underset{H}{|}}{\overset{\overset{H}{|}}{C}}-\underset{\underset{H-\overset{|}{\underset{H-\overset{|}{\underset{H}{C}}-H}{C}}-H}{|}}{\overset{\overset{H}{|}}{C}}-\underset{\underset{H}{|}}{\overset{\overset{H}{|}}{C}}-\underset{\underset{H}{|}}{\overset{\overset{H}{|}}{C}}-\underset{\underset{H}{|}}{\overset{\overset{H}{|}}{C}}-H$$

1.14   Convert each of the following expanded structural formulas into a condensed structural formula.

a.
$$H-\underset{\underset{H}{|}}{\overset{\overset{H}{|}}{C}}-\underset{\underset{H}{|}}{\overset{\overset{H}{|}}{C}}-\underset{\underset{H}{|}}{\overset{\overset{H}{|}}{C}}-\underset{\underset{H}{|}}{\overset{\overset{H}{|}}{C}}-\underset{\underset{H}{|}}{\overset{\overset{H}{|}}{C}}-\underset{\underset{H}{|}}{\overset{\overset{H}{|}}{C}}-H$$

b.
$$H-\underset{\underset{H}{|}}{\overset{\overset{H}{|}}{C}}-\underset{\underset{H-\overset{|}{\underset{H}{C}}-H}{|}}{\overset{\overset{H}{|}}{C}}-\underset{\underset{H}{|}}{\overset{\overset{H}{|}}{C}}-\underset{\underset{H}{|}}{\overset{\overset{H}{|}}{C}}-H$$

c.
$$H-\underset{\underset{H}{|}}{\overset{\overset{H}{|}}{C}}-\underset{\underset{H-\overset{|}{\underset{H}{C}}-H}{|}}{\overset{\overset{H}{|}}{C}}-\underset{\underset{H}{|}}{\overset{\overset{H}{|}}{C}}-\underset{\underset{H-\overset{|}{\underset{H}{C}}-H}{|}}{\overset{\overset{H}{|}}{C}}-\underset{\underset{H}{|}}{\overset{\overset{H}{|}}{C}}-H$$

d.
$$H-\underset{\underset{H}{|}}{\overset{\overset{H}{|}}{C}}-\underset{\underset{H}{|}}{\overset{\overset{H}{|}}{C}}-\underset{\underset{H-\overset{|}{\underset{H-\overset{|}{\underset{H}{C}}-H}{C}}-H}{|}}{\overset{\overset{H}{|}}{C}}-\underset{\underset{H}{|}}{\overset{\overset{H}{|}}{C}}-\underset{\underset{H}{|}}{\overset{\overset{H}{|}}{C}}-\underset{\underset{H}{|}}{\overset{\overset{H}{|}}{C}}-H$$

1.15   The following skeletal structural formulas for alkanes are incomplete in that the hydrogen atoms attached to each carbon are not shown. Complete each of these formulas by writing in the correct number of hydrogen atoms attached to each carbon atom. That is, rewrite each of these formulas as a condensed structural formula such as $CH_3-CH_2-CH_3$.

a. $C-\underset{\underset{C}{|}}{C}-C-C$

b. $C-C-\underset{\underset{C}{|}}{\overset{}{C}}-\underset{\underset{C}{|}}{\overset{}{C}}-\underset{\underset{C}{|}}{\overset{}{C}}-C$

c. $C-C-C-C-C-C$

d. $C-\underset{\underset{C}{|}}{\overset{\overset{C}{|}}{C}}-C-C$

1.16 The following skeletal structural formulas for alkanes are incomplete in that the hydrogen atoms attached to each carbon are not shown. Complete each of these formulas by writing in the correct number of hydrogen atoms attached to each carbon atom. That is, rewrite each of these formulas as a condensed structural formula such as CH_3—CH_2—CH_3.

a. C—C—C—C—C   b.   C
       | | |
      C   C   C—C—C
                           |
                           C

c. C—C—C—C—C   d. C—C—C—C—C
                           | |
                           C   C
                                 |
                                 C

1.17 Draw the indicated type of formula for the following alkanes.
   a. The expanded structural formula for a continuous-chain alkane with the formula C_5H_{12}
   b. The expanded structural formula for CH_3—$(CH_2)_6$—CH_3
   c. The condensed structural formula, using parentheses for the —CH_2— groups, for the continuous-chain alkane $C_{10}H_{22}$
   d. The molecular formula for the alkane CH_3—$(CH_2)_4$—CH_3

1.18 Draw the indicated type of formula for the following alkanes.
   a. The expanded structural formula for a continuous-chain alkane with the molecular formula C_6H_{14}
   b. The condensed structural formula, using parentheses for the —CH_2— groups, for the straight-chain alkane $C_{12}H_{26}$
   c. The molecular formula for the alkane CH_3—$(CH_2)_6$—CH_3
   d. The expanded structural formula for CH_3—$(CH_2)_3$—CH_3

■ Constitutional Isomers and Molecular Conformations (Sections 1.6 and 1.7)

1.19 For each of the following pairs of structures, determine whether they are
   1. Different conformations of the same molecule
   2. Different compounds that are constitutional isomers
   3. Different compounds that are not constitutional isomers

a. CH_3—CH_2—CH_2—CH—CH_3
                             |
                           CH_3
   and CH_3—CH—CH_2—CH_3
                         |
                      CH_3

b. CH_3—CH_2—CH_2—CH_2—CH_3
   and CH_3—CH—CH_3
                        |
                      CH_2
                      |
                      CH_3

c. CH_3—CH_2—CH_2 and CH_3—CH_2
                 | |
             CH_3 CH_2—CH_3

d. CH_3
           |
   CH_3—C—CH_3 and CH_3—CH—CH_2—CH_3
          | |
        CH_3 CH_3

1.20 For each of the following pairs of structures, determine whether they are
   1. Different conformations of the same molecule
   2. Different compounds that are constitutional isomers
   3. Different compounds that are not constitutional isomers

a. CH_3—CH—CH_3
                |
           CH_2—CH_3
   and CH_3—CH—CH_2—CH_3
                       |
                    CH_3

b. CH_3—CH—CH_2—CH_3
            |
         CH_3
   and CH_3—CH_2—CH—CH_3
                             |
                        CH_3

c. CH_3—CH—CH_2—CH_3
            |
         CH_2
            |
         CH_3
   and CH_3—CH—CH—CH_3
                      | |
                    CH_3 CH_3

d. CH_3—CH_2—CH_2—CH_2—CH_2—CH_3
               CH_3
                  |
   and CH_3—C—CH_3
                    |
                  CH_3

■ IUPAC Nomenclature for Alkanes (Section 1.8)

1.21 The first step in naming an alkane is to identify the longest continuous chain of carbon atoms. For each of the following skeletal structural formulas, how many carbon atoms are present in the longest continuous chain?

a.   C
                           |
   C—C—C—C—C—C—C
                |
                C
                |
                C

b. C—C—C—C—C—C
        | |
        C   C
        |
        C

c.   C—C—C
                         |
   C—C—C—C
               |
               C—C—C—C

d. C—C—C
            |
   C—C—C—C
        |
        C
        |
        C—C

1.22 The first step in naming an alkane is to identify the longest continuous chain of carbon atoms. For each of the following skeletal structural formulas, how many carbon atoms are present in the longest continuous chain?

a.   C
                   |
   C—C—C—C—C—C
            | |
            C   C
            |
            C

b.
$$\text{C}-\text{C}-\text{C}-\text{C}-\text{C}$$
with C branches

c.
$$\text{C}-\text{C}-\text{C}-\text{C}$$
branched structure

d.
$$\text{C}-\text{C}-\text{C}-\text{C}-\text{C}-\text{C}-\text{C}$$
branched structure

1.23 Give the IUPAC name for each of the following alkanes.

a. $\text{CH}_3-\text{CH}_2-\text{CH}_2-\text{CH}-\text{CH}_3$ with CH_3 branch

b. $\text{CH}_3-\text{CH}-\text{CH}-\text{CH}_2-\text{CH}-\text{CH}_3$ with CH_2, CH_3, CH_3 branches and CH_3

c. $\text{CH}_3-\text{CH}-\text{C}-\text{CH}_2-\text{CH}_3$ with CH_3, CH_3, CH_2, CH_3 branches

d. $\text{CH}_3-\text{CH}-\text{CH}-\text{CH}-\text{CH}_2-\text{CH}_3$ with CH_3, CH_2, CH_3, CH_3 branches

e. $\text{CH}_3-(\text{CH}_2)_8-\text{CH}_3$

f. $\text{CH}_2-\text{CH}_2-\text{CH}-\text{CH}_2-\text{CH}_2$ with CH_3, CH_2, CH_3, CH_2, CH_3 branches

1.24 Give the IUPAC name for each of the following alkanes.

a. $\text{CH}_3-\text{CH}-\text{CH}-\text{CH}_2-\text{CH}_3$ with CH_3, CH_3 branches

b. $\text{CH}_3-\text{C}-\text{CH}_2-\text{C}-\text{CH}_3$ with CH_3, CH_3, CH_3, CH_3 branches

c. $\text{CH}_3-\text{CH}-\text{CH}_2-\text{CH}-\text{CH}_3$ with CH_3, CH_2, CH_3 branches

d. $\text{CH}_3-\text{CH}_2-\text{CH}_2-\text{CH}-\text{CH}_3$ with $\text{CH}_3-\text{CH}_2-\text{CH}_3$ branch

e. $\text{CH}_3-(\text{CH}_2)_7-\text{CH}_3$

f.
$$\text{CH}_3-(\text{CH}_2)_3-\text{CH}-\text{CH}-(\text{CH}_2)_3-\text{CH}_3$$
with CH_3 on one CH and $\text{CH}_2-\text{CH}_2-\text{CH}_3$ chain

1.25 Two different carbon chains of eight atoms can be located in the following alkane.

$$\text{CH}_3-\text{CH}_2-\text{CH}_2-\text{CH}_2-\text{CH}-\text{CH}-\text{CH}_2-\text{CH}_3$$
with $\text{CH}_2-\text{CH}_2-\text{CH}_3$ and CH_3 branches

Which of these chains should be used in determining the IUPAC name for the alkane? Explain your answer.

1.26 Two different carbon chains of seven atoms can be located in the following alkane.

$$\text{CH}_3-\text{CH}-\text{CH}_2-\text{CH}-\text{CH}_2-\text{CH}-\text{CH}_3$$
with CH_3, CH_2, CH_3 branches and CH_2-CH_3

Which of these chains should be used in determining the IUPAC name for the alkane? Explain your answer.

1.27 Draw a condensed structural formula for each of the following alkanes.

a. 2-Methylbutane   b. 3,4-Dimethylhexane
c. 3-Ethyl-3-methylpentane   d. 2,3,4,5-Tetramethylheptane
e. 3,5-Diethyloctane   f. 4-Propylnonane

1.28 Draw a condensed structural formula for each of the following alkanes.

a. 3-Methylhexane   b. 2,4-Dimethylhexane
c. 5-Propyldecane   d. 2,3,4-Trimethyloctane
e. 3-Ethyl-3-methylheptane   f. 3,3,4,4-Tetramethylheptane

1.29 For each of the alkanes in Problem 1.27 determine (a) the number of alkyl groups present and (b) the number of substituents present.

1.30 For each of the alkanes in Problem 1.28 determine (a) the number of alkyl groups present and (b) the number of substituents present.

1.31 Explain why the name given for each of the following alkanes is not the correct IUPAC name. Then give the correct IUPAC name for the compound.

a. 4-Methylpentane
b. 2-Ethyl-2-methylpropane
c. 2,3,3-Trimethylbutane
d. 1,2,2-Trimethylpentane
e. 3-Methyl-4-ethylhexane
f. 2-Methyl-4-methylhexane

1.32 Explain why the name given for each of the following alkanes is not the correct IUPAC name. Then give the correct IUPAC name for the compound.

a. 1,6-Dimethylhexane
b. 2-Ethylpentane
c. 3,3,4-Trimethylpentane
d. 2-Ethyl-4-methylhexane
e. 3-Ethyl-4-ethylhexane
f. 3,4,5,5,6-Pentamethylhexane

■ **Line-Angle Formulas for Alkanes (Section 1.9)**

1.33   Convert each of the following line-angle formulas to a skeletal structural formula.

a.   b.

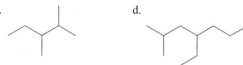

c.   d.

1.34   Convert each of the following line-angle formulas to a skeletal structural formula.

a.   b.

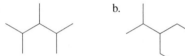

c.   d.

1.35   Convert each of the following line-angle formulas to a condensed structural formula.

a.   b.

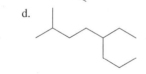

c.   d.

1.36   Convert each of the following line-angle formulas to a condensed structural formula.

a.   b.

c.   d.

1.37   Do the line-angle formulas in each of the following sets represent (1) the same compound, (2) constitutional isomers, or (3) different compounds that are not constitutional isomers?

a.

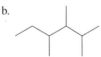   and

b.

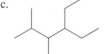   and

1.38   Do the line-angle formulas in each of the following sets represent (1) the same compound, (2) constitutional isomers, or (3) different compounds that are not constitutional isomers?

a.

                    and

b.

                    and

1.39   Convert each of the condensed structural formulas in Problem 1.23 to a line-angle formula.

1.40   Convert each of the condensed structural formulas in Problem 1.24 to a line-angle formula.

1.41   Assign an IUPAC name to each of the compounds in Problem 1.33.

1.42   Assign an IUPAC name to each of the compounds in Problem 1.34.

1.43   *Calculate* the molecular formula for each of the compounds in Problem 1.35.

1.44   *Calculate* the molecular formula for each of the compounds in Problem 1.36.

■ **Classification of Carbon Atoms (Section 1.10)**

1.45   For each of the alkane structures in Problem 1.23, give the number of (a) primary, (b) secondary, (c) tertiary, and (d) quaternary carbon atoms present.

1.46   For each of the alkane structures in Problem 1.24, give the number of (a) primary, (b) secondary, (c) tertiary, and (d) quaternary carbon atoms present.

■ **Branched-Chain Alkyl Groups (Section 1.11)**

1.47   Give the name of the branched alkyl group attached to each of the following carbon chains, where the carbon chain is denoted by a horizontal line.

a. _____   b. _____

$CH-CH_3$ CH_2
| |
CH_3 $CH_3-CH-CH_3$

c. _____   d. _____

$CH_3-CH-CH_3$ $CH-CH_3$
                                   |
                                   CH_2
                                   |
                                   CH_3

1.48   Give the name of the branched alkyl group attached to each of the following carbon chains, where the carbon chain is denoted by a horizontal line.

a. _____   b. _____

CH_2 CH_3-C-CH_3
| |
$CH-CH_3$ CH_3
|
CH_3

c. _____   d. _____

$CH_3-CH_2-CH-CH_3$

1.49 Draw condensed structural formulas for the following branched alkanes.
- a. 5-(*sec*-Butyl)decane
- b. 4,4-Diisopropyloctane
- c. 5-Isobutyl-2,3-dimethylnonane
- d. 4-(1,1-Dimethylethyl)octane

1.50 Draw condensed structural formulas for the following branched alkanes.
- a. 5-Isobutylnonane
- b. 4,4-Di(*sec*-butyl)decane
- c. 4-(*tert*-Butyl)-3,3-diethylheptane
- d. 5-(2-Methylpropyl)nonane

■ Cycloalkanes (Sections 1.12 through 1.14)

1.51 Using the general formula for a cycloalkane, derive the following for specific cycloalkanes.
- a. Number of hydrogen atoms present when 8 carbon atoms are present
- b. Number of carbon atoms present when 12 hydrogen atoms are present
- c. Number of carbon atoms present when a total of 15 atoms are present
- d. Number of covalent bonds present when 5 carbon atoms are present

1.52 Using the general formula for a cycloalkane, derive the following for specific cycloalkanes.
- a. Number of hydrogen atoms present when 4 carbon atoms are present
- b. Number of carbon atoms present when 6 hydrogen atoms are present
- c. Number of hydrogen atoms present when a total of 18 atoms are present
- d. Number of covalent bonds present when 8 hydrogen atoms are present

1.53 What is the molecular formula for each of the following cycloalkane molecules?

a. b. c. d.

1.54 What is the molecular formula for each of the following cycloalkane molecules?

a. b. c. d.

1.55 Assign an IUPAC name to each of the cycloalkanes in Problem 1.53.

1.56 Assign an IUPAC name to each of the cycloalkanes in Problem 1.54.

1.57 What is wrong with each of the following attempts to name a cycloalkane using IUPAC rules?
- a. Dimethylcyclohexane
- b. 3,4-Dimethylcyclohexane
- c. 1-Ethylcyclobutane
- d. 2-Ethyl-1-methylcyclopentane

1.58 What is wrong with each of the following attempts to name a cycloalkane using IUPAC rules?
- a. Dimethylcyclopropane
- b. 1-Methylcyclohexane
- c. 2,5-Dimethylcyclobutane
- d. 1-Propyl-2-ethylcyclohexane

1.59 Draw line-angle formulas for the following cycloalkanes.
- a. Propylcyclobutane
- b. Isopropylcyclobutane
- c. *cis*-1,2-Diethylcyclohexane
- d. *trans*-1-Ethyl-3-propylcyclopentane

1.60 Draw line-angle formulas for the following cycloalkanes.
- a. Butylcyclopentane
- b. Isobutylcyclopentane
- c. *cis*-1,3-Diethylcyclopentane
- d. *trans*-1-Ethyl-4-methylcyclohexane

1.61 Determine whether *cis–trans* isomerism is possible for each of the following cycloalkanes. If it is, then draw structural formulas for the *cis* and *trans* isomers.
- a. Isopropylcyclobutane
- b. 1,2-Diethylcyclopropane
- c. 1-Ethyl-1-propylcyclopentane
- d. 1,3-Dimethylcyclohexane

1.62 Determine whether *cis–trans* isomerism is possible for each of the following cycloalkanes. If it is, then draw structural formulas for the *cis* and *trans* isomers.
- a. *sec*-Butylcyclohexane
- b. 1-Ethyl-3-methylcyclobutane
- c. 1,1-Dimethylcyclohexane
- d. 1,3-Dipropylcyclopentane

■ Sources of Alkanes and Cycloalkanes (Section 1.15)

1.63 What physical property of hydrocarbons is the basis for the fractional distillation process for separating hydrocarbons?

1.64 Describe the process by which crude petroleum is separated into simpler mixtures (fractions).

■ Physical Properties of Alkanes and Cycloalkanes (Section 1.16)

1.65 Which member in each of the following pairs of compounds has the higher boiling point?
- a. Hexane and octane
- b. Cyclobutane and cyclopentane
- c. Pentane and 1-methylbutane
- d. Pentane and cyclopentane

1.66 Which member in each of the following pairs of compounds has the higher boiling point?
- a. Methane and ethane
- b. Cyclohexane and hexane
- c. Butane and methylpropane
- d. Pentane and 2,2-dimethylpropane

1.67 For which of the following pairs of compounds do both members of the pair have the same physical state (solid, liquid, or gas) at room temperature and pressure?
- a. Ethane and hexane
- b. Cyclopropane and butane
- c. Octane and 3-methyloctane
- d. Pentane and decane

1.68 For which of the following pairs of compounds do both members of the pair have the same physical state (solid, liquid, or gas) at room temperature and pressure?
- a. Methane and butane
- b. Cyclobutane and cyclopentane
- c. Hexane and 2,3-dimethylbutane
- d. Pentane and octane

▪ Chemical Properties of Alkanes and Cycloalkanes (Section 1.17)

1.69 Write the formulas of the products from the complete combustion of each of the following alkanes or cycloalkanes.
  a. C_3H_8 b. Butane
  c. Cyclobutane   d. $CH_3-(CH_2)_{15}-CH_3$

1.70 Write the formulas of the products from the complete combustion of each of the following alkanes or cycloalkanes.
  a. C_4H_{10} b. 2-Methylpentane
  c. Cyclopentane   d. $CH_3-(CH_2)_7-CH_3$

1.71 Write molecular formulas for all the possible halogenated hydrocarbon products from the bromination of methane.

1.72 Write molecular formulas for all the possible halogenated hydrocarbon products from the fluorination of methane.

1.73 Write structural formulas for all the possible halogenated hydrocarbon products from the monochlorination of the following alkanes or cycloalkanes.
  a. Ethane   b. Butane
  c. 2-Methylpropane   d. Cyclopentane

1.74 Write structural formulas for all the possible halogenated hydrocarbon products from the monobromination of the following alkanes or cycloalkanes.
  a. Propane   b. Pentane
  c. 2-Methylbutane   d. Cyclohexane

▪ Nomenclature of Halogenated Alkanes (Section 1.18)

1.75 Give both IUPAC and common names to each of the following halogenated hydrocarbons.

  a. CH_3-I b. $CH_3-CH_2-CH_2-Cl$

  c. $CH_3-\underset{\underset{F}{|}}{CH}-CH_2-CH_3$ d. (cyclobutane with Cl)

1.76 Give both IUPAC and common names to each of the following halogenated hydrocarbons.

  a. $CH_3-CH_2-CH_2-CH_2-Br$ b. $CH_3-\underset{\underset{CH_3}{|}}{CH}-Cl$

  c. $CH_3-\underset{\underset{Cl}{|}}{\overset{\overset{CH_3}{|}}{C}}-CH_3$ d. (cyclohexane with $-Br$)

1.77 Draw structural formulas for the following halogenated hydrocarbons.
  a. Trichloromethane
  b. 1,2-Dichloro-1,1,2,2-tetrafluoroethane
  c. Isopropyl bromide
  d. *trans*-1-Bromo-3-chlorocyclopentane

1.78 Draw structural formulas for the following halogenated hydrocarbons.
  a. Trifluorochloromethane
  b. Pentafluoroethane
  c. Isobutyl chloride
  d. *cis*-1,2-Dichlorocyclohexane

ADDITIONAL PROBLEMS

1.79 Answer the following questions about the unbranched alkane with seven carbon atoms.
  a. How many hydrogen atoms are present?
  b. How many carbon–carbon bonds are present?
  c. How many carbon atoms have two hydrogen atoms bonded to them?
  d. How many total covalent bonds are present?
  e. Is the alkane a solid, a liquid, or a gas at room temperature?
  f. Is the alkane less dense or more dense than water?
  g. Is the alkane soluble or insoluble in water?
  h. Is the alkane flammable or nonflammable in air?

1.80 Indicate whether the members of each of the following pairs of compounds are constitutional isomers.
  a. Hexane and 2-methylhexane
  b. Hexane and 2,2-dimethylbutane
  c. Hexane and methylcyclopentane
  d. Hexane and cyclohexane

1.81 Draw structural formulas for the following compounds.
  a. *trans*-1,4-Difluorocyclohexane
  b. *cis*-1-Chloro-2-methylcyclobutane
  c. *tert*-Butyl bromide
  d. Isobutyl iodide

1.82 What is the IUPAC name of the compound obtained by attaching a *tert*-butyl group to carbon 4 and a *sec*-butyl group to carbon 5 of the alkane 2,2-dimethylheptane?

1.83 Give the molecular formula for each of the following compounds.
  a. an 18-carbon alkane
  b. a 7-carbon cycloalkane
  c. a 7-carbon difluorinated alkane
  d. a 6-carbon dibrominated cycloalkane

1.84 Draw the structural formula of each of the following compounds.
  a. Neooctane
  b. Isobutane
  c. Methylene chloride
  d. Freon-12

1.85 Classify each of the following molecular formulas as representing an alkane, a cycloalkane, a halogenated alkane, or a halogenated cycloalkane.
  a. C_6H_{14} b. $C_6H_{10}Cl_2$
  c. $C_4H_8Cl_2$ d. C_4H_8

1.86 Write skeletal structural formulas and assign IUPAC names to all saturated hydrocarbon constitutional isomers (ignore *cis–trans* isomers) with the following molecular formulas.
  a. C_6H_{12} (twelve isomers are possible)
  b. C_6H_{14} (five isomers are possible)
  c. $C_3H_6Br_2$ (four isomers are possible)

1.87 Assign an IUPAC name to each of the following hydrocarbons, whose line-angle formulas are:

a.

b.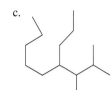

c.

d.

1.88 Assign an IUPAC name to the following alkane, which has a five-carbon branched alkyl group.

$$C-C-C-C-C$$
$$C-C-C-C-\overset{|}{C}-C-C-C-C$$

1.89 Which of the following statements concerning saturated hydrocarbons is *incorrect*?
   a. Every carbon atom present must have four bonds.
   b. All bonds present must be single bonds.
   c. Every carbon atom present must be bonded to at least one hydrogen atom.
   d. This classification includes both alkanes and cycloalkanes.

1.90 Which of the following gives the generalized molecular formulas for alkanes and cycloalkanes, respectively?
   a. C_nH_{2n+2} and C_nH_{2n} b. C_nH_{2n+2} and C_nH_{2n+4}
   c. C_nH_{2n} and C_nH_{2n-2} d. C_nH_{2n} and C_nH_{2n+2}

1.91 The formula $CH_3-CH_2-CH_2-CH_2-CH_3$ is an example of which of the following?
   a. An expanded structural formula
   b. A condensed structural formula
   c. A skeletal structural formula
   d. A line-angle formula

1.92 Which of the following compounds is a constitutional isomer of $CH_3-CH_2-CH_2-CH_2-CH_3$?
   a. 2-Methylpentane   b. 2-Methylbutane
   c. 2,2-Dimethylpentane   d. 2,2-Dimethylbutane

1.93 One of the three five-carbon alkane constitutional isomers has the molecular formula C_5H_{12}. Which of the following gives the molecular formulas for the other two isomers respectively?
   a. C_5H_{11} and C_5H_{10} b. C_5H_{13} and C_5H_{14}
   c. C_4H_{12} and C_6H_{12} d. C_5H_{12} and C_5H_{12}

1.94 Which of the following statements concerning alkanes and alkyl groups is *incorrect*?
   a. Isobutane and 2-methylpropane are two names for the same compound.
   b. 2-Methylpentane and 2-methylbutane contain the same number of alkyl groups.
   c. Butane and cyclobutane contain the same number of hydrogen atoms.
   d. Secondary-butyl group and (1,1-dimethylethyl) group are two names for the same alkyl group.

1.95 In which of the following alkanes are both secondary and tertiary carbon atoms present?
   a. $CH_3-CH_2-CH_2-CH_3$
   b. $CH_3-\overset{\overset{\displaystyle CH_3}{|}}{CH}-CH_3$
   c. $CH_3-CH_2-\overset{\overset{\displaystyle CH_3}{|}}{CH}-CH_3$
   d. $CH_3-\overset{\overset{\displaystyle CH_3}{|}}{CH}-\overset{\overset{\displaystyle CH_3}{|}}{CH}-CH_3$

1.96 For which of the following halogenated cycloalkanes is *cis–trans* isomerism possible?
   a. 1,1-Dibromocyclobutane
   b. 1-Bromo-1-chlorocyclobutane
   c. 1-Bromo-2-chlorocyclobutane
   d. 1,1-Dichlorocyclobutane

1.97 Which of the following statements concerning the boiling points of specific alkanes is *correct*?
   a. Hexane has a higher boiling point than heptane.
   b. Pentane has a higher boiling point than 2-methylbutane.
   c. Butane has a higher boiling point than cyclobutane.
   d. Butane and pentane have approximately the same boiling point.

1.98 Which of the following statements concerning the chemical properties of alkanes and cycloalkanes is *correct*?
   a. Alkanes undergo combustion reactions but cycloalkanes do not.
   b. Neither alkanes nor cycloalkanes undergo combustion reactions.
   c. Both alkanes and cycloalkanes undergo combustion reactions.
   d. Alkanes undergo combustion reactions but do not undergo halogenation reactions.

2 Unsaturated Hydrocarbons

When actylene, an unsaturated hydrocarbon, is burned with oxygen in an oxyacetylene welding torch, a temperature high enough to cut metals is produced.

Two general types of hydrocarbons exist, *saturated* and *unsaturated*. Saturated hydrocarbons, discussed in the previous chapter, include the alkanes and cycloalkanes. All bonds in saturated hydrocarbons are single bonds. Unsaturated hydrocarbons, the topic of this chapter, contain one or more carbon–carbon multiple bonds. There are three classes of unsaturated hydrocarbons: the *alkenes*, the *alkynes*, and the *aromatic hydrocarbons*, all of which we will consider.

2.1 Unsaturated Hydrocarbons

An **unsaturated hydrocarbon** *is a hydrocarbon in which one or more carbon–carbon multiple bonds (double bonds, triple bonds, or both) are present.* Unsaturated hydrocarbons have *physical* properties similar to those of saturated hydrocarbons. However, their *chemical* properties are much different. Unsaturated hydrocarbons are chemically more reactive than their saturated counterparts. The increased reactivity of unsaturated hydrocarbons is related to the presence of the carbon–carbon multiple bond(s) in such compounds. These multiple bonds serve as locations where chemical reactions can occur.

Whenever a specific portion of a molecule governs its chemical properties, that portion of the molecule is called a functional group. A **functional group** *is the part of an organic molecule where most of its chemical reactions occur.* Carbon–carbon multiple bonds are the functional group for an unsaturated hydrocarbon.

The study of various functional groups and their respective reactions provides the organizational structure for organic chemistry. Each of the organic chemistry chapters

> The field of organic chemistry is organized in terms of functional groups.

that follow introduces new functional groups that characterize families of hydrocarbon derivatives.

Unsaturated hydrocarbons are subdivided into three groups on the basis of the type of multiple bond(s) present: (1) *alkenes,* which contain one or more carbon–carbon double bonds, (2) *alkynes,* which contain one or more carbon–carbon triple bonds, and (3) *aromatic hydrocarbons,* which exhibit a special type of "delocalized" bonding that involves a six-membered carbon ring (to be discussed in Section 2.11).

We begin our consideration of unsaturated hydrocarbons with a discussion of alkenes. Information about alkynes and aromatic hydrocarbons then follows.

2.2 Characteristics of Alkenes and Cycloalkenes

An **alkene** *is an acyclic unsaturated hydrocarbon that contains one or more carbon–carbon double bonds.* The alkene functional group is, thus, a $C{=}C$ group. Note the close similarity between the family names *alkene* and *alkane* (Section 1.4); they differ only in their endings: *-ene* versus *-ane.* The *-ene* ending means a double bond is present.

The simplest type of alkene contains only one carbon–carbon double bond. Such compounds have the general formula C_nH_{2n}. Thus alkenes with one double bond have two fewer hydrogen atoms than do alkanes (C_nH_{2n+2}).

The two simplest alkenes are ethene (C_2H_4) and propene (C_3H_6).

$$CH_2{=}CH_2 \qquad CH_2{=}CH{-}CH_3$$
Ethene   Propene

Comparing the geometrical shape of ethene with that of methane (the simplest alkane) reveals a major difference. The arrangement of bonds about the carbon atom in methane is tetrahedral (Section 1.4), whereas the carbon atoms in ethene have a trigonal planar arrangement of bonds; that is, they form a flat, triangle-shaped arrangement (see Figure 2.1). The two carbon atoms participating in a double bond and the four other atoms attached to these two carbon atoms always lie in a plane with a trigonal planar arrangement of atoms about each carbon atom of the double bond. Such an arrangement of atoms is consistent with the principles of VSEPR theory.

A **cycloalkene** *is a cyclic unsaturated hydrocarbon that contains one or more carbon–carbon double bonds within the ring system.* Cycloalkenes in which there is only one double bond have the general formula C_nH_{2n-2}. This general formula reflects the loss of four hydrogen atoms from that of an alkane (C_nH_{2n+2}). Note that two hydrogen atoms are lost because of the double bond and two because of the ring structure.

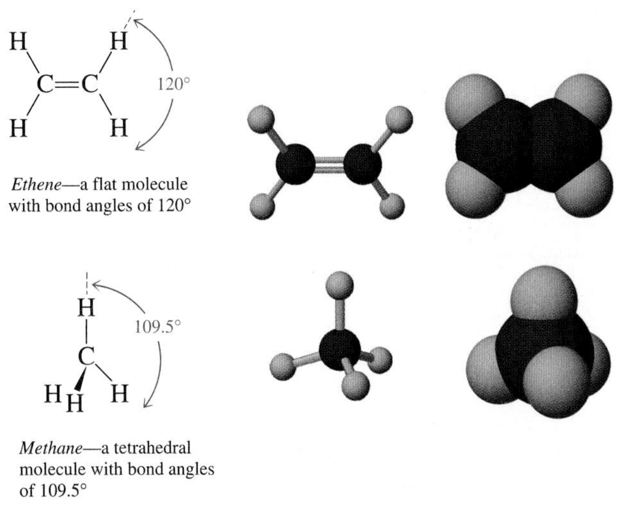

Alkanes and cycloalkanes (Chapter 1) lack functional groups; as a result, they are relatively unreactive.

The general formula for an alkene with one double bond, C_nH_{2n}, is the same as that for a cycloalkane (Section 1.12). Thus such alkenes and cycloalkanes with the same number of carbon atoms are isomeric with one another.

An older but still widely used name for alkenes is *olefins,* pronounced "oh-la-fins." The term *olefin* means "oil-forming." Many alkenes react with Cl_2 to form "oily"compounds.

FIGURE 2.1 Three-dimensional representations of the structures of ethene and methane. In ethene, the atoms are in a flat (planar) rather than a tetrahedral arrangement. Bond angles are 120°.

Ethene—a flat molecule with bond angles of 120°

Methane—a tetrahedral molecule with bond angles of 109.5°

The simplest cycloalkene is the compound cyclopropene (C_3H_4), a three-membered carbon ring system containing one double bond.

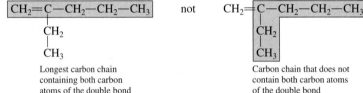

Cyclopropene

Alkenes with more than one carbon–carbon double bond are relatively common. When two double bonds are present, the compounds are often called *dienes*. Cycloalkenes that contain more than one double bond are possible but are not common.

2.3 Names for Alkenes and Cycloalkenes

The rules previously presented for naming alkanes and cycloalkanes (Sections 1.8 and 1.13) can be used, with some modification, to name alkenes and cycloalkenes.

1. Replace the alkane suffix *-ane* with the suffix *-ene,* which is used to indicate the presence of a carbon–carbon double bond.

2. Select as the parent carbon chain the longest continuous chain of carbon atoms that *contains both carbon atoms of the double bond.* For example, select

$$CH_2\!\!=\!\!C\!-\!CH_2\!-\!CH_2\!-\!CH_3 \quad\quad not \quad\quad CH_2\!\!=\!\!C\!-\!CH_2\!-\!CH_2\!-\!CH_3$$
$$\overset{|}{CH_2} \quad\quad\quad\quad\quad\quad\quad\quad \overset{|}{CH_2}$$
$$\overset{|}{CH_3} \quad\quad\quad\quad\quad\quad\quad\quad \overset{|}{CH_3}$$

Longest carbon chain containing both carbon atoms of the double bond

Carbon chain that does not contain both carbon atoms of the double bond

> Carbon–carbon double bonds take precedence over alkyl groups and halogen atoms in determining the direction in which the parent carbon chain is numbered.

3. Number the parent carbon chain beginning at the end nearest the double bond.

$$\overset{1}{C}H_3\!-\!\overset{2}{C}H\!\!=\!\!\overset{3}{C}H\!-\!\overset{4}{C}H_2\!-\!\overset{5}{C}H_3 \quad not \quad \overset{5}{C}H_3\!-\!\overset{4}{C}H\!\!=\!\!\overset{3}{C}H\!-\!\overset{2}{C}H_2\!-\!\overset{1}{C}H_3$$

If the double bond is equidistant from both ends of the parent chain, begin numbering from the end closer to a substituent.

$$\overset{4}{C}H_3\!-\!\overset{3}{C}H\!\!=\!\!\overset{2}{C}H\!-\!\overset{1}{C}H_2 \quad not \quad \overset{1}{C}H_3\!-\!\overset{2}{C}H\!\!=\!\!\overset{3}{C}H\!-\!\overset{4}{C}H_2$$
$$\quad\quad\quad\quad\overset{|}{Cl} \quad\quad\quad\quad\quad\quad\quad\quad\quad\quad\overset{|}{Cl}$$

4. Give the position of the double bond in the chain as a *single* number, which is the lower-numbered carbon atom participating in the double bond. This number is placed immediately before the name of the parent carbon chain.

> A number is not needed to specify double bond position in ethene and propene because there is only one way of positioning the double bond in these molecules.

$$\overset{1}{C}H_3\!-\!\overset{2}{C}H\!\!=\!\!\overset{3}{C}H\!-\!\overset{4}{C}H_3 \quad\quad \overset{1}{C}H_2\!\!=\!\!\overset{2}{C}H\!-\!\overset{3}{C}H\!-\!\overset{4}{C}H_3$$
$$\quad\quad\quad\quad\quad\quad\quad\quad\quad\quad\quad\quad\overset{|}{CH_3}$$

2-Butene   3-Methyl-1-butene

5. Use the suffixes *-diene, -triene, -tetrene,* and so on when more than one double bond is present in the molecule. A separate number must be used to locate each double bond.

$$\overset{1}{C}H_2\!\!=\!\!\overset{2}{C}H\!-\!\overset{3}{C}H\!\!=\!\!\overset{4}{C}H_2 \quad\quad \overset{1}{C}H_2\!\!=\!\!\overset{2}{C}H\!-\!\overset{3}{C}H\!-\!\overset{4}{C}H\!\!=\!\!\overset{5}{C}H_2$$
$$\quad\quad\quad\quad\quad\quad\quad\quad\quad\quad\quad\quad\quad\overset{|}{CH_3}$$

1,3-Butadiene   3-Methyl-1,4-pentadiene

6. Do not use a number to locate the double bond in unsubstituted cycloalkenes with only one double bond because that bond is assumed to be between carbons 1 and 2.

7. In substituted cycloalkenes with only one double bond, the double-bonded carbon atoms are numbered 1 and 2 in the direction (clockwise or counterclockwise) that gives the first-encountered substituent the lower number. Again, no number is used in the name to locate the double bond.

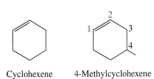

Cyclohexene   4-Methylcyclohexene

8. In cycloalkenes with more than one double bond within the ring, assign one double bond the numbers 1 and 2 and the other double bonds the lowest numbers possible.

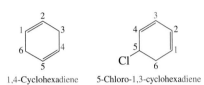

1,4-Cyclohexadiene   5-Chloro-1,3-cyclohexadiene

<hr>

EXAMPLE 2.1

Assigning IUPAC Names to Alkenes and Cycloalkenes

■ Assign IUPAC names to the following alkenes and cycloalkenes.

a. $CH_3—CH{=}CH—CH_2—CH_2—CH_3$ **b.** $CH_3—CH_2—\underset{\underset{\displaystyle CH_3}{\overset{\displaystyle |}{CH_2}}}{C}{=}CH_2$

c.   **d.**

Solution

a. The carbon chain in this hexene is numbered from the end closest to the double bond.

$$\overset{1}{CH_3}—\overset{2}{CH}{=}\overset{3}{CH}—\overset{4}{CH_2}—\overset{5}{CH_2}—\overset{6}{CH_3}$$

The complete IUPAC name is 2-hexene.

b. The longest carbon chain containing *both* carbons of the double bond has four carbon atoms. Thus we have a butene.

$$\boxed{CH_3—CH_2—C{=}CH_2}$$
$$\underset{\underset{\displaystyle CH_3}{\overset{\displaystyle |}{CH_2}}}{|}$$

The chain is numbered from the end closest to the double bond. The complete IUPAC name is 2-ethyl-1-butene.

c. This compound is a methylcyclobutene. The numbers 1 and 2 are assigned to the carbon atoms of the double bond, and the ring is numbered clockwise, which results in a carbon 3 location for the methyl group. (Counterclockwise numbering would have placed the methyl group on carbon 4.) The complete IUPAC name of the cycloalkene is 3-methylcyclobutene. The double bond is understood to involve carbons 1 and 2.

d. A ring system containing five carbon atoms, two double bonds, and a methyl substituent on the ring is called a methylcyclopentadiene. Two different numbering systems produce the same locations (carbons 1 and 3) for the double bonds.

The counterclockwise numbering system assigns the lower number to the methyl group. The complete IUPAC name of the compound is 2-methyl-1,3-cyclopentadiene.

Practical Exercise 2.1

Assign IUPAC names to the following alkenes and cycloalkenes.

a. CH_3—CH=CH—CH_2—CH—CH_3
                                                    |
                                                   CH_3

b.

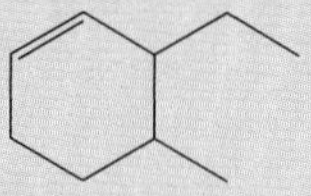

c. CH_2=CH—CH=CH_2

d.

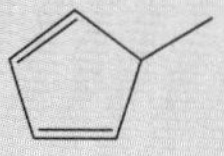

Common Names (Non-IUPAC Names)

The simpler members of most families of organic compounds, including alkenes, have common names in addition to IUPAC names. In many cases these common (non-IUPAC) names are used almost exclusively for the compounds. It would be nice if such common names did not exist, but they do. We have no choice but to memorize these names; fortunately, there are not many of them.

The two simplest alkenes, ethene and propene, have common names you should be familiar with. They are ethylene and propylene, respectively.

$$CH_2=CH_2 \qquad CH_2=CH—CH_3$$
Ethylene   Propylene

Alkenes as Substituents

Just as there are *alkanes* and *alkyl groups* (Section 1.8), there are *alkenes* and *alkenyl groups*. An **alkenyl group** *is a noncyclic hydrocarbon substituent in which a carbon–carbon double bond is present.* The three most frequently encountered alkenyl groups are the one-, two-, and three-carbon entities, which may be named using IUPAC nomenclature (methylidene, ethenyl, and 2-propenyl) or with common names (methylene, vinyl, and allyl).

CH_2= CH_2=CH—— CH_2=CH—CH_2——
Methylene group   Vinyl group   Allyl group
(IUPAC name: methylidene group) (IUPAC name: ethenyl group) (IUPAC name: 2-propenyl group)

The use of these alkenyl group names in actual compound nomenclature is illustrated in the following examples.

CH_2=⬠ CH_2=CH—Cl CH_2=CH—CH_2—Br

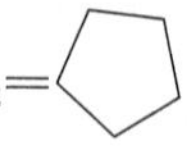

Methylene cyclopentane   Vinyl chloride   Allyl bromide
(IUPAC name: methylidenecyclopentane)   (IUPAC name: chloroethene)   (IUPAC name: 3-bromopropene)

2.4 Line-Angle Formulas for Alkenes

Line-angle formulas for the three- to six-carbon acyclic 1-alkenes are as follows.

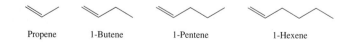

Propene   1-Butene   1-Pentene   1-Hexene

Despite the universal acceptance and precision of the IUPAC nomenclature system, some alkenes (those of low molecular mass) are known almost exclusively by common names.

Representative line-angle formulas for substituent-bearing alkenes include

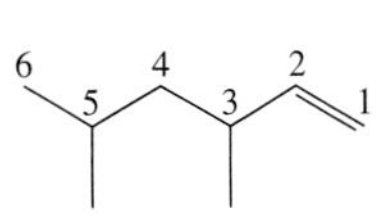

3,5-Dimethyl-1-hexene 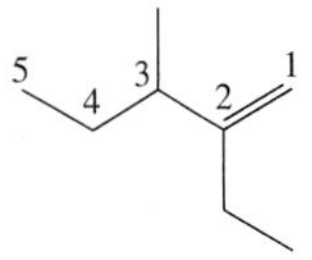   2-Ethyl-3-methyl-1-pentene

Diene representations in terms of line-angle formulas include

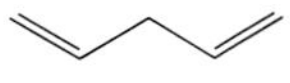

1,4-Pentadiene 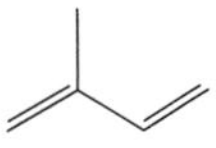   2-Methyl-1,3-butadiene

2.5 Isomerism in Alkenes

Constitutional isomerism is possible for alkenes, just as it was for alkanes (Section 1.6). In general, there are more alkene isomers for a given number of carbon atoms than there are alkane isomers. This is because there is more than one location where a double bond can be placed in systems containing four or more carbon atoms. Figure 2.2 compares constitutional isomer possibilities for C_4 and C_5 alkanes and their counterpart alkenes with one double bond.

Two different subtypes of constitutional isomerism are represented among the alkene isomers shown in Figure 2.2: *positional* isomers and *skeletal* isomers. **Positional isomers** *are constitutional isomers with the same carbon-chain arrangement but different hydrogen atom arrangements as the result of differing location of the functional group present.* Positional isomer sets found in Figure 2.2 are:

1-butene and 2-butene

1-pentene and 2-pentene

2-methyl-1-butene, 3-methyl-1-butene, and 2-methyl-2-butene

FIGURE 2.2 A comparison of structural isomerism possibilities for four- and five-carbon alkane and alkene systems

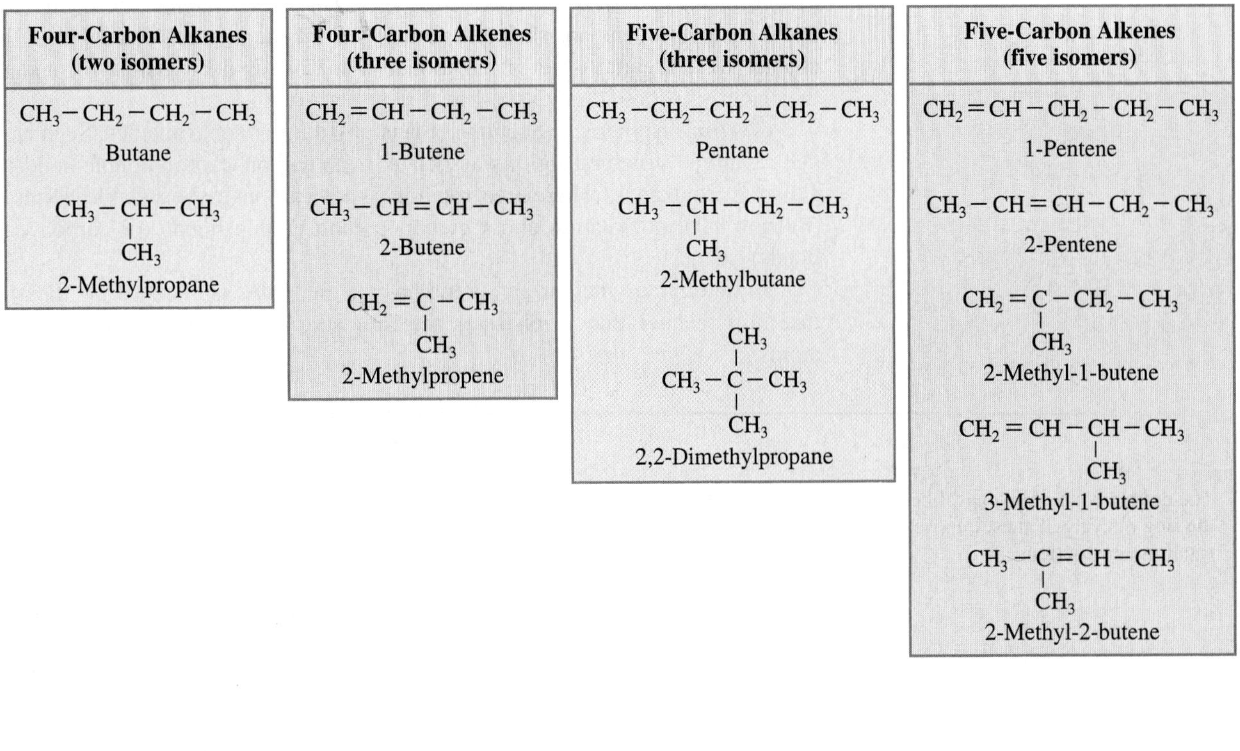

Four-Carbon Alkanes (two isomers)	Four-Carbon Alkenes (three isomers)	Five-Carbon Alkanes (three isomers)	Five-Carbon Alkenes (five isomers)
$CH_3-CH_2-CH_2-CH_3$ Butane	$CH_2=CH-CH_2-CH_3$ 1-Butene	$CH_3-CH_2-CH_2-CH_2-CH_3$ Pentane	$CH_2=CH-CH_2-CH_2-CH_3$ 1-Pentene

$CH_3-CH-CH_3$
       |
     CH_3
 2-Methylpropane
	$CH_3-CH=CH-CH_3$
	 2-Butene
		$CH_3-CH-CH_2-CH_3$
		     |
		     CH_3
		 2-Methylbutane
			$CH_3-CH=CH-CH_2-CH_3$
			 2-Pentene

	$CH_2=C-CH_3$
	     |
	    CH_3
	 2-Methylpropene
		CH_3-C-CH_3
		   CH_3
		   |
		   |
		   CH_3
		 2,2-Dimethylpropane
			$CH_2=C-CH_2-CH_3$
			     |
			    CH_3
			 2-Methyl-1-butene

			$CH_2=CH-CH-CH_3$
			       |
			      CH_3
			 3-Methyl-1-butene

			$CH_3-C=CH-CH_3$
			    |
			   CH_3
			 2-Methyl-2-butene

CHEMICAL CONNECTIONS

Ethene: A Plant Hormone and High-Volume Industrial Chemical

Ethene (ethylene), the simplest unsaturated hydrocarbon (C_2H_4), is a colorless, flammable gas with a slightly sweet odor. It occurs naturally in *small* amounts in plants, where it functions as a plant hormone. A few parts per million ethene (less than 10 parts per million) stimulates the fruit-ripening process.

The commercial fruit industry uses ethene's ripening property to advantage. Bananas, tomatoes, and some citrus fruits are picked green to prevent spoiling and bruising during transportation to markets. At their destinations, the fruits are

Ethene is the hormone that causes tomatoes to ripen.

exposed to small amounts of ethene gas, which stimulates the ripening process.

Despite having no large natural source, ethene is an exceedingly important industrial chemical. Indeed, industrial production of ethene exceeds that of every other organic compound. Petrochemicals (substances found in natural gas and petroleum) are the starting materials for ethene production.

In one process, ethane (from natural gas) is *dehydrogenated* at a high temperature to produce ethene.

$$CH_3-CH_3 \xrightarrow{750°C} CH_2=CH_2 + H_2$$
$$\text{Ethane} \qquad\qquad \text{Ethene}$$

In another process, called *thermal cracking,* hydrocarbons from petroleum are heated to a high temperature in the absence of air (to prevent combustion), which causes the cleavage of carbon–carbon bonds. Ethene is one of the smaller molecules produced by this process.

Industrially produced ethene serves as a starting material for the production of many plastics and fibers. Almost one-half of ethene production is used in the production of the well-known plastic, polyethylene (Section 2.9). Polyvinyl chloride (PVC) and polystyrene (styrofoam) are two other important ethene-based materials. About one-sixth of ethene production is converted to ethylene glycol, the principal component of most brands of antifreeze for automobile radiators (Section 3.5).

Skeletal isomers *are constitutional isomers that have different carbon-chain arrangements as well as different hydrogen atom arrangements.* The C_4 alkenes 1-butene and 2-methylpropene are skeletal isomers. All alkane isomers discussed in the previous chapter were skeletal isomers; positional isomerism is not possible for alkanes because they lack a functional group.

Cis–trans isomerism (Section 1.14) is possible for some alkenes. Such isomerism results from the structural rigidity associated with carbon–carbon double bonds: Unlike the situation in alkanes, where free rotation about carbon–carbon single bonds is possible (Section 1.7), no rotation about carbon–carbon double bonds (or carbon–carbon triple bonds) can occur.

To determine whether an alkene has *cis* and *trans* isomers, draw the alkene structure in a manner that emphasizes the four attachments to the double-bonded carbon atoms.

$$\diagup C=C \diagdown$$

The double bond of alkenes, like the ring of cycloalkanes, imposes rotational restrictions.

If *each* of the two carbons of the double bond has two *different* groups attached to it, *cis* and *trans* isomers exist.

Two groups are different $\diagup CH_3 \qquad CH_2-CH_3 \diagdown$ Two groups are different
$$C=C$$
$\diagdown H \qquad H \diagup$

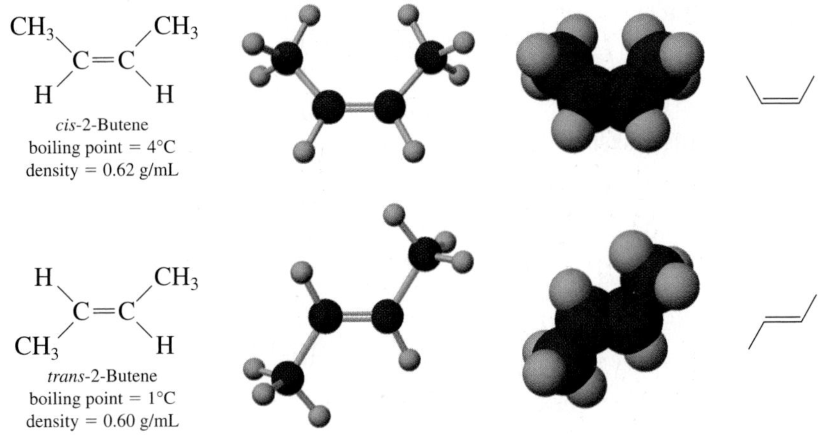

FIGURE 2.3 *Cis–trans* isomers: Different representations of the *cis* and *trans* isomers of 2-butene.

cis-2-Butene
boiling point = 4°C
density = 0.62 g/mL

trans-2-Butene
boiling point = 1°C
density = 0.60 g/mL

The simplest alkene for which *cis* and *trans* isomers exist is 2-butene.

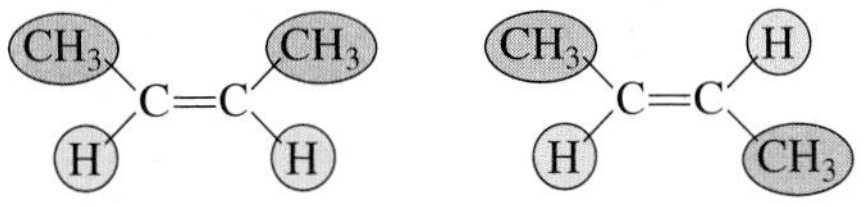

Structure A
(*cis*-2-butene)

Structure B
(*trans*-2-butene)

Recall from Section 1.14 that *cis* means "on the same side" and *trans* means "across from." Structure A is the *cis* isomer; both methyl groups are on the same side of the double bond. Structure B is the *trans* isomer; the methyl groups are on opposite sides of the double bond. The only way to convert structure A to structure B is to break the double bond. At room temperature, such bond breaking does not occur. Hence these two structures represent two different compounds (*cis–trans* isomers) that differ in boiling point, density, and so on. Figure 2.3 shows three-dimensional representations of the *cis* and *trans* isomers of 2-butene.

Cis–trans isomerism is not possible when one of the double-bonded carbons bears two identical groups. Thus neither 1-butene nor 2-methylpropene is capable of existing in *cis* and *trans* forms.

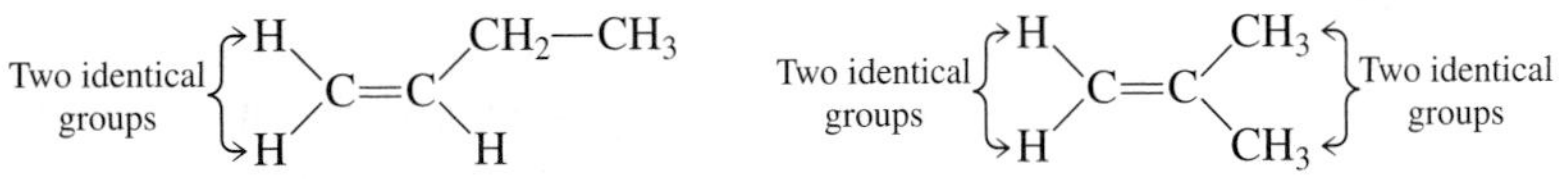

Two identical groups

1-Butene

Two identical groups

Two identical groups

2-Methylpropene

When alkenes contain more than one double bond, *cis–trans* considerations are more complicated. Orientation about each double bond must be considered independently of that at other sites. For example, for the molecule 2,4-heptadiene (two double bonds) there are four different *cis–trans* isomers (*trans–trans, trans–cis, cis–trans,* and *cis–cis*). The structures of two of these isomers are

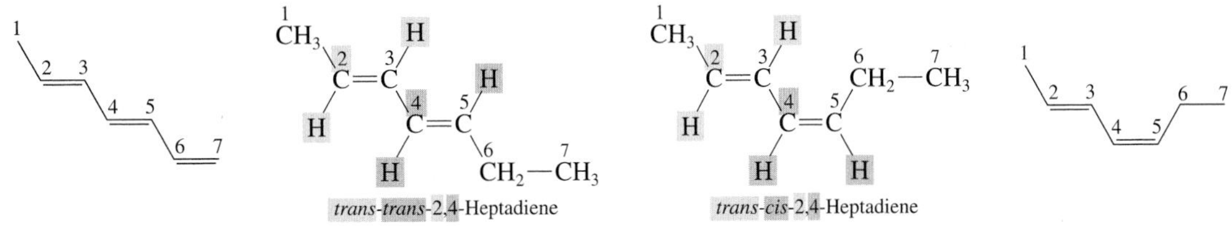

trans-trans-2,4-Heptadiene

trans-cis-2,4-Heptadiene

EXAMPLE 2.2

Determining Whether *Cis–Trans* Isomerism Is Possible in Substituted Alkenes

■ Determine whether each of the following substituted alkenes can exist in *cis–trans* isomeric forms.

a. 1-Bromo-1-chloroethene **b.** 2-Chloro-2-butene

Solution

a. The condensed structural formula for this compound is

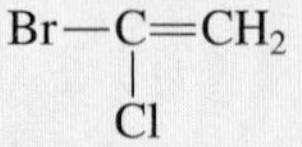

$$Br-\underset{\underset{Cl}{|}}{C}=CH_2$$

Redrawing this formula to emphasize the four attachments to the double-bonded carbon atoms gives

$$\underset{Cl}{\overset{Br}{\diagdown}}C=C\underset{H}{\overset{H}{\diagup}}$$

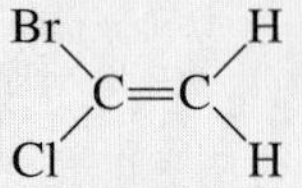

The carbon atom on the right has two identical attachments. Hence *cis–trans* isomerism is not possible.

b. The condensed structural formula for this compound is

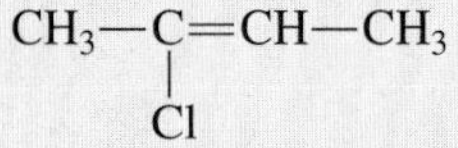

$$CH_3-\underset{\underset{Cl}{|}}{C}=CH-CH_3$$

Redrawing this formula to emphasize the four attachments to the double-bonded carbon atoms gives

$$\underset{Cl}{\overset{CH_3}{\diagdown}}C=C\underset{H}{\overset{CH_3}{\diagup}}$$

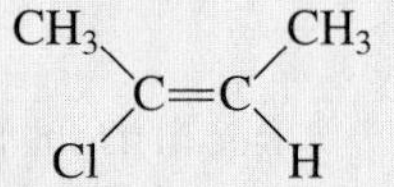

Because both carbon atoms of the double bond bear two different attachments, *cis–trans* isomers are possible.

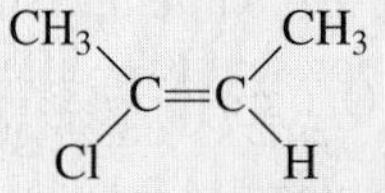

$$\underset{Cl}{\overset{CH_3}{\diagdown}}C=C\underset{H}{\overset{CH_3}{\diagup}} \qquad \underset{CH_3}{\overset{Cl}{\diagdown}}C=C\underset{H}{\overset{CH_3}{\diagup}}$$

cis-2-Chloro-2-butene 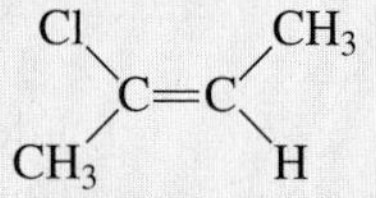   *trans*-2-Chloro-2-butene

Practice Exercise 2.2

Determine whether each of the following substituted alkenes can exist in *cis–trans* isomeric forms.

a. 1-Chloropropene   **b.** 2-Chloropropene

2.6 Naturally Occurring Alkenes

Alkenes are abundant in nature. Many important biological molecules are characterized by the presence of carbon–carbon double bonds within their structure. Two important types of naturally occurring substances to which alkenes contribute are pheromones and terpenes.

■ Pheromones

A **pheromone** *is a compound used by insects (and some animals) to transmit a message to other members of the same species.* Pheromones are often alkenes or alkene derivatives. The biological activity of alkene-type pheromones is usually highly dependent on whether the double bonds present are in a *cis* or a *trans* arrangement (Section 2.5).

CHEMICAL CONNECTIONS

Cis–Trans Isomerism and Vision

Cis–trans isomerism plays an important role in many biochemical processes, including the reception of light by the retina of the eye. Within the retina, microscopic structures called rods and cones contain a compound called *retinal,* which absorbs light. Retinal contains a carbon chain with five carbon–carbon double bonds, four in a *trans* configuration and one in a *cis* configuration. This arrangement of double bonds gives retinal a shape that fits the protein *opsin,* to which it is attached, as shown in the accompanying diagram.

When light strikes retinal, the *cis* double bond is converted to a *trans* double bond. The resulting *trans*-retinal no longer fits the protein opsin and is subsequently released. Accompanying this release is an electrical impulse, which is sent to the brain. Receipt of such impulses by the brain is what enables us to see.

In order to trigger nerve impulses again, *trans*-retinal must be converted back to *cis*-retinal. This occurs in the membranes of the rods and cones, where enzymes change *trans*-retinal back into *cis*-retinal.

The sex attractant of the female silkworm is a 16-carbon alkene derivative containing an —OH group. Two double bonds are present, *trans* at carbon 10 and *cis* at carbon 12.

This compound is 10 billion times more effective in eliciting a response from the male silkworm than the 10-*cis*–12-*trans* isomer and 10 trillion times more effective than the isomer wherein both bonds are in a *trans* configuration.

Insect sex pheromones are useful in insect control. A small amount of synthetically produced sex pheromone is used to lure male insects of a single species into a trap (see Figure 2.4). The trapped males are either killed or sterilized. Releasing sterilized males has proved effective in some situations. A sterile male can mate many times, preventing fertilization in many females, who usually mate only once. Sex attractant pheromones are now used to control the gypsy moth and the Mediterranean fruit fly.

FIGURE 2.4 The application of sex pheromones in insect control involves using a small amount of synthetically produced pheromone to lure a particular insect into a trap. This is accomplished without harming other "beneficial" insects.

■ Terpenes

A **terpene** *is an organic compound whose carbon skeleton is composed of two or more 5-carbon isoprene structural units.* Isoprene (2-methyl-1,3-butadiene) is a five-carbon diene.

2-Methyl-1,3-butadiene
(isoprene)

FIGURE 2.5 Selected terpenes containing two, three, and eight isoprene units. Dashed lines in the structures separate the individual isoprene units.

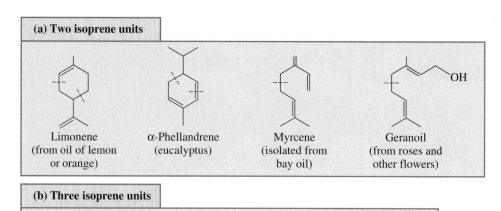

(a) Two isoprene units

Limonene
(from oil of lemon
or orange)

α-Phellandrene
(eucalyptus)

Myrcene
(isolated from
bay oil)

Geranoil
(from roses and
other flowers)

(b) Three isoprene units

Zingiberene
(from oil of ginger)

α-Farnesene
(from natural coating of apples)

Menthol
(mint)

(c) Eight isoprene units

β-Carotene
(present in carrots and other vegetables)

In later chapters, we will encounter additional isoprene-based molecules important in the functioning of the human body. They include vitamin K (Section 10.14), coenzyme Q (Section 12.7), and cholesterol (Section 8.9).

Terpenes are formed by joining the tail of one isoprene structural unit to the head of another unit.

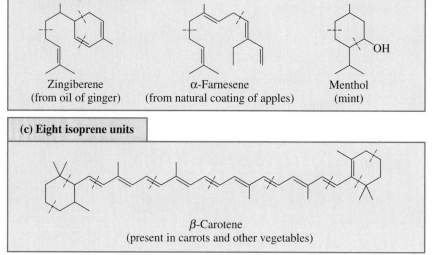

Isoprene structural unit

FIGURE 2.6 The molecule β-carotene is responsible for the yellow-orange color of carrots, apricots, and yams.

The isoprene structural unit maintains its isopentyl structure (Section 1.8) in a terpene, usually with modification of the isoprene double bonds.

Terpenes are among the most widely distributed compounds in the biological world, with over 22,000 structures known. Such compounds are responsible for the odors of many trees and for many characteristic plant fragrances.

The number of carbon atoms present in a terpene is always a multiple of the number 5 (10, 15, and so on). Parts (a) and (b) of Figure 2.5 give the structures of selected 10- and 15-carbon terpenes found in plants. Beta-carotene is a terpene whose structure has 40 carbon atoms present in 8 isoprene units (Figure 2.5c).

In the human body, dietary beta-carotene (obtained by eating yellow-colored vegetables) serves as a precursor for vitamin A (see Figure 2.6); splitting of a beta-carotene molecule produces two vitamin A molecules (Section 10.14). An additional role of beta-carotene in the body, independent of its vitamin A function, is that of antioxidant. An antioxidant is a substance that helps protect cells from damage from reactive oxygen-derived species called free radicals (Section 12.11).

Unbranched 1-Alkenes		
╳ C_3	C_5	C_7
C_2	C_4	C_6	C_8

Unsubstituted Cycloalkenes		
╳	C_3*	C_5	C_7
╳	C_4*	C_6	C_8

☐ Gas ☐ Liquid

*Cyclopropene and cyclobutene are relatively unstable compounds, readily converting to other hydrocarbons because of the severe bond angle strain associated with a small ring containing a double bond.

FIGURE 2.7 A physical-state summary for unbranched 1-alkenes and unsubstituted cycloalkenes with one double bond at room temperature and pressure.

2.7 Physical Properties of Alkenes

The general physical properties of alkenes include insolubility in water, solubility in nonpolar solvents, and densities lower than that of water. Thus alkenes have physical properties similar to those of alkanes (Section 1.16). The melting point of an alkene is usually lower than that of the alkane with the same number of carbon atoms.

Alkenes with 2 to 4 carbon atoms are gases at room temperature. Unsubstituted alkenes with 5 to 17 carbon atoms and one double bond are liquids, and those with still more carbon atoms are solids. Figure 2.7 is a physical-state summary for unbranched 1-alkenes and unsubstituted cycloalkenes with 8 or fewer carbon atoms.

2.8 Chemical Reactions of Alkenes

Alkenes, like alkanes, are very flammable. The combustion products, as with any hydrocarbon, are carbon dioxide and water.

$$C_2H_4 + 3O_2 \longrightarrow 2CO_2 + 2H_2O$$
Ethene

Pure alkenes are, however, too expensive to be used as fuel.

Aside from combustion, nearly all other reactions of alkenes take place at the carbon–carbon double bond(s). These reactions are called *addition reactions* because a substance is *added* to the double bond. This behavior contrasts with that of alkanes, where the most common reaction type, aside from combustion, is *substitution* (Section 1.17).

An **addition reaction** *is a reaction in which atoms or groups of atoms are added to each carbon atom of a carbon–carbon multiple bond in a hydrocarbon or hydrocarbon derivative.* A general equation for an alkene addition reaction is

$$\diagdown C = C \diagup + A—B \longrightarrow —\underset{A}{\overset{|}{C}}=\underset{B}{\overset{|}{C}}—$$

In this reaction, the A part of the reactant A—B becomes attached to one carbon atom of the double bond, and the B part to the other carbon atom (see Figure 2.8). As this occurs, the carbon–carbon double bond simultaneously becomes a carbon–carbon single bond.

Addition reactions can be classified as symmetrical or unsymmetrical. A **symmetrical addition reaction** *is an addition reaction in which identical atoms (or groups of atoms) are added to each carbon of a carbon–carbon multiple bond.* An **unsymmetrical**

FIGURE 2.8 In an alkene addition reaction, the atoms provided by an incoming molecule are attached to the carbon atoms originally joined by a double bond. In the process, the double bond becomes a single bond.

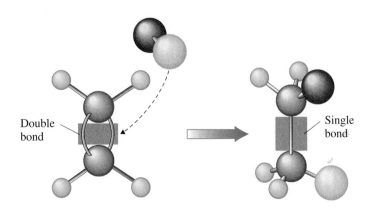

Double bond

Single bond

CHEMICAL CONNECTIONS   Carotenoids: A Source of Color

Carotenoids are the most widely distributed of the substances that give color to our world; they occur in flowers, fruits, plants, insects, and animals. These compounds are terpenes (Section 2.6) in which eight isoprene units are present. Structural formulas for two members of the carotenoid family, β-carotene and lycopene, follow.

β-Carotene

Lycopene

Present in both of these carotenoid structures is a *conjugated* system of 11 double bonds. (Conjugated double bonds are double bonds separated from each other by one single bond.) Color is frequently caused by the presence of compounds that contain extended conjugated-double-bond systems. When *visible* light strikes these compounds, certain wavelengths of the visible light are absorbed by the electrons in the conjugated-bond system. The unabsorbed wavelengths of visible light are reflected and are perceived as color.

The molecule β-carotene is responsible for the yellow-orange color in carrots, apricots, and yams. The yellow-orange color of autumn leaves comes from β-carotene. Leaves contain chlorophyll (green pigment) and β-carotene (yellow-pigment) in a ratio of approximately 3 to 1. The yellow-orange β-carotene color is masked by the chlorophyll until autumn, when the chlorophyll molecules decompose as a result of lower temperatures and less sunlight and are not replaced.

The molecule lycopene is the red pigment in tomatoes, paprika, and watermelon. Lycopene's structure differs from that of β-carotene in that the two rings in β-carotene have been opened. The ripening of a green tomato involves the gradual decomposition of chlorophyll with an associated unmasking of the red color of the lycopene present. A green pepper becomes red after ripening for the same reason.

Research studies indicate that lycopene has anticancer properties. One study comparing a group of men on a lycopene-rich diet with another group on a low-lycopene diet showed the incidence of prostate cancer was one-third lower in the lycopene-rich diet group.

Heat-processed tomatoes are a good source of dietary lycopene, with concentrated juice containing the highest levels of this substance. The lycopene in cooked tomatoes is absorbed more readily during digestion than is the lycopene in raw tomatoes. Recent studies indicate that red seedless watermelon contains as much lycopene as cooked tomatoes.

The anticancer properties of lycopene relate to its ability to react with highly reactive oxygen-containing molecules in the body, thereby preventing these molecules from oxidizing cellular components and creating new substances that might negatively affect cellular activity. Thus, lycopene has *antioxidant* properties. (See Section 3.14 for further information about antioxidants.)

Carotenoids such as β-carotene and lycopene are synthesized only by plants. They can, however, reach animal tissues via feed and can be modified and deposited therein. The yellowish tint of animal fat comes from β-carotene present in animal diets. The chicken egg yolk is another example of color imparted by dietary carotenoids.

Carotenoids, molecules that contain eight isoprene units, are responsible for the yellow-orange color of autumn leaves.

addition reaction *is an addition reaction in which different atoms (or groups of atoms) are added to the carbon atoms of a carbon–carbon multiple bond.*

Symmetrical Addition Reactions

The two most common examples of symmetrical addition reactions are hydrogenation and halogenation.

The following word associations are important to remember:

alkane — substitution reaction

alkene — addition reaction

An analogy can be drawn to a basketball team. When a *substitution* is made, one player leaves the game as another enters. The number of players on the court remains at five per team. If *addition* were allowed during a basketball game, two players could enter the game and no one would leave; there would be seven players per team on the court rather than five.

Hydrogenation of an alkene requires a catalyst. No reaction occurs if the catalyst is not present.

The Chemical Connections feature "*Trans* Fatty Acids and Blood Cholesterol Levels" in Chapter 8 addresses health issues relative to consumption of partially hydrogenated products.

A **hydrogenation reaction** *is an addition reaction in which H$_2$ is incorporated into molecules of an organic compound.* In alkene hydrogenation a hydrogen atom is added to each carbon atom of a double bond. It is accomplished by heating the alkene and H$_2$ in the presence of a catalyst (usually Ni or Pt).

$$CH_2{=}CH{-}CH_3 + H_2 \xrightarrow[\substack{150°C \\ 12–15\ atm \\ pressure}]{Ni\ or\ Pt} \underset{\text{Propane}}{\overset{\overset{H\quad H}{|\quad |}}{CH_2{-}CH{-}CH_3}}$$

Propene

The identity of the catalyst used in hydrogenation is specified by writing it above the arrow in the chemical equation for the hydrogenation. In general terms, hydrogenation of an alkene can be written as

$$\underset{\text{Alkene}}{\Large{>}C{=}C{<}} + H_2 \xrightarrow[\substack{Heat, \\ pressure}]{Ni\ or\ Pt} \underset{\text{Alkane}}{{-}\overset{\overset{H\ \ H}{|\ \ |}}{C}{-}C{-}}$$

The hydrogenation of vegetable oils is a very important commercial process today. Vegetable oils from sources such as soybeans and cottonseeds are composed of long-chain organic molecules that contain several double bonds. When these oils are hydrogenated, they are converted to low-melting solids that are used in margarines and shortenings.

A **halogenation reaction** *is an addition reaction in which a halogen is incorporated into molecules of an organic compound.* In alkene halogenation a halogen atom is added to each carbon atom of a double bond. Chlorination (Cl$_2$) and bromination (Br$_2$) are the two halogenation processes most commonly encountered. No catalyst is needed.

$$\underset{\text{2-Butene}}{CH_3{-}CH{=}CH{-}CH_3} + Cl_2 \longrightarrow \underset{\text{2,3-Dichlorobutane}}{CH_3{-}\overset{\overset{Cl}{|}}{CH}{-}\overset{\overset{Cl}{|}}{CH}{-}CH_3}$$

In general terms, halogenation of an alkene can be written as

$$\underset{\text{Alkene}}{\Large{>}C{=}C{<}} + \underset{\text{Halogen}}{X_2} \longrightarrow \underset{\text{Dihalogenated alkane}}{{-}\overset{\overset{X\ \ X}{|\ \ |}}{C}{-}C{-}} \quad (X = Cl, Br)$$

Bromination is often used to test for the presence of carbon–carbon double bonds in organic substances. Bromine in water or carbon tetrachloride is reddish brown. The dibromo compound(s) formed from the symmetrical addition of bromine to an organic compound is(are) colorless. Thus the decolorization of a Br$_2$ solution indicates the presence of carbon–carbon double bonds (see Figure 2.9).

FIGURE 2.9 A bromine in water solution is reddish brown (left). When a small amount of such a solution is added to an unsaturated hydrocarbon, the added solution is decolorized as the bromine adds to the hydrocarbon to form colorless dibromo compounds (right).

■ Unsymmetrical Addition Reactions

Two important types of unsymmetrical addition reactions are hydrohalogenation and hydration.

A **hydrohalogenation reaction** *is an addition reaction in which a hydrogen halide (HCl, HBr, or HI) is incorporated into molecules of an organic compound.* In alkene hydrohalogenation one carbon atom of a double bond receives a halogen atom and the other carbon atom receives a hydrogen atom. Hydrohalogenation reactions require no catalyst. For *symmetrical* alkenes, such as ethene, only one product results from hydrohalogenation.

$$CH_2\!\!=\!\!CH_2 + H\!-\!Cl \longrightarrow \underset{\text{Chloroethane}}{\overset{\displaystyle \overset{H}{|}\quad\overset{Cl}{|}}{CH_2\!-\!CH_2}}$$

Ethene   Chloroethane

A **hydration reaction** *is an addition reaction in which H₂O is incorporated into molecules of an organic compound.* In alkene hydration one carbon atom of a double bond receives a hydrogen atom and the other carbon atom receives an —OH group. Alkene hydration requires a small amount of H_2SO_4 (sulfuric acid) as a catalyst. For *symmetrical* alkenes, only one product results from hydration.

$$CH_2\!\!=\!\!CH_2 + H\!-\!OH \xrightarrow{H_2SO_4} \underset{\text{An alcohol}}{\overset{\displaystyle \overset{H}{|}\quad\overset{OH}{|}}{CH_2\!-\!CH_2}}$$

Ethene

In this equation, the water (H_2O) is written as H—OH to emphasize how this molecule adds to the double bond. Note also that the product of this hydration reaction contains an —OH group. Hydrocarbon derivatives of this type are called *alcohols.* Such compounds are the subject of Chapter 3.

When the alkene involved in a hydrohalogenation or hydration reaction is itself *unsymmetrical,* more than one product is possible. (An unsymmetrical alkene is one in which the two carbon atoms of the double bond are not equivalently substituted.) For example, the addition of HCl to propene (an unsymmetrical alkene) could produce either 1-chloropropane or 2-chloropropane, depending on whether the H from the HCl attaches itself to carbon 2 or carbon 1.

$$\underset{\text{Propene}}{CH_2\!\!=\!\!CH\!-\!CH_3} + HCl \longrightarrow \underset{\text{1-Chloropropane}}{\overset{\displaystyle \overset{Cl}{|}\quad\overset{H}{|}}{CH_2\!-\!CH\!-\!CH_3}}$$

or

$$\underset{\text{Propene}}{CH_2\!\!=\!\!CH\!-\!CH_3} + HCl \longrightarrow \underset{\text{2-Chloropropane}}{\overset{\displaystyle \overset{H}{|}\quad\overset{Cl}{|}}{CH_2\!-\!CH\!-\!CH_3}}$$

When two isomeric products are possible, one product often predominates. The dominant product can be predicted by using Markovnikov's rule, named after the Russian chemist Vladimir V. Markovnikov (see Figure 2.10). **Markovnikov's rule** states that *when an unsymmetrical molecule of the form* HQ *adds to an unsymmetrical alkene, the hydrogen atom from the* HQ *becomes attached to the unsaturated carbon atom that already has the most hydrogen atoms.* Thus the major product in our example involving propene is 2-chloropropane.

The addition of water to carbon–carbon double bonds occurs in many biochemical reactions that take place in the human body — for example, in the citric acid cycle (Section 12.6) and in the oxidation of fatty acids (Section 14.4).

FIGURE 2.10 Vladimir Vasilevich Markovnikov (1837–1904). A professor of chemistry at several Russian universities, Markovnikov (pronounced Mar-cove-na-coff) synthesized rings containing four carbon atoms and seven carbon atoms, thereby disproving the notion of the day that carbon could form only five- and six-membered rings.

Two catchy summaries of Markovnikov's rule are "Hydrogen goes where hydrogen is" and "The rich get richer" (in terms of hydrogen).

EXAMPLE 2.3

Predicting Products in Alkene Addition Reactions Using Markovnikov's Rule

■ Using Markovnikov's rule, predict the predominant product in each of the following addition reactions.

a. $CH_3\!-\!CH_2\!-\!CH_2\!-\!CH\!\!=\!\!CH_2 + HBr \rightarrow$

b. ⬠ + HCl →

c. $CH_3\!-\!CH\!\!=\!\!CH\!-\!CH_2\!-\!CH_3 + HBr \rightarrow$

Solution

a. The hydrogen atom will add to carbon 1, because carbon 1 already contains more hydrogen atoms than carbon 2. The predominant product of the addition will be 2-bromopentane.

$$CH_3-CH_2-CH_2-\overset{\textcircled{2}}{CH}=\overset{\textcircled{1}}{CH_2} + HBr \longrightarrow CH_3-CH_2-CH_2-\overset{Br}{\underset{|}{CH}}-\overset{H}{\underset{|}{CH_2}}$$

b. Carbon 1 of the double bond does not have any H atoms directly attached to it. Carbon 2 of the double bond has one H atom (H atoms are not shown in the structure but are implied) attached to it. The H atom from the HCl will add to carbon 2, giving 1-chloro-1-methylcyclopentane as the product.

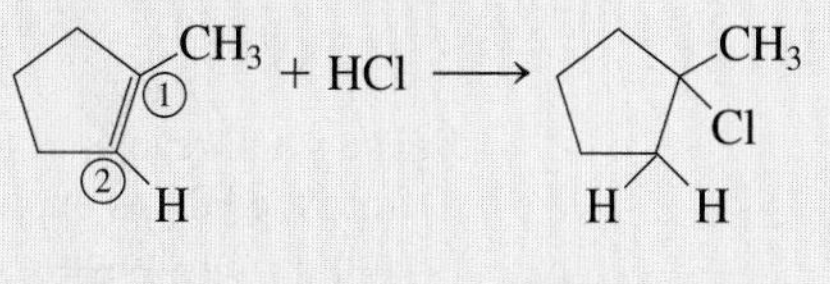

c. Each carbon atom of the double bond in this molecule has one hydrogen atom. Thus Markovnikov's rule does not favor either carbon atom. The result is two isomeric products that are formed in almost equal quantities.

$$CH_3-\overset{}{\underset{\underset{Br}{|}}{CH}}-CH_2-CH_2-CH_3 \qquad and \qquad CH_3-CH_2-\overset{}{\underset{\underset{Br}{|}}{CH}}-CH_2-CH_3$$

2-Bromopentane   3-Bromopentane

Practice Exercise 2.3

Using Markovnikov's rule, predict the predominant product in each of the following addition reactions.

a. $CH_2=CH-CH_2-CH_3 + HCl \rightarrow$

b. 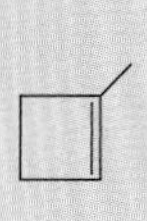 $+ HBr \rightarrow$

In compounds that contain more than one carbon–carbon double bond, such as dienes and trienes, addition can occur at each of the double bonds. In the complete hydrogenation of a diene and in that of a triene, the amounts of hydrogen needed are twice as much and three times as much, respectively, as that needed for the hydrogenation of an alkene with one double bond.

$$CH_2=CH-CH_2-CH_2-CH_2-CH_3 + H_2 \xrightarrow{Ni} CH_3-(CH_2)_4-CH_3$$

$$CH_2=CH-CH=CH-CH_2-CH_3 + 2H_2 \xrightarrow{Ni} CH_3-(CH_2)_4-CH_3$$

$$CH_2=CH-CH=CH-CH=CH_2 + 3H_2 \xrightarrow{Ni} CH_3-(CH_2)_4-CH_3$$

EXAMPLE 2.4

Predicting Reactants and Products in Alkene Addition Reactions

■ Supply the structural formula of the missing substance in each of the following addition reactions.

a. $CH_3-CH_2-CH=CH_2 + H_2O \xrightarrow{H_2SO_4} ?$

b. 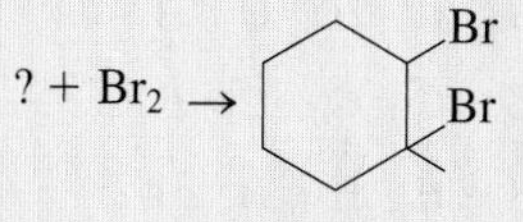
$? + Br_2 \rightarrow$

c. 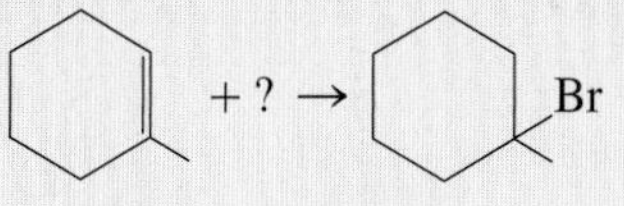
$+ ? \rightarrow$

d. $CH_3-CH=CH-CH=CH_2 + 2H_2 \xrightarrow{Ni} ?$

Solution

a. This is a hydration reaction. Using Markovnikov's rule, we determine that the H will become attached to carbon 1, which has more hydrogen atoms than carbon 2, and that the —OH group will be attached to carbon 2.

$$\underset{\substack{| \\ OH}}{CH_3-CH_2-CH-CH_3}$$

b. The reactant alkene will have to have a double bond between the two carbon atoms that bromine atoms are attached to in the product.

c. The small reactant molecule that adds to the double bond is HBr. The added Br atom from the HBr is explicity shown in the product's structural formula, but the added H atom is not shown.

d. Hydrogen will add at each of the double bonds. The product hydrocarbon is pentane.

$$CH_3-CH_2-CH_2-CH_2-CH_3$$

Practice Exercise 2.4

Supply the structural formula of the missing substance in each of the following addition reactions.

a. $CH_3-CH_2-CH=CH_2 + HBr \rightarrow$?

b. ? + H_2O $\xrightarrow{H_2SO_4}$ (cyclopentane with OH)

c. (cyclohexene) + 2? $\xrightarrow{Ni}$ (cyclohexane)

d. $\underset{\substack{| \\ CH_3}}{CH_3-C=CH-CH_3} + Cl_2 \longrightarrow$?

2.9 Polymerization of Alkenes: Addition Polymers

The word *polymer* comes from the Greek *poly,* which means "many," and *meros,* which means "parts."

A **polymer** *is a large molecule formed by the repetitive bonding together of many smaller molecules.* The smaller repeating units of a polymer are called *monomers.* A **monomer** *is the small molecule that is the structural repeating unit in a polymer.* The process by which a polymer is made is called *polymerization.* A **polymerization reaction** *is a chemical reaction in which the repetitious combining of many small molecules (monomers) produces a very large molecule (the polymer).* With appropriate catalysts, simple alkenes and simple substituted alkenes readily undergo polymerization.

The type of polymer that alkenes and substituted alkenes form is an *addition polymer.* An **addition polymer** *is a polymer in which the monomers simply "add together" with no other products formed besides the polymer.* Addition polymerization is similar to the addition reactions described in Section 2.8 except that there is no reactant other than the alkene or substituted alkene.

We will consider polymer types other than addition polymers in Sections 4.11 and 5.18.

The simplest alkene addition polymer has ethylene (ethene) as the monomer. With appropriate catalysts, ethylene readily adds to itself to produce polyethylene.

$$\underset{\substack{| \\ H}}{\overset{\substack{H \\ |}}{C}}=\underset{\substack{| \\ H}}{\overset{\substack{H \\ |}}{C}} + \underset{\substack{| \\ H}}{\overset{\substack{H \\ |}}{C}}=\underset{\substack{| \\ H}}{\overset{\substack{H \\ |}}{C}} + \underset{\substack{| \\ H}}{\overset{\substack{H \\ |}}{C}}=\underset{\substack{| \\ H}}{\overset{\substack{H \\ |}}{C}} \xrightarrow{Catalyst} -\underset{\substack{| \\ H}}{\overset{\substack{H \\ |}}{C}}-\underset{\substack{| \\ H}}{\overset{\substack{H \\ |}}{C}}-\underset{\substack{| \\ H}}{\overset{\substack{H \\ |}}{C}}-\underset{\substack{| \\ H}}{\overset{\substack{H \\ |}}{C}}-\underset{\substack{| \\ H}}{\overset{\substack{H \\ |}}{C}}-\underset{\substack{| \\ H}}{\overset{\substack{H \\ |}}{C}}-$$

Polyethylene

An *exact* formula for a polymer such as polyethylene cannot be written because the length of the carbon chain varies from polymer molecule to polymer molecule. In recognition of this "inexactness" of formula, the notation used for denoting polymer formulas is independent of carbon chain length. We write the formula of the simplest repeating unit (the monomer with the double bond changed to a single bond) in parentheses and then add the subscript n after the parentheses, with n being understood to represent a very large number. Using this notation, we have, for the formula of polyethylene,

$$\left(\begin{array}{cc} H & H \\ | & | \\ -C - C- \\ | & | \\ H & H \end{array}\right)_n$$

This notation clearly identifies the basic repeating unit found in the polymer.

■ Substituted-ethene Addition Polymers

Many substituted alkenes undergo polymerization similar to that of ethene when they are treated with the proper catalyst. For a monosubstituted-ethene monomer, the general polymerization equation is

$$H_2C = \overset{\displaystyle \overset{Z}{|}}{CH} \xrightarrow{\text{Polymerization}} \left(CH_2 - \overset{\displaystyle \overset{Z}{|}}{CH}\right)_n$$

Variation in the substituent group Z can change polymer properties dramatically, as is shown by the entries in Table 2.1, a listing of monomers for the five ethene-based polymers *polyethylene, polypropylene, poly(vinyl chloride) (PVC), Teflon,* and *polystyrene,* along with several uses of each. Figure 2.11 depicts the preparation of polystyrene. Figure 2.12 contrasts the structures of polyethylene, polypropylene, and poly (vinyl chloride) as depicted in space-filling models.

The properties of an ethene-based polymer depend not only on monomer identity but also on the average size (length) of polymer molecules and on the extent of polymer branching. For example, there are three major types of polyethylene: high-density polyethylene (HDPE), low-density polyethylene (LDPE) and linear low-density polyethylene (LLDPE). The major difference among these three materials is the degree of branching of the polymer chain. HDPE and LLDPE are composed of linear, unbranched carbon chains, while LDPE chains are branched. The strong and thick plastic bags from a shopping mall are LLDPE, the thin and flimsy grocery store plastic bags are HDPE, and the very wispy garment bags dry cleaners use are LDPE.

In general, HDPE materials are rigid or semi-rigid with uses such as threaded bottle caps, toys, bottles, and milk jugs whereas LDPE materials are more flexible with uses such as plastic film and squeeze bottles (see Figure 2.13). Objects made of HDPE hold their shape in boiling water, whereas objects made of LDPE become severely deformed at this temperature.

FIGURE 2.11 Preparation of polystyrene. When styrene, C_6H_5—CH=CH$_2$, is heated with a catalyst (benzoyl peroxide), it yields a viscous liquid. After some time, this liquid sets to a hard plastic (sample shown at left).

FIGURE 2.12 Line-angle formulas and space-filling models of segments of the ethene-based polymers (a) polyethylene, (b) polypropylene, and (c) poly(vinyl chloride).

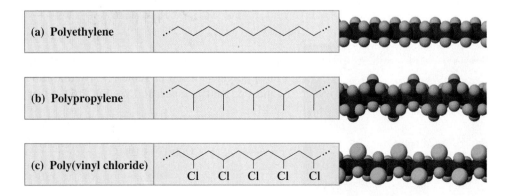

TABLE 2.1

Some Common Polymers Obtained from Ethene-Based Monomers

Polymer Formula and Name	Monomer Formula and Name	Uses of Polymer
polyethylene	ethylene	bottles, plastic bags, toys, electrical insulation
polypropylene	propylene	indoor–outdoor carpeting, bottles, molded parts (including heart valves)
poly(vinyl chloride) (PVC)	vinyl chloride	plastic wrap, bags for intravenous drugs, garden hose, plastic pipe, simulated leather (Naugahyde)
Teflon	tetrafluoroethylene	cooking utensil coverings, electrical insulation, component of artificial joints in body parts replacement
polystyrene	styrene	toys, Styrofoam packaging, cups, simulated wood furniture

FIGURE 2.13 Examples of objects made of polyethylene. Polyethylene objects that are strong and rigid (bottles, toys, covering for wire) contain HDPE (high-density polyethylene). Polyethylene objects that are very flexible (plastic bags and packaging materials) contain LDPE (low-density polyethylene).

■ Butadiene-Based Addition Polymers

When dienes such as 1,3-butadiene are used as the monomers in addition polymerization reactions, the resulting polymers contain double bonds and are thus still unsaturated.

$$CH_2{=}CH{-}CH{=}CH_2 \xrightarrow{\text{Polymerization}} {\left(\!CH_2{-}CH{=}CH{-}CH_2\!\right)}_n$$

1,3-Butadiene   Polybutadiene

In general, unsaturated polymers are much more flexible than the ethene-based saturated polymers listed in Table 2.1. Natural rubber is a flexible addition polymer whose repeating unit is isoprene (Section 2.6)—that is, 2-methyl-1,3-butadiene (see Figure 2.14).

$$CH_2{=}C{-}CH{=}CH_2 \xrightarrow{\text{Polymerization}} {\left(\!CH_2{-}C{=}CH{-}CH_2\!\right)}_n$$

           | |
          CH_3 CH_3

Isoprene   Polyisoprene
(2-methyl-1,3-butadiene)   (natural rubber)

Addition Copolymers

Saran Wrap is a polymer in which two different monomers are present: chloroethene (vinyl chloride) and 1,1-dichloroethene.

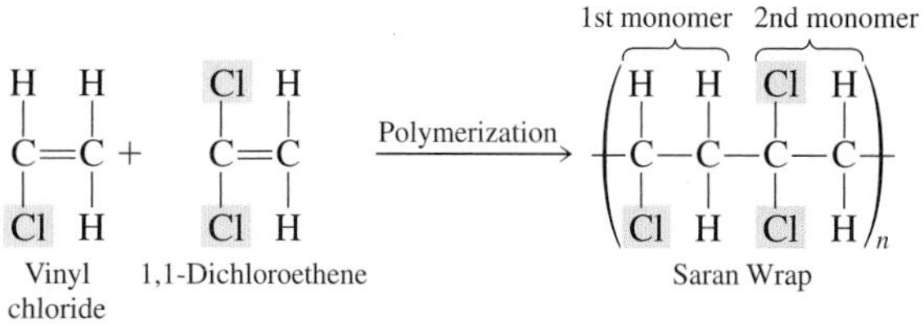

$$
\underset{\substack{\text{Vinyl}\\\text{chloride}}}{\underset{\text{Cl}}{\overset{\text{H}}{\text{C}}}=\underset{\text{H}}{\overset{\text{H}}{\text{C}}}} + \underset{\text{1,1-Dichloroethene}}{\underset{\text{Cl}}{\overset{\text{Cl}}{\text{C}}}=\underset{\text{H}}{\overset{\text{H}}{\text{C}}}} \xrightarrow{\text{Polymerization}} \underset{\text{Saran Wrap}}{\left(\underset{\text{Cl}}{\overset{\text{H}}{\text{C}}}-\underset{\text{H}}{\overset{\text{H}}{\text{C}}}-\underset{\text{Cl}}{\overset{\text{Cl}}{\text{C}}}-\underset{\text{H}}{\overset{\text{H}}{\text{C}}}\right)_n}
$$

1st monomer 2nd monomer

Such a polymer is an example of a *copolymer.* A **copolymer** *is a polymer in which two different monomers are present.* Another important copolymer is styrene–butadiene rubber, the leading synthetic rubber in use today. It contains the monomers 1,3-butadiene and styrene in a 3:1 ratio. It is a major ingredient in automobile tires

The Chemistry at a Glance feature on page 56 summarizes the reaction chemistry of alkenes presented in this and the previous section.

2.10 Alkynes

Alkynes represent a second type of unsaturated hydrocarbon. An **alkyne** *is an acyclic unsaturated hydrocarbon that contains one or more carbon–carbon triple bonds.* The alkyne functional group is, thus, a C≡C group. As the family name *alkyne* indicates, the characteristic "ending" associated with a triple bond is -*yne*.

The general formula for an alkyne with one triple bond is C_nH_{2n-2}. Thus the simplest member of this type of alkyne has the formula C_2H_2, and the next member, with $n = 3$, has the formula C_3H_4.

$$
\underset{\text{Ethyene}}{\text{CH}\equiv\text{CH}} \qquad \underset{\text{Propyne}}{\text{CH}\equiv\text{C}-\text{CH}_3}
$$

> Very few biological molecules are known that contain a carbon–carbon triple bond.

The presence of a carbon–carbon triple bond in a molecule always results in a linear arrangement for the two atoms attached to the carbons of the triple bond. Thus ethyne is a linear molecule (see Figure 2.15).

The simplest alkyne, ethyne (C_2H_2), is the most important alkyne from an industrial standpoint. A colorless gas, it goes by the common name *acetylene* and is used in oxyacetylene torches, high-temperature torches used for cutting and welding materials.

IUPAC Nomenclature for Alkynes

> *Cycloalkynes,* molecules that contain a triple bond as part of a ring structure, are known, but they are not common. Because of the 180° angle associated with a triple bond, a ring system containing a triple bond has to be quite large. The smallest cycloalkyne that has been isolated is cyclooctyne.

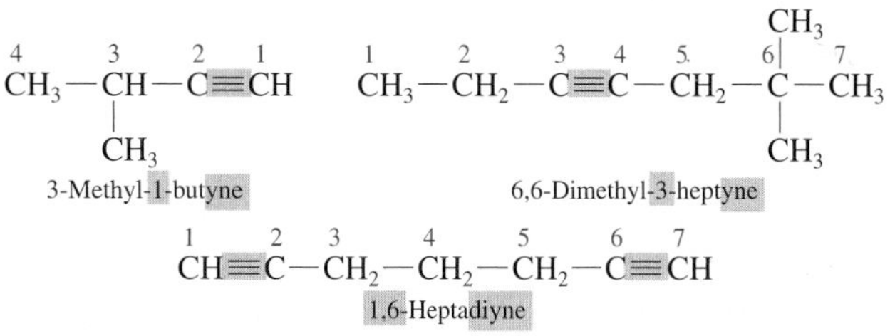

The rules for naming alkynes are identical to those used to name alkenes (Section 2.3), except the ending -*yne* is used instead of -*ene*. Consider the following structures and their IUPAC names.

$$
\overset{4}{\text{CH}_3}-\overset{3}{\underset{\underset{\text{CH}_3}{|}}{\text{CH}}}-\overset{2}{\text{C}}\equiv\overset{1}{\text{CH}} \qquad \overset{1}{\text{CH}_3}-\overset{2}{\text{CH}_2}-\overset{3}{\text{C}}\equiv\overset{4}{\text{C}}-\overset{5}{\text{CH}_2}-\overset{6}{\underset{\underset{\text{CH}_3}{|}}{\overset{\overset{\text{CH}_3}{|}}{\text{C}}}}-\overset{7}{\text{CH}_3}
$$

3-Methyl-1-butyne   6,6-Dimethyl-3-heptyne

$$
\overset{1}{\text{CH}}\equiv\overset{2}{\text{C}}-\overset{3}{\text{CH}_2}-\overset{4}{\text{CH}_2}-\overset{5}{\text{CH}_2}-\overset{6}{\text{C}}\equiv\overset{7}{\text{CH}}
$$

1,6-Heptadiyne

$$
\text{H}-\text{C}\equiv\text{C}-\text{H}\text{-----}
$$
180°

Ethyne—a linear molecule with bond angles of 180°

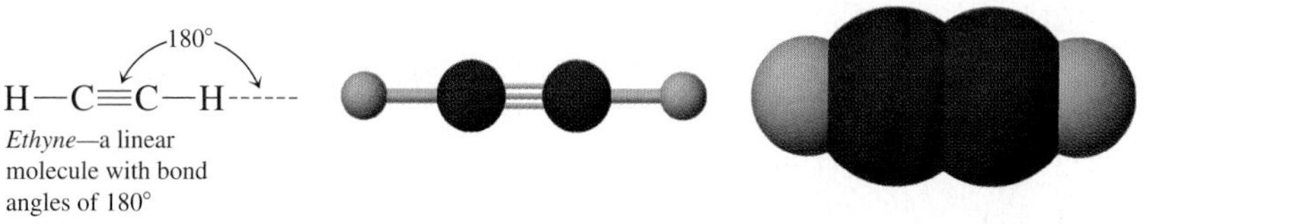

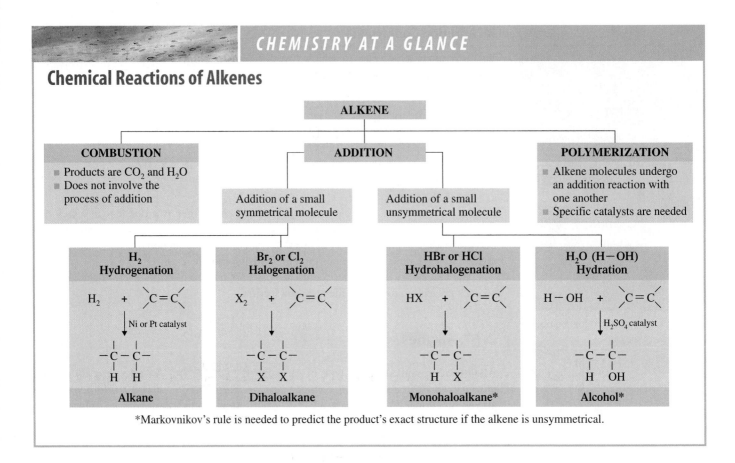

CHEMISTRY AT A GLANCE

Chemical Reactions of Alkenes

*Markovnikov's rule is needed to predict the product's exact structure if the alkene is unsymmetrical.

Early cars had carbide headlights that produced acetylene by the action of slowly dripping water on calcium carbide. This same type of lamp, which was also used by miners, is still often used by spelunkers (cave explorers).

Common names for simple alkynes are based on the name *acetylene,* as shown in the following examples.

$CH \equiv CH$ $CH_3 - C \equiv CH$ $CH_3 - C \equiv C - CH_3$
Acetylene   Methylacetylene   Dimethylacetylene

■ Isomerism and Alkynes

Because of the linearity (180° angles) about an alkyne's triple bond, *cis–trans* isomerism, such as that found in alkenes, is not possible for alkynes because there are no "up" and "down" positions. However, constitutional isomers are possible—both relative to the carbon chain (skeletal isomers) and to the position of the triple bond (positional isomers).

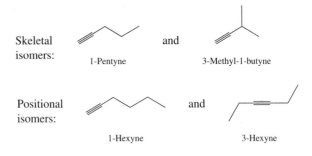

Skeletal isomers: 1-Pentyne   and   3-Methyl-1-butyne

Positional isomers: 1-Hexyne   and   3-Hexyne

FIGURE 2.16 A physical-state summary for unbranched 1-alkynes at room temperature and pressure.

Unbranched 1-Alkynes			
✕	C_3	C_5	C_7
C_2	C_4	C_6	C_8

☐ Gas ☐ Liquid

■ Physical and Chemical Properties of Alkynes

The physical properties of alkynes are similar to those of alkenes and alkanes. In general, alkynes are insoluble in water but soluble in organic solvents, have densities less than that of water, and have boiling points that increase with molecular mass. Low-molecular-mass alkynes are gases at room temperature. Figure 2.16 is a physical-state summary for unbranched 1-alkynes with eight or fewer carbon atoms.

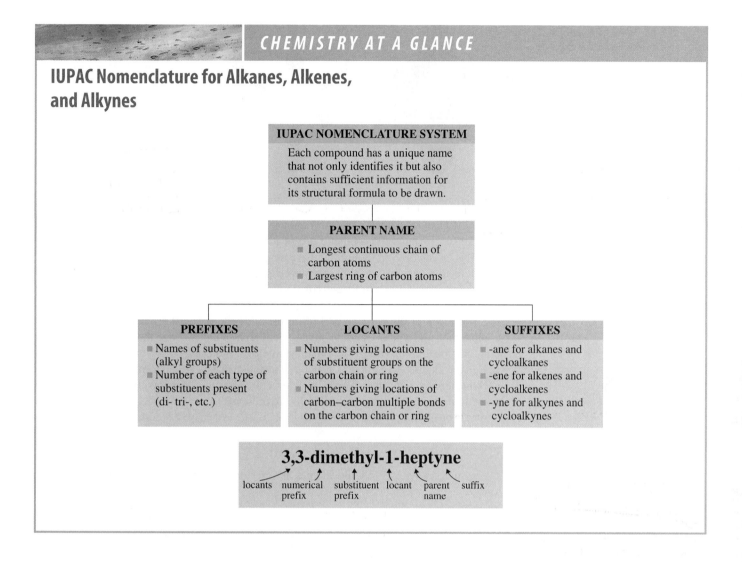

CHEMISTRY AT A GLANCE

IUPAC Nomenclature for Alkanes, Alkenes, and Alkynes

IUPAC NOMENCLATURE SYSTEM

Each compound has a unique name that not only identifies it but also contains sufficient information for its structural formula to be drawn.

PARENT NAME

- Longest continuous chain of carbon atoms
- Largest ring of carbon atoms

PREFIXES

- Names of substituents (alkyl groups)
- Number of each type of substituents present (di- tri-, etc.)

LOCANTS

- Numbers giving locations of substituent groups on the carbon chain or ring
- Numbers giving locations of carbon–carbon multiple bonds on the carbon chain or ring

SUFFIXES

- -ane for alkanes and cycloalkanes
- -ene for alkenes and cycloalkenes
- -yne for alkynes and cycloalkynes

3,3-dimethyl-1-heptyne

locants   numerical   substituent   locant   parent   suffix
prefix   prefix   name

The triple-bond functional group of alkynes behaves chemically quite similarly to the double-bond functional group of alkenes. Thus there are many parallels between alkene chemistry and alkyne chemistry. The same substances that add to double bonds (H_2, HCl, Cl_2, and so on) also add to triple bonds. However, two molecules of a specific reactant can add to a triple bond, as contrasted to the addition of one molecule of reactant to a double bond. In triple-bond addition, the first molecule converts the triple bond into a double bond, and the second molecule then converts the double bond into a single bond. For example, propyne reacts with H_2 to form propene first and then to form propane.

$$CH\equiv C-CH_3 \xrightarrow[Ni]{H_2} CH_2\!=\!CH-CH_3 \xrightarrow[Ni]{H_2} CH_3\!-\!CH_2-CH_3$$

An alkyne   An alkene   An alkane
(propyne)   (propene)   (propane)

Alkynes, like alkenes and alkanes, are flammable, that is, they readily undergo combustion reactions.

2.11 Aromatic Hydrocarbons

Aromatic hydrocarbons are the third class of unsaturated hydrocarbons; the alkenes and alkynes (previously considered) are the other two classes. An **aromatic hydrocarbon** *is an unsaturated cyclic hydrocarbon that does not readily undergo addition reactions.* This

---

Students often ask whether it is possible to have hydrocarbons in which both double and triple bonds are present. The answer is yes. Immediately, another question is asked. How do we name such compounds? Such compounds are called *alkenynes*. An example is

$$CH\equiv C-CH\!=\!CH_2$$
1-Buten-3-yne

A double bond has priority over a triple bond in numbering the chain when numbering systems are equivalent. Otherwise, the chain is numbered from the end closest to a multiple bond.

FIGURE 2.17 Space-filling and ball-and-stick models for the structure of benzene.

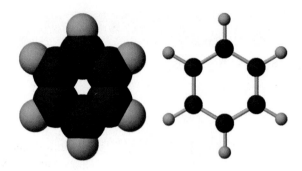

reaction behavior, which is very different from that of alkenes and alkynes, explains the separate classification for aromatic hydrocarbons.

The fact that, even though they are unsaturated compounds, aromatic hydrocarbons do not readily undergo addition reactions suggests that the bonding present in this type of compound must differ significantly from that in alkenes and alkynes. Such is indeed the case.

Let's look at the bonding present in *benzene,* the simplest aromatic hydrocarbon, to explore this new type of bonding situation and to also characterize the aromatic hydrocarbon functional group. Benzene, a flat, symmetrical molecule with a molecular formula of C_6H_6 (see Figure 2.17), has a structural formula that is often formalized as that of a cyclohexatriene—in other words, as a structural formula that involves a six-membered carbon ring in which three double bonds are present.

This structure is one of two equivalent structures that can be drawn for benzene that differ only in the locations of the double bonds (1,3,5 positions versus 2,4,6 positions):

Neither of these conventional structures, however, is totally correct. Experimental evidence indicates that all of the carbon–carbon bonds in benzene are equivalent (identical), and these structures imply three bonds of one type (double bonds) and three bonds of a different type (single bonds).

The equivalent nature of the carbon–carbon bonds in benzene is addressed by considering the correct bonding structure for benzene to be an *average* of the two "triene" structures. Related to this "average"-structure situation is the concept that electrons associated with the ring double bonds are not held between specific carbon atoms; instead, they are free to move "around" the carbon ring. Thus the true structure for benzene, an intermediate between that represented by the two "triene" structures, is a situation in which all carbon–carbon bonds are equivalent; they are neither single nor double bonds but something in between. Placing a double-headed arrow between the conventional structures that are averaged to obtain the true structure is one way to denote the average structure.

An alternative notation for denoting the bonding in benzene—a notation that involves a single structure—is

In this "circle-in-the-ring" structure for benzene, the circle denotes the electrons associated with the double bond that move "around" the ring. Each carbon atom in the ring can be considered to participate in three conventional (localized) bonds (two C—C bonds and one C—H bond) and in one *delocalized bond* (the circle) that involves all six carbon atoms. A **delocalized bond** *is a covalent bond in which electrons are shared among more than two atoms.* This delocalized bond is what causes benzene and its derivatives to be resistant to addition reactions, a property normally associated with unsaturation in a molecule.

The structure represented by the notation

is called an *aromatic ring system,* and it is the functional group present in aromatic compounds. An **aromatic ring system** *is a highly unsaturated carbon ring system in which both localized and delocalized bonds are present.*

2.12 Names for Aromatic Hydrocarbons

Replacement of one or more of the hydrogen atoms on benzene with other groups produces benzene derivatives. Compounds with alkyl groups or halogen atoms attached to the benzene ring are commonly encountered. We consider first the naming of benzene derivatives with one substituent, then the naming of those with two substituents, and finally the naming of those with three or more substituents.

■ Benzene Derivatives with One Substituent

The IUPAC system of naming monosubstituted benzene derivatives uses the name of the substituent as a prefix to the name benzene. Examples of this type of nomenclature include

F	Cl	CH_3 $CH—CH_3$	CH_3 CH_2
Fluorobenzene	Chlorobenzene	Isopropylbenzene	Ethylbenzene

A few monosubstituted benzenes have names wherein the substituent and the benzene ring taken together constitute a new parent name. Two important examples of such nomenclature with hydrocarbon substituents are

CH_3	$CH=CH_2$
Toluene (not methylbenzene)	Styrene (not vinylbenzene)

Both of these compounds are industrially important chemicals.

Monosubstituted benzene structures are often drawn with the substituent at the "12 o'clock" position, as in the previous structures. However, because all the hydrogen atoms in benzene are equivalent, it does not matter at which carbon of the ring the substituted group is located. Each of the following formulas represents toluene.

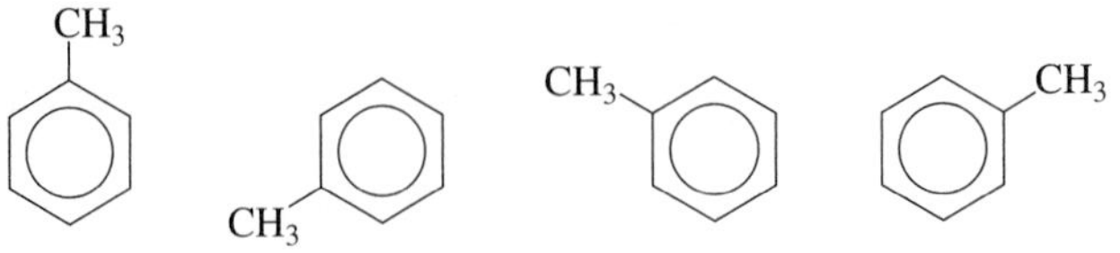

For monosubstituted benzene rings that have a group attached that is not easily named as a substituent, the benzene ring is often treated as a group attached to this substituent. In this reversed approach, the benzene ring attachment is called a *phenyl* group, and the compound is named according to the rules for naming alkanes, alkenes, and alkynes.

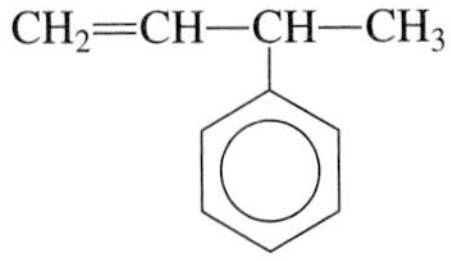

$$CH_2\!=\!CH\!-\!CH\!-\!CH_3$$

3-Phenyl-1-butene

The word *phenyl* comes from "phene," a European term used during the 1800s for benzene. The word is pronounced *fen*-nil.

■ Benzene Derivatives with Two Substituents

When two substituents, either the same or different, are attached to a benzene ring, three isomeric structures are possible.

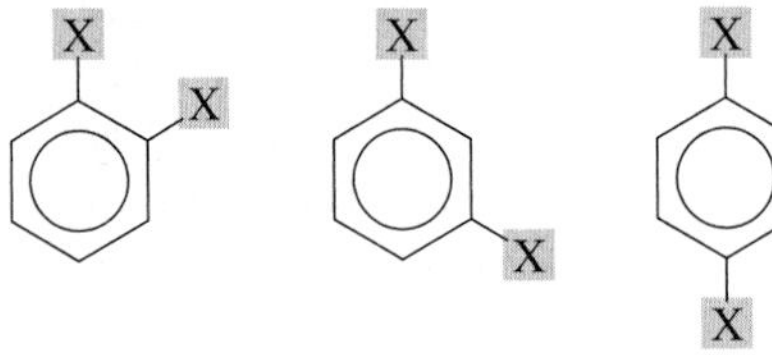

To distinguish among these three isomers, we must specify the positions of the substituents relative to one another. This can be done in either of two ways: by using numbers or by using nonnumerical prefixes.

When numbers are used, the three isomeric dichlorobenzenes have the first-listed set of names:

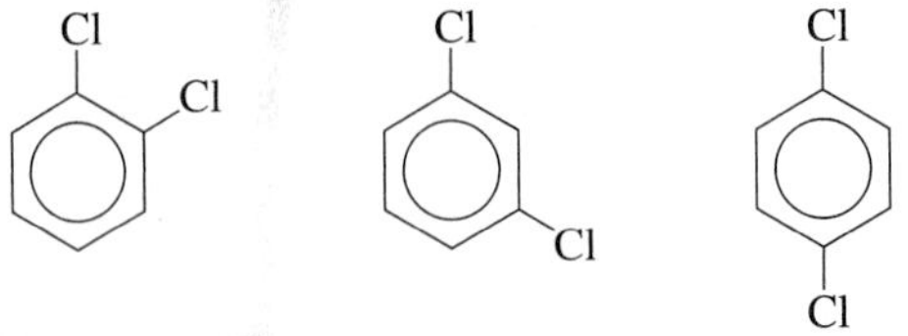

Cis–trans isomerism is not possible for disubstituted benzenes. All 12 atoms of benzene are in the same plane—that is, benzene is a flat molecule. When a substituent group replaces an H atom, the atom that bonds the group to the ring is also in the plane of the ring.

| 1,2-Dichlorobenzene | 1,3-Dichlorobenzene | 1,4-Dichlorobenzene |
| (*ortho*-dichlorobenzene) | (*meta*-dichlorobenzene) | (*para*-dichlorobenzene) |

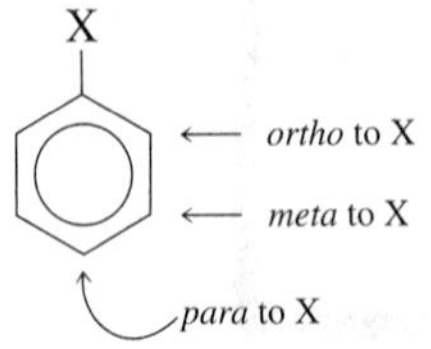

The prefix system uses the prefixes *ortho-*, *meta-*, and *para-* (abbreviated *o-*, *m-*, and *p-*).

Learn the meaning of the prefixes *ortho-*, *meta-*, and *para-*. These prefixes are extensively used in naming disubstituted benzenes.

Ortho- means 1,2 disubstitution; the substituents are on adjacent carbon atoms.
Meta- means 1,3 disubstitution; the substituents are one carbon removed from each other.
Para- means 1,4 disubstitution; the substituents are two carbons removed from each other (on opposite sides of the ring).

When prefixes are used, the three isomeric dichlorobenzenes have the second-listed set of names above.

The use of *ortho-, meta-,* and *para-* in place of position numbers is reserved for disubstituted benzenes. The system is never used with cyclohexanes or other ring systems.

When one of the two substituents in a disubstituted benzene imparts a special name to the compound (as, for example, toluene), the compound is named as a derivative of that parent molecule. The special substituent is assumed to be at ring position 1.

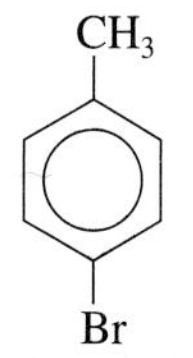

CH₃

Br

4-Bromotoluene
(not 1-bromo-4-methylbenzene)

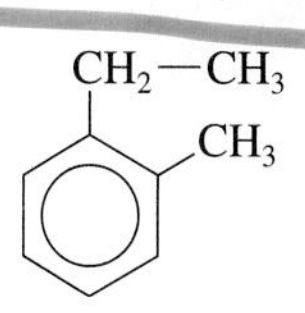

CH₂—CH₃
CH₃

2-Ethyltoluene
(not 1-ethyl-2-methylbenzene)

When neither substituent group imparts a special name, the substituents are cited in alphabetical order before the ending *-benzene.* The carbon of the benzene ring bearing the substituent with alphabetical priority becomes carbon 1.

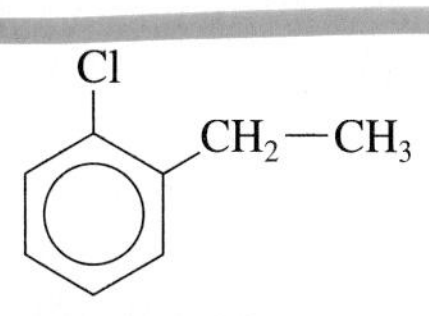

Cl
CH₂—CH₃

1-Chloro-2-ethylbenzene
(not 2-chloro-1-ethylbenzene)

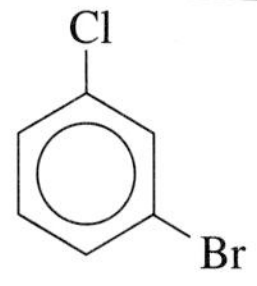

Cl

Br

1-Bromo-3-chlorobenzene
(not 3-bromo-1-chlorobenzene)

When parent names such as *toluene* and *xylene* are used, additional substituents present cannot be the same as those included in the parent name. If such is the case, name the compound as a substituted benzene. The compound

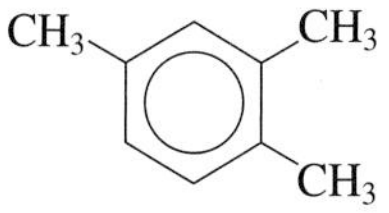

CH₃   CH₃

CH₃

is named as a trimethylbenzene and not as a methylxylene or a dimethyltoluene.

A benzene ring bearing two methyl groups is a situation that generates a new special base name. Such compounds (there are three isomers) are not named as dimethylbenzenes or as methyl toluenes. They are called xylenes.

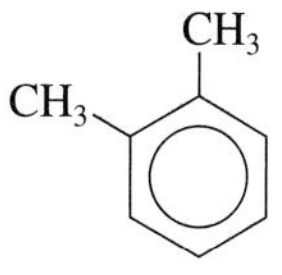

CH₃
CH₃

o-Xylene

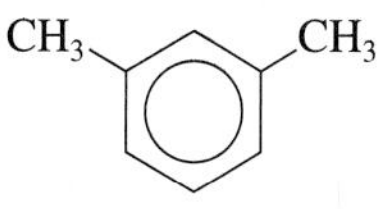

CH₃   CH₃

m-Xylene

CH₃

CH₃

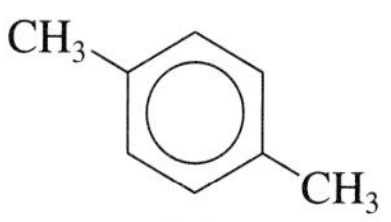

p-Xylene

The xylenes are good solvents for grease and oil and are used for cleaning microscope slides and optical lenses and for removing wax from skis.

Benzene Derivatives with Three or More Substituents

When more than two groups are present on the benzene ring, their positions are indicated with *numbers.* The ring is numbered in such a way as to obtain the lowest possible numbers for the carbon atoms that have substituents. If there is a choice of numbering systems (two systems give the same lowest set), then the group that comes first alphabetically is given the lower number.

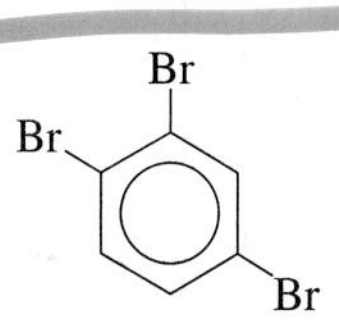

Br
Br

Br

1,2,4-Tribromobenzene

Cl

Br   Cl

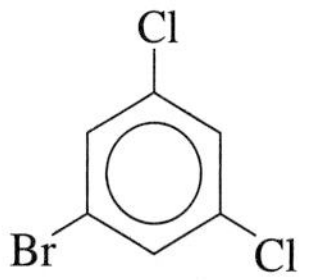

1-Bromo-3,5-dichlorobenzene

EXAMPLE 2.5

Assigning IUPAC Names to Benzene Derivatives

■ Assign IUPAC names to the following benzene derivatives.

a.   Cl

CH₂—CH₃

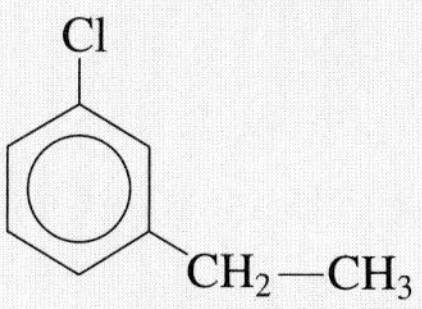

b.   Br

Cl   CH₂—CH₃

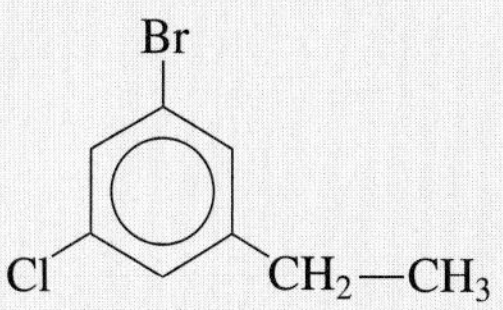

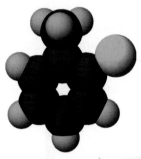

FIGURE 2.18 Space-filling model for the compound 2-chlorotoluene.

c.
$$CH_3-\underset{\displaystyle \bigcirc}{\overset{\displaystyle Br}{CH}}-CH-CH_3$$

d.
CH₃, Cl on benzene ring

Solution

a. No substituents that will change the parent name from benzene are present on the ring. Alphabetical priority dictates that the chloro group is on carbon 1 and the ethyl group on carbon 3. The compound is named 1-chloro-3-ethylbenzene (or *m*- chloroethylbenzene).

b. Again, no substituents that will change the parent name from benzene are present on the ring. Alphabetical priority among substituents dictates that the bromo group is on carbon 1, the chloro group on carbon 3, and the ethyl group on carbon 5. The compound is named 1-bromo-3-chloro-5-ethylbenzene.

c. This compound is named with the benzene ring treated as a substituent—that is, as a phenyl group. The compound is named 2-bromo-3-phenylbutane.

d. The methyl group present on the benzene ring changes the parent name from benzene to toluene. Carbon 1 bears the methyl group. Numbering clockwise, we obtain the name 2-chlorotoluene. (See Figure 2.18)

Practice Exercise 2.5

Assign IUPAC names to the following benzene derivatives.

a. Br, CH₂—CH₂—CH₃ on benzene ring

b. CH₂—CH₂—CH₃, Cl on benzene ring

c. CH₃—CH₂—CH₂—CH—CH₂—CH₃ on benzene ring

d. Br, Cl, Cl on benzene ring

2.13 Aromatic Hydrocarbons: Physical Properties and Sources

In general, aromatic hydrocarbons resemble other hydrocarbons in physical properties. They are insoluble in water, are good solvents for other nonpolar materials, and are less dense than water.

Benzene, monosubstituted benzenes, and many disubstituted benzenes are liquids at room temperature. Benzene itself is a colorless, flammable liquid that burns with a sooty flame because of incomplete combustion.

At one time, coal tar was the main source of aromatic hydrocarbons. Petroleum is now the primary source of such compounds. At high temperatures, with special catalysts, saturated hydrocarbons obtained from petroleum can be converted to aromatic hydrocarbons. The production of toluene from heptane is representative of such a conversion.

$$CH_3-CH_2-CH_2-CH_2-CH_2-CH_2-CH_3 \xrightarrow[\text{High temperature}]{\text{Catalyst}} \underset{\displaystyle \bigcirc}{\overset{\displaystyle CH_3}{}} + 4H_2$$

Two common situations in which a person can be exposed to low-level benzene vapors are

1. Inhaling gasoline vapors while refueling an automobile. Gasoline contains about 2% (v/v) benzene.
2. Being around a cigarette smoker. Benzene is a combustion product present in cigarette smoke. For smokers themselves, inhaled cigarette smoke is a serious benzene-exposure source.

Benzene was once widely used as an organic solvent. Such use has been discontinued because benzene's short- and long-term toxic effects are now recognized. Benzene inhalation can cause nausea and respiratory problems.

2.14 Chemical Reactions of Aromatic Hydrocarbons

We have noted that aromatic hydrocarbons do not readily undergo the addition reactions characteristic of other unsaturated hydrocarbons. An addition reaction would require breaking up the delocalized bonding (Section 2.11) present in the ring system.

If benzene is so unresponsive to addition reactions, what reactions does it undergo? Benzene undergoes *substitution* reactions. As you recall from Section 1.17, substitution reactions are characterized by different atoms or groups of atoms replacing hydrogen atoms in a hydrocarbon molecule. Two important types of substitution reactions for benzene and other aromatic hydrocarbons are alkylation and halogenation.

1. *Alkylation:* An alkyl group (R—) from an alkyl chloride (R—Cl) substitutes for a hydrogen atom on the benzene ring. A catalyst, $AlCl_3$, is needed for alkylation.

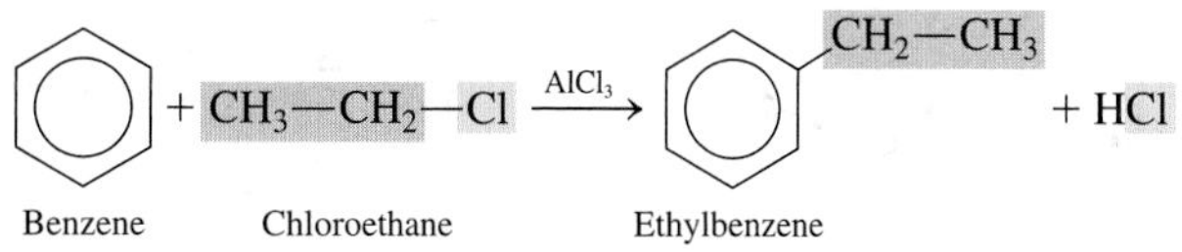

In general terms, the alkylation of benzene can be written as

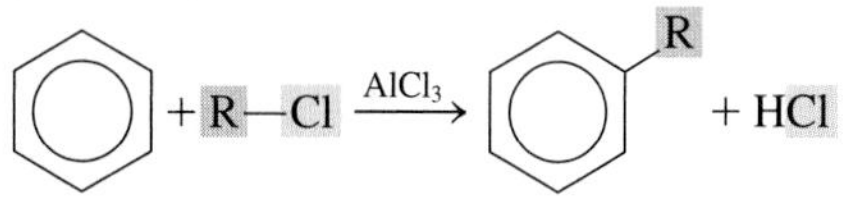

Alkylation, the reaction that attaches an alkyl group to an aromatic ring, is also known as a *Friedel–Crafts reaction,* named after Charles Friedel and James Mason Crafts, the French and American chemists responsible for its discovery in 1877.

Alkylation is the most important industrial reaction of benzene.

2. *Halogenation* (bromination or chlorination): A hydrogen atom on a benzene ring can be replaced by bromine or chlorine if benzene is treated with Br_2 or Cl_2 in the presence of a catalyst. The catalyst is usually $FeBr_3$ for bromination and $FeCl_3$ for chlorination.

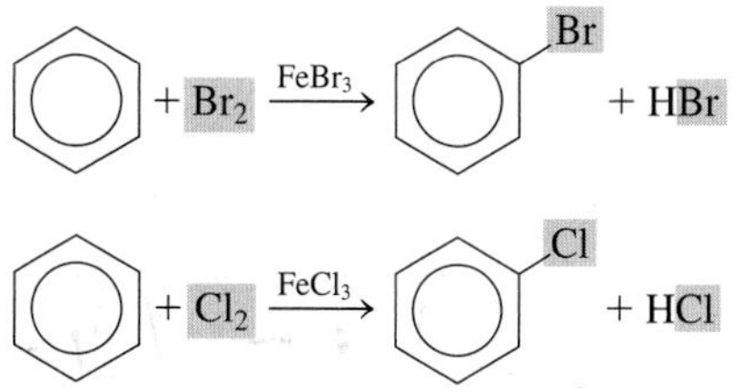

Aromatic halogenation differs from alkane halogenation (Section 1.17) in that light is not required to initiate aromatic halogenation.

2.15 Fused-Ring Aromatic Compounds

Benzene and its substituted derivatives are not the only type of aromatic hydrocarbon that exists. Another large class of aromatic hydrocarbons is the fused-ring aromatic hydrocarbons. A **fused-ring aromatic hydrocarbon** is *an aromatic hydrocarbon whose structure contains two or more rings fused together.* Two carbon rings that share a pair of carbon atoms are said to be *fused.*

64   Chapter 2   Unsaturated Hydrocarbons

CHEMICAL CONNECTIONS   Fused-Ring Aromatic Hydrocarbons and Cancer

A number of fused-ring aromatic hydrocarbons are known to be carcinogens—that is, to cause cancer. Three of the most potent carcinogens are

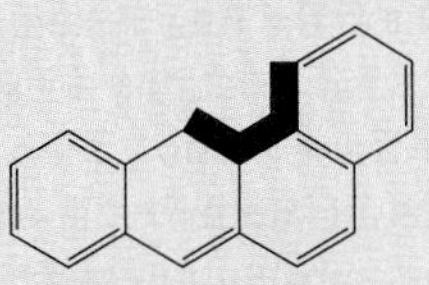

1,2-Benzanthracene 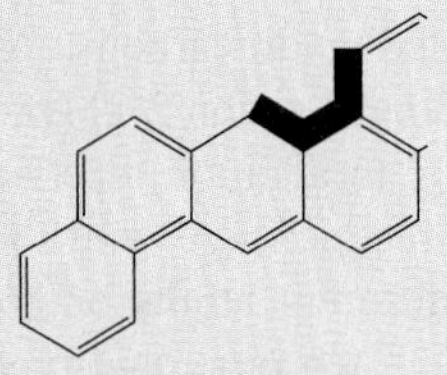   1,2,5,6-Dibenzanthracene

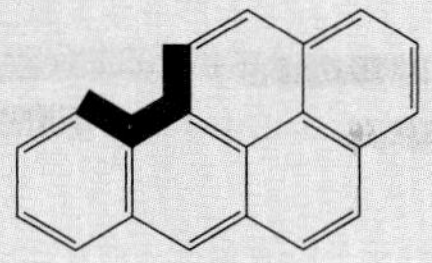

3,4-Benzpyrene

Very small amounts of these substances, when applied to the skin of mice, cause cancer.

Carcinogenic fused-ring aromatic hydrocarbons share some structural features. They all contain four or more fused rings, and they all have the same "angle" in the series of rings (the dark area in the structures shown). Fused-ring aromatic hydrocarbons are often formed when hydrocarbon materials are heated to high temperatures. These resultant compounds are present in low concentrations in tobacco smoke, in automobile exhaust, and sometimes in burned (charred) food. The charred portions of a well-done steak cooked over charcoal are a likely source.

Angular, fused-ring hydrocarbon systems are believed to be partially responsible for the high incidence of lung and lip cancer among cigarette smokers because tobacco smoke contains 3,4-benzpyrene. The more a person smokes, the greater his or her risk of developing cancer.

We now know that the high incidence of lung cancer in British chimney sweeps (documented over 200 years ago) was caused by fused-ring hydrocarbon compounds present in the chimney soot that the sweeps inhaled regularly.

The three simplest fused-ring aromatic compounds are naphthalene, anthracene, and phenanthrene. All three are solids at room temperature.

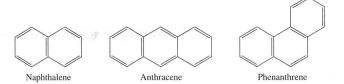

Naphthalene   Anthracene   Phenanthrene

CONCEPTS TO REMEMBER

Unsaturated hydrocarbons. An unsaturated hydrocarbon is a hydrocarbon that contains one or more carbon–carbon multiple bonds. Three main classes of unsaturated hydrocarbons exist: alkenes, alkynes, and aromatic hydrocarbons (Section 2.1).

Alkenes and cycloalkenes. An alkene is an acyclic unsaturated hydrocarbon in which one or more carbon–carbon double bonds are present. A cycloalkene is a cyclic unsaturated hydrocarbon that contains one or more carbon–carbon double bonds within the ring system (Section 2.2).

Alkene nomenclature. Alkenes and cycloalkenes are given IUPAC names using rules similar to those for alkanes and cycloalkanes, except that the ending -ene is used. Also, the double bond takes precedence both in selecting and in numbering the main chain or ring (Section 2.3).

Isomerism in alkenes. Two subtypes of constitutional isomers are possible for alkenes: skeletal isomers and positional isomers. Positional isomers differ in the location of the functional group (double bond)

present. *Cis–trans* isomerism is also possible for some alkenes because there is restricted rotation about a carbon–carbon double bond (Section 2.5).

Physical properties of alkenes. Alkenes and alkanes have similar physical properties. They are nonpolar, insoluble in water, less dense than water, and soluble in nonpolar solvents (Section 2.7).

Addition reactions of alkanes. Numerous substances, including H_2, Cl_2, Br_2, HCl, HBr, and H_2O, add to an alkene carbon–carbon double bond. When both the alkene and an alkene reactant are unsymmetrical, the addition proceeds according to Markovnikov's rule: The carbon atom of the double bond that already has the greater number of H atoms gets one more (Section 2.8).

Addition polymers. Addition polymers are formed from alkene monomers that undergo repeated addition reactions with each other. Many familiar and widely used materials, such as fibers and plastics, are addition polymers. (Section 2.9).

Alkynes and cycloalkynes. Alkynes and cycloalkynes are unsaturated hydrocarbons that contain one or more carbon–carbon triple bonds. They are named in the same way as alkenes and cycloalkenes, except that their parent names end in -*yne*. Like alkenes, alkynes undergo addition reactions. These occur in two steps, an alkene forming first and then an alkane (Section 2.10).

Aromatic hydrocarbons. Benzene, the simplest aromatic hydrocarbon, and other members of this family of compounds contain a six-membered ring with a cyclic, delocalized bond. This aromatic ring is often drawn as a hexagon containing a circle, which represents the six electrons that move freely around the ring (Section 2.11).

Nomenclature of aromatic hydrocarbons. Monosubstituted benzene compounds are named by adding the substituent name to the word *benzene*. Positions of substituents in disubstituted benzenes are indicated by using a numbering system or the *ortho-* (1,2), *meta-* (1,3), and *para-* (1,4) prefix system (Section 2.12).

Chemical reactions of aromatic hydrocarbons. Aromatic hydrocarbons undergo substitution reactions rather than addition reactions. Important substitution reactions are alkylation and halogenation (Section 2.14).

KEY REACTIONS AND EQUATIONS

1. Halogenation of an alkene (Section 2.8)

$$\text{C=C} + \text{Br—Br} \longrightarrow -\overset{|}{\underset{Br}{C}}-\overset{|}{\underset{Br}{C}}-$$

2. Hydrogenation of an alkene (Section 2.8)

$$\text{C=C} + \text{H—H} \xrightarrow{\text{Ni or Pt}} -\overset{|}{\underset{H}{C}}-\overset{|}{\underset{H}{C}}-$$

3. Hydrohalogenation of an alkene (Section 2.8)

$$\text{C=C} + \text{H—Cl} \longrightarrow -\overset{|}{\underset{H}{C}}-\overset{|}{\underset{Cl}{C}}-$$

4. Hydration of an alkene (Section 2.8)

$$\text{C=C} + \text{H—OH} \xrightarrow{\text{H}_2\text{SO}_4} -\overset{|}{\underset{H}{C}}-\overset{|}{\underset{OH}{C}}-$$

5. Hydrogenation of an alkyne (Section 2.10)

$$-\text{C}\equiv\text{C}- + \text{H}_2 \xrightarrow{\text{Ni}} -\overset{H}{\underset{H}{C}}=\overset{H}{\underset{H}{C}}- \xrightarrow[\text{Ni}]{\text{H}_2} -\overset{H}{\underset{H}{C}}-\overset{H}{\underset{H}{C}}-$$

6. Halogenation of an alkyne (Section 2.10)

$$-\text{C}\equiv\text{C}- + \text{Br}_2 \longrightarrow -\overset{Br}{\underset{}{C}}=\overset{Br}{\underset{}{C}}- \xrightarrow{\text{Br}_2} -\overset{Br}{\underset{Br}{C}}-\overset{Br}{\underset{Br}{C}}-$$

7. Hydrohalogenation of an alkyne (Section 2.10)

$$-\text{C}\equiv\text{C}- + \text{HBr} \longrightarrow -\overset{H}{\underset{}{C}}=\overset{Br}{\underset{}{C}}- \xrightarrow{\text{HBr}} -\overset{H}{\underset{H}{C}}-\overset{Br}{\underset{Br}{C}}-$$

8. Alkylation of benzene (Section 2.14)

benzene + R—Cl $\xrightarrow{\text{AlCl}_3}$ R-substituted benzene + HCl

9. Halogenation of benzene (Section 2.14)

benzene + Br$_2$ $\xrightarrow{\text{FeBr}_3}$ Br-substituted benzene + HBr

EXERCISES AND PROBLEMS

The members of each pair of problems in this section test similar material.

■ Unsaturated Hydrocarbons with Double Bonds (Section 2.2)

2.1   Classify each of the following hydrocarbons as saturated or unsaturated. Further classify any unsaturated hydrocarbons as alkenes with one double bond, dienes, or trienes.

   a. $\text{CH}_3-\text{CH}_2-\text{CH}=\text{CH}-\text{CH}_3$

   b. $\text{CH}_3-\text{CH}_2-\text{CH}_2-\text{CH}_2-\text{CH}_3$

   c.
$$\text{CH}_2=\overset{\overset{\displaystyle \text{CH}_3}{|}}{\text{C}}-\overset{\overset{\displaystyle \text{CH}_3}{|}}{\underset{\underset{\displaystyle \text{CH}_3}{|}}{\text{C}}}-\text{CH}_3$$

   d. $\text{CH}_2=\text{CH}-\text{CH}_2-\overset{\overset{\displaystyle }{\|}}{\underset{\underset{\displaystyle \text{CH}_2}{}}{\text{C}}}-\text{CH}_3$

   e. $\text{CH}_2=\text{CH}-\text{CH}=\text{CH}-\text{CH}=\text{CH}_2$

   f. $\text{CH}_3-\text{CH}=\text{C}=\text{CH}-\text{CH}_3$

2.2   Classify each of the following hydrocarbons as saturated or unsaturated. Further classify any unsaturated hydrocarbons as alkenes with one double bond, dienes, or trienes.

   a. $\text{CH}_3-\text{CH}=\text{CH}-\text{CH}=\text{CH}_2$

   b. $\text{CH}_3-\text{CH}=\text{CH}-\text{CH}_3$

c. $CH_2{=}C{-}CH_2{-}CH_3$
     |
     CH_3

d. $CH_2{=}C{-}CH{=}CH{-}CH{=}CH_2$
     |
     CH_3

e. $CH_3{-}CH_2{-}CH_2{-}CH{=}CH_2$

f. $CH_3{-}C{-}CH_2{-}CH_2{-}C{-}CH_3$
     ‖   ‖
     CH_2 CH_2

2.3 Write the *molecular formula* for hydrocarbons with each of the following structural features.
  a. Acylic, four carbon atoms, no multiple bonds
  b. Acylic, five carbon atoms, one double bond
  c. Cyclic, five carbon atoms, one double bond
  d. Cyclic, seven carbon atoms, two double bonds

2.4 Write the *molecular formula* for hydrocarbons with each of the following structural features.
  a. Acyclic, six carbon atoms, two double bonds
  b. Acyclic, six carbon atoms, three double bonds
  c. Cyclic, five carbon atoms, no multiple bonds
  d. Cyclic, eight carbon atoms, four double bonds

2.5 Write the *general* molecular formula (C_nH_{2n} and so on) for each of the following families of compounds.
  a. Cycloalkene with one double bond
  b. Alkadiene
  c. Diene
  d. Cycloalkatriene

2.6 Write the *general* molecular formula (C_nH_{2n} and so on) for each of the following families of compounds.
  a. Cycloalkadiene   b. Alkene with one double bond
  c. Triene   d. Alkatriene

▉ Names for Hydrocarbons Containing Double Bonds (Sections 2.3 and 2.4)

2.7 Assign an IUPAC name to each of the following unsaturated hydrocarbons.
  a. $CH_3{-}CH{=}CH{-}CH_3$
  b. $CH_3{-}C{=}CH{-}CH{-}CH_3$
           | |
           CH_3 CH_3

  c.

  d.

  e. $CH_2{=}C{-}CH_2{-}CH_2{-}CH_3$
           |
     $CH_3{-}CH_2$

  f.

2.8 Assign an IUPAC name to each of the following unsaturated hydrocarbons.
  a. $CH_3{-}CH_2{-}CH{=}CH{-}CH_3$
  b. $CH_3{-}CH_2{-}C{=}CH{-}CH_3$
                   |
                   CH_3
  c.

2.9 Assign an IUPAC name to each of the hydrocarbons in Problem 2.1.

2.10 Assign an IUPAC name to each of the hydrocarbons in Problem 2.2.

2.11 Draw a condensed structural formula for each of the following unsaturated hydrocarbons.
  a. 3-Methyl-1-pentene
  b. 3-Methylcyclopentene
  c. 1,3-Butadiene
  d. 3-Ethyl-1,4-pentadiene
  e. 4-Propyl-2-heptene
  f. 3,6-Diethyl-1,4-cyclohexadiene

2.12 Draw a condensed structural formula for each of the following unsaturated hydrocarbons.
  a. 4-Methly-1-hexene
  b. 4-methylcyclohexene
  c. 1,3-Pentadiene
  d. 2-Ethyl-1,4-pentadiene
  e. 1,3-Cyclohexadiene
  f. 4,4,5-Trimethyl-2-heptene

2.13 The following names are *incorrect* by IUPAC rules. Determine the correct IUPAC name for each compound.
  a. 2-Ethyl-2-pentene
  b. 4,5-Dimethyl-4-hexene
  c. 3,5-Cyclopentadiene
  d. 1,2-Dimethyl-4-cyclohexene

2.14 The following names are *incorrect* by IUPAC rules. Determine the correct IUPAC name for each compound.
  a. 2-Methyl-4-pentene
  b. 3-Methyl-2,4-pentadiene
  c. 3-Methyl-3-cyclopentene
  d. 1,2-Dimethyl-3-cyclohexene

▉ Isomerism in Alkenes (Section 2.5)

2.15 Draw skeletal structural formulas and give the IUPAC names for the 13 possible alkene constitutional isomers with the formula C_6H_{12}. (Three of the constitutional isomers are hexenes, six are methylpentenes, three are dimethylbutenes, and one is an ethylbutene.)

2.16 Draw skeletal structural formulas and give the IUPAC names for the 16 possible alkadiene constitutional isomers with the formula C_6H_{10}. (Six of the constitutional isomers are hexadienes, eight are methylpentadienes, one is a dimethylbutadiene, and one is an ethylbutadiene.)

2.17 For each molecule, tell whether *cis–trans* isomers exist. If they do, draw the two isomers and label them as *cis* and *trans*.
  a. $CH_2{=}CH{-}CH_3$ b. $CH_2{=}CH{-}CH_2$
                                              |
                                              Cl
  c. $CH_3{-}C{=}CH{-}CH_3$ d. 3-Hexene
           |
           CH_3
  e. 4-Methyl-2-pentene   f. 1,2-Dimethylcyclopentane

2.18   For each molecule, tell whether *cis–trans* isomers exist. If they do, draw the two isomers and label them as *cis* and *trans*.

a. $CH_3—CH_2—CH=CH_2$ b. $CH_3—CH_2—C=CH_2$
                                              |
                                              Cl

c. $CH_3—CH_2—CH=CH$ d. 2-Pentene
                          |
                          Cl

e. 1,2-Dichloroethene   f. 1,3-Dichlorocyclobutane

2.19   Assign an IUPAC name to each of the following molecules. Include the prefix *cis-* or *trans-* when appropriate.

a.
$$\begin{array}{c} CH_3 \quad\quad CH_2—CH_3 \\ \diagdown\quad\quad\diagup \\ C=C \\ \diagup\quad\quad\diagdown \\ H \quad\quad\quad H \end{array}$$

b.
$$\begin{array}{c} I \quad\quad\quad H \\ \diagdown\quad\quad\diagup \\ C=C \\ \diagup\quad\quad\diagdown \\ H \quad\quad\quad Br \end{array}$$

c.
$$\begin{array}{c} F \quad\quad\quad F \\ \diagdown\quad\quad\diagup \\ C=C \\ \diagup\quad\quad\diagdown \\ F \quad\quad\quad F \end{array}$$

d.
$$\begin{array}{c} H \quad\quad\quad CH_3 \\ \diagdown\quad\quad\diagup \\ C=C \\ \diagup\quad\quad\diagdown \\ CH_3 \quad\quad CH_3 \end{array}$$

2.20   Assign an IUPAC name to each of the following molecules. Include the prefix *cis-* or *trans-* when appropriate.

a.
$$\begin{array}{c} H \quad\quad\quad Cl \\ \diagdown\quad\quad\diagup \\ C=C \\ \diagup\quad\quad\diagdown \\ H \quad\quad\quad H \end{array}$$

b.
$$\begin{array}{c} Br \quad\quad\quad Br \\ \diagdown\quad\quad\diagup \\ C=C \\ \diagup\quad\quad\diagdown \\ H \quad\quad\quad H \end{array}$$

c.
$$\begin{array}{c} H \quad\quad\quad CH_3 \\ \diagdown\quad\quad\diagup \\ C=C \\ \diagup\quad\quad\diagdown \\ CH_3 \quad\quad H \end{array}$$

d.
$$\begin{array}{c} H \quad\quad\quad Br \\ \diagdown\quad\quad\diagup \\ C=C \\ \diagup\quad\quad\diagdown \\ H \quad\quad\quad Br \end{array}$$

2.21   Draw a structural formula for each of the following compounds.
a. *trans*-3-Methyl-3-hexene   b. *cis*-2-Pentene
c. *trans*-5-Methyl-2-heptene   d. *trans*-1,3-Pentadiene

2.22   Draw a structural formula for each of the following compounds.
a. *trans*-2-hexene   b. *cis*-4-Methyl-2-pentene
c. *cis*-1-Chloro-1-pentene   d. *cis*-1,3-Pentadiene

■ **Naturally Occurring Alkenes (Section 2.6)**

2.23   What is a pheromone?

2.24   What is a terpene?

2.25   Why is the number of carbon atoms in a terpene always a multiple of the number 5?

2.26   What is the structural relationship between β-carotene and vitamin A?

■ **Physical Properties of Alkenes (Section 2.7)**

2.27   Indicate whether each of the following alkenes would be expected to be a solid, a liquid, or a gas at room temperature and pressure.
a. Propene   b. 1-Pentene
c. 1-Octene   d. Cyclopentene

2.28   Indicate whether each of the following statements is true or false.
a. 1-Butene has a density greater than that of water.
b. 1-Butene has a higher boiling point than 1-hexene.
c. 1-Butene is flammable but 1-hexene is not.
d. Both 1-pentene and cyclopentene are gases at room temperature and pressure.

■ **Alkene Addition Reactions (Section 2.8)**

2.29   Which of the following reactions are addition reactions?
a. $C_4H_8 + Cl_2 \rightarrow C_4H_8Cl_2$
b. $C_6H_6 + Cl_2 \rightarrow C_6H_5Cl + HCl$
c. $C_3H_6 + HCl \rightarrow C_3H_7Cl$
d. $C_7H_{16} \rightarrow C_7H_8 + 4H_2$

2.30   Which of the following reactions are addition reactions?
a. $C_3H_6 + Cl_2 \rightarrow C_3H_6Cl_2$
b. $C_8H_{10} \rightarrow C_8H_8 + H_2$
c. $C_6H_6 + C_2H_5Cl \rightarrow C_8H_{10} + HCl$
d. $C_4H_8 + HCl \rightarrow C_4H_9Cl$

2.31   Write a chemical equation showing reactants, products, and catalysts needed (if any) for the reaction of ethene with each of the following substances.
a. Cl_2 b. HCl   c. H_2 d. HBr

2.32   Write a chemical equation showing reactants, products, and catalysts needed (if any) for the reaction of ethene with each of the following substances.
a. H_2O b. Br_2 c. HI   d. I_2

2.33   Write a chemical equation showing reactants, products, and catalysts needed (if any) for the reaction of propene with each of the reactants in Problem 2.31. Use Markovnikov's rule as needed.

2.34   Write a chemical equation showing reactants, products, and catalysts needed (if any) for the reaction of propene with each of the reactants in Problem 2.32. Use Markovnikov's rule as needed.

2.35   Supply the structural formula of the product in each of the following alkene addition reactions.

a. $CH_3—CH=CH—CH_3 + Cl_2 \rightarrow$?

b. $CH_3—C=CH_2 + HBr \rightarrow$?
                |
                CH_3

c. $CH_3—CH_2—CH=CH_2 + HCl \rightarrow$?

d.
$$\text{(cyclopentane)} + H_2 \xrightarrow[\text{catalyst}]{\text{Ni}} ?$$

e.
$$\text{(cyclopentene)} + H_2 \xrightarrow[\text{catalyst}]{\text{no}} ?$$

f.
$$\text{(cyclobutane)} + H_2O \xrightarrow{H_2SO_4} ?$$

2.36   Supply the structural formula of the product in each of the following alkene addition reactions.

a. $CH_3—CH_2—CH=CH_2 + Cl_2 \rightarrow$?

b. $CH_3—CH—CH=CH_2 + HBr \rightarrow$?
               |
               CH_3

c. $CH_3—C=C—CH_3 + HCl \rightarrow$?
            | |
           CH_3 CH_3

d.
$$\text{(cyclohexane)} + H_2 \xrightarrow[\text{catalyst}]{\text{Ni}} ?$$

e.
$$\text{(cyclohexene)} + H_2 \xrightarrow[\text{catalyst}]{\text{no}} ?$$

f. $CH_3—CH=CH_2 + H_2O \xrightarrow{H_2SO_4} ?$

2.37 What reactant would you use to prepare each of the following compounds from cyclohexene?

a.

b.

c.   Cl

d.   OH

2.38 What reactant would you use to prepare each of the following compounds from cyclopentene?

a.

b.   OH

c.   Cl Cl

d.   Br

2.39 How many molecules of H_2 gas will react with 1 molecule of each of the following unsaturated hydrocarbons?

a. $CH_3—CH{=}CH—CH{=}CH—CH_3$

b.

c. $CH{=}CH_2$

d. $CH_3—CH{=}C{=}C—CH{=}CH_2$
                          |
                        CH_3

2.40 How many molecules of H_2 gas will react with 1 molecule of each of the following unsaturated hydrocarbons?

a. $CH_3—CH{=}CH—CH_3$

b.

c. $CH{=}CH_2$

d. $CH_2{=}CH—C—CH{=}CH_2$
                     ||
                   CH_2

▪ Polymerization of Alkenes (Section 2.9)

2.41 Draw the structural formula of the monomer(s) from which each of the following polymers was made.

a. $\left(\begin{array}{c} F \; F \\ -C-C- \\ F \; F \end{array}\right)_n$

b. $\left(\begin{array}{c} H \quad\quad H \\ -C-C{=}C-C- \\ H \; Cl \; H \; H \end{array}\right)_n$

c. $\left(\begin{array}{c} H \; H \\ -C-C- \\ H \; Cl \end{array}\right)_n$

d. $\left(\begin{array}{c} H \quad H \\ -C-\;-C- \\ H \end{array}\right)_n$ (phenyl group)

2.42 Draw the structural formula of the monomer(s) from which each of the following polymers was made.

a. $\left(\begin{array}{c} H \; F \\ -C-C- \\ H \; F \end{array}\right)_n$

b. $\left(\begin{array}{c} H \quad\quad H \\ -C-C{=}C-C- \\ H \; Cl \; Cl \; H \end{array}\right)_n$

c. $\left(\begin{array}{c} H \; H \\ -C-C- \\ Cl \; CH_3 \end{array}\right)_n$

d. $\left(\begin{array}{c} H \quad H \\ -C-\;-C- \\ Cl \end{array}\right)_n$ (phenyl group)

2.43 Draw the "start" (the first three repeating units) of the structural formula of the addition polymers made from the following monomers.
a. Ethylene   b. Vinyl chloride
c. 1,2-Dichloroethene   d. 1-Chloroethene

2.44 Draw the "start" (the first three repeating units) of the structural formula of the addition polymers made from the following monomers.
a. Propylene   b. 1,1,2,2-Tetrafluoroethene
c. 2-Methyl-1-propene   d. 1,2-Dichloroethylene

▪ Alkynes (Section 2.10)

2.45 Assign an IUPAC name to each of the following unsaturated hydrocarbons.

a. $CH_3—CH_2—CH_2—CH_2—C{\equiv}CH$

b. $CH_3—C{\equiv}C—CH—CH_3$
                             |
                           CH_3

c.
    $\quad\quad CH_3$
    $\quad\quad |$
    $CH_3—C—C{\equiv}C—CH_2—CH_2—CH_3$
    $\quad\quad |$
    $\quad\quad CH_3$

d.

e. $CH{\equiv}C—CH—C{\equiv}C—CH_3$
                          |
                        CH_3

f.

2.46 Assign an IUPAC name to each of the following unsaturated hydrocarbons.

a. $CH_3—CH—C{\equiv}CH$
                |
              CH_3

b.

c. $CH_3-CH-C\equiv C-CH-CH_3$
        | |
       CH_3 CH_3

d. $CH_3-CH-CH_2-C$
        | ‖
       CH_2 CH
        |
       CH_3

e. $CH\equiv C-C\equiv CH$

f.

2.47 Draw skeletal structural formulas and give the IUPAC names for the three possible alkyne isomers with the molecular formula C_5H_8.

2.48 Draw skeletal structural formulas and give the IUPAC names for the seven possible alkyne isomers with the molecular formula C_6H_{10}. (Three of the constitutional isomers are hexynes, three are pentynes, and one is a butyne.)

2.49 Supply the condensed structural formula of the product in each of the following alkyne addition reactions.

a. $CH\equiv CH + 2H_2 \xrightarrow{Ni}$?

b. $CH_3-C\equiv CH + 2Br_2 \longrightarrow$?

c. $CH_3-C\equiv CH + 2HBr \longrightarrow$?

d. $CH\equiv CH + 1HCl \longrightarrow$?

e.

$+ 3H_2 \xrightarrow{Ni}$?

f. $CH_3-CH_2-C\equiv CH + 1HBr \longrightarrow$?

2.50 Supply the condensed structural formula of the product in each of the following alkyne addition reactions.

a. $CH_3-C\equiv C-CH_3 + 2Br_2 \longrightarrow$?

b. $CH_3-C\equiv C-CH_3 + 2HBr \longrightarrow$?

c. $CH\equiv C-CH_2-CH_3 + 1H_2 \xrightarrow{Ni}$?

d. $CH\equiv C-CH_3 + 1HCl \longrightarrow$?

e.

$+ 4H_2 \xrightarrow{Ni}$?

f. $CH_3-CH_2-C\equiv CH + 1HBr \longrightarrow$?

■ **Nomenclature for Aromatic Compounds (Section 2.12)**

2.51 Assign an IUPAC name to each of the following disubstituted benzenes. Use numbers rather than prefixes to locate the substituents on the benzene ring.

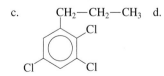

2.52 Assign an IUPAC name to each of the following disubstituted benzenes. Use numbers rather than prefixes to locate the substituents on the benzene ring.

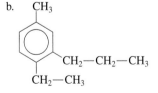

2.53 Assign each of the compounds in Problem 2.51 an IUPAC name in which the substituents on the benzene ring are located using the *ortho-*, *meta-*, *para-* prefix system.

2.54 Assign each of the compounds in Problem 2.52 an IUPAC name in which the substituents on the benzene ring are located using the *ortho-*, *meta-*, *para-* prefix system.

2.55 Assign an IUPAC name to each of the following substituted benzenes.

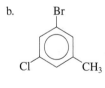

2.56 Assign an IUPAC name to each of the following substituted benzenes.

2.57   Assign an IUPAC name to each of the following compounds, in which the benzene ring is treated as a substituent.

a. 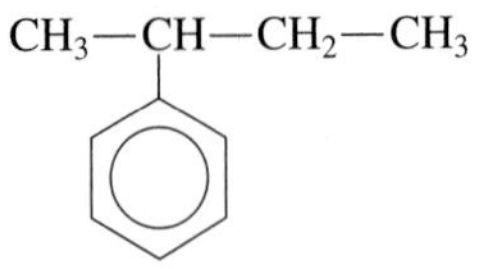 CH_3—CH—CH_2—CH_3

b. 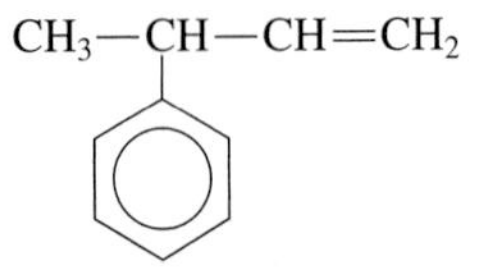 CH_3—CH—CH=CH_2

c. 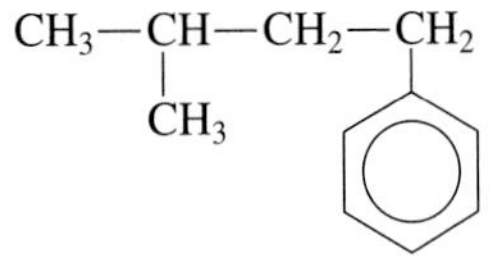 CH_3—CH—CH_2—CH_2
   |
   CH_3

d. 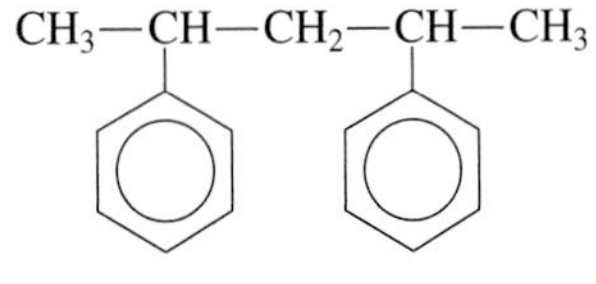 CH_3—CH—CH_2—CH—CH_3

2.58   Assign an IUPAC name to each of the following compounds in which the benzene ring is treated as a substituent.

a. 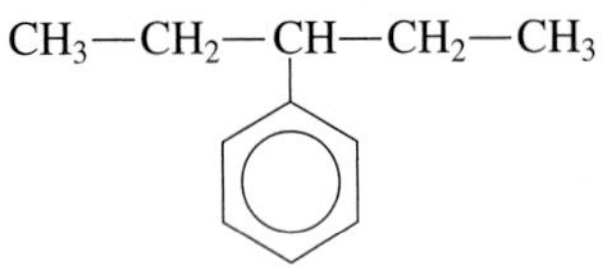 CH_3—CH_2—CH—CH_2—CH_3

b. 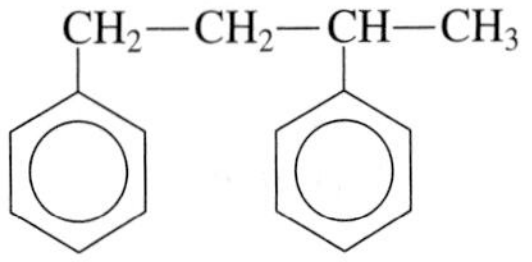 CH_2—CH_2—CH—CH_3

c. 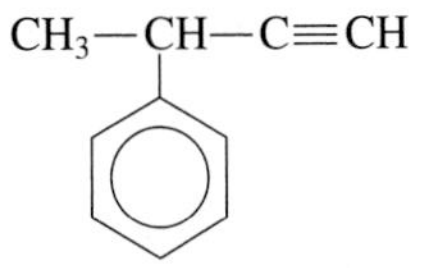 CH_3—CH—C≡CH

d. 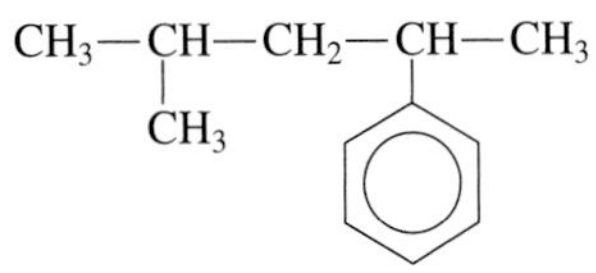 CH_3—CH—CH_2—CH—CH_3
   |
   CH_3

2.59   Write a structural formula for each of the following compounds.
a. 1,3-Diethylbenzene
b. o-Xylene
c. p-Ethyltoluene
d. Phenylbenzene
e. 1,2-Diphenylethane
f. 3-Methyl-3-phenylpentane

2.60   Write a structural formula for each of the following compounds.
a. o-Ethylpropylbenzene   b. m-Xylene
c. 2-Bromotoluene   d. 2-Phenylpropane
e. Isopropylbenzene   f. Triphenylmethane

■ **Chemical Reactions of Aromatic Hydrocarbons (Section 2.14)**

2.61   For each of the following classes of compounds, indicate whether addition or substitution is the most characteristic reaction for the class.
a. Alkanes   b. Dienes
c. Alkylbenzenes   d. Cycloalkenes

2.62   For each of the following classes of compounds, indicate whether addition or substitution is the most characteristic reaction for the class.
a. Alkynes   b. Cycloalkanes
c. Aromatic hydrocarbons   d. Saturated hydrocarbons

2.63   Complete the following reaction equations by supplying the formula of the missing reactant or product.

a. 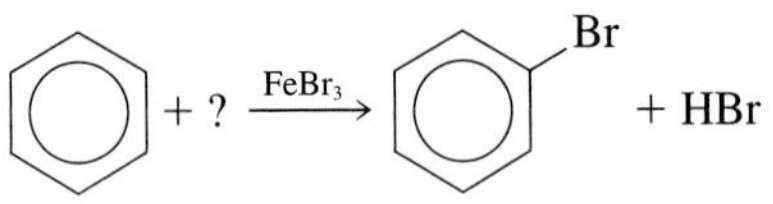 ⬡ + ? $\xrightarrow{FeBr_3}$ ⬡—Br + HBr

b. 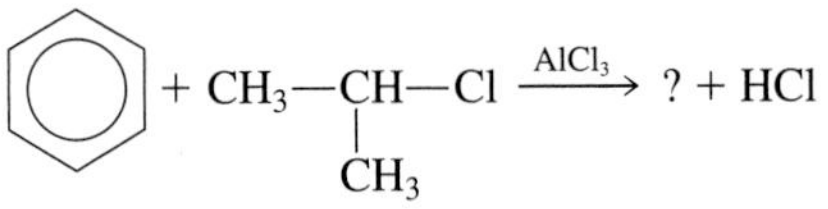 ⬡ + CH_3—CH—Cl $\xrightarrow{AlCl_3}$? + HCl
                   |
                   CH_3

c. 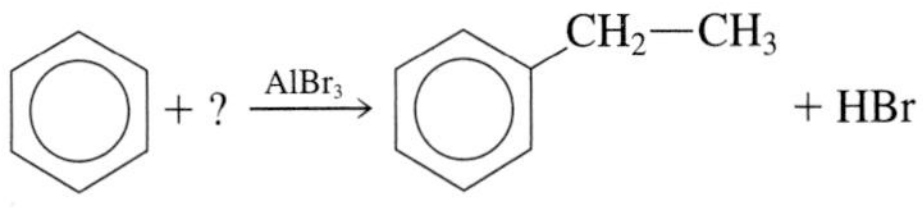 ⬡ + ? $\xrightarrow{AlBr_3}$ ⬡—CH_2—CH_3 + HBr

2.64   Complete the following reaction equations by supplying the formula of the missing reactant, product, or catalyst.

a. 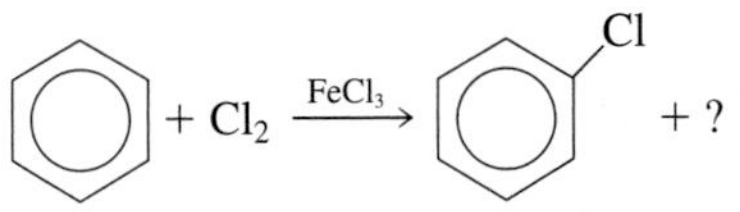 ⬡ + Cl_2 $\xrightarrow{FeCl_3}$ ⬡—Cl + ?

b. 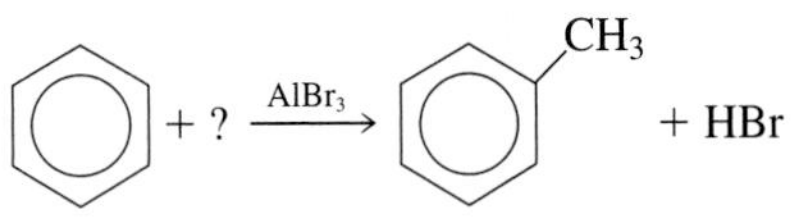 ⬡ + ? $\xrightarrow{AlBr_3}$ ⬡—CH_3 + HBr

c. ⬡ + CH_3—C—Br $\xrightarrow{AlBr_3}$? + HBr
              |
              CH_3
    (with CH_3 groups above and below C)

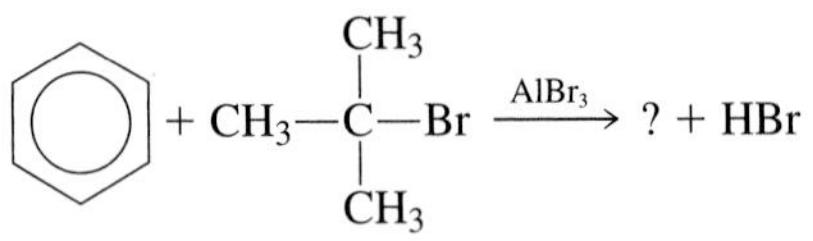

ADDITIONAL PROBLEMS

2.65   What is the molecular formula for the simplest compound of each of the following types?
a. Alkene with one multiple bond
b. Cycloalkene with one multiple bond
c. Alkyne with one multiple bond
d. Alkane

2.66   Indicate whether the hydrocarbon listed first in each of the following pairs of hydrocarbons contains (1) more hydrogen atoms, (2) the same number of hydrogen atoms, or (3) fewer hydrogen atoms than the hydrocarbon listed second.
a. Propane and propene   b. Propene and propyne
c. Propene and cyclopropene d. Propyne and cyclopropene

2.67 Indicate whether each of the following pairs of hydrocarbons are *constitutional* isomers.
   a. Propene and cyclopropene
   b. 1-Pentene and 2-pentene
   c. *cis*-2-Butene and *trans*-2-butene
   d. Cyclobutene and 2-butyne

2.68 Contrast the compounds cyclohexane, cyclohexene, and benzene in terms of each of the following:
   a. Number of carbon atoms present
   b. Number of hydrogen atoms present
   c. Whether they undergo substitution or addition reactions
   d. Whether they are a solid, a liquid, or a gas at room temperature and pressure

2.69 Draw a condensed structural formula for each of the following unsaturated hydrocarbons or hydrocarbon derivatives.
   a. 5-Methyl-2-hexyne   b. 1-Chloro-2-butene
   c. 5,6-Dimethyl-2-heptyne   d. 3-Isopropyl-1-hexene
   e. 1,6-Heptadiene   f. 3-Methyl-1,4-pentadiene

2.70 How many molecules of H_2 will react with 1 molecule of each of the compounds in Problem 2.69 when the appropriate catalyst is present?

2.71 Draw a condensed structural formula for each of the following compounds.

   a. Vinylbenzene   b. Allyl chloride
   c. Propylacetylene   d. Dipropylacetylene
   e. *o*-Xylene   f. *m*-Phenyltoluene

2.72 The compound 2-methyl-1-propene is a well-known substance. The compound 2,2-dimethyl-1-propene does not exist. Explain why this is so.

2.73 The compound 1,2-dichlorocyclohexane exists in *cis–trans* forms. However, *cis–trans* isomerism is not possible for the compound 1,2-dichlorobenzene. Explain why this is so.

2.74 Hydrocarbons with the formula C_5H_{10} can be either alkenes or cycloalkanes. Draw the ten possible constitutional isomers that fit this formula; five are alkenes and five are cycloalkanes. Then indicate which of these ten isomers can exist in *cis–trans* forms.

2.75 There are eight isomeric substituted benzenes that have the formula C_9H_{12}. What are the IUPAC names for these eight constitutional isomers?

2.76 How many different compounds are there that fit each of the following descriptions?
   a. Bromochlorobenzenes
   b. Trichlorobenzenes
   c. Dibromodichlorobenzenes
   d. Monobromoanthracenes

MULTIPLE-CHOICE PRACTICE TEST

2.77 All of the following compounds are unsaturated hydrocarbons except which one?
   a. 2-Butene   b. 3-Heptyne
   c. Cyclopropane   d. 1,3-Dimethylbenzene

2.78 What is the correct IUPAC name for the compound

$CH_3—CH—CH = CH_2$
         |
        CH_3

   a. 2-Methylbutene
   b. 2-Methyl-3, 4-butene
   c. 2-Methyl-3-butene
   d. 3-Methyl-1-butene

2.79 What is the number of carbon atoms present in a vinyl group?
   a. One   b. Two   c. Three   d. Four

2.80 Which of the following types of unsaturated hydrocarbons does *not* have the general formula C_nH_{2n-2}?
   a. Alkenes with one double bond
   b. Cycloalkenes with one double bond
   c. Alkenes with two double bonds
   d. Alkenes with one triple bond

2.81 For which of the following halogenated hydrocarbons is *cis–trans* isomerism possible?
   a. 1,1-Dichloro-1-propene
   b. 1,3-Dichloro-1-propene
   c. 2,3-Dichloro-1-propene
   d. 3,3-Dichloro-1-propene

2.82 Which of the following reactions can be used to convert an alkene to an alkane?
   a. Hydrogenation   b. Halogenation
   c. Hydrohalogenation   d. Hydration

2.83 In which of the following addition polymers are methyl groups present as attachments to the carbon chain?
   a. Polyethylene   b. Polypropylene
   c. Teflon   d. PVC

2.84 Which of the following statements concerning alkynes is *incorrect*?
   a. Alkynes are generally insoluble in water.
   b. Alkynes generally have densities less than that of water.
   c. Alkynes do not undergo halogenation reactions.
   d. Alkynes undergo combustion reactions.

2.85 Which of the following is a correct pairing of "prefix" and "numbers"?
   a. *Para-* and 1,2-
   b. *Ortho-* and 1,4-
   c. *Meta-* and 1,3-
   d. *Iso-* and 2,3-

2.86 Which of the following aromatic compounds contains 7 carbon atoms?
   a. Toluene
   b. 1,2-Dichlorobenzene
   c. *o*-Xylene
   d. 3-Phenylbutane

3 Alcohols, Phenols, and Ethers

The physiological effects of poison ivy are caused by certain phenol compounds present in the leaves.

This chapter is the first of three that consider hydrocarbon derivatives with *oxygen-containing functional groups*. Many biochemically important molecules contain carbon atoms bonded to oxygen atoms.

In this chapter we consider hydrocarbon derivatives whose functional groups contain one oxygen atom participating in two single bonds (alcohols, phenols, and ethers). Chapter 4 focuses on derivatives whose functional groups have one oxygen atom participating in a double bond (aldehydes and ketones), and in Chapter 5 we examine functional groups that contain two oxygen atoms, one participating in single bonds and the other in a double bond (carboxylic acids, esters, and other acid derivatives).

3.1 Bonding Characteristics of Oxygen Atoms in Organic Compounds

An understanding of the bonding characteristics of the oxygen atom is a prerequisite to our study of compounds with oxygen-containing functional groups. Normal bonding behavior for oxygen atoms in such functional groups is the formation of two covalent bonds. Oxygen is a member of Group VIA of the periodic table and thus possesses six valence electrons. To complete its octet by electron sharing, an oxygen atom can form either two single bonds or a double bond.

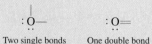

Two single bonds   One double bond

FIGURE 3.1 Space-filling models for the three simplest unbranched-chain alcohols: methyl alcohol, ethyl alcohol, and propyl alcohol.

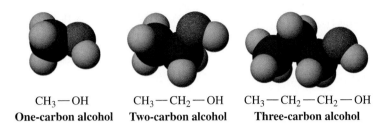

$CH_3—OH$ $CH_3—CH_2—OH$ $CH_3—CH_2—CH_2—OH$
One-carbon alcohol   **Two-carbon alcohol**   **Three-carbon alcohol**

Thus, in organic chemistry, carbon forms four bonds, hydrogen forms one bond, and oxygen forms two bonds.

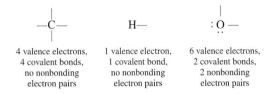

4 valence electrons,   1 valence electron,   6 valence electrons,
4 covalent bonds,   1 covalent bond,   2 covalent bonds,
no nonbonding   no nonbonding   2 nonbonding
electron pairs   electron pairs   electron pairs

3.2 Structural Characteristics of Alcohols

We begin our discussion of hydrocarbon derivatives containing a single oxygen atom by considering *alcohols,* substances with the generalized formula

R—OH

An **alcohol** *is an organic compound in which an —OH group is bonded to a saturated carbon atom.* A *saturated* carbon atom is a carbon atom that is bonded to four other atoms.

Saturated
carbon atom   —C—OH   Alcohol
                                        functional group

The —OH group, the functional group that is characteristic of an alcohol, is called a *hydroxyl group.* A **hydroxyl group** *is the —OH functional group.*

Examples of structural formulas for alcohols include

$CH_3—OH$ $CH_3—CH_2—OH$ $CH_3—CH_2—CH_2—OH$

The hydroxyl group (—OH) should not be confused with the *hydroxide* ion (OH⁻) that we have encountered previously. Alcohols are not *hydroxides.* Hydroxides are ionic compounds that contain the OH⁻ polyatomic ion. Alcohols are not ionic compounds. In an alcohol, the —OH group, which is not an ion, is *covalently* bonded to a saturated carbon atom.

Space-filling models for these three alcohols, the simplest alcohols possible that have unbranched carbon chains, are given in Figure 3.1.

Alcohols may be viewed structurally as being alkyl derivatives of water in which a hydrogen atom has been replaced by an alkyl group.

H—O—H   R—O—H
Water   An alcohol

Figure 3.2 shows the similarity in oxygen bond angles for water and $CH_3—OH$, the simplest alcohol.

Alcohols may also be viewed structurally as hydroxyl derivatives of alkanes in which a hydrogen atom has been replaced by a hydroxyl group

R—H   R—OH
An alkane   An alcohol

FIGURE 3.2 The similar shapes of water and methanol. Methyl alcohol may be viewed structurally as an alkyl derivative of water.

~105°

Water (HOH)

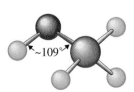

~109°

Methyl alcohol (CH_3OH)

3.3 Nomenclature for Alcohols

Common names exist for alcohols with simple (generally C_1 through C_4) alkyl groups. The word *alcohol,* as a separate word, is placed after the name of the alkyl or cycloalkyl group present.

Line-angle formulas for selected simple alcohols:

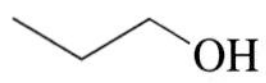

OH

Propyl alcohol
(1-propanol)

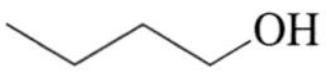

OH

Butyl alcohol
(1-butanol)

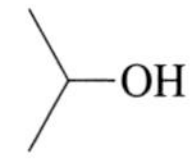

OH

Isopropyl alcohol
(2-propanol)

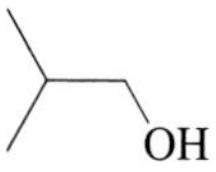

OH

Isobutyl alcohol
(2-methyl-1-propanol)

CH_3—OH
Methyl alcohol

CH_3—CH_2—OH
Ethyl alcohol

CH_3—CH_2—CH_2—OH
Propyl alcohol

CH_3—CH—OH
         |
        CH_3
Isopropyl alcohol

OH

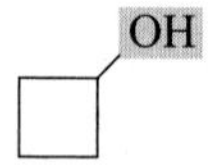

Cyclobutyl alcohol

IUPAC rules for naming alcohols that contain a single hydroxyl group follow.

Rule 1: *Name the longest carbon chain to which the hydroxyl group is attached.* The chain name is obtained by dropping the final *-e* from the alkane name and adding the suffix *-ol.*

Rule 2: *Number the chain starting at the end nearest the hydroxyl group, and use the appropriate number to indicate the position of the* —*OH group.* (In numbering of the longest carbon chain, the hydroxyl group has priority over double and triple bonds, as well as over alkyl, cycloalkyl, and halogen substituents.)

Rule 3: *Name and locate any other substituents present.*

Rule 4: *In alcohols where the* —*OH group is attached to a carbon atom in a ring, the hydroxyl group is assumed to be on carbon 1.*

Table 3.1 gives both IUPAC and common names for monohydroxy alcohols that contain four or fewer carbon atoms.

EXAMPLE 3.1

Determining IUPAC Names for Alcohols

■ Name the following alcohols, utilizing IUPAC nomenclature rules.

a.
          CH_3
          |
CH_3—CH_2—C—CH_2—CH_2—CH_3
          |
          OH

b. CH_3—CH_2—CH—CH_2—CH_3
                |
                CH_2—OH

c.
      CH_3

CH_3—⟨ ⟩—OH

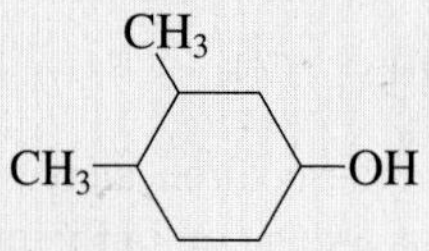

d.

OH

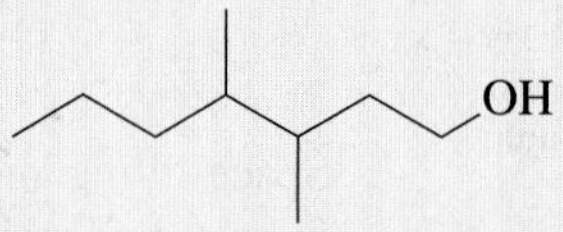

Solution

a. The longest carbon chain that contains the alcohol functional group has six carbons. When we change the *-e* to *-ol*, hexane becomes *hexanol*. Numbering the chain from the end nearest the —OH group identifies carbon number 3 as the location of both the —OH group and a methyl group. The complete name is 3-methyl-3-hexanol.

In the naming of alcohols with *unsaturated* carbon chains, two endings are needed: one for the double or triple bond and one for the hydroxyl group. The *-ol* suffix always comes last in the name; that is, unsaturated alcohols are named as *alkenols* or *alkynols.*

  3   2   1
CH_2=CH—CH_2—OH

2-Propen-1-ol
(common name: allyl alcohol)

                    CH_3
  1   2   3| 4   5   6
CH_3—CH_2—C—CH_2—CH_2—CH_3
                    |
                    OH

b. The longest carbon chain containing the —OH group has four carbon atoms. It is numbered from the end closest to the —OH group as follows:

                  2   3   4
CH_3—CH_2—CH—CH_2—CH_3
                  |
                  CH_2—OH

The base name is 1-butanol. The complete name is 2-ethyl-1-butanol.

The contrast between IUPAC and common names for alcohols is as follows:

IUPAC (one word)

$\boxed{\text{alkanol}}$

ethanol

Common (two words)

$\boxed{\text{alkyl alcohol}}$

ethyl alcohol

c. This alcohol is a cyclohexanol. The carbon to which the —OH group is attached is assigned the number 1. The complete name for this alcohol is 3,4-dimethylcyclohexanol. Note that the number 1 is not part of the name.

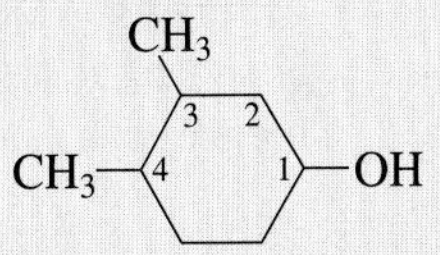

d. This alcohol is a dimethylheptanol. Numbering from right to left, the location of the hydroxyl group is 1, and locants for the methyl groups are 3 and 4. The complete IUPAC name is 3,4-dimethyl-1-heptanol.

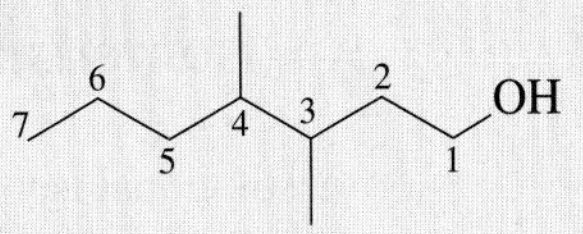

Practice Exercise 3.1

Name the following alcohols utilizing IUPAC nomenclature rules.

a. $CH_3-CH-CH-CH_2-CH-CH_3$
         $\quad\quad\; |\quad\;\; |\quad\quad\quad\quad\; |$
         $\quad\quad CH_3\; OH\quad\quad\; CH_3$

b. $CH_3-CH_2-CH-CH_3$
         $\quad\quad\quad\quad\quad\; |$
         $\quad\quad\quad\quad CH_2-CH_2-OH$

c.

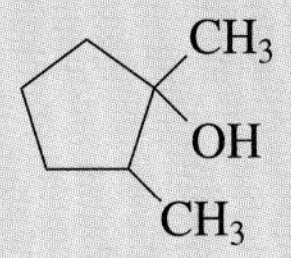

d.

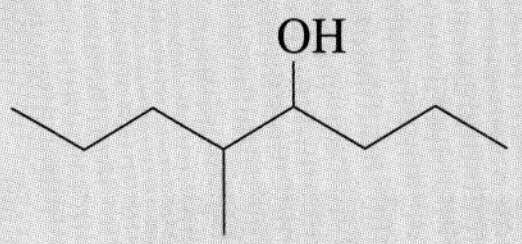

TABLE 3.1
IUPAC and Common Names of Monohydroxy Alcohols That Contain Up to Four Carbon Atoms

Formula	IUPAC Name	Common Name
One carbon atom (CH_3OH)		
CH_3-OH	methanol	methyl alcohol
Two carbon atoms (C_2H_5OH)		
CH_3-CH_2-OH	ethanol	ethyl alcohol
Three carbon atoms (C_3H_7OH); two constitutional isomers exist		
$CH_3-CH_2-CH_2-OH$	1-propanol	propyl alcohol
$CH_3-CH-CH_3$ $\quad\quad\; |$ $\quad\quad OH$	2-propanol	isopropyl alcohol
Four carbon atoms (C_4H_9OH); four constitutional isomers exist		
$CH_3-CH_2-CH_2-CH_2-OH$	1-butanol	butyl alcohol
$CH_3-CH-CH_2-OH$ $\quad\quad\; |$ $\quad\quad CH_3$	2-methyl-1-propanol	isobutyl alcohol
$CH_3-CH_2-CH-OH$ $\quad\quad\quad\quad\; |$ $\quad\quad\quad\quad CH_3$	2-butanol	*sec*-butyl alcohol
$CH_3-\overset{\displaystyle CH_3}{\underset{\displaystyle CH_3}{C}}-OH$	2-methyl-2-propanol	*tert*-butyl alcohol

A hydroxyl group as a substituent in a molecule is called a hydroxy group; an *-oxy* rather than an *-oxyl* ending is used.

Alcohols with More Than One Hydroxyl Group

Polyhydroxy alcohols—alcohols that possess more than one hydroxyl group—can be named with only a slight modification of the preceding IUPAC rules. An alcohol in which two hydroxyl groups are present is named as a *diol*, one containing three hydroxyl groups is named as a *triol*, and so on. In these names for diols, triols, and so forth, the final *-e* of the parent alkane name is retained for pronunciation reasons.

$$\underset{\underset{\text{1,2-Ethanediol}}{\overset{|\qquad\quad|}{OH\quad\ OH}}}{CH_2-CH_2} \qquad \underset{\underset{\text{1,2-Propanediol}}{\overset{\quad\ |\quad\ |}{OH\quad OH}}}{CH_3-CH-CH_3} \qquad \underset{\underset{\text{1,2,3-Propanetriol}}{\overset{|\qquad\ |\qquad\ |}{OH\quad OH\quad OH}}}{CH_2-CH-CH_2}$$

3.4 Isomerism for Alcohols

Constitutional isomerism is possible for alcohols containing three or more carbon atoms. As with alkenes (Section 2.5), both *skeletal* isomers and *positional* isomers are possible. For monohydroxy saturated alcohols, there are two C_3 isomers, four C_4 isomers, and eight C_5 isomers. Structures for the C_3 and C_4 isomers are found in Table 3.1. The C_5 isomers are

Addition of a functional group greatly increases constitutional isomer possibilities. There are 75 alkane isomers with the formula $C_{10}H_{22}$ and 507 alcohol isomers with the formula $C_{10}H_{21}OH$.

$$\underset{\underset{\text{1-Pentanol}}{\overset{|}{OH}}}{C-C-C-C-C} \qquad \underset{\underset{\text{2-Pentanol}}{\overset{\quad\ |}{OH}}}{C-C-C-C-C} \qquad \underset{\underset{\text{3-Pentanol}}{\overset{\quad\quad\ |}{OH}}}{C-C-C-C-C} \qquad \underset{\underset{\text{2,2-Dimethyl-1-propanol}}{\overset{|}{\underset{OH\ C}{C-C-C}}}}{\overset{C}{|}}$$

$$\underset{\underset{\text{2-Methyl-2-butanol}}{\overset{OH}{\underset{C}{\overset{|}{C-C-C-C}}}}}{} \qquad \underset{\underset{\text{3-Methyl-2-butanol}}{\overset{}{\underset{C\ \ OH}{C-C-C-C}}}}{} \qquad \underset{\underset{\text{3-Methyl-1-butanol}}{\overset{}{\underset{C\ \ OH}{C-C-C-C}}}}{} \qquad \underset{\underset{\text{2,2-Dimethyl-1-propanol}}{\overset{C}{\underset{OH\ C}{C-C-C}}}}{}$$

The three pentanols are positional isomers as are the four methylbutanols.

3.5 Important Commonly Encountered Alcohols

In this section we consider the properties and uses of six commonly encountered alcohols: methyl, ethyl, and isopropyl alcohols (all monohydroxy alcohols), ethylene glycol and propylene glycol (both diols), and glycerin (a triol).

Methyl Alcohol (Methanol)

Methyl alcohol, with one carbon atom and one —OH group, is the simplest alcohol. This colorless liquid is a good fuel for internal combustion engines. Since 1965 all racing cars at the Indianapolis Speedway have been fueled with methyl alcohol (Figure 3.3). (Methyl alcohol fires are easier to put out than gasoline fires because water mixes with and dilutes methyl alcohol.) Methyl alcohol also has excellent solvent properties, and it is the solvent of choice for paints, shellacs, and varnishes.

Methyl alcohol is sometimes called *wood alcohol*, terminology that draws attention to an early method for its preparation—the heating of wood to a high temperature in the absence of air. Today, nearly all methyl alcohol is produced via the reaction between H_2 and CO.

$$CO + 2H_2 \xrightarrow[\text{300°C − 400°C, 200 atm}]{\text{ZnO—Cr}_2\text{O}_3} CH_3\text{—OH}$$

Drinking methyl alcohol is very dangerous. Within the human body, methyl alcohol is oxidized by the liver enzyme *alcohol dehydrogenase* to the toxic metabolites formaldehyde and formic acid.

FIGURE 3.3 Racing cars at the Indianapolis Speedway are fueled with methyl alcohol.

Methyl alcohol poisoning is treated with ethyl alcohol, which ties up the enzyme that oxidizes methyl alcohol to its toxic metabolites. Ethyl alcohol has 10 times the affinity for the alcohol dehydrogenase enzyme that methyl alcohol has. This situation is considered further in Section 10.7.

$$CH_3-OH \xrightarrow[\text{dehydrogenase}]{\text{Alcohol}} \underset{\text{Formaldehyde}}{H-\overset{\displaystyle O}{\overset{\|}{C}}-H} \xrightarrow[\text{oxidation}]{\text{Further}} \underset{\text{Formic acid}}{H-\overset{\displaystyle O}{\overset{\|}{C}}-OH}$$

Formaldehyde can cause blindness (temporary or permanent). Formic acid causes acidosis. Ingesting as little as 1 oz (30 mL) of methyl alcohol can cause optic nerve damage.

■ Ethyl Alcohol (Ethanol)

Ethyl alcohol, the two-carbon monohydroxy alcohol, is the alcohol present in alcoholic beverages and is commonly referred to simply as alcohol or *drinking alcohol*. Like methyl alcohol, ethyl alcohol is oxidized in the human body by the liver enzyme *alcohol dehydrogenase*.

$$CH_3-CH_2-OH \xrightarrow[\text{dehydrogenase}]{\text{Alcohol}} \underset{\text{Acetaldehyde}}{CH_3-\overset{\displaystyle O}{\overset{\|}{C}}-H} \xrightarrow[\text{oxidation}]{\text{Further}} \underset{\text{Acetic acid}}{CH_3-\overset{\displaystyle O}{\overset{\|}{C}}-OH}$$

Acetaldehyde, the first oxidation product, is largely responsible for the symptoms of hangover. The odors of both acetaldehyde and acetic acid are detected on the breath of someone who has consumed a large amount of alcohol. Ethyl alcohol oxidation products are less toxic than those of methyl alcohol.

Long-term excessive use of ethyl alcohol may cause undesirable effects such as cirrhosis of the liver, loss of memory, and strong physiological addiction. Links have also been established between certain birth defects and the ingestion of ethyl alcohol by women during pregnancy (fetal alcohol syndrome).

Ethyl alcohol can be produced by yeast fermentation of sugars found in plant extracts (see Figure 3.4). The synthesis of ethyl alcohol in this manner, from grains such as corn, rice, and barley, is the reason why ethyl alcohol is often called *grain alcohol*.

Fermentation is the process by which ethyl alcohol for alcoholic beverages is produced. The maximum concentration of ethyl alcohol obtainable by fermentation is about 18% (v/v),

> Many people imagine ethanol to be relatively nontoxic and methanol to be extremely toxic. Actually, their toxicities differ by a factor of only 2. Typical fatal doses for adults are about 100 mL for methanol and about 200 mL for ethanol, although smaller doses of methanol may damage the optic nerve.

> The alcohol content of strong alcoholic beverages is often stated in terms of proof. *Proof* is twice the percentage of alcohol. This system dates back to the seventeenth century and is based on the fact that a 50% (v/v) alcohol–water mixture will burn. Its flammability was *proof* that a liquor had not been watered down.

FIGURE 3.4 An experimental setup for preparing ethyl alcohol by fermentation. (a) A small amount of yeast has been added to the aqueous sugar solution in the flask. Yeast enzymes catalyze the decomposition of sugar to ethanol and carbon dioxide, CO_2. The CO_2 is bubbling through lime water, $Ca(OH)_2$, producing calcium carbonate, $CaCO_3$. (b) More concentrated ethanol is produced from the solution in the flask by collecting the fraction that boils at about 78°C. (c) Concentrated ethanol (50% v/v) burns when it is ignited.

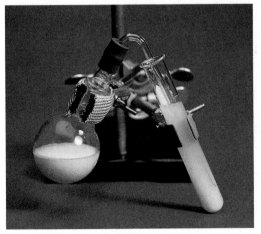

(a)

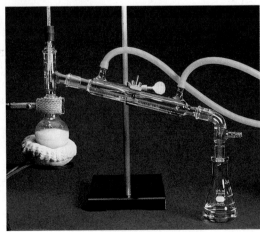

(b)

(c)

because yeast enzymes cannot function in stronger alcohol solutions. Alcoholic beverages with a higher concentration of alcohol than this are prepared by either distillation or fortification with alcohol obtained by the distillation of another fermentation product. Table 3.2 lists the alcohol content of common alcoholic beverages and of selected common household products and over-the-counter drug products.

Denatured alcohol is ethyl alcohol that has been rendered unfit to drink by the addition of small amounts of toxic substances (denaturing agents). Almost all of the ethyl alcohol used for industrial purposes is denatured alcohol.

Most ethyl alcohol used in industry is prepared from ethene via a hydration reaction (Section 2.8).

$$CH_2{=}CH_2 + H_2O \xrightarrow{\text{Catalyst}} CH_3{-}CH_2{-}OH$$

The reaction produces a product that is 95% alcohol and 5% water. In applications where water does interfere with its use, the mixture is treated with a dehydrating agent to produce 100% ethyl alcohol. Such alcohol, with all traces of water removed, is called *absolute alcohol*.

■ Isopropyl Alcohol (2-Propanol)

Isopropyl alcohol is one of two three-carbon monohydroxy alcohols; the other is propyl alcohol. A 70% isopropyl alcohol–30% water solution is marketed as *rubbing alcohol.* Isopropyl alcohol's rapid evaporation rate creates a dramatic cooling effect when it is applied to the skin, hence its use for alcohol rubs to combat high body temperature. It also finds use in cosmetics formulations such as after-shave lotion and hand lotions.

Isopropyl alcohol has a bitter taste. Its toxicity is twice that of ethyl alcohol, but it causes few fatalities because it often induces vomiting and thus doesn't stay down long enough to be fatal. In the body it is oxidized to acetone.

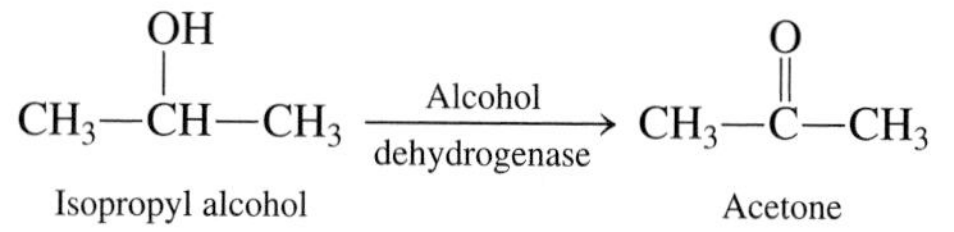

$$\underset{\text{Isopropyl alcohol}}{CH_3{-}\underset{\underset{\displaystyle OH}{|}}{CH}{-}CH_3} \xrightarrow[\text{dehydrogenase}]{\text{Alcohol}} \underset{\text{Acetone}}{CH_3{-}\overset{\overset{\displaystyle O}{\|}}{C}{-}CH_3}$$

Large amounts (about 150 mL) of ingested isopropyl alcohol can be fatal; death occurs from paralysis of the central nervous system.

The "medicinal" odor associated with doctors' offices is usually that of isopropyl alcohol.

TABLE 3.2
Ethyl Alcohol Content (volume percent) of Common Alcoholic Beverages, Household Products, and Over-the-Counter Drugs

Product Type	Product	Volume Percent Ethyl Alcohol
Alcoholic Beverages	Beer	3.2–9
	Wine (unfortified)	12
	Brandy	40–45
	Whiskey	45–55
	Rum	45
Flavorings	Vanilla extract	35
	Almond extract	50
Cough and Cold Remedies	Pertussin Plus	25
	Nyquil	25
	Dristan	12
	Vicks 44	10
	Robitussin, DM	1.4
Mouthwashes	Listerine	25
	Scope	18
	Colgate 100	17
	Cepacol	14
	Lavoris	5

■ Ethylene Glycol (1,2-Ethanediol) and Propylene Glycol (1,2-Propanediol)

The ethylene glycol and propylene glycol used in antifreeze formulations are colorless and odorless; the color and odor of antifreezes come from additives for rust protection and the like.

Ethylene glycol and propylene glycol are synthesized from ethylene and propylene, respectively, hence their common names.

Ethylene glycol and propylene glycol are the two simplest alcohols possessing two —OH groups. Besides being diols, they are also classified as glycols. A **glycol** *is a diol in which the two —OH groups are on adjacent carbon atoms.*

$$
\begin{array}{cc}
\mathrm{CH_2{-}CH_2} & \mathrm{CH_3{-}CH{-}CH_2} \\
\,|\quad\;\;| & \qquad\;\;|\quad\;\;| \\
\mathrm{OH}\;\;\mathrm{OH} & \;\;\;\;\mathrm{OH}\;\;\mathrm{OH} \\
\text{Ethylene glycol} & \text{Propylene glycol}
\end{array}
$$

Both of these glycols are colorless, odorless, high-boiling liquids that are completely miscible with water. Their major uses are as the main ingredient in automobile "year-round" antifreeze and airplane "de-icers" (Figure 3.5) and as a starting material for the manufacture of polyester fibers (Section 5.18).

Ethylene glycol is extremely toxic when ingested. In the body, liver enzymes oxidize it to oxalic acid.

$$
\mathrm{HO{-}CH_2{-}CH_2{-}OH} \xrightarrow[\text{enzymes}]{\text{Liver}} \mathrm{HO{-}\underset{\displaystyle \overset{\|}{O}}{C}{-}\underset{\displaystyle \overset{\|}{O}}{C}{-}OH}
$$

Ethylene glycol   Oxalic acid

Oxalic acid, as a calcium salt, crystallizes in the kidneys, which leads to renal problems.

Propylene glycol, on the other hand, is essentially nontoxic and has been used as a solvent for drugs. Like ethylene glycol, it is oxidized by liver enzymes; however, pyruvic acid, its oxidation product, is a compound normally found in the human body, being an intermediate in carbohydrate metabolism (Chapter 13).

$$
\mathrm{CH_3{-}CH{-}CH_2} \xrightarrow[\text{enzymes}]{\text{Liver}} \mathrm{CH_3{-}\underset{\displaystyle \overset{\|}{O}}{C}{-}\underset{\displaystyle \overset{\|}{O}}{C}{-}OH}
$$

Propylene glycol   Pyruvic acid

■ Glycerol (1,2,3-Propanetriol)

Glycerol is a clear, thick liquid that has the consistency of honey. Its molecular structure involves three —OH groups on three different carbon atoms.

$$
\begin{array}{c}
\mathrm{CH_2{-}CH{-}CH_2} \\
\;|\quad\;\;\;|\quad\;\;\;| \\
\mathrm{OH}\;\;\mathrm{OH}\;\;\mathrm{OH}
\end{array}
$$

Glycerol is normally present in the human body because it is a product of fat metabolism. It is present, in combined form, in all animal fats and vegetable oils (Section 8.4). In some Arctic species, glycerol functions as a "biological antifreeze" (see Figure 3.6).

Because glycerol has a great affinity for water vapor (moisture), it is often added to pharmaceutical preparations such as skin lotions and soap. Florists sometimes use glycerol on cut flowers to help retain water and maintain freshness. Its lubricative properties also make it useful in shaving creams and in applications such as glycerin suppositories for rectal administration of medicines. It is used in candies and icings as a retardant for preventing sugar crystallization.

3.6 Physical Properties of Alcohols

Alcohol molecules have both polar and nonpolar character. The hydroxyl groups present are polar, and the alkyl (R) group present is nonpolar.

Nonpolar portion↘   ↙Polar portion

$$
\overline{\mathrm{CH_3{-}CH_2{-}CH_2}}\;\overline{\mathrm{OH}}
$$

FIGURE 3.5 Ethylene glycol is the major ingredient in airplane "de-icers."

FIGURE 3.6 Glycerol is often called biological antifreeze. For survival in Arctic and northern winters, many fish and insects, including the common housefly, produce large amounts of glycerol that dissolve in their blood, thereby lowering the freezing point of the blood.

CHEMICAL CONNECTIONS   Menthol: A Useful Naturally Occurring Terpene Alcohol

Menthol is a naturally occurring terpene (Section 2.6) alcohol with a pleasant, minty odor. Its IUPAC name is 2-isopropyl-5-methylcyclohexanol.

In the pure state, menthol is a white crystalline solid with a melting point of 41°C to 43°C. It can be obtained from peppermint oil and can also be prepared synthetically.

Topical application of menthol to the skin causes a refreshing, cooling sensation followed by a slight burning-and-prickling sensation. Its mode of action is that of a *differential* anesthetic. It stimulates the receptor cells in the skin that normally respond to cold to give a sensation of coolness that is unrelated to body temperature. (This cooling sensation is particularly noticeable in the respiratory tract when low concentrations of menthol are inhaled.) At the same time as cooling is perceived, menthol can depress the nerves for pain reception.

Numerous products contain menthol.

- Throat sprays and lozenges containing menthol temporarily soothe inflamed mucous surfaces of the nose and throat. Lozenges contain 2–20 milligrams of menthol per wafer.
- Cough drops and cigarettes of the "mentholated" type use menthol for its counterirritant effect.

- Pre-electric shave preparations and aftershave lotions often contain menthol. A concentration of only 0.1% (m/v) gives ample cooling to allay the irritation of a "close" shave.
- Many dermatologic preparations contain menthol as an antipruritic (anti-itching agent).
- Chest-rub preparations containing menthol include Ben Gay [7% (m/v)] and Mentholatum [6% (m/v)].
- Artificial mint flavors have menthol as an ingredient. Several toothpastes and mouthwashes use menthol as a flavoring agent.

The physical properties of an alcohol depend on whether the polar or the nonpolar portion of its structure "dominates." Factors that determine this include the *length* of the nonpolar carbon chain present and the *number* of polar hydroxyl groups present (see Figure 3.7).

Boiling Points and Water Solubilities

Figure 3.8a shows that the boiling point for 1-alcohols, unbranched-chain alcohols with an —OH group on an end carbon, increases as the length of the carbon chain increases. This trend results from increasing London forces (Section 1.16) with increasing carbon chain length. Alcohols with more than one hydroxyl group present have significantly higher boiling points (bp) than their monohydroxy counterparts.

$$CH_3-CH_2-CH_2 \quad CH_3-CH-CH_2 \quad CH_2-CH-CH_2$$

bp = 97°C   bp = 188°C   bp = 290°C

FIGURE 3.7 Space-filling molecular models showing the nonpolar (green) and polar (pink) parts of methanol and 1-octanol. (a) The polar hydroxyl functional group dominates the physical properties of methanol. The molecule is completely soluble in water (polar) but only partially so in hexane (nonpolar). (b) Conversely, the nonpolar portion of 1-octanol dominates its physical properties; it is infinitely soluble in hexane and has limited solubility in water.

CH_3 — OH
Nonpolar   Polar

(a) Methanol

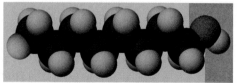

$CH_3CH_2CH_2CH_2CH_2CH_2CH_2CH_2$—OH
Nonpolar   Polar

(b) 1- Octanol

FIGURE 3.8 (a) Boiling points and (b) solubilities in water of selected 1-alcohols.

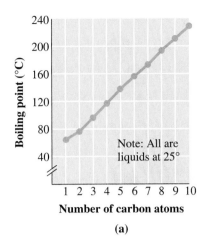
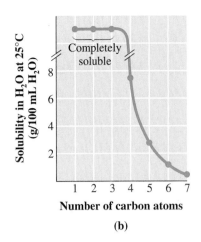

(a)   (b)

FIGURE 3.9 A physical-state summary for unbranched 1-alcohols and unsubstituted cycloalcohols at room temperature and pressure.

Unbranched 1-Alcohols			
C_1	C_3	C_5	C_7
C_2	C_4	C_6	C_8

Unsubstituted Cycloalcohols			
✕	C_3	C_5	C_7
✕	C_4	C_6	C_8

☐ Liquid

This boiling-point trend is related to increased hydrogen bonding between alcohol molecules (to be discussed shortly). Figure 3.9 is a physical-state summary for unbranched 1-alcohols and unsubstituted cycloalcohols with eight or fewer carbon atoms.

Small monohydroxy alcohols are soluble in water in all proportions. As carbon chain length increases beyond three carbons, solubility in water rapidly decreases (Figure 3.8b) because of the increasingly nonpolar character of the alcohol. Alcohols with two —OH groups present are more soluble in water than their counterparts with only one —OH group. Increased hydrogen bonding is responsible for this. Diols containing as many as seven carbon atoms show appreciable solubility in water.

■ Alcohols and Hydrogen Bonding

A comparison of the properties of alcohols with their alkane counterparts (Table 3.3) shows that

1. Alcohols have *higher* boiling points than alkanes of similar molecular mass.
2. Alcohols have much *higher* solubility in water than alkanes of similar molecular mass.

The differences in physical properties between alcohols and alkanes are related to hydrogen bonding. Because of their hydroxyl group(s), alcohols can participate in hydrogen bonding, whereas alkanes cannot. Hydrogen bonding between alcohol molecules (see Figure 3.10) is similar to that which occurs between water molecules.

Extra energy is needed to overcome alcohol–alcohol hydrogen bonds before alcohol molecules can enter the vapor phase. Hence alcohol boiling points are higher than those for the corresponding alkanes (where no hydrogen bonds are present).

TABLE 3.3
A Comparison of Selected Physical Properties of Alcohols with Alkane Counterparts of Similar Molecular Mass

Type of Compound	Compound	Structure	Molecular Mass	Boiling Point (°C)	Solubility in Water
alkane	ethane	$CH_3—CH_3$	30	−89	slight solubility
alcohol	methanol	$CH_3—OH$	32	65	unlimited solubility
alkane	propane	$CH_3—CH_2—CH_3$	44	−42	slight solubility
alcohol	ethanol	$CH_3—CH_2—OH$	46	78	unlimited solubility
alkane	butane	$CH_3—CH_2—CH_2—CH_3$	58	−1	slight solubility
alcohol	1-propanol	$CH_3—CH_2—CH_2—OH$	60	97	unlimited solubility
alcohol	2-propanol	$CH_3—\overset{\displaystyle |}{\underset{\displaystyle CH_3}{CH}}—OH$	60	83	unlimited solubility

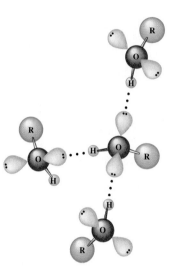

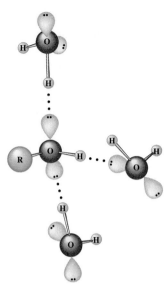

FIGURE 3.10 Alcohol boiling points are higher than those of the corresponding alkanes because of alcohol–alcohol hydrogen bonding.

FIGURE 3.11 Because of hydrogen bonding between alcohol molecules and water molecules, alcohols of small molecular mass have unlimited solubility in water.

Alcohol molecules can also hydrogen-bond to water molecules (see Figure 3.11). The formation of such hydrogen bonds explains the solubility of small alcohol molecules in water. As the alcohol chain length increases, alcohols become more alkane-like (nonpolar), and solubility decreases.

3.7 Preparation of Alcohols

A general method for preparing alcohols—the hydration of alkenes—was discussed in the previous chapter (Section 2.8). Alkenes react with water (an unsymmetrical addition agent) in the presence of sulfuric acid (the catalyst) to form an alcohol. Markovnikov's rule is used to determine the predominant alcohol product.

$$\text{C=C} + \text{H—OH} \xrightarrow{\text{H}_2\text{SO}_4} \underset{\underset{\text{H}\quad\text{OH}}{\mid\quad\mid}}{\overset{\overset{\mid\quad\mid}{}}{-\text{C}-\text{C}-}}$$

> Alcohols are intermediate products in the metabolism of both carbohydrates (Chapter 13) and fats (Chapter 14). In these metabolic processes, both addition of water to a carbon–carbon double bond and addition of hydrogen to a carbon–oxygen double bond lead to the introduction of the alcohol functional group into a biomolecule.

Another method of synthesizing alcohols involves the addition of H_2 to a carbon–oxygen double bond (a carbonyl group, C=O). (The carbonyl group is a functional group that will be discussed in detail in Chapter 4.) A carbonyl group behaves very much like a carbon–carbon double bond when it reacts with H_2 under the proper conditions. As a result of H_2 addition, the oxygen of the carbonyl group is converted to an —OH group.

$$\text{R}-\overset{\overset{\displaystyle O}{\parallel}}{\text{C}}-\text{H} + \text{H}_2 \xrightarrow{\text{Catalyst}} \text{R}-\overset{\overset{\displaystyle OH}{\mid}}{\underset{\underset{\displaystyle H}{\mid}}{\text{C}}}-\text{H}$$

Aldehyde (Section 15.2)   Alcohol

$$\text{R}-\overset{\overset{\displaystyle O}{\parallel}}{\text{C}}-\text{R}' + \text{H}_2 \xrightarrow{\text{Catalyst}} \text{R}-\overset{\overset{\displaystyle OH}{\mid}}{\underset{\underset{\displaystyle H}{\mid}}{\text{C}}}-\text{R}'$$

Ketone (Section 15.2)   Alcohol

3.8 Classification of Alcohols

Prior to considering chemical reactions of alcohols (Section 3.9), we consider a classification system for alcohols that is often needed when predicting the products in a chemical reaction that involves an alcohol.

Alcohols are classified as primary (1°), secondary (2°), or tertiary (3°) depending on the number of carbon atoms bonded to the carbon atom that bears the hydroxyl group. A **primary alcohol** *is an alcohol in which the hydroxyl-bearing carbon atom is bonded to only one other carbon atom.* A **secondary alcohol** *is an alcohol in which the hydroxyl-bearing carbon atom is bonded to two other carbon atoms.* A **tertiary alcohol** *is an alcohol in which the hydroxyl-bearing carbon atom is bonded to three other carbon atoms.* Chemical reactions of alcohols often depend on alcohol class (1°, 2°, or 3°).

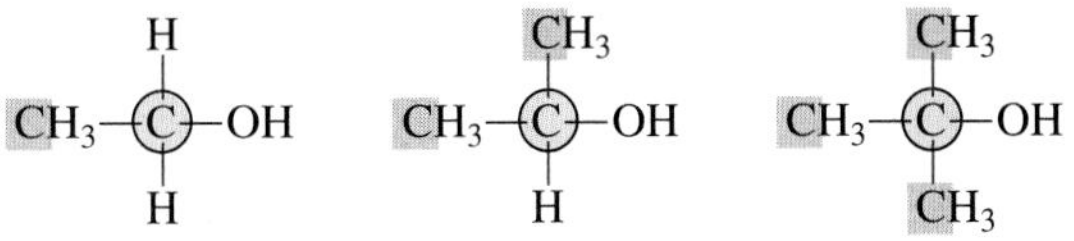

> Pronounce 1° as "primary," 2° as "secondary," and 3° as "tertiary."

> Methyl alcohol, CH_3—OH, an alcohol in which the hydroxyl-bearing carbon atom is attached to three hydrogen atoms, does not fit any of the alcohol classification definitions. It is usually grouped with the primary alcohols because its reactions are similar to theirs.

$$CH_3-\overset{\overset{\displaystyle H}{|}}{\underset{\underset{\displaystyle H}{|}}{C}}-OH \qquad CH_3-\overset{\overset{\displaystyle CH_3}{|}}{\underset{\underset{\displaystyle H}{|}}{C}}-OH \qquad CH_3-\overset{\overset{\displaystyle CH_3}{|}}{\underset{\underset{\displaystyle CH_3}{|}}{C}}-OH$$

$$\text{1° Alcohol} \qquad\qquad \text{2° Alcohol} \qquad\qquad \text{3° Alcohol}$$

EXAMPLE 3.2

Classifying Alcohols as Primary, Secondary, or Tertiary Alcohols

■ Classify each of the following alcohols as a primary, secondary, or tertiary alcohol.

a. $CH_3-CH_2-CH_2-OH$

b. 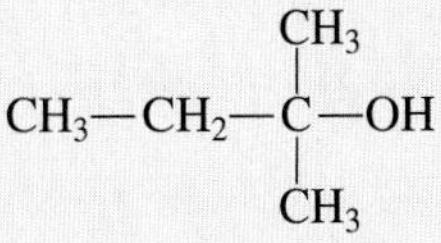 $CH_3-CH_2-\overset{\overset{\displaystyle CH_3}{|}}{\underset{\underset{\displaystyle CH_3}{|}}{C}}-OH$

c. $CH_3-\overset{\overset{\displaystyle CH_3}{|}}{CH}-\underset{\underset{\displaystyle OH}{|}}{CH}-\overset{\overset{\displaystyle CH_3}{|}}{CH}-CH_3$

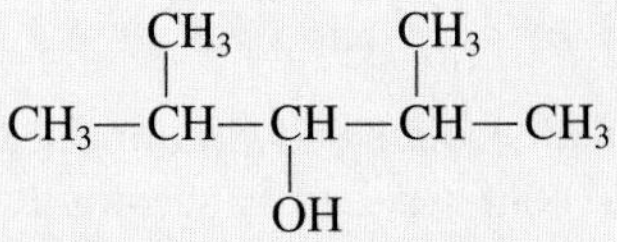

d. 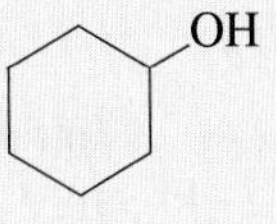 (cyclohexane ring with OH)

Solution

a. This is a primary alcohol. The carbon atom to which the —OH group is attached is bonded to only one other carbon atom.

b. This is a tertiary alcohol. The carbon atom bearing the —OH group is bonded to three other carbon atoms.

c. This is a secondary alcohol. The hydroxyl-bearing carbon atom is bonded to two other carbon atoms.

d. This is a secondary alcohol. The ring carbon atom to which the —OH group is attached is bonded to two other ring carbon atoms.

Practice Exercise 3.2

Classify each of the following alcohols as a primary, secondary, or tertiary alcohol.

a. 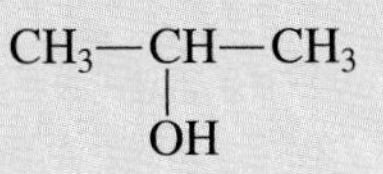 $CH_3-\underset{\underset{\displaystyle OH}{|}}{CH}-CH_3$

b. 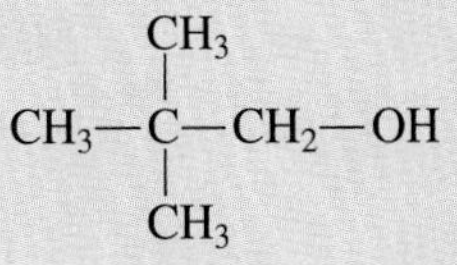 $CH_3-\overset{\overset{\displaystyle CH_3}{|}}{\underset{\underset{\displaystyle CH_3}{|}}{C}}-CH_2-OH$

c. $CH_3-\underset{\underset{\displaystyle CH_3}{|}}{CH}-\underset{\underset{\displaystyle CH_3}{|}}{CH}-OH$

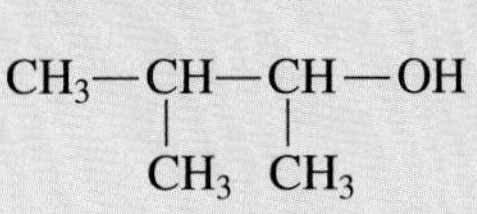

d. 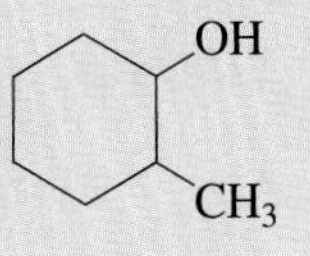 (cyclohexane ring with OH and CH_3)

3.9 Chemical Reactions of Alcohols

Of the many chemical reactions that alcohols undergo, we consider four in this section: (1) combustion, (2) dehydration, (3) oxidation, and (4) halogenation.

■ Combustion

As we have seen in the previous two chapters, hydrocarbons of all types undergo combustion in air to produce carbon dioxide and water. Alcohols are also flammable; as with hydrocarbons, the combustion products are carbon dioxide and water. Methyl alcohol is the fuel of choice for racing cars (Section 3.5). Oxygenated gasoline, which is used in winter in many areas of the United States because it burns "cleaner," contains ethyl alcohol as one of the "oxygenates."

■ Intramolecular Alcohol Dehydration

A **dehydration reaction** *is a chemical reaction in which the components of water (H and OH) are removed from a single reactant or from two reactants (H from one and OH from the other).* In *intramolecular* dehydration, both water components are removed from the same molecule.

Reaction conditions for the intramolecular dehydration of an alcohol are a temperature of 180°C and the presence of sulfuric acid (H_2SO_4) as a catalyst. The dehydration product is an alkene.

$$-\overset{|}{\underset{H}{C}}-\overset{|}{\underset{OH}{C}}- \xrightarrow[180°C]{H_2SO_4} \enspace \overset{}{C}{=}\overset{}{C} + H-OH$$

$$CH_3-\overset{|}{\underset{H}{CH}}=\overset{|}{\underset{OH}{CH_2}} \xrightarrow[180°C]{H_2SO_4} CH_3-CH{=}CH_2 + H_2O$$

Intramolecular alcohol dehydration is an example of an *elimination reaction* (see Figure 3.12), as contrasted to a substitution reaction (Section 1.17) and an addition reaction (Section 2.8). An **elimination reaction** *is a reaction in which two groups or two atoms on neighboring carbon atoms are removed, or eliminated, from a molecule, leaving a multiple bond between the carbon atoms.*

$$-\overset{|}{\underset{A}{C}}=\overset{|}{\underset{B}{C}}- \longrightarrow \enspace C{=}C + A-B$$

What occurs in an elimination reaction is the reverse of what occurs in an addition reaction.

FIGURE 3.12 In an intramolecular alcohol dehydration, the components of water (H and OH) are removed from neighboring carbon atoms with the resultant introduction of a double bond into the molecule.

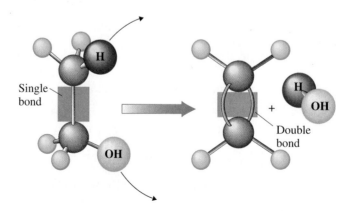

Dehydration of an alcohol can result in the production of more than one alkene product. This happens when there is more than one neighboring carbon atom from which hydrogen loss can occur. Dehydration of 2-butanol produces two alkenes.

$$CH_2-CH-CH-CH_3 \xrightarrow[180°C]{H_2SO_4}$$

H   OH   H

Removal produces 1-butene   2-Butanol   Removal produces 2-butene

①   ②   ③   ④   ①   ②   ③   ④
$$CH_2=CH-CH-CH_3 \ + \ CH_2-CH=CH-CH_3 \ + \ H_2O$$

H   H

1-Butene   2-Butene

The dominant product can be predicted using Zaitsev's rule, named after the Russian chemist Alexander Zaitsev. **Zaitsev's rule** states that *the major product in an intramolecular alcohol dehydration reaction is the alkene that has the greatest number of alkyl groups attached to the carbon atoms of the double bond.* In the preceding reaction, 2-butene (with two alkyl groups) is favored over 1-butene (with one alkyl group).

Two alkyl groups on double-bonded carbons $CH_3-CH=CH-CH_3$ 2-Butene

$CH_2=CH-CH_2-CH_3$ 1-Butene   One alkyl group on double-bonded carbons

Alkene formation via intramolecular alcohol *dehydration* is the "reverse reaction" to the reaction for preparing an alcohol through *hydration* of an alkene (Section 3.7). This relationship can be diagrammed as follows:

Hydration

An alkene   **An alcohol**

Dehydration

This "reverse reaction" situation illustrates the fact that many organic reactions can go both forward or backward, depending on reaction conditions. Noting relationships such as this helps in keeping track of the numerous reactions that hydrocarbon derivatives undergo. These two "reverse reactions" actually involve an equilibrium situation.

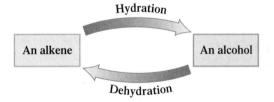

$$\text{C=C} \ + \ H_2O \ \underset{\text{Dehydration}}{\overset{\text{Hydration}}{\rightleftharpoons}} \ -\overset{|}{\underset{H}{C}}-\overset{|}{\underset{OH}{C}}-$$

An alkene   An alcohol

Whether the forward reaction (alcohol formation) or the reverse reaction (alkene formation) is favored depends on experimental conditions. The favored direction for the reaction can be predicted using Le Châtelier's principle.

1. The addition of water favors alcohol formation.
2. The removal of water favors alkene formation.

Experimental conditions for alcohol formation involve the use of a *dilute* sulfuric acid solution as a catalyst. *Concentrated* sulfuric acid (a dehydrating agent) as well as higher temperatures are used for alkene formation. Dilute acid solutions are mainly water; concentrated acid solutions have less water and heat also removes water.

Dehydration of alcohols to form carbon–carbon double bonds occurs in several metabolic pathways in living systems, such as the citric acid cycle (Section 12.6) and the fatty acid spiral (Section 14.4). In these biochemical dehydrations, enzymes serve as catalysts instead of acids, and the reaction temperature is 37°C instead of the elevated temperatures required in the laboratory.

Alexander Zaitsev (1841–1910), a nineteenth-century Russian chemist, studied at the University of Paris and then returned to his native Russia to become a professor of chemistry at the University of Kazan. His surname is pronounced "zait-zeff."

An alternative way of expressing Zaitsev's rule is "Hydrogen atom loss, during intramolecular alcohol dehydration to form an alkene, will occur preferentially from the carbon atom (adjacent to the hydroxyl-bearing carbon) that already has the fewest hydrogen atoms."

■ Intermolecular Alcohol Dehydration

At a lower temperature (140°C) than that required for alkene formation (180°C), an *inter*molecular rather than an *intra*molecular alcohol dehydration process can occur to produce an ether—a compound with the general structure R—O—R (Section 3.3). In ether formation, two alcohol molecules interact, an H atom being lost from one and an —OH group from the other. The resulting "leftover" portions of the two alcohol molecules join to form the ether. This reaction, which gives useful yields only for primary alcohol reactants (2° and 3° alcohols yield predominantly alkenes), can be written as

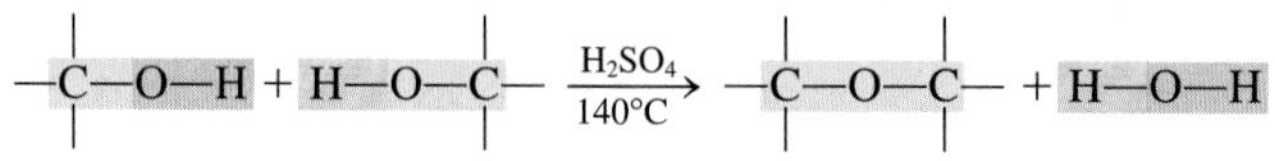

$$-\overset{|}{\underset{|}{C}}-O-H + H-O-\overset{|}{\underset{|}{C}}- \xrightarrow[140°C]{H_2SO_4} -\overset{|}{\underset{|}{C}}-O-\overset{|}{\underset{|}{C}}- + H-O-H$$

$$CH_3-CH_2-O-H + H-O-CH_2-CH_3 \xrightarrow[140°C]{H_2SO_4}$$
$$\text{Ethanol} \qquad\qquad \text{Ethanol}$$

$$CH_3-CH_2-O-CH_2-CH_3 + H_2O$$

The preceding reaction is an example of *condensation*. A **condensation reaction** *is a chemical reaction in which two molecules combine to form a larger one while liberating a small molecule, usually water.* In this case, two alcohol molecules combine to give an ether and water.

EXAMPLE 3.3

Predicting the Reactant in an Alcohol Dehydration Reaction When Given the Product

The following is a summary of products obtained from alcohol dehydration reactions using H_2SO_4 as a catalyst.

Primary alcohol	$\xrightarrow{180°C}$	alkene
	$\xrightarrow{140°C}$	ether
Secondary alcohol	$\xrightarrow{180°C}$	alkene
	$\xrightarrow{140°C}$	alkene
Tertiary alcohol	$\xrightarrow{180°C}$	alkene
	$\xrightarrow{140°C}$	alkene

■ Identify the alcohol reactant needed to produce each of the following compounds as the *major product* of an alcohol dehydration reaction.

a. Alcohol $\xrightarrow[180°C]{H_2SO_4}$ $CH_3-CH=CH-CH_3$

b. Alcohol $\xrightarrow[180°C]{H_2SO_4}$ $CH_2=CH-\underset{\underset{CH_3}{|}}{CH}-CH_3$

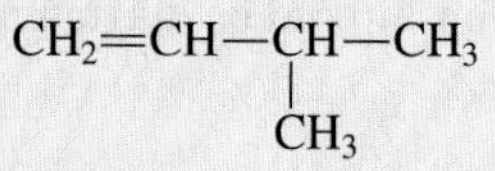

c. Alcohol $\xrightarrow[140°C]{H_2SO_4}$ $CH_3-\underset{\underset{CH_3}{|}}{CH}-CH_2-O-CH_2-\underset{\underset{CH_3}{|}}{CH}-CH_3$

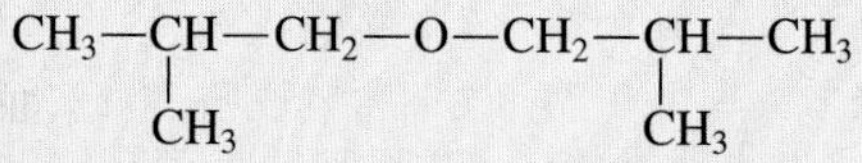

Solution

a. Both carbon atoms of the double bond are equivalent to each other. Add an H atom to one carbon atom of the double bond and an OH group to the other carbon atom of the double bond. It does not matter which goes where; you get the same molecule either way.

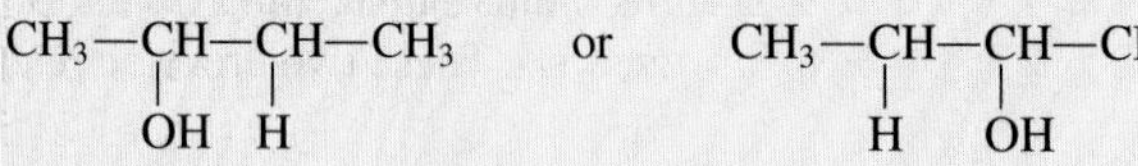

$$CH_3-\underset{\underset{OH}{|}}{CH}-\underset{\underset{H}{|}}{CH}-CH_3 \qquad \text{or} \qquad CH_3-\underset{\underset{H}{|}}{CH}-\underset{\underset{OH}{|}}{CH}-CH_3$$

b. There are two possible parent alcohols: one with an —OH group on carbon 1 and the other with an —OH group on carbon 2.

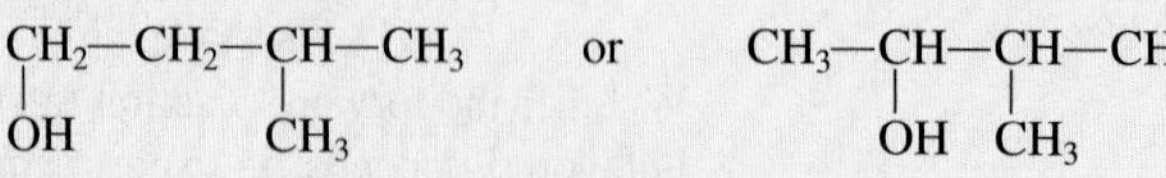

$$\underset{\underset{OH}{|}}{CH_2}-CH_2-\underset{\underset{CH_3}{|}}{CH}-CH_3 \qquad \text{or} \qquad CH_3-\underset{\underset{OH}{|}}{CH}-\underset{\underset{CH_3}{|}}{CH}-CH_3$$

Using the reverse of Zaitsev's rule, we find that the hydrogen atom will go back on the double-bonded carbon that bears the most alkyl groups.

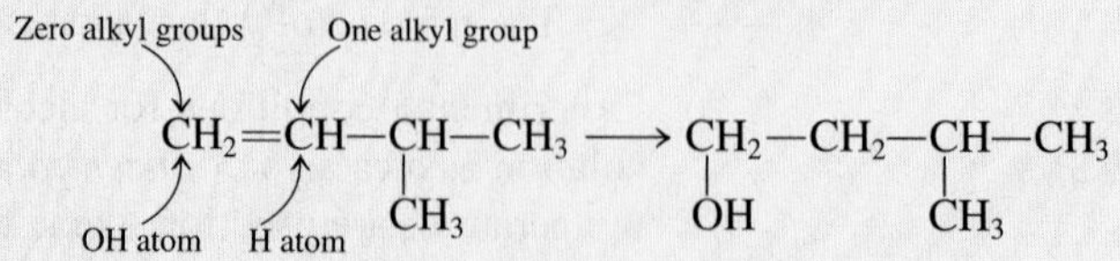

Zero alkyl groups   One alkyl group

$$CH_2=CH-\underset{\underset{CH_3}{|}}{CH}-CH_3 \longrightarrow \underset{\underset{OH}{|}}{CH_2}-CH_2-\underset{\underset{CH_3}{|}}{CH}-CH_3$$

OH atom   H atom

c. This is an ether. The primary alcohol from which the ether was formed will have the same alkyl group present as is in the ether. Thus the alcohol is

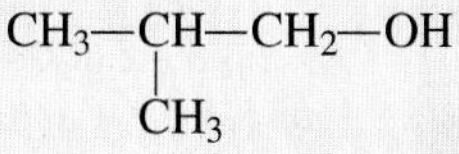

$$CH_3-CH-CH_2-OH$$
$$\quad\quad\ |$$
$$\quad\quad CH_3$$

Practice Exercise 3.3

Identify the starting alcohol from which each of the following products was obtained by an alcohol dehydration reaction.

a. Alcohol $\xrightarrow[140°C]{H_2SO_4}$ $CH_2=CH-CH_2-CH_3$

b. Alcohol $\xrightarrow[140°C]{H_2SO_4}$ $CH_3-C=C-CH_3$
$\quad\quad\quad\quad\quad\quad\quad\quad\quad | \quad |$
$\quad\quad\quad\quad\quad\quad\quad\quad CH_3 \ CH_3$

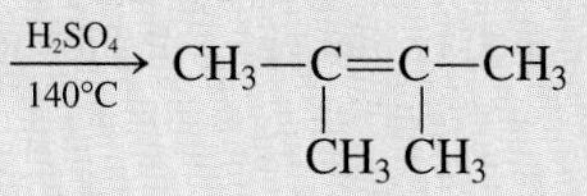

c. Alcohol $\xrightarrow[140°C]{H_2SO_4}$ $CH_3-CH_2-CH_2-O-CH_2-CH_2-CH_3$

■ Oxidation

Before discussing alcohol oxidation reactions, we consider a new method for recognizing when oxidation and reduction have occurred in a chemical reaction.

Oxidation numbers are used to characterize oxidation–reduction processes in inorganic chemistry. This same technique could be used in characterizing oxidation–reduction processes involving organic compounds, but it is not. Formal use of the oxidation number rules with organic compounds is usually cumbersome because of the many carbon and hydrogen atoms present; often, fractional oxidation numbers for carbon result.

A better approach for organic redox reactions is to use the following set of operational rules instead of oxidation numbers.

1. A carbon atom in an organic compound is considered *oxidized* if it *loses hydrogen atoms* or *gains oxygen atoms* in a redox reaction.
2. A carbon atom in an organic compound is considered *reduced* if it *gains hydrogen atoms* or *loses oxygen atoms* in a redox reaction.

Note that these operational definitions for oxidation and reduction are "opposites." This is just as it should be; oxidation and reduction are "opposite" processes.

Some alcohols readily undergo oxidation with mild oxidizing agents; others are resistant to oxidation with these same oxidizing agents. Primary and secondary alcohols, but not tertiary alcohols, readily undergo oxidation in the presence of mild oxidizing agents to produce compounds that contain a carbon–oxygen double bond (aldehydes, ketones, and carboxylic acids). A number of different oxidizing agents can be used for the oxidation, including potassium permanganate ($KMnO_4$), potassium dichromate ($K_2Cr_2O_7$), and chromic acid (H_2CrO_4).

The net effect of the action of a mild oxidizing agent on a primary or secondary alcohol is the removal of two hydrogen atoms from the alcohol. One hydrogen comes from the —OH group, the other from the carbon atom to which the —OH group is attached. This H removal generates a carbon–oxygen double bond.

$$\begin{array}{ccc} O-H & & O \\ | & & \| \\ -C-H & \xrightarrow[\text{agent}]{\text{Mild oxidizing}} & -C + 2H \\ | & & | \end{array}$$

An alcohol   Compound containing
a carbon–oxygen
double bond

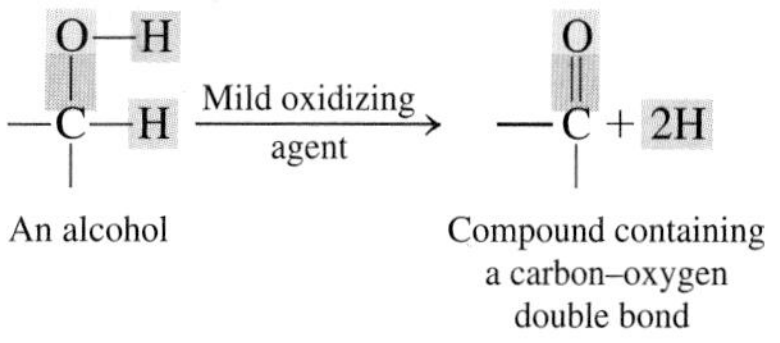

The two "removed" hydrogen atoms combine with oxygen supplied by the oxidizing agent to give H_2O.

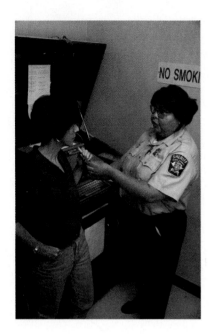

FIGURE 3.13 The oxidation of ethanol is the basis for the "breathalyzer test" that law enforcement officers use to determine whether an individual suspected of driving under the influence (DUI) has a blood alcohol level exceeding legal limits.

The DUI suspect is required to breathe into an apparatus containing a solution of potassium dichromate ($K_2Cr_2O_7$). The unmetabolized alcohol in the person's breath is oxidized by the dichromate ion ($Cr_2O_7^{2-}$), and the extent of the reaction gives a measure of the amount of alcohol present.

The dichromate ion is a yellow-orange color in solution. As oxidation of the alcohol proceeds, the dichromate ions are converted to Cr^{3+} ions, which have a green color in solution. The intensity of the green color that develops is measured and is proportional to the amount of ethanol in the suspect's breath, which in turn has been shown to be proportional to the person's blood alcohol level.

Primary and secondary alcohols, the two types of oxidizable alcohols, yield different products upon oxidation. A 1° alcohol produces an *aldehyde* that is often then further oxidized to a *carboxylic acid,* and a 2° alcohol produces a *ketone.*

$$\text{Primary alcohol} \xrightarrow[\text{ox. agent}]{\text{Mild}} \text{aldehyde} \xrightarrow[\text{ox. agent}]{\text{Mild}} \text{carboxylic acid}$$

$$\text{Secondary alcohol} \xrightarrow[\text{ox. agent}]{\text{Mild}} \text{ketone}$$

$$\text{Tertiary alcohol} \xrightarrow[\text{ox. agent}]{\text{Mild}} \text{no reaction}$$

The general reaction for the oxidation of a primary alcohol is

In this equation, the symbol [O] represents the mild oxidizing agent. The immediate product of the oxidation of a primary alcohol is an aldehyde. Because aldehydes themselves are readily oxidized by the same oxidizing agents that oxidize alcohols, aldehydes are further converted to carboxylic acids. A specific example of a primary alcohol oxidation reaction is

This specific oxidation reaction — that of ethanol — is the basis for the "breathalyzer test" used by law enforcement officers to determine whether an automobile driver is "drunk" (see Figure 3.13).

The general reaction for the oxidation of a secondary alcohol is

As with primary alcohols, oxidation involves the removal of two hydrogen atoms. Unlike aldehydes, ketones are resistant to further oxidation. A specific example of the oxidation of a secondary alcohol is

Tertiary alcohols do not undergo oxidation with mild oxidizing agents. This is because they do not have hydrogen on the —OH-bearing carbon atom.

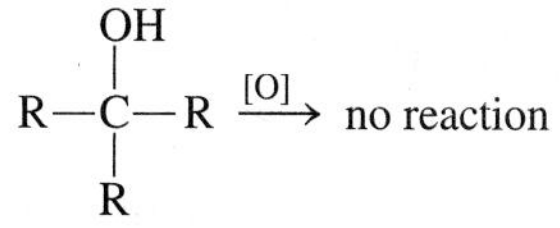

$$
\begin{array}{c}
\text{OH} \\
| \\
\text{R}\!-\!\text{C}\!-\!\text{R} \xrightarrow{\text{[O]}} \text{no reaction} \\
| \\
\text{R}
\end{array}
$$

3° Alcohol

EXAMPLE 3.4

Predicting Products in Alcohol Oxidation Reactions

■ Draw the structural formula(s) for the product(s) formed by oxidation of the following alcohols with a mild oxidizing agent. If no reaction occurs, write "no reaction."

a. CH₃—CH₂—CH₂—CH—CH₃
  |
  OH

b. CH₃—CH—CH₂—OH
  |
  CH₃

c. CH₃—CH₂—CH—OH
  |
  CH₃

d. (cyclohexane ring with OH and CH₃ on same carbon)

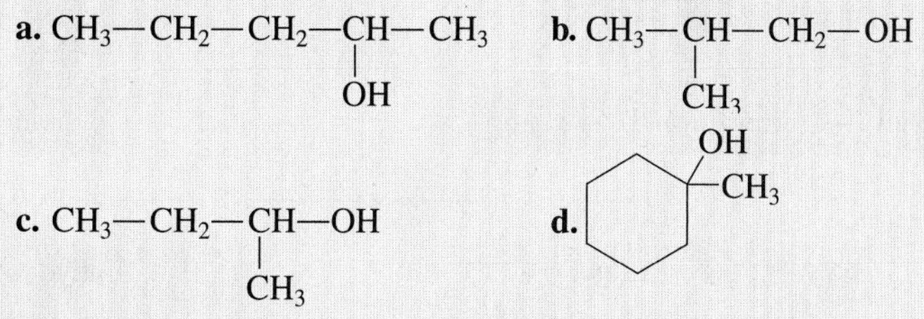

Solution

a. The oxidation product will be a ketone, as this is a 2° alcohol.

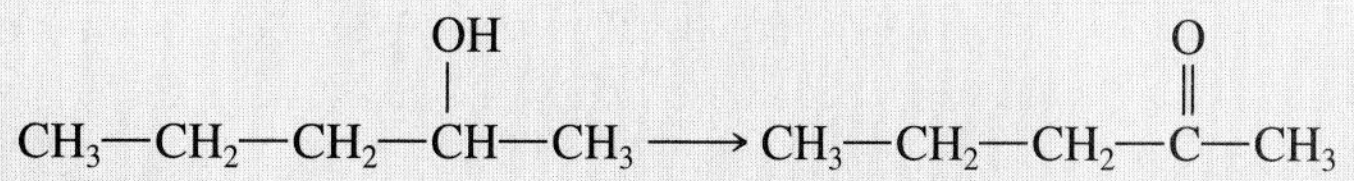

$$
\begin{array}{ccc}
\text{OH} & & \text{O} \\
| & & \| \\
\text{CH}_3\!-\!\text{CH}_2\!-\!\text{CH}_2\!-\!\text{CH}\!-\!\text{CH}_3 \longrightarrow & & \text{CH}_3\!-\!\text{CH}_2\!-\!\text{CH}_2\!-\!\text{C}\!-\!\text{CH}_3
\end{array}
$$

b. A 1° alcohol undergoes oxidation first to an aldehyde and then to a carboxylic acid.

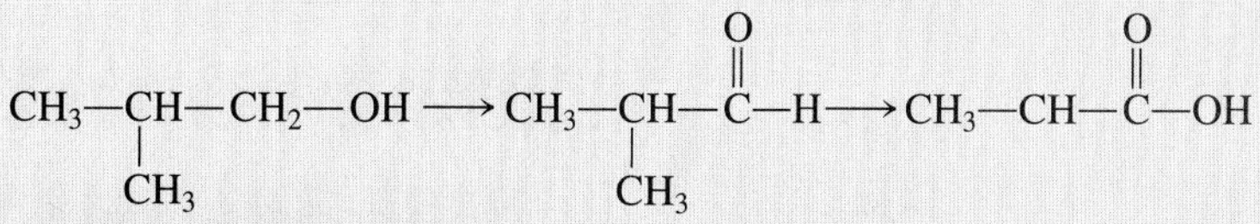

$$
\begin{array}{ccc}
 & \text{O} & \text{O} \\
 & \| & \| \\
\text{CH}_3\!-\!\text{CH}\!-\!\text{CH}_2\!-\!\text{OH} \longrightarrow \text{CH}_3\!-\!\text{CH}\!-\!\text{C}\!-\!\text{H} \longrightarrow \text{CH}_3\!-\!\text{CH}\!-\!\text{C}\!-\!\text{OH} \\
| & | & | \\
\text{CH}_3 & \text{CH}_3 & \text{CH}_3
\end{array}
$$

c. A ketone is the product from the oxidation of a 2° alcohol.

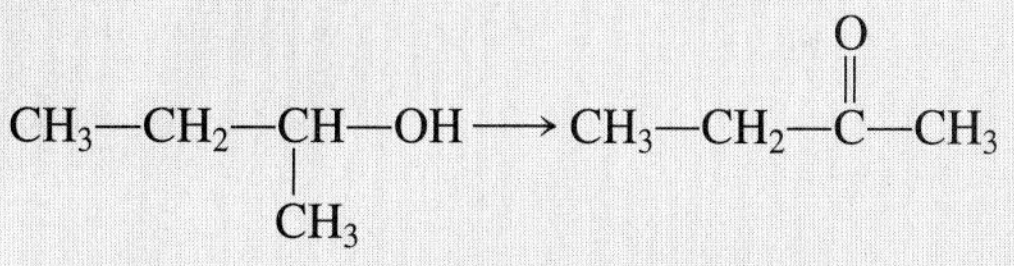

$$
\begin{array}{cc}
 & \text{O} \\
 & \| \\
\text{CH}_3\!-\!\text{CH}_2\!-\!\text{CH}\!-\!\text{OH} \longrightarrow \text{CH}_3\!-\!\text{CH}_2\!-\!\text{C}\!-\!\text{CH}_3 \\
| \\
\text{CH}_3
\end{array}
$$

d. This cyclic alcohol is a tertiary alcohol. The hydroxyl-bearing carbon atom is attached to two ring carbon atoms and a methyl group. Tertiary alcohols do not undergo oxidation with mild oxidizing agents. Therefore, "no reaction."

Practice Exercise 3.4

Draw the structural formula(s) for the product(s) formed by oxidation of the following alcohols with a mild oxidizing agent. If no reaction occurs, write "no reaction."

a. CH₃—CH₂—CH₂—OH

b. CH₃—C—OH
  with CH₃ above and CH₃ below the central C

c. CH₃—CH—CH₂—CH₃
  |
  OH

d. (cyclohexane ring with OH and CH₃ substituents)

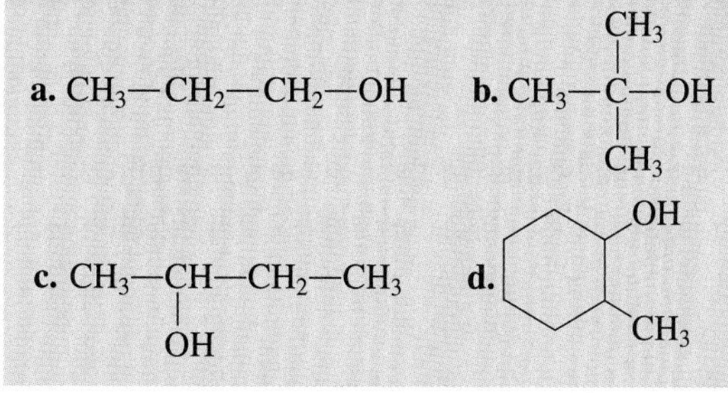

■ **Halogenation**

Alcohols undergo halogenation reactions in which a halogen atom is substituted for the hydroxyl group, producing an alkyl halide. Alkyl halide production in this manner is

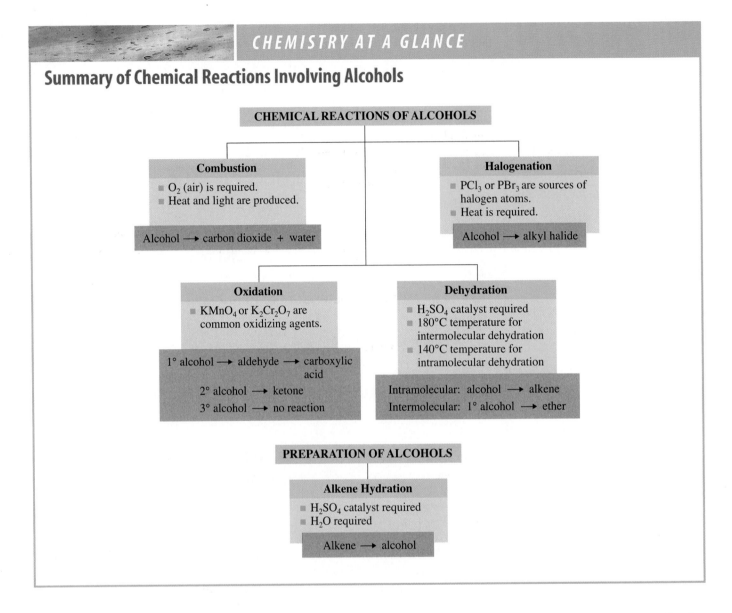

CHEMISTRY AT A GLANCE

Summary of Chemical Reactions Involving Alcohols

CHEMICAL REACTIONS OF ALCOHOLS

Combustion
- O_2 (air) is required.
- Heat and light are produced.

Alcohol ⟶ carbon dioxide + water

Halogenation
- PCl_3 or PBr_3 are sources of halogen atoms.
- Heat is required.

Alcohol ⟶ alkyl halide

Oxidation
- $KMnO_4$ or $K_2Cr_2O_7$ are common oxidizing agents.

1° alcohol ⟶ aldehyde ⟶ carboxylic acid

2° alcohol ⟶ ketone

3° alcohol ⟶ no reaction

Dehydration
- H_2SO_4 catalyst required
- 180°C temperature for intermolecular dehydration
- 140°C temperature for intramolecular dehydration

Intramolecular: alcohol ⟶ alkene

Intermolecular: 1° alcohol ⟶ ether

PREPARATION OF ALCOHOLS

Alkene Hydration
- H_2SO_4 catalyst required
- H_2O required

Alkene ⟶ alcohol

superior to alkyl halide production through halogenation of an alkane (Section 1.18) because mixtures of products are *not* obtained. A single product is produced in which the halogen atom is found only where the —OH group was originally located.

Several different halogen-containing reactants, including phosphorus trihalides (PX_3; X is Cl or Br), are useful in producing alkyl halides from alcohols.

$$3R\text{—}OH + PX_3 \xrightarrow{\text{heat}} 3R\text{—}X + H_3PO_3$$

Note that heating of the reactants is required.

The Chemistry at a Glance feature summarizes the reaction chemistry of alcohols.

3.10 Polymeric Alcohols

It is possible to synthesize polymeric alcohols with structures similar to those of substituted polyethylenes (Section 2.9). Two of the simplest such compounds are poly(vinyl alcohol) (PVA) and poly(ethylene glycol) (PEG).

Poly(vinyl alcohol) is a tough, whitish polymer that can be formed into strong films, tubes, and fibers that are highly resistant to hydrocarbon solvents. Unlike most organic polymers, PVA is water-soluble. Water-soluble films and sheetings are important PVA entities. PVA has oxygen-barrier properties under dry conditions that are superior to those of any other polymer. PVA can be rendered insoluble in water, if needed, by use of chemical agents that cross-link individual polymer strands.

Aqueous solutions of PEG are very viscous (thick) because of the great solubility of PEG in water. PEG is used as an additive in many shampoos. It contributes little to the cleansing action of the shampoo, but it gives the shampoo texture or richness.

3.11 Structural Characteristics of Phenols

A **phenol** *is an organic compound in which an —OH group is attached to a carbon atom that is part of an aromatic carbon ring system.*

The general formula for phenols is Ar–OH, where Ar represents an *aryl group.* An **aryl group** *is an aromatic carbon ring system from which one hydrogen atom has been removed.*

> The generic term *aryl group* (Ar) is the aromatic counterpart of the nonaromatic general term *alkyl group* (R).

A hydroxyl group is thus the functional group for both phenols and alcohols. The reaction chemistry for phenols is sufficiently different from that for nonaromatic alcohols (Section 3.9) to justify discussing these compounds separately. Remember that phenols contain a "benzene ring" and that the chemistry of benzene is much different from that of other unsaturated hydrocarbons (Section 2.14).

The following are examples of compounds classified as phenols.

3.12 Nomenclature for Phenols

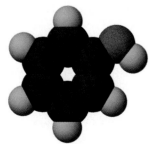

FIGURE 3.14 A space-filling model for *phenol*, a compound that has an —OH group bonded directly to a benzene (aromatic) ring.

Besides being the name for a family of compounds, *phenol* is also the IUPAC-approved name for the simplest member of the phenol family of compounds.

Phenol

A space-filling model for the compound *phenol* is shown in Figure 3.14. The name *phenol* is derived from a combination of the terms *phen*yl and alcoh*ol*.

The IUPAC rules for naming phenols are simply extensions of the rules used to name benzene derivatives with hydrocarbon or halogen substituents (Section 2.12). The parent name is phenol. Ring numbering always begins with the hydroxyl group and proceeds in the direction that gives the lower number to the next carbon atom bearing a substituent. The numerical position of the hydroxyl group is not specified in the name because it is 1 by definition.

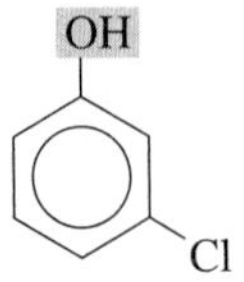

3-Chlorophenol
(or *meta*-Chlorophenol)

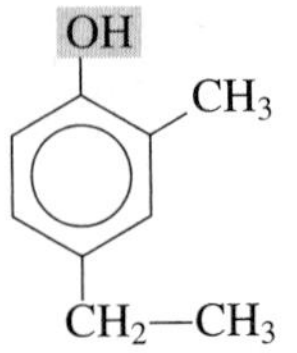

4-Ethyl-2-methylphenol

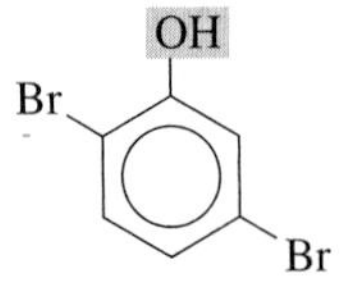

2,5-Dibromophenol

Methyl and hydroxy derivatives of phenol have IUPAC-accepted common names. Methylphenols are called cresols. The name *cresol* applies to all three isomeric methylphenols.

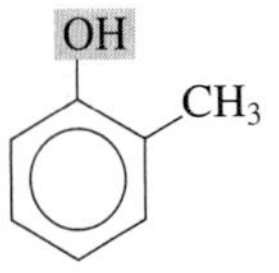

ortho-Cresol

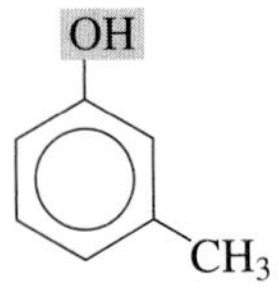

meta-Cresol

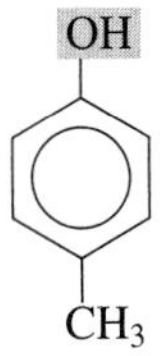

para-Cresol

For hydroxyphenols, each of the three isomers has a different common name.

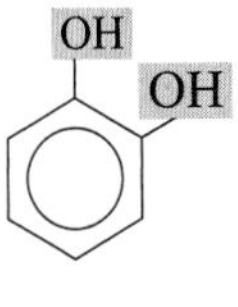

Catechol

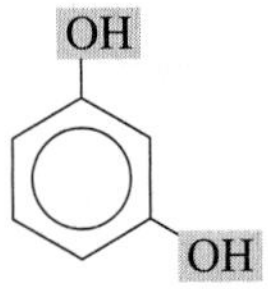

Resorcinol

Hydroquinone

Several neurotransmitters in the human body (Section 6.10), including norepinephrine, epinephrine (adrenaline), and dopamine, are catechol derivatives.

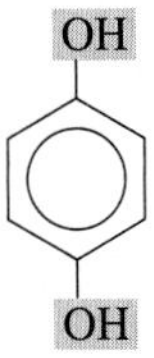

3.13 Physical and Chemical Properties of Phenols

Phenols are generally low-melting solids or oily liquids at room temperature. Most of them are only slightly soluble in water. Many phenols have antiseptic and disinfectant properties. The simplest phenol, phenol itself, is a colorless solid with a medicinal odor. Its melting point is 41°C, and it is more soluble in water than are most other phenols.

We have previously noted that the chemical properties of phenols are significantly different from those of alcohols (Section 3.11). The similarities and differences between these two reaction chemistries are as follows:

An *antiseptic* is a substance that kills microorganisms on living tissue. A *disinfectant* is a substance that kills microorganisms on inanimate objects.

1. Both alcohols and phenols are flammable.
2. Dehydration is a reaction of alcohols but not of phenols; phenols cannot be dehydrated.
3. Both 1° and 2° alcohols are oxidized by mild oxidizing agents. Tertiary (3°) alcohols and phenols do not react with the oxidizing agents that cause 1° and 2° alcohol oxidation. Phenols can be oxidized by stronger oxidizing agents.
4. Both alcohols and phenols undergo halogenation in which the hydroxyl group is replaced by a halogen atom in a substitution reaction.

■ Acidity of Phenols

One of the most important properties of phenols is their acidity. Unlike alcohols, phenols are weak acids in solution. As acids, phenols have K_a values of about 10^{-10}. Such K_a values are lower than those of most weak inorganic acids (10^{-5} to 10^{-10}). The acid ionization reaction for phenol itself is

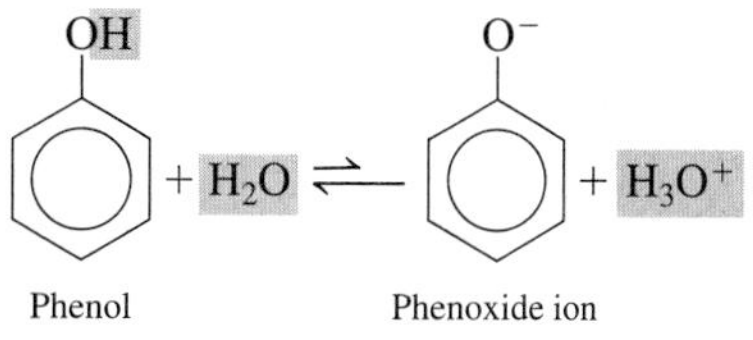

Phenol   Phenoxide ion

Note that the negative ion produced from the ionization is called the phenoxide ion. When phenol itself is reacted with sodium hydroxide (a base), the salt sodium phenoxide is produced.

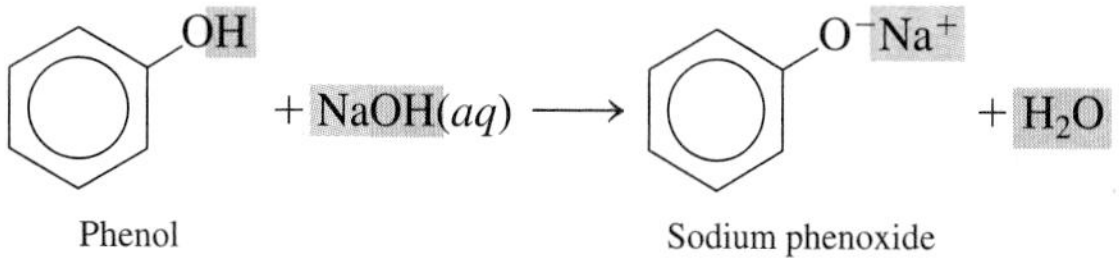

Phenol   Sodium phenoxide

3.14 Occurrence of and Uses for Phenols

Dilute (2%) solutions of phenol have long been used as antiseptics. Concentrated phenol solutions, however, can cause severe skin burns. Today, phenol has been largely replaced by more effective phenol derivatives such as 4-hexylresorcinol. The compound 4-hexylresorcinol is an ingredient in many mouthwashes and throat lozenges.

The "parent" name for a benzene ring bearing two hydroxyl groups "meta" to each other is resorcinol (Section 3.12).

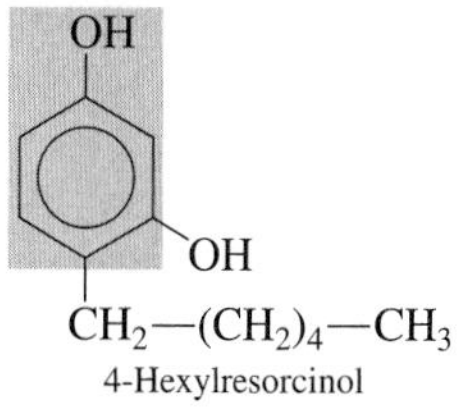

4-Hexylresorcinol

The phenol derivatives *o*-phenylphenol and 2-benzyl-4-chlorophenol are the active ingredients in Lysol, a disinfectant for walls, floors, and furniture in homes and hospitals.

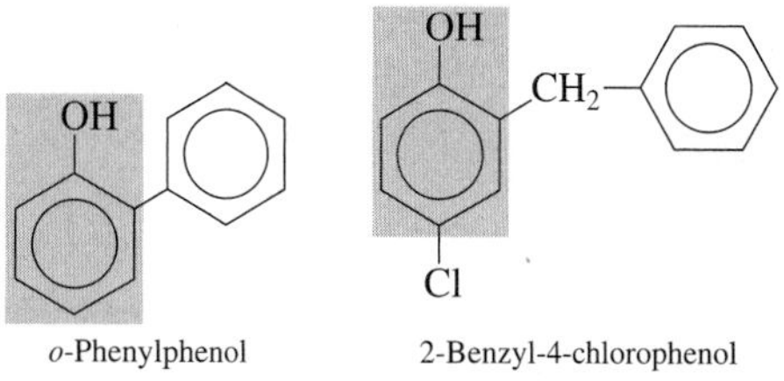

o-Phenylphenol   2-Benzyl-4-chlorophenol

A number of phenols possess antioxidant activity. An **antioxidant** *is a substance that protects other substances from being oxidized by being oxidized itself in preference to the other substances.* An antioxidant has a greater affinity for a particular oxidizing agent than do the substances the antioxidant is "protecting"; the antioxidant, therefore, reacts with the oxidizing agent first. Many foods sensitive to air are protected from oxidation through the

FIGURE 3.15 Many commercially baked goods contain the antioxidants BHA and BHT to help prevent spoilage.

Within the human body, natural dietary antioxidants also offer protection against undesirable oxidizing agents. They include vitamin C (section 10.13), beta-carotene (Section 10.14), vitamin E (Section 10.14), and flavonoids (Sec. 12.11).

use of phenolic antioxidants. Two commercial phenolic antioxidant food additives are BHA (butylated hydroxy anisole) and BHT (butylated hydroxy toluene) (see Figure 3.15).

BHA (2 isomers)

BHT

A naturally occurring phenolic antioxidant that is important in the functioning of the human body is vitamin E (Section 10.14).

Vitamin E

A number of phenols found in plants are used as flavoring agents and/or antibacterials. Included among these phenols are

Thymol

Eugenol

Isoeugenol

Vanillin

FIGURE 3.16 Nutmeg tree fruit. A phenolic compound, isoeugenol, is responsible for the odor associated with nutmeg.

Thymol, obtained from the herb thyme, possesses both flavorant and antibacterial properties. It is used as an ingredient in several mouthwash formulations.

Eugenol is responsible for the flavor of cloves. Dentists traditionally used clove oil as an antiseptic because of eugenol's presence; they use it to a limited extent even today.

Isoeugenol, which differs in structure from eugenol only in the location of the double bond in the hydrocarbon side chain, is responsible for the odor associated with nutmeg (see Figure 3.16).

Vanillin, which gives vanilla its flavor, is extracted from the dried seed pods of the vanilla orchid. Natural supplies of vanillin are inadequate to meet demand for this flavoring agent. Synthetic vanillin is produced by oxidation of eugenol. Vanillin is an unusual substance in that even though its odor can be perceived at extremely low concentrations, the strength of its odor does not increase greatly as its concentration is increased.

Certain phenols exert profound physiological effects. For example, the irritating constituents of poison ivy and poison oak are derivatives of catechol (Section 3.12). These skin irritants have 15-carbon alkyl side chains with varying degrees of unsaturation (zero to three double bonds).

Catechol   Poison ivy irritants

3.15 Structural Characteristics of Ethers

An **ether** *is an organic compound in which an oxygen atom is bonded to two carbon atoms by single bonds.* In an ether, the carbon atoms that are attached to the oxygen atom can be part of alkyl, cycloalkyl, or aryl groups. Examples of ethers include

CH_3—O—CH_3 CH_3—CH_2—O—⬠ CH_3—O—⬡

The two groups attached to the oxygen atom of an ether can be the same (first structure), but they need not be so (second and third structures).

All ethers contain a C—O—C unit, which is the ether functional group.

Ether functional group

---C—O—C---

FIGURE 3.17 The similar shapes of water and dimethyl ether molecules. Dimethyl ether may be viewed structurally as a dialkyl derivative of water.

Water (HOH)

~105°

Dimethyl ether (CH_3OCH_3)

~111°

Generalized formulas for ethers, which depend on the types of groups attached to the oxygen atom (alkyl or aryl), include R—O—R, R—O—R′ (where R′ is an alkyl group different from R), R—O—Ar, and Ar—O—Ar.

Structurally, an ether can be visualized as a derivative of water in which both hydrogen atoms have been replaced by hydrocarbon groups (see Figure 3.17). Note that unlike alcohols and phenols, ethers do not possess a hydroxyl (—OH) group.

H—O—H   R—O—R

Water   An ether

3.16 Nomenclature for Ethers

Common names for ethers are formed by naming the two hydrocarbon groups attached to the oxygen atom and adding the word *ether.* The hydrocarbon groups are listed in alphabetical order. When both hydrocarbon groups are the same, the prefix *di-* is placed before the name of the hydrocarbon group. In this system, ether names consist of two or three separate words.

CH_3—O—CH_2—CH_3 CH_3—CH_2—O—⬡ CH_3—O—CH_3

Ethyl methyl ether   Ethyl phenyl ether   Dimethyl ether

In the IUPAC nomenclature system, ethers are named as substituted hydrocarbons. The smaller hydrocarbon attachment and the oxygen atom are called an *alkoxy group,* and

Line-angle drawings for selected simple ethers:

Methyl ethyl ether
(methoxyethane)

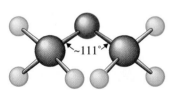

Diethyl ether
(ethoxyethane)

Dipropyl ether
(1-propoxypropane)

this group is considered a substituent on the larger hydrocarbon group. An **alkoxy group** *is an —OR group, an alkyl (or aryl) group attached to an oxygen atom.* Simple alkoxy groups include the following:

$$CH_3-O- \qquad CH_3-CH_2-O- \qquad CH_3-CH_2-CH_2-O-$$
Methoxy group Ethoxy group Propoxy group

The general symbol for an alkoxy group is —O—R (or —OR).

The steps in naming an ether using the IUPAC system are

1. Select the longest carbon chain and use its name as the base name.
2. Change the *-yl* ending of the other hydrocarbon group to *-oxy* to obtain the alkoxy group name; *methyl* becomes *methoxy, ethyl* becomes *ethoxy,* etc.
3. Place the alkoxy name, with a locator number, in front of the base chain name.

Here are two examples of IUPAC ether nomenclature, with the alkoxy groups present highlighted in each structure:

$$CH_3-O-CH_2-CH_2-CH_2-CH_3$$
1-Methoxybutane

$$CH_3-CH-CH_2-O-CH_2-CH_3$$
$$\quad CH_3$$
1-Ethoxy-2-methylpropane

The simplest aromatic ether involves a methoxy group attached to a benzene ring. This ether goes by the common name *anisole.*

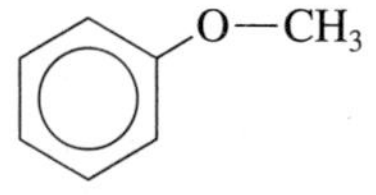

Anisole

Derivatives of anisole are named as substituted anisoles, in a manner similar to that for substituted phenols (Section 3.12). Anisole derivatives were encountered in Section 3.14 when considering antioxidant food additives: BHAs are both a phenol and an anisole.

It is possible to have compounds that contain both ether and alcohol functional groups such as

$$CH_3-CH-CH_2-CH_2-O-CH_3$$
$$\quad OH$$
4-Methoxy-2-butanol

(The alcohol functional group has higher priority in IUPAC nomenclature, so the compound is named as an alcohol rather than as an ether.)

The compound responsible for the characteristic odor of anise and fennel is *anethole,* an allyl derivative of anisole.

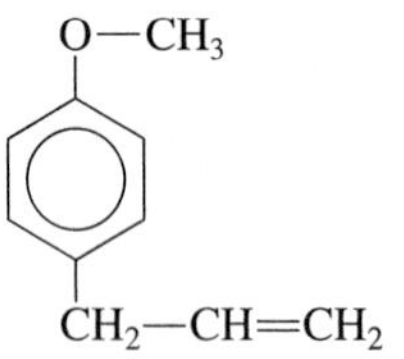

EXAMPLE 3.5

Determining IUPAC Names for Ethers

■ Name the following ethers utilizing IUPAC nomenclature rules.

a. $CH_3-CH_2-O-CH_2-CH_2-CH_3$

b. $CH_3-O-CH-CH_2-CH_3$
$\qquad CH_3$

c. 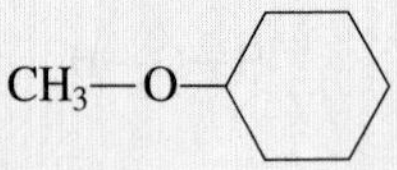 $CH_3-O-\langle hexagon \rangle$

d. Ethyl methyl ether

Solution

a. The base name is propane. An ethoxy group is attached to carbon-1 of the propane chain.

$$CH_3-CH_2-O-CH_2-CH_2-CH_3$$
The IUPAC name is 1-ethoxypropane

b. The base name is butane, as the longest carbon chain contains four carbon atoms.

$$CH_3-O-CH-CH_2-CH_3$$
$$\qquad CH_3$$
The IUPAC name is 2-methoxybutane.

CHEMICAL CONNECTIONS   Ethers as General Anesthetics

For many people, the word *ether* evokes thoughts of hospital operating rooms and anesthesia. This response derives from the *former* large-scale use of diethyl ether as a general anesthetic. In 1846, the Boston dentist William Morton was the first to demonstrate publicly the use of diethyl ether as a surgical anesthetic.

In many ways, diethyl ether is an ideal general anesthetic. It is relatively easy to administer, it is readily made in pure form, and it causes excellent muscle relaxation. There is less danger of an overdose with diethyl ether than with almost any other anesthetic because there is a large gap between the effective level for anesthesia and the lethal dose.

Despite these ideal properties, diethyl ether is rarely used today because of two drawbacks: (1) It causes nausea and irritation of the respiratory passage, and (2) it is a highly flammable substance, forming explosive mixtures with air, which can be set off by a spark.

By the 1930s, nonether anesthetics had been developed that solved the problems of nausea and irritation. They also, however, were extremely flammable compounds. The simple hydrocarbon cyclopropane was the most widely used of these newer compounds.

It was not until the late 1950s and early 1960s that nonflammable general anesthetics became available. Anesthetic nonflammability was achieved by incorporating halogen atoms into anesthetic molecules. Three of the most used of these "halogenated" anesthetics are enflurane, isoflurane, and halothane.

Enflurane and isoflurane, which are constitutional isomers, are hexahalogenated ethers.

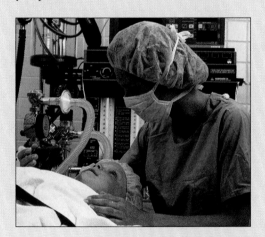

Enflurane   Isoflurane

With these compounds, induction of anesthesia can be achieved in less than 10 minutes with an inhaled concentration of 3% in oxygen.

Halothane, which is potent at relatively low doses, and whose effects wear off quickly, is a pentahalogenated alkane derivative rather than an ether.

Halothane

---

c. The base name is cyclohexane. The complete IUPAC name is methoxycyclohexane. No number is needed to locate the methoxy group since all ring carbon atoms are equivalent to each other.

d. The ether structure is CH_3—CH_2—O—CH_3, and the IUPAC name is methoxyethane.

Practice Exercise 3.5

Name the following ethers utilizing IUPAC nomenclature rules.

a. CH_3—CH_2—CH_2—O—CH_2—CH_2—CH_3 **b.** CH_3—O—CH_2—CH—CH_3
                                                                             |
                                                                          CH_3

c.   **d.** Dimethyl ether

The contrast between IUPAC and common names for ethers is as follows:

IUPAC (one word)

[alkoxyalkane]

2-methoxybutane

Common (three or two words)

[alkyl alkyl ether]

ethyl methyl ether

or

[dialkyl ether]

dipropyl ether

The ether MTBE (methyl *tert*-butyl ether) has been a widely used gasoline additive since the early 1980s.

$$CH_3-O-\underset{\underset{CH_3}{|}}{\overset{\overset{CH_3}{|}}{C}}-CH_3$$

Methyl *tert*-butyl ether
(MTBE)

As an additive, MTBE not only raises octane levels but also functions as a clean-burning "oxygenate" in EPA-mandated reformulated gasolines used to improve air quality in polluted areas. The amount of MTBE used in gasoline is now decreasing in response to a growing problem: contamination of water supplies by small amounts of MTBE from leaking gasoline tanks and from spills. MTBE in the water supplies is not a health-and-safety issue at this time, but its presence does affect taste and odor in contaminated supplies.

Compounds with ether functional groups occur in a variety of plants. The phenolic flavoring agents eugenol, isoeugenol, and vanillin (Section 3.11) are also ethers; each has a methoxy substituent on the ring.

> Technically, the name methyl *tert*-butyl ether (MTBE) is incorrect because the convention for naming ethers dictates an alphabetical ordering of alkyl groups (*tert*-butyl methyl ether). However, the compound is called MTBE rather than TBME by those in the petroleum industry and by environmental scientists.

3.17 Isomerism for Ethers

Ethers contain two carbon chains (two alkyl groups), unlike the one carbon chain found in alcohols. Constitutional isomerism possibilities in ethers depend on (1) the partitioning of carbon atoms between the two alkyl groups and (2) isomerism possibilities for the individual alkyl groups present. Isomerism is not possible for a C_2 ether (two methyl groups) or a C_3 ether (a methyl and an ethyl group). For C_4 ethers, isomerism arises not only from carbon atom partitioning between the alkyl groups (C_1-C_3 and C_2-C_2) but also from isomerism within a C_3 group (propyl and isopropyl). There are three C_4 ether constitutional isomers.

$$CH_3-CH_2-O-CH_2-CH_3 \qquad CH_3-O-CH_2-CH_2-CH_3 \qquad CH_3-O-\underset{\underset{CH_3}{|}}{CH}-CH_3$$

Diethyl ether   Methyl propyl ether

Isopropyl methyl ether

For C_5 ethers, carbon partitioning possibilities are C_2-C_3 and C_1-C_4. For C_4 groups there are four isomeric variations: butyl, isobutyl, *sec*-butyl, and *tert*-butyl (Section 1.11).

■ Functional Group Isomerism

Ethers and alcohols with the same number of carbon atoms and the same degree of saturation have the same molecular formula. The simplest manifestation of this phenomenon involves dimethyl ether, the C_2 ether, and ethyl alcohol, the C_2 alcohol. Both have the molecular formula C_2H_6O.

$$CH_3-O-CH_3 \qquad CH_3-CH_2-OH$$

Dimethyl ether   Ethyl alcohol

With the same molecular formula and different structural formulas, these two compounds are constitutional isomers. This type of constitutional isomerism is the subtype called *functional group isomerism.* **Functional group isomers** *are constitutional isomers that contain different functional groups.* When three carbon atoms are present the ether–alcohol functional group isomerism possibilities are

> Later in this chapter (Section 3.21) and in each of the next two chapters we will encounter other pairs of functional groups for which functional group isomerism is possible.

$$CH_3-CH_2-O-CH_3 \qquad CH_3-CH_2-CH_2-OH \qquad CH_3-\underset{\underset{CH_3}{|}}{CH}-OH$$

Ethyl methyl ether   Propyl alcohol

Isopropyl alcohol

FIGURE 3.18 Alcohols and ethers with the same number of carbon atoms and the same degree of saturation are functional group isomers, as is illustrated here for propyl alcohol and ethyl methyl ether.

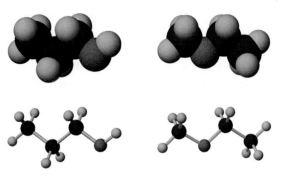

Propyl alcohol (C$_3$H$_8$O)   **Ethyl methyl ether (C$_3$H$_8$O)**

FIGURE 3.19 A physical-state summary for unbranched alkyl alkyl ethers at room temperature and pressure.

Unbranched Alkyl Alkyl Ethers			
C$_1$–C$_1$			
C$_1$–C$_2$	C$_2$–C$_2$		
C$_1$–C$_3$	C$_2$–C$_3$	C$_3$–C$_3$	
C$_1$–C$_4$	C$_2$–C$_4$	C$_3$–C$_4$	C$_4$–C$_4$

☐ Gas   ☐ Liquid

The term *ether* comes from the Latin *aether*, which means "to ignite." This name is given to these compounds because of their high vapor pressure at room temperature, which makes them very flammable.

FIGURE 3.20 Although ether molecules cannot hydrogen-bond to one another, they can hydrogen-bond to water molecues. Such hydrogen bonding causes ethers to be more soluble in water than alkanes of similar molecular mass.

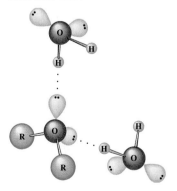

All three compounds have the molecular formula C$_3$H$_8$O. Figure 3.18 shows molecular models for the isomeric propyl alcohol and ethyl methyl ether molecules.

3.18 Physical and Chemical Properties of Ethers

The boiling points of ethers are similar to those of alkanes of comparable molecular mass and are much lower than those of alcohols of comparable molecular mass.

Alkane   CH$_3$—CH$_2$—CH$_2$—CH$_2$—CH$_3$ Mol. mass = 72 amu
                                                                            bp = 36°C

Ether   CH$_3$—CH$_2$—O—CH$_2$—CH$_3$ Mol. mass = 74 amu
                                                                        bp = 35°C

Alcohol   CH$_3$—CH$_2$—CH$_2$—CH$_2$—OH   Mol. mass = 74 amu
                                                                            bp = 117°C

The much higher boiling point of the alcohol results from hydrogen bonding between alcohol molecules. Ether molecules, like alkanes, cannot hydrogen-bond to one another. Ether oxygen atoms have no hydrogen atom attached directly to them. Figure 3.19 is a physical-state summary for unbranched alkyl alkyl ethers where the alkyl groups range in size from C$_1$ to C$_4$.

Ethers, in general, are more soluble in water than are alkanes of similar molecular mass because ether molecules are able to form hydrogen bonds with water (Figure 3.20).

Ethers have water solubilities similar to those of alcohols of the same molecular mass. For example, diethyl ether and butyl alcohol have the same solubility in water. Because ethers can also hydrogen-bond to alcohols, alcohols and ethers tend to be mutually soluble. Nonpolar substances tend to be more soluble in ethers than in alcohols because ethers have no hydrogen-bonding network that has to be broken up for solubility to occur.

Two chemical properties of ethers are especially important.

1. *Ethers are flammable.* Special care must be exercised in laboratories where ethers are used. Diethyl ether, whose boiling point of 35°C is only a few degrees above room temperature, is a particular flash-fire hazard.

2. *Ethers react slowly with oxygen from the air to form unstable hydroperoxides and peroxides.*

R—O—O—H   R—O—O—R
Hydroperoxide   Peroxide

Such compounds, when concentrated, represent an explosion hazard and must be removed before *stored* ethers are used.

Ethers are unreactive toward acids, bases, and oxidizing agents. Like alkanes, they do undergo halogenation reactions.

The general chemical unreactivity of ethers, coupled with the fact that most organic compounds are ether-soluble, makes ethers excellent solvents in which to carry out organic reactions. Their relatively low boiling points simplify their separation from the reaction products.

A chemical reaction for the preparation of ethers has been previously considered. In Section 3.9 we noted that the intermolecular dehydration of a primary alcohol will produce an ether. Although additional methods exist for ether preparation, we will not consider them in this text.

3.19 Cyclic Ethers

Cyclic ethers contain ether functional groups as part of a ring system. Some examples of such cyclic ethers, along with their common names, follow.

Ethylene oxide   Tetrahydrofuran (THF)   Furan   Pyran

Ethylene oxide has few direct uses. Its importance is as a starting material for the production of ethylene glycol (Section 3.5), a major component of automobile antifreeze. THF is a particularly useful solvent in that it dissolves many organic compounds and yet is miscible with water. In carbohydrate chemistry (Chapter 7), we will encounter many cyclic structures that are polyhydroxy derivatives of the five-membered (furan) and six-membered (pyran) cyclic ether systems. These carbohydrate derivatives are called *furanoses* and *pyranoses,* respectively (Section 7.10).

Vitamin E (Section 3.14) and THC (the active ingredient in marijuana; page 101) have structures in which a cyclic ether component is present.

Cyclic ethers are our first encounter with heterocyclic organic compounds. A **heterocyclic organic compound** *is a cyclic organic compound in which one or more of the carbon atoms in the ring have been replaced with atoms of other elements.* The hetero atom is usually oxygen or nitrogen.

We have just seen that *cyclic ethers*—compounds in which the ether functional group is part of a ring system—exist. In contrast, cyclic alcohols—compounds in which the alcohol functional group is part of a ring system—do not exist. To incorporate an alcohol functional group into a ring system would require an oxygen atom with three bonds, and oxygen atoms form only two bonds.

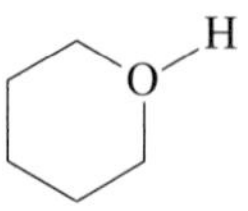

The oxygen atom in this structure has three bonds, which is not possible.

Compounds such as

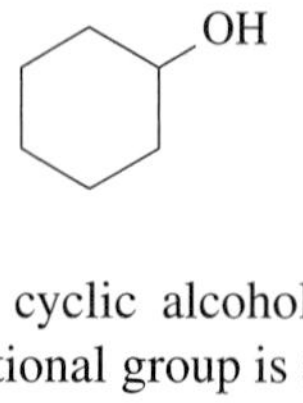

and

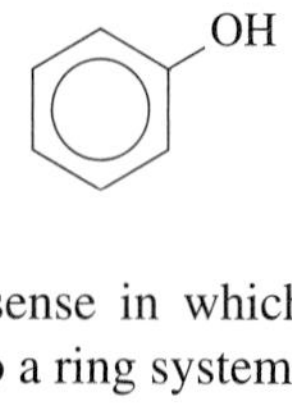

which do exist, are not cyclic alcohols in the sense in which we are using the term because the alcohol functional group is attached to a ring system rather than part of it.

CHEMICAL CONNECTIONS   Marijuana: The Most Commonly Used Illicit Drug

Prepared from the leaves, flowers, seeds, and small stems of a hemp plant called *Cannabis sativa*, marijuana, which is also called pot or grass, is the most commonly used illicit drug in the United States. The most active ingredient of the many in marijuana is the molecule *tetrahydrocannabinol*, called THC for short. Three different functional groups are present in a THC molecule; it is a phenol, a cyclic ether, and a cycloalkene. The THC content of marijuana varies considerably. Most marijuana sold in the North American illegal drug market has a THC content of 1% to 2%.

Marijuana has a pharmacology unlike that of any other drug. A marijuana "high" is a combination of sedation, tranquilization, and mild hallucination. THC readily penetrates the brain. The portions of the brain that involve memory and motor control contain the receptor sites where THC molecules interact. Even moderate doses of marijuana cause short-term memory loss. Marijuana unquestionably impairs driving ability, even after ordinary social use. THC readily crosses the placental barrier and reaches the fetus. Heavy marijuana users experience inflammation of the bronchi, sore throat, and inflamed sinuses. Increased heart rate, to as high as a dangerous 160 beats per minute, can occur with marijuana use.

The onset of action of THC is usually within minutes after smoking begins, and peak concentration in plasma occurs in 10 to 30 minutes. Unless more is smoked, the effects seldom last longer than 2 to 3 hours. Because THC is only slightly soluble in water, it tends to be deposited in fatty tissues. Unlike alcohol, THC persists in the bloodstream for several days, and the products of its breakdown remain in the blood for as long as 8 days.

New research indicates that physical dependence on THC can develop. Drug withdrawal symptoms are seen in some individuals who have been exposed repeatedly to high doses.

Tetrahydrocannabinol

3.20 Sulfur Analogs of Alcohols

Many organic compounds containing oxygen have sulfur analogs, in which a sulfur atom has replaced an oxygen atom. Sulfur is in the same group of the periodic table as oxygen, so the two elements have similar electron configurations.

Thiols, the sulfur analogs of alcohols, contain —SH functional groups instead of —OH functional groups. The thiol functional group is called a *sulfhydryl group.* A **sulfhydryl group** *is the —SH functional group.* A **thiol** *is an organic compound in which a sulfhydryl group is bonded to a saturated carbon atom.* An older term used for thiols is *mercaptans.* Contrasting the general structures for alcohols and thiols, we have

> The root *thio-* indicates that a sulfur atom has replaced an oxygen atom in a compound. It originates from the Greek *theion,* meaning "brimstone," which is an older name for the element sulfur.

$$R—OH \qquad \text{and} \qquad R—SH$$

An alcohol (Hydroxyl group)   A thiol (Sulfhydryl group)

■ Nomenclature for Thiols

Thiols are named in the same way as alcohols in the IUPAC system, except that the *-ol* becomes *-thiol.* The prefix *thio-* indicates the substitution of a sulfur atom for an oxygen atom in a compound.

$$CH_3—CH—CH_2—CH_3 \qquad CH_3—CH—CH_2—CH_3$$
$$\quad\ |\qquad\qquad\qquad\qquad\qquad |$$
$$\quad OH \qquad\qquad\qquad\qquad\quad SH$$
2-Butanol   2-Butanethiol

As in the case of diols and triols, the *-e* at the end of the alkane name is also retained for thiols.

Even though thiols have a higher molecular mass than alcohols with the same number of carbon atoms, they have much lower boiling points because they do not exhibit hydrogen bonding as alcohols do.

Common names for thiols are based on use of the term *mercaptan,* the older name for thiols. The name of the alkyl group present (as a separate word) precedes the word *mercaptan.*

$$CH_3-CH_2-SH$$
Ethyl mercaptan

$$CH_3-CH-SH$$
$$| $$
$$CH_3$$
Isopropyl mercaptan

EXAMPLE 3.6

Determining IUPAC and Common Names for Thiols

■ Convert each of the following common names for thiols to IUPAC names or vice versa.

a. Propyl mercaptan   **b.** Isobutyl mercaptan
c. 1-Butanethiol   **d.** 2-Propanethiol

Solution

a. The structural formula for propyl mercaptan is $CH_3-CH_2-CH_2-SH$. In the IUPAC system the name base is propane; the complete name is 1-propanethiol.
b. The structural formula for isobutyl mercaptan is

$$CH_3-CH-CH_2-SH$$
$$|$$
$$CH_3$$

The longest carbon chain has three carbon atoms (propane), and both a methyl group and a sulfhydryl group are attached to the chain. The IUPAC name is 2-methyl-1-propanethiol.
c. The structure of this thiol is $CH_3-CH_2-CH_2-CH_2-SH$. The alkyl group is a butyl group, giving a common name of butyl mercaptan for this thiol.
d. The thiol structural formula is

$$CH_3-CH-CH_3$$
$$|$$
$$SH$$

The sulfhydryl group is attached to an isopropyl group; the common name is isopropyl mercaptan.

Practice Exercise 3.6

Convert each of the following common names for thiols to IUPAC names or vice versa.

a. Methyl mercaptan   **b.** *sec*-Butyl mercaptan
c. 2-Methyl-2-propanethiol   **d.** 1-Pentanethiol

FIGURE 3.21 Thiols are responsible for the strong odor of "essence of skunk." Their odor is an effective defense mechanism.

Properties of Thiols

Two important properties of thiols are low boiling points (because of lack of hydrogen bonding) and a strong, disagreeable odor. The familiar odor of natural gas results from the addition of a low concentration of methanethiol (CH_3-SH) to the gas. The exceptionally low threshold of detection for this thiol enables consumers to smell a gas leak long before the gas, which is itself odorless, reaches dangerous levels. The scent of skunks (Figure 3.21) is due primarily to two thiols.

$$CH_2-CH_2-CH-CH_3$$
$$| \qquad\qquad |$$
$$SH \qquad\qquad CH_3$$
3-Methyl-1-butanethiol

$$SH-CH_2 \overset{H}{\underset{}{\diagdown}} C=C \overset{CH_3}{\underset{H}{\diagup}}$$
trans-2-Butene-1-thiol

A major contributor to the typical smell of the human armpit is a compound that contains both an alcohol and a thiol functional group.

$$CH_2-CH_2-\underset{\underset{SH}{|}}{\overset{\overset{CH_3}{|}}{C}}-CH_2-CH_2-CH_3$$

$$\underset{OH}{|}$$

3-Methyl-3-sulfanyl-1-hexanol

Thiols are easily oxidized but yield different products than their alcohol analogs. Thiols form *disulfides*. Each of two thiol groups loses a hydrogen atom, thus linking the two sulfur atoms together via a disulfide group, —S—S—.

$$R-SH + HS-R \xrightarrow{\text{Oxidation}} R-S-S-R + 2H$$
$$\text{A disulfide}$$

Reversal of this reaction, a reduction process, is also readily accomplished. Breaking of the disulfide bond regenerates two thiol molecules.

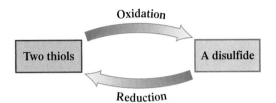

These two "opposite reactions" are of biological importance in the area of protein chemistry. Disulfide bonds formed from the interaction of two —SH groups contribute in a major way to protein structure (Chapter 9).

3.21 Sulfur Analogs of Ethers

Sulfur analogs of ethers are known as thioethers (or sulfides). A **thioether** *is an organic compound in which a sulfur atom is bonded to two carbon atoms by single bonds.* The generalized formula for a thioether is R—S—R. Like thiols, thioethers (or sulfides) have strong characteristic odors.

Thioethers are named in the same way as ethers, with *sulfide* used in place of *ether* in common names and *alkylthio* used in place of *alkoxy* in IUPAC names.

CH₃—S—CH₃
Dimethyl sulfide
(methylthiomethane)

Methyl phenyl sulfide
(methylthiobenzene)

4-(ethylthio)-2-Methyl-2-pentene

Bacteria in the mouth interact with saliva and leftover food to produce such compounds as hydrogen sulfide, methanethiol (a thioalcohol), and dimethyl sulfide (a thioether). These compounds, which have odors detectable in air at concentrations of parts per billion, are responsible for "morning breath."

In general, thiols are more reactive than their alcohol counterparts, and thioethers are more reactive than their corresponding ethers. The larger size of a sulfur atom compared to an oxygen atom (see Figure 3.22) results in a carbon–sulfur covalent bond that is weaker than a carbon–oxygen covalent bond. An added factor is that sulfur's electronegativity (2.5) is significantly lower than that of oxygen (3.5).

Thiols and thioethers are functional group isomers in the same manner that alcohols and ethers are functional group isomers (Section 3.17). For example, the thiol 1-propanethiol and the thioether methylthioethane both have the molecular formula C_3H_8S.

CH₃—CH₂—CH₂—SH   CH₃—S—CH₂—CH₃
1-Propanethiol   Methylthioethane

FIGURE 3.22 A comparison involving space-filling models for dimethyl ether and dimethyl sulfide (dimethyl thioether). A sulfur atom is much larger than an oxygen atom. This results in a carbon–sulfur bond being weaker than a carbon–oxygen bond.

Dimethyl ether   **Dimethyl sulfide**

CHEMICAL CONNECTIONS   Garlic and Onions: Odiferous Medicinal Plants

Garlic and onions, which botanically belong to the same plant genus, are vegetables known for the bad breath—and perspiration odors—associated with their consumption. These effects are caused by organic sulfur-containing compounds, produced when garlic and onions are cut, that reach the lungs and sweat glands via the bloodstream. The total sulfur content of garlic and onions amounts to about one percent of their dry weight.

Less well known about garlic and onions are the numerous studies showing that these same "bad breath" sulfur-containing compounds are health-promoting substances that have the capacity to prevent or at least ameliorate a host of ailments in humans and animals. The list of beneficial effects associated with garlic use is longer than that for any other medicinal plant. Only onions come close to having the same kind of efficacy. Garlic has been shown to function as an antibacterial, antiviral, antifungal, antiprotozal, and antiparasitic agent. In the area of heart and circulatory problems, garlic contains vasodilative compounds that improve blood fluidity and reduce platelet aggregation. The health-promoting role of onions has not been explored as thoroughly as that of garlic, but the studies undertaken so far seem to confirm that onions are second only to garlic in their "healing powers."

Whole garlic bulbs and whole onions that remain undisturbed and intact do not contain any strongly odiferous compounds and display virtually no physiological activity. The act of cutting or crushing these vegetables causes a cascade of reactions to occur in damaged plant cells. Exposure to oxygen in the air is an important facet of these reactions. Over one hundred sulfur-containing organic compounds are formed in garlic and probably a similar number are produced in the less-studied onion. Many of the compounds so produced are common to both garlic and onions. The compounds associated with garlic ingestion that contribute to bad breath include allyl methyl sulfide, allyl methyl disulfide, diallyl sulfide, and diallyl disulfide. Their structures are given in the accompanying table.

Not all of the strongly odiferous compounds associated with garlic and onions elicit negative responses from the human olfactory system. For example, the smell of fried onions is considered a pleasant odor by most people. Compounds contributing to the "fried onion smell" include methyl propyl disulfide, methyl propyl trisulfide, allyl propyl disulfide, and

dipropyl trisulfide. Structures for these compounds are also given in the accompanying table.

In addition to physiologically active sulfur compounds, garlic and onions also contain a variety of other healthful ingredients. Among these are the B vitamins thiamine and riboflavin and vitamin C. Almost all of the trace elements are also present, including manganese, iron, phosphorus, selenium, and chromium. The actual amount of a given trace element depends on the soil in which the garlic or onion was grown.

Garlic Breath

$CH_2 = CH - CH_2 - S - CH_3$
Allyl methyl sulfide

$CH_2 = CH - CH_2 - S - S - CH_3$
Allyl methyl disulfide

$CH_2 = CH - CH_2 - S - CH_2 - CH = CH_2$
Diallyl sulfide

$CH_2 = CH - CH_2 - S - S - CH_2 - CH = CH_2$
Diallyl disulfide

Fried Onions

$CH_3 - S - S - CH_2 - CH_2 - CH_3$
Methyl propyl disulfide

$CH_3 - S - S - S - CH_2 - CH_2 - CH_3$
Methyl propyl trisulfide

$CH_2 = CH - CH_2 - S - S - CH_2 - CH_2 - CH_3$
Allyl propyl disulfide

$CH_3 - CH_2 - CH_2 - S - S - S - CH_2 - CH_2 - CH_3$
Dipropyl trisulfide

CONCEPTS TO REMEMBER

Alcohols. Alcohols are organic compounds that contain an —OH group attached to a saturated carbon atom. The general formula for an alcohol is R—OH, where R is an alkyl group (Section 3.2).

Nomenclature of alcohols. The IUPAC names of simple alcohols end in -ol, and their carbon chains are numbered to give precedence to the location of the —OH group. Alcohol common names contain the word *alcohol* preceded by the name of the alkyl group (Section 3.3).

Isomerism for alcohols. Constitutional isomerism is possible for alcohols containing three or more carbon atoms. Both skeletal and positional isomers are possible (Section 3.4).

Physical properties of alcohols. Alcohol molecules hydrogen-bond to each other and to water molecules. They thus have higher-than-normal boiling points, and the low-molecular-mass alcohols are soluble in water (Section 3.6).

Classifications of alcohols. Alcohols are classified on the basis of the number of carbon atoms bonded to the carbon attached to the —OH group. In primary alcohols, the —OH group is bonded to a carbon atom bonded to only one other C atom. In secondary alcohols, the —OH-containing C atom is attached to two other C atoms. In tertiary alcohols, it is attached to three other C atoms (Section 3.8).

Alcohol dehydration. Alcohols can be dehydrated in the presence of sulfuric acid to form alkenes or ethers. At 180°C, an alkene is produced; at 140°C, primary alcohols produce an ether (Section 3.9).

Alcohol oxidation. Oxidation of primary alcohols first produces an aldehyde, which is then further oxidized to a carboxylic acid. Secondary alcohols are oxidized to ketones, and tertiary alcohols are resistant to oxidation (Section 3.9).

Phenols. Phenols have the general formula Ar—OH, where Ar represents an aryl group derived from an aromatic compound. Phenols are named as derivatives of the parent compound phenol, using the conventions for aromatic hydrocarbon nomenclature (Sections 3.11 and 3.12).

Properties of phenols. Phenols are generally low-melting solids; most are only slightly soluble in water. The chemical reactions of phenols are significantly different from those of alcohols, even though both types of compounds possess hydroxyl groups. Phenols are more resistant to oxidation and do not undergo dehydration. Phenols have acidic properties, whereas alcohols do not (Section 3.13).

Ethers. The general formula for an ether is R—O—R′, where R and R′ are alkyl, cycloalkyl, or aryl groups. In the IUPAC system, ethers are named as alkoxy derivatives of alkanes. Common names are obtained by giving the R group names in alphabetical order and adding the word *ether* (Sections 3.15 and 3.16).

Functional group isomerism. Ethers and alcohols with the same number of carbon atoms and the same degree of saturation have the same molecular formula and are thus isomers of each other. This type of constitutional isomerism is known as functional group isomerism (Section 3.17).

Properties of ethers. Ethers have lower boiling points than alcohols because ether molecules do not hydrogen-bond to each other. Ethers are slightly soluble in water because water forms hydrogen bonds with ethers (Section 3.18).

Thiols and disulfides. Thiols are the sulfur analogs of alcohols. They have the general formula R—SH. The —SH group is called the sulfhydryl group. Oxidation of thiols forms disulfides, which have the general formula R—S—S—R. The most distinctive physical property of thiols is their foul odor (Section 3.20).

Thioethers (sulfides). Thioethers (sulfides) are the sulfur analogs of ethers. They have the general formula R—S—R (Section 3.21).

KEY REACTIONS AND EQUATIONS

1. Intramolecular dehydration of alcohols to give alkenes (Section 3.9)

$$-\overset{|}{\underset{H}{C}}-\overset{|}{\underset{OH}{C}}- \xrightarrow[180°C]{H_2SO_4} \ \ \overset{}{>}C=C\overset{}{<} + H_2O$$

2. Intermolecular dehydration of primary alcohols to give ethers (Section 3.9)

$$R—O—H + H—O—R \xrightarrow[140°C]{H_2SO_4} R—O—R + H_2O$$

3. Oxidation of a primary alcohol to give an aldehyde and then a carboxylic acid (Section 3.9)

$$R—\overset{OH}{\underset{H}{\overset{|}{C}}}—H \xrightarrow{[O]} R—\overset{O}{\overset{||}{C}}—H + H_2O$$

Aldehyde

$$\xrightarrow[\text{oxidation}]{[O] \ \text{Further}} R—\overset{O}{\overset{||}{C}}—OH$$

Carboxylic acid

4. Oxidation of a secondary alcohol to give a ketone (Section 3.9)

$$R—\overset{OH}{\underset{H}{\overset{|}{C}}}—R′ \xrightarrow{[O]} R—\overset{O}{\overset{||}{C}}—R′ + 2H$$

5. Attempted oxidation of a tertiary alcohol, which gives no reaction (Section 3.9)

$$R—\overset{OH}{\underset{R''}{\overset{|}{C}}}—R′ \xrightarrow{[O]} \text{no reaction}$$

6. Production of an alkyl halide from an alcohol by substitution using PX$_3$ (X is Cl or Br) (Section 3.9)

$$R—OH \xrightarrow[\text{heat}]{PX_3} R—X$$

7. Oxidation of a thiol to give a disulfide (Section 3.19)

$$R—SH + HS—R \xrightarrow{[O]} R—S—S—R + 2H$$

8. Reduction of a disulfide to give a thiol (Section 3.19)

$$R—S—S—R + 2H \longrightarrow R—SH + HS—R$$

EXERCISES AND PROBLEMS

The members of each pair of problems in this section test similar material.

■ **Bonding Characteristics of Oxygen (Section 3.1)**

3.1   In organic compounds, how many covalent bonds does each of the following types of atoms form?
　　　a. Oxygen   b. Hydrogen
　　　c. Carbon   d. A halogen

3.2   In organic compounds, which of the following bonding behaviors does an oxygen atom exhibit?

　　　a. One single bond   b. Two single bonds
　　　c. One double bond   d. Two double bonds

■ **Structural Characteristics of Alcohols (Section 3.2)**

3.3   What is the generalized formula for an alcohol?

3.4   What is the name of the functional group that characterizes an alcohol?

3.5   Contrast, in general terms, the structures of an alcohol and water.

3.6   Contrast, in general terms, the structures of an alcohol and an alkane.

■ Nomenclature for Alcohols (Section 3.3)

3.7   Assign an IUPAC name to each of the following alcohols.

a.
$$\text{OH}$$
$$\text{CH}_3{-}\text{CH}_2{-}\text{CH}_2{-}\overset{|}{\text{CH}}{-}\text{CH}_3$$

b. $\text{CH}_3{-}\text{CH}_2{-}\text{OH}$

c.
$$\text{CH}_3 \quad \text{OH}$$
$$\text{CH}_3{-}\overset{|}{\text{CH}}{-}\overset{|}{\text{CH}}{-}\text{CH}_3$$

d. $\text{CH}_3{-}\text{CH}_2{-}\text{CH}_2{-}\overset{|}{\text{CH}}{-}\text{CH}_2{-}\text{CH}_3$
$$\overset{|}{\text{CH}_2}$$
$$\overset{|}{\text{OH}}$$

e. $\text{CH}_3{-}\text{CH}_2{-}\overset{|}{\text{CH}}{-}\text{OH}$
$$\overset{|}{\text{CH}_3}$$

f.
$$\text{CH}_3$$
$$\overset{|}{}$$
$$\text{CH}_3{-}\overset{|}{\text{C}}{-}\text{CH}_2{-}\text{CH}_2{-}\text{OH}$$
$$\overset{|}{\text{CH}_3}$$

3.8   Assign an IUPAC name to each of the following alcohols.

a. $\text{CH}_3{-}\text{CH}_2{-}\overset{|}{\text{CH}}{-}\text{CH}_2{-}\text{CH}_3$
$$\overset{|}{\text{OH}}$$

b. $\text{CH}_3{-}\text{CH}_2{-}\overset{|}{\text{CH}}{-}\overset{|}{\text{CH}}{-}\text{CH}_3$
$$\overset{|}{\text{OH}} \quad \overset{|}{\text{CH}_3}$$

c. $\text{CH}_3{-}\text{CH}_2{-}\overset{|}{\text{CH}}{-}\text{CH}_2{-}\text{CH}_2{-}\text{CH}_3$
$$\overset{|}{\text{CH}_2{-}\text{CH}_2{-}\text{CH}_2{-}\text{OH}}$$

d. $\text{CH}_3{-}\overset{|}{\text{CH}}{-}\overset{|}{\text{CH}}{-}\text{CH}_2{-}\text{OH}$
$$\overset{|}{\text{CH}_3} \quad \overset{|}{\text{CH}_3}$$

e.
$$\text{OH}$$
$$\overset{|}{}$$
$$\text{CH}_3{-}\overset{|}{\text{C}}{-}\text{CH}_2{-}\text{CH}_3$$
$$\overset{|}{\text{CH}_3}$$

f.
$$\text{CH}_3$$
$$\overset{|}{}$$
$$\text{CH}_3{-}\text{CH}_2{-}\overset{|}{\text{CH}}{-}\overset{|}{\text{C}}{-}\overset{|}{\text{CH}}{-}\text{CH}_3$$
$$\overset{|}{\text{OH}} \quad \overset{|}{\text{CH}_3} \quad \overset{|}{\text{CH}_2{-}\text{CH}_3}$$

3.9   Assign an IUPAC name to each of the following alcohols.

a. 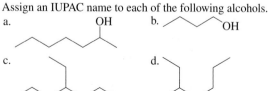   b.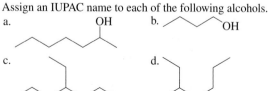

c.   d.

3.10   Assign an IUPAC name to each of the following alcohols.

a.   b.

c.   d.

3.11   Write a condensed structural formula for each of the following alcohols.

  a. 3-Pentanol   b. 3-Ethyl-3-hexanol
  c. 2-Methyl-1-propanol d. 4-Methyl-2-pentanol
  e. 2-Phenyl-2-propanol f. 2-Methylcyclobutanol

3.12   Write a condensed structural formula for each of the following alcohols.

  a. 2-Heptanol   b. 2,2-Dimethyl-1-hexanol
  c. 2-Methyl-2-heptanol d. 3-Ethyl-2-pentanol
  e. 3-Phenyl-1-butanol f. 3,5-Dimethylcyclohexanol

3.13   Write a condensed structural formula for, and assign an IUPAC name to, each of the following alcohols.

  a. Pentyl alcohol   b. Propyl alcohol
  c. Isobutyl alcohol   d. *sec*-Butyl alcohol

3.14   Write a condensed structural formula for, and assign an IUPAC name to, each of the following alcohols.

  a. Butyl alcohol   b. Hexyl alcohol
  c. Isopropyl alcohol   d. *tert*-Butyl alcohol

3.15   Assign an IUPAC name to each of the following polyhydroxy alcohols.

a. $\text{CH}_2{-}\text{CH}{-}\text{CH}_3$
$$\overset{|}{\text{OH}} \quad \overset{|}{\text{OH}}$$

b. $\text{CH}_2{-}\text{CH}_2{-}\text{CH}_2{-}\text{CH}{-}\text{CH}_3$
$$\overset{|}{\text{OH}} \qquad\qquad \overset{|}{\text{OH}}$$

c. $\text{CH}_3{-}\text{CH}_2{-}\text{CH}{-}\text{CH}_2{-}\text{CH}_2$
$$\qquad\qquad\;\; \overset{|}{\text{OH}} \qquad\;\; \overset{|}{\text{OH}}$$

d. $\text{CH}_2{-}\text{CH}{-}\text{CH}{-}\text{CH}_2$
$$\overset{|}{\text{OH}} \quad \overset{|}{\text{OH}} \quad \overset{|}{\text{CH}_3} \quad \overset{|}{\text{OH}}$$

3.16   Assign an IUPAC name to each of the following polyhydroxy alcohols.

a. $\text{CH}_2{-}\text{CH}{-}\text{CH}_2$
$$\overset{|}{\text{OH}} \quad \overset{|}{\text{CH}_3} \quad \overset{|}{\text{OH}}$$

b. $\text{CH}_2{-}\text{CH}{-}\text{CH}_2$
$$\overset{|}{\text{OH}} \quad \overset{|}{\text{OH}} \quad \overset{|}{\text{CH}_3}$$

c. $\text{CH}_3{-}\text{CH}_2{-}\text{CH}{-}\text{CH}{-}\text{CH}_3$
$$\qquad\qquad\;\; \overset{|}{\text{OH}} \quad \overset{|}{\text{OH}}$$

d. $\text{CH}_2{-}\text{CH}{-}\text{CH}{-}\text{CH}_3$
$$\overset{|}{\text{OH}} \quad \overset{|}{\text{OH}} \quad \overset{|}{\text{OH}}$$

3.17   Utilizing IUPAC rules, name each of the following compounds. Don't forget to use *cis-* and *trans-* prefixes (Section 1.14) where needed.

a.   b.

c.   d.

3.18   Utilizing IUPAC rules, name each of the following compounds. Don't forget to use *cis-* and *trans-* prefixes (Section 1.14) where needed.

a.   b.

c.   d.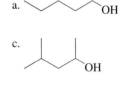

<hr/>

3.19 Write a condensed structural formula for each of the following *unsaturated* alcohols.
  a. 4-Penten-2-ol   b. 1-Pentyn-3-ol
  c. 3-Methyl-3-buten-2-ol   d. *cis*-2-Buten-1-ol

3.20 Write a condensed structural formula for each of the following *unsaturated* alcohols.
  a. 1-Penten-3-ol   b. 3-Butyn-1-ol
  c. 2-Methyl-3-buten-1-ol   d. *trans*-3-Penten-1-ol

3.21 Each of the following alcohols is named incorrectly. However, the names give correct structural formulas. Draw structural formulas for the compounds, and then write the correct IUPAC name for each alcohol.
  a. 2-Ethyl-1-propanol   b. 2,4-Butanediol
  c. 2-Methyl-3-butanol   d. 1,4-Cyclopentanediol

3.22 Each of the following alcohols is named incorrectly. However, the names give correct structural formulas. Draw structural formulas for the compounds, and then write the correct IUPAC name for each alcohol.
  a. 3-Ethyl-2-butanol   b. 3,4-Pentanediol
  c. 3-Methyl-3-butanol   d. 1,1-Dimethyl-1-butanol

Isomerism for Alcohols (Section 3.4)

3.23 Indicate whether each of the following compounds is or is not a constitutional isomer of 1-hexanol.
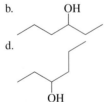

3.24 Indicate whether each of the following compounds is or is not a constitutional isomer of 2-pentanol.

3.25 Give IUPAC names for all isomeric C_7 monohydroxy alcohols in which the carbon chain is unbranched.

3.26 Give IUPAC names for all isomeric C_8 monohydroxy alcohols in which the carbon chain is unbranched.

3.27 For which values of x is the alcohol name 2-methyl-x-pentanol a correct IUPAC name?

3.28 For which values of x is the alcohol name 3-methyl-x-pentanol a correct IUPAC name?

Important Common Alcohols (Section 3.5)

3.29 What does each of the following terms mean?

a. Absolute alcohol   b. Grain alcohol
c. Rubbing alcohol   d. Drinking alcohol

3.30 What does each of the following terms mean?
  a. Wood alcohol   b. Denatured alcohol
  c. 70-Proof alcohol   d. "Alcohol"

3.31 Give the common name of the alcohol that fits each of the following descriptions.
  a. Thick liquid that has the consistency of honey
  b. Often produced via a fermentation process
  c. Used as a race car fuel
  d. Industrially produced from CO and H_2

3.32 Give the common name of the alcohol that fits each of the following descriptions.
  a. Sometimes used as a skin coolant for the human body
  b. Antifreeze ingredient
  c. Active ingredient in alcoholic beverages
  d. Moistening agent in many cosmetics

Physical Properties of Alcohols (Section 3.6)

3.33 Explain why the boiling points of alcohols are much higher than those of alkanes with similar molecular masses.

3.34 Explain why the water solubilities of alcohols are much higher than those of alkanes with similar molecular masses.

3.35 Which member of each of the following pairs of compounds would you expect to have the higher boiling point?
  a. 1-Butanol and 1-heptanol
  b. Butane and 1-propanol
  c. Ethanol and 1,2-ethanediol

3.36 Which member of each of the following pairs of compounds would you expect to have the higher boiling point?
  a. 1-Octanol and 1-pentanol
  b. Pentane and 1-butanol
  c. 1,3-Propanediol and 1-propanol

3.37 Which member of each of the following pairs of compounds would you expect to be more soluble in water?
  a. Butane and 1-butanol
  b. 1-Octanol and 1-pentanol
  c. 1,2-Butanediol and 1-butanol

3.38 Which member of each of the following pairs of compounds would you expect to be more soluble in water?
  a. 1-Pentanol and 1-butanol
  b. 1-Propanol and 1-hexanol
  c. 1,2,3-Propanetriol and 1-hexanol

3.39 Determine the maximum number of hydrogen bonds that can form between an ethanol molecule and
  a. other ethanol molecules   b. water molecules
  c. methanol molecules   d. 1-propanol molecules

3.40 Determine the maximum number of hydrogen bonds that can form between a methanol molecule and
  a. other methanol molecules b. water molecules
  c. 1-propanol molecules   d. 2-propanol molecules

Preparation of Alcohols (Section 3.7)

3.41 Write the structure of the expected predominant organic product formed in each of the following reactions.

a. $CH_2{=}CH_2 + H_2O \xrightarrow{H_2SO_4}$

b.
$$CH_3{-}CH_2{-}\overset{\overset{\displaystyle O}{\|}}{C}{-}H + H_2 \xrightarrow{Catalyst}$$

c. $CH_3-CH_2-\underset{\underset{CH_3}{|}}{C}=CH_2 + H_2O \xrightarrow{H_2SO_4}$

d.

$CH_3-CH_2-\overset{\overset{O}{||}}{C}-CH_2-CH_3 + H_2 \xrightarrow{Catalyst}$

3.42 Write the structure of the expected predominant organic product formed in each of the following reactions.

a. $CH_3-CH=CH-CH_3 + H_2O \xrightarrow{H_2SO_4}$

b.

$CH_3-CH_2-\overset{\overset{O}{||}}{C}-CH_3 + H_2 \xrightarrow{Catalyst}$

c.

$CH_3-\overset{\overset{O}{||}}{C}-H + H_2 \xrightarrow{Catalyst}$

d. $CH_3-\underset{\underset{CH_3}{|}}{CH}-CH=CH-CH_3 + H_2O \xrightarrow{H_2SO_4}$

Classification of Alcohols (Section 3.8)

3.43 Classify each of the alcohols in Problem 3.7 as a primary, secondary, or tertiary alcohol.

3.44 Classify each of the alcohols in Problem 3.8 as a primary, secondary, or tertiary alcohol.

Chemical Reactions of Alcohols (Section 3.9)

3.45 Draw the structure of the organic product expected to be predominant when each of the following alcohols is dehydrated using sulfuric acid at the temperature indicated.

a. $CH_3-\underset{\underset{OH}{|}}{CH}-CH_3 \xrightarrow[180°C]{H_2SO_4}$

b. $CH_3-CH_2-\underset{\underset{CH_3}{|}}{CH}-CH_2-OH \xrightarrow[180°C]{H_2SO_4}$

c. $CH_3-\underset{\underset{CH_3}{|}}{CH}-OH \xrightarrow[140°C]{H_2SO_4}$

d. $CH_3-CH_2-CH_2-OH \xrightarrow[140°C]{H_2SO_4}$

3.46 Draw the structure of the organic product expected to be predominant when each of the following alcohols is dehydrated using sulfuric acid at the temperature indicated.

a. $CH_3-\underset{\underset{CH_3}{|}}{CH}-CH_2-OH \xrightarrow[140°C]{H_2SO_4}$

b. $CH_3-\underset{\underset{CH_3}{|}}{CH}-CH_2-OH \xrightarrow[180°C]{H_2SO_4}$

c. $CH_3-\underset{\underset{OH}{|}}{CH}-CH_2-CH_3 \xrightarrow[180°C]{H_2SO_4}$

d. $CH_3-\underset{\underset{OH}{|}}{CH}-CH_2-CH_3 \xrightarrow[140°C]{H_2SO_4}$

3.47 Identify the alcohol reactant from which each of the following products was obtained by an alcohol dehydration reaction.

a. Alcohol $\xrightarrow[180°C]{H_2SO_4} CH_3-CH=\underset{\underset{CH_3}{|}}{C}-CH_3$

b. Alcohol $\xrightarrow[180°C]{H_2SO_4} CH_3-CH=CH_2$

c. Alcohol $\xrightarrow[140°C]{H_2SO_4} CH_3-CH_2-O-CH_2-CH_3$

d. Alcohol $\xrightarrow[140°C]{H_2SO_4} CH_3-\underset{\underset{CH_3}{|}}{CH}-CH_2-O-CH_2-\underset{\underset{CH_3}{|}}{CH}-CH_3$

3.48 Identify the alcohol reactant from which each of the following products was obtained by an alcohol dehydration reaction.

a. Alcohol $\xrightarrow[180°C]{H_2SO_4} CH_2=\underset{\underset{CH_3}{|}}{C}-CH_2-CH_3$

b. Alcohol $\xrightarrow[180°C]{H_2SO_4} CH_3-CH_2-CH=CH_2$

c. Alcohol $\xrightarrow[140°C]{H_2SO_4} CH_3-O-CH_3$

d. Alcohol $\xrightarrow[140°C]{H_2SO_4} CH_3-CH_2-O-CH_2-CH_3$

3.49 Draw the structure of the alcohol that could be used to prepare each of the following compounds in an oxidation reaction.

a.

$CH_3-CH_2-\overset{\overset{O}{||}}{C}-CH_3$

b.

$CH_3-CH_2-\overset{\overset{O}{||}}{C}-OH$

c.

$CH_3-CH_2-\overset{\overset{O}{||}}{C}-H$

d.

3.50 Draw the structure of the alcohol that could be used to prepare each of the following compounds in an oxidation reaction.

a.

$CH_3-\underset{\underset{CH_3}{|}}{CH}-\overset{\overset{O}{||}}{C}-H$

b.

$CH_3-\underset{\underset{CH_3}{|}}{CH}-\overset{\overset{O}{||}}{C}-CH_3$

c.

$CH_3-\underset{\underset{CH_3}{|}}{CH}-CH_2-\overset{\overset{O}{||}}{C}-OH$

d.

3.51 Draw the structure of the expected predominant organic product formed in each of the following reactions.

a. $CH_3-CH_2-CH_2-OH \xrightarrow[heat]{PCl_3}$

b.

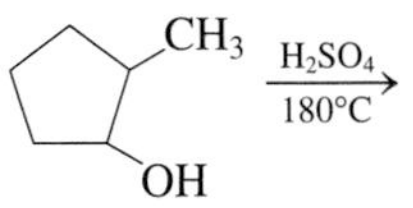

c. $CH_3\!-\!CH\!-\!CH_2\!-\!CH_3 \xrightarrow{K_2Cr_2O_7}$
    |
    OH

d. $CH_3\!-\!CH_2\!-\!CH\!-\!CH_2\!-\!CH_3 \xrightarrow[\text{heat}]{PCl_3}$
    |
    OH

e. $CH_3\!-\!CH_2\!-\!OH \xrightarrow[140°C]{H_2SO_4}$

f. $CH_2\!-\!CH_2 \xrightarrow[\text{heat}]{PBr_3}$
    | |
    OH   OH

3.52 Draw the structure of the expected predominant organic product formed in each of the following reactions.

a. $CH_3\!-\!CH_2\!-\!CH_2\!-\!OH \xrightarrow[\text{heat}]{PBr_3}$

b.

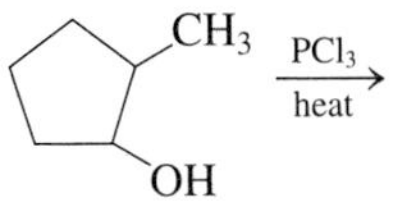

c. $CH_2\!-\!CH_2\!-\!CH_2\!-\!CH_3 \xrightarrow[140°C]{H_2SO_4}$
    |
    OH

d. $CH_3\!-\!CH_2\!-\!CH\!-\!CH_2\!-\!CH_3 \xrightarrow{K_2Cr_2O_7}$
    |
    OH

e. $CH_3\!-\!CH\!-\!OH \xrightarrow[180°C]{H_2SO_4}$
    |
    CH_3

f. $CH_2\!-\!CH_2 \xrightarrow[\text{heat}]{PCl_3}$
    | |
    OH   OH

■ **Polymeric Alcohols (Section 3.10)**

3.53 Draw a structural representation for the polymeric alcohol PEG [poly(ethylene glycol)].

3.54 Draw a structural representation for the polymeric alcohol PVA [poly(vinyl alcohol)].

■ **Structural Characteristics of and Nomenclature for Phenols (Sections 3.11 and 3.12)**

3.55 Explain why the first of the following two compounds is a phenol, and the second is not.

$CH_3\!-\!\bigcirc\!-\!OH$ $CH_3\!-\!\bigcirc\!-\!CH_2\!-\!OH$

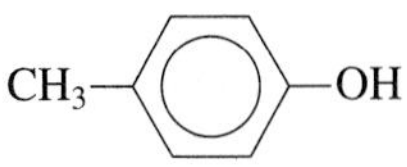

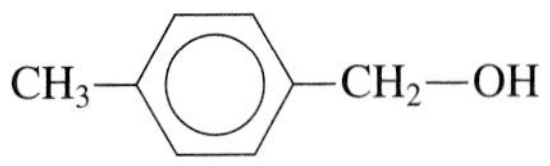

3.56 Explain why the first of the following two compounds is a phenol, and the second is not.

$CH_3\!-\!\bigcirc\!-\!OH$ $CH_3\!-\!\bigcirc\!-\!OH$

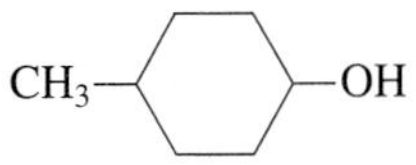

3.57 Name the following phenols.

a. 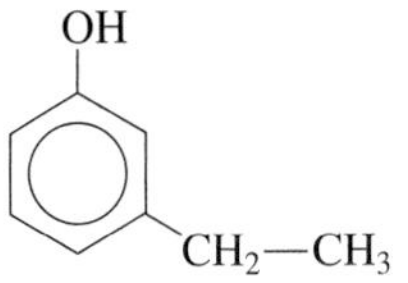 OH
   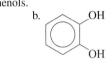
   CH_2—CH_3

b. 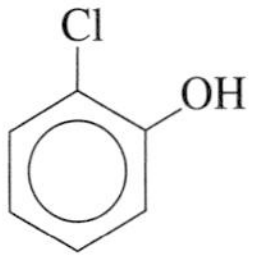 Cl
   OH

c. 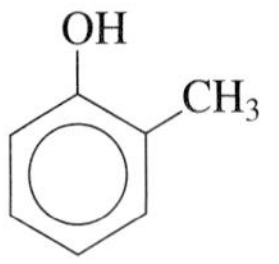 OH
   CH_3

d. 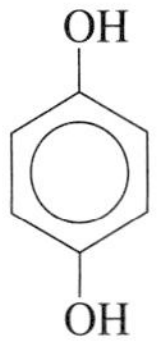 OH
   OH

e. 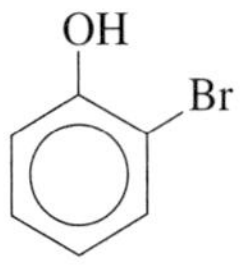 OH
   Br

f. 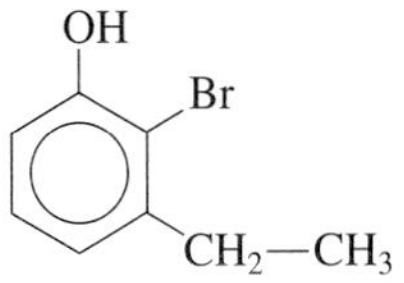 OH
   Br
   CH_2—CH_3

3.58 Name the following phenols.

a. 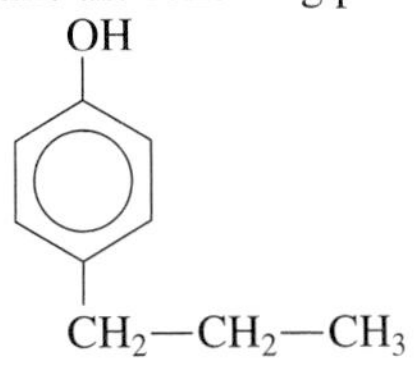 OH

   CH_2—CH_2—CH_3

b. OH
   OH

c. 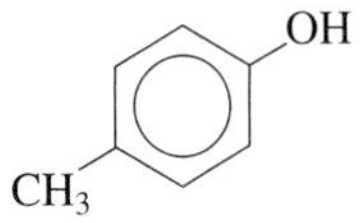 OH
   CH_3

d. 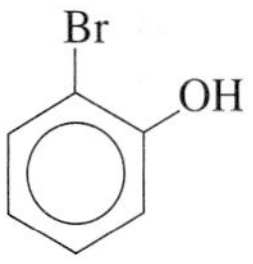 Br
   OH

e. OH
   Cl

f. OH
   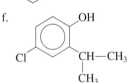
   Cl   CH—CH_3
         CH_3

3.59 Draw a structural formula for each of the following phenols.
   a. 4-Chlorophenol   b. 2-Ethylphenol
   c. 2,4-Dibromophenol   d. *m*-Cresol
   e. Resorcinol   f. 2,6-Diethyl-4-methylphenol

3.60 Draw a structural formula for each of the following phenols.
   a. 3-Bromophenol   b. *m*-Ethylphenol
   c. Hydroquinone   d. *o*-Cresol
   e. Catechol   f. 2,6-Dichlorophenol

■ **Properties and Uses of Phenols (Sections 3.13 and 3.14)**

3.61 Phenolic compounds are frequently used as antiseptics and disinfectants. What is the difference between an antiseptic and a disinfectant?

3.62 Phenolic compounds are frequently used as antioxidants. What is an antioxidant?

3.63 Phenols are weak acids. Write an equation for the acid ionization of the compound phenol.

3.64 How does the acidity of phenols compare with that of inorganic weak acids?

■ **Structural Characteristics of and Nomenclature for Ethers (Sections 3.15 and 3.16)**

3.65 Indicate whether each of the following structural notations denotes an ether.
   a. R—O—R   b. R—O—H
   c. Ar—O—R   d. Ar—O—Ar

3.66   What is the difference in meaning associated with each of the following pairs of notations?
a. R—O—R and R—O—H
b. Ar—O—R and Ar—O—Ar
c. Ar—O—H and Ar—O—R
d. R—O—R and Ar—O—Ar

3.67   Assign an IUPAC name to each of the following ethers.
a. CH₃—O—CH₂—CH₂—CH₃
b. CH₃—CH₂—CH₂—O—CH₂—CH₃
c. CH₃—CH—CH₃
              |
              O—CH₃
d. 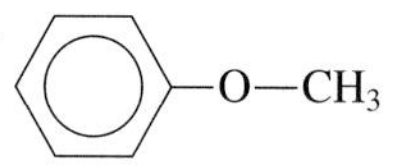
e. 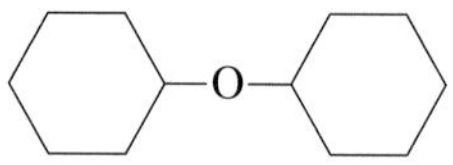
f. 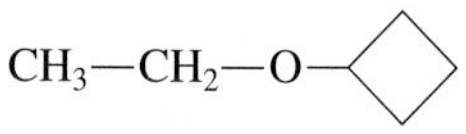 CH₃—CH₂—O—

3.68   Assign an IUPAC name to each of the following ethers.
a. CH₃—CH₂—O—CH₂—CH₃
b. CH₃—CH—O—CH₃
         |
         CH₃
c. CH₃—CH₂—CH—CH₃
                 |
                 O—CH₂—CH₃
d. 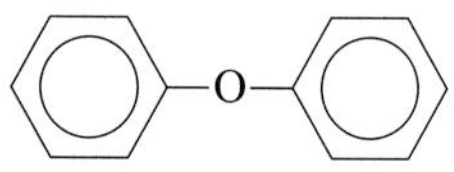
e. 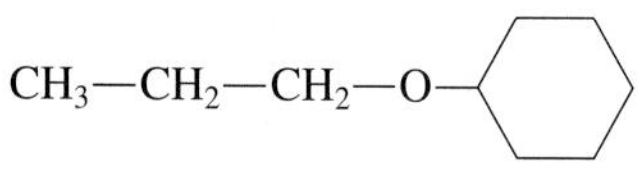 CH₃—CH₂—CH₂—O—
f. 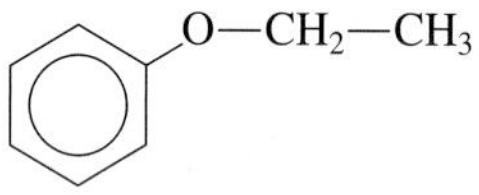 O—CH₂—CH₃

3.69   Assign a common name to each of the ethers in Problem 3.67.

3.70   Assign a common name to each of the ethers in Problem 3.68.

3.71   Assign an IUPAC name to each of the following ethers.
a.    b. 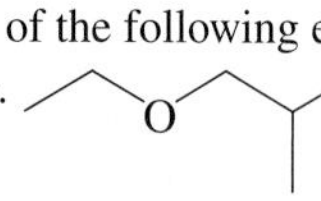
c. 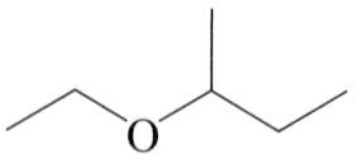   d.

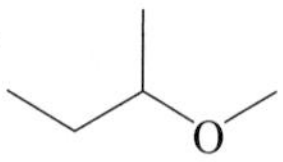

3.72   Assign an IUPAC name to each of the following ethers.
a. 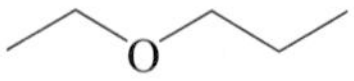   b. 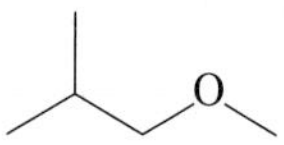
c. 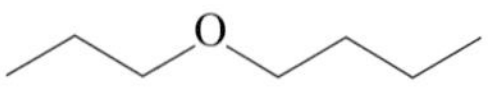   d.

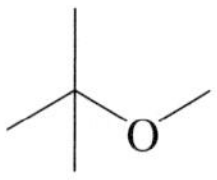

3.73   Draw the structure of each of the following ethers.
a. Isopropyl propyl ether
b. Ethyl phenyl ether

c. 3-Methylanisole
d. 2-Ethoxypentane
e. Ethoxycyclobutane
f. 1-Methoxy-2,2-dimethylpropane

3.74   Draw the structure of each of the following ethers.
a. Butyl methyl ether
b. Anisole
c. Phenyl propyl ether
d. 3-Propoxyheptane
e. 1,3-Dimethoxybenzene
f. 3-Methoxy-2-methylhexane

■ **Isomerism for Ethers (Section 3.17)**

3.75   Indicate whether each of the following ethers is or is not a constitutional isomer of ethyl propyl ether.
a.    b. 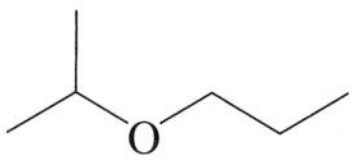
c. 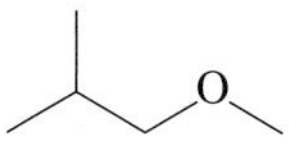   d.

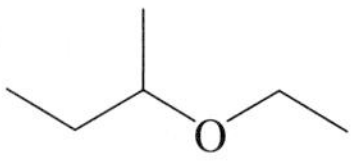

3.76   Indicate whether each of the following ethers is or is not a constitutional isomer of dipropyl ether.
a. 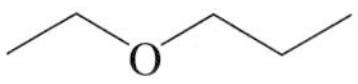   b. 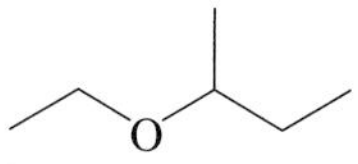
c. 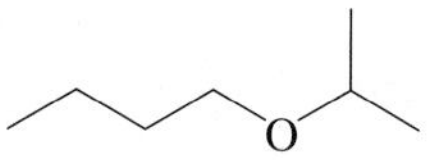   d.

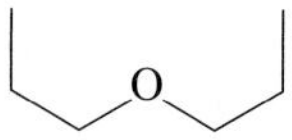

3.77   Give common names for all ethers that are constitutional isomers of ethyl propyl ether.

3.78   Give common names for all ethers that are constitutional isomers of butyl methyl ether.

3.79   Draw condensed structural formulas for the following.
a. All ethers that are functional group isomers of 1-butanol
b. All alcohols that are functional group isomers of 2-methoxypropane

3.80   Draw condensed structural formulas for the following.
a. All ethers that are functional group isomers of 2-methyl-1-propanol
b. All alcohols that are functional group isomers of 1-ethoxyethane

3.81   For which values of x is the ether name x-methoxy-3-methylpentane a correct IUPAC name?

3.82   For which values of x is the ether name x-methoxy-2-methylpentane a correct IUPAC name?

■ **Properties of Ethers (Section 3.18)**

3.83   Dimethyl ether and ethanol have the same molecular mass. Dimethyl ether is a gas at room temperature, and ethanol is a liquid at room temperature. Explain these observations.

3.84   Compare the solubility in water of ethers and alcohols that have similar molecular masses.

3.85   What are the two chemical hazards associated with ether use?

3.86   How do the chemical reactivities of ethers compare with those of
a. alkanes   b. alcohols

3.87 Explain why ether molecules cannot hydrogen-bond to each other.

3.88 How many hydrogen bonds can form between a single ether molecule and water molecules?

Cyclic Ethers (Section 3.19)

3.89 Classify each of the following molecular structures as that of a cyclic ether, a noncyclic ether, or a nonether.

a. 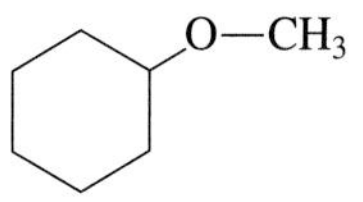 O—CH₃

b.

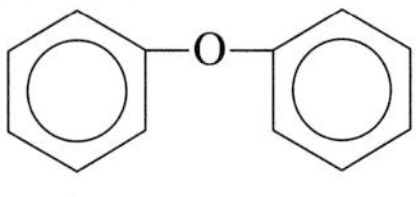

c. 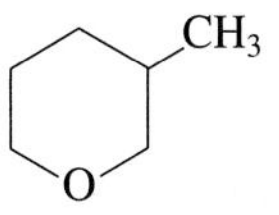 CH₃

d. 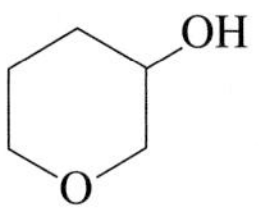 OH

e. 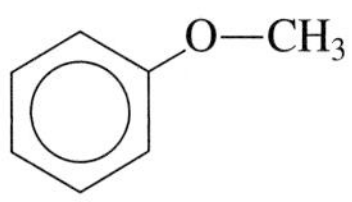 O—CH₃

f. 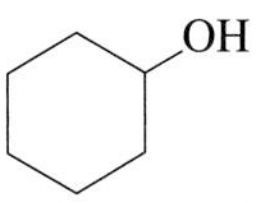 OH

3.90 Classify each of the following molecular structures as that of a cyclic ether, a noncyclic ether, or a nonether.

a. 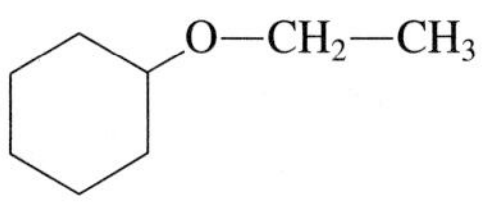 O—CH₂—CH₃

b.

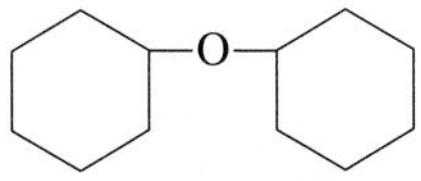

c.

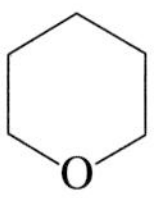

d. 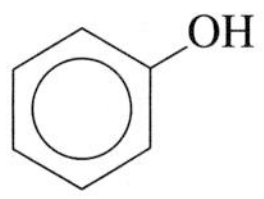 OH

e. 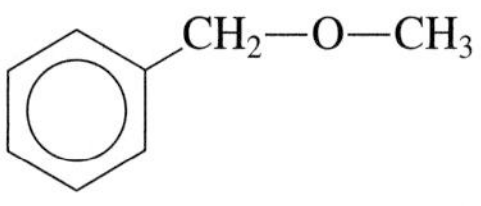 CH₂—O—CH₃

f. 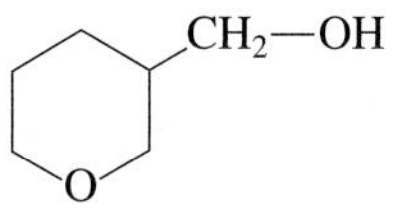 CH₂—OH

Sulfur Analogs of Alcohols and Ethers (Sections 3.20 and 3.21)

3.91 Contrast the general structural formulas for a thioalcohol and an alcohol.

3.92 Contrast the general structural formulas for a thioether and an ether.

3.93 Draw a condensed structural formula for each of the following thiols.
a. Methanethiol
b. 2-Propanethiol
c. 1-Butanethiol
d. 3-Methyl-1-pentanethiol
e. Cyclopentanethiol
f. 1,2-Ethanedithiol

3.94 Draw a condensed structural formula for each of the following thiols.
a. 1-Propanethiol
b. Ethanethiol
c. 1,3-Pentanedithiol
d. 3-Methyl-3-pentanethiol
e. 2-Methylcyclopentanethiol
f. 2,2-Dimethyl-1-hexanethiol

3.95 Assign a common name to each of the following thiols.
a. CH_3—SH
b. CH_3—CH_2—CH_2—SH
c. CH_3—CH_2—CH—SH
              |
              CH_3
d. CH_3—CH—CH_2—SH
         |
         CH_3

3.96 Assign a common name to each of the following thiols.
a. CH_3—CH_2—SH
b. CH_3—CH—SH
         |
         CH_3
c. CH_3—CH_2—CH_2—CH_2—SH
d. CH_3—C—SH
         |
         CH_3

3.97 Contrast the products that result from the oxidation of an alcohol and the oxidation of a thiol.

3.98 Write the formulas for the sulfur-containing organic products of the following reactions.

a. $2CH_3$—CH_2—SH $\xrightarrow[\text{agent}]{\text{Oxidizing}}$

b. CH_3—CH_2—S—S—CH_2—CH_3 $\xrightarrow[\text{agent}]{\text{Reducing}}$

3.99 Assign both an IUPAC name and a common name to each of the following thioethers.
a. CH_3—CH_2—S—CH_3
b. CH_3—CH—S—CH_3
         |
         CH_3

c. 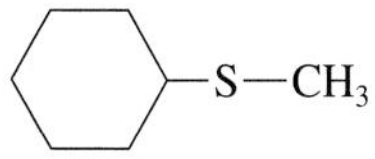 S—CH₃

d. 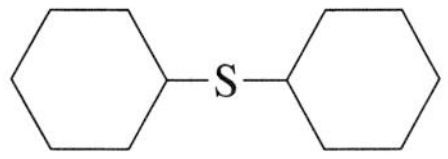 S

e. CH_2=CH—CH_2—S—CH_3
f. CH_3—CH_2—CH—CH_3
            |
            S
            |
          CH_3

3.100 Assign both an IUPAC name and a common name to each of the following thioethers.
a. CH_3—CH_2—S—CH_2—CH_3
b. CH_3—CH—S—CH_2—CH_3
         |
         CH_3

c. 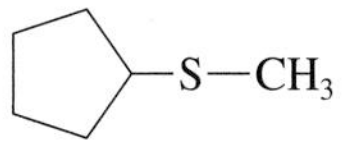 S—CH₃

d. 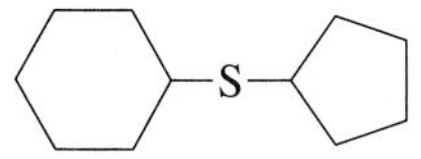 S

e. CH_2=CH—CH_2—S—CH_2—CH_3
f. CH_3—CH—CH_3
        |
        S
        |
       CH_3

ADDITIONAL PROBLEMS

3.101 Assign an IUPAC name to each of the following compounds.

a. [structure: OH on hexane chain]

b. [structure: OH on pentane chain]

c. [structure: benzene ring with O—allyl group]

d. [structure: OH with isobutyl chain]

e. [structure: tert-butyl with OH]

f. [structure: ethyl ether CH₃CH₂—O—CH₂CH₃]

3.102 Draw structural formulas for the eight isomeric alcohols and six isomeric ethers that have the molecular formula $C_5H_{10}O$.

3.103 Three isomeric pentanols with unbranched carbon chains exist. Which of these, upon dehydration at 180°C, yields only 1-pentene as a product?

3.104 A mixture of methanol, 1-propanol, and H_2SO_4 (catalyst) is heated to 140°C. After reaction, the solution contains three different ethers. Draw a structural formula for each of the ethers.

3.105 Which of the terms *ether, alcohol, diol, thiol, thioether, thioalcohol, disulfide, sulfide,* and *peroxide* characterize(s) each of the following compounds? Note that more than one term may apply to a given compound.

a. CH_3—S—S—CH_3

b. CH_3—CH_2—SH

c. CH_3—CH—CH_2—CH_3
　　　　　|
　　　　　OH

d. CH_3—CH_2—O—O—CH_3

e. HO—CH_2—CH_2—SH

f. CH_3—O—CH_2—S—CH_3

3.106 Assign IUPAC names to the following compounds.

a. HS—CH_2—CH_2—SH

b. CH_3—O—CH_2—CH_2—CH_2—OH

c. HO—CH_2—CH_2—CH_3

d. CH_3—CH_2—O—CH_2—CH_2—O—CH_2—CH_3

e. CH_3—S—CH_2—CH_3

f. CH_3—O—CH_2—CH_2—S—CH_2—CH_3

MULTIPLE-CHOICE PRACTICE TEST

3.107 What is the correct IUPAC name for the alcohol whose structural formula is

$$CH_3—CH—CH—CH_3$$
　　　　|　　|
　　　CH_3　OH

a. 2-Methylbutanol b. 3-Methylbutanol
c. 2-Methyl-3-butanol d. 3-Methyl-2-butanol

3.108 Which of the following statements concerning common alcohols is *incorrect*?

a. Wood alcohol and methyl alcohol are two names for the same compound.

b. Denatured alcohol is drinking alcohol rendered unfit to drink.

c. Rubbing alcohol is a 70% solution of ethyl alcohol.

d. Glycerin and ethylene glycol are both polyhydroxy alcohols.

3.109 What is the organic product formed by the oxidation of a secondary alcohol?

a. Aldehyde b. Ketone
c. Carboxylic acid d. Alkene

3.110 How many constitutional isomeric alcohols are there that have the molecular formula $C_4H_{10}O$?

a. Two b. Three
c. Four d. Five

3.111 Which of the following statements concerning the physical properties of alcohols is *incorrect*?

a. Alcohol solubility in water decreases as the carbon chain length increases.

b. Alcohol solubility in water decreases as the number of —OH groups present increases.

c. Alcohol boiling points increase as carbon chain length increases.

d. C_1 to C_4 straight-chain alcohols are liquids at room temperature.

3.112 What is the molecular formula for the compound called phenol?

a. C_6H_6O b. $C_6H_{12}O$
c. $C_6H_6O_2$ d. $C_6H_{12}O_2$

3.113 Simple ethers may be viewed as derivatives of water in which both hydrogen atoms have been replaced with which of the following?

a. Alkyl groups b. Alkoxy groups
c. Hydroxyl groups d. Sulfhydryl groups

3.114 What is the common name for the compound 2-ethoxypropane?

a. Diethyl ether b. Diisopropyl ether
c. Ethyl propyl ether d. Ethyl isopropyl ether

3.115 Which of the following is a characteristic property of thiols?

a. Extremely strong odors

b. Abnormally high boiling points

c. Extensive intermolecular hydrogen bonding

d. Strong resistance to oxidation

3.116 In which of the following pairs of compounds are the two members of the pair constitutional isomers?

a. Methoxymethane and methoxyethane

b. 2-Propanol and isopropyl alcohol

c. Ethanol and ethanediol

d. Propyl alcohol and ethyl methyl ether

Aldehydes and Ketones

Benzaldehyde is the main flavor component in almonds. Aldehydes and ketones are responsible for the odor and taste of numerous nuts and spices.

In this chapter, we continue our discussion of hydrocarbon derivatives that contain the element oxygen. The functional groups we considered in the previous chapter (alcohols, phenols, and ethers) have the common feature of carbon–oxygen *single* bonds. Carbon–oxygen *double* bonds are also possible in hydrocarbon derivatives. We will now consider the simplest types of compounds that contain this structural feature: aldehydes and ketones.

4.1 The Carbonyl Group

Both aldehydes and ketones contain a carbonyl functional group. A **carbonyl group** *is a carbon atom double-bonded to an oxygen atom.* The structural representation for a carbonyl group is

Carbonyl group

Carbon–oxygen and carbon–carbon double bonds differ in a major way. A carbon–oxygen double bond is *polar,* and a carbon–carbon double bond is *nonpolar.* The electronegativity of oxygen (3.5) is much greater than that of carbon (2.5). Hence the

> The word *carbonyl* is pronounced "carbon-EEL."

> The difference in electronegativity between oxygen and carbon causes a carbon–oxygen double bond to be polar.

carbon–oxygen double bond is polarized, the oxygen atom acquiring a partial negative charge (δ^-) and the carbon atom acquiring a partial positive charge (δ^+).

$$\overset{\delta^+}{C}=\overset{\delta^-}{O} \quad \text{or} \quad C=O$$

Polar nature of carbon–oxygen double bond

4.2 Structure of Aldehydes and Ketones

The word *aldehyde* is pronounced "AL-da-hide."

The identity of the atoms directly bonded to the carbon atom of the carbonyl group is what distinguishes aldehydes from ketones. **An aldehyde** *is a carbonyl-containing organic compound in which the carbonyl carbon atom has at least one hydrogen atom directly attached to it.* The remaining group attached to the carbonyl carbon atom can be hydrogen, an alkyl group (R), a cycloalkyl group, or an aryl group (Ar).

In interpreting general *condensed* functional group structures such as RCHO, remember that carbon always has four bonds and hydrogen always has only one. In RCHO, you know one of carbon's bonds goes to the R group and one to H; therefore, two bonds must go to O.

The aldehyde functional group, the structural feature common to all the preceding compounds, is

Linear notations for an aldehyde functional group and for an aldehyde itself are —CHO and RCHO, respectively. Note that the ordering of the symbols H and O in these notations is HO, not OH (which denotes a hydroxyl group).

The word *ketone* is pronounced. "KEY-tone."

    A **ketone** *is a carbonyl-containing organic compound in which the carbonyl carbon atom has two other carbon atoms directly attached to it.* The groups containing these bonded carbon atoms may be alkyl, cycloalkyl, or aryl.

The ketone functional group, the structural feature common to all the preceding compounds, is

In an aldehyde, the carbonyl group is always located at the end of a hydrocarbon chain.

$$CH_3—CH_2—CH_2—CH_2—\boxed{\overset{O}{\underset{\|}{C}}}—H$$

In a ketone, the carbonyl group is always at a nonterminal (interior) position on the hydrocarbon chain.

$$CH_3—CH_2—CH_2—\boxed{\overset{O}{\underset{\|}{C}}}—CH_2—CH_3$$

The general condensed formula for a ketone is RCOR, in which the oxygen atom is understood to be double-bonded to the carbonyl carbon at the left of it in the formula.

    An aldehyde functional group can be bonded to only one carbon atom because three of the four bonds from an aldehyde carbonyl carbon must go to oxygen and hydrogen. Thus, an aldehyde functional group is always found at the end of a carbon chain. A ketone functional group, by contrast, is always found within a carbon chain, as it must be bonded to two other carbon atoms.

    Cyclic aldehydes are not possible. For an aldehyde carbonyl carbon atom to be part of a ring system it would have to form two bonds to ring atoms, which would give it five bonds. Unlike aldehydes, ketones can form cyclic structures, such as

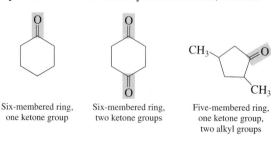

Six-membered ring, one ketone group

Six-membered ring, two ketone groups

Five-membered ring, one ketone group, two alkyl groups

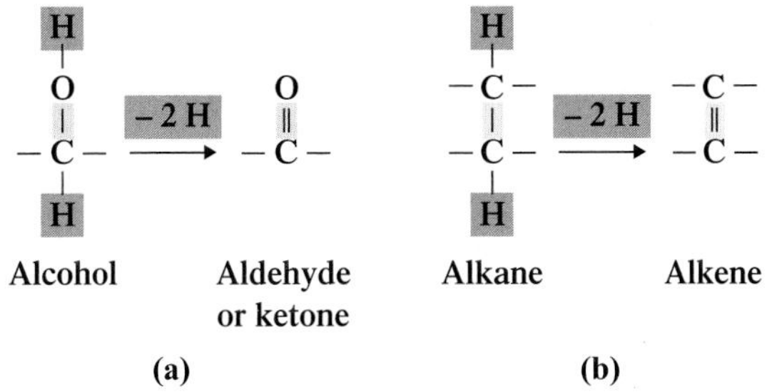

FIGURE 4.1 Aldehydes and ketones are related to alcohols in the same manner that alkenes are related to alkanes; removal of two hydrogen atoms produces a double bond.

Other families of organic compounds are known in which a carbonyl group is present, but they differ from aldehydes and ketones in having a *heteroatom*—any atom other than carbon or hydrogen—directly attached to the carbonyl carbon atom. *Carboxylic acids* and *esters,* to be discussed in Chapter 5, have an oxygen heteroatom and *amides,* to be discussed in Chapter 6, have a nitrogen heteroatom.

Cyclic ketones are *not* heterocyclic ring systems as were cyclic ethers (Section 3.19).

Aldehydes and ketones are related to alcohols in the same manner that alkenes are related to alkanes. Removal of hydrogen atoms from each of two adjacent carbon atoms in an alkane produces an alkene. In a like manner, removal of a hydrogen atom from the —OH group of an alcohol and from the carbon atom to which the hydroxyl group is attached produces a carbonyl group (see Figure 4.1)

4.3 Nomenclature for Aldehydes

The IUPAC rules for naming aldehydes are as follows:

1. Select as the parent carbon chain the longest chain that *includes* the carbon atom of the carbonyl group.
2. Name the parent chain by changing the *-e* ending of the corresponding alkane name to *-al*.
3. Number the parent chain by assigning the number 1 to the carbonyl carbon atom of the aldehyde group.
4. Determine the identity and location of any substituents, and append this information to the front of the parent chain name.

The carbonyl carbon atom in an aldehyde cannot have any number but 1, so we do not have to include this number in the aldehyde's IUPAC name.

EXAMPLE 4.1

Determining IUPAC Names for Aldehydes

Line-angle formulas for the simpler unbranched-chain aldehydes:

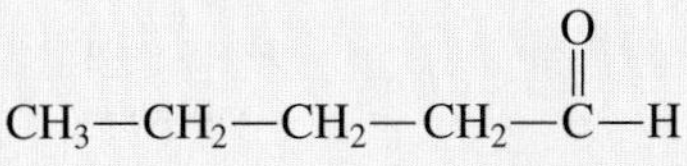

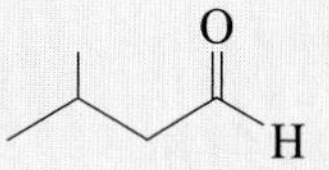

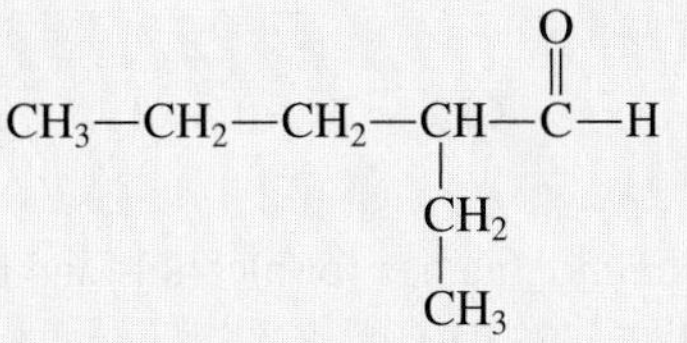

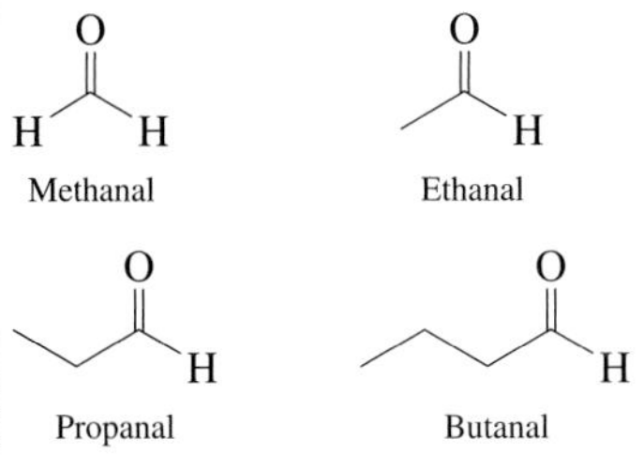

Be careful about the endings *-al* and *-ol.* They are easily confused. The suffix *-al* (pronounced like the man's name Al) denotes an aldehyde; the suffix *-ol* (pronounced like the *ol* in old) denotes an alcohol.

■ Assign IUPAC names to the following aldehydes.

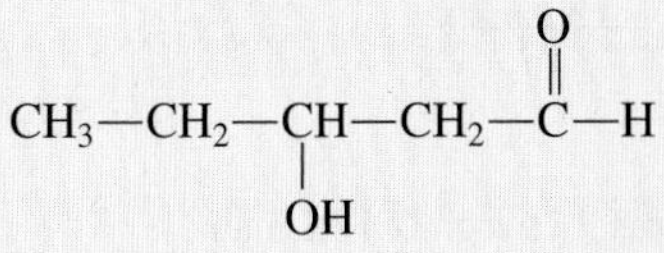

Solution

a. The parent chain name comes from pentane. Remove the *-e* ending and add the aldehyde suffix *-al*. The name becomes *pentanal*. The location of the carbonyl carbon atom need not be specified because this carbon atom is always number 1. The complete name is simply *pentanal*.

b. The parent chain name is *butanal*. To locate the methyl group, we number the carbon chain beginning with the carbonyl carbon atom. The complete name of the aldehyde is *3-methylbutanal*

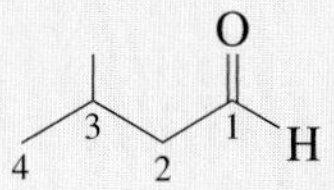

(*continued*)

When a compound contains more than one type of functional group, the suffix for only one of them can be used as the ending of the name. The IUPAC rules establish priorities that specify which suffix is used. For the functional groups we have discussed up to this point in the text, the IUPAC priority system is

aldehyde
ketone
alcohol
Increasing priority — alkene
alkyne
alkoxy ⎤
alkyl   ⎬ Equal-priority substituents (listed in alphabetical order)
halogen ⎦

aldehyde–ether

$$CH_3-O-CH_2-\overset{\overset{\displaystyle O}{\|}}{C}-H$$

2-methoxyethanal

aldehyde–alkene

$$CH_2{=}CH-\overset{\overset{\displaystyle O}{\|}}{C}-H$$

2-propenal

aldehyde–alcohol

$$HO-CH_2-CH_2-\overset{\overset{\displaystyle O}{\|}}{C}-H$$

3-hydroxypropanal

The common names for simple aldehydes illustrate a second method for counting from one to four: *form-, acet-, propion-,* and *butyr-*. We will use this method again in the next chapter when we consider the common names for carboxylic acids and esters. (The first method for counting from one to four, with which you are now thoroughly familiar, is *meth-, eth-, prop-,* and *but-,* as in methane, ethane, propane, and butane.)

The contrast between IUPAC names and common names for aldehydes is as follows:

IUPAC (one word)
⎡ alkanal ⎤
pentanal

Common (one word)
⎡ (prefix) aldehyde* ⎤
butyraldehyde

*The common-name prefixes are related to natural sources for carboxylic acids with the same number of carbon atoms (see Section 5.3).

c. The longest chain containing the carbonyl atom is five carbons long, giving a parent chain name of *pentanal*. An ethyl group is present on carbon 2. Thus the complete name is *2-ethylpentanal*.

$$\overset{5}{C}H_3-\overset{4}{C}H_2-\overset{3}{C}H_2-\overset{2}{C}H-\overset{1}{\overset{\overset{\displaystyle O}{\|}}{C}}-H$$
$$\underset{CH_3}{\overset{|}{\underset{|}{CH_2}}}$$

d. This is a hydroxyaldehyde, with the hydroxyl group located on carbon 3.

$$\overset{5}{C}H_3-\overset{4}{C}H_2-\overset{3}{\underset{\underset{OH}{|}}{C}}H-\overset{2}{C}H_2-\overset{1}{\overset{\overset{\displaystyle O}{\|}}{C}}-H$$

The complete name of the compound is *3-hydroxypentanal*. An aldehyde functional group has priority over an alcohol functional group in IUPAC nomenclature. An alcohol group named as a substituent is a *hydroxy* group.

Practice Exercise 4.1

Assign IUPAC names to the following aldehydes.

a.
$$CH_3-\overset{\underset{\underset{CH_3}{|}}{}}{C}H-\overset{\overset{\displaystyle O}{\|}}{C}-H$$

b.
$$CH_3-CH_2-\overset{\underset{\underset{CH_3-CH_2-CH_2}{|}}{}}{C}H-\overset{\overset{\displaystyle O}{\|}}{C}-H$$

c.
$$CH_3-\overset{\underset{\underset{Cl}{|}}{}}{C}H-\overset{\underset{\underset{Cl}{|}}{}}{C}H-\overset{\overset{\displaystyle O}{\|}}{C}-H$$

d.

Unbranched aldehydes with a small number of carbon atoms have common names:

$$H-\overset{\overset{\displaystyle O}{\|}}{C}-H \qquad CH_3-\overset{\overset{\displaystyle O}{\|}}{C}-H$$
Formaldehyde   Acetaldehyde

$$CH_3-CH_2-\overset{\overset{\displaystyle O}{\|}}{C}-H \qquad CH_3-CH_2-CH_2-\overset{\overset{\displaystyle O}{\|}}{C}-H$$
Propionaldehyde   Butyraldehyde

Unlike the common names for alcohols and ethers, the common names for aldehydes are *one* word rather than two or three.

In the IUPAC system, aromatic aldehydes—compounds in which an aldehyde group is attached to a benzene ring—are named as derivatives of benzaldehyde, the parent compound (see Figure 4.2).

Benzaldehyde   3-Chloro-5-methylbenzaldehyde   4-Hydroxybenzaldehyde

Aromatic aldehydes are not cyclic aldehydes (which do not exist). The carbonyl carbon atom in an aromatic aldehyde is not part of the ring system.

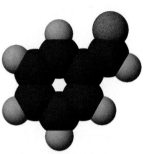

FIGURE 4.2 Space-filling model for benzaldehyde, the simplest aromatic aldehyde.

The last of these compounds is named as a benzaldehyde rather than as a phenol because the aldehyde group has priority over the hydroxyl group in the IUPAC naming system.

4.4 Nomenclature for Ketones

Assigning IUPAC names to ketones is similar to naming aldehydes except that the ending *-one* is used instead of *-al*. The rules for IUPAC ketone nomenclature follow.

1. Select as the parent carbon chain the longest carbon chain that *includes* the carbon atom of the carbonyl group.
2. Name the parent chain by changing the *-e* ending of the corresponding alkane name to *-one*.
3. Number the carbon chain such that the carbonyl carbon atom receives the lowest possible number. The position of the carbonyl carbon atom is noted by placing a number immediately before the name of the parent chain.
4. Determine the identity and location of any substituents, and append this information to the front of the parent chain name.
5. Cyclic ketones are named by assigning the number 1 to the carbon atom of the carbonyl group. The ring is then numbered to give the lowest number(s) to the atom(s) bearing substituents.

EXAMPLE 4.2

Determining IUPAC Names for Ketones

Propanone is the simplest possible ketone. One- and two-carbon ketones cannot exist. A minimum of three carbon atoms is required for a ketone: one C atom for the carbonyl group and one C atom for each of the groups attached to the carbonyl carbon atom. No locator number is needed in the name propanone, because there is only one possible location for the double bond.

In IUPAC nomenclature, the ketone functional group has precedence over all groups we have discussed so far except the aldehyde group. When both aldehyde and ketone groups are present in the same molecule, the ketone group is named as a substituent (the *oxo-* group).

$$CH_3-\overset{\overset{\displaystyle O}{\|}}{C}-CH_2-CH_2-\overset{\overset{\displaystyle O}{\|}}{C}-H$$

4-Oxopentanal

■ Assign IUPAC names to the following ketones.

a.

$$CH_3-\overset{\overset{\displaystyle O}{\|}}{C}-CH_2-CH_2-CH_3$$

b.

c.

d.

Solution

a. The parent chain name is *pentanone*. We number the chain from the end closest to the carbonyl carbon atom. Locating the carbonyl carbon at carbon 2 completes the name. *2-pentanone*.

b. The longest carbon chain of which the carbonyl carbon is a member is four carbons long. The parent chain name is *butanone*. There is one methyl group attached, and the numbering system is from right to left.

The complete name for the compound is *3-methyl-2-pentanone*.

c. The base name is *cyclohexanone*. The methyl group is bonded to carbon 2 because we begin numbering at the carbonyl carbon. The name is *2-methylcyclohexanone*.

d. This ketone has a base name of *cyclopentanone*. Numbering clockwise from the carbonyl carbon atom locates the bromo group on carbon 3. The complete name is *3-bromocyclopentanone*.

(*continued*)

Line-angle formulas for the simpler unbranched-chain 2-ketones:

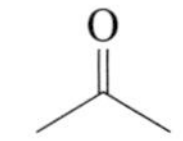

2-Propanone

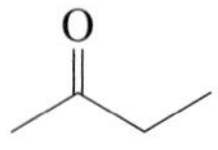

2-Butanone

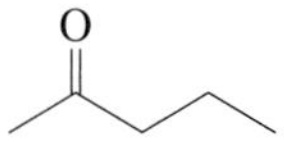

2-Pentanone

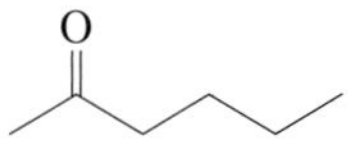

2-Hexanone

Practice Exercise 4.2

Assign IUPAC names to the following ketones.

a.

b. 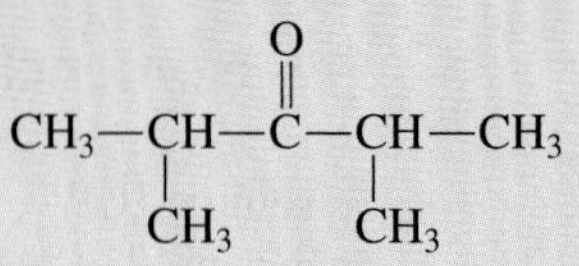 $CH_3-CH-C-CH-CH_3$
         | || |
        CH_3 O CH_3

c.

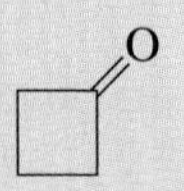

d.

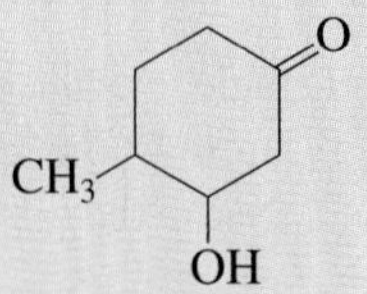

The procedure for coining common names for ketones is the same as that used for ether common names (Section 3.16). They are constructed by giving, in alphabetical order, the names of the alkyl or aryl groups attached to the carbonyl functional group and then adding the word *ketone*. Unlike aldehyde common names, which are one word, those for ketones are two or three words.

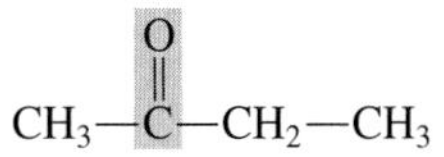

$CH_3-C-CH_2-CH_3$ CH_3-CH_2-C-⬡ $CH_3-CH-C-CH_3$
                                                                   CH_3

Ethyl methyl ketone 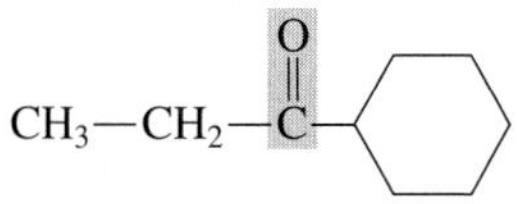   Cyclohexyl ethyl ketone 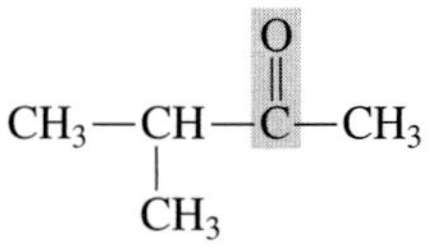   Isopropyl methyl ketone

Three ketones have additional common names besides those obtained with the preceding procedures. These three ketones are

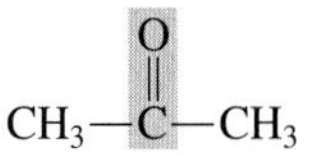

CH_3-C-CH_3 ⬡$-C-CH_3$ ⬡$-C-$⬡

Acetone   Acetophenone   Benzophenone
(dimethyl ketone)   (methyl phenyl ketone)   (diphenyl ketone)

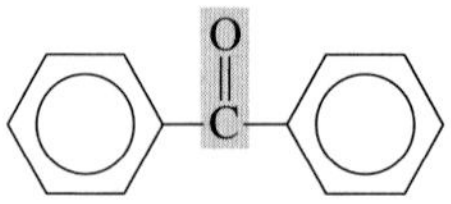

Acetophenone is the simplest aromatic ketone.

The contrast between IUPAC names and common names for ketones is as follows:

IUPAC (one word)

| alkanone |

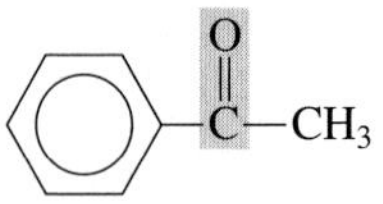

2-butanone

Common (three or two words)

| alkyl alkyl ketone |

ethyl methyl ketone

or

| dialkyl ketone |

dipropyl ketone

4.5 Isomerism for Aldehydes and Ketones

Like the classes of organic compounds previously discussed (alkanes, alkenes, alkynes, alcohols, ethers, etc.), constitutional isomers exist for aldehydes and for ketones, and *between* aldehydes and ketones (functional group isomerism). The compounds butanal and 2-methylpropanal are examples of skeletal aldehyde isomers; the compounds 2-pentanone and 3-pentanone are examples of positional ketone isomers.

Aldehyde
skeletal isomers: $CH_3-CH_2-CH_2-C-H$ and $CH_3-CH-C-H$ (with CH_3 and O)

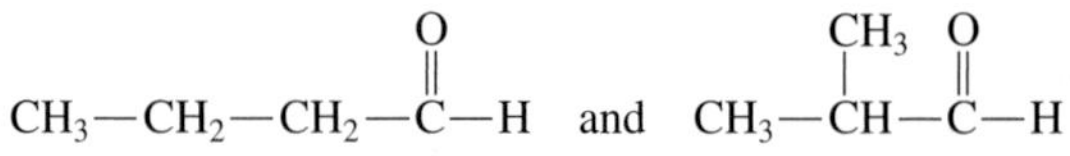

Ketone
positional isomers: $CH_3-C-CH_2-CH_2-CH_3$ and $CH_3-CH_2-C-CH_2-CH_3$

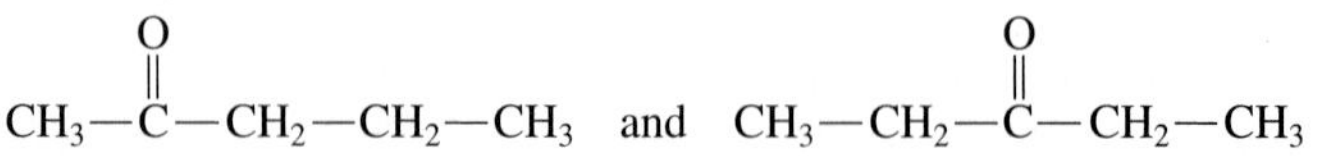

This is the third time we have encountered functional group isomerism. Alcohols and ethers (Section 3.17) and thiols and thioethers (Section 3.21) also exhibit this type of isomerism.

Aldehydes and ketones with the same number of carbon atoms and the same degree of saturation are functional group isomers. Molecular models for the isomeric C_3 compounds propanal and propanone, which both have the molecular formula C_3H_6O, are shown in Figure 4.3.

CHEMICAL CONNECTIONS

Lachrymatory Aldehydes and Ketones

A lachrymator, pronounced "lack-ra-mater," is a compound that causes the production of tears. A number of aldehydes and ketones have lachrymatic properties.

Two lachrymal ketones are 2-chloroacetophenone and bromoacetone.

2-Chloroacetophenone
(2-chloro-1-phenylethanone)

Bromoacetone
(1-bromopropanone)

2-Chloroacetophenone is a component of the tear gas used by police and the military. It is also the active ingredient in MACE canisters now marketed for use by individuals to protect themselves from attackers. The compound bromoacetone has been used as a chemical war gas.

Smoke contains compounds that cause the eyes to tear. A predominant lachrymator in wood smoke is formaldehyde, the one-carbon aldehyde. The smoke associated with an outdoor barbecue contains the unsaturated aldehyde *acrolein*.

$$CH_2=CH-\overset{\overset{\displaystyle O}{\|}}{C}-H$$

Acrolein
(propenal)

Acrolein is formed as fats that are present in meat break down when heated. (Besides being a lachrymator, acrolein is responsible for the "pleasant" odor associated with the process of barbecuing meat.)

The lachrymatory compound associated with onions is a derivative of thiopropionaldehyde.

Thiopropionaldehyde
(propanethial)

Thiopropionaldehyde-S-oxide
(propanethial-S-oxide)
[lachrymator in chopped onions]

Onions do not cause tear production until they are chopped or sliced. The onion cells damaged by these actions release the enzyme *allinase*, which converts an odorless compound naturally present in onions to the lachrymatic compound.

Scientists are not sure why thiopropionaldehyde-S-oxide causes tear production, but it is known that it is an unstable molecule that is readily broken up by water into propanal, hydrogen sulfide, and sulfuric acid.

$$CH_3-CH_2-\overset{\overset{\displaystyle S \nearrow O}{\|}}{C}-H \xrightarrow{H_2O}$$

$$CH_3-CH_2-\overset{\overset{\displaystyle O}{\|}}{C}-H + H_2S + H_2SO_4$$

Sulfuric acid may be responsible for the eye irritation.

Many people can peel onions under water, or peel cold ones from the refrigerator, without crying. Water washes away the soluble lachrymator and also breaks it down chemically. If the onion is cold, the enzymatic reaction making the lachrymator is slower, so less is formed. Also the vapor pressure of the lachrymator is greatly reduced at the lower temperature, so its concentration in air is reduced.

FIGURE 4.3 Aldehydes and ketones with the same number of carbon atoms and the same degree of saturation are functional group isomers, as is illustrated here for the three-carbon aldehyde (propanal) and the three-carbon ketone (propanone). Both have the molecular formula C_3H_6O.

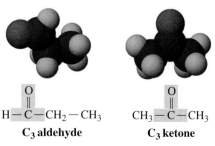

C₃ aldehyde   **C₃ ketone**

4.6 Selected Common Aldehydes and Ketones

Formaldehyde, the simplest aldehyde, with only one carbon atom, is manufactured on a large scale by the oxidation of methanol.

$$CH_3-OH \xrightarrow[600°C-700°C]{\text{Ag catalyst}} H-\overset{\displaystyle O}{\overset{\|}{C}}-H + H_2$$

Its major use is in the manufacture of polymers (Section 4.11). At room temperature and pressure, formaldehyde is an irritating gas. Bubbling this gas through water produces *formalin*, an aqueous solution containing 37% formaldehyde by mass or 40% by volume. (This represents the solubility limit of formaldehyde gas in water.) Very little free formaldehyde gas is actually present in formalin; most of it reacts with water, producing methylene glycol.

$$H-\overset{\displaystyle O}{\overset{\|}{C}}-H + H-O-H \longrightarrow HO-CH_2-OH$$

Formalin is used for preserving biological specimens (see Figure 4.4); anyone who has experience in a biology laboratory is familiar with the pungent odor of formalin. Formalin is also the most widely used preservative chemical in embalming fluids used by morticians. Its mode of action involves reaction with protein molecules in a manner that links the protein molecules together; the result is a "hardening" of the protein.

Acetone, a colorless, volatile liquid with a pleasant, mildly "sweet" odor, is the simplest ketone and is also the ketone used in largest volume in industry. Acetone is an excellent solvent because it is miscible with both water and nonpolar solvents. Acetone is the main ingredient in gasoline treatments that are designed to solubilize water in the gas tank and allow it to pass through the engine in miscible form. Acetone can also be used to remove water from glassware in the laboratory. And it is a major component of some nail polish removers.

Small amounts of acetone are produced in the human body in reactions related to obtaining energy from fats. Normally, such acetone is degraded to CO_2 and H_2O. Diabetic people produce larger amounts of acetone, not all of which can be degraded. The presence of acetone in urine is a sign of diabetes. In severe diabetes, the odor of acetone can be detected on the person's breath.

The oxidation of methanol to formaldehyde has been previously mentioned (Section 3.5). Ingested methanol is oxidized in the human body to formaldehyde, and it is the formaldehyde that causes blindness.

■ Naturally Occurring Aldehydes and Ketones

Aldehydes and ketones occur widely in nature. Naturally occurring compounds of these types, with higher molecular masses, usually have pleasant odors and flavors and are often used for these properties in consumer products (perfumes, air fresheners, and the like). Table 4.1 gives the structures and uses for selected naturally occurring aldehydes and ketones. The unmistakable odor of melted butter is largely due to the four-carbon diketone butanedione (see Figure 4.5).

Many important steroid hormones (Section 8.12) are ketones, including testosterone, the hormone that controls the development of male sex characteristics; progesterone, the hormone secreted at the time of ovulation in females; and cortisone, a hormone from the adrenal glands that is used medicinally to relieve inflammation.

FIGURE 4.5 The delightful odor of melted butter is largely due to butanedione.

Testosterone

Progesterone

Cortisone

TABLE 4.1
Selected Aldehydes and Ketones Whose Uses Are Based on Their Odor or Flavor

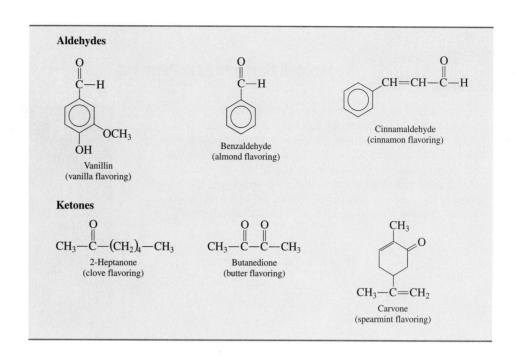

4.7 Physical Properties of Aldehydes and Ketones

FIGURE 4.6 A physical-state summary for unbranched aldehydes and unbranched 2-ketones at room temperature and pressure.

Unbranched Aldehydes			
C_1	C_3	C_5	C_7
C_2	C_4	C_6	C_8

Unbranched 2-Ketones			
✕	C_3	C_5	C_7
✕	C_4	C_6	C_8

☐ Gas   ☐ Liquid

The C_1 and C_2 aldehydes are gases at room temperature (Figure 4.6). The C_3 through C_{11} straight-chain saturated aldehydes are liquids, and the higher aldehydes are solids. The presence of alkyl groups tends to lower both boiling points and melting points, as does the presence of unsaturation in the carbon chain. Lower-molecular-mass ketones are colorless liquids at room temperature (Figure 4.6).

The boiling points of aldehydes and ketones are intermediate between those of alcohols and alkanes of similar molecular mass. Aldehydes and ketones have higher boiling points than alkanes because of dipole–dipole attractions between molecules. Carbonyl group polarity (Section 4.1) makes these dipole–dipole interactions possible. Aldehydes and ketones have lower boiling points than the corresponding alcohols because no hydrogen bonding occurs as it does with alcohols. Dipole–dipole attractions are weaker forces than hydrogen bonds. Table 4.2 provides boiling-point information for selected aldehydes, alcohols, and alkanes.

Water molecules can hydrogen-bond with aldehyde and ketone molecules (Figure 4.7). This hydrogen bonding causes low-molecular-mass aldehydes and ketones to be water soluble. As the hydrocarbon portions get larger, the water solubility of aldehydes and ketones decreases. Table 4.3 gives data on solubility in water for selected aldehydes and ketones.

TABLE 4.2
Boiling Points of Some Alkanes, Aldehydes, and Alcohols of Similar Molecular Mass

Type of Compound	Compound	Structure	Molecular Mass	Boiling Point (8°C)
alkane	ethane	$CH_3—CH_3$	30	−89
aldehyde	methanal	$H—CHO$	30	−21
alcohol	methanol	$CH_3—OH$	32	65
alkane	propane	$CH_3—CH_2—CH_3$	44	−42
aldehyde	ethanal	$CH_3—CHO$	44	20
alcohol	ethanol	$CH_3—CH_2—OH$	46	78
alkane	butane	$CH_3—CH_2—CH_2—CH_3$	58	−1
aldehyde	propanal	$CH_3—CH_2—CHO$	58	49
alcohol	1-propanol	$CH_3—CH_3—CH_2—OH$	60	97

CHEMICAL CONNECTIONS   Melanin: A Hair and Skin Pigment

Human hair as well as human skin is colored naturally by the pigment melanin—a polymeric substance involving many interconnected *cyclic ketone* units. The more melanin a person produces, which is genetically controlled, the darker his or her hair and skin. The number of melanin-producing cells is essentially the same in dark-skinned and light-skinned people, but they are more active in dark-skinned people. The following is a representation of a portion of the structure of polymeric melanin pigment.

Hair pigmentation (hair color) results from biosynthesis of melanin within hair follicles. The melanin molecules so produced are incorporated into the growing hair shaft and distributed throughout the hair cortex. The melanin tends to accumulate within hair protein as granules. Hair, once it exits the scalp, is no longer alive and any damage to the melanin or the hair itself cannot be repaired by the body.

For most people starting sometime in their thirties, the production of melanin in hair follicles begins to gradually decrease. Once a hair follicle stops producing melanin, the hair will be colorless but will appear white due to light scattering. A proportion of white hair in colored hair will make a head of hair appear gray.

Suntan and sunburn both involve melanin (as a skin pigment) and ultraviolet (UV) radiation. Sudden high-level exposure to UV radiation causes the skin to burn, and steady low-level exposure to UV radiation can cause tanning. Melanin molecules in the skin constitute a built-in defense system to protect the skin against UV radiation. The melanin molecules act as a protective barrier by absorbing and scattering the UV radiation. A dark-skinned person has more melanin molecules in the upper layers of the skin (and more protection against sunburn) than a light-skinned person.

When melanin-producing cells deep in the skin are exposed to UV radiation, melanin production increases. The presence of this extra melanin in the skin gives the skin an appearance that we call a "tan." The larger the melanin molecules so produced, the deeper the tan. People who tan readily have skin that can produce a large amount of melanin.

When a person experiences a sunburn, the skin peels. When peeling occurs, any tan that has been built up (excess melanin) sloughs off with the dead skin. Thus the tanning process must begin anew.

The ordering of boiling points for carbonyl compounds (aldehydes and ketones), alcohols, and alkanes of similar molecular mass is

Alcohols > carbonyl compounds > alkanes

Although low-molecular-mass aldehydes have pungent, penetrating, unpleasant odors, higher-molecular-mass aldehydes (above C_8) are more fragrant, especially benzaldehyde derivatives. Ketones generally have pleasant odors, and several are used in perfumes and air fresheners.

FIGURE 4.7 Low-molecular-mass aldehydes and ketones are soluble in water because of hydrogen bonding.

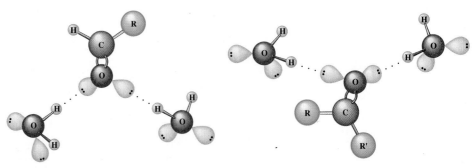

(a) Aldehyde–Water Hydrogen Bonding   **(b) Ketone–Water Hydrogen Bonding**

TABLE 4.3
Water Solubility (g/100 g H₂O) for Various Aldehydes and Ketones

Number of Carbon Atoms	Aldehyde	Water Solubility of Aldehyde	Ketone	Water Solubility of Ketone
1	methanal	very soluble		
2	ethanal	infinite		
3	propanal	16	propanone	infinite
4	butanal	7	2-butanone	26
5	pentanal	4	2-pentanone	5
6	hexanal	1	2-hexanone	1.6
7	heptanal	0.1	2-heptanone	0.4
8	octanal	insoluble	2-octanone	insoluble

4.8 Preparation of Aldehydes and Ketones

Aldehydes and ketones can be produced by the oxidation of primary and secondary alcohols, respectively, using mild oxidizing agents such as $KMnO_4$ or $K_2Cr_2O_7$ (Section 3.9).

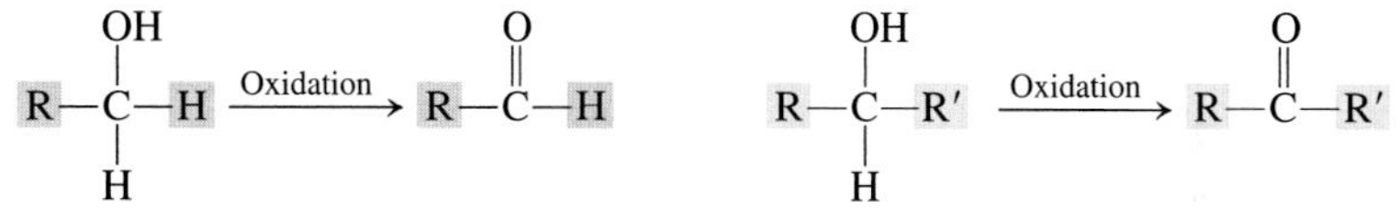

$$
\begin{array}{cc}
\underset{\substack{| \\ H}}{\overset{\substack{OH \\ |}}{R-C-H}} \xrightarrow{\text{Oxidation}} R-\overset{\overset{O}{\|}}{C}-H & \underset{\substack{| \\ H}}{\overset{\substack{OH \\ |}}{R-C-R'}} \xrightarrow{\text{Oxidation}} R-\overset{\overset{O}{\|}}{C}-R' \\
\text{Primary alcohol} \qquad \text{Aldehyde} & \text{Secondary alcohol} \qquad \text{Ketone}
\end{array}
$$

The term *aldehyde* stems from *alc*ohol *dehyd*rogenation, indicating that aldehydes are related to alcohols by the loss of hydrogen.

When this type of reaction is used for aldehyde preparation, reaction conditions must be sufficiently mild to avoid further oxidation of the aldehyde to a carboxylic acid (Section 3.9). Ketones do not undergo the further oxidation that aldehydes do.

In the oxidation of an alcohol to an aldehyde or a ketone, the alcohol molecule loses H atoms. Recall that the *loss of H atoms* by an organic molecule is one of the operational definitions for the process of *oxidation* (Section 3.9).

EXAMPLE 4.3

Predicting Products in Alcohol Oxidation Reactions

■ Draw the structure of the aldehyde or ketone formed from the oxidation of each of the following alcohols. Assume that reaction conditions are sufficiently mild that any aldehydes produced are not oxidized further.

a. $CH_3-CH_2-CH_2-OH$

b. $\underset{\substack{| \\ OH}}{CH_3-CH-CH_3}$

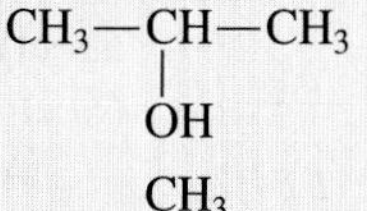

c. ⬡—OH

d. $\underset{\substack{| \\ CH_3}}{\overset{\substack{CH_3 \\ |}}{CH_3-C-OH}}$

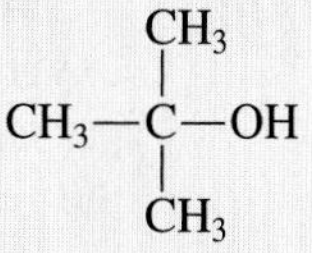

Solution

a. This is a primary alcohol that will give the aldehyde *propanal* as the oxidation product.

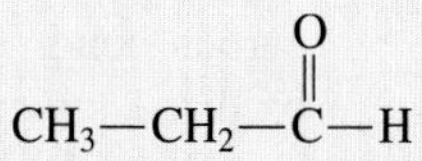

$$CH_3-CH_2-\overset{\overset{O}{\|}}{C}-H$$

Propanal

b. This is a secondary alcohol. Upon oxidation, secondary alcohols are converted to ketones.

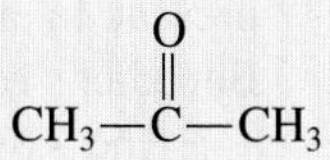

$$CH_3-\overset{\overset{O}{\|}}{C}-CH_3$$

Propanone

(*continued*)

c. This cyclic alcohol is a secondary alcohol; hence a ketone is the oxidation product.

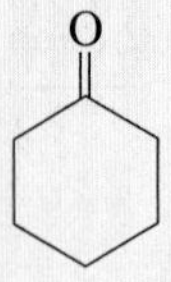

Cyclohexanone

d. This is a tertiary alcohol. Tertiary alcohols do not undergo oxidation (Section 3.9).

Practice Exercise 4.3

Draw the structure of the aldehyde or ketone formed from the oxidation of each of the following alcohols. Assume that reaction conditions are sufficiently mild that any aldehydes produced are not oxidized further to carboxylic acids.

a. 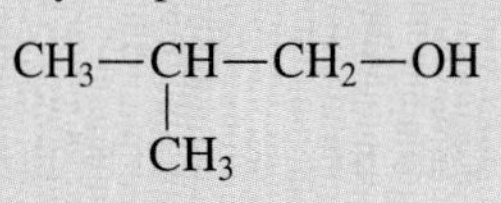 $CH_3-CH-CH_2-OH$
                    |
                   CH_3

b. 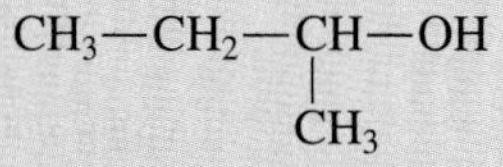 $CH_3-CH_2-CH-OH$
                       |
                      CH_3

c. 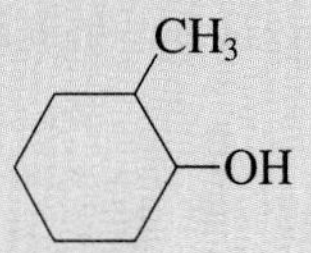 CH_3
         |
    —OH

d. 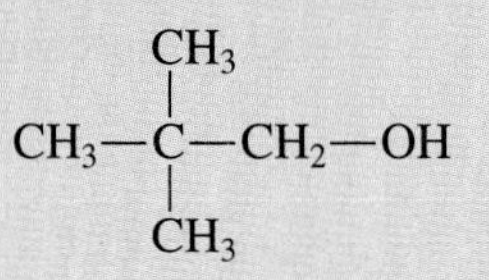 CH_3
              |
      CH_3-C-CH_2-OH
              |
             CH_3

4.9 Oxidation and Reduction of Aldehydes and Ketones

■ Oxidation of Aldehydes and Ketones

Aldehydes readily undergo oxidation to carboxylic acids (Section 4.8), and ketones are resistant to oxidation.

In aldehyde oxidation, the aldehyde gains an oxygen atom (supplied by the oxidizing agent). *Gain of oxygen* is one of the operational definitions for the process of *oxidation* (Section 3.9).

Among the mild oxidizing agents that convert aldehydes into carboxylic acids is oxygen in air. Thus aldehydes must be protected from air. When an aldehyde is prepared from oxidation of a primary alcohol (Section 4.8), it is usually removed from the reaction mixture immediately to prevent it from being further oxidized to a carboxylic acid.

Because both aldehydes and ketones contain carbonyl groups, we might expect similar oxidation reactions for the two types of compounds. Oxidation of an aldehyde involves breaking a carbon–hydrogen bond, and oxidation of a ketone involves breaking a carbon–carbon bond. The former is much easier to accomplish than the latter. For ketones to be oxidized, strenuous reaction conditions must be employed.

Several tests, based on the ease with which aldehydes are oxidized, have been developed for distinguishing between aldehydes and ketones, for detecting the presence of aldehyde groups in sugars (carbohydrates), and for measuring the amounts of sugars present in a solution. The most widely used of these tests are the Tollens test and Benedict's test.

The Tollens test, also called the silver mirror test, involves a solution that contains silver nitrate ($AgNO_3$) and ammonia (NH_3) in water. When Tollens solution is added

CHEMICAL CONNECTIONS — **Diabetes, Aldehyde Oxidation, and Glucose Testing**

Diabetes mellitus is a disease that involves the hormone insulin, a substance necessary to control blood-sugar (glucose) levels. There are two forms of diabetes. In one form, the pancreas does not produce insulin at all. Patients with this condition require injections of insulin to control glucose levels. In the second form, the body cannot make proper use of insulin. Patients with this form of diabetes can often control glucose levels through their diet but may require medication. If the blood-sugar level of a diabetic becomes too high, serious kidney damage can result.

Normal urine does not contain glucose. When the kidneys become overloaded with glucose (the blood-glucose level is too high), glucose is excreted in the urine. Benedict's test (Section 4.9) can be used to detect glucose in urine, because glucose has an aldehyde group present in its structure.

$$CH_2-CH-CH-CH-CH-\overset{\displaystyle O}{\overset{\|}{C}}-H$$
$$\quad\;\, |\quad\;\, |\quad\;\, |\quad\;\, |\quad\;\, |$$
$$OH\quad OH\quad OH\quad OH\quad OH$$
Glucose

A urine glucose test is carried out using either plastic test strips coated with Benedict's solution or Clinitest tablets (a convenient solid form of Benedict's reagent). A few drops of urine are added to the plastic strip or tablet, and the degree of coloration is used to estimate the blood-glucose level. The solution turns greenish at a low glucose level, then yellow-orange, and finally a dark orange-red.

Tests are also available for directly measuring glucose concentration in blood. These tests involve placing a drop of blood (from a finger prick) on a plastic strip containing a dye and an enzyme that will oxidize glucose's aldehyde group. A two-step reaction sequence occurs. First, the enzyme causes glucose oxidation to a carboxylic acid, with hydrogen peroxide (H_2O_2) being another product of the reaction.

$$R-\overset{\displaystyle O}{\overset{\|}{C}}-H + O_2 \xrightarrow{\text{Enzyme}} R-\overset{\displaystyle O}{\overset{\|}{C}}-OH + H_2O_2$$
Aldehyde   Carboxylic acid   Hydrogen
(glucose)   peroxide

Then the H_2O_2 reacts with the dye to produce a colored product. The amount of color produced, measured by comparison with a color chart or by an electronic monitor, is proportional to the blood-glucose concentration.

to an aldehyde, Ag^+ ion (the oxidizing agent) is reduced to silver metal, which deposits on the inside of the test tube, forming a silver mirror. The appearance of this silver mirror (see Figure 4.8) is a positive test for the presence of the aldehyde group.

$$R-\overset{\displaystyle O}{\overset{\|}{C}}-H + Ag^+ \xrightarrow[\text{heat}]{NH_3,\, H_2O} R-\overset{\displaystyle O}{\overset{\|}{C}}-OH + Ag$$
Aldehyde   Carboxylic acid   Silver metal

The Ag^+ ion will not oxidize ketones.

FIGURE 4.8 A positive Tollens test for aldehydes involves the formation of a silver mirror. (a) An aqueous solution of ethanal is added to a solution of silver nitrate in aqueous ammonia and stirred. (b) The solution darkens as ethanal is oxidized to ethanoic acid, and Ag^+ ion is reduced to silver. (c) The inside of the beaker becomes coated with metallic silver.

(a)   (b)   (c)

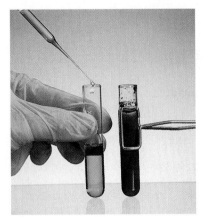

FIGURE 4.9 Benedict's solution, which is blue in color, turns brick red when an aldehyde reacts with it.

Benedict's test is similar to the Tollens test in that a metal ion is the oxidizing agent. With this test, Cu^{2+} ion is reduced to Cu^+ ion, which precipitates from solution as Cu_2O (a brick-red solid; Figure 4.9).

$$R-\overset{\overset{O}{\|}}{C}-H + Cu^{2+} \longrightarrow R-\overset{\overset{O}{\|}}{C}-OH + Cu_2O$$

Aldehyde   Carboxylic acid   Brick-red solid

Benedict's solution is made by dissolving copper sulfate, sodium citrate, and sodium carbonate in water.

■ Reduction of Aldehydes and Ketones

Aldehydes and ketones are easily reduced by hydrogen gas (H_2), in the presence of a catalyst (Ni, Pt, or Cu), to form alcohols. The reduction of aldehydes produces primary alcohols, and the reduction of ketones yields secondary alcohols.

$$\text{Aldehyde} \xrightarrow{\text{Reduction}} \text{primary alcohol}$$

$$\text{Ketone} \xrightarrow{\text{Reduction}} \text{secondary alcohol}$$

Specific examples of such reactions follow.

Aldehyde reduction: $CH_3-\overset{\overset{O}{\|}}{C}-H + H_2 \xrightarrow{Ni} CH_3-\overset{\overset{OH}{|}}{\underset{\underset{H}{|}}{C}}-H$

       Ethanal   Ethanol

Ketone reduction: $CH_3-\overset{\overset{O}{\|}}{C}-CH_3 + H_2 \xrightarrow{Ni} CH_3-\overset{\overset{OH}{|}}{\underset{\underset{H}{|}}{C}}-CH_3$

       Propanone   2-Propanol

Recall that *gain of H atoms* by an organic molecule is one of the operational definitions for the process of *reduction* (Section 3.9).

It is the addition of hydrogen atoms to the carbon–oxygen double bond that produces the alcohol in each of these reactions.

$$\overset{}{\underset{}{C}}=O + H_2 \xrightarrow{\text{Catalyst}} \underset{\underset{H}{|}}{C}-\underset{\underset{H}{|}}{O}$$

This hydrogen addition process is very similar to the addition of hydrogen to the carbon–carbon double bond of an alkene to produce an alkane, which we encountered in Section 2.8.

$$\overset{}{\underset{}{C}}=\overset{}{\underset{}{C}} + H_2 \xrightarrow{\text{Catalyst}} \underset{\underset{H}{|}}{C}-\underset{\underset{H}{|}}{C}$$

Aldehyde reduction and ketone reduction to produce alcohols are the "opposite" of the oxidation of alcohols to produce aldehydes and ketones (Section 4.8). These "opposite" relationships can be diagrammed as follows:

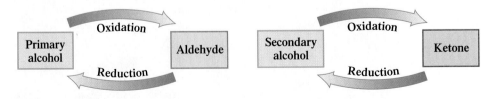

As we noted in Section 3.9, "keeping track" of such relationships is an aid in remembering organic chemistry reaction schemes.

4.10 Reaction of Aldehydes and Ketones with Alcohols

Aldehydes and ketones react with alcohols to form hemiacetals and acetals. Reaction with *one* molecule of alcohol produces a hemiacetal, which is then converted to an acetal by reaction with a second alcohol molecule.

$$\text{Aldehyde or ketone} + \text{alcohol} \xrightarrow[\text{catalyst}]{\text{Acid}} \text{hemiacetal}$$

$$\text{Hemiacetal} + \text{alcohol} \xrightarrow[\text{catalyst}]{\text{Acid}} \text{acetal}$$

The Greek prefix *hemi-* means "half." When one alcohol molecule has reacted with the aldehyde or ketone, the compound is halfway to the final acetal.

Further information about these two reactions follows.

■ **Hemiacetal Formation**

Hemiacetal formation is an addition reaction in which a molecule of alcohol adds to the carbonyl group of an aldehyde or ketone. The H portion of the alcohol adds to the carbonyl oxygen atom, and the R—O portion of the alcohol adds to the carbonyl carbon atom.

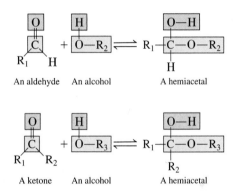

Formally defined, a **hemiacetal** *is an organic compound in which a carbon atom is bonded to both a hydroxyl group (—OH) and an alkoxy group (—OR).* The functional group for a hemiacetal is thus

$$-\overset{\displaystyle \text{OH}}{\underset{\displaystyle |}{\overset{\displaystyle |}{C}}}-\text{OR}$$

The carbon atom of the hemiacetal functional group is often referred to as the *hemiacetal carbon atom;* it was the carbonyl carbon atom of the aldehyde or ketone that reacted.

A reaction mixture containing a hemiacetal is always in equilibrium with the alcohol and carbonyl compound from which it was made, and the equilibrium lies to the carbonyl compound side of the reaction.

$$\text{Alcohol} + \text{aldehyde} \rightleftharpoons \text{hemiacetal}$$
$$\text{Alcohol} + \text{ketone} \rightleftharpoons \text{hemiacetal}$$

This situation makes isolation of the hemiacetal difficult; in practice, it usually cannot be done.

An important exception to this difficulty with isolation is the case where the —OH and $\text{C}{=}\text{O}$ functional groups that react to form the hemiacetal come from the *same* molecule.

This produces a *cyclic* hemiacetal rather than a noncyclic one, and cyclic acetals are more stable than the noncyclic ones and can be isolated.

Hemiacetal and acetal formation are very important biochemical reactions; they are crucial to understanding the chemistry of carbohydrates, which is considered in Chapter 7.

Hemiacetals contain an alcohol group (hydroxyl group) and an ether group (alkoxy group) on the same carbon atom.

Illustrative of intramolecular hemiacetal formation is the reaction

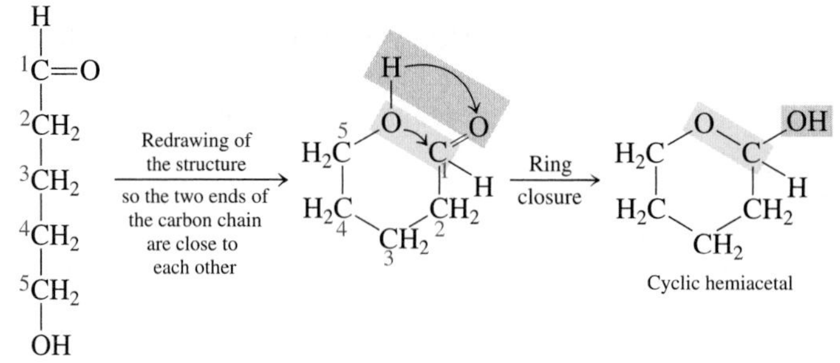

Cyclic hemiacetals are very important compounds in carbohydrate chemistry, the topic of Chapter 7.

EXAMPLE 4.4

Recognizing Hemiacetal Structures

■ Indicate whether each of the following compounds is a hemiacetal.

a. CH₃—CH—O—CH₃
       |
      OH

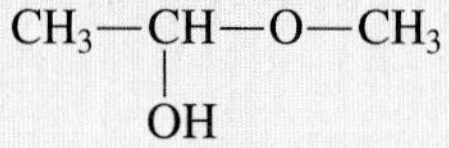

b.
           OH
           |
    CH₃—C—CH₃
           |
        O—CH₃

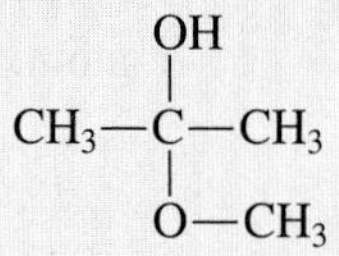

c.
           CH₃
           |
CH₃—CH—CH—O—CH₃
     |
   OH

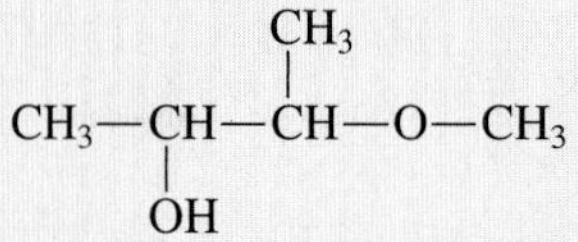

d.

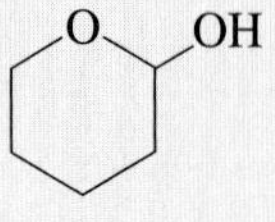

Solution

In each part, we will be looking for the following structural feature: the presence of an —OH group and an —OR group attached to the same carbon atom.

a. We have an —OH group and an —OR group attached to the same carbon atom. The compound is a *hemiacetal*.
b. We have an —OH group and an —OR group attached to the same carbon atom. The compound is a *hemiacetal*.
c. The —OH and —OR groups present in this molecule are attached to *different* carbon atoms. Therefore, the molecule is *not a hemiacetal*.
d. We have a ring carbon atom bonded to two oxygen atoms: one oxygen atom in an —OH substituent and the other oxygen atom bonded to the rest of the ring (the same as an R group). This is a *hemiacetal*.

Practice Exercise 4.4

Indicate whether each of the following compounds is a hemiacetal.

a. OH
   |
   CH₂
   |
   O—CH₃

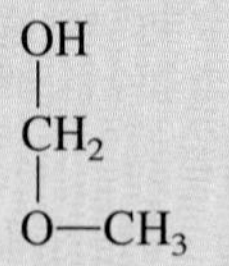

b.
           CH₃
           |
           CH₂
           |
   CH₃—O—C—OH
           |
           CH₃

c.
          CH₃
          |
  CH₃—O—CH
         |
     HO—CH
         |
        CH₃

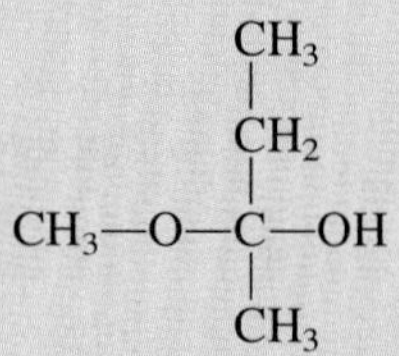

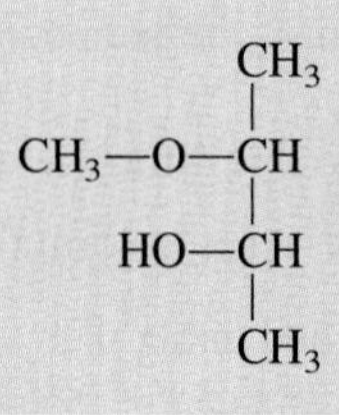

d.

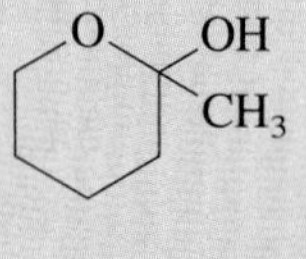

■ Acetal Formation

This is our second encounter with condensation reactions. The first encounter involved intermolecular alcohol dehydration (Section 3.9).

If a small amount of acid catalyst is added to a hemiacetal reaction mixture, the hemiacetal reacts with a second alcohol molecule, in a condensation reaction, to form an acetal.

$$R_1 - \underset{\underset{H}{|}}{\overset{\overset{OH}{|}}{C}} - OR_2 \ + \ R_3 - OH \ \underset{}{\overset{H^+}{\rightleftharpoons}} \ R_1 - \underset{\underset{H}{|}}{\overset{\overset{OR_3}{|}}{C}} - OR_2 \ + \ H - OH$$

A hemiacetal   An alcohol   An acetal

Acetals have two alkoxy groups (—OR) attached to the same carbon atom.

An **acetal** *is an organic compound in which a carbon atom is bonded to two alkoxy groups (—OR).* The functional group for an acetal is thus

$$-\underset{\underset{|}{|}}{\overset{\overset{OR}{|}}{C}} - OR$$

A specific example of acetal formation from a hemiacetal is

$$CH_3 - \underset{\underset{O-CH_3}{|}}{\overset{\overset{OH}{|}}{CH}} \ + \ CH_3 - CH_2 - OH \ \overset{H^+}{\rightleftharpoons} \ CH_3 - \underset{\underset{O-CH_3}{|}}{\overset{\overset{O-CH_2-CH_3}{|}}{CH}} \ + \ H - OH$$

Note that acetal formation does not involve addition to a carbon–oxygen double bond as hemiacetal formation does; no double bond is present in either of the reactants involved in acetal formation. Acetal formation involves a substitution reaction; the —OR group of the alcohol replaces the —OH group on the hemiacetal.

Figure 4.10 shows molecular models for acetaldehyde (the two-carbon aldehyde) and the hemiacetal and acetal formed when this aldehyde reacts with ethyl alcohol.

■ Acetal Hydrolysis

In Section 13.1, we will find that the enzyme-catalyzed hydrolysis of acetals is an important step in the digestion of carbohydrates.

Acetals, unlike hemiacetals, are easily isolated from reaction mixtures. They are stable in basic solution but undergo *hydrolysis* in acidic solution. A **hydrolysis reaction** *is the reaction of a compound with H_2O, in which the compound splits into two or more fragments as the elements of water (H— and —OH) are added to the compound.* The products of acetal hydrolysis are the aldehyde or ketone and alcohols that originally reacted to form the acetal.

$$-\underset{\underset{|}{|}}{\overset{\overset{O-R_1}{|}}{C}} - O - R_2 \ + \ H - OH \ \underset{}{\overset{\overset{Acid}{catalyst}}{\rightleftharpoons}} \ \underset{\underset{ketone}{Aldehyde\ or}}{\overset{\overset{O}{\|}}{C}} \ + \ R_1 - OH \ + \ R_2 - O - H$$

Acetal

FIGURE 4.10 Molecular models for acetaldehyde and its hemiacetal and acetal formed by reaction with ethyl alcohol.

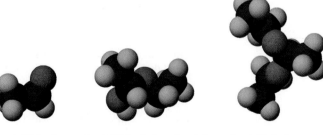

Acetaldehyde   **Acetaldehyde hemiacetal with ethyl alcohol**   **Acetaldehyde acetal with ethyl alcohol**

For example,

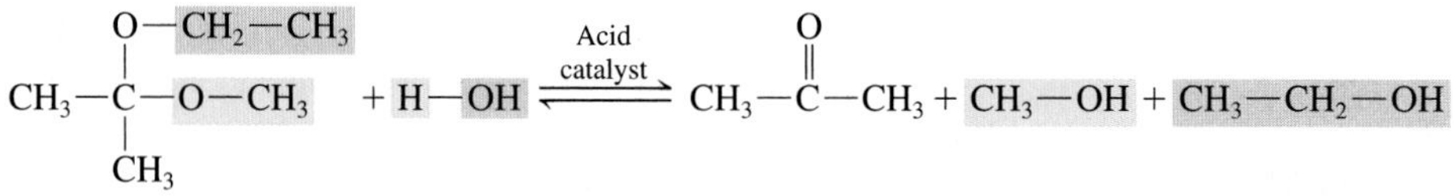

$$CH_3 - \underset{\underset{CH_3}{|}}{\overset{\overset{O-CH_2-CH_3}{|}}{C}} - O-CH_3 \quad + \text{ H}-\text{OH} \xrightleftharpoons[]{\substack{\text{Acid} \\ \text{catalyst}}} CH_3 - \overset{\overset{O}{\|}}{C} - CH_3 + CH_3 - OH + CH_3 - CH_2 - OH$$

The carbonyl hydrolysis product is an aldehyde if the acetal carbon atom has a hydrogen atom attached directly to it, and it is a ketone if no hydrogen attachment is present. In the preceding example, the carbonyl product is a ketone because the two additional acetal carbon atom attachments are methyl groups.

EXAMPLE 4.5

Predicting Products in Acetal
Hydrolysis Reactions

■ Draw the structures of the aldehyde (or ketone) and the two alcohols produced when the following acetals undergo hydrolysis in acidic solution.

a.

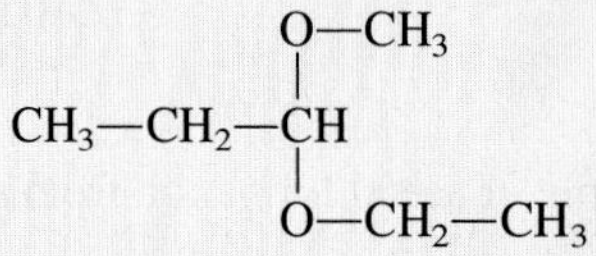

$$CH_3-CH_2-\underset{\underset{O-CH_2-CH_3}{|}}{\overset{\overset{O-CH_3}{|}}{CH}}$$

b.

$$CH_3-\underset{\underset{O}{|}}{\overset{\overset{CH_3}{|}}{C}}-O-\underset{\underset{CH_3}{|}}{CH}-CH_3$$
$$\underset{\underset{CH_3}{|}}{\overset{|}{CH_3-C-CH_3}}$$

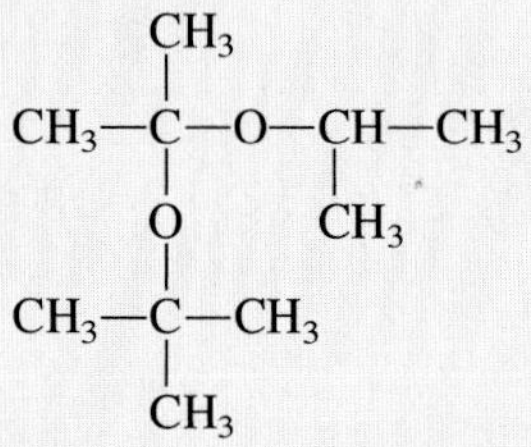

Solution

a. Each of the alkoxy (—OR) groups present will be converted into an alcohol during the hydrolysis. Because the acetal carbon atom has a H attachment, the remainder of the molecule becomes an aldehyde, with the carbon atom to which the alkoxy groups were attached becoming the carbonyl carbon atom.

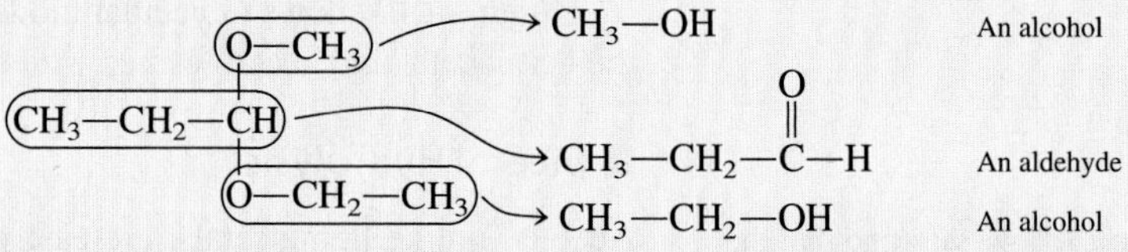

CH_3—OH   An alcohol

CH_3—CH_2—C—H   An aldehyde

CH_3—CH_2—OH   An alcohol

b. Again, each of the alkoxy groups present will be converted into an alcohol during the hydrolysis. Because the acetal carbon atoms lacks a H attachment, the remainder of the molecule becomes a ketone.

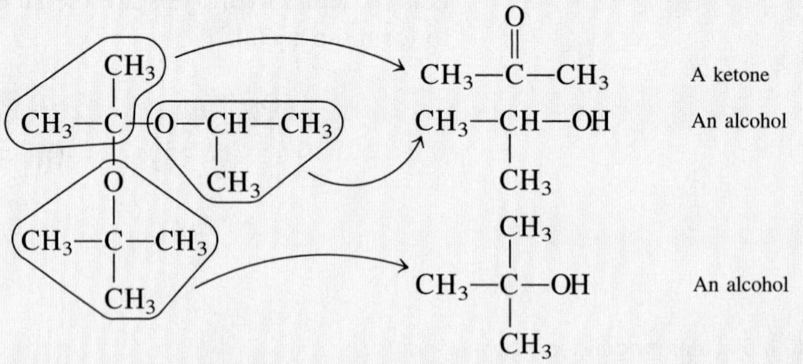

CH_3—C—CH_3   A ketone

CH_3—CH—OH   An alcohol
        |
       CH_3

CH_3—C—OH   An alcohol
        |
       CH_3

Practice Exercise 4.5

Draw the structures of the aldehyde (or ketone) and the two alcohols produced when the following acetal undergoes hydrolysis in acidic solution.

$$CH_3-CH_2-CH_2-O-\underset{\underset{CH_3}{|}}{\overset{\overset{CH_3}{|}}{C}}-O-CH_2-CH_3$$

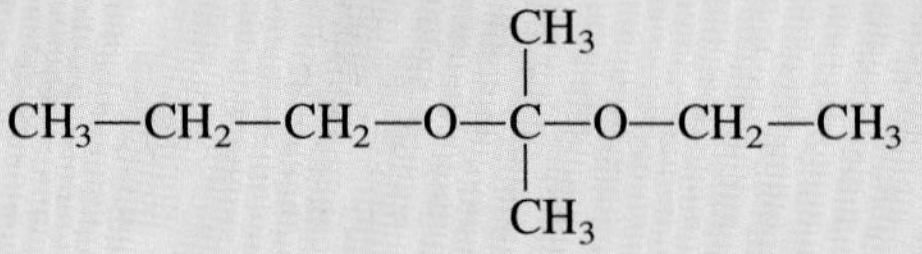

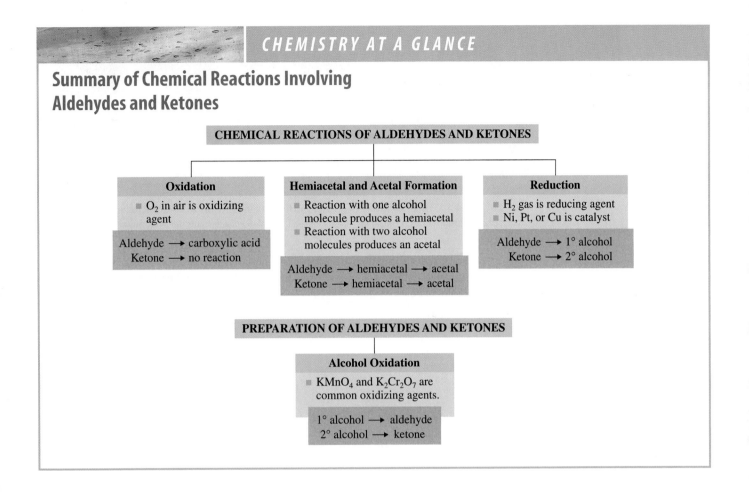

Summary of Chemical Reactions Involving Aldehydes and Ketones

CHEMICAL REACTIONS OF ALDEHYDES AND KETONES

Oxidation
- O_2 in air is oxidizing agent

Aldehyde ⟶ carboxylic acid
Ketone ⟶ no reaction

Hemiacetal and Acetal Formation
- Reaction with one alcohol molecule produces a hemiacetal
- Reaction with two alcohol molecules produces an acetal

Aldehyde ⟶ hemiacetal ⟶ acetal
Ketone ⟶ hemiacetal ⟶ acetal

Reduction
- H_2 gas is reducing agent
- Ni, Pt, or Cu is catalyst

Aldehyde ⟶ 1° alcohol
Ketone ⟶ 2° alcohol

PREPARATION OF ALDEHYDES AND KETONES

Alcohol Oxidation
- $KMnO_4$ and $K_2Cr_2O_7$ are common oxidizing agents.

1° alcohol ⟶ aldehyde
2° alcohol ⟶ ketone

■ Nomenclature for Hemiacetals and Acetals

A "descriptive" type of common nomenclature that includes the terms *hemiacetal* and *acetal* as well as the name of the carbonyl compound (aldehyde or ketone) produced in the hydrolysis of the hemiacetal or acetal is commonly used in describing such compounds. Two examples of such nomenclature are

Methyl hemiacetal of propanal   Diethyl acetal of propanone

The Chemistry at a Glance feature summarizes reactions that involve aldehydes and ketones.

4.11 Formaldehyde-Based Polymers

Many types of organic compounds can serve as reactants (monomers) for polymerization reactions, including ethylenes (Section 2.9), alcohols (Section 3.10), and carbonyl compounds.

Formaldehyde, the simplest aldehyde, is a prolific "polymer former." As representative of its polymer reactions, let us consider the reaction between formaldehyde and phenol to form a phenol–formaldehyde network polymer (see Figure 4.11). A **network polymer** *is a polymer in which monomers are connected in a three-dimensional cross-linked network.*

FIGURE 4.11 When a mixture of phenol and formaldehyde dissolved in acetic acid is treated with concentrated hydrochloric acid, a cross-linked phenol–formaldehyde network polymer is formed.

When excess formaldehyde is present, the polymerization proceeds via mono-, di-, and trisubstituted phenols that are formed as intermediates in the reaction between phenol and formaldehyde.

The substituted phenols then interact with each other by splitting out water molecules. The final product is a complex, large, three-dimensional network polymer:

FIGURE 4.12 Bakelite jewelry in use during the 1930–1950 time period.

The first synthetic plastic, Bakelite, produced in 1907, was a phenol–formaldehyde polymer. Early uses of Bakelite were in the manufacture of billiard balls and "plastic" jewelry (Figure 4.12). Modern phenol–formaldehyde polymers, called phenolics, are adhesives used in the production of plywood and particle board.

4.12 Sulfur-Containing Carbonyl Groups

The introduction of sulfur into a carbonyl group produces two different classes of compounds depending on whether the sulfur atom replaces the carbonyl oxygen atom or the carbonyl carbon atom.

Replacement of the carbonyl *oxygen* atom with sulfur produces *thiocarbonyl compounds*—thioaldehydes (thials) and thioketones (thiones)—the simplest of which are

Thioformaldehyde
(Methanethial)

Thioacetone
(Propanethione)

Thiocarbonyl compounds such as these are unstable and readily decompose.

Replacement of the carbonyl *carbon* atom with sulfur produces *sulfoxides,* compounds that are much more stable than thiocarbonyl compounds. The oxidation of a thioether (sulfide) [Section 3.21] constitutes the most common route to a sulfoxide.

$$R-S-R \xrightarrow{[O]} R-\overset{\overset{\textstyle O}{\|}}{S}-R$$

Thioether   Sulfoxide

A highly interesting sulfoxide is DMSO (dimethyl sulfoxide), a sulfur analog of acetone, the simplest ketone.

$$CH_3-\overset{\overset{\textstyle O}{\|}}{S}-CH_3 \qquad CH_3-\overset{\overset{\textstyle O}{\|}}{C}-CH_3$$

DMSO   Acetone

DMSO is an odorless liquid with unusual properties. Because of the presence of the polar sulfur–oxide bond, DMSO is miscible with water and also quite soluble in less polar organic solvents. When rubbed on the skin, DMSO has remarkable penetrating power and is quickly absorbed into the body, where it relieves pain and inflammation. For many years it has been heralded as a "miracle drug" for arthritis, sprains, burns, herpes, infections, and high blood pressure. However, the FDA has steadfastly refused to approve it for general medical use. For example, the FDA says that DMSO's powerful penetrating action could cause an insecticide on a gardener's skin to be carried accidentally into his or her bloodstream. Another complication is that DMSO is reduced in the body to dimethyl sulfide, a compound with a strong garlic-like odor that soon appears on the breath.

$$CH_3-\overset{\overset{\textstyle O}{\|}}{S}-CH_3 \xrightarrow{\text{Reduction}} CH_3-S-CH_3$$

The FDA has approved DMSO for use in certain bladder conditions and as a veterinary drug for topical use in nonbreeding dogs and horses. For example, DMSO is used as an anti-inflammatory rub for race horses.

CONCEPTS TO REMEMBER

The carbonyl group. A carbonyl group consists of a carbon atom bonded to an oxygen atom through a double bond. Aldehydes and ketones are compounds that contain a carbonyl functional group. The carbonyl carbon in an aldehyde has at least one hydrogen attached to it, and the carbonyl carbon in a ketone has no hydrogens attached to it (Sections 4.1 and 4.2).

Nomenclature of aldehydes and ketones. The IUPAC names of aldehydes and ketones are based on the longest carbon chain that contains the carbonyl group. The chain numbering is done from the end that results in the lowest number for the carbonyl group. The names of aldehydes end in -*al,* those of ketones in -*one* (Sections 4.3 and 4.4).

Isomerism for aldehydes and ketones. Constitutional isomerism is possible for aldehydes and for ketones when four or more carbon atoms are present. Aldehydes and ketones with the same number of carbon atoms and the same degree of saturation have the same molecular formula and thus are functional group isomers of each other (Section 4.5).

Physical properties of aldehydes and ketones. The boiling points of aldehydes and ketones are intermediate between those of alcohols and alkanes. The polarity of the carbonyl groups enables aldehyde and ketone molecules to interact with each other through dipole–dipole interactions. They cannot, however, hydrogen-bond to each other. Lower-molecular-mass aldehydes and ketones are soluble in water (Section 4.7).

Preparation of aldehydes and ketones. Oxidation of primary and secondary alcohols, using mild oxidizing agents, produces aldehydes and ketones, respectively (Section 4.8).

Oxidation and reduction of aldehydes and ketones. Aldehydes are easily oxidized to carboxylic acids; ketones do not readily undergo oxidation. Reduction of aldehydes and ketones produces primary and secondary alcohols, respectively (Section 4.9).

Hemiacetals and acetals. A characteristic reaction of aldehydes and ketones is the addition of an alcohol across the carbonyl double bond to produce hemiacetals. The reaction of a second alcohol molecule with a hemiacetal produces an acetal (Section 4.10).

KEY REACTIONS AND EQUATIONS

1. Oxidation of an aldehyde to give a carboxylic acid (Section 4.9)

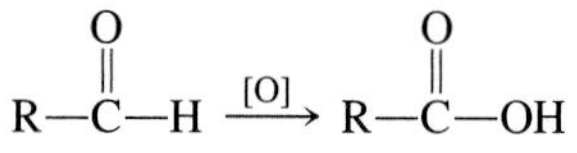

$$R-\overset{O}{\overset{\|}{C}}-H \xrightarrow{[O]} R-\overset{O}{\overset{\|}{C}}-OH$$

2. Attempted oxidation of a ketone (Section 4.9)

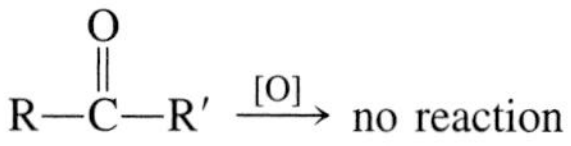

$$R-\overset{O}{\overset{\|}{C}}-R' \xrightarrow{[O]} \text{no reaction}$$

3. Reduction of an aldehyde to give a primary alcohol (Section 4.9)

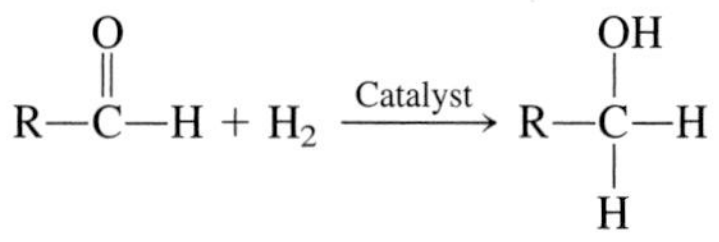

$$R-\overset{O}{\overset{\|}{C}}-H + H_2 \xrightarrow{\text{Catalyst}} R-\overset{OH}{\underset{H}{\overset{|}{C}}}-H$$

4. Reduction of a ketone to give a secondary alcohol (Section 4.9)

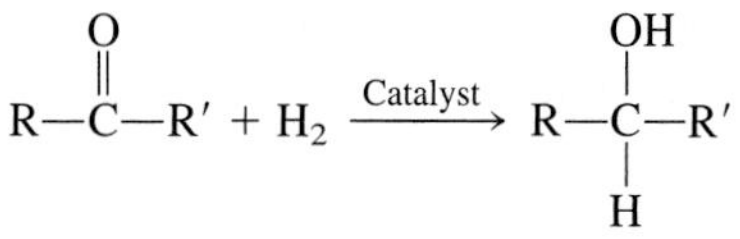

$$R-\overset{O}{\overset{\|}{C}}-R' + H_2 \xrightarrow{\text{Catalyst}} R-\overset{OH}{\underset{H}{\overset{|}{C}}}-R'$$

5. Addition of an alcohol to an aldehyde to form a hemiacetal and then an acetal (Section 4.10)

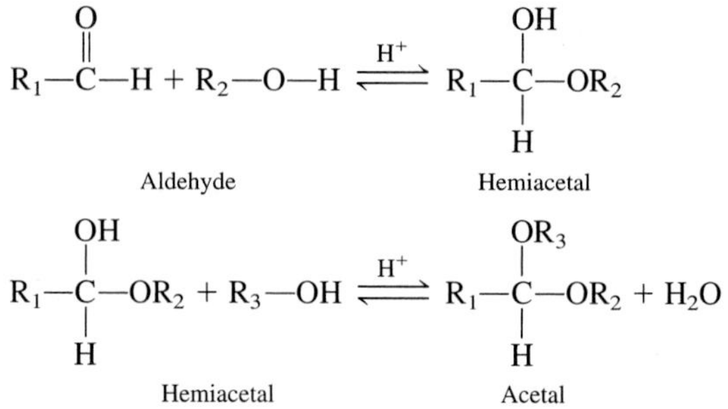

$$R_1-\overset{O}{\overset{\|}{C}}-H + R_2-O-H \overset{H^+}{\rightleftharpoons} R_1-\underset{H}{\overset{OH}{\overset{|}{\underset{|}{C}}}}-OR_2$$

Aldehyde   Hemiacetal

$$R_1-\underset{H}{\overset{OH}{\overset{|}{\underset{|}{C}}}}-OR_2 + R_3-OH \overset{H^+}{\rightleftharpoons} R_1-\underset{H}{\overset{OR_3}{\overset{|}{\underset{|}{C}}}}-OR_2 + H_2O$$

Hemiacetal   Acetal

6. Hydrolysis of an acetal to yield an aldehyde and two alcohols (Section 4.10)

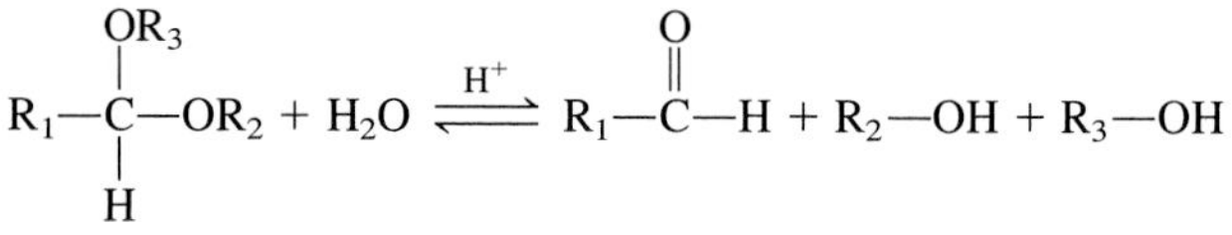

$$R_1-\underset{H}{\overset{OR_3}{\overset{|}{\underset{|}{C}}}}-OR_2 + H_2O \overset{H^+}{\rightleftharpoons} R_1-\overset{O}{\overset{\|}{C}}-H + R_2-OH + R_3-OH$$

EXERCISES AND PROBLEMS

The members of each pair of problems in this section test similar material.

■ The Carbonyl Group (Section 4.1)

4.1   Indicate which of the following compounds contain a carbonyl group.

a. 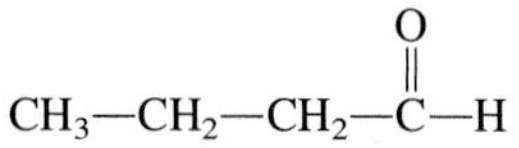 $CH_3-CH_2-CH_2-\overset{O}{\overset{\|}{C}}-H$ b. CH_3-O-CH_3

c. 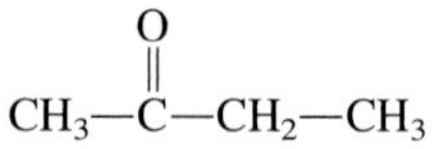 $CH_3-\overset{O}{\overset{\|}{C}}-CH_2-CH_3$ d. 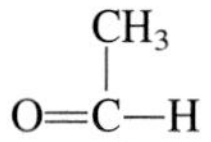 $O=\overset{CH_3}{\overset{|}{C}}-H$

e. $CH_3-O-CH_2-O-CH_3$ f. 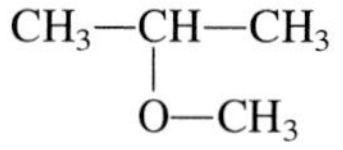 $CH_3-\overset{CH_3}{\underset{O-CH_3}{\overset{|}{\underset{|}{CH}}}}-CH_3$

4.2   Indicate which of the following compounds contain a carbonyl group.

a. 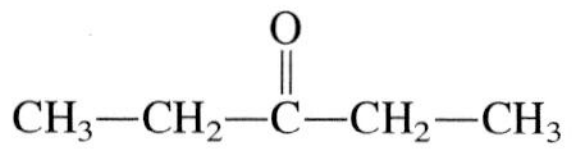 $CH_3-CH_2-\overset{O}{\overset{\|}{C}}-CH_2-CH_3$

b. $CH_3-CH_2-O-CH_2-CH_3$

c. $CH_3-CH_2-CH_2-OH$ d. 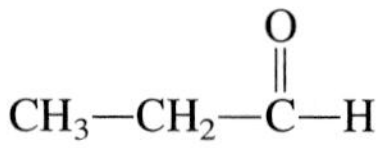 $CH_3-CH_2-\overset{O}{\overset{\|}{C}}-H$

e. $CH_3-O-CH_2-CH_2-OH$

f. 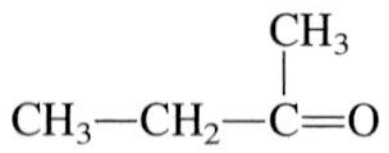 $CH_3-CH_2-\overset{CH_3}{\overset{|}{C}}=O$

4.3   What are the similarities and differences between the bonding in a carbon–oxygen double bond and that in a carbon–carbon double bond?

4.4   Use δ^+ and δ^- notation to show the polarity in a carbon–oxygen double bond.

■ Structure of Aldehydes and Ketones (Section 4.2)

4.5   Classify each of the following structures as an aldehyde, a ketone, or neither.

a. 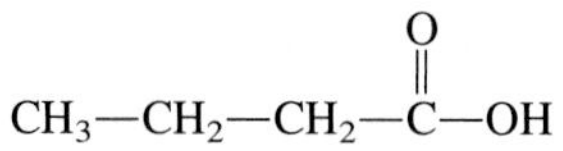 $CH_3-CH_2-CH_2-\overset{O}{\overset{\|}{C}}-OH$

b. $CH_3-CH_2-CH_2-CH_2-\overset{O}{\overset{\|}{C}}-H$

c. $CH_3-\overset{O}{\overset{\|}{C}}-CH_3$

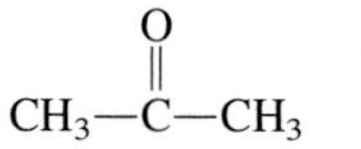

d. $CH_3-O-CH_2-CH_3$

e. 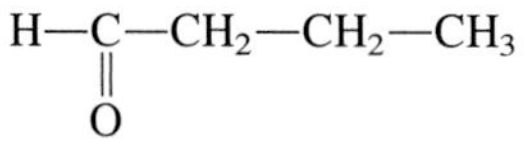 $H-\underset{O}{\overset{\|}{C}}-CH_2-CH_2-CH_3$

f. CH_3-CHO

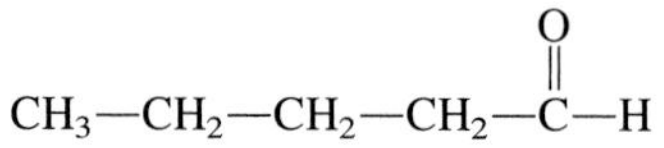

4.6   Classify each of the following structures as an aldehyde, a ketone, or neither.

a. 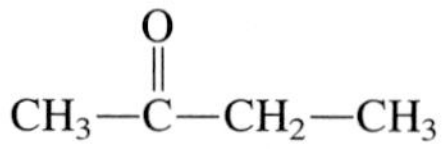 $CH_3-\overset{O}{\overset{\|}{C}}-CH_2-CH_3$

b. $CH_3-CH_2-CH_2-\overset{O}{\overset{\|}{C}}-O-CH_3$

c. $CH_3-CH_2-\overset{CH_3}{\underset{CH_3}{\overset{|}{C}}}=O$

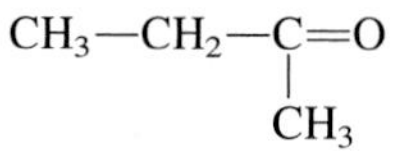

d. 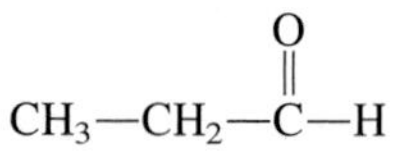 $CH_3-CH_2-\overset{O}{\overset{\|}{C}}-H$

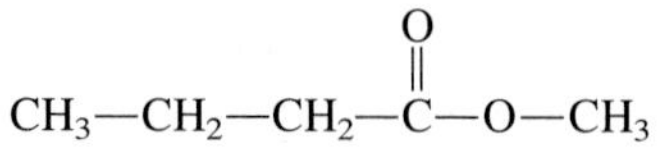

e.

$$CH_3-CH_2-CH-\overset{\displaystyle O}{\overset{\|}{C}}-H$$
$$\quad\quad\quad\quad | $$
$$\quad\quad\quad\quad CH_3$$

f.

$$\quad\quad\quad CH_3$$
$$\quad\quad\quad | $$
$$CH_3-\overset{|}{\underset{|}{C}}-CH_2-CHO$$
$$\quad\quad\quad CH_3$$

4.7 Draw the structures of the two simplest aldehydes and the two simplest ketones.

4.8 One- and two-carbon ketones do not exist. Explain why.

4.9 Classify each of the following structures as an aldehyde, a ketone, or neither.

a.

OH
CH_3 (cyclohexane ring)

b.

$$\overset{\displaystyle O}{\overset{\|}{C}}-H$$ (benzene ring)

c.

$$\overset{\displaystyle O}{\overset{\|}{C}}-O-CH_3$$ (benzene ring)

d.

O (cyclohexanone ring)

e.

$$\overset{\displaystyle O}{\overset{\|}{C}}-CH_3$$ (benzene ring)

f.

$$\overset{\displaystyle O}{\overset{\|}{C}}-H$$
CH_3 (cyclohexane ring)

4.10 Classify each of the following structures as an aldehyde, a ketone, or neither.

a.

$$\overset{\displaystyle O}{\overset{\|}{C}}-CH_2-CH_3$$ (cyclohexane ring)

b.

$$\overset{\displaystyle O}{\overset{\|}{C}}-OH$$ (benzene ring)

c.

$$\overset{\displaystyle O}{\overset{\|}{C}}-H$$ (cyclohexane ring)

d.

O (cyclohexane ring with two O)

e.

$$\overset{\displaystyle O}{\overset{\|}{C}}-O-CH_3$$ (benzene ring)

f.

$$\overset{\displaystyle O}{\overset{\|}{C}}-CH_3$$
CH_3   CH_3 (benzene ring)

▨ Nomenclature for Aldehydes (Section 4.3)

4.11 Assign an IUPAC name to each of the following aldehydes.

a.

$$CH_3-CH_2-CH_2-\overset{\displaystyle O}{\overset{\|}{C}}-H$$

b.

$$CH_3-CH_2-CH-\overset{\displaystyle O}{\overset{\|}{C}}-H$$
$$\quad\quad\quad\quad | $$
$$\quad\quad\quad\quad CH_3$$

c.

$$CH_3-CH-CH_2-CH_2-\overset{\displaystyle O}{\overset{\|}{C}}-H$$
$$\quad\quad | $$
$$\quad\quad CH_2-CH_2-CH_3$$

d.

$$-CH_2-CH_2-\overset{\displaystyle O}{\overset{\|}{C}}-H$$ (benzene ring)

e. CH_3-CH_2-CHO

f.

$$\quad\quad\quad CH_3\quad\quad O$$
$$\quad\quad\quad | \quad\quad\quad \|$$
$$CH_3-\overset{|}{\underset{|}{C}}-CH_2-\overset{}{C}-H$$
$$\quad\quad\quad CH_3$$

4.12 Assign an IUPAC name to each of the following aldehydes.

a.

$$CH_3-CH-CH_2-CH_2-\overset{\displaystyle O}{\overset{\|}{C}}-H$$
$$\quad\quad | $$
$$\quad\quad CH_3$$

b.

$$CH_3-CH_2-CH-\overset{\displaystyle O}{\overset{\|}{C}}-H$$
$$\quad\quad\quad\quad | $$
$$\quad\quad\quad\quad CH_2$$
$$\quad\quad\quad\quad | $$
$$\quad\quad\quad\quad CH_3$$

c.

$$CH_2-CH_2-CH-CH_2-\overset{\displaystyle O}{\overset{\|}{C}}-H$$
$$| \quad\quad\quad\quad\quad | $$
$$CH_3\quad\quad\quad CH_2-CH_2-CH_3$$

d.

$$-CH_2-\overset{\displaystyle O}{\overset{\|}{C}}-H$$ (benzene ring)

e.

$$\quad\quad\quad\quad CH_3\quad\quad O$$
$$\quad\quad\quad\quad | \quad\quad\quad \|$$
$$CH_3-CH_2-\overset{|}{\underset{|}{C}}-CH_2-\overset{}{C}-H$$
$$\quad\quad\quad\quad CH_3$$

f. $CH_3-CH_2-CH_2-CHO$

4.13 Assign an IUPAC name to each of the following aldehydes.

a.   b.

c.   d.

4.14 Assign an IUPAC name to each of the following aldehydes.

a.   b.

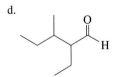

c.   d.

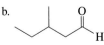

4.15 Draw a structural formula for each of the following aldehydes.
a. 3-Methylpentanal
b. 2-Ethylhexanal
c. 3,4-Dimethylheptanal
d. 2,2-Dichloropropanal
e. 2,4,5-Trimethylheptanal
f. 4-Hydroxy-2-methyloctanal

4.16 Draw a structural formula for each of the following aldehydes.
a. 2-Methylpentanal
b. 4-Ethylhexanal
c. 3,3-Dimethylhexanal
d. 2,3-Dibromopropanal
e. 2-Bromo-4-methylhexanal
f. 2,4-Dichloroheptanal

4.17 Draw a structural formula for each of the following aldehydes.
a. Formaldehyde
b. Propionaldehyde
c. Chloroacetaldehyde
d. 2-Chlorobenzaldehyde
e. o-Methylbenzaldehyde
f. 2,4-Dimethylbenzaldehyde

4.18 Draw a structural formula for each of the following aldehydes.
a. Acetaldehyde
b. Butyraldehyde
c. Dichloroacetaldehyde
d. Benzaldehyde
e. 2-Methylbenzaldehyde
f. p-Bromobenzaldehyde

4.19 Assign a common name to each of the following aldehydes.

a.
$$CH_3-CH_2-\overset{\displaystyle O}{\overset{\|}{C}}-H$$

b.
$$CH_3-CH_2-CHO$$

c.
$$\underset{\displaystyle CH_3}{\overset{}{\underset{|}{CH_2}}}-CH_2-\overset{\displaystyle O}{\overset{\|}{C}}-H$$

d.
$$Cl-\underset{\displaystyle Cl}{\underset{|}{CH}}-\overset{\displaystyle O}{\overset{\|}{C}}-H$$

e.

f.
(benzaldehyde ring with Cl and HO substituents, $\overset{O}{\overset{\|}{C}}-H$)

4.20 Assign a common name to each of the following aldehydes.

a.
$$CH_3-CH_2-CH_2-\overset{\displaystyle O}{\overset{\|}{C}}-H$$

b.
$$\underset{\displaystyle CH_2}{\overset{\displaystyle CH_3}{\underset{|}{\overset{|}{}}}}-\overset{\displaystyle O}{\overset{\|}{C}}-H$$

c.
$$CH_3-CH_2-CH_2-CHO$$

d.
$$Cl-\underset{\displaystyle Cl}{\overset{\displaystyle Cl}{\underset{|}{\overset{|}{C}}}}-\overset{\displaystyle O}{\overset{\|}{C}}-H$$

e.
(benzaldehyde ring with Br substituent, $\overset{O}{\overset{\|}{C}}-H$)

f.
(benzaldehyde ring with Br and OH substituents, $\overset{O}{\overset{\|}{C}}-H$)

■ **Nomenclature for Ketones (Section 4.4)**

4.21 Using IUPAC nomenclature, name each of the following ketones.

a.
$$CH_3-CH_2-\overset{\displaystyle O}{\overset{\|}{C}}-CH_3$$

b.
$$CH_3-\underset{\displaystyle CH_3}{\underset{|}{CH}}-\underset{\displaystyle CH_3}{\underset{|}{CH}}-\overset{\displaystyle O}{\overset{\|}{C}}-\underset{\displaystyle CH_3}{\underset{|}{CH}}-CH_3$$

c.
$$CH_3-\underset{\displaystyle CH_3}{\underset{|}{CH}}-CH_2-CH_2-\overset{\displaystyle O}{\overset{\|}{C}}-CH_2-CH_3$$

d.
$$CH_3-CH_2-CH_2 \quad \underset{\displaystyle CH_2-CH_2}{\underset{|}{CH_2}}-\overset{\displaystyle O}{\overset{\|}{C}}-CH_3$$

e.
$$Cl-CH_2-CH_2-\overset{\displaystyle O}{\overset{\|}{C}}-CH_2-CH_2-Cl$$

f.
$$CH_3-CH_2-\overset{\displaystyle O}{\overset{\|}{C}}-\underset{\displaystyle Cl}{\underset{|}{CH}}-Cl$$

4.22 Using IUPAC nomenclature, name each of the following ketones.

a.
$$CH_3-\overset{\displaystyle O}{\overset{\|}{C}}-CH_2-CH_2-CH_2-CH_3$$

b.
$$CH_3-\underset{\displaystyle CH_3}{\underset{|}{CH}}-\overset{\displaystyle O}{\overset{|}{C}}-CH_3$$

c.
$$CH_3-CH_2-\overset{\displaystyle O}{\overset{\|}{C}}-CH_2-\underset{\displaystyle CH_2}{\underset{|}{CH}}-CH_3 \quad (\text{with } CH_2-CH_3 \text{ branch})$$

d.
$$CH_3-\underset{\displaystyle Cl}{\underset{|}{CH}}-\overset{\displaystyle O}{\overset{\|}{C}}-\underset{\displaystyle Br}{\underset{|}{CH}}-CH_3$$

e.
$$CH_3-\underset{\displaystyle Cl}{\underset{|}{CH}}-\overset{\displaystyle O}{\overset{\|}{C}}-CH_2-\underset{\displaystyle Cl}{\underset{|}{CH_2}}$$

f.
$$\underset{\displaystyle CH_3}{\underset{|}{CH_2}}-CH_2-\underset{\displaystyle CH_3}{\underset{|}{CH}}-\overset{\displaystyle O}{\overset{\|}{C}}-\underset{\displaystyle CH_3}{\underset{|}{CH_2}}$$

4.23 Assign an IUPAC name to each of the following ketones.

a.
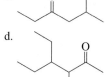

b.
(skeletal ketone structure)

c.
(skeletal ketone structure)

d.
(skeletal ketone structure)

4.24 Assign an IUPAC name to each of the following ketones.

a.
(skeletal ketone structure)

b.
(skeletal ketone structure)

c.
(skeletal ketone structure)

d.
(skeletal ketone structure)

4.25 Using IUPAC nomenclature, name each of the following ketones.

a.

b.

c.

d.

4.26 Using IUPAC nomenclature, name each of the following ketones.

a.

b.

c.

d.

4.27 Draw a structural formula for each of the following ketones.
a. 3-Methyl-2-pentanone   b. 3-Hexanone
c. Cyclobutanone   d. 2,4-Dimethyl-3-pentanone
e. Chloropropanone   f. 1,3-Dichloropropanone

4.28 Draw a structural formula for each of the following ketones.
a. 2-Methyl-3-pentanone   b. 2-Pentanone
c. 2,2-Dimethyl-4-octanone   d. Bromopropanone
e. 1,1-Dibromopropanone   f. Cyclopentanone

4.29 Draw a structural formula for each of the following ketones.
a. Diethyl ketone   b. Acetone
c. Isopropyl propyl ketone   d. Chloromethyl methyl ketone
e. Acetophenone   f. Methyl phenyl ketone

4.30 Draw a structural formula for each of the following ketones.
a. Dimethyl ketone
b. Phenyl propyl ketone
c. Methyl *tert*-butyl ketone
d. Dichloromethyl ethyl ketone
e. Benzophenone
f. Diphenyl ketone

■ **Isomerism for Aldehydes and Ketones (Section 4.5)**

4.31 Give IUPAC names for all saturated unbranched-chain compounds that are named as the following.
a. Heptanals   b. Heptanones

4.32 Give IUPAC names for all saturated unbranched-chain compounds that are named as the following.
a. Hexanals   b. Hexanones

4.33 How many aldehydes and how many ketones exist with each of the following molecular formulas?
a. CH_2O b. C_3H_6O

4.34 How many aldehydes and how many ketones exist with each of the following molecular formulas?
a. C_2H_4O b. C_4H_8O

4.35 For which values of x is the ketone name x-methyl-3-hexanone a correct IUPAC name?

4.36 For which values of x is the ketone name x-methyl-3-pentanone a correct IUPAC name?

4.37 Draw skeletal structural formulas for the four aldehydes and three ketones that have the molecular formula $C_5H_{10}O$.

4.38 Draw skeletal structural formulas for the eight aldehydes and six ketones that have the molecular formula $C_6H_{12}O$.

■ **Physical Properties of Aldehydes and Ketones (Section 4.7)**

4.39 Aldehydes and ketones have higher boiling points than alkanes of similar molecular mass. Explain why.

4.40 Aldehydes and ketones have lower boiling points than alcohols of similar molecular mass. Explain why.

4.41 How many hydrogen bonds can form between an acetone molecule and water molecules?

4.42 How many hydrogen bonds can form between an acetaldehyde molecule and water molecules?

4.43 Would you expect ethanal or octanal to be more soluble in water? Explain your answer.

4.44 Would you expect ethanal or octanal to have the more fragrant odor? Explain your answer.

■ **Preparation of Aldehydes and Ketones (Section 4.8)**

4.45 Draw the structure of the aldehyde or ketone formed from oxidation of each of the following alcohols. Assume that reaction conditions are sufficiently mild that any aldehydes produced are not oxidized further to carboxylic acids.

a. $CH_3-CH_2-CH_2-CH_2-CH_2-OH$

b. $CH_3-CH_2-\overset{\displaystyle CH-OH}{\underset{\displaystyle CH_3}{|}}$

c. $CH_3-\overset{\displaystyle CH_3}{\underset{\displaystyle CH_3}{\overset{|}{\underset{|}{C}}}}-CH_2-CH_2-OH$

d. $CH_3-CH_2-\overset{\displaystyle CH-CH_2-CH_3}{\underset{\displaystyle OH}{|}}$

e.

f. $CH_3 OH$

4.46 Draw the structure of the aldehyde or ketone formed from oxidation of each of the following alcohols. Assume that reaction conditions are sufficiently mild that any aldehydes formed are not oxidized further to carboxylic acids.

a. $CH_3-CH_2-\overset{\displaystyle CH-CH_2-OH}{\underset{\displaystyle CH_3}{|}}$

b. $CH_3-\overset{\displaystyle CH-CH-OH}{\underset{\displaystyle CH_3 CH_3}{| |}}$

c. $CH_3-\overset{\displaystyle CH_3}{\underset{\displaystyle CH_3}{\overset{|}{\underset{|}{C}}}}-OH$

d. $CH_3-CH_2-CH_2-\underset{\underset{OH}{|}}{CH}-CH_3$

e. (cyclohexane ring)—OH

f. (cyclohexane ring with CH_2-CH_3 substituent)—OH

4.47   Draw the structure of the alcohol needed to prepare each of the following aldehydes or ketones by alcohol oxidation.
   a. Ethanal   b. Diethyl ketone
   c. Phenylpropanone   d. Acetaldehyde
   e. Acetone   f. 2-Ethylhexanal

4.48   Draw the structure of the alcohol needed to prepare each of the following aldehydes or ketones by alcohol oxidation.
   a. Propanal   b. Dipropyl ketone
   c. 3-Phenyl-2-butanone   d. Chloroacetone
   e. Formaldehyde   f. Cyclohexanone

■ **Oxidation and Reduction of Aldehydes and Ketones (Section 4.9)**

4.49   Draw the structural formula of the organic product when each of the following aldehydes is oxidized to a carboxylic acid.
   a. Ethanal   b. Pentanal
   c. Formaldehyde   d. 3,4-Dichlorohexanal

4.50   Draw the structural formula of the organic product when each of the following aldehydes is oxidized to a carboxylic acid.
   a. Butanal   b. 2-Methylpentanal
   c. Acetaldehyde   d. Benzaldehyde

4.51   What are the characteristics of a positive Tollens test for aldehydes?

4.52   What are the characteristics of a positive Benedict's test for aldehydes?

4.53   What is the oxidizing agent in Benedict's solution?

4.54   What is the oxidizing agent in Tollens solution?

4.55   Which of the following compounds would react with Tollens solution?

   a. $CH_3-CH_2-CH_2-\overset{\overset{O}{\|}}{C}-CH_3$

   b. $CH_3-CH_2-CH_2-\overset{\overset{O}{\|}}{C}-H$

   c. $CH_3-\underset{\underset{OH}{|}}{CH}-CH_2-\overset{\overset{O}{\|}}{C}-H$ d. $CH_3-\underset{\underset{OH}{|}}{CH}-\overset{\overset{O}{\|}}{C}-CH_3$

4.56   Which of the following compounds would react with Benedict's solution?

   a. $CH_3-CH_2-\overset{\overset{O}{\|}}{C}-H$ b. $CH_3-\overset{\overset{O}{\|}}{C}-CH_3$

   c. $CH_3-CH_2-\underset{\underset{OH}{|}}{CH}-\overset{\overset{O}{\|}}{C}-CH_2-CH_3$

d. $CH_3-\underset{\underset{CH_3}{|}}{CH}-\underset{\underset{CH_3}{|}}{CH}-CH_2-\overset{\overset{O}{\|}}{C}-H$

4.57   Draw the structure of the major organic compound produced when each of the following compounds is reduced using molecular H_2 and a Ni catalyst.

   a. $CH_3-CH_2-CH_2-\overset{\overset{O}{\|}}{C}-H$

   b. $CH_3-CH_2-\overset{\overset{O}{\|}}{C}-CH_2-CH_3$

   c. $CH_3-\underset{\underset{CH_3}{|}}{CH}-CH_2-\overset{\overset{O}{\|}}{C}-H$

   d. $CH_3-\underset{\underset{CH_3}{|}}{CH}-\overset{\overset{O}{\|}}{C}-CH_2-CH_2-CH_3$

4.58   Draw the structure of the major organic compound produced when each of the following compounds is reduced using molecular H_2 and a Ni catalyst.

   a. $CH_3-CH_2-CH_2-\overset{\overset{O}{\|}}{C}-CH_3$

   b. $CH_3-CH_2-CH_2-CH_2-\overset{\overset{O}{\|}}{C}-H$

   c. $CH_3-\underset{\underset{CH_3}{|}}{\overset{\overset{CH_3}{|}}{C}}-CH_2-CH_2-\overset{\overset{O}{\|}}{C}-H$

   d. $CH_3-CH_2-\overset{\overset{O}{\|}}{C}-CH_3$

■ **Hemiacetal Formation (Section 4.10)**

4.59   When an alcohol molecule (R—O—H) adds across a carbon–oxygen double bond, into what "fragments" is the alcohol split?

4.60   When an alcohol molecule (R—O—H) adds across a carbon–oxygen double bond, which part of the alcohol molecule adds to the carbonyl oxygen atom?

4.61   Indicate whether each of the following compounds is a hemiacetal.
   a. $CH_3-CH_2-O-CH_3$
   b. CH_3-O
      $\quad\ \ CH_3-CH-OH$
   c. $CH_3-O-CH_2-CH_2-OH$
   d. $\qquad OH$
      $CH_3-\underset{\underset{O-CH_3}{|}}{\overset{\overset{|}{C}}{C}}-CH_3$

   e. (tetrahydrofuran ring)—OH   f. (tetrahydrofuran ring)—O—CH_3

4.62   Indicate whether each of the following compounds is a hemiacetal.

a.

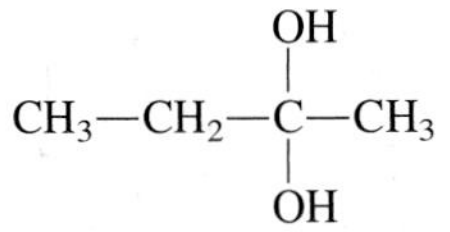

$$CH_3-CH_2-\overset{\overset{\displaystyle OH}{|}}{\underset{\underset{\displaystyle OH}{|}}{C}}-CH_3$$

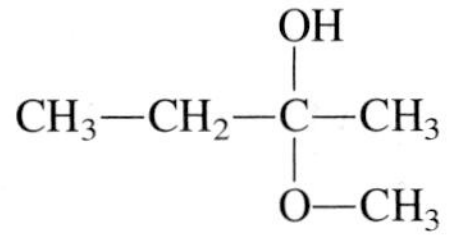

b.

$$CH_3-CH_2-\overset{\overset{\displaystyle OH}{|}}{\underset{\underset{\displaystyle O-CH_3}{|}}{C}}-CH_3$$

c.

$$CH_3-CH_2-\overset{\overset{\displaystyle O-CH_3}{|}}{\underset{\underset{\displaystyle O-CH_3}{|}}{C}}-CH_3$$

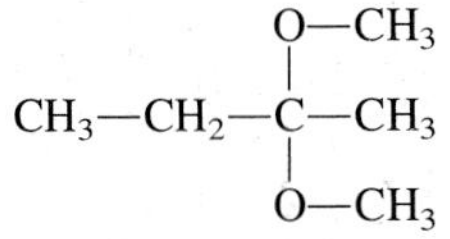

d. $CH_3-CH_2-O-CH_3$

e. 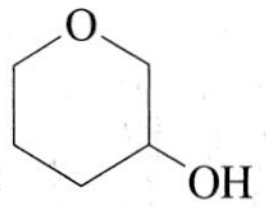 (tetrahydropyran ring with O in ring and OH substituent)

f. 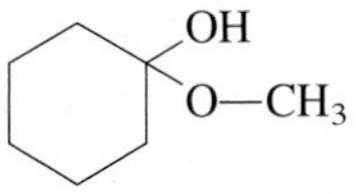 (cyclohexane ring with OH and O—CH₃ substituents)

4.63 Draw the structural formula of the hemiacetal formed from each of the following pairs of reactants.
   a. Acetaldehyde and ethyl alcohol
   b. 2-Pentanone and methanol
   c. Butanal and ethanol
   d. Acetone and isopropyl alcohol

4.64 Draw the structural formula of the hemiacetal formed from each of the following pairs of reactants.
   a. Acetaldehyde and methanol
   b. 2-Pentanone and ethyl alcohol
   c. Butanal and isopropyl alcohol
   d. Acetone and ethanol

4.65 Draw the structural formula of the missing compound in each of the following reactions.

a. 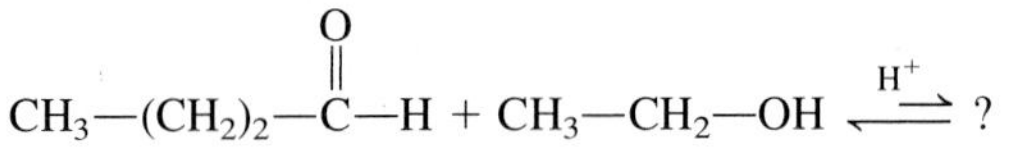
$$CH_3-(CH_2)_2-\overset{\overset{\displaystyle O}{||}}{C}-H + CH_3-CH_2-OH \overset{H^+}{\rightleftharpoons} ?$$

b.
$$? + CH_3-OH \overset{H^+}{\rightleftharpoons} CH_3-CH_2-\overset{\overset{\displaystyle OH}{|}}{\underset{\underset{\displaystyle O-CH_3}{|}}{C}}H$$

c.
$$CH_3-CH_2-\overset{\overset{\displaystyle O}{||}}{C}-CH_3 + CH_3-OH \overset{H^+}{\rightleftharpoons} ?$$

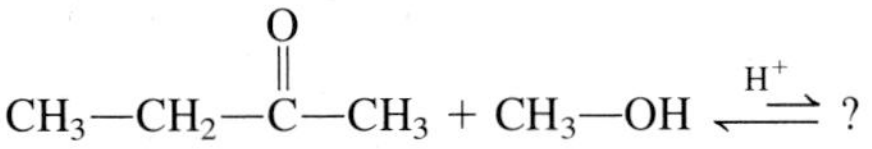

d. 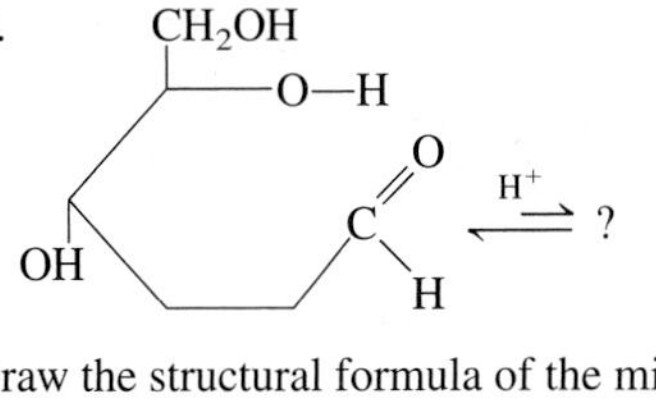 (cyclic structure with CH₂OH, O—H, OH groups and C=O aldehyde) $\overset{H^+}{\rightleftharpoons} ?$

4.66 Draw the structural formula of the missing compound in each of the following reactions.

a. 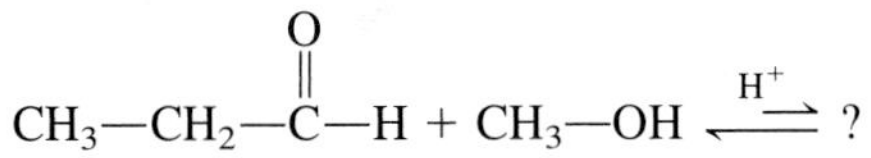
$$CH_3-CH_2-\overset{\overset{\displaystyle O}{||}}{C}-H + CH_3-OH \overset{H^+}{\rightleftharpoons} ?$$

b.
$$? + CH_3-CH_2-OH \overset{H^+}{\rightleftharpoons} CH_3-CH_2-\overset{\overset{\displaystyle OH}{|}}{\underset{\underset{\displaystyle O-CH_2-CH_3}{|}}{C}}H$$

c.
$$CH_3-CH_2-CH_2-\overset{\overset{\displaystyle O}{||}}{C}-CH_3 + CH_3-CH_2-OH \overset{H^+}{\rightleftharpoons} ?$$

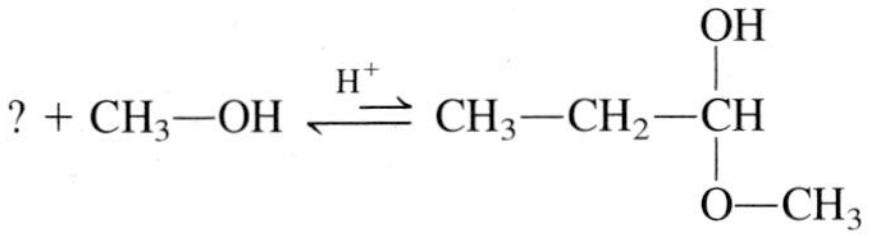

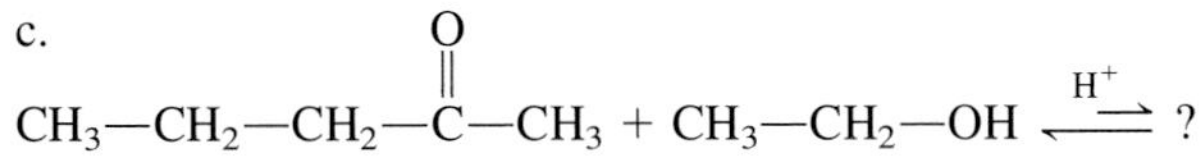

d. 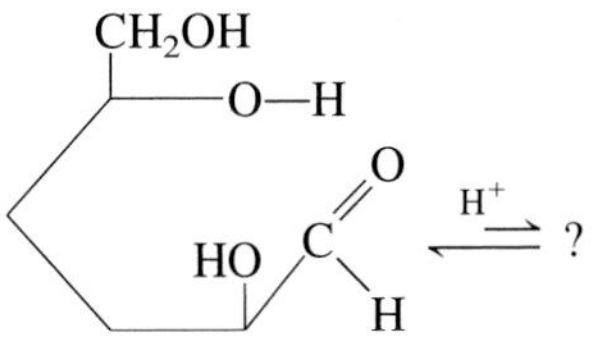 (cyclic structure with CH₂OH, O—H, and HO—C—H aldehyde) $\overset{H^+}{\rightleftharpoons} ?$

■ Acetal Formation (Section 4.10)

4.67 Indicate whether each of the following compounds is an acetal.

a. 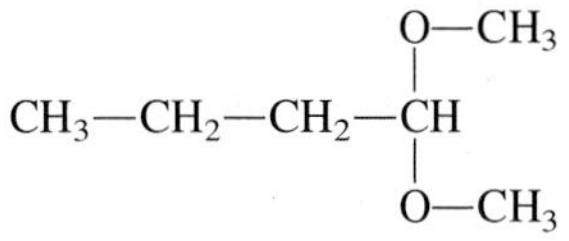
$$CH_3-CH_2-CH_2-\overset{\overset{\displaystyle O-CH_3}{|}}{\underset{\underset{\displaystyle O-CH_3}{|}}{C}}H$$

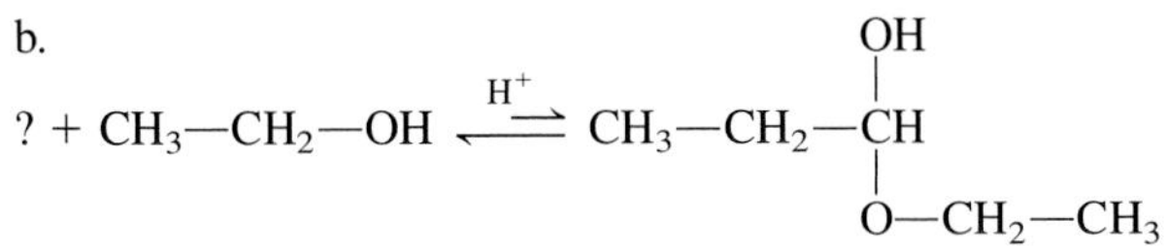

b.
$$CH_3-CH_2-CH_2-O-\overset{\overset{\displaystyle O-CH_3}{|}}{\underset{\underset{\displaystyle CH_3}{|}}{C}}-CH_3$$

c.
$$CH_3-CH_2-CH_2-O-\overset{\overset{\displaystyle CH_3}{|}}{\underset{\underset{\displaystyle CH_3}{|}}{C}}-OH$$

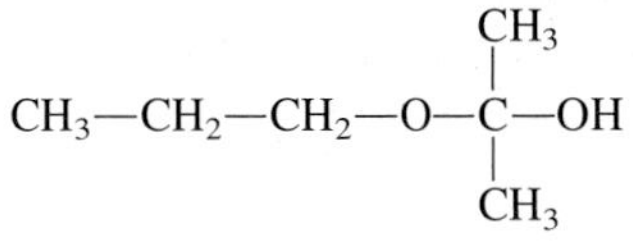

d. 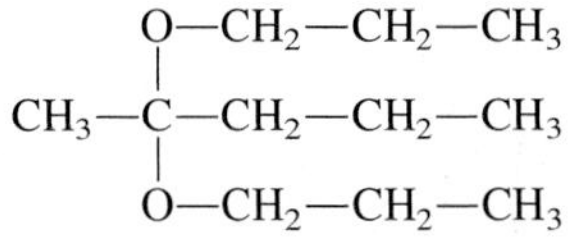
$$CH_3-\overset{\overset{\displaystyle O-CH_2-CH_2-CH_3}{|}}{\underset{\underset{\displaystyle O-CH_2-CH_2-CH_3}{|}}{C}}-CH_2-CH_2-CH_3$$

4.68 Indicate whether each of the following compounds is an acetal.

a. 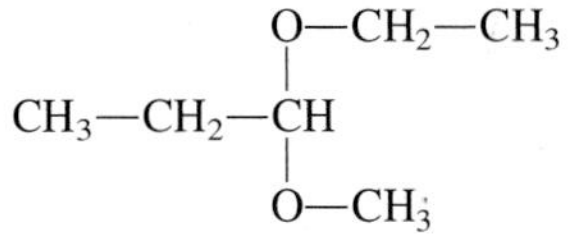
$$CH_3-CH_2-\overset{\overset{\displaystyle O-CH_2-CH_3}{|}}{\underset{\underset{\displaystyle O-CH_3}{|}}{C}}H$$

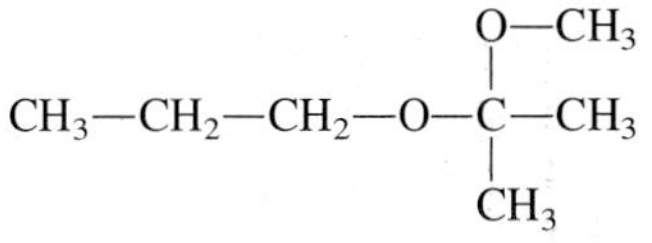

b.
$$CH_3-CH_2-O-\overset{\overset{\displaystyle O-CH_3}{|}}{\underset{\underset{\displaystyle CH_3}{|}}{C}}-CH_3$$

c.
$$CH_3-CH_2-CH_2-\overset{\overset{\displaystyle H}{|}}{\underset{\underset{\displaystyle OH}{|}}{C}}-O-CH_3$$

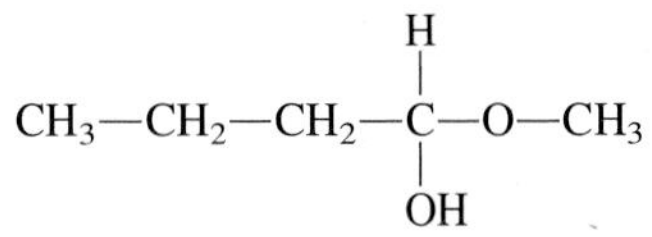

d. 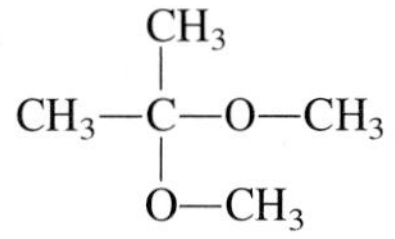
$$CH_3-\overset{\overset{\displaystyle CH_3}{|}}{\underset{\underset{\displaystyle O-CH_3}{|}}{C}}-O-CH_3$$

4.69 Draw the structural formula of the missing compound(s) in each of the following reactions.

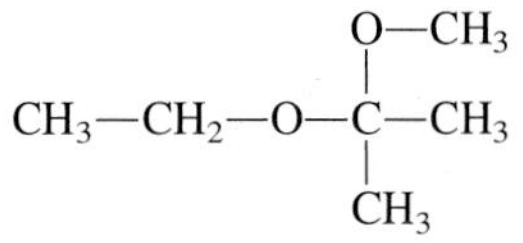

a. 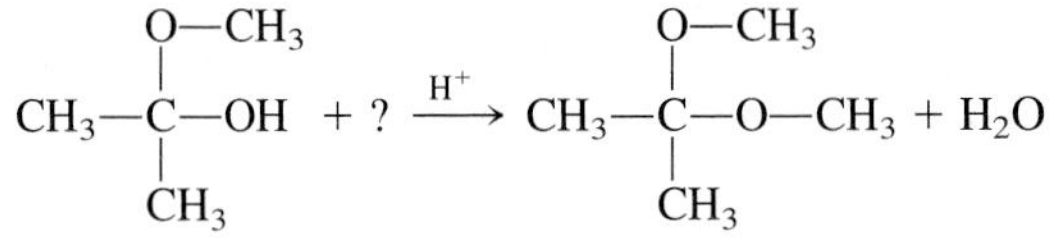
$$CH_3-\overset{\overset{\displaystyle O-CH_3}{|}}{\underset{\underset{\displaystyle CH_3}{|}}{C}}-OH + ? \overset{H^+}{\longrightarrow} CH_3-\overset{\overset{\displaystyle O-CH_3}{|}}{\underset{\underset{\displaystyle CH_3}{|}}{C}}-O-CH_3 + H_2O$$

b.
$$? + CH_3-CH_2-OH \overset{H^+}{\longrightarrow} CH_3-\overset{\overset{\displaystyle H}{|}}{\underset{\underset{\displaystyle O-CH_2-CH_3}{|}}{C}}-O-CH_3 + H_2O$$

c.
$$CH_3-CH_2-\overset{\overset{\displaystyle OH}{|}}{\underset{\underset{\displaystyle H}{|}}{C}}-O-CH_3 + CH_3-\overset{\overset{\displaystyle}{}}{\underset{\underset{\displaystyle CH_3}{|}}{CH}}-OH \overset{H^+}{\longrightarrow}$$
$$? + H_2O$$

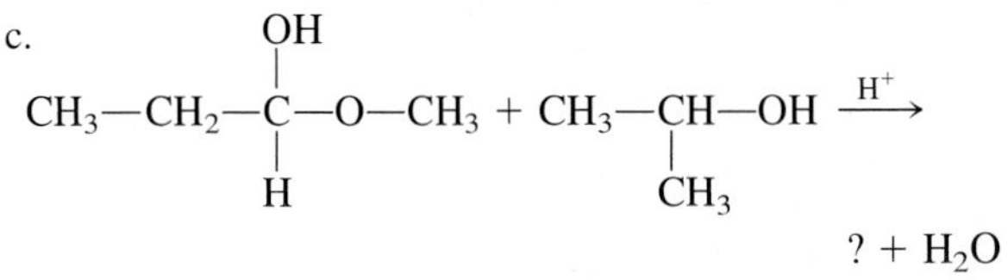

d. 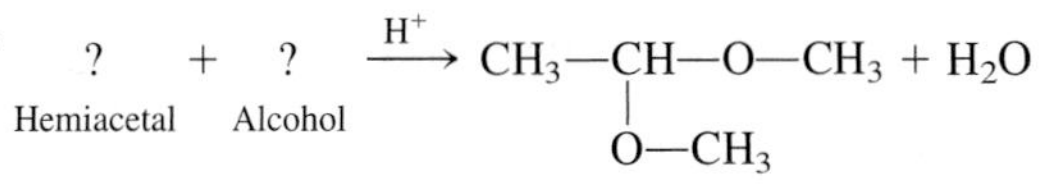
$$\underset{\text{Hemiacetal}}{?} + \underset{\text{Alcohol}}{?} \overset{H^+}{\longrightarrow} CH_3-\overset{\overset{\displaystyle}{}}{\underset{\underset{\displaystyle O-CH_3}{|}}{CH}}-O-CH_3 + H_2O$$

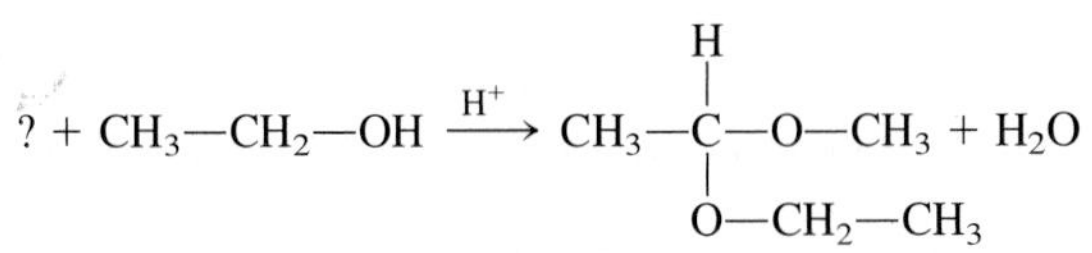

4.70 Draw the structural formula of the missing compound(s) in each of the following reactions.

a.

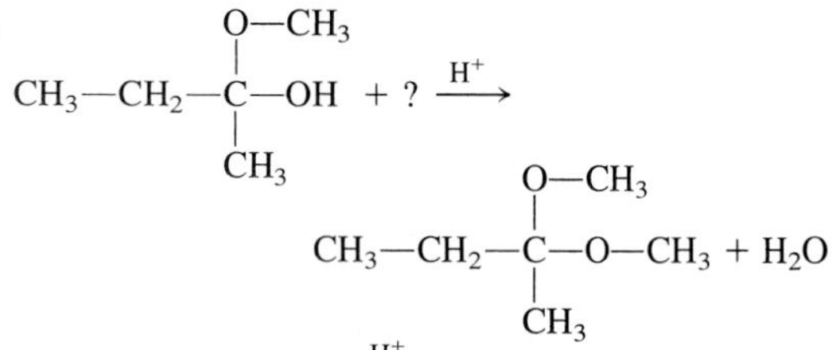

$$CH_3-CH_2-\overset{\overset{\displaystyle O-CH_3}{|}}{\underset{\underset{\displaystyle CH_3}{|}}{C}}-OH \ + \ ? \ \xrightarrow{H^+}$$

$$CH_3-CH_2-\overset{\overset{\displaystyle O-CH_3}{|}}{\underset{\underset{\displaystyle CH_3}{|}}{C}}-O-CH_3 \ + \ H_2O$$

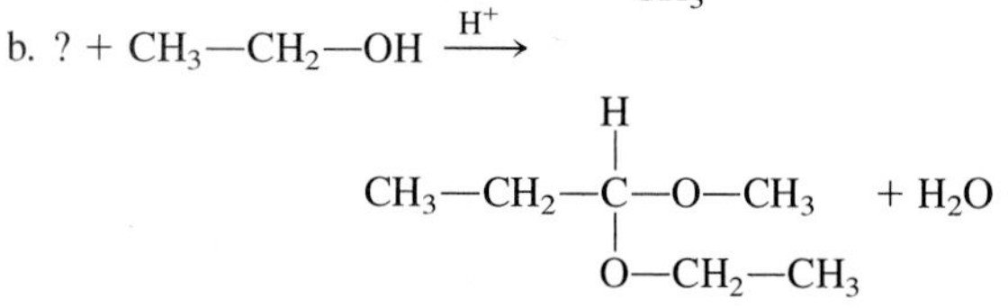

b. $? + CH_3-CH_2-OH \xrightarrow{H^+}$

$$CH_3-CH_2-\overset{\overset{\displaystyle H}{|}}{\underset{\underset{\displaystyle O-CH_2-CH_3}{|}}{C}}-O-CH_3 \ + \ H_2O$$

c.

$$CH_3-CH_2-\overset{\overset{\displaystyle OH}{|}}{\underset{\underset{\displaystyle H}{|}}{C}}-O-CH_3 + CH_3-\overset{\overset{\displaystyle }{}}{\underset{\underset{\displaystyle CH_3}{|}}{CH}}-OH \xrightarrow{H^+}$$

$$? + H_2O$$

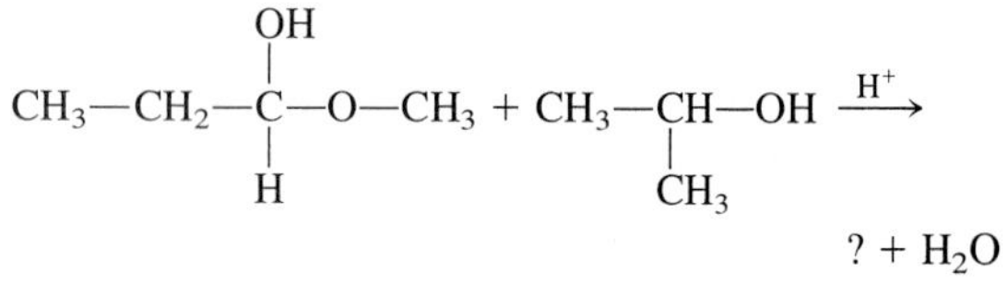

d. $? \ + \ ? \xrightarrow{H^+}$

Hemiacetal   Alcohol

$$CH_3-CH_2-\overset{\overset{\displaystyle }{}}{\underset{\underset{\displaystyle O-CH_3}{|}}{CH}}-O-CH_3 + H_2O$$

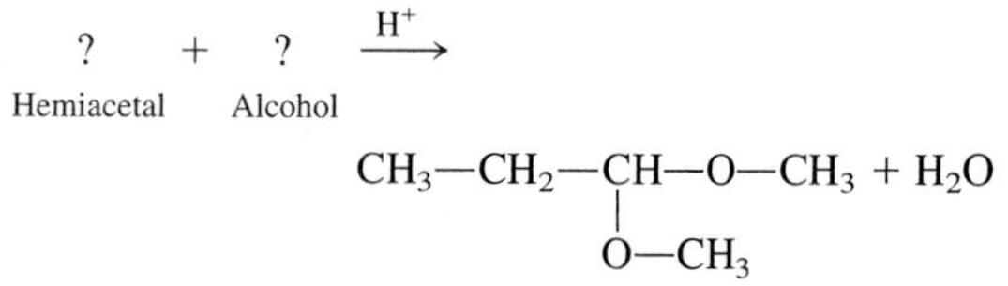

4.71 Draw the structural formulas of the aldehyde (or ketone) and the two alcohols produced when the following acetals undergo hydrolysis in acidic solution.

a.

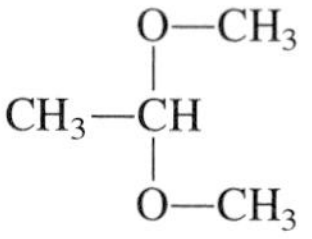

$$CH_3-\overset{\overset{\displaystyle O-CH_3}{|}}{\underset{\underset{\displaystyle O-CH_3}{|}}{CH}}$$

b.

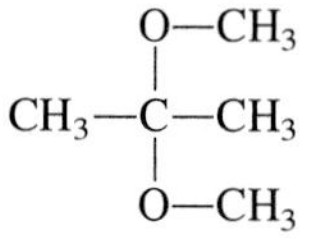

$$CH_3-\overset{\overset{\displaystyle O-CH_3}{|}}{\underset{\underset{\displaystyle O-CH_3}{|}}{C}}-CH_3$$

c.

$$CH_3-O-\overset{\overset{\displaystyle CH_2-CH_3}{|}}{\underset{\underset{\displaystyle CH_2-CH_3}{|}}{C}}-O-CH_2-CH_3$$

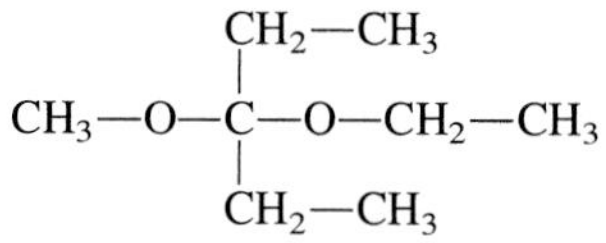

d.

$$CH_3-CH_2-CH_2-CH_2-\overset{\overset{\displaystyle O-CH_3}{|}}{\underset{\underset{\displaystyle H}{|}}{C}}-O-CH_3$$

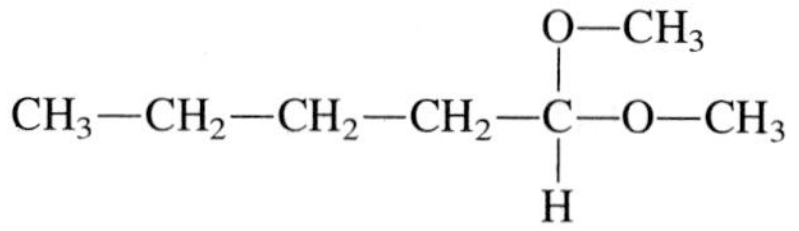

4.72 Draw the structural formulas of the aldehyde (or ketone) and the two alcohols produced when the following acetals undergo hydrolysis in acidic solution.

a.

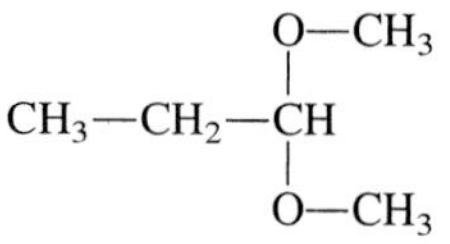

$$CH_3-CH_2-\overset{\overset{\displaystyle O-CH_3}{|}}{\underset{\underset{\displaystyle O-CH_3}{|}}{CH}}$$

b.

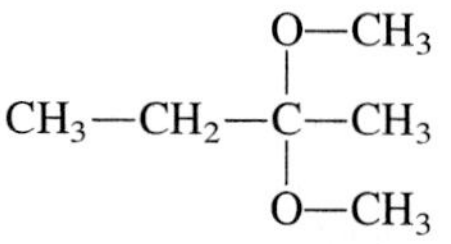

$$CH_3-CH_2-\overset{\overset{\displaystyle O-CH_3}{|}}{\underset{\underset{\displaystyle O-CH_3}{|}}{C}}-CH_3$$

c.

$$CH_3-CH_2-O-\overset{\overset{\displaystyle H}{|}}{\underset{\underset{\displaystyle CH_3}{|}}{C}}-O-CH_2-CH_3$$

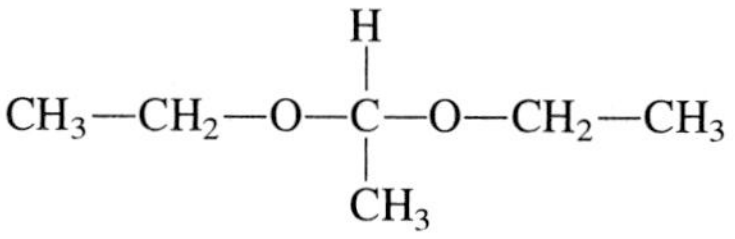

d.

$$CH_3-CH_2-\overset{\overset{\displaystyle O-CH_2-CH_3}{|}}{\underset{\underset{\displaystyle O-CH_2-CH_3}{|}}{C}}-CH_2-CH_2-CH_3$$

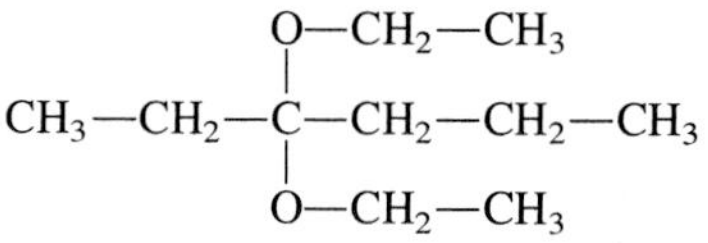

4.73 Name each of the compounds in Problem 4.71 in the manner described in Section 4.10.

4.74 Name each of the compounds in Problem 4.72 in the manner described in Section 4.10.

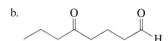

ADDITIONAL PROBLEMS

4.75 Explain each of the following.
a. The IUPAC name for the three-carbon aldehyde is propanal rather than 1-propanal.
b. The IUPAC name for the three-carbon ketone is propanone rather than 2-propanone.

4.76 Each of the following compound names represents an impossible structure. In each case, explain why.
a. Methanone
b. 1-Chlorobutanal
c. 3-Methyl-3-pentanone
d. Cyclohexanal

4.77 What is the characteristic structural feature of each of the following?
a. Hemiacetal   b. Acetal

4.78 Draw the structural formula of the hemiacetal formed and then the acetal formed when each of the following compounds reacts with an excess of the reactant alcohol.
a. Propanal and ethanol
b. Cyclohexanone and methanol

4.79 The compound 4-hydroxybutanal can form an intramolecular cyclic hemiacetal. Draw the structural formula of this cyclic hemiacetal.

4.80 Name the functional groups present in each of the following polyfunctional compounds.
a. 4-Octen-2-one
b. 2-Methoxy-4-hydroxypentanal
c. 3-Hexyn-2-one
d. 4-Oxohexanal

4.81 Indicate whether each of the following compounds would be named as an alcohol, an aldehyde, or a ketone.

a.

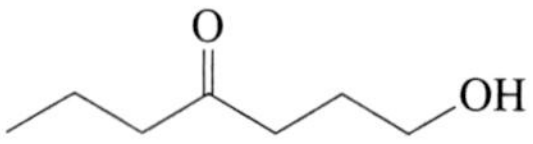

b.

c.

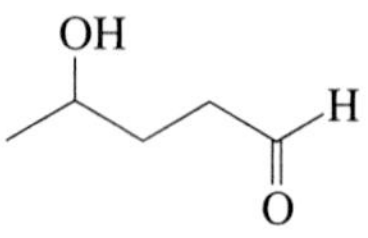

d.

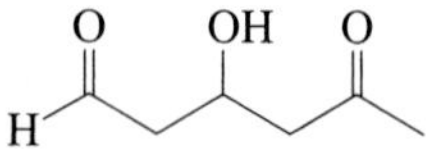

MULTIPLE-CHOICE PRACTICE TEST

4.82 Which of the following statements concerning aldehydes and ketones is correct?
a. Aldehydes contain a carbonyl group but ketones do not.
b. Ketones contain a carbonyl group but aldehydes do not.
c. Both aldehydes and ketones contain a carbonyl group.
d. Neither aldehydes nor ketones contain a carbonyl group.

4.83 Which is the IUPAC name for the ketone *ethyl propyl ketone*?
a. 3-Pentanone   b. 4-Pentanone
c. 3-Hexanone   d. 4-Hexanone

4.84 Which of the following compounds is a constitutional isomer of acetone?
a. Formaldehyde   b. Acetaldehyde
c. Propionaldehyde   d. Butyraldehyde

4.85 The physical state, at room temperature and pressure, for the simplest aldehyde and the simplest ketone is, respectively, which of the following?
a. Gas and gas   b. Gas and liquid
c. Liquid and gas   d. Liquid and liquid

4.86 For which of the following molecular combinations is hydrogen bonding possible?
a. Aldehyde–aldehyde   b. Ketone–ketone
c. Aldehyde–ketone   d. Water–ketone

4.87 A general method for the preparation of ketones is oxidation of which of the following?
a. 1° alcohols   b. 2° alcohols
c. 3° alcohols   d. Aldehydes

4.88 Which of the following reactions is classified as a reduction reaction?
a. Alcohol to ketone   b. Alcohol to aldehyde
c. Aldehyde to alcohol d. Aldehyde to carboxylic acid

4.89 In a hemiacetal, the hemiacetal carbon atom is bonded to
a. Two hydroxyl groups
b. Two alkoxy groups
c. One hydroxyl group and one alkoxy group
d. Two hydroxyl groups and one alkoxy group

4.90 To produce an acetal from a ketone, the ketone must react with which of the following?
a. One alcohol molecule
b. Two identical alcohol molecules
c. Two different alcohol molecules
d. Two alcohol molecules, which may or may not be identical

4.91 What is the number of organic product molecules produced from the complete hydrolysis of an acetal molecule?
a. Two   b. Three
c. Four   d. Five

5 Carboxylic Acids, Esters, and Other Acid Derivatives

The term *carboxyl* is a contraction of the words *carb*onyl and hydr*oxyl*.

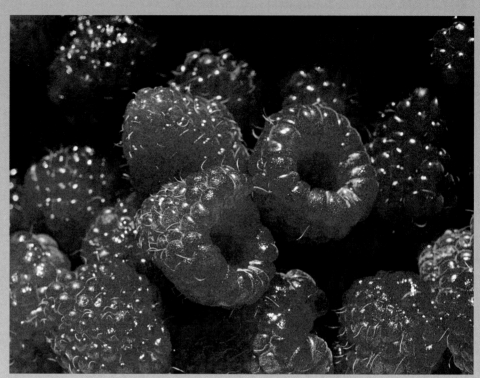

Esters, a type of carboxylic acid derivative, are largely responsible for the flavors and fragrances of ripe fruits such as red raspberries.

In Chapter 4, we discussed the carbonyl group and two families of compounds—aldehydes and ketones—that contain this group. In this chapter, we discuss four more families of compounds in which the carbonyl group is present: carboxylic acids, esters, acid chlorides, and acid anhydrides.

5.1 Structure of Carboxylic Acids and Their Derivatives

A **carboxylic acid** *is an organic compound whose functional group is the carboxyl group.* What is a carboxyl group? A **carboxyl group** *is a carbonyl group (C═O) with a hydroxyl group (—OH) bonded to the carbonyl carbon atom.* A general structural representation for a carboxyl group is

$$\begin{matrix} & O \\ & \| \\ — & C—OH \end{matrix}$$

Abbreviated linear designations for the carboxyl group are

$$—COOH \quad \text{and} \quad —CO_2H$$

Although we see within a carboxyl group both a carbonyl group (C═O) and a hydroxyl group (—OH), the carboxyl group does not show characteristic behavior of either an alcohol or a carbonyl compound (aldehyde or ketone). Rather, it is a unique functional group with a set of characteristics different from those of its component parts.

The simplest carboxylic acid has a hydrogen atom attached to the carboxyl group carbon atom.

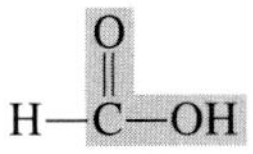

$$H-\underset{\underset{\|}{\overset{O}{\|}}}{C}-OH$$

Structures for the next two simplest carboxylic acids, those with methyl and ethyl alkyl groups, are

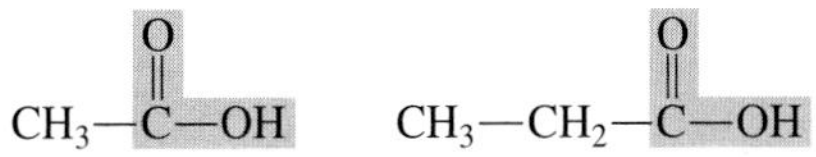

$$CH_3-\underset{\underset{\|}{\overset{O}{\|}}}{C}-OH \qquad CH_3-CH_2-\underset{\underset{\|}{\overset{O}{\|}}}{C}-OH$$

The structure of the simplest aromatic carboxylic acid involves a benzene ring to which a carboxyl group is attached.

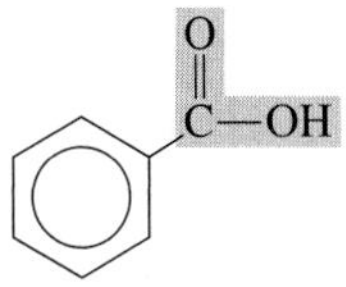

General formulas for carboxylic acids containing alkyl and aryl groups, respectively, are

R—COOH   and   Ar—COOH

Cyclic carboxylic acids do not exist; having the carboxyl carbon atom as part of a ring system creates a situation where the carboxyl carbon atom would have five bonds. The nonexistence of cyclic carboxylic acids parallels the nonexistence of cyclic aldehydes (Section 4.2).

A carboxylic acid derivative *is an organic compound that can be synthesized from or converted into a carboxylic acid.* Four important families of carboxylic acid derivatives are esters, acid chlorides, acid anhydrides, and amides. The group attached to the carbonyl carbon atom distinguishes these derivative types from each other and also from carboxylic acids.

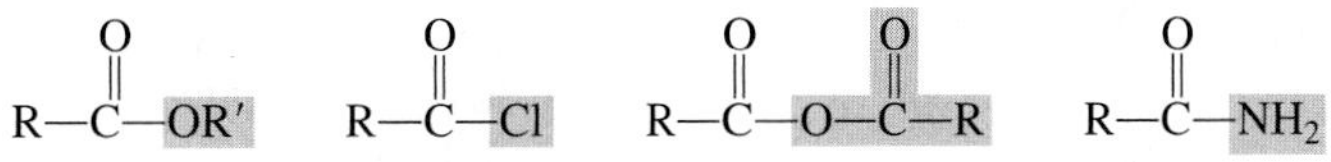

R—C—OR'	R—C—Cl	R—C—O—C—R	R—C—NH₂
Ester	Acid chloride	Acid anhydride	Amide

Further information about the first three of these four families of carboxylic acid derivatives is found in later sections of this chapter. Consideration of amides, which are nitrogen-containing compounds, will be found in Chapter 6.

5.2 IUPAC Nomenclature for Carboxylic Acids

IUPAC rules for naming carboxylic acids resemble those for naming aldehydes (Section 4.3).

■ Monocarboxylic Acids

A **monocarboxylic acid** *is a carboxylic acid in which one carboxyl group is present.* IUPAC rules for naming such compounds are:

1. Select as the parent carbon chain the longest carbon chain that *includes* the carbon atom of the carboxyl group.
2. Name the parent chain by changing the *-e* ending of the corresponding alkane to *-oic acid.*
3. Number the parent chain by assigning the number 1 to the carboxyl carbon atom.
4. Determine the identity and location of any substituents in the usual manner, and append this information to the front of the parent chain name.

A carboxyl group must occupy a terminal (end) position in a carbon chain because there can be only one other bond to it.

Space-filling models for the three simplest carboxylic acids—methanoic acid, ethanoic acid, and propanoic acid—are shown in Figure 5.1.

FIGURE 5.1 Space-filling models for the three simplest carboxylic acids: methanoic acid, ethanoic acid, and propanoic acid.

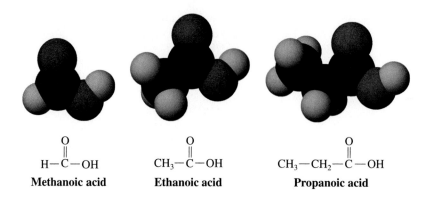

$$H-\overset{\overset{\displaystyle O}{\|}}{C}-OH \qquad CH_3-\overset{\overset{\displaystyle O}{\|}}{C}-OH \qquad CH_3-CH_2-\overset{\overset{\displaystyle O}{\|}}{C}-OH$$

Methanoic acid   **Ethanoic acid**   **Propanoic acid**

EXAMPLE 5.1

Determining IUPAC Names for Carboxylic Acids

Line-angle formulas for the simpler unbranched-chain carboxylic acids:

Methanoic acid

Ethanoic acid

Propanoic acid

Butanoic acid

The carboxyl functional group has the highest priority in the IUPAC naming system of all functional groups considered so far. When both a carboxyl group and a carbonyl group (aldehyde, ketone) are present in the same molecule, the prefix *oxo-* is used to denote the carbonyl group.

4-Oxobutanoic acid

■ Assign IUPAC names to the following carboxylic acids.

a. $CH_3-CH_2-CH_2-CH_2-\overset{\overset{\displaystyle O}{\|}}{C}-OH$

b.

c. $CH_3-\overset{}{CH}-\overset{}{CH}-\overset{\overset{\displaystyle O}{\|}}{C}-OH$ with Br and CH₂CH₃

Solution

a. The parent chain name is based on pentane. Removing the *-e* ending from pentane and replacing it with the ending *-oic acid* gives *pentanoic acid*. The location of the carboxyl group need not be specified, because by definition the carboxyl carbon atom is always carbon 1.

b. The parent chain name is *butanoic acid*. To locate the methyl group substituent, we number the carbon chain beginning with the carboxyl carbon atom. The complete name of the acid is *3-methylbutanoic acid*.

c. The longest carboxyl-carbon-containing chain has four carbon atoms. The parent chain name is thus *butanoic acid*. There are two substituents present, an ethyl group on carbon 2 and a bromo group on carbon 3. The complete name is *3-bromo-2-ethylbutanoic acid*.

Practice Exercise 5.1

Assign IUPAC names to the following carboxylic acids.

a.   **b.**   **c.**

■ Dicarboxylic Acids

A **dicarboxylic acid** *is a carboxylic acid that contains two carboxyl groups, one at each end of a carbon chain.* Saturated acids of this type are named by appending the suffix *-dioic acid* to the corresponding alkane name (the *-e* is retained to facilitate pronunciation). Both carboxyl carbon atoms must be part of the parent carbon chain, and the carboxyl locations need not be specified with numbers because they will always be at the two ends of the chain.

Pentanedioic acid

2-Methylbutanedioic acid

■ Aromatic Carboxylic Acids

The simplest aromatic carboxylic acid is called benzoic acid (Figure 5.2).

Benzoic acid

Other simple aromatic acids are named as derivatives of benzoic acid.

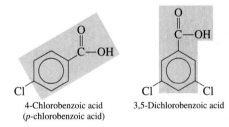

4-Chlorobenzoic acid
(*p*-chlorobenzoic acid)

3,5-Dichlorobenzoic acid

In substituted benzoic acids, the ring carbon atom bearing the carboxyl group is always carbon 1.

5.3 Common Names for Carboxylic Acids

The use of common names is more prevalent for carboxylic acids than for any other family of organic compounds. Because of their abundance in nature, carboxylic acids were among the earliest classes of organic compounds to be studied, and they acquired names before the advent of the IUPAC naming system. These common names are usually derived from some Latin or Greek word that is related to a source for the acid.

■ Monocarboxylic Acids

Table 5.1 gives the common names for the first six unbranched monocarboxylic acids. The stinging sensation associated with red ant bites is due in part to formic acid (Latin, *formica,* "ant"). Acetic acid gives vinegar its tartness (sour taste); vinegar is a 4%–8% (v/v) acetic acid solution (Latin, *acetum,* "sour"). Propionic acid is the smallest acid that can be obtained from fats (Greek, *protos,* "first," and *pion,* "fat"). Rancid butter contains

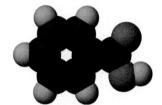

FIGURE 5.2 Space-filling model for benzoic acid, the simplest aromatic carboxylic acid.

Methyl benzoic acids go by the name *toluic acid.* (This situation parallels methyl benzene being called toluene.)

o-Toluic acid

The common names of monocarboxylic acids are the basis for aldehyde common names (Section 4.3).

C_1:   formic acid and formaldehyde
C_2:   acetic acid and acetaldehyde
C_3:   propionic acid and propionaldehyde
C_4:   butyric acid and butyraldehyde

TABLE 5.1
Common Names for the First Six Monocarboxylic Acids

Length of Carbon Chain	Structural Formula	Common Name[a]	IUPAC Name
C_1 monoacid	H—COOH	formic acid	methanoic acid
C_2 monoacid	CH_3—COOH	acetic acid	ethanoic acid
C_3 monoacid	CH_3—CH_2—COOH	propionic acid	propanoic acid
C_4 monoacid	CH_3—$(CH_2)_2$—COOH	butyric acid	butanoic acid
C_5 monoacid	CH_3—$(CH_2)_3$—COOH	valeric acid	pentanoic acid
C_6 monoacid	CH_3—$(CH_2)_4$—COOH	caproic acid	hexanoic acid

[a] The mnemonic "*Frogs are polite, being very courteous*" is helpful in remembering, in order, the first letters of the common names of these six simple saturated monocarboxylic acids.

There is a connection between acetic acid and sourdough bread. The yeast used in leavening the dough for this bread is a type that cannot metabolize the sugar maltose as most yeasts do. Consequently, bacteria that thrive on maltose become abundant in the dough. These bacteria produce acetic acid and lactic acid from the maltose, and the dough becomes *sour* (acidic); hence the name *sourdough* bread.

The alpha-carbon atom in a carboxylic acid is the carbon atom to which the carboxyl group is attached. It is never the carboxyl carbon atom itself.

Greek letters are *never* used for specifying substituent location in the IUPAC system.

FIGURE 5.3 "Drug-sniffing" dogs used by narcotics agents can find hidden heroin by detecting the odor of acetic acid (vinegar odor). Acetic acid is a by-product of the final step in illicit heroin production, and trace amounts remain in the heroin.

butyric acid (Latin, *butyrum,* "butter"). Valeric acid, found in valerian root (an herb), has a strong odor (Latin, *valere,* "to be strong"). The skin secretions of goats contain caproic acid, which contributes to the odor associated with these animals (Latin, *caper,* "goats").

Acetic acid is the most widely used of all carboxylic acids. Its primary use is as an *acidulant*—a substance that gives the proper acidic conditions for a chemical reaction. In the pure state, acetic acid is a colorless liquid with a sharp odor (see Figure 5.3). Vinegar is a 4%–8% (v/v) acetic acid solution; its characteristic odor comes from the acetic acid present. Pure acetic acid is often called *glacial* acetic acid because it freezes on a moderately cold day (f.p. = 17°C), producing icy-looking crystals.

When using common names for carboxylic acids, we designate the positions (locations) of substituents by using letters of the Greek alphabet rather than numbers. The first four letters of the Greek alphabet are alpha (α), beta (β), gamma (γ), and delta (δ). The alpha-carbon atom is carbon 2, the beta-carbon atom is carbon 3, and so on.

$$\cdots\cdots C-C-C-C-\overset{\displaystyle O}{\overset{\displaystyle \|}{C}}-OH$$

IUPAC: 5   4   3   2   1
Greek letter: δ γ β α

With the Greek-letter system, the compound

$$CH_3-CH_2-\underset{\underset{\displaystyle CH_3}{|}}{CH}-CH_2-\overset{\displaystyle O}{\overset{\displaystyle \|}{C}}-OH$$

β carbon α carbon

would be called β-methylvaleric acid.

■ Dicarboxylic Acids

Common names for the first six dicarboxylic acids are given in Table 5.2. Oxalic acid, the simplest dicarboxylic acid, is found in plants of the genus *Oxalis,* which includes

TABLE 5.2
Common Names for the First Six Dicarboxylic Acids

Length of Carbon Chain	Structural Formula	Common Name[a]	IUPAC Name
C_2 diacid	HOOC—COOH	oxalic acid	ethanedioic acid
C_3 diacid	HOOC—CH_2—COOH	malonic acid	propanedioic acid
C_4 diacid	HOOC—$(CH_2)_2$—COOH	succinic acid	butanedioic acid
C_5 diacid	HOOC—$(CH_2)_3$—COOH	glutaric acid	pentanedioic acid
C_6 diacid	HOOC—$(CH_2)_4$—COOH	adipic acid	hexanedioic acid
C_7 diacid	HOOC—$(CH_2)_5$—COOH	pimelic acid	heptanedioic acid

[a] The mnemonic "*Oh my, such good apple pie*" is helpful in remembering, in order, the first letters of the common names of these six simple dicarboxylic acids.

rhubarb and spinach, and in cabbage (see Figure 5.4). This acid and its salts are poisonous in *high* concentrations. The amount of oxalic acid present in spinach, cabbage, and rhubarb is not harmful. Oxalic acid is used to remove rust, bleach straw and leather, and remove ink stains. Succinic and glutaric acid and their derivatives play important roles in biochemical reactions that occur in the human body (Section 12.6).

EXAMPLE 5.2

Generating the Structural Formulas of Carboxylic Acids from Their Names

The contrast between IUPAC names and common names for mono- and dicarboxylic acids is as follows:

Monocarboxylic Acids

    IUPAC (two words)

        alkanoic acid

    Common (two words)

        (prefix)ic acid*

Dicarboxylic Acids

    IUPAC (two words)

        alkanedioic acid

    Common (two words)

        (prefix)ic acid*

*The common-name prefixes are related to natural sources for the acids.

FIGURE 5.4 The C_2 dicarboxylic acid, oxalic acid, contributes to the tart taste of rhubarb stalks.

■ Draw a structural formula for each of the following carboxylic acids.

  a. Caproic acid   **b.** Glutaric acid

  c. α-Phenylsuccinic acid   **d.** β-Chlorobutyric acid

Solution

a. Caproic acid is the six-carbon unsubstituted monocarboxylic acid. Its structural formula is

$$CH_3-CH_2-CH_2-CH_2-CH_2-\overset{\displaystyle O}{\overset{\displaystyle \|}{C}}-OH$$

b. Glutaric acid is the five-carbon unsubstituted dicarboxylic acid, with a carboxyl group at each end of the carbon chain.

$$HO-\overset{\displaystyle O}{\overset{\displaystyle \|}{C}}-CH_2-CH_2-CH_2-\overset{\displaystyle O}{\overset{\displaystyle \|}{C}}-OH$$

c. Succinic acid is the four-carbon unsubstituted dicarboxylic acid. A phenyl group (Section 2.12) is present on the alpha-carbon atom.

$$HO-\overset{\displaystyle O}{\overset{\displaystyle \|}{C}}-\overset{\alpha}{CH}-\overset{\beta}{CH_2}-\overset{\displaystyle O}{\overset{\displaystyle \|}{C}}-OH$$

d. Butyric acid is the four-carbon unsubstituted monocarboxylic acid. A chloro group is attached to the beta-carbon atom (carbon 3).

$$\overset{\gamma}{CH_3}-\overset{\beta}{\underset{\underset{\displaystyle Cl}{|}}{CH}}-\overset{\alpha}{CH_2}-\overset{\displaystyle O}{\overset{\displaystyle \|}{C}}-OH$$

Practice Exercise 5.2

Draw a structural formula for each of the following carboxylic acids.

  a. Adipic acid   **b.** β-Chlorovaleric acid

  c. Malonic acid   **d.** Phenylacetic acid

5.4 Polyfunctional Carboxylic Acids

Polyfunctional carboxylic acids are carboxylic acids that contain one or more additional functional groups besides the carboxyl group. Such acids occur naturally in many fruits, are important in the normal functioning of the human body (metabolism), and find use in over-the-counter skin-care products and in prescription drugs. Three commonly encountered types of polyfunctional carboxylic acids are *unsaturated* acids, *hydroxy* acids, and *keto* acids.

$$\underset{\text{An unsaturated acid}}{C-C=C-COOH} \qquad \underset{\text{A hydroxy acid}}{C-C-\overset{\overset{\displaystyle OH}{\displaystyle |}}{C}-COOH} \qquad \underset{\text{A keto acid}}{C-\overset{\displaystyle O}{\overset{\displaystyle \|}{C}}-C-COOH}$$

More information about these types of polyfunctional acids follows.

CHEMICAL CONNECTIONS

Nonprescription Pain Relievers Derived from Propanoic Acid

Consumers are faced with a shelf-full of choices when looking for an over-the-counter medicine to treat aches, pains, and fever. The vast majority of brands available, however, represent only four chemical formulations. Besides the long-available aspirin and acetaminophen, consumers can now purchase products that contain ibuprofen and naproxen.

These two newer entrants into the over-the-counter pain-reliever market are derivatives of propanoic acid, the three-carbon monocarboxylic acid.

Ibuprofen, marketed under the brand names Advil, Motrin-IB, and Nuprin, was cleared by the FDA in 1984 for nonprescription sales. Numerous studies have shown that nonprescription-strength ibuprofen relieves minor pain and fever as well as aspirin or acetaminophen. Like aspirin, ibuprofen reduces inflammation. (Prescription-strength ibuprofen has extensive use as an anti-inflammatory agent for the treatment of rheumatoid arthritis.) There is evidence that ibuprofen is more effective than either aspirin or acetaminophen in reducing dental pain and menstrual pain. Both aspirin and ibuprofen can cause stomach bleeding in some people, although ibuprofen seems to cause fewer problems. Ibuprofen is more expensive than either aspirin or acetaminophen.

Naproxen, marketed under the brand names Aleve and Anaprox, was cleared by the FDA in 1994 for nonprescription

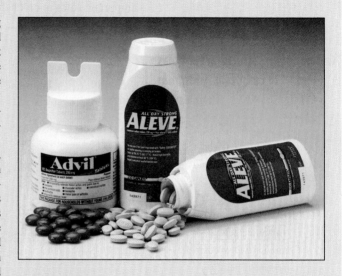

use. The effects of naproxen last longer in the body (8–12 hr per dose) than the effects of ibuprofen (4–6 hr per dose) and of aspirin and acetaminophen (4 hr per dose). Naproxen is more likely to cause slight intestinal bleeding and stomach upset than is ibuprofen. It is also not recommended for use by children under 12.

Propanoic acid

Ibuprofen

Naproxen

■ Unsaturated Acids

An unsaturated monocarboxylic acid with the structure

$$CH_3-CH_2-CH_2-C=CH-COOH$$
$$|$$
$$CH_3$$

3-Methyl-2-hexenoic acid

has been found to be largely responsible for "body odor." It is produced by skin bacteria, particularly those found in armpits.

The simplest *unsaturated mono*carboxylic acid is propenoic acid (acrylic acid), a substance used in the manufacture of several polymeric materials. Two forms exist for the simplest *unsaturated di*carboxylic acid, butenedioic acid. The two isomers have separate common names, fumaric acid (*trans*) and maleic acid (*cis*), a naming procedure seldom encountered.

$$CH_2=CH-COOH$$

Acrylic acid

Maleic acid
(*cis* isomer)

Fumaric acid
(*trans* isomer)

Some antihistamines (Section 6.10) are salts of maleic acid. The addition of small amounts of maleic acid to fats and oils prevents them from becoming rancid. Fumaric acid is a *metabolic acid*. Metabolic acids are intermediate compounds in the metabolic reactions (Section 12.1) that occur in the human body. More information about metabolic acids is presented in the next section.

FIGURE 5.5 Tartaric acid, the dihydroxy derivative of succinic acid, is particularly abundant in ripe grapes.

The IUPAC name for citric acid is 2-hydroxy-1,2,3-propanetrioic acid.

■ Hydroxy Acids

Four of the simpler *hydroxy* acids are

CH$_2$—COOH   CH$_3$—CH—COOH   HOOC—CH—CH$_2$—COOH   HOOC—CH—CH—COOH
|
OH   OH   OH   OH   OH
Glycolic acid   Lactic acid   Malic acid   Tartaric acid

Malic and tartaric acids are derivatives of succinic acid, the four-carbon unsubstituted diacid (Section 5.3).

Hydroxy acids occur naturally in many foods. Glycolic acid is present in the juice from sugar cane and sugar beets. Lactic acid is present in sour milk, sauerkraut, and dill pickles. Both malic acid and tartaric acid occur naturally in fruits. The sharp taste of apples (fruit of trees of the genus *Malus*) is due to malic acid. Tartaric acid is particularly abundant in grapes (Figure 5.5). It is also a component of tartar sauce and an acidic ingredient in many baking powders. Lactic and malic acids are also *metabolic acids* (Section 5.5).

Citric acid, perhaps the best known of all carboxylic acids, is a hydroxy acid with a structural feature we have not previously encountered. It is a hydroxy *tri*carboxylic acid. Besides there being acid groups at both ends of a carbon chain, a third acid group is present as a substituent on the chain. An acid group as a substituent is called a *carboxy* group. Thus citric acid is a hydroxycarboxy diacid.

$$\underset{\text{Citric acid}}{\text{HOOC—CH}_2\text{—}\underset{\underset{\text{COOH}}{|}}{\overset{\overset{\text{OH}}{|}}{\text{C}}}\text{—CH}_2\text{—COOH}}$$

Citric acid gives citrus fruits their "sharp" taste; lemon juice contains 4%–8% citric acid, and orange juice is about 1% citric acid. Citric acid is used widely in beverages and in foods. In jams, jellies, and preserves, it produces tartness and pH adjustment to optimize conditions for gelation. In fresh salads, citric acid prevents enzymatic browning reactions, and in frozen fruits it prevents deterioration of color and flavor. Addition of citric acid to seafood retards microbial growth by lowering pH. Citric acid is also a metabolic acid (Section 5.5).

■ Keto Acids

Keto acids, as the designation implies, contain a carboxyl group within a carbon chain. Pyruvic acid, with three carbon atoms, is the simplest keto acid that can exist.

$$\underset{\text{Pyruvic acid}}{\text{CH}_3\text{—}\overset{\overset{\text{O}}{||}}{\text{C}}\text{—COOH}}$$

In the pure state, pyruvic acid is a liquid with an odor resembling that of vinegar (acetic acid; Section 5.3). Pyruvic acid is a metabolic acid (Section 5.5).

5.5 "Metabolic" Acids

Numerous polyfunctional acids, including some mentioned in the previous section, are intermediates in the metabolic reactions that occur in the human body as food is processed. There are eight such acids that will appear repeatedly in the biochemistry chapters of this text.

CHEMICAL CONNECTIONS — Carboxylic Acids and Skin Care

A number of carboxylic acids are used as "skin-care acids." Heavily advertised at present are cosmetic products that contain *alpha-hydroxy* acids, carboxylic acids in which a hydroxyl group is attached to the acid's alpha-carbon atom (the carbon atom attached to the carboxyl group carbon atom). Such cosmetic products address problems such as dryness, flaking, and itchiness of the skin and are highly promoted for removing wrinkles.

Alpha-hydroxy acids "work" by loosening the cells of the outer layer of skin (the epidermis) and by accelerating the flaking off of dead skin. The result is healthier-looking skin.

The alpha-hydroxy acids most commonly found in cosmetic products are glycolic acid and lactic acid, the two simplest alpha-hydroxy acids.

Alpha-carbon atom

$$CH_2—COOH$$
$$OH$$
Glycolic acid

Alpha-carbon atom

$$CH_3—CH—COOH$$
$$OH$$
Lactic acid

Both acids are naturally occurring substances. Glycolic acid occurs in sugar cane and sugar beets, and lactic acid occurs in sour milk.

The use of alpha-hydroxy acids in cosmetics is considered safe at acid concentrations of less than 10%; higher concentrations can cause skin irritation, burning, and stinging. (Lactic acid becomes a prescription drug at concentrations of 12% or more.) One drawback of the cosmetic use of alpha-hydroxy products is that such use can increase the skin's sensitivity to the ultraviolet light component of sunlight; it is this component that causes sunburn. Individuals who are using "alpha-hydroxys" should apply a sunscreen whenever they go outside for an extended period of time.

Glycolic acid, at higher concentrations than that found in cosmetics, is used by dermatologists for the "spot" removal of *keratoses* (precancerous lesions and/or patches of darker, thickened skin).

Polyunsaturated carboxylic acids are used extensively in the treatment of severe acne. The prescription drugs Tretinoin and Accutane are such compounds.

Tretinoin

Accutane

Tretinoin has an all-*trans* double-bond configuration. Accutane has a *cis* double bond in the position closest to the carboxyl group, with the rest of the double bonds in *trans* configurations.

Two skin-care products containing alpha-hydroxy acids.

Interestingly, these eight key metabolic intermediates are derived from only three of the simple carboxylic acids. These three simple acids and the metabolic acids related to them are

propionic acid (3-carbon monoacid): lactic, glyceric, and pyruvic acids
succinic acid (4-carbon diacid): fumaric, oxaloacetic, and malic acids
glutaric acid (5-carbon diacid): α-ketoglutaric and citric acids

Metabolic acids derived from the diacids succinic and glutaric are encountered in the citric acid cycle (Section 12.6), a series of reactions in which C_2 units obtained from all types of foods are further processed for the purpose of obtaining energy. Glyceric and pyruvic acid (propionic acid derivatives) are encountered in glycolysis (Section 13.2), a series of reactions in which glucose is processed. Lactic acid (a propionic derivative) is a by-product of strenuous exercise (Section 13.3). Figure 5.6 gives further details about the eight "metabolic" acids.

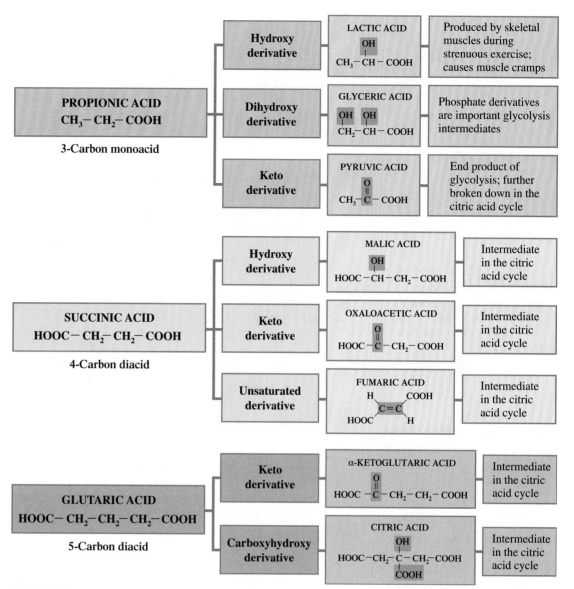

FIGURE 5.6 Structural characteristics and functions of several polyfunctional carboxylic acids that are important in metabolic reactions in the human body.

5.6 Physical Properties of Carboxylic Acids

Carboxylic acids are the most *polar* organic compounds we have discussed so far. Both the carbonyl part ($\diagdown$C=O) and the hydroxyl part (—OH) of the carboxyl functional group are polar. The result is very high melting and boiling points for carboxylic acids, the highest of any type of organic compound yet considered (Figure 5.7).

Unsubstituted saturated monocarboxylic acids containing up to nine carbon atoms are liquids that have strong, sharp odors (Figure 5.8). Acids with 10 or more carbon atoms in an unbranched chain are waxy solids that are odorless (because of low volatility). Aromatic carboxylic acids, as well as dicarboxylic acids, are also odorless solids.

The high boiling points of carboxylic acids indicate the presence of strong intermolecular attractive forces. A unique hydrogen-bonding arrangement, shown in Figure 5.9, contributes to these attractive forces. A given carboxylic acid molecule forms two hydrogen bonds to another carboxylic acid molecule, producing a "complex" known as a

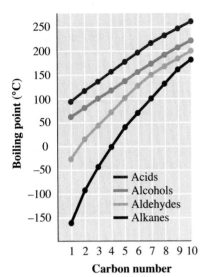

FIGURE 5.7 The boiling points of monocarboxylic acids compared to those of other types of compounds. All compounds in the comparison have unbranched carbon chains.

FIGURE 5.8 A physical-state summary for unbranched mono- and dicarboxylic acids at room temperature and pressure.

Unbranched Monocarboxylic Acids			
C_1	C_3	C_5	C_7
C_2	C_4	C_6	C_8

Unbranched Dicarboxylic Acids			
✕	C_3	C_5	C_7
C_2	C_4	C_6	C_8

☐ Liquid ☐ Solid

FIGURE 5.9 A given carboxylic acid molecule can form two hydrogen bonds to another carboxylic acid molecule, producing a "dimer"—a complex with a mass twice that of a single molecule.

dimer. Because dimers have twice the mass of a single molecule, a higher temperature is needed to boil a carboxylic acid than would be needed for similarly sized aldehyde and alcohol molecules where dimerization does not occur.

Carboxylic acids readily hydrogen-bond to water molecules. Such hydrogen bonding contributes to water solubility for short-chain carboxylic acids. The unsubstituted C_1 to C_4 monocarboxylic acids are completely miscible with water. Solubility then rapidly decreases with carbon number, as shown in Figure 5.10. Short-chain dicarboxylic acids are also water-soluble. In general, aromatic acids are not water-soluble.

5.7 Preparation of Carboxylic Acids

Oxidation of primary alcohols or aldehydes, using an oxidizing agent such as CrO_3 or $K_2Cr_2O_7$, produces carboxylic acids, a process that we examined in Sections 3.9 and 4.8.

$$\text{Primary alcohol} \xrightarrow{[O]} \text{aldehyde} \xrightarrow{[O]} \text{carboxylic acid}$$

Aromatic acids can be prepared by oxidizing a carbon side chain (alkyl group) on a benzene derivative. In this process, all the carbon atoms of the alkyl group except the one attached to the ring are lost. The remaining carbon becomes part of a carboxyl group.

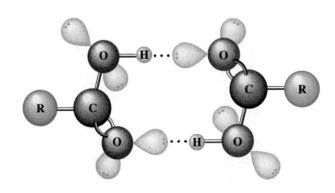

5.8 Acidity of Carboxylic Acids

Carboxylic acids, as the name implies, are *acidic.* When a carboxylic acid is placed in water, hydrogen ion transfer (proton transfer) occurs to produce hydronium ion (the acidic species in water) and carboxylate ion.

$$R{-}COOH + H_2O \longrightarrow H_3O^+ + R{-}COO^-$$
$$\text{Hydronium} \qquad \text{Carboxylate}$$
$$\text{ion} \qquad \text{ion}$$

A **carboxylate ion** *is the negative ion produced when a carboxylic acid loses one or more acidic hydrogen atoms.*

Carboxylate ions formed from monocarboxylic acids always carry a -1 charge; only one acidic hydrogen atom is present in such molecules. Dicarboxylic acids, which possess two acidic hydrogen atoms (one in each carboxyl group), can produce carboxylate ions bearing a -2 charge.

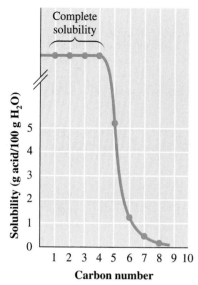

FIGURE 5.10 The solubility in water of saturated unbranched-chain carboxylic acids.

At normal human body pH values (pH = 7.35 to 7.45), most carboxylic acids exist as carboxylate ions. Acetic acid is in the form of acetate ion, pyruvic acid is in the form of pyruvate ion, lactic acid is in the form of lactate ion, and so on.

Carboxylic acid salt formation involves an acid–base neutralization reaction.

Carboxylate ions are named by dropping the -ic acid ending from the name of the parent acid and replacing it with -ate.

$$CH_3-\overset{O}{\overset{\|}{C}}-OH + H_2O \longrightarrow H_3O^+ + CH_3-\overset{O}{\overset{\|}{C}}-O^-$$

Acetic acid
(ethanoic acid)   Acetate ion
(ethanoate ion)

$$HO-\overset{O}{\overset{\|}{C}}-\overset{O}{\overset{\|}{C}}-OH + 2H_2O \longrightarrow 2H_3O^+ + {}^-O-\overset{O}{\overset{\|}{C}}-\overset{O}{\overset{\|}{C}}-O^-$$

Oxalic acid
(ethanedioic acid)   Oxalate ion
(ethanedioate ion)

Carboxylic acids are weak acids. The extent of proton transfer is usually less than 5%; that is, an equilibrium situation exists in which the equilibrium lies far to the left.

$$R-COOH + H_2O \rightleftharpoons H_3O^+ + R-COO^-$$

More than 95%
of molecules in this form   Less than 5%
of molecules in this form

Table 5.3 gives K_a values and percent ionization in 0.100 M solution for selected monocarboxylic acids.

5.9 Carboxylic Acid Salts

In a manner similar to that of inorganic acids, carboxylic acids react completely with strong bases to produce water and a carboxylic acid salt.

$$CH_3-\overset{O}{\overset{\|}{C}}-OH + NaOH \longrightarrow CH_3-\overset{O}{\overset{\|}{C}}-O^-Na^+ + H_2O$$

Carboxylic acid   Strong base   Carboxylic
acid salt   Water

A **carboxylic acid salt** is an ionic compound in which the negative ion is a carboxylate ion.

Carboxylic acid salts are named similarly to other ionic compounds: *The positive ion is named first, followed by a separate word giving the name of the negative ion.* The salt formed in the preceding reaction contains sodium ions and acetate ions (from acetic acid); hence the salt's name is sodium acetate.

EXAMPLE 5.3

Writing Equations for the Formation of Carboxylic Acid Salts

■ Using an acid–base neutralization reaction, write a chemical equation for the formation of each of the following carboxylic acid salts.

a. Sodium propionate   **b.** Potassium oxalate

(continued)

TABLE 5.3
Acid Strength for Selected Monocarboxylic Acids

Acid	K_a	Percent Ionization (0.100 M Solution)
Formic	1.8×10^{-4}	4.2%
Acetic	1.8×10^{-5}	1.3%
Propionic	1.3×10^{-5}	1.2%
Butyric	1.5×10^{-5}	1.2%
Valeric	1.5×10^{-5}	1.2%
Caproic	1.4×10^{-5}	1.2%

Solution

a. This salt contains sodium ion (Na^+) and propionate ion, the three-carbon monocarboxylate ion.

$$CH_3-CH_2-\overset{\displaystyle O}{\overset{\displaystyle \|}{C}}-O^-Na^+$$

From a neutralization standpoint, the sodium ion's source is the base sodium hydroxide, NaOH, and the negative ion's source is the acid propanoic acid. The acid–base neutralization equation is

$$CH_3-CH_2-\overset{\displaystyle O}{\overset{\displaystyle \|}{C}}-OH + NaOH \longrightarrow CH_3-CH_2-\overset{\displaystyle O}{\overset{\displaystyle \|}{C}}-O^-Na^+ + H_2O$$

Propionic acid   Sodium hydroxide   Sodium propionate   Water

b. This salt contains potassium ions (K^+) whose source would be the base potassium hydroxide, KOH. The salt also contains oxalate ions, whose source would be the acid oxalic acid.

$$HO-\overset{\displaystyle O}{\overset{\displaystyle \|}{C}}-\overset{\displaystyle O}{\overset{\displaystyle \|}{C}}-OH + 2KOH \longrightarrow K^+{}^-O-\overset{\displaystyle O}{\overset{\displaystyle \|}{C}}-\overset{\displaystyle O}{\overset{\displaystyle \|}{C}}-O^-K^+ + 2H_2O$$

Oxalic acid   Potassium hydroxide   Potassium oxalate   Water

Note that two molecules of base are needed to react completely with one molecule of acid because the acid is a dicarboxylic acid.

Practice Exercise 5.3

Using an acid–base neutralization reaction, write a chemical equation for the formation of each of the following carboxylic acid salts.

a. Sodium formate   **b.** Potassium malonate

Converting a carboxylic acid salt back to a carboxylic acid is very simple. React the salt with a solution of a strong acid such as hydrochloric acid (HCl) or sulfuric acid (H_2SO_4).

$$CH_3-\overset{\displaystyle O}{\overset{\displaystyle \|}{C}}-O^-\ Na^+ + HCl \longrightarrow CH_3-\overset{\displaystyle O}{\overset{\displaystyle \|}{C}}-OH + NaCl$$

Sodium acetate   Hydrochloric acid   Acetic acid   Sodium chloride

The interconversion reactions between carboxylic acid salts and their "parent" carboxylic acids are so easy to carry out that organic chemists consider these two types of compounds interchangeable.

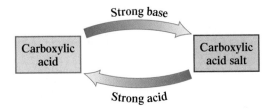

Strong base

Carboxylic acid   Carboxylic acid salt

Strong acid

■ Uses for Carboxylic Acid Salts

The solubility of carboxylic acid salts in water is much greater than that of the carboxylic acids from which they are derived. Drugs and medicines that contain acid groups are usually marketed as the sodium or potassium salt of the acid. This greatly enhances the solubility of the medication, increasing the ease of its absorption by the body.

Many *antimicrobials,* compounds used as food preservatives, are carboxylic acid salts. Particularly important are the salts of benzoic, sorbic, and propionic acids.

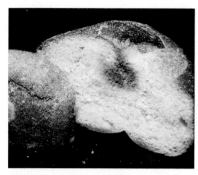

FIGURE 5.11 Propionates, salts of propionic acid, extend the shelf life of bread by preventing the formation of mold.

Benzoic acid

CH_3—CH=CH—CH=CH—COOH

Sorbic acid
(2,4-hexadienoic acid)

CH_3—CH_2—COOH

Propionic acid

The benzoate salts of sodium and potassium are effective against yeast and mold in beverages, jams and jellies, pie fillings, ketchup, and syrups. Concentrations of up to 0.1% (m/m) benzoate are found in such products.

> The solubility of benzoic acid in water at 25°C is 3.4 g/L. The solubility of sodium benzoate, the sodium salt of benzoic acid in water at 25°C is 550 g/L.

Sodium benzoate

Potassium benzoate

Sodium and potassium sorbates inhibit mold and yeast growth in dairy products, dried fruits, sauerkraut, and some meat and fish products. Sorbate preservative concentrations range from 0.02% to 0.2% (m/m).

CH_3—CH=CH—CH=CH—C—O^- Na^+

Sodium sorbate

CH_3—CH=CH—CH=CH—C—O^- K^+

Potassium sorbate

Calcium and sodium propionates are used in baked products and also in cheese foods and spreads (see Figure 5.11). Benzoates and sorbates cannot be used in yeast-leavened baked goods because they affect the activity of the yeast.

$(CH_3$—CH_2—C—$O^-)_2$ Ca^{2+}

Calcium propionate

CH_3—CH_2—C—O^- Na^+

Sodium propionate

Carboxylate salts do not directly kill microorganisms present in food. Rather, they prevent further growth and proliferation of these organisms by increasing the pH of the foods in which they are used.

5.10 Structure of Esters

An **ester** *is a carboxylic acid derivative in which the —OH portion of the carboxyl group has been replaced with an —OR group.*

R—C—O—H

Carboxylic acid

R—C—O—R

Ester

The ester functional group is thus

—C—O—R

In linear form, the ester functional group can be represented as —COOR or —CO_2R.

The simplest ester, which has two carbon atoms, has a hydrogen atom attached to the ester functional group.

$$H-\overset{\overset{\displaystyle O}{\|}}{C}-O-CH_3$$

Note that the two carbon atoms present are not bonded to each other.

There are two three-carbon esters.

$$H-\overset{\overset{\displaystyle O}{\|}}{C}-O-CH_2-CH_3 \quad \text{and} \quad CH_3-\overset{\overset{\displaystyle O}{\|}}{C}-O-CH_3$$

The structure of the simplest aromatic ester is derived from the structure of benzoic acid, the simplest aromatic carboxylic acid.

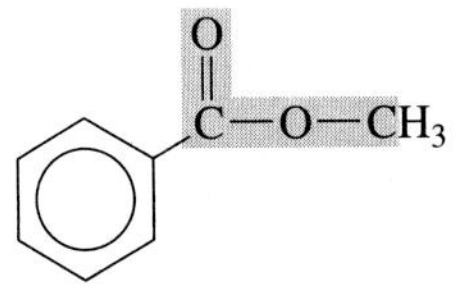

$$\overset{\overset{\displaystyle O}{\|}}{C}-O-CH_3$$

Note that the difference between a carboxylic acid and an ester is a "H versus R" relationship.

$$R-\overset{\overset{\displaystyle O}{\|}}{C}-O-H \quad \text{and} \quad R-\overset{\overset{\displaystyle O}{\|}}{C}-O-R$$
$$\text{Acid} \qquad\qquad\qquad \text{Ester}$$

We have encountered this "H versus R" relationship several times before in our study of hydrocarbon derivatives. The Chemistry at a Glance feature on page 157 summarizes the "H versus R" relationships we have encountered so far.

5.11 Preparation of Esters

Esters are produced through *esterification*. An **esterification reaction** *is the reaction of a carboxylic acid with an alcohol (or phenol) to produce an ester.* A strong acid catalyst (generally H_2SO_4) is needed for esterification.

$$R-\overset{\overset{\displaystyle O}{\|}}{C}-O-H + H-O-R' \overset{H^+}{\rightleftharpoons} R-\overset{\overset{\displaystyle O}{\|}}{C}-O-R' + H_2O$$
$$\text{Carboxylic acid} \qquad \text{Alcohol} \qquad\qquad \text{Ester} \qquad \text{Water}$$

> Esterification is a *condensation reaction*. This is the third time we have encountered this type of reaction. The first encounter involved intermolecular alcohol dehydration (Section 3.9) and the second encounter involved the preparation of acetals (Section 4.10).

In the esterification process, a —OH group is lost from the carboxylic acid, a —H atom is lost from the alcohol, and water is formed as a by-product. The net effect of this reaction is substitution of the —OR group of the alcohol for the —OH group of the acid.

$$R-\overset{\overset{\displaystyle O}{\|}}{C}-O-H + H-O-R' \overset{H^+}{\rightleftharpoons} R-\overset{\overset{\displaystyle O}{\|}}{C}-O-R' + H_2O$$

> Studies show that in ester formation, the hydroxyl group of the acid (not of the alcohol) becomes part of the water molecule.

A specific example of esterification is the reaction of acetic acid with methyl alcohol.

$$CH_3-\overset{\overset{\displaystyle O}{\|}}{C}-O-H + H-O-CH_3 \overset{H^+}{\rightleftharpoons} CH_3-\overset{\overset{\displaystyle O}{\|}}{C}-O-CH_3 + H_2O$$

CHEMISTRY AT A GLANCE

Summary of the "H Versus R" Relationship for Pairs of Hydrocarbon Derivatives

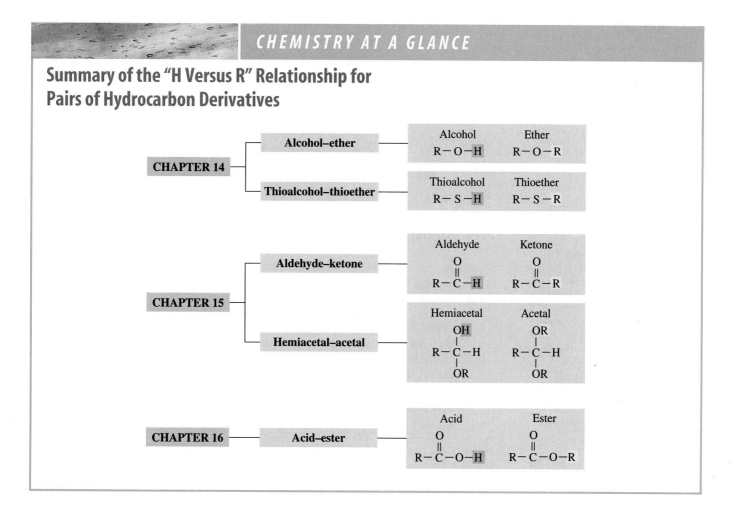

Esterification reactions are equilibrium processes, with the position of equilibrium usually favoring products only slightly. That is, at equilibrium, substantial amounts of both reactants and products are present. The amount of ester formed can be increased by using an excess of alcohol or by constantly removing one of the products. According to Le Châtelier's principle, either of these techniques will shift the position of equilibrium to the right (the product side of the equation). This equilibrium problem explains the use of the "double-arrow notation" in all the esterification equations in this section.

It is often useful to think of the structure of an ester in terms of its "parent" alcohol and acid molecules; the ester has an acid part and an alcohol part.

In this context, it is easy to identify the acid and alcohol from which a given ester can be produced; just add a —OH group to the acid part of the ester and a —H atom to the alcohol part to generate the parent molecules.

Hydroxy acids—compounds which contain both a hydroxyl and a carboxyl group (Section 5.4)—have the capacity to undergo intermolecular esterification to form *lactones* (cyclic esters). Such internal esterification easily takes place in situations where a five- or six-membered ring can be formed.

Cyclic esters formed from hydroxyacids are called *lactones*.

$$
\underset{\underset{OH}{|}}{\overset{4}{CH_2}}-\overset{3}{CH_2}-\overset{2}{CH_2}-\overset{O}{\overset{\|}{C}}-OH \longrightarrow \underset{\overset{3}{CH_2}-\overset{4}{CH_2}}{\overset{\overset{O}{\|}}{\underset{2}{\overset{\|}{C}}-\boxed{OH}}}{\overset{\boxed{OH}}{}} \longrightarrow \overset{O}{\underset{\text{Cyclic ester}}{\text{(ring)}}} + H_2O
$$

5.12 Nomenclature for Esters

Salts and esters of carboxylic acids are named in the same way. The name of the positive ion (in the case of a salt) or the name of the organic group attached to the single-bonded oxygen of the carbonyl group (in the case of an ester) precedes the name of the acid. The *-ic acid* part of the name of the acid is converted to *-ate*.

$$
CH_3-CH_2-CH_2-\overset{\overset{O}{\|}}{C}-O^-\,Na^+
$$

IUPAC: Sodium butanoate
Common: Sodium butyrate

$$
CH_3-CH_2-CH_2-\overset{\overset{O}{\|}}{C}-O-CH_3
$$

IUPAC: Methyl butanoate
Common: Methyl butyrate

FIGURE 5.12 Space-filling models for the methyl and ethyl esters of acetic acid.

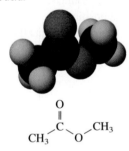

$$
\underset{CH_3}{}\overset{\overset{O}{\|}}{C}\underset{O}{}{}^{CH_3}
$$

Methyl acetate

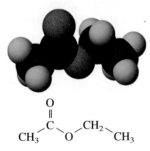

$$
\underset{CH_3}{}\overset{\overset{O}{\|}}{C}\underset{O}{}{}^{CH_2}{}^{CH_3}
$$

Ethyl acetate

Visualizing esters as having an "alcohol part" and an "acid part" (Section 5.11) is the key to naming them in both the common and the IUPAC systems of nomenclature. The rules are as follows:

1. The name for the alcohol part of the ester appears first and is followed by a *separate word* giving the name for the acid part of the ester.
2. The name for the alcohol part of the ester is simply the name of the R group (alkyl, cycloalkyl, or aryl) present in the —OR portion of the ester.
3. The name for the acid part of the ester is obtained by dropping the *-ic acid* ending for the acid's name and adding the suffix *-ate*.

Consider the ester derived from ethanoic acid (acetic acid) and methanol (methyl alcohol). Its name will be *methyl ethanoate* (IUPAC) or *methyl acetate* (common); see Figure 5.12.

$$
CH_3-\overset{\overset{O}{\|}}{C}-OH + HO-CH_3 \longrightarrow CH_3-\overset{\overset{O}{\|}}{C}-O-CH_3 + H_2O
$$

IUPAC: Ethanoic acid   Methanol   Methyl ethanoate
Common: Acetic acid   Methyl alcohol   Methyl acetate

Dicarboxylic acids can form diesters, with each of the carboxyl groups undergoing esterification. An example of such a molecule and how it is named is

$$
CH_3-O-\overset{\overset{O}{\|}}{C}-CH_2-CH_2-\overset{\overset{O}{\|}}{C}-O-CH_3
$$

IUPAC: Dimethyl butanedioate
Common: Dimethyl succinate

Further examples of ester nomenclature, for compounds in which substituents are present, are

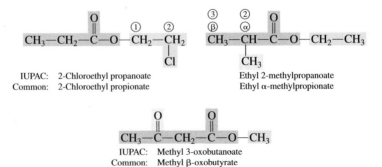

IUPAC: 2-Chloroethyl propanoate
Common: 2-Chloroethyl propionate

Ethyl 2-methylpropanoate
Ethyl α-methylpropionate

$$
CH_3-\overset{\overset{O}{\|}}{C}-CH_2-\overset{\overset{O}{\|}}{C}-O-CH_3
$$

IUPAC: Methyl 3-oxobutanoate
Common: Methyl β-oxobutyrate

EXAMPLE 5.4

Determining IUPAC and Common Names for Esters

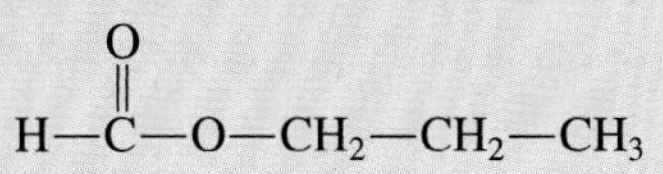

Line-angle formulas for the simpler unbranched-chain methyl esters:

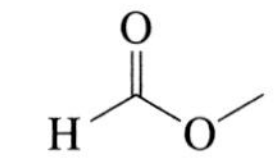

Methyl methanoate

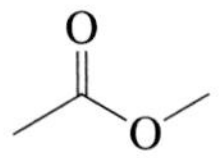

Methyl ethanoate

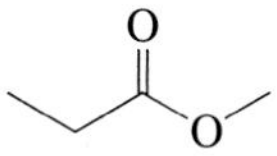

Methyl propanoate

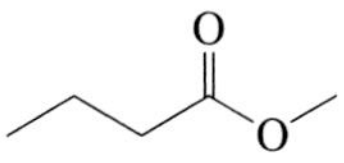

Methyl butanoate

The contrast between IUPAC names and common names for unbranched esters of carboxylic acids is as follows:

IUPAC (two words)

alkyl alkanoate

methyl propanoate

Common (two words)

alkyl (prefix)ate*

methyl acetate

*The common-name prefixes are related to natural sources for the "parent" carboxylic acids.

■ Assign both IUPAC and common names to the following esters.

a.
$$CH_3-CH_2-\overset{\overset{\displaystyle O}{\|}}{C}-O-CH_2-CH_3$$

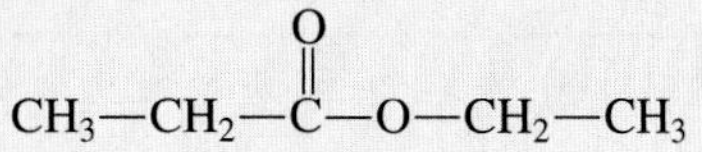

b.

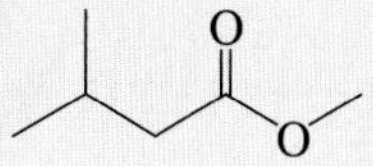

c.

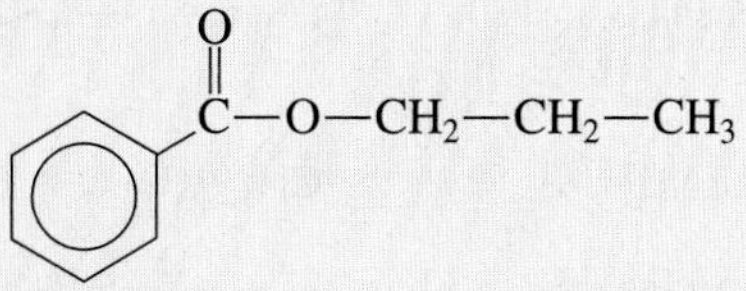

Solution

a. The name *ethyl* characterizes the alcohol part of the molecule. The name of the acid is propanoic acid (IUPAC) or propionic acid (common). Deleting the -*ic acid* ending and adding -*ate* gives the name *ethyl propanoate* (IUPAC) or *ethyl propionate* (common).

b. The name of the alcohol part of the molecule is methyl (from methanol or methyl alcohol). The name of the five-carbon acid is 3-methylbutanoic acid or β-methylbutyric acid. Hence the ester name is *methyl 3-methylbutanoate* (IUPAC) or *methyl β-methylbutyrate* (common).

c. The name *propyl* characterizes the alcohol part of the molecule. The acid part of the molecule is derived from benzoic acid (both IUPAC and common name). Hence the ester name in both systems is *propyl benzoate*.

Practice Exercise 5.4

Assign both IUPAC and common names to the following esters.

a.
$$CH_3-\overset{\overset{\displaystyle O}{\|}}{C}-O-CH_2-CH_3$$

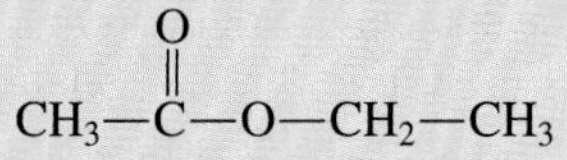

b.

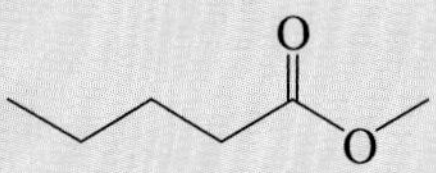

c.
$$H-\overset{\overset{\displaystyle O}{\|}}{C}-O-CH_2-CH_2-CH_3$$

Lactones (cyclic esters) are named by replacing the -*oic* ending of the parent hydroxy-carboxylic acid name with -*olide* and identifying the hydroxyl-bearing carbon by number.

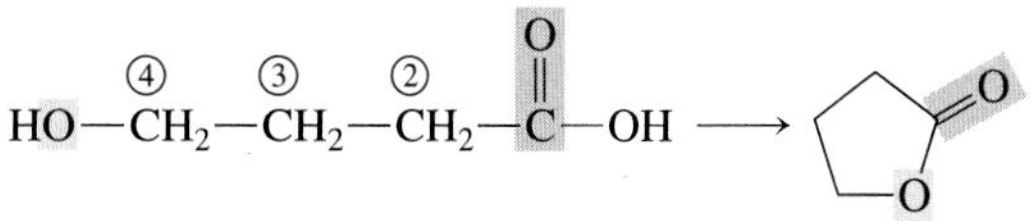

$$HO-\overset{④}{CH_2}-\overset{③}{CH_2}-\overset{②}{CH_2}-\overset{\overset{\displaystyle O}{\|}}{C}-OH \longrightarrow$$

4-Hydroxybutanoic acid   4-Butanolide

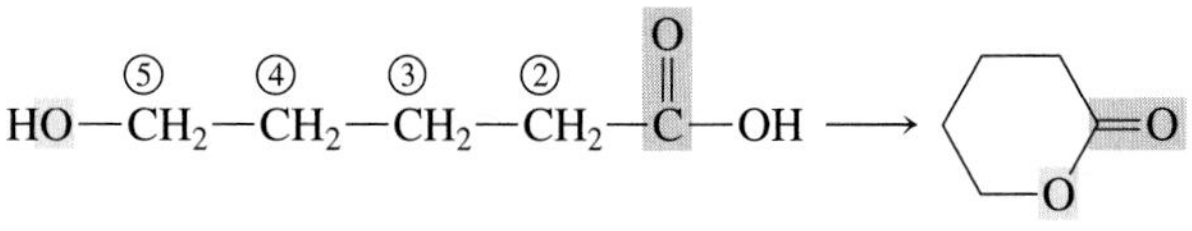

$$HO-\overset{⑤}{CH_2}-\overset{④}{CH_2}-\overset{③}{CH_2}-\overset{②}{CH_2}-\overset{\overset{\displaystyle O}{\|}}{C}-OH \longrightarrow$$

5-Hydroxypentanoic acid   5-Pentanolide

5.13 Selected Common Esters

In this section we consider selected esters that function as flavoring agents, pheromones, and medications.

■ Flavor/Fragrance Agents

Esters are largely responsible for the flavor and fragrance of fruits and flowers. Generally, a natural flavor or odor is caused by a mixture of esters, with one particular compound

Fats and oils, substances that are part of our dietary intake, are triesters— molecules containing three ester functional groups. Such compounds are considered in Chapter 8.

TABLE 5.4

Selected Esters That Are Used as Flavoring Agents

IUPAC Name	Structural Formula	Characteristic Flavor and Odor
isobutyl methanoate	$H-\overset{\overset{\displaystyle O}{\|}}{C}-O-CH_2-\overset{\overset{\displaystyle CH_3}{\|}}{CH}-CH_3$	raspberry
propyl ethanoate	$CH_3-\overset{\overset{\displaystyle O}{\|}}{C}-O-(CH_2)_2-CH_3$	pear
pentyl ethanoate	$CH_3-\overset{\overset{\displaystyle O}{\|}}{C}-O-(CH_2)_4-CH_3$	banana
octyl ethanoate	$CH_3-\overset{\overset{\displaystyle O}{\|}}{C}-O-(CH_2)_7-CH_3$	orange
pentyl propanoate	$CH_3-CH_2-\overset{\overset{\displaystyle O}{\|}}{C}-O-(CH_2)_4-CH_3$	apricot
methyl butanoate	$CH_3-(CH_2)_2-\overset{\overset{\displaystyle O}{\|}}{C}-O-CH_3$	apple
ethyl butanoate	$CH_3-(CH_2)_2-\overset{\overset{\displaystyle O}{\|}}{C}-O-CH_2-CH_3$	pineapple

being dominant. The synthetic production of these "dominant" compounds is the basis for the flavoring agents used in ice cream, gelatins, soft drinks, and so on. Table 5.4 gives the structures of selected esters used as flavoring agents. What is surprising about the structures in Table 5.4 is how closely some of them resemble each other. For example, the apple and pineapple flavoring agents differ by one carbon atom (methyl versus ethyl); a five-carbon chain versus an eight-carbon chain makes the difference between banana and orange flavor.

Numerous lactones are common in plants. Two examples are 4-decanolide, a compound partially responsible for the taste and odor of ripe peaches, and coumarin (common name), the compound responsible for the pleasant odor of newly mown hay.

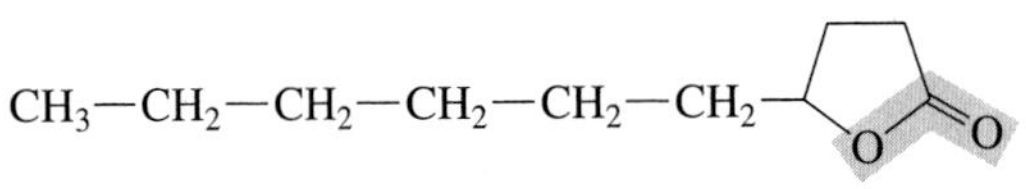

4-Decanolide
(peach odor)

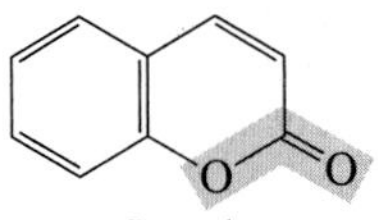

Coumarin
(newly mown hay odor)

■ Pheromones

A number of pheromones (Section 2.6) contain ester functional groups. The compound isoamyl acetate,

is an alarm pheromone for the honey bee. The compound methyl *p*-hydroxybenzoate,

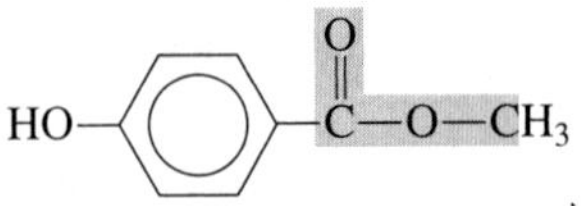

is a sexual attractant for canine species. It is secreted by female dogs in heat and evokes attraction and sexual arousal in male dogs.

The compound nepetalactone, a lactone present in the catnip plant, is an attractant for cats of all types. It is not considered a pheromone, however, because different species are involved (see Figure 5.13).

■ Medications

Numerous esters have medicinal value, including benzocaine (a local anesthetic), aspirin, and oil of wintergreen (a counterirritant). The structure of benzocaine is

$$H_2N-\text{⟨ ⟩}-\overset{\overset{O}{\|}}{C}-O-CH_2-CH_3$$

Both aspirin and oil of wintergreen are esters of salicylic acid, an aromatic hydroxy-acid.

$$\overset{\overset{O}{\|}}{C}-OH$$
OH
Salicylic acid

Because this acid has both an acid group and a hydroxyl group, it can form two different types of esters: one by reaction of its acid group with an alcohol, the other by reaction of its alcohol group with a carboxylic acid.

Reaction of acetic acid with the alcohol group of salicylic acid produces aspirin.

Salicylic acid + Acetic acid $\xrightarrow[\text{Heat}]{H^+}$ Aspirin + H_2O

Aspirin's mode of action in the human body is considered in the Chemical Connections feature on page 162.

Reaction of methanol with the acid group of salicylic acid produces oil of wintergreen.

Salicylic acid + Methanol $\xrightarrow[\text{Heat}]{H^+}$ Oil of wintergreen + H_2O

Oil of wintergreen, also called methyl salicylate, is used in skin rubs and liniments to help decrease the pain of sore muscles. It is absorbed through the skin, where it is hydrolyzed to produce salicylic acid. Salicylic acid, as with aspirin, is the actual pain reliever.

5.14 Isomerism for Carboxylic Acids and Esters

As with the other families of organic compounds previously discussed, constitutional isomers based on different carbon skeletons and on different positions for the functional group are possible for carboxylic acids and esters as well as other types of carboxylic acid derivatives. The following two examples illustrate carboxylic acid skeletal isomerism and ester positional isomerism.

FIGURE 5.13 Cats of all types (from lions to house cats) are strongly attracted to the catnip plant. The attractant in the catnip plant is nepetalactone, a cyclic ester.

CH₃ O
Nepetalactone

CHEMICAL CONNECTIONS   Aspirin

Aspirin, an ester of salicylic acid (Section 5.13), is a drug that has the ability to decrease pain (analgesic properties), to lower body temperature (antipyretic properties), and to reduce inflammation (anti-inflammatory properties). It is most frequently taken in tablet form, and the tablet usually contains 325 mg of aspirin held together with an inert starch binder.

After ingestion, aspirin undergoes hydrolysis to produce salicylic acid and acetic acid. Salicylic acid is the active ingredient of aspirin—the substance that has analgesic, antipyretic, and anti-inflammatory effects.

Salicylic acid is capable of irritating the lining of the stomach, inducing a small amount of bleeding. Breaking (or chewing) an aspirin tablet, rather than taking it whole, reduces the chance of bleeding by eliminating drug concentration on

one part of the stomach lining. Buffered aspirin products contain alkaline chemicals (such as aluminum glycinate or aluminum hydroxide) to neutralize the acidity of the aspirin when it contacts the stomach lining.

Aspirin—that is, salicylic acid—inhibits the synthesis of a class of hormones called prostaglandins (Section 8.13), molecules that cause pain, fever, and inflammation when present in the bloodstream in higher-than-normal levels. Salicylic acid's mode of action is irreversible inhibition (Section 10.7) of *cyclooxygenase,* an enzyme necessary for the production of prostaglandins.

Recent studies show that aspirin also increases the time it takes blood to coagulate (clot). For blood to coagulate, platelets must first be able to aggregate, and prostaglandins (which aspirin inhibits) appear to be necessary for platelet aggregation to occur. One study suggests that healthy men can cut their risk of heart attacks nearly in half by taking one baby aspirin per day (81 mg compared to the 325 mg in a regular tablet). Aspirin acts by making the blood less likely to clot. Heart attacks usually occur when clots form in the coronary arteries, cutting off blood supply to the heart.

Aspirin manufacturers indicate that "low dose" (81 mg) aspirin tablets now represent 23% of the total market for aspirin tablets. In 2001, about 26 million people in America regularly took aspirin for "heart health"—up from 7 million in 1997.

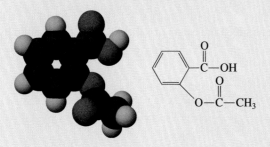

Esters and carboxylic acid functional group isomerism represents the fourth time we have encountered this type of isomerism. Previous examples are alcohol–ether, thiol–thioether, and aldehyde–ketone isomers.

Carboxylic acids and esters with the same number of carbon atoms and the same degree of saturation are functional group isomers. The ester ethyl propanoate and the carboxylic acid pentanoic acid both have the molecular formula, $C_5H_{10}O_2$, and are thus functional group isomers.

TABLE 5.5
Boiling Points of Compounds of Similar Molecular Mass That Contain Different Functional Groups

Name	Functional-Group Class	Molecular Mass	Boiling Point (°C)
diethyl ether	ether	74	34
ethyl formate	ester	74	54
methyl acetate	ester	74	57
butanal	aldehyde	72	76
1-butanol	alcohol	74	118
propionic acid	acid	74	141

5.15 Physical Properties of Esters

Ester molecules cannot form hydrogen bonds to each other because they do not have a hydrogen atom bonded to an oxygen atom. Consequently, the boiling points of esters are much lower than those of alcohols and carboxylic acids of comparable molecular mass. Esters are more like ethers in their physical properties. Table 5.5 gives boiling-point data for compounds of similar molecular mass that contain different functional groups.

Water molecules can hydrogen-bond to esters through the oxygen atoms present in the ester functional group (Figure 5.14). Because of such hydrogen bonding, low-molecular-mass esters are soluble in water. Solubility rapidly decreases with increasing carbon chain length; borderline solubility situations are reached when three to five carbon atoms are in a chain.

Low- and intermediate-molecular-mass esters are usually colorless liquids at room temperature (see Figure 5.15). Most have pleasant odors (Section 5.13).

5.16 Chemical Reactions of Esters

The most important reaction of esters involves breaking the carbon–oxygen single bond that holds the "alcohol part" and the "acid part" of the ester together. This reaction process is called either ester hydrolysis or ester saponification, depending on reaction conditions.

■ Ester Hydrolysis

This is our second encounter with hydrolysis reactions. The first encounter involved the hydrolysis of acetals (Section 4.10).

The breaking of a bond within a molecule and the attachment of the components of water to the fragments are characteristics of all hydrolysis reactions.

In ester hydrolysis, an ester reacts with water, producing the carboxylic acid and alcohol from which the ester was formed.

$$R-\overset{\overset{\text{O}}{\|}}{C}-O-R' + H-OH \xrightarrow{H^+} R-\overset{\overset{\text{O}}{\|}}{C}-OH + R'-O-H$$

$$CH_3-\overset{\overset{\text{O}}{\|}}{C}-O-CH_3 + H-OH \xrightarrow{H^+} CH_3-\overset{\overset{\text{O}}{\|}}{C}-OH + CH_3-O-H$$

Methyl acetate   Water   Acetic acid   Methyl alcohol

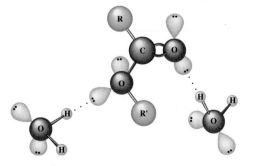

FIGURE 5.14 Low-molecular-mass esters are soluble in water because of ester–water hydrogen bonding.

Methyl Esters		
⨉	C_3	C_5	C_7
C_2	C_4	C_6	C_8

Ethyl Esters		
⨉	C_3	C_5	C_7
⨉	C_4	C_6	C_8

☐ Liquid

FIGURE 5.15 A physical-state summary for methyl and ethyl esters of unbranched-chain carboxylic acids at room temperature and pressure.

Ester hydrolysis requires the presence of a strong-acid catalyst or enzymes. Ester hydrolysis is the reverse of esterification (Section 5.11), the formation of an ester from a carboxylic acid and an alcohol.

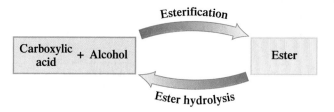

Ester Saponification

A **saponification reaction** *is the hydrolysis of an organic compound, under basic conditions, in which a carboxylic acid salt is one of the products.* Esters, amides (Section 6.17), and fats and oils (Section 8.6) all undergo saponification reactions.

In ester saponification either NaOH or KOH is used as the base and the saponification products are an alcohol and a carboxylic acid salt. (Any carboxylic acid product formed is converted to its salt because of the basic reaction conditions.)

$$R-\overset{\overset{\displaystyle O}{\|}}{C}-O-R' + \text{NaOH} \xrightarrow{H_2O} R-\overset{\overset{\displaystyle O}{\|}}{C}-O^- \text{Na}^+ + R'-OH$$

An ester   A strong base   A carboxylate salt   An alcohol

A specific example of ester saponification is

Methyl benzoate   Sodium hydroxide   Sodium benzoate   Methyl alcohol

In both ester hydrolysis and ester saponification, an alcohol is produced. Under acidic conditions (ester hydrolysis), the other product is a carboxylic acid. Under basic conditions (ester saponification), the other product is a carboxylic acid salt.

EXAMPLE 5.5

Structural Equations for Reactions That Involve Esters

■ Write structural equations for each of the following reactions.

a. Hydrolysis, with an acidic catalyst, of ethyl acetate
b. Saponification, with NaOH, of methyl formate
c. Esterification of propionic acid using isopropyl alcohol

Solution

a. Hydrolysis, under acidic conditions, cleaves an ester to produce its "parent" carboxylic acid and alcohol.

$$CH_3-\overset{\overset{\displaystyle O}{\|}}{C}-O-CH_2-CH_3 + H_2O \xrightarrow{H^+} CH_3-\overset{\overset{\displaystyle O}{\|}}{C}-OH + CH_3-CH_2-OH$$

Ethyl acetate   Acetic acid   Ethyl alcohol

b. Saponification cleaves an ester to produce its "parent" alcohol and the *salt* of its "parent" carboxylic acid.

$$H-\overset{\overset{\displaystyle O}{\|}}{C}-O-CH_3 + \text{NaOH} \xrightarrow{H_2O} H-\overset{\overset{\displaystyle O}{\|}}{C}-O^- \text{Na}^+ + CH_3-OH$$

Methyl formate   Sodium hydroxide   Sodium formate   Methyl alcohol

c. Esterification is the reaction in which a carboxylic acid and an alcohol react to produce an ester.

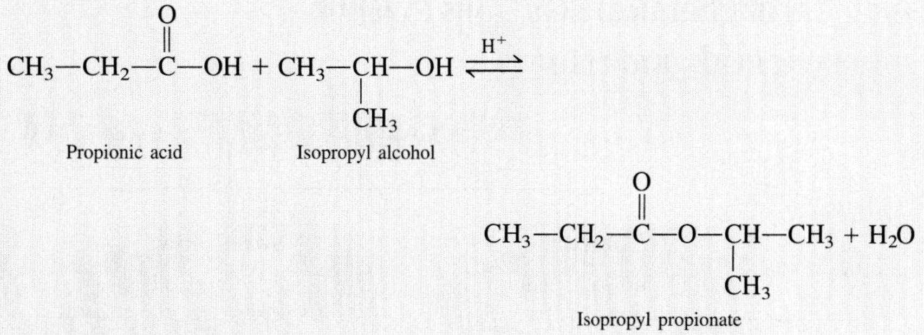

$$CH_3-CH_2-\overset{\overset{\textstyle O}{\|}}{C}-OH + CH_3-\underset{\underset{\textstyle CH_3}{|}}{CH}-OH \underset{}{\overset{H^+}{\rightleftharpoons}}$$

Propionic acid   Isopropyl alcohol

$$CH_3-CH_2-\overset{\overset{\textstyle O}{\|}}{C}-O-\underset{\underset{\textstyle CH_3}{|}}{CH}-CH_3 + H_2O$$

Isopropyl propionate

Practice Exercise 5.5

Write structural equations for each of the following reactions.

a. Hydrolysis, with an acidic catalyst, of propyl propanoate
b. Saponification, with KOH, of ethyl propanoate
c. Esterification of acetic acid with propyl alcohol

The Chemistry at a Glance feature on page 166 summarizes reactions that involve carboxylic acids and esters.

5.17 Sulfur Analogs of Esters

Just as alcohols react with carboxylic acids to produce esters, thiols (Section 3.20) react with carboxylic acids to produce thioesters. A **thioester** *is a sulfur-containing analog of an ester in which an —SR group has replaced the —OR group.*

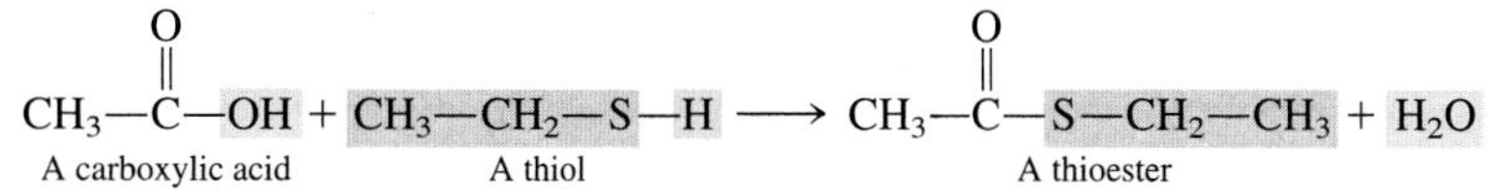

$$CH_3-\overset{\overset{\textstyle O}{\|}}{C}-OH + CH_3-CH_2-S-H \longrightarrow CH_3-\overset{\overset{\textstyle O}{\|}}{C}-S-CH_2-CH_3 + H_2O$$

A carboxylic acid   A thiol   A thioester

The thioester methyl thiobutanoate is used as an artificial flavoring agent. It generates the taste we call strawberry.

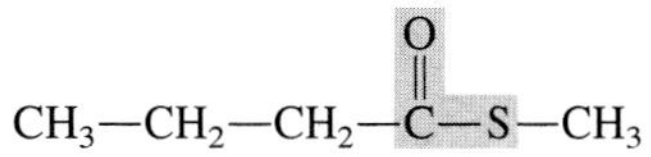

$$CH_3-CH_2-CH_2-\overset{\overset{\textstyle O}{\|}}{C}-S-CH_3$$

Methyl thiobutanoate

The most important naturally occurring thioester is acetyl coenzyme A, whose abbreviated structure is

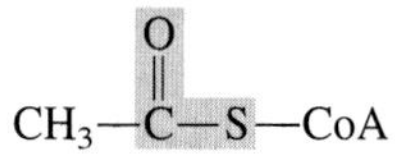

$$CH_3-\overset{\overset{\textstyle O}{\|}}{C}-S-CoA$$

Acetyl coenzyme A

Coenzyme A, the parent molecule for acetyl coenzyme A, is a large, complex *thiol* whose structure, for simplicity, is usually abbreviated as CoA—S—H. The formation of acetyl coenzyme A (acetyl CoA) from coenzyme A can be envisioned as a thioesterification reaction between acetic acid and coenzyme A.

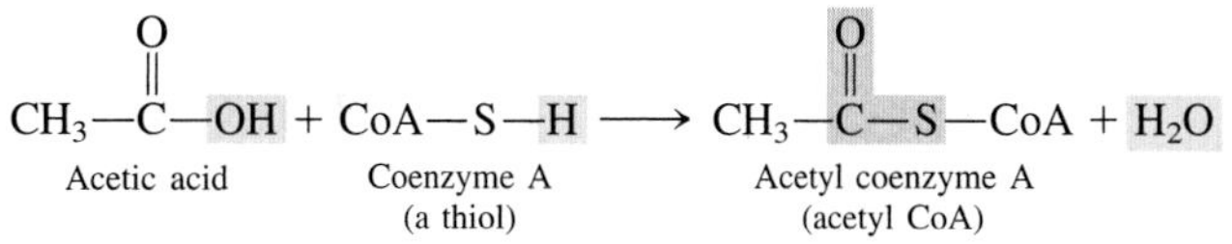

$$CH_3-\overset{\overset{\textstyle O}{\|}}{C}-OH + CoA-S-H \longrightarrow CH_3-\overset{\overset{\textstyle O}{\|}}{C}-S-CoA + H_2O$$

Acetic acid   Coenzyme A   Acetyl coenzyme A
                    (a thiol)   (acetyl CoA)

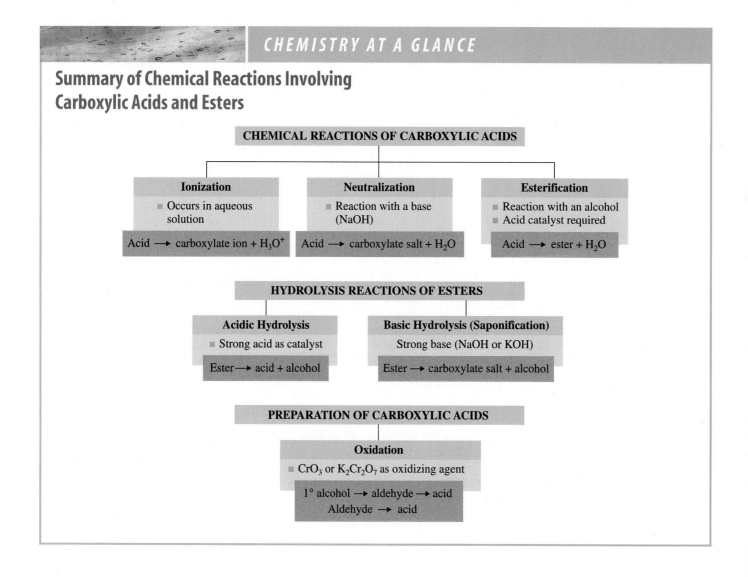

CHEMISTRY AT A GLANCE

Summary of Chemical Reactions Involving Carboxylic Acids and Esters

CHEMICAL REACTIONS OF CARBOXYLIC ACIDS

Ionization
- Occurs in aqueous solution

Acid ⟶ carboxylate ion + H_3O^+

Neutralization
- Reaction with a base (NaOH)

Acid ⟶ carboxylate salt + H_2O

Esterification
- Reaction with an alcohol
- Acid catalyst required

Acid ⟶ ester + H_2O

HYDROLYSIS REACTIONS OF ESTERS

Acidic Hydrolysis
- Strong acid as catalyst

Ester ⟶ acid + alcohol

Basic Hydrolysis (Saponification)
- Strong base (NaOH or KOH)

Ester ⟶ carboxylate salt + alcohol

PREPARATION OF CARBOXYLIC ACIDS

Oxidation
- CrO_3 or $K_2Cr_2O_7$ as oxidizing agent

1° alcohol ⟶ aldehyde ⟶ acid
Aldehyde ⟶ acid

Acetyl coenzyme A plays a central role in the metabolic cycles through which the body obtains energy to "run itself" (Section 12.6).

The complete structure of acetyl CoA is given in Section 12.3.

5.18 Polyesters

A **condensation polymer** *is a polymer formed by reacting difunctional monomers to give a polymer and some small molecule (such as water) as a by-product of the process.* Polyesters are an important type of condensation polymer. A **polyester** *is a condensation polymer in which the monomers are joined through ester linkages.* Dicarboxylic acids and dialcohols are the monomers generally used in forming polyesters.

The best known of the many polyesters now marketed is *poly(ethylene terephthalate)*, which is also known by the acronym *PET*. The monomers used to produce PET are terephthalic acid (a diacid) and ethylene glycol (a dialcohol).

Condensation polymerization reactions produce two products: the polymer and a small molecule. This contrasts with addition polymerization reactions (Section 2.9) where the polymer is the only product.

$$HO-\underset{\underset{O}{\|}}{C}-\underset{}{\bigcirc}-\underset{\underset{O}{\|}}{C}-OH \qquad\qquad HO-CH_2-CH_2-OH$$

Terephthalic acid   Ethylene glycol

FIGURE 5.16 Space-filling model of a segment of the polyester condensation polymer known as poly(ethylene terephthalate), or PET.

The reaction of one acid group of the diacid with one alcohol group of the dialcohol initially produces an ester molecule, with an acid group left over on one end and an alcohol group left over on the other end.

Leftover acid group that can react further

Ester linkage

Leftover alcohol group that can react further

This species can react further. The remaining acid group can react with an alcohol group from another monomer, and the alcohol group can react with an acid group from another monomer. This process continues until an extremely long polymer molecule called a *polyester* is produced (see Figure 5.16).

Ester linkage

Ester linkage

Poly(ethylene terephthalate), a polyester

About 80% of PET production goes into textile products, including clothing fibers, curtain and upholstery materials, and tire cord. The trade name for PET as a clothing fiber is *Dacron.* The other 20% of PET production goes into plastics applications. As a film-like material, it is called *Mylar.* Mylar products include the plastic backing for audio and video tapes and computer diskettes. Its chemical name PET is applied when this polyester is used in clear, flexible soft-drink bottles and as the wrapping material for frozen foods and boil-in bags for foods.

PET is also used in medicine. Because it is physiologically inert, PET is used in the form of a mesh to replace diseased sections of arteries. It has also been used in synthetic heart valves.

A variation of the diacid–dialcohol monomer formulation for polyesters involves using hydroxyacids as monomers. In this situation, both of the functional groups required are present in the same molecule.

A polymerization reaction in which lactic acid and glycolic acid (both hydroxyacids, Section 5.4) are monomers produces a biodegradable material (trade name *Lactomer*) that is used as surgical staples in several types of surgery. Traditional suture materials must be removed later on, after they have served their purpose. Lactomer staples start to dissolve (hydrolyze) after a period of several weeks. The hydrolysis products are the starting monomers, lactic acid and glycolic acid, both of which are normally present in the human body. By the time the tissue has fully healed, the staples have fully degraded.

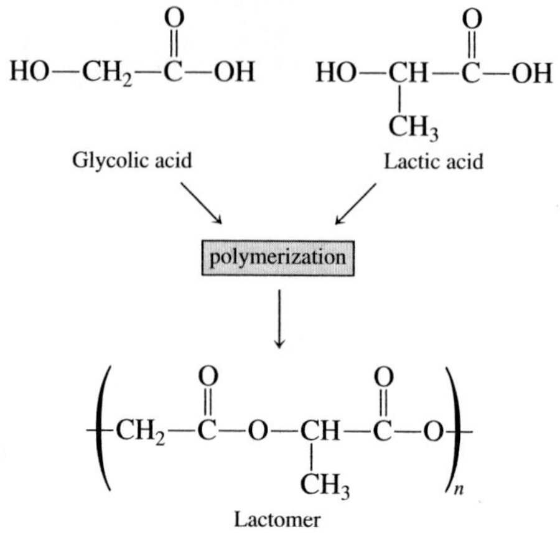

5.19 Acid Chlorides and Acid Anhydrides

Sections 5.11 through 5.18 have focused on the carboxylic acid derivatives called esters. We now consider two more of the carboxylic acid derivatives types listed in Section 5.1, namely carboxylic acid chlorides and carboxylic acid anhydrides.

Acid Chlorides

An **acid chloride** *is a carboxylic acid derivative in which the —OH portion of the carboxyl group has been replaced with a —Cl atom.* Thus, acid chlorides have the general formula

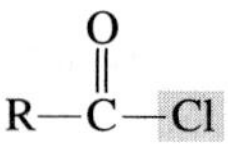

Acid chlorides are named in either of two ways:

1. Replace the *-ic acid* ending of the common name of the parent carboxylic acid with *-yl chloride.*

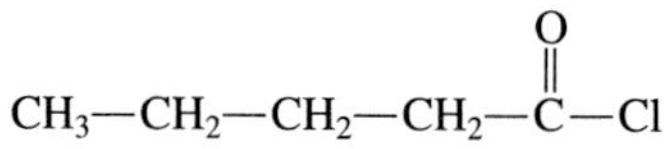

Butyr*ic acid* becomes butyr*yl chloride.*

2. Replace the *-oic acid* ending of the IUPAC name of the parent carboxylic acid with *-oyl chloride.*

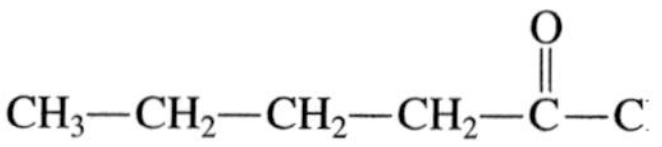

3-Methylpentan*oic acid* becomes 3-methylpentan*oyl chloride.*

Preparation of an acid chloride from its parent carboxylic acid involves reacting the acid with one of several inorganic chlorides (PCl$_3$, PCl$_5$, or SOCl$_2$). The general reaction is

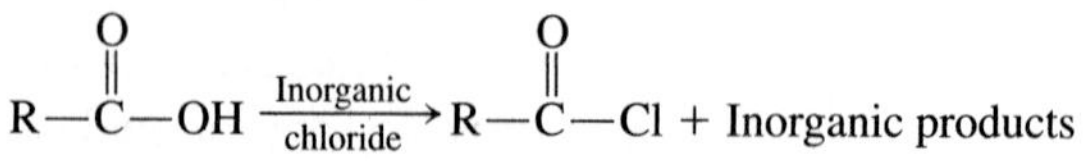

Acid chlorides react rapidly with water, in a hydrolysis reaction, to regenerate the parent carboxylic acid.

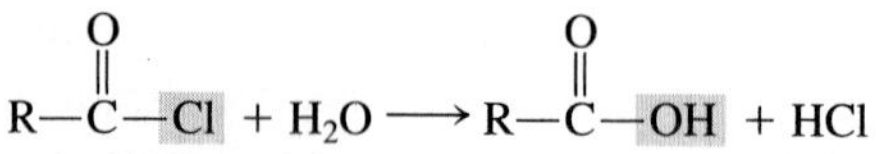

This reactivity with water means that acid chlorides cannot exist in biological systems.

Acid chlorides are useful starting materials for the synthesis of other carboxylic acid derivatives, particularly esters and amides. Synthesis of esters and amides using acid chlorides is a more efficient process than ester and amide synthesis using a carboxylic acid.

■ Acid Anhydrides

An **acid anhydride** *is a carboxylic acid derivative in which the* —*OH portion of the carboxyl*

group *has been replaced with a* —O—C̈—R *group*. Thus, acid anhydrides have the general formula

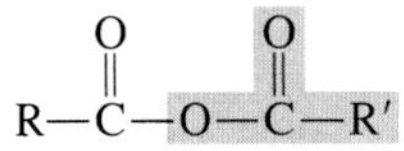

$$R-\overset{O}{\underset{\|}{C}}-O-\overset{O}{\underset{\|}{C}}-R'$$

The word *anhydride* means "without water." Structurally, acid anhydrides can be visualized as two carboxylic acid molecules bonded together after removal of a water molecule from the acid molecules.

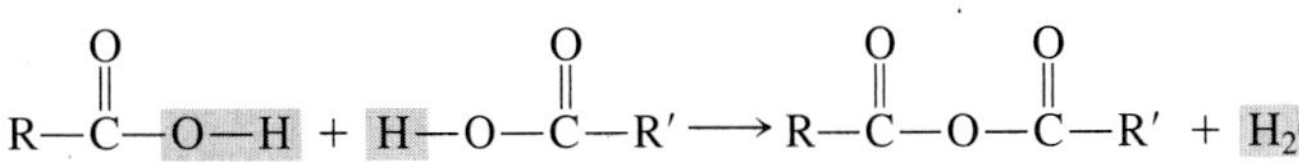

$$R-\overset{O}{\underset{\|}{C}}-O-H + H-O-\overset{O}{\underset{\|}{C}}-R' \longrightarrow R-\overset{O}{\underset{\|}{C}}-O-\overset{O}{\underset{\|}{C}}-R' + H_2O$$

Symmetrical acid anhydrides (both R groups are the same) are named by replacing the *acid* ending of the parent carboxylic acid name with the word *anhydride*.

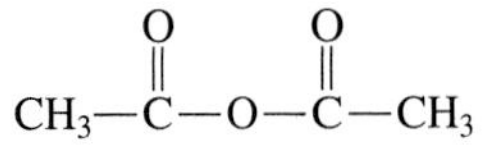

$$CH_3-\overset{O}{\underset{\|}{C}}-O-\overset{O}{\underset{\|}{C}}-CH_3$$

IUPAC name: Ethanoic anhydride
Common name: Acetic anhydride

Mixed acid anhydrides (different R groups present) are named by using the names of the individual parent carboxylic acids (in alphabetic order) followed by the word *anhydride*.

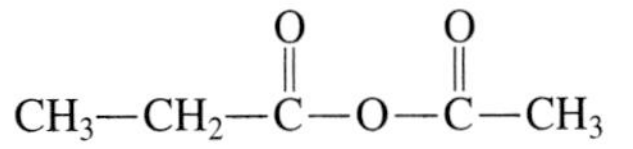

$$CH_3-CH_2-\overset{O}{\underset{\|}{C}}-O-\overset{O}{\underset{\|}{C}}-CH_3$$

IUPAC name: Ethanoic propanoic anhydride
Common name: Acetic propionic anhydride

In general, acid anhydrides cannot be formed by directly reacting the parent carboxylic acids together, Instead, an acid chloride is reacted with a carboxylate ion to produce the acid anhydride.

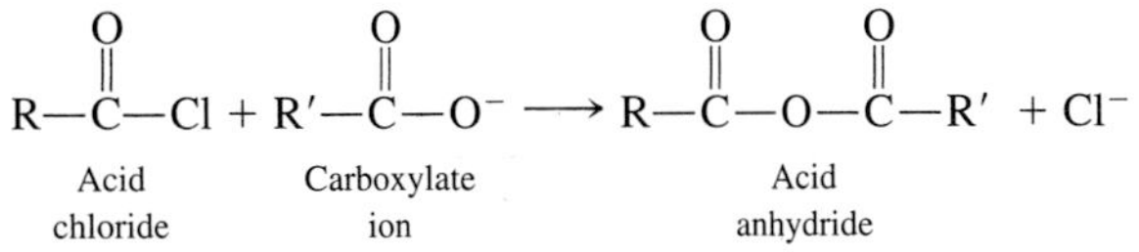

$$R-\overset{O}{\underset{\|}{C}}-Cl + R'-\overset{O}{\underset{\|}{C}}-O^- \longrightarrow R-\overset{O}{\underset{\|}{C}}-O-\overset{O}{\underset{\|}{C}}-R' + Cl^-$$

Acid   Carboxylate   Acid
chloride   ion   anhydride

Acid anhydrides are very reactive compounds, although generally not as reactive as the acid chlorides. Like acid chlorides, they cannot exist in biological systems, as they undergo hydrolysis to regenerate the parent carboxylic acids.

$$R-\overset{O}{\underset{\|}{C}}-O-\overset{O}{\underset{\|}{C}}-R' + H_2O \xrightarrow{\text{Heat}} R-\overset{O}{\underset{\|}{C}}-OH + R'-\overset{O}{\underset{\|}{C}}-OH$$

Acid   Acid   Acid
anhydride

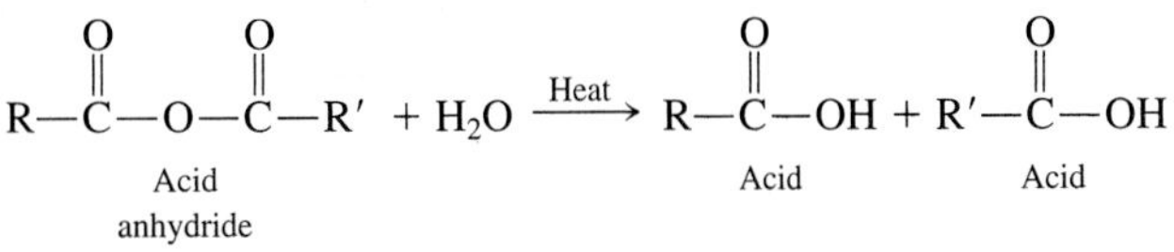

Reaction of an alcohol with an acid anhydride is a useful method for synthesizing esters.

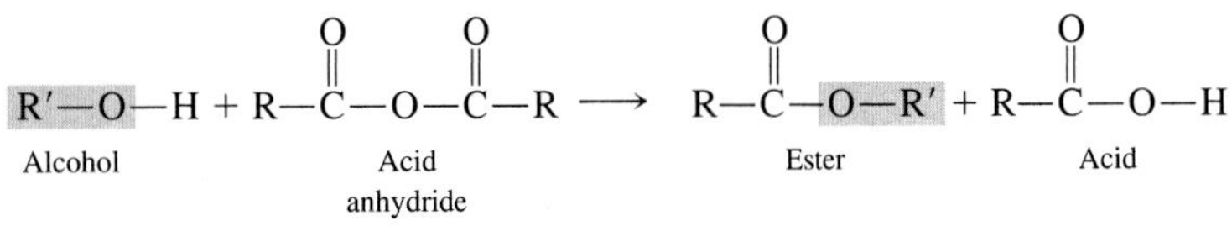

$$R'-O-H + R-\overset{O}{\underset{\|}{C}}-O-\overset{O}{\underset{\|}{C}}-R \longrightarrow R-\overset{O}{\underset{\|}{C}}-O-R' + R-\overset{O}{\underset{\|}{C}}-O-H$$

Alcohol   Acid   Ester   Acid
            anhydride

5.20 Esters and Anhydrides of Inorganic Acids

Inorganic acids such as sulfuric, phosphoric, and nitric acids react with alcohols to form esters in a manner similar to that for carboxylic acids.

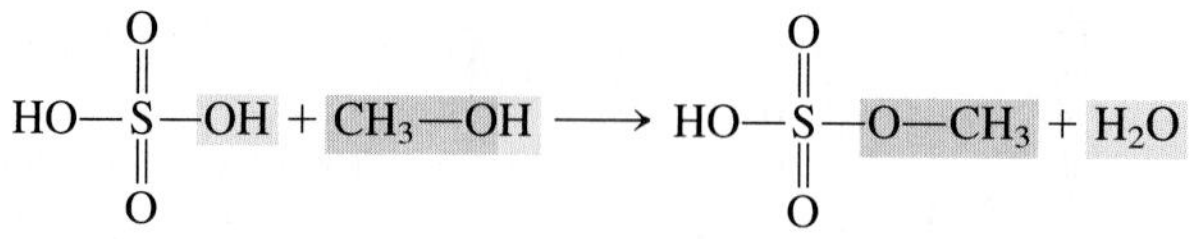

Sulfuric acid
(H_2SO_4)

Methyl ester of
sulfuric acid

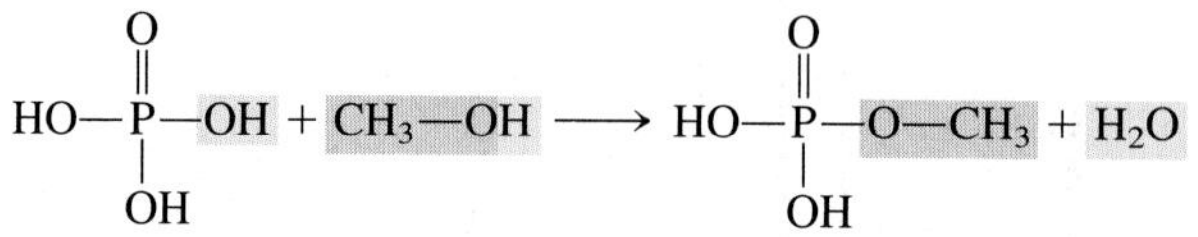

Phosphoric acid
(H_3PO_4)

Methyl ester of
phosphoric acid

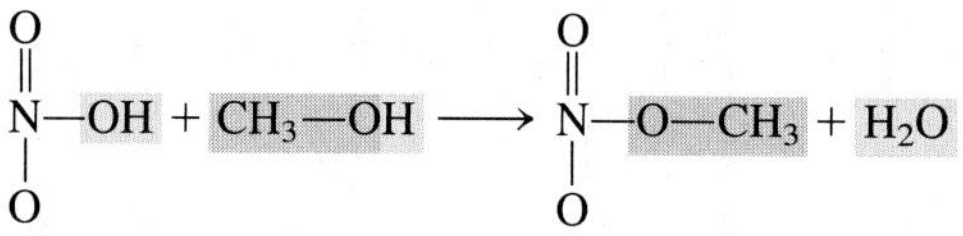

Nitric acid
(HNO_3)

Methyl ester of
nitric acid

Esters of inorganic acids undergo hydrolysis reactions in a manner similar to that for esters of carboxylic acids (Section 5.16).

The most important inorganic esters, from a biochemical standpoint, are those of phosphoric acid—that is, phosphate esters. A **phosphate ester** *is an organic compound formed by reaction of an alcohol with phosphoric acid.* Because phosphoric acid has three hydroxyl groups, it can form mono-, di-, and triesters by reaction with one, two, and three molecules of alcohol, respectively.

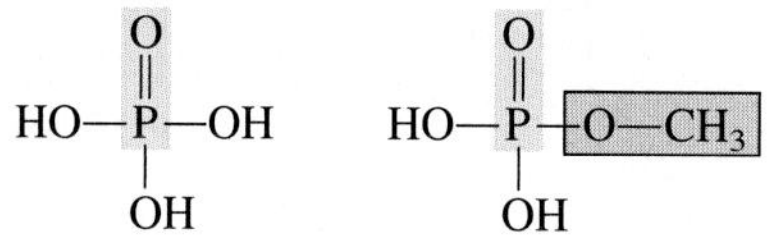

Phosphoric acid   Monoester (one —OR group)

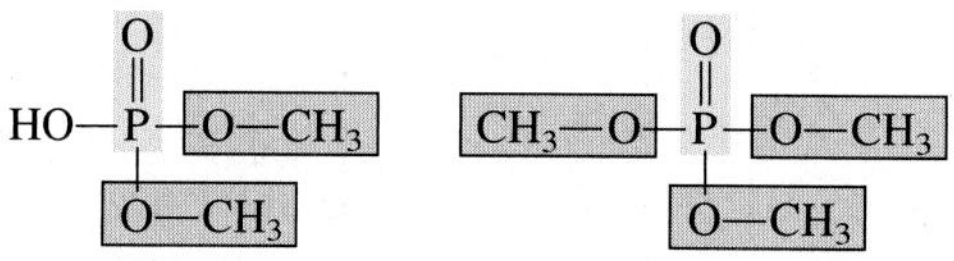

Diester (two —OR groups)   Triester (three —OR groups)

■ Phosphoric Acid Anhydrides

Three biologically important phosphoric acids exist: phosphoric acid, diphosphoric acid, and triphosphoric acid. Phosphoric acid, the simplest of the three acids, undergoes intermolecular dehydration to produce diphosphoric acid.

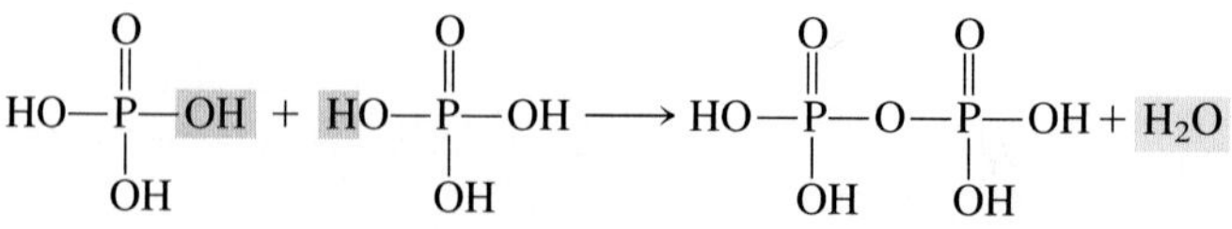

Phosphoric acid   Phosphoric acid   Diphosphoric acid

Another intermolecular dehydration, involving diphosphoric acid and phosphoric acid, produces triphosphoric acid.

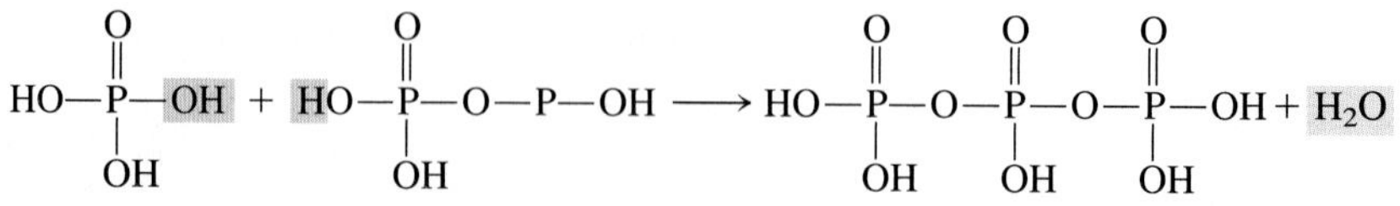

Triphosphoric acid

CHEMICAL CONNECTIONS Nitroglycerin: An Inorganic Triester

The reaction of one molecule of glycerol (a trihydroxy-alcohol) with three molecules of nitric acid produces the trinitrate ester called nitroglycerin.

$$
\begin{array}{l}
CH_2{-}OH \\
| \\
CH{-}OH \quad + 3HO{-}NO_2 \\
| \\
CH_2{-}OH
\end{array}
\longrightarrow
\begin{array}{l}
CH_2{-}O{-}NO_2 \\
| \\
CH{-}O{-}NO_2 \quad + 3H_2O \\
| \\
CH_2{-}O{-}NO_2
\end{array}
$$

Besides being a component of dynamite explosives, nitroglycerin has medicinal value. It is used in treating patients with angina pectoris—sharp chest pains caused by an insufficient supply of oxygen reaching heart muscle. Its effect on the human body is that of a vasodilator, a substance that increases blood flow by relaxing constricted muscles around blood vessels.

Nitroglycerin medication is available in several forms: (1) as a liquid diluted with alcohol to render it nonexplosive, (2) as a liquid adsorbed to a tablet for convenience of sublingual (under the tongue) administration, (3) in ointments for topical use, and (4) as "skin patches" that release the drug continuously through the skin over a 24-hr period. Nitroglycerin is rapidly absorbed through the skin, enters the bloodstream, and finds its way to heart muscle within seconds.

In the pure state, nitroglycerin is a shock-sensitive liquid that can decompose to produce large volumes of gases (N_2, CO_2, H_2O, and O_2). When used in dynamite, it is adsorbed on clay-like materials, giving products that will not explode without a formal ignition system.

Another compound used for the same medicinal purposes as nitroglycerin is isopentyl nitrite. It is a monoester involving nitrous acid (HNO_2) and isopentyl alcohol (3-methyl-1-butanol).

$$
\begin{array}{l}
CH_3{-}CH{-}CH_2{-}CH_2{-}OH + HO{-}NO \longrightarrow \\
\qquad\quad | \\
\qquad\quad CH_3
\end{array}
$$

$$
\begin{array}{l}
CH_3{-}CH{-}CH_2{-}CH_2{-}O{-}NO + H_2O \\
\qquad\quad | \\
\qquad\quad CH_3
\end{array}
$$

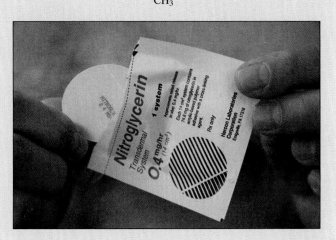

In the same manner that carboxylic acids are acidic (Section 5.8), phosphoric acid, diphosphoric acid, and triphosphoric acid are also acidic. The phosphoric acids are, however, polyprotic rather than monoprotic acids. The hydrogen atom in each of the —OH groups possesses acidic properties. All three phosphoric acids undergo esterification reactions with alcohols, producing species such as

$$
\underset{\text{Diphosphate monoester}}{
R{-}O{-}\overset{\displaystyle O}{\underset{\displaystyle OH}{\overset{\displaystyle \|}{P}}}{-}O{-}\overset{\displaystyle O}{\underset{\displaystyle OH}{\overset{\displaystyle \|}{P}}}{-}OH
}
\qquad \text{and} \qquad
\underset{\text{Triphosphate monoester}}{
R{-}O{-}\overset{\displaystyle O}{\underset{\displaystyle OH}{\overset{\displaystyle \|}{P}}}{-}O{-}\overset{\displaystyle O}{\underset{\displaystyle OH}{\overset{\displaystyle \|}{P}}}{-}O{-}\overset{\displaystyle O}{\underset{\displaystyle OH}{\overset{\displaystyle \|}{P}}}{-}OH
}
$$

We will encounter esters of these types in Chapter 12 when we consider the biochemical production of energy in the human body. Adenosine diphosphate (ADP) and adenosine triphosphate (ATP) are important examples of such compounds.

Diphosphoric acid and triphosphoric acid are phosphoric acid anhydrides as well as acids. Note the structural similarities between a carboxylic acid anhydride and diphosphoric acid.

$$
\underset{\text{Carboxylic acid anhydride}}{
R{-}\overset{\displaystyle O}{\overset{\displaystyle \|}{C}}{-}O{-}\overset{\displaystyle O}{\overset{\displaystyle \|}{C}}{-}R
}
\qquad\qquad
\underset{\text{Diphosphoric acid}}{
HO{-}\overset{\displaystyle O}{\underset{\displaystyle OH}{\overset{\displaystyle \|}{P}}}{-}O{-}\overset{\displaystyle O}{\underset{\displaystyle OH}{\overset{\displaystyle \|}{P}}}{-}OH
}
$$

Phosphoric acid anhydride systems play important roles in cellular processes through which biochemical energy is produced. The presence of phosphoric anhydride systems in biological settings contrasts markedly with carboxylic acid anhydride systems (Section 5.19), which are not found in biological settings because of their reactivity with water.

CONCEPTS TO REMEMBER

The carboxyl group. The functional group present in carboxylic acids is the carboxyl group. A carboxyl group is composed of a hydroxyl group bonded to a carbonyl carbon atom. It thus contains two oxygen atoms directly bonded to the same carbon atom (Section 5.1).

Carboxylic acid derivatives. Four important families of carboxylic acid derivatives are esters, acid chlorides, acid anhydrides, and amides. The group attached to the carbonyl carbon atom distinguishes these derivatives from each other and also from carboxylic acids (Section 5.1).

Nomenclature of carboxylic acids. The IUPAC name for a monocarboxylic acid is formed by replacing the final *-e* of the hydrocarbon parent name with *-oic acid*. As with previous IUPAC nomenclature, the longest carbon chain containing the functional group is identified, and it is numbered starting with the carboxyl carbon atom. Common-name usage is more prevalent for carboxylic acids than for any other type of organic compound (Sections 5.2 and 5.3).

Types of carboxylic acids. Carboxylic acids are classified by the number of carboxyl groups present (monocarboxylic, dicarboxylic, etc.), by the degree of saturation (saturated, unsaturated, aromatic), and by additional functional groups present (hydroxy, keto, etc.) (Sections 5.4 and 5.5).

Physical properties of carboxylic acids. Low-molecular-mass carboxylic acids are liquids at room temperature and have sharp or unpleasant odors. Long-chain acids are waxlike solids. The carboxyl group is polar and forms hydrogen bonds to other carboxyl groups or other molecules. Thus carboxylic acids have relatively high boiling points, and those with lower molecular masses are soluble in water (Section 5.6).

Preparation of carboxylic acids. Carboxylic acids are synthesized through oxidation of primary alcohols or aldehydes using strong oxidizing agents. Aromatic carboxylic acids can be prepared by oxidizing a carbon side chain on a benzene derivative using a strong oxidizing agent (Section 5.7).

Acidity of carboxylic acids. Soluble carboxylic acids behave as weak acids, donating protons to water molecules. The portion of the acid molecule left after proton loss is called a carboxylate ion (Section 5.8).

Carboxylic acid salts. Carboxylic acids are neutralized by bases to produce carboxylic acid salts. Such salts are usually more soluble in water than are the acids from which they were derived. Carboxylic acid salts are named by changing the *-ic* ending of the acid to *-ate* (Section 5.9).

Esters. Esters are formed by the reaction of an acid with an alcohol. In such reactions, the —OR group from the alcohol replaces the —OH group in the carboxylic acid. Esters are polar compounds, but they cannot form hydrogen bonds to each other. Therefore, their boiling points are lower than those of alcohols and acids of similar molecular mass (Sections 5.10 and 5.11).

Nomenclature of esters. An ester is named as an alkyl (from the name of the alcohol reactant) carboxylate (from the name of the acid reactant) (Section 5.12).

Chemical reactions of esters. Esters can be converted back to carboxylic acids and alcohols under either acidic or basic conditions. Under acidic conditions, the process is called hydrolysis, and the products are the acid and alcohol. Under basic conditions, the process is called saponification, and the products are the acid salt and alcohol (Section 5.16).

Thioesters. Thioesters are sulfur-containing analogs of esters in which a —SR group has replaced the —OR group (Section 5.17).

Polyesters. Polyesters are polymers in which the monomers (diacids and dialcohols) are joined through ester linkages (Section 5.18).

Acid chlorides and acid anhydrides. An acid chloride is a carboxylic acid derivative in which the —OH portion of the carboxyl group has been replaced with a —Cl atom. An acid anhydride involves two carboxylic acid molecules bonded together after intermolecular dehydration has occurred. Both acid chlorides and acid anhydrides are very reactive molecules (Section 5.19).

Esters and anhydrides of inorganic acids. Alcohols can react with inorganic acids, such as nitric, sulfuric, and phosphoric acids, to form esters. Phosphate esters are an important class of biochemical compounds. Anhydrides of phosphoric acid (diphosphoric acid and triphosphoric acid) and their esters are also important types of biochemical molecules (Section 5.20).

KEY REACTIONS AND EQUATIONS

1. Oxidation of a primary alcohol to an acid (Section 5.7)

$$R-CH_2-OH \xrightarrow{[O]} R-\overset{\overset{\displaystyle O}{\|}}{C}-H \xrightarrow{[O]} R-\overset{\overset{\displaystyle O}{\|}}{C}-OH$$

2. Oxidation of an alkylbenzene to an acid (Section 5.7)

3. Ionization of a carboxylic acid to give a carboxylate ion and a hydronium ion (Section 5.8)

$$R-\overset{\overset{\displaystyle O}{\|}}{C}-OH + H_2O \rightleftharpoons R-\overset{\overset{\displaystyle O}{\|}}{C}-O^- + H_3O^+$$

4. Reaction of a carboxylic acid with a base to produce a carboxylic acid salt plus water (Section 5.9)

$$R-\overset{\overset{\displaystyle O}{\|}}{C}-OH + NaOH \longrightarrow R-\overset{\overset{\displaystyle O}{\|}}{C}-O^-Na^+ + H_2O$$

5. Preparation of an ester from an acid and an alcohol
(Section 5.11)

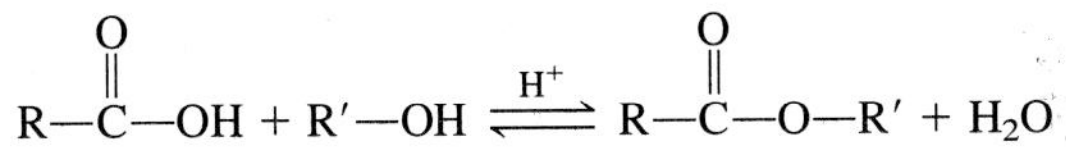

$$R-\overset{O}{\underset{\|}{C}}-OH + R'-OH \underset{}{\overset{H^+}{\rightleftharpoons}} R-\overset{O}{\underset{\|}{C}}-O-R' + H_2O$$

6. Ester hydrolysis to produce a carboxylic acid and an alcohol
(Section 5.16)

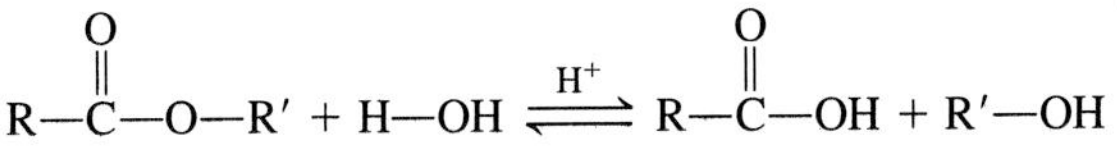

$$R-\overset{O}{\underset{\|}{C}}-O-R' + H-OH \underset{}{\overset{H^+}{\rightleftharpoons}} R-\overset{O}{\underset{\|}{C}}-OH + R'-OH$$

7. Ester saponification to give a carboxylic acid salt and alcohol
(Section 5.16)

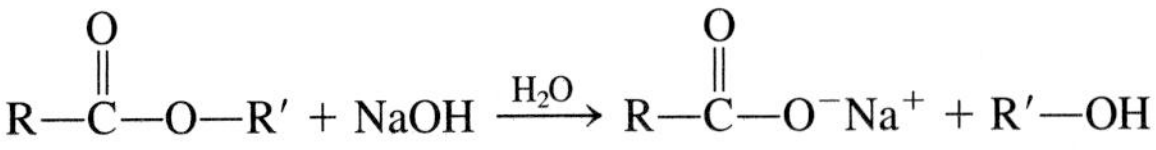

$$R-\overset{O}{\underset{\|}{C}}-O-R' + NaOH \xrightarrow{H_2O} R-\overset{O}{\underset{\|}{C}}-O^-Na^+ + R'-OH$$

8. Preparation of a thioester (Section 5.17)

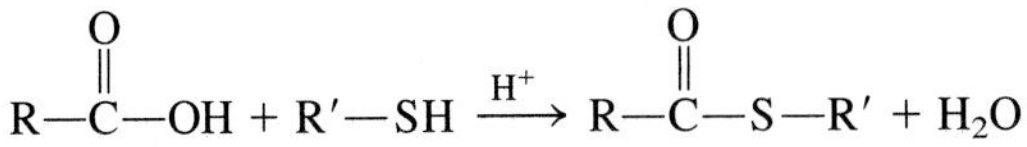

$$R-\overset{O}{\underset{\|}{C}}-OH + R'-SH \xrightarrow{H^+} R-\overset{O}{\underset{\|}{C}}-S-R' + H_2O$$

9. Phosphate ester formation (Section 5.20)

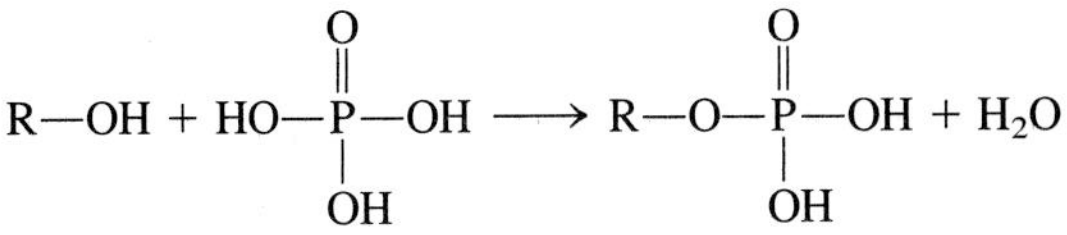

$$R-OH + HO-\overset{O}{\underset{\underset{OH}{|}}{\overset{\|}{P}}}-OH \longrightarrow R-O-\overset{O}{\underset{\underset{OH}{|}}{\overset{\|}{P}}}-OH + H_2O$$

EXERCISES AND PROBLEMS

The members of each pair of problems in this section test similar material.

The Carboxyl Functional Group (Section 5.1)

5.1 In which of the following compounds is a carboxyl group present?

a. 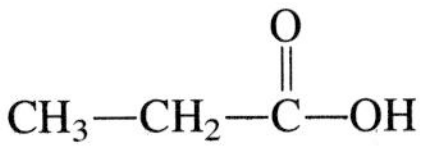
$$CH_3-CH_2-\overset{O}{\underset{\|}{C}}-OH$$

b.
$$CH_3-CH_2-CH_2-\overset{O}{\underset{\|}{C}}-CH_3$$

c.
phenyl-$\overset{O}{\underset{\|}{C}}$-OH

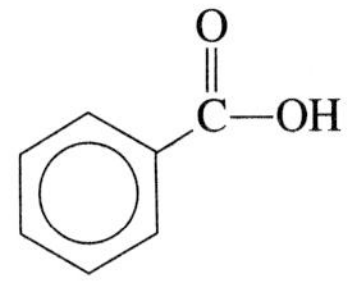

d. 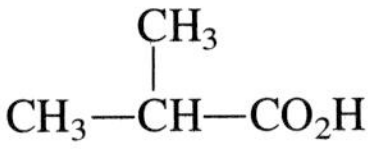
$$CH_3-\overset{\overset{CH_3}{|}}{C}H-CO_2H$$

e. 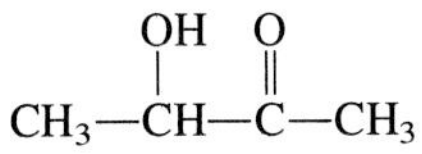
$$CH_3-\overset{\overset{OH}{|}}{C}H-\overset{O}{\underset{\|}{C}}-CH_3$$

f. CH_3-CH_2-COOH

5.2 In which of the following compounds is a carboxyl group present?

a. 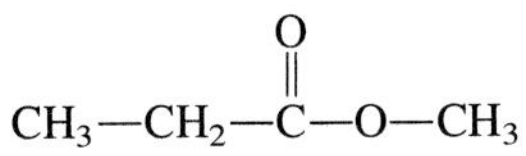
$$CH_3-CH_2-\overset{O}{\underset{\|}{C}}-O-CH_3$$

b. 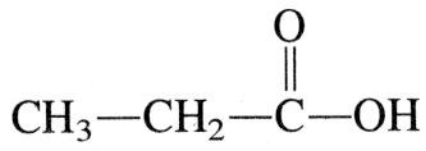
$$CH_3-CH_2-\overset{O}{\underset{\|}{C}}-OH$$

c. $HOOC-CH_2-CH_3$

d. 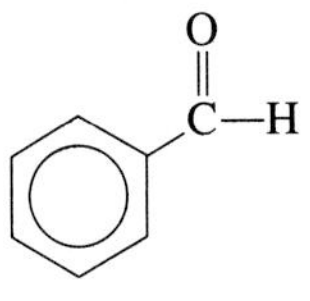
phenyl-$\overset{O}{\underset{\|}{C}}$-H

e. 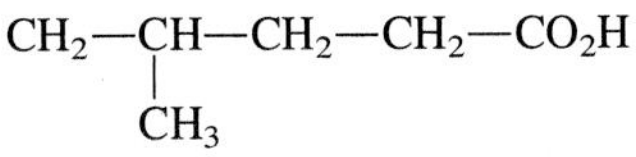 $CH_2-CH-CH_2-CH_2-CO_2H$ with CH_3 branch

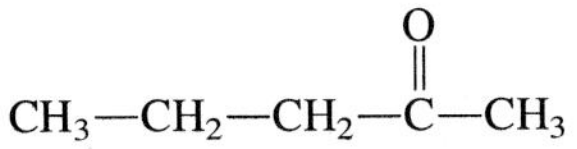

f. 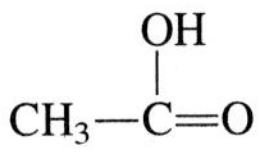
$$CH_3-\overset{OH}{\underset{}{C}}=O$$

IUPAC Nomenclature for Carboxylic Acids (Section 5.2)

5.3 Give the IUPAC name for each of the following carboxylic acids.

a. 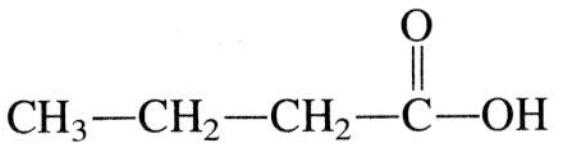
$$CH_3-CH_2-CH_2-\overset{O}{\underset{\|}{C}}-OH$$

b.
$$CH_3-CH_2-CH_2-CH_2-CH_2-CH_2-\overset{O}{\underset{\|}{C}}-OH$$

c.
$$CH_3-CH_2-\overset{\overset{CH_3}{|}}{C}H-\overset{\overset{CH_3}{|}}{C}H-\overset{O}{\underset{\|}{C}}-OH$$

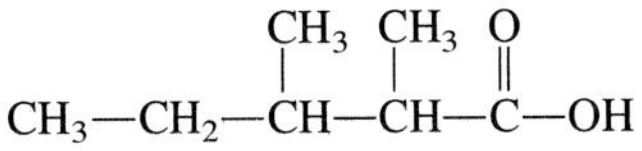

d. 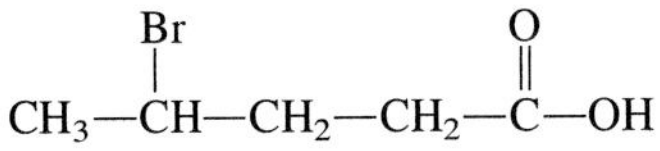
$$CH_3-\overset{\overset{Br}{|}}{C}H-CH_2-CH_2-\overset{O}{\underset{\|}{C}}-OH$$

e. 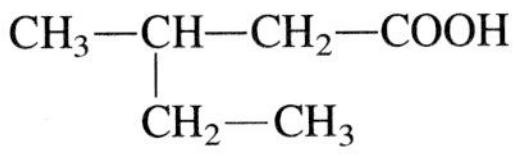 $CH_3-CH-CH_2-COOH$ with CH_2-CH_3 branch

f. 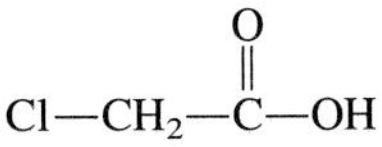
$Cl-CH_2-\overset{O}{\underset{\|}{C}}-OH$

5.4 Give the IUPAC name for each of the following carboxylic acids.

a. 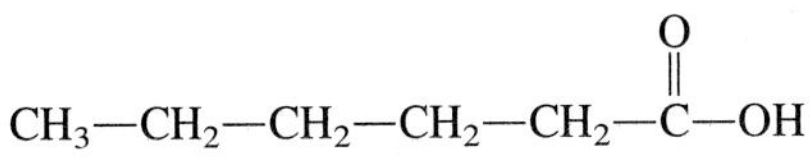
$$CH_3-CH_2-CH_2-CH_2-CH_2-\overset{O}{\underset{\|}{C}}-OH$$

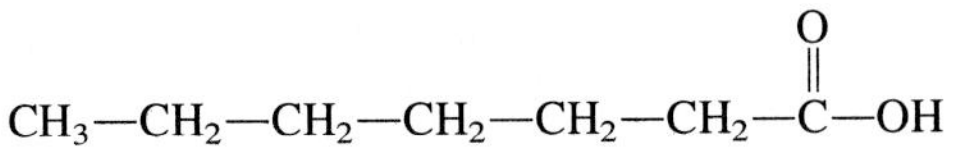

b.
$$CH_2-CH-\overset{O}{\underset{\|}{C}}-OH$$ with CH_3, CH_3 branches

c.
$$CH_3-\overset{\overset{CH_3}{|}}{\underset{\underset{Cl}{|}}{C}}-\overset{O}{\underset{\|}{C}}-OH$$

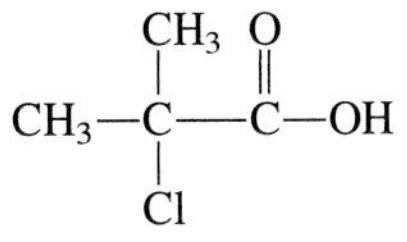

d. 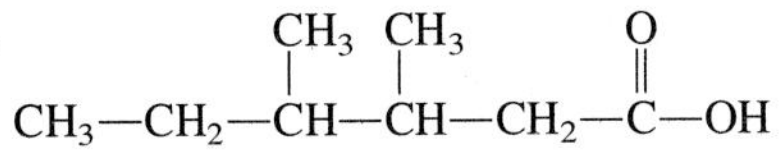
$$CH_3-CH_2-\overset{\overset{CH_3}{|}}{C}H-\overset{\overset{CH_3}{|}}{C}H-CH_2-\overset{O}{\underset{\|}{C}}-OH$$

e. $HOOC-CH_3$

f. 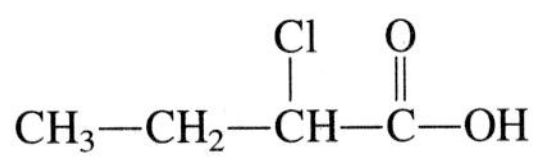
$$CH_3-CH_2-\overset{\overset{Cl}{|}}{C}H-\overset{O}{\underset{\|}{C}}-OH$$

5.5 Assign an IUPAC name to each of the following carboxylic acids.

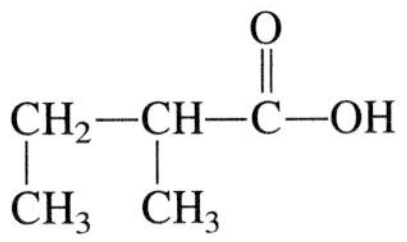

a. 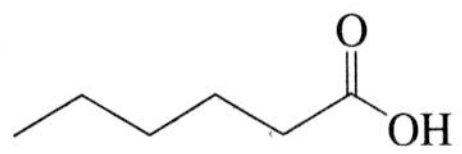 structure with OH
b. 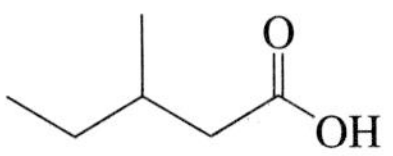 structure with OH
c. 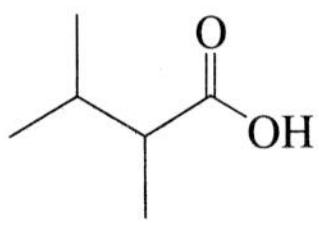 structure with OH
d. 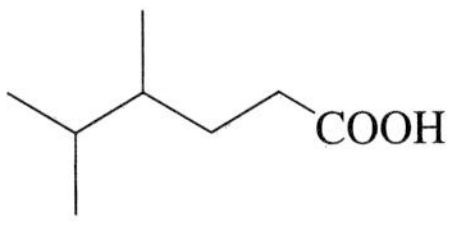 structure with COOH

5.6 Assign an IUPAC name to each of the following carboxylic acids.

a.

b.

c.

d.

5.7 Draw a condensed structural formula that corresponds to each of the following carboxylic acids.
a. 2-Ethylbutanoic acid
b. 2,5-Dimethylhexanoic acid
c. Methylpropanoic acid
d. Dichloroethanoic acid
e. 3-Bromo-5-chlorooctanoic acid
f. 2,3-Dimethylbutanoic acid

5.8 Draw a condensed structural formula that corresponds to each of the following carboxylic acids.
a. 3,3-Dimethylheptanoic acid
b. 4-Methylpentanoic acid
c. 3-Chloropropanoic acid
d. Trichloroethanoic acid
e. 3-Isopropylhexanoic acid
f. 4-Ethyl-3,5-dimethylhexanoic acid

5.9 Give the IUPAC name for each of the following carboxylic acids.

a.
$$HO-\overset{O}{\overset{||}{C}}-CH_2-CH_2-\overset{O}{\overset{||}{C}}-OH$$

b.

c.
$$HO-\overset{O}{\overset{||}{C}}-CH_2-\overset{CH_3}{\overset{|}{CH}}-CH_2-\overset{O}{\overset{||}{C}}-OH$$

d.

e.

f.

5.10 Give the IUPAC name for each of the following carboxylic acids.

a.
$$HO-\overset{O}{\overset{||}{C}}-CH_2-CH_2-CH_2-\overset{O}{\overset{||}{C}}-OH$$

b.

c.
$$HO-\overset{O}{\overset{||}{C}}-\overset{Cl}{\overset{|}{CH}}-CH_2-CH_2-\overset{O}{\overset{||}{C}}-OH$$

d.

---

e.

f.

5.11 Draw a condensed structural formula that corresponds to each of the following carboxylic acids.
a. 2,2-Dimethylbutanoic acid
b. 2,2-Dimethylbutanedioic acid
c. 2,2-Dimethylpentanedioic acid
d. *o*-Bromobenzoic acid
e. 2,4-Dichlorobenzoic acid
f. *p*-Toluic acid

5.12 Draw a condensed structural formula that corresponds to each of the following carboxylic acids.
a. 2,3-Dichlorohexanoic acid
b. 2,3-Dichlorohexanedioic acid
c. 2,3-Dichloroheptanedioic acid
d. *m*-Bromobenzoic acid
e. 3,5-Dichlorobenzoic acid
f. *o*-Toluic acid

■ Common Names for Carboxylic Acids (Section 5.3)

5.13 Draw a condensed structural formula that corresponds to each of the following carboxylic acids.
a. Valeric acid
b. Propionic acid
c. Acetic acid
d. α-Chlorobutyric acid
e. β-Bromocaproic acid
f. γ-Chloro-α-methylvaleric acid

5.14 Draw a condensed structural formula that corresponds to each of the following carboxylic acids.
a. Butyric acid
b. Caproic acid
c. Formic acid
d. Chloroacetic acid
e. α-Methylpropionic acid
f. β-Chloro-β-iodocaproic acid

5.15 Draw a condensed structural formula that corresponds to each of the following carboxylic acids.
a. Malonic acid
b. Succinic acid
c. Adipic acid
d. γ-Bromopimelic acid
e. α-Methylglutaric acid
f. α-Bromo-α-chlorosuccinic acid

5.16 Draw a condensed structural formula that corresponds to each of the following carboxylic acids.
a. Oxalic acid
b. Glutaric acid
c. Pimelic acid
d. Chloromalonic acid
e. α-Methyladipic acid
f. α,β-Dichlorosuccinic acid

5.17 Classify the two carboxylic acids in each of the following pairs as (1) both dicarboxylic acids, (2) both monocarboxylic acids, or (3) one dicarboxylic and one monocarboxylic acid.
a. Glutaric acid and valeric acid
b. Adipic acid and oxalic acid
c. Caproic acid and formic acid
d. Succinic acid and malonic acid

5.18 Classify the two carboxylic acids in each of the following pairs as (1) both dicarboxylic acids, (2) both monocarboxylic acids, or (3) one dicarboxylic and one monocarboxylic acid.
a. Formic acid and acetic acid
b. Butyric acid and succinic acid
c. Pimelic acid and caproic acid
d. Malonic acid and adipic acid

■ Polyfunctional Carboxylic Acids (Section 5.4)

5.19   Each of the following acids contains an additional type of functional group besides the carboxyl group. For each acid, specify the noncarboxyl functional group present.
   a. Acrylic acid   b. Lactic acid
   c. Maleic acid   d. Glycolic acid

5.20   Each of the following acids contains an additional type of functional group besides the carboxyl group. For each acid, specify the noncarboxyl functional group present.
   a. Fumaric acid   b. Pyruvic acid
   c. Malic acid   d. Tartaric acid

5.21   Give the IUPAC name for each of the acids in Problem 5.19.

5.22   Give the IUPAC name for each of the acids in Problem 5.20.

5.23   Draw a structural formula for each of the following acids.
   a. 3-Oxopentanoic acid
   b. 2-Hydroxybutanoic acid
   c. *trans*-4-Hexenoic acid
   d. α,β-Dihydroxyglutaric acid

5.24   Draw a structural formula for each of the following acids.
   a. 3-Hydroxypentanoic acid
   b. α,γ-Dihydroxyvaleric acid
   c. 2-Oxobutanoic acid
   d. *cis*-3-Heptenoic acid

■ "Metabolic" Acids (Section 5.5)

5.25   Classify each of the following polyfunctional acids as a derivative of (1) propionic acid, (2) succinic acid, or (3) glutaric acid.
   a. Lactic acid   b. Glyceric acid
   c. Oxaloacetic acid   d. Citric acid

5.26   Classify each of the following polyfunctional acids as a derivative of (1) propionic acid, (2) succinic acid, or (3) glutaric acid
   a. Pyruvic acid   b. Malic acid
   c. Fumaric acid   d. α-Ketoglutaric acid

5.27   For each of the acids in Problem 5.25, list the functional groups that are present.

5.28   For each of the acids in Problem 5.26, list the functional groups that are present.

■ Physical Properties of Carboxylic Acids (Section 5.6)

5.29   Determine the maximum number of hydrogen bonds that can form between an acetic acid molecule and
   a. another acetic acid molecule
   b. water molecules

5.30   Determine the maximum number of hydrogen bonds that can form between a butanoic acid molecule and
   a. another butanoic acid molecule
   b. water molecules

5.31   What is the physical state (solid, liquid, or gas) of each of the following carboxylic acids at room temperature?
   a. Oxalic acid   b. Decanoic acid
   c. Hexanoic acid   d. Benzoic acid

5.32   What is the physical state (solid, liquid, or gas) of each of the following carboxylic acids at room temperature?
   a. Succinic acid   b. Octanoic acid
   c. Pentanoic acid   d. *p*-Chlorobenzoic acid

■ Preparation of Carboxylic Acids (Section 5.7)

5.33   Draw a structural formula for the carboxylic acid expected to be formed when each of the following substances is oxidized using a strong oxidizing agent.
   a. CH_3—CH_2—OH

   b.
$$CH_3-\overset{\overset{\textstyle O}{\|}}{C}-H$$

   c.
$$CH_3-CH_2-\overset{\overset{\textstyle CH_3}{|}}{CH}-CH_2-\overset{\overset{\textstyle O}{\|}}{C}-H$$

   d.

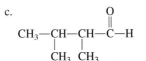

5.34   Draw a structural formula for the carboxylic acid expected to be formed when each of the following substances is oxidized using a strong oxidizing agent.
   a.
$$CH_3-CH_2-\overset{\overset{\textstyle O}{\|}}{C}-H$$
   b. CH_3—CH_2—CH_2—OH

   c.
$$CH_3-\overset{\overset{\textstyle }{|}}{\underset{\underset{\textstyle CH_3}{|}}{CH}}-\overset{\overset{\textstyle }{|}}{\underset{\underset{\textstyle CH_3}{|}}{CH}}-\overset{\overset{\textstyle O}{\|}}{C}-H$$
   d. (benzene ring with CH_3)

■ Acidity of Carboxylic Acids (Section 5.8)

5.35   How many acidic hydrogen atoms are present in each of the following carboxylic acids?
   a. Pentanoic acid   b. Citric acid
   c. Succinic acid   d. Oxalic acid

5.36   How many acidic hydrogen atoms are present in each of the following carboxylic acids?
   a. Acetic acid   b. Benzoic acid
   c. Propanoic acid   d. Glutaric acid

5.37   What is the charge on the carboxylate ion formed when each of the acids in Problem 5.35 ionizes in water?

5.38   What is the charge on the carboxylate ion formed when each of the acids in Problem 5.36 ionizes in water?

5.39   What is the name of the carboxylate ion that forms when each of the acids in Problem 5.35 ionizes in water? (Use an IUPAC carboxylate name if the acid name is IUPAC; use a common name if the acid name is common.)

5.40   What is the name of the carboxylate ion that forms when each of the acids in Problem 5.36 ionizes in water? (Use an IUPAC carboxylate name if the acid name is IUPAC; use a common name if the acid name is common.)

5.41   Write a chemical equation for the formation of each of the following carboxylate ions, in aqueous solution, from its parent acid.
   a. Acetate   b. Citrate
   c. Ethanoate   d. 2-Methylbutanoate

5.42   Write a chemical equation for the formation of each of the following carboxylate ions, in aqueous solution, from its parent acid.
   a. Butanoate   b. Succinate
   c. Benzoate   d. α-Methylbutyrate

■ Carboxylic Acid Salts (Section 5.9)

5.43 Give the IUPAC name for each of the following carboxylic acid salts.

a.
$$CH_3-\overset{\overset{\displaystyle O}{\|}}{C}-O^- \, K^+$$

b.
$$\left(CH_3-CH_2-\overset{\overset{\displaystyle O}{\|}}{C}-O^-\right)_2 Ca^{2+}$$

c.
$$K^+ \, {}^-O-\overset{\overset{\displaystyle O}{\|}}{C}-CH_2-CH_2-\overset{\overset{\displaystyle O}{\|}}{C}-O^- \, K^+$$

d. 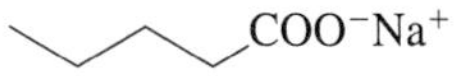
$$\diagup\!\!\diagdown\!\!\diagup\!\!\diagdown COO^- Na^+$$

5.44 Give the IUPAC name for each of the following carboxylic acid salts.

a.
$$\left(H-\overset{\overset{\displaystyle O}{\|}}{C}-O^-\right)_2 Ca^{2+}$$

b.
$$CH_3-CH_2-\overset{\overset{\displaystyle O}{\|}}{C}-O^- \, Na^+$$

c. 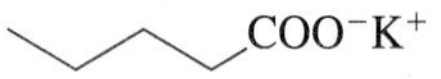
$$\diagup\!\!\diagdown\!\!\diagup\!\!\diagdown COO^- K^+$$

d. 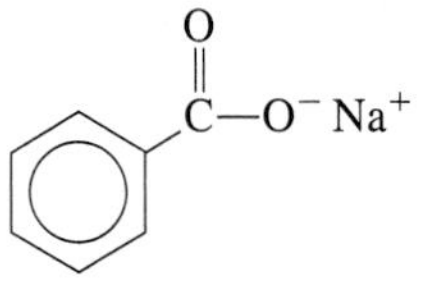
$$\text{(benzene ring)}-\overset{\overset{\displaystyle O}{\|}}{C}-O^- Na^+$$

5.45 Write a chemical equation for the preparation of each of the salts in Problem 5.43 using an acid–base neutralization reaction.

5.46 Write a chemical equation for the preparation of each of the salts in Problem 5.44 using an acid–base neutralization reaction.

5.47 Write a chemical equation for the conversion of each of the following carboxylic acid salts to its parent carboxylic acid. Let hydrochloric acid (HCl) be the source of the needed hydronium ions.
a. Sodium butanoate   b. Potassium oxalate
c. Calcium malonate   d. Sodium benzoate

5.48 Write a chemical equation for the conversion of each of the following carboxylic acid salts to its parent carboxylic acid. Let hydrochloric acid (HCl) be the source of the needed hydronium ions.
a. Calcium propanoate   b. Sodium lactate
c. Magnesium succinate   d. Potassium benzoate

■ Structure of Esters (Section 5.10)

5.49 Which of the following structures represent esters?

a.
$$CH_3-CH_2-CH_2-\overset{\overset{\displaystyle O}{\|}}{C}-O-CH_3$$

b.
$$CH_3-O-\overset{\overset{\displaystyle O}{\|}}{C}-CH_3$$

c.
$$CH_3-O-CH_2-\overset{\overset{\displaystyle O}{\|}}{C}-CH_3$$

d.
$$CH_3-\overset{\overset{\displaystyle O}{\|}}{C}-O-CH_2-CH_3$$

e.      f.

5.50 Which of the following structures represent esters?

a.
$$CH_3-CH_2-CH_2-\overset{\overset{\displaystyle O}{\|}}{C}-OH$$

b.
$$CH_3-\overset{\overset{\displaystyle CH_3}{|}}{CH}-\overset{\overset{\displaystyle O}{\|}}{C}-O-CH_3$$

c. CH_3-O-CH_3

d.
$$CH_3-CH_2-O-\overset{\overset{\displaystyle O}{\|}}{C}-CH_2-CH_3$$

e. 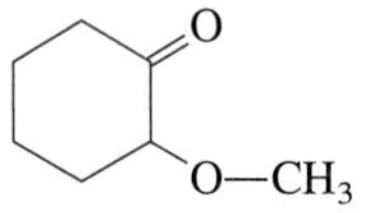
(cyclohexane ring with C=O and O—CH₃)

f. 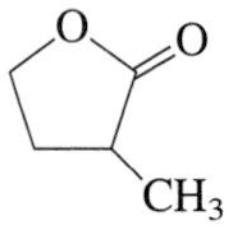
(cyclic lactone with O, =O, and CH₃)

■ Preparation of Esters (Section 5.11)

5.51 Draw the structure of the ester produced when each of the following pairs of carboxylic acid and alcohol react.
a. Propanoic acid and methanol
b. Acetic acid and 1-propanol
c. 2-Methylbutanoic acid and 2-propanol
d. Valeric acid and sec-butyl alcohol

5.52 Draw the structure of the ester produced when each of the following pairs of carboxylic acid and alcohol react.
a. Methanoic acid and 1-propanol
b. Propanoic acid and ethanol
c. 2-Methylpropanoic acid and 2-butanol
d. Valeric acid and isobutyl alcohol

5.53 For each of the following esters, draw the structural formula of the "parent" acid and the "parent" alcohol.

a.
$$CH_3-CH_2-\overset{\overset{\displaystyle O}{\|}}{C}-O-CH_2-CH_3$$

b.
$$CH_3-CH_2-CH_2-\overset{\overset{\displaystyle O}{\|}}{C}-O-CH_3$$

c.
$$CH_3-O-\overset{\overset{\displaystyle O}{\|}}{C}-CH_2-CH_2-CH_3$$

d. 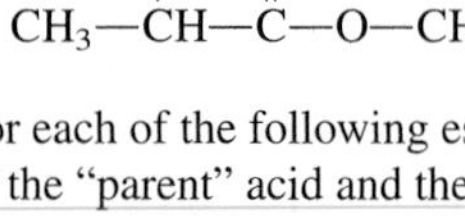
$$CH_3-\overset{\overset{\displaystyle O}{\|}}{C}-O-\text{(benzene ring)}$$

e. 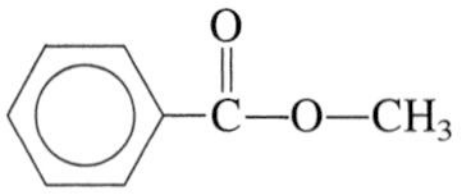
$$\text{(benzene ring)}-\overset{\overset{\displaystyle O}{\|}}{C}-O-CH_3$$

f.
$$CH_3-\overset{\overset{\displaystyle Cl}{|}}{CH}-\overset{\overset{\displaystyle O}{\|}}{C}-O-CH_2-CH_3$$

5.54 For each of the following esters, draw the structural formula of the "parent" acid and the "parent" alcohol.

a.
$$CH_3-\overset{\overset{\displaystyle O}{\|}}{C}-O-CH_2-CH_3$$

b.

CH₃—CH₂—C(=O)—O—CH₃

c.

CH₃—O—C(=O)—CH₂—CH₃

d.

(benzene ring)—C(=O)—O—CH₃

e.

CH₃—CH(CH₃)—CH₂—C(=O)—O—(benzene ring)

f.

CH₃—CH—CH₂—C(=O)—O—CH₃ (with phenyl substituent on CH)

■ Nomenclature for Esters (Section 5.12)

5.55 Assign an IUPAC name to each of the following esters.

a.

CH₃—CH₂—C(=O)—O—CH₃

b.

H—C(=O)—O—CH₃

c.

CH₃—C(=O)—O—CH₃

d.

CH₃—CH₂—CH₂—O—C(=O)—CH₃

e.

CH₃—CH₂—C(=O)—O—CH(CH₃)—CH₃

f.

(benzene ring)—C(=O)—O—CH₂—CH₃

5.56 Assign an IUPAC name to each of the following esters.

a.

CH₃—C(=O)—O—CH₂—CH₂—CH₂—CH₃

b.

CH₃—CH₂—CH₂—C(=O)—O—CH₃

c.

CH₃—CH₂—C(=O)—O—CH₂—CH₂—CH₃

d.

CH₃—CH₂—O—C(=O)—H

e.

CH₃—CH(CH₃)—CH(CH₃)—C(=O)—O—CH₂—CH₃

f.

CH₃—CH₂—C(=O)—O—(benzene ring)

5.57 Assign a common name to each of the esters in Problem 5.55.

5.58 Assign a common name to each of the esters in Problem 5.56.

5.59 Assign an IUPAC name to each of the following esters.

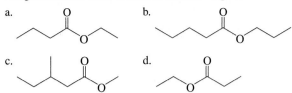

a. b.

c. d.

5.60 Assign an IUPAC name to each of the following esters.

a. b.

c. d.

5.61 Draw a structural formula for each of the following esters.
a. Methyl formate
b. Propyl acetate
c. Octyl decanoate
d. Ethyl phenylacetate
e. Isopropyl acetate
f. 2-Bromopropyl ethanoate

5.62 Draw a structural formula for each of the following esters.
a. Ethyl butyrate
b. Butyl ethanoate
c. 2-Methylpropyl formate
d. Ethyl α-methylpropanoate
e. Methyl valerate
f. Phenyl benzoate

5.63 Assign IUPAC names to the esters that are produced from the reaction of the following carboxylic acids and alcohols.
a. Acetic acid and ethanol
b. Ethanoic acid and methanol
c. Butyric acid and ethyl alcohol
d. Lactic acid and propyl alcohol
e. 1-Pentanol and pentanoic acid
f. 2-Butanol and caproic acid

5.64 Assign IUPAC names to the esters that are produced from the reaction of the following carboxylic acids and alcohols.
a. Ethanoic acid and propyl alcohol
b. Acetic acid and 1-pentanol
c. Acetic acid and 2-pentanol
d. Methyl alcohol and butyric acid
e. Ethanol and benzoic acid
f. Pyruvic acid and methyl alcohol

■ Isomerism for Carboxylic Acids and Esters (Section 5.14)

5.65 Give IUPAC names for the four isomeric C_5 monocarboxylic acids with saturated carbon chains.

5.66 Give IUPAC names for the eight isomeric C_6 monocarboxylic acids with saturated carbon chains.

5.67 Give IUPAC names for the four isomeric methyl esters that contain six carbon atoms and saturated carbon chains.

5.68 Give IUPAC names for the two isomeric ethyl esters that contain six carbon atoms and saturated carbon chains.

5.69 How many esters exist that are isomeric with 2-methylbutanoic acid?

5.70 How many esters exist that are isomeric with 2-methylpropanoic acid?

5.71 Draw condensed structural formulas for all carboxylic acids and all esters that have the molecular formula $C_3H_6O_2$.

5.72 Draw condensed structural formulas for all carboxylic acids and all esters that have the molecular formula $C_4H_8O_2$.

■ Physical Properties of Esters (Section 5.15)

5.73 Explain why ester molecules cannot form hydrogen bonds to each other.

5.74 How many hydrogen bonds can form between a methyl acetate molecule and two water molecules?

5.75 Explain why esters have lower boiling points than carboxylic acids of comparable molecular mass.

5.76 Explain why esters are less soluble in water than carboxylic acids of comparable molecular mass.

■ Chemical Reactions of Esters (Section 5.16)

5.77 Write the structural formulas of the reaction products when each of the following esters is hydrolyzed under acidic conditions.

a.
$$CH_3-CH_2-\overset{\overset{\displaystyle O}{\|}}{C}-O-CH_2-CH_3$$

b.
$$CH_3-\overset{\overset{\displaystyle O}{\|}}{C}-O-CH_2-CH_3$$

c.
$$CH_3-\overset{\overset{\displaystyle CH_3}{|}}{CH}-\overset{\overset{\displaystyle O}{\|}}{C}-O-\bigcirc$$

d. Methyl butanoate
e. Ethyl formate
f. Isopropyl benzoate

5.78 Write the structural formulas of the reaction products when each of the following esters is hydrolyzed under acidic conditions.

a.
$$H-\overset{\overset{\displaystyle O}{\|}}{C}-O-CH_2-CH_2-CH_3$$

b.
$$CH_3-\overset{\overset{\displaystyle CH_3}{|}}{CH}-\overset{\overset{\displaystyle O}{\|}}{C}-O-\overset{\overset{\displaystyle CH_3}{|}}{CH}-CH_3$$

c.
$$\bigcirc-\overset{\overset{\displaystyle O}{\|}}{C}-O-\bigcirc$$

d. Ethyl valerate
e. Butyl butyrate
f. Pentyl benzoate

5.79 Write the structural formulas of the reaction products when each of the esters in Problem 5.77 is saponified using sodium hydroxide.

5.80 Write the structural formulas of the reaction products when each of the esters in Problem 5.78 is saponified using sodium hydroxide.

5.81 Draw structures of the reaction products in the following chemical reactions.

a.
$$CH_3-\overset{\overset{\displaystyle CH_3}{|}}{CH}-\overset{\overset{\displaystyle O}{\|}}{C}-O-CH_2-CH_3 + H_2O \xrightarrow{H^+}$$

b.
$$CH_3-\overset{\overset{\displaystyle CH_3}{|}}{CH}-\overset{\overset{\displaystyle O}{\|}}{C}-O-CH_2-CH_3 + NaOH \xrightarrow{H_2O}$$

c.
$$H-\overset{\overset{\displaystyle O}{\|}}{C}-O-CH_2-CH_2-CH_2-CH_3 + H_2O \xrightarrow{H^+}$$

d.
$$CH_3-\overset{\overset{\displaystyle O}{\|}}{C}-O-CH_2-\overset{\overset{\displaystyle CH_3}{|}}{CH}-\overset{\overset{\displaystyle CH_3}{|}}{CH}-CH_3 + NaOH \xrightarrow{H_2O}$$

5.82 Draw structures of the reaction products in the following chemical reactions.

a.
$$CH_3-\overset{\overset{\displaystyle CH_3}{|}}{CH}-CH_2-\overset{\overset{\displaystyle O}{\|}}{C}-O-CH_3 + H_2O \xrightarrow{H^+}$$

b.
$$CH_3-\overset{\overset{\displaystyle CH_3}{|}}{CH}-CH_2-\overset{\overset{\displaystyle O}{\|}}{C}-O-CH_3 + NaOH \xrightarrow{H_2O}$$

c.
$$CH_3-CH_2-\overset{\overset{\displaystyle O}{\|}}{C}-O-(CH_2)_5-CH_3 + H_2O \xrightarrow{H^+}$$

d.
$$CH_3-(CH_2)_5-\overset{\overset{\displaystyle O}{\|}}{C}-O-CH_2-CH_3 + NaOH \xrightarrow{H_2O}$$

■ Sulfur Analogs of Esters (Section 5.17)

5.83 Draw the structures of the thioesters formed as a result of each of the following reactions between carboxylic acids and thiols.

a.
$$CH_3-\overset{\overset{\displaystyle O}{\|}}{C}-OH + CH_3-CH_2-SH \rightarrow$$

b.
$$CH_3-(CH_2)_8-\overset{\overset{\displaystyle O}{\|}}{C}-OH + CH_3-SH \rightarrow$$

c.
$$\bigcirc-COOH + CH_3-\overset{\overset{\displaystyle CH_3}{|}}{CH}-SH \rightarrow$$

d.
$$H-\overset{\overset{\displaystyle O}{\|}}{C}-OH + CH_3-CH_2-CH_2-SH \rightarrow$$

5.84 Draw the structures of the thioesters formed as a result of each of the following reactions between carboxylic acids and thiols.

a.
$$CH_3-CH_2-\overset{\overset{\displaystyle O}{\|}}{C}-OH + CH_3-CH_2-SH \rightarrow$$

b. $CH_3-CH_2-CH_2-COOH + CH_3-SH \rightarrow$

c.
$$CH_3-\overset{\overset{\displaystyle O}{\|}}{C}-OH + CH_3-CH_2-\overset{\overset{\displaystyle |}{CH}}{\underset{\overset{\displaystyle |}{CH_3}}{}}-SH \rightarrow$$

d.
$$\bigcirc-COOH + \bigcirc-SH \rightarrow$$

■ Polyesters (Section 5.18)

5.85 Write the structure (two repeating units) of the polyester polymer formed from oxalic acid and 1,3-propanediol.

5.86 Write the structure (two repeating units) of the polyester polymer formed from malonic acid and ethylene glycol.

5.87 Draw the structural formulas of the monomers needed to form the following polyester.

$$\left(\!O\!-\!(CH_2)_3\!-\!O\!-\!\overset{\displaystyle O}{\overset{\|}{C}}\!-\!(CH_2)_2\!-\!\overset{\displaystyle O}{\overset{\|}{C}}\!-\!O\!\right)_{\!n}$$

5.88 Draw the structural formulas of the monomers needed to form the following polyester.

$$\left(\!O\!-\!\overset{\displaystyle O}{\overset{\|}{C}}\!-\!(CH_2)_3\!-\!\overset{\displaystyle O}{\overset{\|}{C}}\!-\!O\!-\!(CH_2)_2\!-\!O\!\right)_{\!n}$$

◼ Acid Chlorides and Acid Anhydrides (Section 5.19)

5.89 Draw the condensed structural formula for each of the following compounds.
  a. Propionyl chloride
  b. 3-Methylbutanoyl chloride
  c. Butyric anhydride
  d. Butanoic ethanoic anhydride

5.90 Draw the condensed structural formula for each of the following compounds.
  a. Acetyl chloride
  b. 2-Methylbutanoyl chloride
  c. Propionic anhydride
  d. Ethanoic methanoic anhydride

5.91 Assign an IUPAC name to each of the following compounds.

  a.
  $$CH_3\!-\!\overset{\displaystyle O}{\overset{\|}{C}}\!-\!O\!-\!\overset{\displaystyle O}{\overset{\|}{C}}\!-\!CH_2\!-\!CH_3$$

  b.
  $$CH_3\!-\!CH_2\!-\!CH_2\!-\!CH_2\!-\!\overset{\displaystyle O}{\overset{\|}{C}}\!-\!Cl$$

  c.
  $$CH_3\!-\!\underset{\underset{\displaystyle CH_3}{|}}{CH}\!-\!\underset{\underset{\displaystyle CH_3}{|}}{CH}\!-\!\overset{\displaystyle O}{\overset{\|}{C}}\!-\!Cl$$

  d.
  $$CH_3\!-\!CH_2\!-\!\overset{\displaystyle O}{\overset{\|}{C}}\!-\!O\!-\!\overset{\displaystyle O}{\overset{\|}{C}}\!-\!H$$

5.92 Assign an IUPAC name to each of the following compounds.

  a.
  $$CH_3\!-\!CH_2\!-\!\overset{\displaystyle O}{\overset{\|}{C}}\!-\!O\!-\!\overset{\displaystyle O}{\overset{\|}{C}}\!-\!CH_2\!-\!CH_3$$

  b.
  $$CH_3\!-\!CH_2\!-\!\overset{\displaystyle O}{\overset{\|}{C}}\!-\!Cl$$

  c.
  $$CH_3\!-\!\underset{\underset{\displaystyle CH_3}{|}}{\overset{\overset{\displaystyle CH_3}{|}}{C}}\!-\!CH_2\!-\!\overset{\displaystyle O}{\overset{\|}{C}}\!-\!Cl$$

  d.
  $$CH_3\!-\!\overset{\displaystyle O}{\overset{\|}{C}}\!-\!O\!-\!\overset{\displaystyle O}{\overset{\|}{C}}\!-\!H$$

5.93 Draw a condensed structural formula for the organic product of the reaction of each of the following compounds with water.
  a. Pentanoyl chloride   b. Pentanoic anhydride

5.94 Draw a condensed structural formula for the organic product of the reaction of each of the following compounds with water.
  a. Butanoyl chloride   b. Butanoic anhydride

5.95 Draw the condensed structural formulas for the ester formed and the carboxylic acid formed when acetic anhydride reacts with the following alcohols.
  a. Ethyl alcohol   b. 1-Butanol

5.96 Draw the condensed structural formulas for the ester formed and the carboxylic acid formed when ethanoic anhydride reacts with the following alcohols.
  a. Propyl alcohol   b. 2-Butanol

◼ Esters and Anhydrides of Inorganic Acids (Section 5.20)

5.97 Draw the structures of the esters formed by reacting the following substances.
  a. 1 molecule methanol and 1 molecule phosphoric acid
  b. 2 molecules methanol and 1 molecule phosphoric acid
  c. 1 molecule methanol and 1 molecule nitric acid
  d. 1 molecule ethylene glycol and 2 molecules nitric acid

5.98 Draw the structures of the esters formed by reacting the following substances.
  a. 1 molecule ethanol and 1 molecule phosphoric acid
  b. 2 molecules methanol and 1 molecule sulfuric acid
  c. 1 molecule ethylene glycol and 1 molecule nitric acid
  d. 1 molecule glycerol and 3 molecules nitric acid

5.99 Phosphoric acid can form triesters but sulfuric acid cannot. Explain why.

5.100 Sulfuric acid can form diesters but nitric acid cannot. Explain why.

ADDITIONAL PROBLEMS

5.101 With the help of Figure 5.6 and IUPAC naming rules, specify the number of carbon atoms present and the number of carboxyl groups present in each of the following carboxylic acids.
  a. Oxalic acid
  b. Heptanoic acid
  c. *cis*-3-Heptenoic acid
  d. Citric acid
  e. Pyruvic acid
  f. Dichloroethanoic acid

5.102 Malonic, maleic, and malic acids are dicarboxylic acids with similar-sounding names. How do the structures of these acids differ from each other?

5.103 The general molecular formula for an alkane is C_nH_{2n+2}. What is the general molecular formula for an unsaturated unsubstituted monocarboxylic acid containing one carbon–carbon double bond?

5.104 Assign IUPAC names to the following compounds.

  a.   b.
  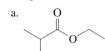

  c.   d.
  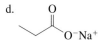

5.105 A sample of ethyl alcohol is divided into two portions. Portion A is added to an aqueous solution of a strong oxidizing agent and allowed to react. The organic product of this reaction is mixed with portion B of the ethyl alcohol. A trace of acid is added and the solution is heated. What is the structure of the final product of this reaction scheme?

5.106 For each of the following reactions, draw the structure(s) of the organic product(s).

a.

$$CH_3-CH_2-\overset{\overset{\displaystyle O}{\|}}{C}-O-CH_3 + NaOH \xrightarrow{H_2O}$$

b.

$$CH_3-CH_2-\overset{\overset{\displaystyle O}{\|}}{C}-OH + CH_3-SH \longrightarrow$$

c.

$$CH_3-\overset{\overset{\displaystyle O}{\|}}{C}-OH + NaOH \longrightarrow$$

d.

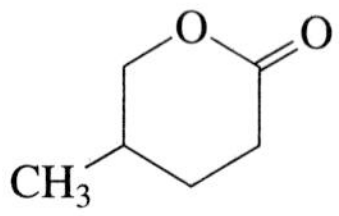

(lactone ring structure with CH₃ substituent) $+ H_2O \xrightarrow{H^+}$

MULTIPLE-CHOICE PRACTICE TEST

5.107 Which of the following statements concerning the carboxylic acid functional group is *correct*?
a. It is called a carboxylate group.
b. It can be denoted using the notation —COOH.
c. An oxygen–oxygen single bond is present.
d. A carbon–hydrogen single bond is present.

5.108 What are the common names for the C_1 and C_2 monocarboxylic acids, respectively?
a. Formic acid and acetic acid
b. Acetic acid and formic acid
c. Oxalic acid and acetic acid
d. Acetic acid and oxalic acid

5.109 In which of the following pairs of carboxylic acids does the first member of the pair have more carbon atoms than the second member of the pair?
a. Malonic acid and succinic acid
b. Glutaric acid and succinic acid
c. Oxalic acid and malonic acid
d. Oxalic acid and glutaric acid

5.110 Which statement is true for the carboxyl carbon atom in the IUPAC nomenclature system for monocarboxylic acids?
a. It is always assigned the number one.
b. It is always assigned the highest number possible.
c. It is always known as the alpha carbon atom.
d. It is always known as the beta carbon atom.

5.111 Which of the following is a C_3 monohydroxy carboxylic acid?
a. Tartaric acid   b. Lactic acid
c. Citric acid   d. Pyruvic acid

5.112 An ester is a carboxylic acid derivative in which the —OH portion of the carboxyl group has been replaced with which of the following?
a. —OR group   b. —OCl group
c. —Cl atom   d. —O⁻Na⁺ group

5.113 Which of the following esters, upon hydrolysis, produces a two-carbon alcohol as one of the products?
a. Methyl methanoate   b. Propyl ethanoate
c. Methyl propanoate   d. Ethyl methanoate

5.114 Which of the following is neither a reactant nor a product in an ester saponification reaction?
a. A strong base   b. An alcohol
c. A carboxylic acid   d. A carboxylic acid salt

5.115 A polyester is a condensation polymer in which the reacting monomers are a dicarboxylic acid and which of the following?
a. Carboxylic acid anhydride
b. Carboxylic acid salt
c. Monoalcohol
d. Dialcohol

5.116 What is the number of oxygen atoms present in a triester of phosphoric acid?
a. One   b. Two
c. Three   d. Four

6

Amines and Amides

Parachutist with a parachute made of the polyamide nylon.

The four most abundant elements in living organisms are carbon, hydrogen, oxygen, and nitrogen. In previous chapters, we have discussed compounds containing the first three of these elements. Alkanes, alkenes, alkynes, and aromatic hydrocarbons are all carbon–hydrogen compounds. The carbon–hydrogen–oxygen compounds we have discussed include alcohols, phenols, ethers, aldehydes, ketones, carboxylic acids, and esters. We now extend our discussion to organic compounds that contain the element nitrogen.

Two types of organic nitrogen-containing compounds are the focus of this chapter: amines and amides. Amines are carbon–hydrogen–nitrogen compounds, and amides contain oxygen in addition to these elements. Amines and amides occur widely in nature in living organisms. Many of these naturally occurring compounds are very active physiologically. In addition, numerous drugs used for the treatment of mental illness, hay fever, heart problems, and other physical disorders are amines or amides.

6.1 Bonding Characteristics of Nitrogen Atoms in Organic Compounds

An understanding of the bonding characteristics of the nitrogen atom is a prerequisite to our study of amines and amides. Nitrogen is a member of Group VA of the periodic table; it has five valence electrons and will form three covalent bonds to complete its octet of electrons. Thus, in organic chemistry, carbon forms four bonds (Section 1.2), nitrogen forms three bonds, and oxygen forms two bonds (Section 3.1).

$$-\overset{|}{\underset{|}{C}}- \qquad -\overset{..}{\underset{|}{N}}- \qquad :\overset{..}{\underset{|}{O}}-$$

4 valence electrons	5 valence electrons	6 valence electrons
4 covalent bonds	3 covalent bonds	2 covalent bonds
no nonbonding	1 nonbonding	2 nonbonding
electron pairs	electron pair	electron pairs

6.2 Structure and Classification of Amines

> Amines bear the same relationship to ammonia that alcohols and ethers bear to water (Sections 3.2 and 3.15).

An **amine** *is an organic derivative of ammonia (NH₃) in which one or more alkyl, cycloalkyl, or aryl groups are attached to the nitrogen atom.* Amines are classified as primary (1°), secondary (2°), or tertiary (3°) on the basis of how many hydrocarbon groups are bonded to the ammonia nitrogen atom (see Figure 6.1). A **primary amine** *is an amine in which the nitrogen atom is bonded to one hydrocarbon group and two hydrogen atoms.* The generalized formula for a primary amine is RNH_2. A **secondary amine** *is an amine in which the nitrogen atom is bonded to two hydrocarbon groups and one hydrogen atom.* The generalized formula for a secondary amine is R_2NH. A **tertiary amine** *is an amine in which the nitrogen atom is bonded to three hydrocarbon groups and no hydrogen atoms.* The generalized formula for a tertiary amine is R_3N.

The basis for the amine primary-secondary-tertiary classification system differs from that for alcohols (Section 3.8).

1. For alcohols we look at how many R groups are on a *carbon* atom, the hydroxyl-bearing carbon atom.
2. For amines we look at how many R groups are on the *nitrogen* atom.

Tert-butyl alcohol is a *tertiary* alcohol, whereas *tert*-butylamine is a *primary* amine.

Tertiary carbon atom →
$$CH_3-\overset{\overset{CH_3}{|}}{\underset{\underset{CH_3}{|}}{C}}-OH$$
tert-Butyl alcohol
(a tertiary alcohol)

Tertiary carbon atom → ← Primary nitrogen atom
$$CH_3-\overset{\overset{CH_3}{|}}{\underset{\underset{CH_3}{|}}{C}}-NH_2$$
tert-Butylamine
(a primary amine)

FIGURE 6.1 Classification of amines is related to the number of R groups attached to the nitrogen atom.

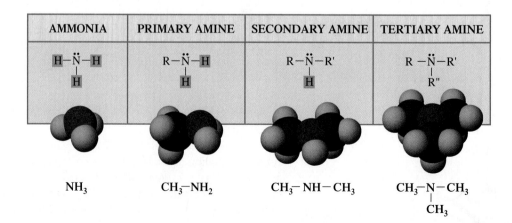

AMMONIA	PRIMARY AMINE	SECONDARY AMINE	TERTIARY AMINE
$H-\overset{..}{\underset{\underset{H}{|}}{N}}-H$	$R-\overset{..}{\underset{\underset{H}{|}}{N}}-H$	$R-\overset{..}{\underset{\underset{H}{|}}{N}}-R'$	$R-\overset{..}{\underset{\underset{R''}{|}}{N}}-R'$
NH_3	CH_3-NH_2	$CH_3-NH-CH_3$	$CH_3-\underset{\underset{CH_3}{|}}{N}-CH_3$

The functional group present in a primary amine, the —NH$_2$ group, is called an *amino* group. An **amino group** *is the* —NH$_2$ *functional group.* Secondary and tertiary amines possess substituted amino groups.

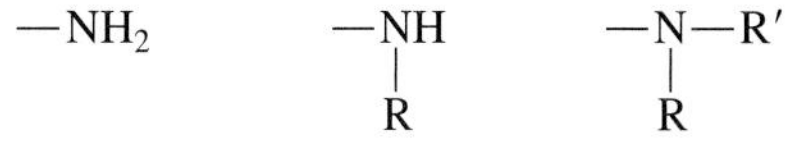

—NH$_2$ —NH   —N—R′
                 | |
                 R   R

Amino group   Monosubstituted   Disubstituted
               amino group   amino group

EXAMPLE 6.1

Classifying Amines as Primary, Secondary, or Tertiary

■ Classify each of the following amines as a primary, secondary, or tertiary amine.

a.
CH$_3$—NH—⬡

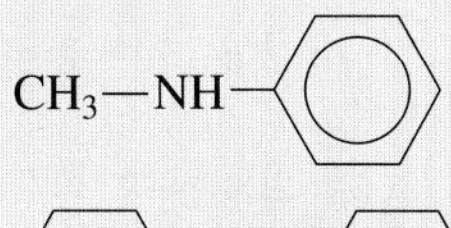

b. CH$_3$—N—CH$_3$
              |
             CH$_3$

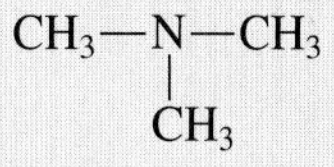

c.
⬡—N—⬡
   |
  CH$_3$

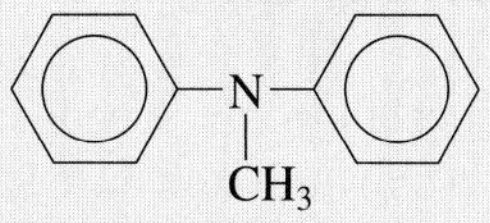

d.
(cyclohexane with CH$_3$ and NH$_2$)

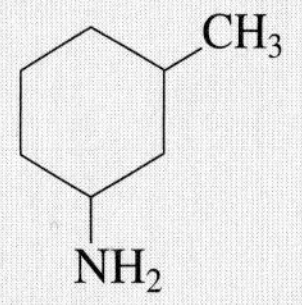

Solution

The number of carbon atoms directly bonded to the nitrogen atom determines the amine classification.

a. This is a secondary amine because the nitrogen is bonded to both a methyl group and a phenyl group.
b. Here we have a tertiary amine because the nitrogen atom is bonded to three methyl groups.
c. This is also a tertiary amine; the nitrogen atom is bonded to two phenyl groups and a methyl group.
d. This is a primary amine. The nitrogen atom is bonded to only one carbon atom.

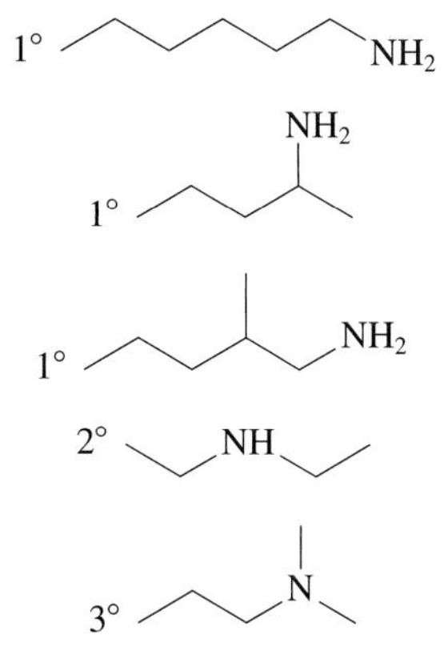

Practice Exercise 6.1

Classify each of the following amines as a primary, secondary, or tertiary amine.

a. CH$_3$—CH$_2$—CH$_2$—NH$_2$
b. CH$_3$—NH—CH$_2$—CH$_3$
c. 
⬡—NH$_2$
d.
CH$_3$—N—⬡
        |
       CH$_3$

Line-angle formulas for selected primary, secondary, and tertiary amines.

1° (line-angle structure) NH$_2$

1° (line-angle structure) NH$_2$

1° (line-angle structure) NH$_2$

2° (line-angle structure) NH

3° (line-angle structure) N

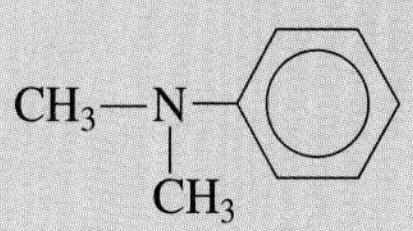

Cyclic amines exist. Such compounds are always either secondary or tertiary amines.

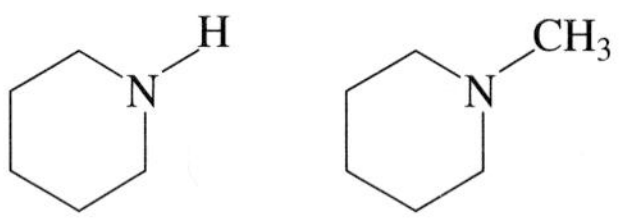

(piperidine with N—H)   (piperidine with N—CH$_3$)

2° Cyclic amine   3° Cyclic amine

Cyclic amines are heterocyclic compounds (Section 3.19). Numerous cyclic amine compounds are found in biochemical systems (Section 6.9).

6.3 Nomenclature for Amines

Both common and IUPAC names are extensively used for amines. In the common system of nomenclature, amines are named by listing the alkyl group or groups attached to the nitrogen atom in alphabetical order and adding the suffix *-amine;* all of this appears as

> The common names of amines, like those of aldehydes, are written as a single word, which is different from the common names of alcohols (two words), ethers (two or three words), ketones (two or three words), acids (two words), and esters (two words).

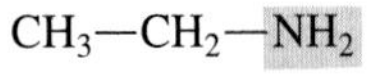

one word. Prefixes such as *di-* and *tri-* are added when identical groups are bonded to the nitrogen atom.

Ethylamine    Dimethylamine 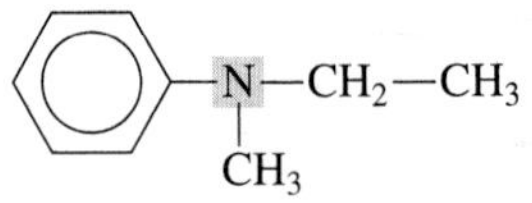   Ethylmethylphenylamine

The IUPAC rules for naming amines are similar to those for alcohols (Section 3.3). Alcohols are named as *alkanols* and amines are named as *alkanamines*. IUPAC rules for naming *primary* amines are as follows:

1. Select as the parent carbon chain the longest chain to which the nitrogen atom is attached.
2. Name the parent chain by changing the *-e* ending of the corresponding alkane name to *-amine*.
3. Number the parent chain from the end nearest the nitrogen atom.
4. The position of attachment of the nitrogen atom is indicated by a number in front of the parent chain name.
5. The identity and location of any substituents are appended to the front of the parent chain name.

> IUPAC nomenclature for primary amines is similar to that for alcohols, except that the suffix is *-amine* rather than *-ol*. An —NH$_2$ group, like an —OH group, has priority in numbering the parent carbon chain.

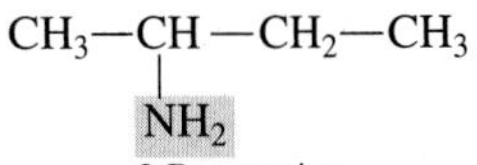

2-Butanamine 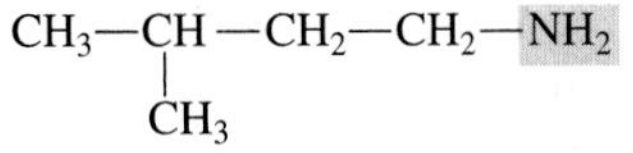   3-Methyl-1-butanamine

In diamines, the final *-e* of the carbon chain name is retained for ease of pronunciation. Thus the base name for a four-carbon chain bearing two amino groups is butane-diamine.

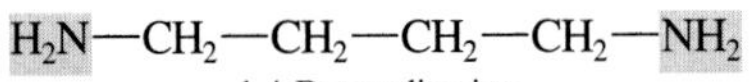

1,4-Butanediamine

Secondary and tertiary amines are named as *N*-substituted primary amines. The largest carbon group bonded to the nitrogen is used as the parent amine name. The names of the other groups attached to the nitrogen are appended to the front of the base name, and *N-* or *N,N-* prefixes are used to indicate that these groups are attached to the nitrogen atom rather than to the base carbon chain.

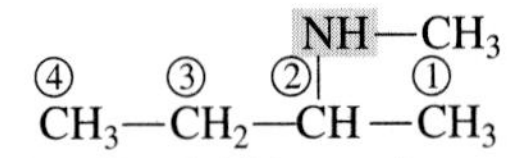

N-methyl-2-butanamine 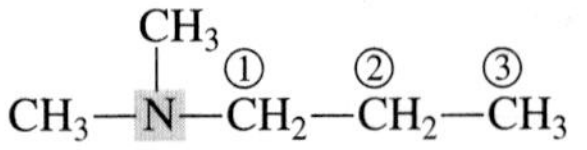   *N,N*-dimethyl-1-propanamine

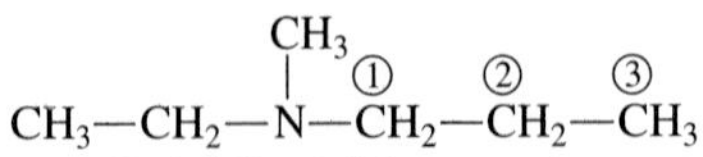

N-ethyl-*N*-methyl-1-propanamine 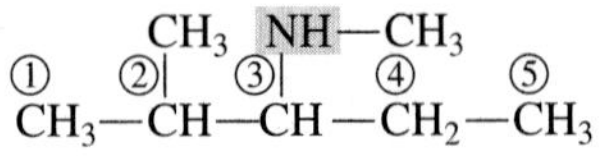   2,*N*-dimethyl-3-pentanamine

> In IUPAC nomenclature, the amino group has a priority just below that of an alcohol. The priority list for functional groups is
>
> carboxylic acid ↑
> aldehyde
> ketone   Increasing
> alcohol   priority
> amine

In amines where additional functional groups are present, the amine group is treated as a substituent. As a substituent, an —NH$_2$ group is called an *amino* group.

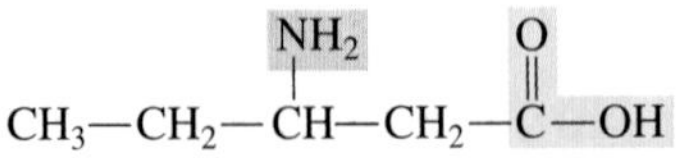

3-Aminopentanoic acid 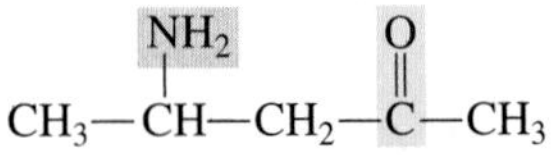   4-Amino-2-pentanone

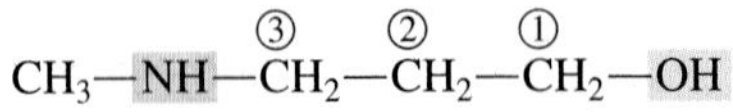

3-(*N*-methylamino)-1-propanol

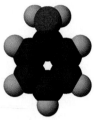

FIGURE 6.2 Space-filling model of aniline, the simplest aromatic amine. Aromatic amines, including aniline, are generally toxic; they are readily absorbed through the skin.

The simplest aromatic amine, a benzene ring bearing an amino group, is called *aniline* (Figure 6.2). Other simple aromatic amines are named as derivatives of aniline.

Aniline   *m*-Chloroaniline   2,3-Dichloroaniline

In secondary and tertiary aromatic amines, the additional group or groups attached to the nitrogen atom are located using a capital *N*-.

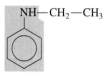

N-ethylaniline   *N,N*-dimethylaniline   3,*N*-dimethylaniline

---

EXAMPLE 6.2

Determining IUPAC Names for Amines

A benzene ring with both an amino group and a methyl group as substituents is called *toluidine*. This name is a combination of the names *toluene* and *aniline*.

The contrast between IUPAC names and common names for primary, secondary, and tertiary amines is as follows:

Primary Amines

    IUPAC (one word)

        | alkanamine |

    Common (one word)

        | alkylamine |

Secondary Amines

    IUPAC (one word)

        | *N*-alkylalkanamine |

    Common (one word)

        | alkylalkylamine |

Tertiary Amines

    IUPAC (one word)

    | *N*-alkyl-*N*-alkylalkanamine |

    Common (one word)

        | alkylalkylalkylamine |

■ Assign IUPAC names to each of the following amines.

a. $CH_3-CH_2-NH-(CH_2)_4-CH_3$

b.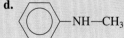

c. $H_2N-CH_2-CH_2-NH_2$

d. [structure: ethyl and two methyl groups attached to N]

Solution

a. The longest carbon chain has five carbons. The name of the compound is *N-ethyl-1-pentanamine*.

b. This compound is named as a derivative of aniline: *4-bromoaniline* (or *p*-bromoaniline). The carbon in the ring to which the —NH_2 is attached is carbon 1.

c. Two —NH_2 groups are present in this molecule. The name is *1,2-ethanediamine*.

d. This is a tertiary amine in which the longest carbon chain has two carbons (ethane). The base name is thus *ethanamine*. We also have two methyl groups attached to the nitrogen atom. The name of the compound is *N,N-dimethylethanamine*.

Practice Exercise 6.2

Assign IUPAC names to each of the following amines.

a. [structure with NH$_2$ group]

b. $CH_3-CH_2-CH_2-NH-CH_2-CH_2-CH_3$

c. CH_3-N-CH_3 **d.** [benzene ring]—$NH-CH_3$
          $|$
         CH_3

6.4 Isomerism for Amines

Constitutional isomerism in amines can arise from several causes. Different carbon atom arrangements produce isomers, as in

$$CH_3-CH_2-CH_2-CH_2-CH_2-NH_2 \quad \text{and} \quad CH_3-CH_2-CH-CH_2-NH_2$$
$$\underset{\qquad\qquad\qquad\qquad\qquad\qquad\qquad\quad CH_3}{}$$

Unbranched Primary Amines			
C_1	C_3	C_5	C_7
C_2	C_4	C_6	C_8

☐ Gas ☐ Liquid

FIGURE 6.3 A physical-state summary for unbranched primary amines at room temperature and room pressure.

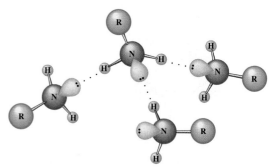

FIGURE 6.4 Hydrogen bonding interactions among amine molecules involves the hydrogen atoms and nitrogen atoms of amino groups.

Different positioning of the nitrogen atom on a carbon chain is another cause for isomerism, illustrated in the following compounds.

$$\underset{NH_2}{CH_2}-CH_2-CH_2-CH_3 \quad \text{and} \quad CH_3-\underset{NH_2}{CH}-CH_2-CH_3$$

For secondary and tertiary amines, different partitioning of carbon atoms among the carbon chains present produces constitutional isomers. There are three C_4 secondary amines; carbon atom partitioning can be two ethyl groups, a propyl group and a methyl group, or an isopropyl group and a methyl group.

$$CH_3-CH_2-NH-CH_2-CH_3 \text{ and } CH_3-CH_2-CH_2-NH-CH_3 \text{ and } CH_3-\underset{CH_3}{CH}-NH-CH_3$$

6.5 Physical Properties of Amines

The methylamines (mono-, di-, and tri-) and ethylamine are gases at room temperature and have ammonia-like odors. Most other amines are liquids (see Figure 6.3), and many have odors resembling that of raw fish. A few amines, particularly diamines, have strong, disagreeable odors. The foul odor arising from dead fish and decaying flesh is due to amines released by the bacterial decomposition of protein. Two of these "odoriferous" compounds are the diamines putrescine and cadaverine.

$$H_2N-(CH_2)_4-NH_2 \qquad H_2N-(CH_2)_5-NH_2$$
$$\text{Putrescine} \qquad\qquad \text{Cadaverine}$$
$$\text{(1,4-butanediamine)} \qquad \text{(1,5-pentanediamine)}$$

The simpler amines are irritating to the skin, eyes, and mucous membranes and are toxic by ingestion. Aromatic amines are generally toxic (see Figure 6.3). Many are readily absorbed through the skin and affect both the blood and the nervous system.

The boiling points of amines are intermediate between those of alkanes and alcohols of similar molecular mass. They are higher than alkane boiling points, because hydrogen bonding is possible between amine molecules but not between alkane molecules. Intermolecular hydrogen bonding of amines involves the hydrogen atoms and nitrogen atoms of the amino groups (Figure 6.4).

The boiling points of amines are lower than those of corresponding alcohols (Figure 6.5), because N···H hydrogen bonds are weaker than O···H hydrogen bonds. (The difference in hydrogen-bond strength results from electronegativity differences; nitrogen is less electronegative than oxygen.)

Amines with fewer than six carbon atoms are infinitely soluble in water. This solubility results from hydrogen bonding between the amines and water. Even tertiary amines are water-soluble, because the amine nitrogen atom has a nonbonding electron pair that can form a hydrogen bond with a hydrogen atom of water (Figure 6.6).

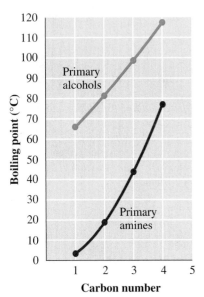

FIGURE 6.5 A comparison of boiling points of unbranched primary amines and unbranched primary alcohols.

FIGURE 6.6 Low-molecular-mass amines are soluble in water because of amine–water hydrogen bonding interactions.

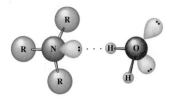

6.6 Basicity of Amines

Amines, like ammonia, are weak bases. Ammonia's weak-base behavior results from its accepting a proton (H^+) from water to produce ammonium ion (NH_4^+) and hydroxide ion (OH^-).

$$\ddot{N}H_3 \; + \; \overset{\frown}{H}OH \;\rightleftharpoons\; NH_4^+ \; + \; OH^-$$

Ammonia   Ammonium   Hydroxide
                    ion   ion

Amines behave in a similar manner.

$$CH_3-\ddot{N}H_2 + \overset{\frown}{H}OH \rightleftharpoons CH_3-\overset{+}{N}H_3 \; + \; OH^-$$

Methylamine   Methylammonium Hydroxide
                       ion   ion

The result of the interaction of an amine with water is a basic solution containing substituted ammonium ions and hydroxide ions. A **substituted ammonium ion** *is an ammonium ion in which one or more alkyl, cycloalkyl, or aryl groups have been substituted for hydrogen atoms.*

Three important generalizations apply to substituted ammonium ions.

1. *Substituted ammonium ions are charged species rather than neutral molecules.*
2. *The nitrogen atom in an ammonium ion or a substituted ammonium ion participates in four bonds.* In a neutral compound, nitrogen atoms form only three bonds. Four bonds about a nitrogen atom are possible, however, when the species is a positive ion.
3. *Substituted ammonium ions have common names derived from the names of the "parent" amines.* Replacement of the word *amine* in the name of the "parent" amine with the words *ammonium ion* generates the name of the substituted ammonium ion. The following two examples illustrate this nomenclature pattern.

$$CH_3-CH_2-NH_2 \xrightarrow{H_2O} CH_3-CH_2-\overset{+}{N}H_3 + OH^-$$

Ethylamine   Ethylammonium ion

$$CH_3-\underset{\underset{CH_3}{|}}{N}-CH_2-CH_3 \xrightarrow{H_2O} CH_3-\underset{\underset{CH_3}{|}}{\overset{+}{N}H}-CH_2-CH_3 + OH^-$$

Ethyldimethylamine   Ethyldimethylammonium ion

Aromatic amines also exhibit basic behavior in water. With such compounds, the positive ion formed is called a substituted *anilinium ion.*

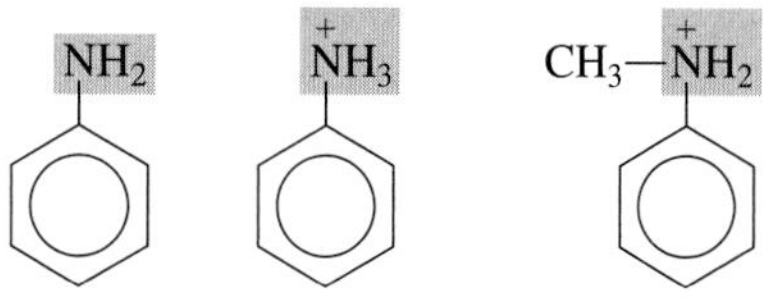

Aniline   Anilinium ion   *N*-methylanilinium ion

EXAMPLE 6.3

Determining Names for Substituted Ammonium and Substituted Anilinium Ions

■ Name the following substituted ammonium or substituted anilinium ions.

a. $CH_3-CH_2-\overset{+}{N}H_2-CH_2-CH_3$

b. $CH_3-\underset{\underset{CH_3}{|}}{CH}-CH_2-\overset{+}{N}H_3$

c. $CH_3-\underset{\underset{CH_3}{|}}{\overset{+}{N}H}-CH_3$

d. $CH_3-\overset{+}{N}H-CH_3$

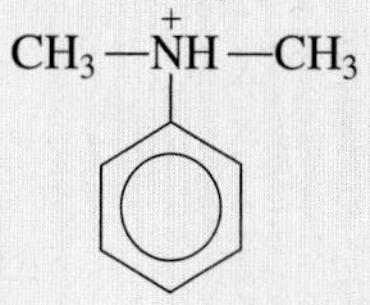

(*continued*)

Solution

a. The parent amine is diethylamine. Replacing the word *amine* in the parent name with *ammonium ion* generates the name of the ion, *diethylammonium ion*.
b. The parent amine is isobutylamine. The name of the ion is *isobutylammonium ion*.
c. The parent amine is trimethylamine. The name of the ion is *trimethylammonium ion*.
d. The parent name is *N,N*-dimethylaniline. Replacing the word *aniline* in the parent name with *anilinium ion* generates the name of the ion, *N,N-dimethylanilinium ion*.

Practice Exercise 6.3

Name the following substituted ammonium or substituted anilinium ions.

a. $CH_3—CH_2—\overset{+}{NH_2}—CH_3$

b. $CH_3—\underset{\underset{CH_3}{|}}{CH}—\overset{+}{NH_3}$

c. $CH_3—CH_2—\underset{\underset{CH_3}{|}}{\overset{+}{NH}}—CH_2—CH_3$

d. 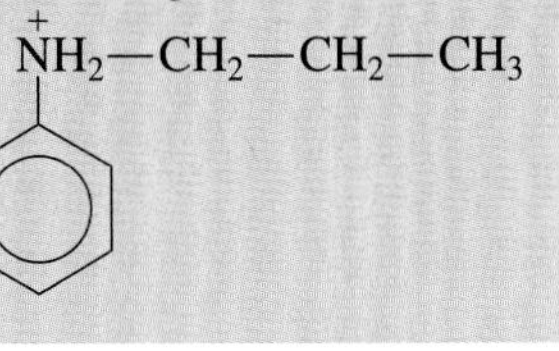 $\overset{+}{NH_2}—CH_2—CH_2—CH_3$

6.7 Amine Salts

The reaction of an acid with a base (neutralization) produces a salt. Because amines are bases (Section 6.6), their reaction with an acid produces a salt, an amine salt.

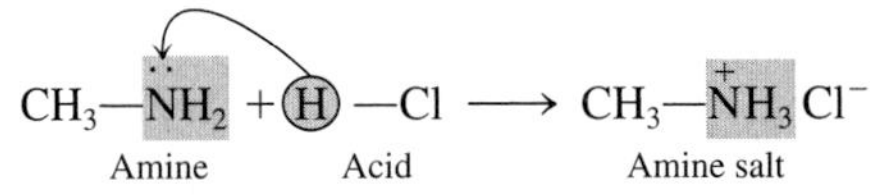

$$CH_3—\overset{\cdot\cdot}{NH_2} + \text{(H)}—Cl \longrightarrow CH_3—\overset{+}{NH_3}\,Cl^-$$

Amine   Acid   Amine salt

Aromatic amines react with acids in a similar manner.

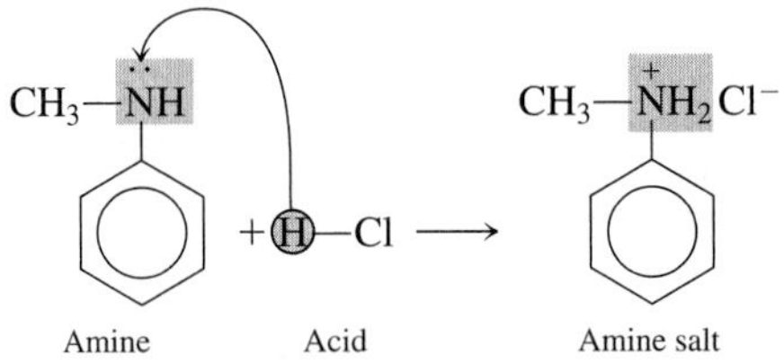

$$CH_3—\overset{\cdot\cdot}{NH} + \text{(H)}—Cl \longrightarrow CH_3—\overset{+}{NH_2}\,Cl^-$$

Amine   Acid   Amine salt

An **amine salt** *is an ionic compound in which the positive ion is a mono-, di-, or trisubstituted ammonium ion (RNH_3^+, $R_2NH_2^+$, or R_3NH^+) and the negative ion comes from an acid.* Amine salts can be obtained in crystalline form (odorless, white crystals) by evaporating the water from the acidic solutions in which amine salts are prepared.

Amine salts are named using standard nomenclature procedures for ionic compounds. The name of the positive ion, the substituted ammonium or anilinium ion, is given first and is followed by a separate word for the name of the negative ion.

$$CH_3—CH_2—\overset{+}{NH_3}\,Cl^- \qquad CH_3—\overset{+}{NH_2}—CH_3\,Br^-$$

Ethylammonium chloride   Dimethylammonium bromide

An older naming system for amine salts, still used in the pharmaceutical industry, treats amine salts as amine–acid complexes rather than as ionic compounds. In this

The cocaine molecule is both an amine and an ester; one amine and two ester functional groups are present. As an illegal street drug, cocaine is consumed as a water-soluble amine salt and in a water-insoluble, free-base form (nonsalt form—freed of the base required to make the salt). Cocaine hydrochloride, the amine salt, is a white powder that is snorted or injected intravenously. Free-base cocaine is heated and its vapors are inhaled. Cocaine users and dealers call the water-insoluble form of the drug "crack." "Snow" and "coke" are street names for the water-soluble form of the drug.

Cocaine
(an amine)

Cocaine hydrochloride
(an amine salt)

system, the amine salt made from dimethylamine and hydrochloric acid is named and represented as

$$CH_3-NH\cdot HCl \atop | \atop CH_3 \qquad \text{rather than as} \qquad CH_3-\overset{+}{N}H_2\ Cl^- \atop | \atop CH_3}$$

Dimethylamine hydrochloride   Dimethylammonium chloride

Many medication labels refer to hydrochlorides or hydrogen sulfates (from sulfuric acid), indicating that the medications are in a water-soluble ionic (salt) form.

Many higher-molecular-mass amines are water-insoluble; however, virtually all amine salts are water-soluble. Thus amine salt formation, like carboxylic acid salt formation (Section 5.9), provides a means for converting water-insoluble compounds into water-soluble compounds. Many drugs that contain amine functional groups are administered to patients in the form of amine salts because of their increased solubility in water in this form.

Many people unknowingly use acids to form amine salts when they put vinegar or lemon juice on fish. Such action converts amines in fish (often smelly compounds) to salts, which are odorless.

The process of forming amine salts with acids is an easily reversed process. Treating an amine salt with a strong base such as NaOH regenerates the "parent" amine.

$$CH_3-\overset{+}{N}H_3\ Cl^- + NaOH \longrightarrow CH_3-NH_2 + NaCl + H_2O$$
Amine salt   Base   Amine

The "opposite nature" of the processes of amine salt formation from an amine and the regeneration of the amine from its amine salt can be diagrammed as follows:

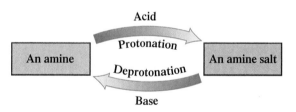

An amine gains a hydrogen ion to produce an amine salt when treated with an acid (a protonation reaction), and an amine salt loses a hydrogen ion to produce an amine when treated with a base (a deprotonation reaction).

EXAMPLE 6.4

Writing Chemical Equations for Reactions That Involve Amine Salts

■ Write the structures of the products that form when each of the following reactions involving amines or amine salts takes place.

a. $CH_3-NH-CH_3 + HCl \longrightarrow$

b.

NH—CH₃

+ $H_2SO_4 \longrightarrow$

c. $CH_3-\overset{+}{N}H_2-CH_3\ Cl^- + NaOH \longrightarrow$

Solution

a. The reactants are an amine and a strong acid. Their interaction produces an amine salt.

$$CH_3-\overset{..}{N}H-CH_3 + \overset{\frown}{\textcircled{H}}Cl \longrightarrow CH_3-\overset{+}{N}H_2-CH_3\ Cl^-$$

(continued)

b. Again, we have the reaction of an amine with a strong acid. A hydrogen ion is transferred from the acid to the amine.

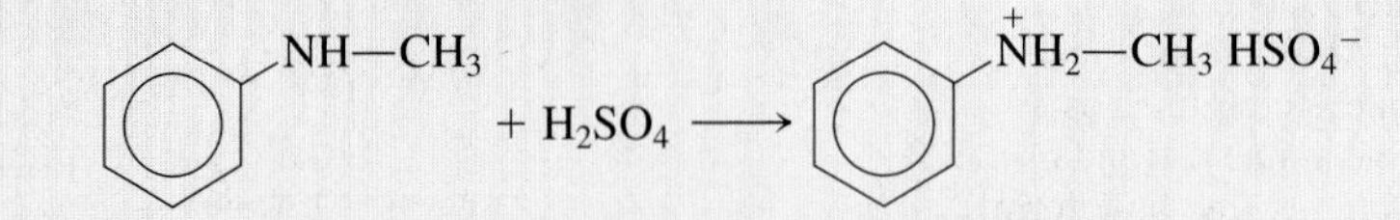

$$\underset{\text{(benzene ring)}}{}\text{NH—CH}_3 + \text{H}_2\text{SO}_4 \longrightarrow \underset{\text{(benzene ring)}}{}\overset{+}{\text{NH}}_2\text{—CH}_3 \; \text{HSO}_4^-$$

c. The reactants are an amine salt and a strong base. Their interaction regenerates the "parent" amine.

$$\text{CH}_3\text{—}\overset{+}{\text{NH}}_2\text{—CH}_3 \; \text{Cl}^- + \text{NaOH} \longrightarrow \text{CH}_3\text{—NH—CH}_3 + \text{NaCl} + \text{H}_2\text{O}$$

Practice Exercise 6.4

Write the structures of the products formed in the following reactions.

a. $\text{CH}_3\text{—CH}_2\text{—}\underset{\underset{\text{CH}_3}{|}}{\text{N}}\text{—CH}_3 + \text{HCl} \longrightarrow$

b. $\text{CH}_3\text{—CH}_2\text{—NH}_2 + \text{H}_2\text{SO}_4 \longrightarrow$

c. $\text{CH}_3\text{—CH}_2\text{—}\underset{\underset{\text{CH}_3}{|}}{\overset{+}{\text{NH}}}\text{—CH}_3 \; \text{Br}^- + \text{NaOH} \longrightarrow$

6.8 Preparation of Amines and Quaternary Ammonium Salts

Several methods exist for preparing amines. We consider only one: alkylation in the presence of base. Generalized equations for the alkylation process are

$$\text{Ammonia} + \text{alkyl halide} \xrightarrow{\text{Base}} 1° \text{ amine}$$

$$1° \text{ Amine} + \text{alkyl halide} \xrightarrow{\text{Base}} 2° \text{ amine}$$

$$2° \text{ Amine} + \text{alkyl halide} \xrightarrow{\text{Base}} 3° \text{ amine}$$

$$3° \text{ Amine} + \text{alkyl halide} \xrightarrow{\text{Base}} \text{quaternary ammonium salt}$$

Alkylation under basic conditions is actually a two-step process. In the first step, using a primary amine preparation as an example, an amine salt is produced.

$$\text{NH}_3 + \text{R—X} \longrightarrow \text{R—}\overset{+}{\text{NH}}_3 \; \text{X}^-$$

The second step, which involves the base present (NaOH), converts the amine salt to free amine.

$$\text{R—}\overset{+}{\text{NH}}_3 \; \text{X}^- + \text{NaOH} \longrightarrow \text{RNH}_2 + \text{NaX} + \text{H}_2\text{O}$$

A specific example of the production of a primary amine from ammonia is the reaction of ethyl bromide with ammonia to produce ethylamine. The chemical equation (with both steps combined) is

$$\text{NH}_3 + \text{CH}_3\text{—CH}_2\text{—Br} + \text{NaOH} \longrightarrow \text{CH}_3\text{—CH}_2\text{—NH}_2 + \text{NaBr} + \text{H}_2\text{O}$$

If the newly formed primary amine produced in an ammonia alkylation reaction is not quickly removed from the reaction mixture, the nitrogen atom of the amine may react with further alkyl halide molecules, giving, in succession, secondary and tertiary amines.

$$\text{NH}_3 \xrightarrow[\text{OH}^-]{\text{RX}} \underset{\substack{\text{Primary} \\ \text{amine}}}{\text{RNH}_2} \xrightarrow[\text{OH}^-]{\text{RX}} \underset{\substack{\text{Secondary} \\ \text{amine}}}{\text{R}_2\text{NH}} \xrightarrow[\text{OH}^-]{\text{RX}} \underset{\substack{\text{Tertiary} \\ \text{amine}}}{\text{R}_3\text{N}}$$

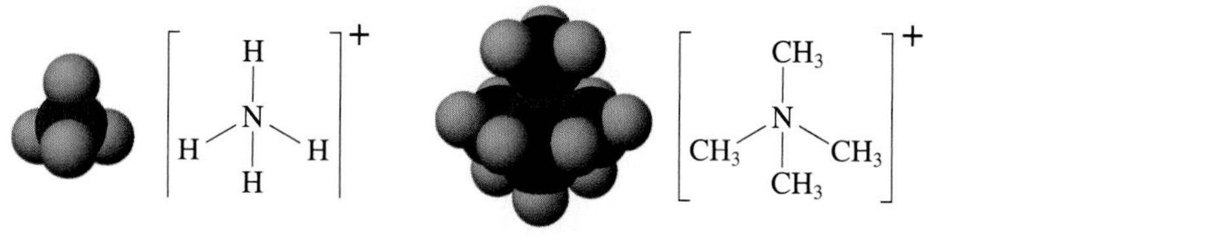

FIGURE 6.7 Space-filling models showing that the ammonium ion (NH_4^+) has a tetrahedral structure, as does the quaternary ammonium ion, in which four methyl groups are present [$(CH_3)_4N^+$].

Examples of the production of a 2° amine and a 3° amine via alkylation are

$$CH_3-CH_2-\boxed{NH_2} + CH_3-Br + NaOH \longrightarrow CH_3-CH_2-\boxed{NH}-CH_3 + NaBr + H_2O$$

Primary amine   Alkyl halide   Base   Secondary amine

$$CH_3-CH_2-\boxed{NH}-CH_3 + CH_3-Br + NaOH \longrightarrow CH_3-CH_2-\underset{\underset{CH_3}{|}}{N}-CH_3 + NaBr + H_2O$$

Secondary amine   Alkyl halide   Base   Tertiary amine

Tertiary amines react with alkyl halides in the presence of a strong base to produce a quaternary ammonium salt. A **quaternary ammonium salt** *is an ammonium salt in which all four groups attached to the nitrogen atom of the ammonium ion are hydrocarbon groups.*

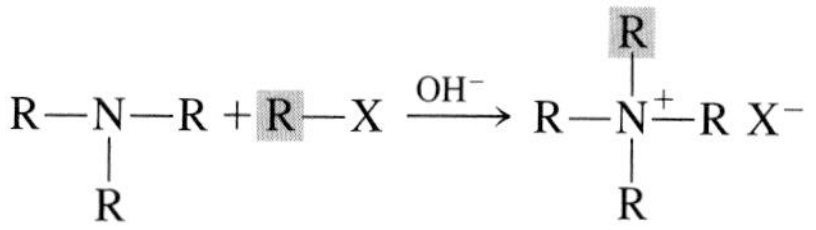

$$R-\underset{\underset{R}{|}}{N}-R + \boxed{R}-X \xrightarrow{OH^-} R-\overset{\overset{\boxed{R}}{|}}{\underset{\underset{R}{|}}{N^+}}-R \ X^-$$

Quaternary ammonium salts differ from amine salts in that addition of strong base does not convert quaternary ammonium salts back to their "parent" amines; there is no hydrogen atom on the nitrogen with which the OH^- can react. Quaternary ammonium salts are colorless, odorless, crystalline solids that have high melting points and are usually water-soluble.

Quaternary ammonium salts are named in the same way as amine salts (Section 6.7), taking into account that four organic groups are attached to the nitrogen atom rather than a lesser number of groups. Figure 6.7 contrasts the structures of an ammonium ion and a tetramethyl ammonium ion.

Compounds that contain quaternary ammonium ions are important in biochemical systems. Choline and acetylcholine are two important quaternary ammonium *ions* present in the human body. Choline has important roles in both fat transport and growth regulation. Acetylcholine is involved in the transmission of nerve impulses.

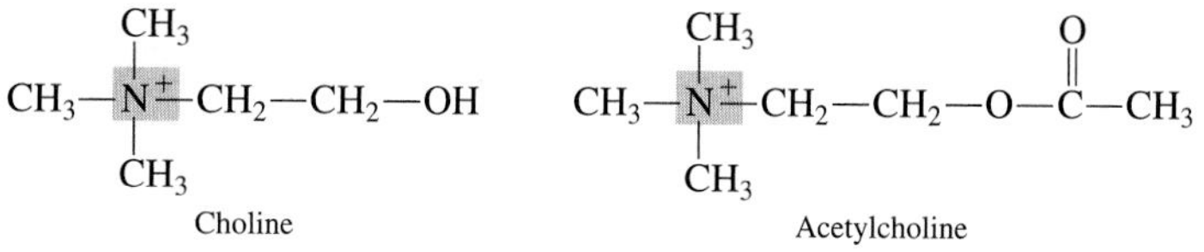

$$CH_3-\overset{\overset{CH_3}{|}}{\underset{\underset{CH_3}{|}}{N^+}}-CH_2-CH_2-OH \qquad CH_3-\overset{\overset{CH_3}{|}}{\underset{\underset{CH_3}{|}}{N^+}}-CH_2-CH_2-O-\overset{\overset{O}{||}}{C}-CH_3$$

Choline   Acetylcholine

6.9 Heterocyclic Amines

Heterocyclic amines are the first heterocyclic compounds we have encountered that have nitrogen heteroatoms. In previous chapters, we have discussed heterocyclic compounds with oxygen as the heteroatom: cyclic ethers (Section 3.19); cyclic ketones (Section 4.2); the cyclic forms of hemiacetals and acetals (Section 4.10); and cyclic esters (Section 5.10).

A **heterocyclic amine** *is an organic compound in which nitrogen atoms of amine groups are part of either an aromatic or a nonaromatic ring system.* Heterocyclic amines are the most common type of heterocyclic organic compound (Section 3.19). Figure 6.8 gives structures for a number of "key" unsubstituted heterocyclic amines. These compounds are the "parent" compounds for numerous derivatives that are important in medicinal, agricultural, food, and industrial chemistry, as well as in the functioning of the human body.

Study of the heterocyclic amine structures in Figure 6.8 shows that (1) ring systems may be saturated, unsaturated, or aromatic, (2) more than one nitrogen atom may be present in a given ring, and (3) fused ring systems often occur.

FIGURE 6.8 Structural formulas for selected heterocyclic amines that serve as "parent" molecules for more complex amine derivatives.

PYRROLIDINE	PYRROLE	IMIDAZOLE	INDOLE
PYRIDINE	PYRIMIDINE	QUINOLINE	PURINE

Heterocyclic amines often have strong odors, some agreeable and others disagreeable. The "pleasant" aroma of many heat-treated foods is caused by heterocyclic amines formed during the heat treatment. The compounds responsible for the pervasive odors of popped popcorn and hot roasted peanuts are heterocyclic amines.

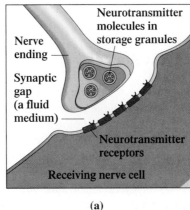

Methyl-2-pyridyl ketone
(odor of popcorn)

2-Methoxy-5-methylpyrazine
(odor of peanuts)

The two most widely used central nervous system stimulants in our society, caffeine and nicotine, are heterocyclic amine derivatives. Caffeine's structure is based on a purine ring system. Nicotine's structure contains one pyridine ring and one pyrrolidine ring.

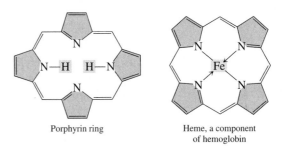

Caffeine   Nicotine

A large cyclic structure built on four pyrrole rings (Figure 6.8), called a porphyrin, is important in the chemistry of living organisms. Porphyrins form metal ion complexes in which the metal ion is located in the middle of the large ring structure. *Heme,* an iron–porphyrin complex present in the red blood pigment hemoglobin, is responsible for oxygen transport in the human body.

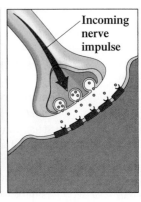
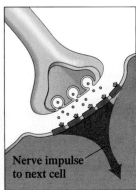

Porphyrin ring   Heme, a component
of hemoglobin

FIGURE 6.9 Neurotransmitters are chemical messengers between nerve cells. Neurotransmitters released from one nerve cell stimulate (activate) an adjacent nerve cell. (a) Before the conduction of a nerve impulse. (b) An incoming nerve impulse triggers the release of neurotransmitter molecules. (c) Neurotransmitters bind to receptor sites, activating the receptor nerve cell.

CHEMICAL CONNECTIONS   Caffeine: The Most Widely Used Central Nervous System Stimulant

Caffeine, which is naturally present in coffee beans and tea leaves and is added to many soft drinks, is both the most widely used and the most frequently used central nervous system (CNS) stimulant in our society. Studies indicate that 82% of the adult U.S. population drink coffee and 52% drink tea. Over one-half of the coffee drinkers consume three or more cups daily. In the soft drink market, the use of caffeine has now spread beyond cola drinks to orange drinks.

Caffeine belongs to a family of compounds called *xanthines*. Its formal chemical name is 1,3,7-trimethylxanthine and its structure is

Caffeine
(1,3,7-trimethylxanthine)

Xanthine

Caffeine is naturally present in coffee beans. Over 60 plants and trees cultivated by humans contain caffeine.

Besides its CNS system effects, caffeine also increases basal metabolic rate, increases heart rate by stimulating heart muscles, promotes secretion of stomach acid, functions as a bronchial tube dilator and increases urine production because of its diuretic properties. The overall effect that most individuals experience from caffeine consumption is interpreted as a "lift."

Used in small quantities, caffeine's effects are temporary; hence it must be consumed on a regular basis throughout the day. It's half-life in the body is 3.0–7.5 hours. Caffeine tolerance develops in regular users of the substance. Over time, larger amounts of caffeine are needed in order for an individual to achieve his or her "lift."

Caffeine is mildly addicting. People who ordinarily consume substantial amounts of caffeine-containing beverages or drugs experience withdrawal symptoms if caffeine is eliminated. Such symptoms include headache and depression for a period of several days. As a result of caffeine dependence, many people need a cup of coffee before they feel good each morning.

In large quantities, caffeine has been shown to cause significant undesirable effects including anxiety, sleeplessness, headaches, and dehydration. The latter occurs because of caffeine's diuretic effects.

Several studies indicate that caffeine has an effect on the child-bearing abilities of women. One study indicates a decreased rate of conception for women who consume more than three cups of coffee daily. Another study indicates that the consumption of this amount of caffeine increases the risk of miscarriage by 30% compared to women who do not drink coffee. Lactating mothers probably should limit their caffeine intake because it appears in mother's milk, thus affecting newborn babies.

Current scientific thought holds that caffeine's mode of action in the body is exerted through a chemical substance called cyclic adenosine monophosphate (cyclic AMP). Caffeine inhibits an enzyme that ordinarily breaks down cyclic AMP to its inactive end product. The resulting increase in cyclic AMP leads to increased glucose production within cells and thus makes available more energy to allow higher rates of cellular activity.

Coffee is the major source of caffeine for most Americans. However, substantial amounts of caffeine may be consumed in soft drinks, tea, and numerous nonprescription medications including combination pain relievers (Anacin, Midol, Empirin), cold remedies (Dristan, Triaminicin), and antisleep agents (No Doz, Vivarin).

6.10 Selected Biochemically Important Amines

■ Neurotransmitters

A **neurotransmitter** *is a chemical substance that is released at the end of a nerve, travels across the synaptic gap between the nerve and another nerve, and then bonds to a receptor site on the other nerve, triggering a nerve impulse.* Figure 6.9 shows schematically how neurotransmitters function.

The most important neurotransmitters in the human body are acetylcholine (Section 6.8) and the amines norepinephrine, dopamine, and serotonin.

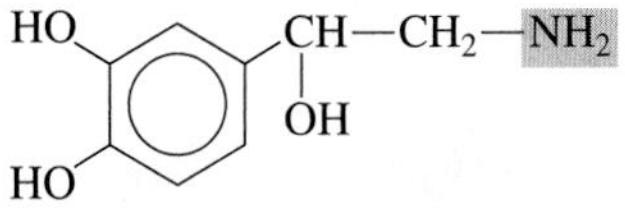

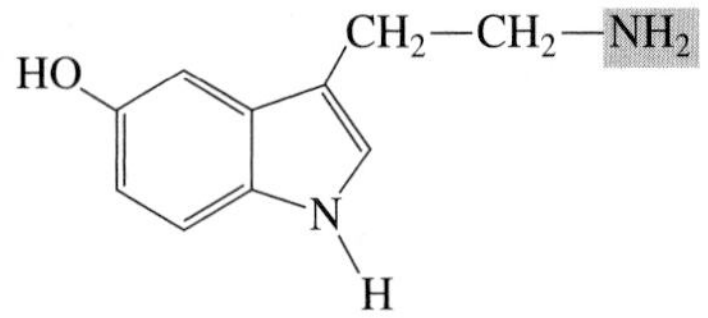

Norepinephrine, a compound secreted by the adrenal glands into the blood, helps maintain muscle tone in the blood vessels.

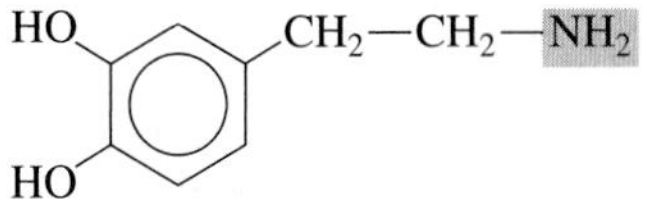

Dopamine is found in the brain. A deficiency of this neurotransmitter results in Parkinson's disease, a degenerative neurological disease. Administration of dopamine to a patient does not relieve the symptoms of this disease because dopamine in the blood cannot cross the blood–brain barrier. The drug L-dopa, which can pass through the blood–brain barrier, does give relief from Parkinson's symptoms. Inside brain cells, enzymes catalyze the conversion of L-dopa to dopamine.

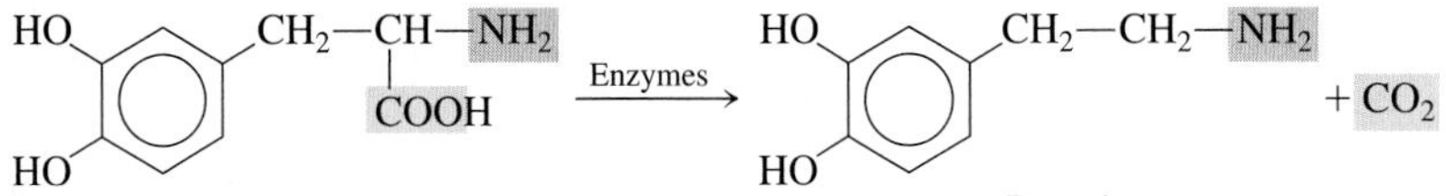

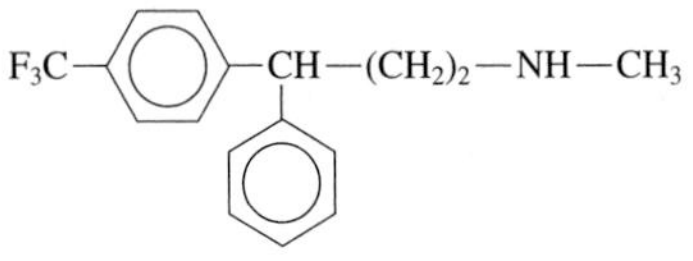

Serotonin, also a brain chemical, is involved in sleep, sensory perception, and the regulation of body temperature. Serotonin deficiency has been implicated in mental illness. Treatment of mental depression can involve the use of drugs that help maintain serotonin at normal levels by preventing its breakdown within the brain.

▇ Epinephrine

Epinephrine, also known as adrenaline, has some neurotransmitter functions but is more important as a central nervous system stimulant. Produced by the adrenal glands, epinephrine differs in structure from norepinephrine in that a methyl group substituent is present on the amine nitrogen atom.

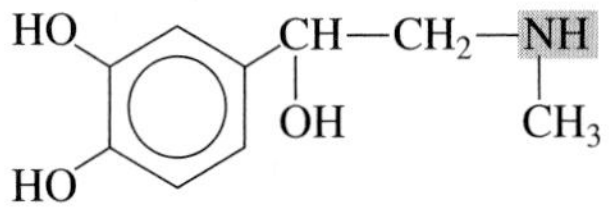

Pain, excitement, and fear trigger the release of large amounts of epinephrine into the bloodstream. The effect is increased blood glucose levels, which in turn increase blood pressure, rate and force of heart contraction, and muscular strength. These changes cause the body to function at a "higher" level. Epinephrine is often called the "fight or flight" hormone.

▇ Histamine

The heterocyclic amine *histamine* is responsible for the unpleasant effects felt by individuals susceptible to hay fever and various pollen allergies.

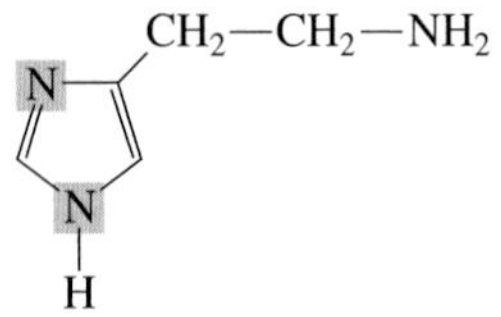

Histamine is naturally present in the human body in a "stored" form; it is part of more complex molecules. A number of situations can trigger the release of the "stored" histamine. Activators include (1) contact with pollen, dust, and other allergens, (2) substances

---

The names of the amine neurotransmitters are pronounced nor-ep-in-NEFF-rin, DOPE-a-mean, and SER-oh-tone-in.

Prozac, the most widely prescribed drug for mental depression, inhibits the reuptake of serotonin, thus maintaining serotonin levels. Chemically, Prozac is a derivative of methyl propyl amine with three fluorine atoms present in the structure. The element fluorine is rarely encountered in biochemical molecules.

Ephedrin, pronounced "eh-FEH-drin," is a substance extracted from the Asian plant ephedra that was used in numerous dietary supplements sold as weight-loss aids and energy boosters at nutrition stores, in supermarkets, and on the Internet until it was banned in the United States by the Food and Drug Administration (FDA) in 2004. Its chemical structure resembles that of epinephrine (adrenaline).

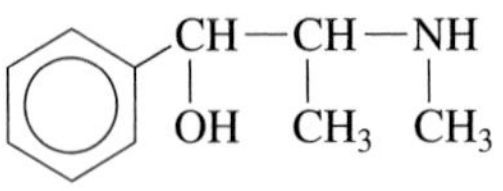

Ephedrin

Within the body, it behaves as a stimulant to the heart and the central nervous system. Concerns about the safety of its use, particularly with people who have hypertension and other cardiovascular problems, led to its ban.

Antihistamines are drugs that counteract, to some extent, the effects of histamine release in the body. Antihistamines share a common structural feature with histamine— an ethanamine chain.

$$-CH_2-CH_2-N\diagup^{\diagdown}$$

This structure allows antihistamines to occupy receptor sites in nerves normally occupied by histamine, thus blocking histamine from occupying the nerve sites.

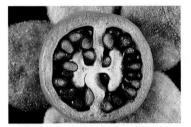

FIGURE 6.10 Fruit of the belladonna plant; the alkaloid atropine is obtained from this plant.

The name *alkaloid,* which means "like a base," reflects the fact that alkaloids react with acids. Such behavior is expected for substances with amine functional groups because amines are weak bases.

FIGURE 6.11 Oriental poppy plants, the source of several narcotic painkillers, including morphine.

released from damaged cells, and (3) contact with chemicals to which an individual has become sensitized.

The presence of "free" histamine causes the symptoms associated with hay fever, such as watery eyes and stuffy nose, and many of the symptoms associated with the common cold. A group of substances called *antihistamines* can be taken as medication to counteract the effects of the histamine.

6.11 Alkaloids

People in various parts of the world have known for centuries that physiological effects can be obtained by eating or chewing the leaves, roots, or bark of certain plants. Over 5000 different compounds that are physiologically active have been isolated from such plants. Nearly all of these compounds, which are collectively called alkaloids, contain amine functional groups. An **alkaloid** *is a nitrogen-containing organic compound extracted from plant material.*

Three well-known compounds that we have considered previously are alkaloids. They are nicotine (tobacco plant), caffeine (coffee beans and tea leaves), and cocaine (coca plant).

A number of alkaloids are currently used in medicine. Quinine, which occurs in cinchona bark, is used to treat malaria. Atropine, which is isolated from the belladonna plant, is used to dilate the pupil of the eye in patients undergoing eye examinations (Figure 6.10). Atropine is also used as a preoperative drug to relax muscles and reduce the secretion of saliva in surgical patients.

Quinine

Atropine

An extremely important family of alkaloids is the narcotic painkillers, a class of drugs derived from the resin (opium) of the oriental poppy plant (Figure 6.11). The most important drugs obtained from opium are morphine and codeine. Synthetic modification of morphine produces the illegal drug heroin.

These three compounds have similar chemical structures.

Morphine

Codeine

Heroin

CHEMICAL CONNECTIONS   Amphetamines: Central Nervous System Stimulants

Amphetamines are a set of powerful *synthetic* amines that function as central nervous system stimulants. Benzedrine, also called simply amphetamine, is the parent compound for the amphetamine family. Important amphetamine derivatives include methamphetamine (prescribed as an antidepressant), methoxyamphetamine (prescrib]ed as a bronchodilator), and isoproterenol (prescribed for emphysema and asthma).

Amphetamine
(benzedrine)

Methamphetamine
(methedrine)

Methoxyamphetamine

Isoproternol

Structurally, these compounds are all related to adrenaline and mimic its stimulant effects.

Generally, amphetamines increase both heart rate and respiratory rate. They also reduce fatigue and diminish hunger by raising the glucose level in blood. At one time, they were widely used as appetite suppressants in the treatment of obesity, but because of many adverse effects, their use in weight control has diminished.

Hyperactive children (so overactive that they cannot sit still or concentrate) can benefit from amphetamines. Paradoxically, these CNS system stimulants calm hyperactive patients. (Recent research suggests that the mode of action involves increasing serotonin levels, with resulting inhibition of aggressive and impulsive behavior.) Ritalin (methylphenidate), an amphetaminelike drug used for treating hyperactive children, is one of the most widely prescribed stimulant drugs in the United States. Its structure is similar enough to that of amphetamine

that, pharmacologically, its effects are virtually identical to those of amphetamine.

Amphetamine

Methylphenidate
(Ritalin)

A great problem with amphetamines is the diversion of large quantities of these relatively inexpensive drugs into the illegal drug market. In this illegal drug market, *speed, splash,* and *crank* are names for powdered forms of methamphetamine; *ice* is a pure, crystallized form of methamphetamine that is smoked. *STP* is methoxyamphetamine. Abuse of these drugs can produce severe physiological reactions. Once the drugs wear off, the user tends to "crash" into a state of physical and mental exhaustion. Withdrawal produces fatigue and profound and prolonged sleep.

Methylenedioxymethamphetamine (also known as MDMA and as Ecstasy) is an illegal hallucinogenic amphetamine derivative whose use is rapidly increasing; it now ranks behind only alcohol and marijuana in use by teens. It is typically sold as a white powder, and it can be inhaled, injected, or swallowed. Animal studies indicate that Ecstasy use can damage the brain's serotonin cells; serotonin is involved in appetite, sleep, mood regulation, memory, and sexual function. One of the most dangerous aspects of the use of this particular drug is that it can damage brain cells without any warning that the damage is taking place.

Methylenedioxymethamphetamine
(MDMA, Ecstasy)

Morphine is one of the most effective painkillers known; its painkilling properties are about a hundred times greater than those of aspirin. Morphine acts by blocking the process in the brain that interprets pain signals coming from the peripheral nervous system. The major drawback to the use of morphine is that it is addictive.

Codeine is a methylmorphine. Almost all codeine used in modern medicine is produced by methylating the more abundant morphine. Codeine is less potent than morphine, having a painkilling effect about one-sixth that of morphine.

Heroin is a synthetic compound, the diacetyl ester of morphine; it is produced from morphine. This chemical modification increases painkilling potency; heroin has more than three times the painkilling effect of morphine. However, heroin is so addictive that it has no accepted medical use in the United States.

Alkaloids Present in Chocolate

Chocolate is a food preparation made from the beans (seeds) of the tropical cacao tree. Growth conditions for cacao trees require a warm, moist climate like that found near the equator. The majority of the world's supply of cacao beans comes from the west coast of Africa—Ivory Coast, Ghana, and Nigeria. (Because of a mistake in spelling, probably made by early English importers, cacao beans are known as cocoa beans in English-speaking countries.)

All chocolate products are manufactured from ground cocoa beans. The heat from the grinding process causes the cocoa bean mixture to melt, forming a free-flowing mixture called *chocolate liquor.* Unsweetened baker's chocolate is simply cooled, hardened chocolate liquor. Semisweet chocolate has added granulated sugar. Milk chocolate has added sugar, milk solids, and vanilla flavoring.

Because of their plant origins, chocolate products contain alkaloids. The dominant alkaloid present is theobromine, with caffeine being present in a smaller amount. The name theobromine comes from the Greek term *theobroma* meaning "food of the Gods." The concentrations of these two alkaloids in cocoa beans varies depending on the origin of the beans. The following table gives theobromine and caffeine content of several finished chocolate products.

The caffeine content of a typical chocolate bar is 30 mg and that of a slice of chocolate cake 20–30 mg. By contrast, a cup of coffee contains 100–150 mg of caffeine and a twelve-ounce cola drink contains 33–52 mg.

Structurally, theobromine and caffeine differ only by a methyl group.

This close structural similarity does not, however, translate into close pharmacological properties. Theobromine's stimulant effects on the central nervous system are minimal compared to those of caffeine. A mild diuretic effect and relaxation of the smooth muscles of the bronchi in the lungs are two other effects of theobromine; caffeine has similar effects in these areas.

Theobromine has been used as a pharmaceutical drug for its diuretic effect. Because of its ability to dilate blood vessels, theobromine also has been used to treat high blood pressure. Research shows that pets, especially dogs, are sensitive to theobromine because the animals metabolize theobromine more slowly than humans. A chocolate bar, inadvertently ingested, is poisonous to dogs and can even be lethal. The same holds true for cats.

Occasionally, chocolate is touted as a "health" food because cocoa beans have relatively high levels of several kinds of antioxidant flavonoids (see the Chemical Connections feature on page 153). Studies show that people with high blood levels of flavonoids are at lower risk of developing heart disease, asthma, and type 2 diabetes. Dark chocolate contains the most cocoa and thus the most flavonoids. As a "health" food, however, chocolate should be consumed only occasionally because the downside of consumption is the high number of calories associated with chocolate.

Theobromine
(3,7-dimethylxanthine)

Caffeine
(1,3,7-trimethylxanthine)

Theobromine and Caffeine Content of Finished Chocolate Products

Product	Theobromine, %	Caffeine, %	Theobromine/ Caffeine Ratio
baking chocolate	1.386	0.164	8.45 to 1
dark sweet chocolate	0.474	0.076	6.3 to 1
milk chocolate	0.197	0.022	9.0 to 1

6.12 Structure and Classification of Amides

An **amide** *is a carboxylic acid derivative in which the carboxyl —OH group has been replaced with an amino or a substituted amino group.* The amide functional group is thus

depending on the degree of substitution.

Amides, like amines, can be classified as primary (1°), secondary (2°), or tertiary (3°), depending on how many hydrogen atoms are attached to the nitrogen atom.

> Primary, secondary, and tertiary amides are also called unsubstituted, monosubstituted, and disubstituted amides, respectively.

$$\underset{\text{Primary amide}}{R-\overset{\displaystyle O}{\overset{\|}{C}}-NH_2} \qquad \underset{\text{Secondary amide}}{R-\overset{\displaystyle O}{\overset{\|}{C}}-NH-R'} \qquad \underset{\text{Tertiary amide}}{R-\overset{\displaystyle O}{\overset{\|}{C}}-\underset{\underset{R''}{|}}{N}-R'}$$

A **primary amide** *is an amide in which two hydrogen atoms are bonded to the amide nitrogen atom.* Such amides are also called *unsubstituted* amides. A **secondary amide** *is an amide in which an alkyl (or aryl) group and a hydrogen atom are bonded to the amide nitrogen atom. Monosubstituted* amide is another name for this type of amide. A **tertiary amide** *is an amide in which two alkyl (or aryl) groups and no hydrogen atoms are bonded to the amide nitrogen atom.* Such amides are *disubstituted* amides.

Note that the difference between a 1° amide and a 2° amide is "H versus R" and that the difference between a 2° amide and a 3° amide is again "H versus R." These "H versus R" relationships are the same relationships that exist between 1° and 2° amines and 2° and 3° amines (Section 6.2), as is summarized in Figure 6.12.

The simplest amide has a hydrogen atom attached to an unsubstituted amide functional group.

$$H-\overset{\displaystyle O}{\overset{\|}{C}}-NH_2$$

Next in complexity are amides in which a methyl group is present. There are two of them, one with the methyl group attached to the carbon atom and the other with the methyl group attached to the nitrogen atom.

FIGURE 6.12 Primary, secondary, and tertiary amines and amides and the "H versus R" relationship.

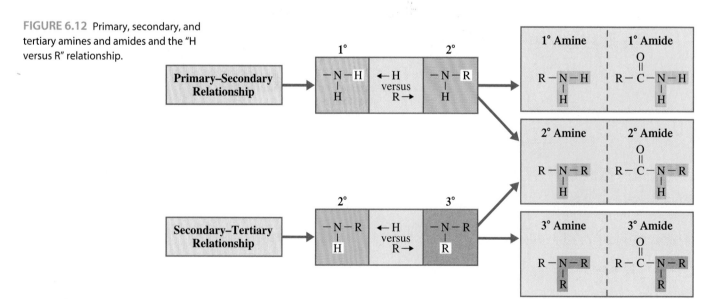

Line-angle formulas for selected primary, secondary, and tertiary amides:

$$CH_3-\overset{\displaystyle O}{\overset{\|}{C}}-NH_2 \quad \text{and} \quad H-\overset{\displaystyle O}{\overset{\|}{C}}-NH-CH_3$$

The first of these structures is a 1° amide, and the second structure is a 2° amide. The structure of the simplest aromatic amide involves a benzene ring to which an unsubstituted amide functional group is attached.

Cyclic amide structures are possible. Examples of such structures include

Cyclic amides are called *lactams,* a term that parallels the use of the term *lactones* for cyclic esters (Section 5.11).

A lactone
(a cyclic ester)

A lactam
(a cyclic amide)

The members of the penicillin family of antibiotics (Section 10.10) have structures that contain a four-membered lactam ring.

6.13 Nomenclature for Amides

For nomenclature purposes (both IUPAC and common), amides are considered to be derivatives of carboxylic acids. Hence their names are based on the name of the parent carboxylic acid. (A similar procedure was used for naming esters; Section 5.12). The rules are as follows:

1. The ending of the name of the carboxylic acid is changed from -*ic acid* (common) or -*oic acid* (IUPAC) to -*amide.* For example, *benzoic acid* becomes *benzamide.*
2. The names of groups attached to the nitrogen (2° and 3° amides) are appended to the front of the base name, using an *N*- prefix as a locator.

Selected primary amide IUPAC names (with the common name in parentheses) are

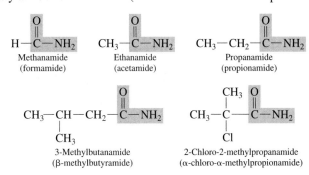

Acrylamide (2-propenamide), the simplest unsaturated amide, has the structure

$$CH_2=CH-\overset{\displaystyle O}{\overset{\|}{C}}-NH_2$$

It is a neurotoxic agent and a possible human carcinogen.

Surprisingly, low concentrations of acrylamide have been found in potato chips, french fries, and other starchy foods prepared at high temperatures (greater than 120°C). Its possible source is the reaction between the amino acid asparagine (present in food proteins; Section 9.2) and carbohydrate sugars (Section 7.8) present in food.

Human risk studies are underway concerning acrylamide presence in fried and some baked foods. No traces of acrylamide have been found in uncooked or boiled foods.

Nomenclature for secondary and tertiary amides, amides with substituted amino groups, involves use of the prefix *N-*, a practice we previously encountered with amine nomenclature (Section 6.3).

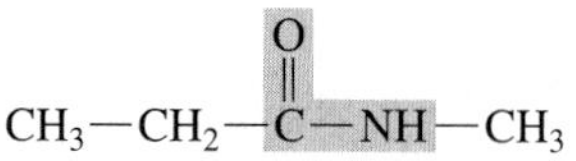

$$CH_3-CH_2-\overset{\overset{\textstyle O}{\|}}{C}-NH-CH_3$$

N-Methylpropanamide
(*N*-methylpropionamide)

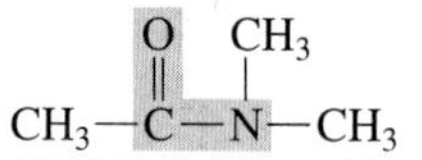

$$CH_3-\overset{\overset{\textstyle O}{\|}}{C}-\overset{\overset{\textstyle CH_3}{|}}{N}-CH_3$$

N,N-Dimethylethanamide
(*N,N*-dimethylacetamide)

Molecular models for methanamide and its *N*-methyl and *N,N*-dimethyl derivatives (the simplest 1°, 2°, and 3° amides, respectively) are given in Figure 6.13.

The simplest aromatic amide, a benzene ring bearing an unsubstituted amide group, is called *benzamide.* Other aromatic amides are named as benzamide derivatives.

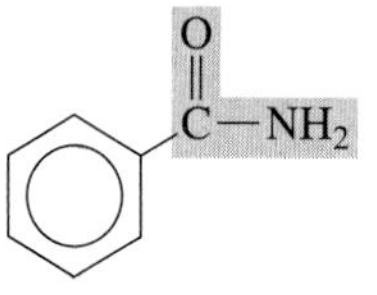

Benzamide 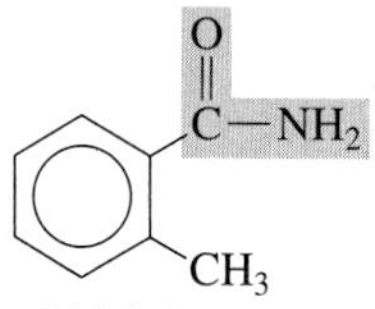   2-Methylbenzamide 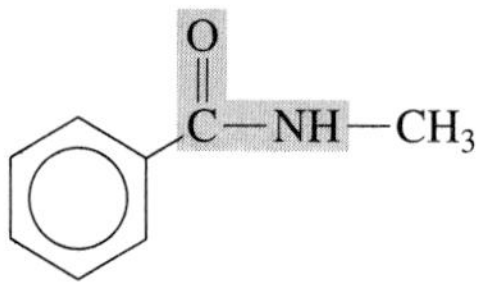   *N*-Methylbenzamide

---

EXAMPLE 6.5

Determining IUPAC and Common Names for Amides

■ Assign both common and IUPAC names to each of the following amides.

a. 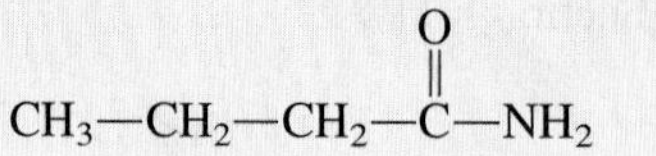
$$CH_3-CH_2-CH_2-\overset{\overset{\textstyle O}{\|}}{C}-NH_2$$

b. 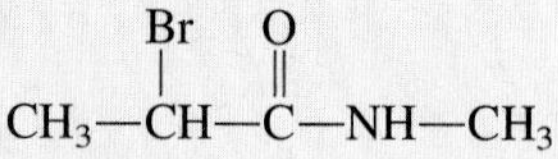
$$CH_3-\overset{\overset{\textstyle Br}{|}}{CH}-\overset{\overset{\textstyle O}{\|}}{C}-NH-CH_3$$

c.

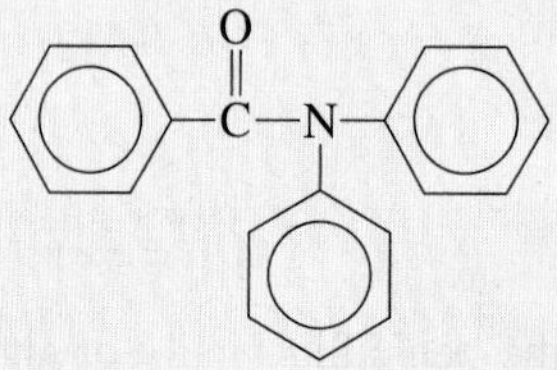

d.

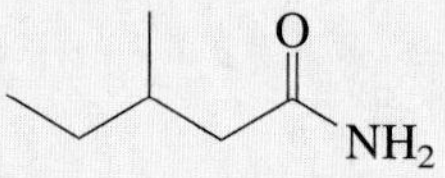

The contrast between IUPAC names and common names for unbranched unsubstituted amides is as follows:

IUPAC (one word)

| alkanamide |

ethanamide

Common (one word)

| (prefix)amide* |

acetamide

*The common-name prefixes are related to natural sources for the acids.

Solution

a. The parent acid for this amide is butyric acid (common) or butanoic acid (IUPAC). The common name for this amide is *butyramide,* and the IUPAC name is *butanamide.*

b. The common and IUPAC names of the acid are very similar; they are propionic acid and propanoic acid, respectively. The common name is *α-bromo-N-methylpropionamide,* and the IUPAC name is *2-bromo-N-methylpropanamide.* The prefix *N-* must be used with the methyl group to indicate that it is attached to the nitrogen atom.

c. In both the common and IUPAC systems of nomenclature, the name of the parent acid is the same: benzoic acid. The name of the amide is *N,N-diphenylbenzamide.*

d. The amide is derived from valeric acid (common name) or pentanoic acid (IUPAC name). The complete name must take into account the presence of the methyl group on the carbon chain. The amide's common name is *β-methylvaleramide* and its IUPAC name is 3-methylpentanamide.

Practice Exercise 6.5

Assign both common and IUPAC names to each of the following amides.

a. 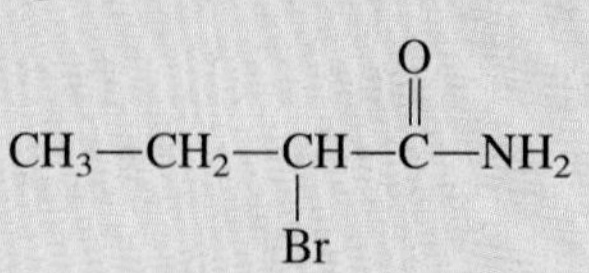
$$CH_3-CH_2-\overset{\overset{\textstyle O}{\|}}{\underset{\overset{\textstyle |}{Br}}{CH}}-\overset{\overset{\textstyle O}{\|}}{C}-NH_2$$

b. 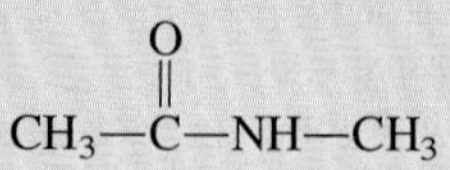
$$CH_3-\overset{\overset{\textstyle O}{\|}}{C}-NH-CH_3$$

c.

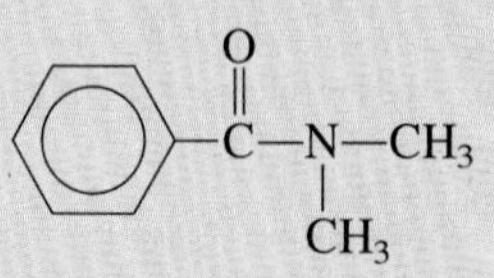

d.

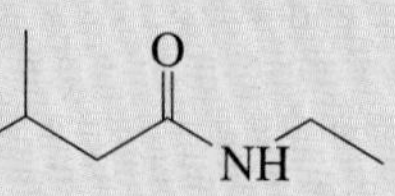

$$\overset{O}{\overset{\|}{H-C-NH_2}}$$

Methanamide
(a primary amide)

$$\overset{O}{\overset{\|}{H-C-NH-CH_3}}$$

N-Methyl methanamide
(a secondary amide)

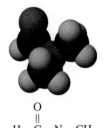

$$\overset{O}{\overset{\|}{H-C-\underset{\underset{CH_3}{|}}{N}-CH_3}}$$

N,N-Dimethyl methanamide
(a tertiary amide)

FIGURE 6.13 Space-filling models for the simplest primary, secondary, and tertiary amides.

Synthetically produced melatonin is under investigation as a drug for treating jet lag. Jet lag is a condition caused by desynchronization of the biological clock. It is usually caused by drastically changing the sleep–wake cycle, as when crossing several time zones during an airline flight or when performing shift work. Symptoms of jet lag include fatigue, early awakening or inability to sleep, and headaches. Studies indicate that melatonin taken in the evening in a new time zone will usually reset a person's biological clock and almost totally alleviate (or prevent) the symptoms of jet lag.

6.14 Selected Amides and Their Uses

The simplest naturally occurring amide is urea, a water-soluble white solid produced in the human body from carbon dioxide and ammonia through a complex series of metabolic reactions (Section 15.4).

$$CO_2 + 2NH_3 \longrightarrow (H_2N)_2CO + H_2O$$

Urea is a one-carbon diamide. Its molecular structure is

$$\overset{O}{\overset{\|}{H_2N-C-NH_2}}$$

Urea formation is the human body's primary method for eliminating "waste" nitrogen. The kidneys remove urea from the blood and provide for its excretion in urine. With malfunctioning kidneys, urea concentrations in the body can build to toxic levels—a condition called *uremia*.

Melatonin is a hormone that is synthesized by the pineal gland and that regulates the sleep–wake cycle in humans. Melatonin levels within the body increase in evening hours and then decrease as morning approaches. High melatonin levels are associated with longer and more sound sleeping. The concentration of this hormone in the blood decreases with age; a six-year-old has a blood melatonin concentration over five times that of an 80-year-old. This is one reason why young children have less trouble sleeping than senior citizens. As a prescription drug, melatonin is used to treat insomnia and jet lag.

Structurally, melatonin is a polyfunctional amide; amine and ether groups are also present.

A number of synthetic amides exhibit physiological activity and are used as drugs in the human body. Foremost among them, in terms of use, is acetaminophen, which in 1992 replaced aspirin as the top-selling over-the-counter pain reliever. Acetaminophen is a derivative of acetamide (see the Chemical Connections feature on page 203).

Barbiturates, which are cyclic amide compounds, are a heavily used group of prescription drugs that cause relaxation (tranquilizers), sleep (sedatives), and death (overdoses). All barbiturates are derivatives of barbituric acid, a cyclic amide that was first synthesized from urea and malonic acid.

Urea   Malonic acid   Barbituric acid

(The researcher who first synthesized this compound named it after his girlfriend Barbara.)

6.15 Properties of Amides

Amides do not exhibit basic properties in solution as amines do (Section 6.6). Although the nitrogen atom present in amides has a nonbonding pair of electrons, as in amines, these electrons are not available for bonding to a H^+ ion. The reason for this is related to the polarity of the carbonyl portion (—C=O) of the amide functional group.

Unbranched Primary Amides			
C_1	C_3	C_5	C_7
C_2	C_4	C_6	C_8

☐ Liquid ☐ Solid

FIGURE 6.14 A physical-state summary for unbranched primary amides at room temperature and pressure.

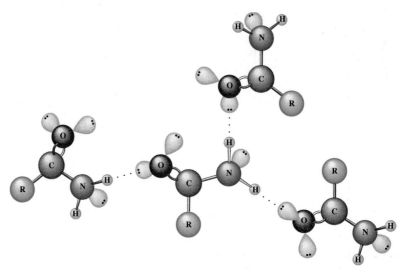

FIGURE 6.15 The high boiling points of amides are related to the numerous amide–amide hydrogen-bonding possibilities that exist.

Lidocaine (xylocaine), a substance commonly administered by injection as a dental anesthetic, is a synthetic molecule that contains both amide and amine functional groups.

Another well known local anesthetic is procaine (novocaine). Its structure contains two amine groups and an ester group but no amide group.

Both lidocaine and procaine share a common structural feature—the presence of a diethyl amino group (on the right side of each structure).

Methanamide and its *N*-methyl and *N,N*-dimethyl derivatives (the simplest 1°, 2°, and 3° amides, respectively), are all liquids at room temperature. All unbranched primary amides, except methanamide, are solids at room temperature (Figure 6.14), as are most other amides. In many cases, the amide melting point is even higher than that of the corresponding carboxylic acid. The high melting points result from the numerous intermolecular hydrogen-bonding possibilities that exist between amide H atoms and carbonyl O atoms. Figure 6.15 shows selected hydrogen-bonding interactions that are possible among several primary amide molecules.

Fewer hydrogen-bonding possibilities exist for 2° amides because the nitrogen atom now has only one hydrogen atom; hence lower melting points are the rule for such amides. Still lower melting points are observed for 3° amides because no hydrogen bonding is possible. The disubstituted *N,N*-dimethylacetamide has a melting point of −20°C, which is about 100°C lower than that of the unsubstituted acetamide.

Amides of low molecular mass, up to five or six carbon atoms, are soluble in water. Again, numerous hydrogen-bonding possibilities exist between water and the amide. Even disubstituted amides can participate in such hydrogen bonding.

Arrows denote sites where hydrogen bonding to water can occur.

6.16 Preparation of Amides

Amides are the least reactive of the common carboxylic acid derivatives and they can be synthesized from an acid chloride, an acid anhydride, an ester, or the carboxylic acid itself.

The reaction of a carboxylic acid with ammonia or a 1° or 2° amine produces an amide, provided that the reaction is carried out at an elevated temperature (greater than 100°C) and a dehydrating agent is present.

$$\text{Ammonia + carboxylic acid} \xrightarrow[\text{Catalyst}]{100°C} 1° \text{ amide}$$

$$1° \text{ Amine + carboxylic acid} \xrightarrow[\text{Catalyst}]{100°C} 2° \text{ amide}$$

$$2° \text{ Amine + carboxylic acid} \xrightarrow[\text{Catalyst}]{100°C} 3° \text{ amide}$$

CHEMICAL CONNECTIONS   Acetaminophen: A Substituted Amide

Often called the aspirin substitute, acetaminophen is the most widely used of all nonprescription pain relievers, accounting for over half of that market. Acetaminophen is a derivative of acetamide in which a hydroxyphenyl group has replaced one of the amide hydrogens.

O
‖
$CH_3—C—NH_2$

Acetamide

O
‖
$CH_3—C—NH$⟨⟩—OH

Acetaminophen

The pharmaceutical designation APAP for this compound comes from its IUPAC name, which is *N-acetyl-p-a*mino*p*henol.

Acetaminophen is the active ingredient in Tylenol, Datril, Tempra, and Anacin-3. Excedrin, which contains both acetaminophen and aspirin, is a combination pain reliever.

Acetaminophen is often used as an aspirin substitute because it has no irritating effect on the intestinal tract and yet has comparable analgesic and antipyretic effects. Unlike aspirin, however, it is not effective against inflammation and is of limited use for the aches and pains of arthritis. Also, acetaminophen does not inhibit platelet aggregation and therefore is not useful for preventing vascular clotting.

Acetaminophen is available in a liquid form that is used extensively for small children and other patients who have difficulty taking solid tablets. The wide use of acetaminophen for children has a drawback; it is the drug most often involved in childhood poisonings.

In *large* doses, acetaminophen can cause liver and kidney damage. Such effects are not found when acetaminophen is taken as directed. For this reason, the maximum adult daily dosage of 4 g should not be exceeded (eight 500 mg tablets) and extra-strength formulations should be used with great caution. Analgesic abuse is a real potential with the heavily-advertised extra-strength formulations.

Acetaminophen's mode of action in the body is similar to that of aspirin—inhibition of prostaglandin synthesis.

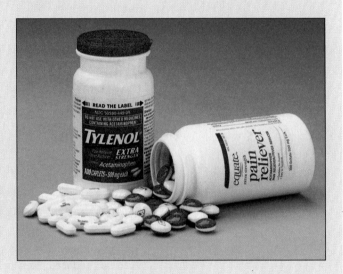

If the preceding reactions are run at room temperature (25°C), no amide formation occurs; instead an acid–base reaction occurs in which a carboxylic acid salt is produced. This acid–base reaction when a 1° amine is the reactant is

O   H   H   O
‖ | | ‖
R—C—OH + H—N—R $\xrightarrow{25°C}$ H—N$^+$—R   R—C—O$^-$
                                                   |
                                                   H

Acid   *Primary amine*   *Carboxylate salt*

General structural equations for 1°, 2°, and 3° amide production from carboxylic acids are

> The reaction of a carboxylic acid with ammonia or an amine to produce an amide is similar to the reaction of a carboxylic acid with an alcohol. In both cases, water is formed as a by-product as the —OH part of the carboxylic acid is replaced.

O   H   O   H
‖ | ‖ |
R—C—OH + H—N—H $\xrightarrow[\text{catalyst}]{100°C}$ R—C—N—H + H₂O
Carboxylic   *Ammonia*   *Primary amide*
acid

O   H   O   H
‖ | ‖ |
R—C—OH + H—N—R $\xrightarrow[\text{catalyst}]{100°C}$ R—C—N—R + H₂O
Carboxylic   *Primary*   *Secondary amide*
acid   *amine*

O   R   O   R
‖ | ‖ |
R—C—OH + H—N—R $\xrightarrow[\text{catalyst}]{100°C}$ R—C—N—R + H₂O
Carboxylic   *Secondary*   *Tertiary amide*
acid   *amine*

This is the fourth time we have encountered *condensation* reactions. Esterification (Section 5.11), acetal formation (Section 4.10), and intermolecular alcohol dehydration (Section 3.9) were the other three condensation situations.

These reactions are called *amidification* reactions. An **amidification reaction** *is the reaction of a carboxylic acid with an amine (or ammonia) to produce an amide.* In amidification, an —OH group is lost from the carboxylic acid, a —H atom is lost from the ammonia or amine, and water is formed as a by-product. Amidification reactions are thus *condensation* reactions.

Two specific amidification reactions, in which a 2° amide and a 3° amide are produced, respectively, are

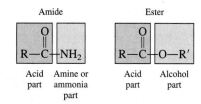

Just as it is useful to think of the structure of an ester (Section 5.10) in terms of an "acid part" and an "alcohol part," it is useful to think of an amide in terms of an "acid part" and an "amine (or ammonia) part."

In this context, it is easy to identify the parent acid and amine from which a given amide can be produced; to generate the parent molecules, just add an —OH group to the acid part of the amide and a H atom to the amine part.

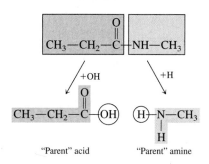

EXAMPLE 6.6

Predicting Reactants Needed to Prepare Specific Amides

■ What carboxylic acid and amine (or ammonia) are needed to prepare each of the following amides?

a.

$$CH_3-\overset{\overset{\displaystyle O}{\|}}{C}-NH-CH_2-CH_3$$

b.

$$CH_3-CH_2-\overset{\overset{\displaystyle O}{\|}}{C}-NH_2$$

c.

$$CH_3-CH_2-\overset{\overset{\displaystyle O}{\|}}{C}-\overset{\displaystyle N}{\underset{\displaystyle |}{}}-CH_3$$
$$\qquad\qquad\qquad CH_3$$

Solution

a. Viewing the molecule as having an acid part and an amine part, we obtain

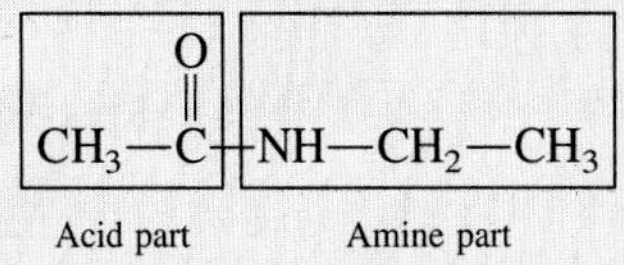

$$\underbrace{CH_3-\overset{\overset{\displaystyle O}{\|}}{C}}_{\text{Acid part}}\Big|\underbrace{NH-CH_2-CH_3}_{\text{Amine part}}$$

Adding an —OH group to the acid part and a H atom to the amine part, we obtain the "parent" molecules, which are

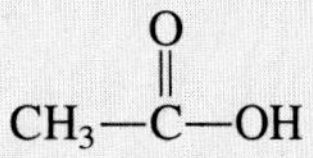

$$CH_3-\overset{\overset{\displaystyle O}{\|}}{C}-OH \quad \text{and} \quad CH_3-CH_2-NH_2$$

b. Proceeding as in part **a,** we find that the "parent" acid and amine molecules are, respectively,

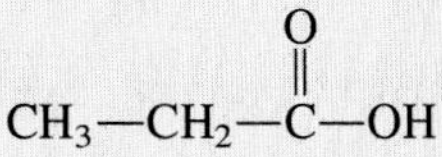

$$CH_3-CH_2-\overset{\overset{\displaystyle O}{\|}}{C}-OH \quad \text{and} \quad NH_3$$

c. Proceeding again as in part **a,** we find that the "parent" acid and amine molecules are, respectively,

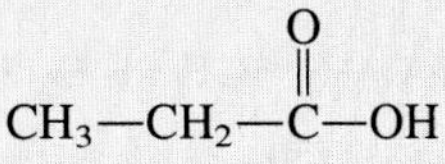

$$CH_3-CH_2-\overset{\overset{\displaystyle O}{\|}}{C}-OH \quad \text{and} \quad CH_3-NH-CH_3$$

Practice Exercise 6.6

What carboxylic acid and amine (or ammonia) are needed to prepare each of the following amides?

a.
$$CH_3-\overset{\overset{\displaystyle O}{\|}}{C}-\underset{\underset{\displaystyle CH_3}{|}}{N}-CH_2-CH_3$$

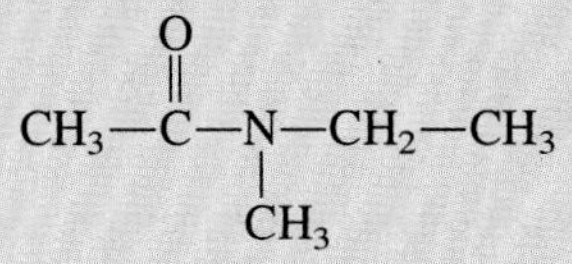

b.
$$CH_3-CH_2-\overset{\overset{\displaystyle O}{\|}}{C}-NH-CH_3$$

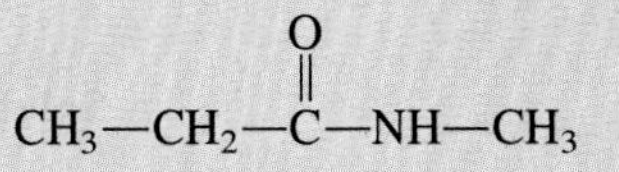

c.
$$\bigcirc\!\!\!\!-\overset{\overset{\displaystyle O}{\|}}{C}-NH_2$$

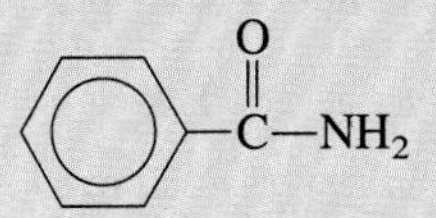

6.17 Hydrolysis of Amides

As was the case with esters (Section 5.16), the most important reaction of amides is hydrolysis. In amide hydrolysis, the bond between the carbonyl carbon atom and the nitrogen is broken, and free acid and free amine are produced. Amide hydrolysis is catalyzed by acids, bases, or certain enzymes; sustained heating is also often required.

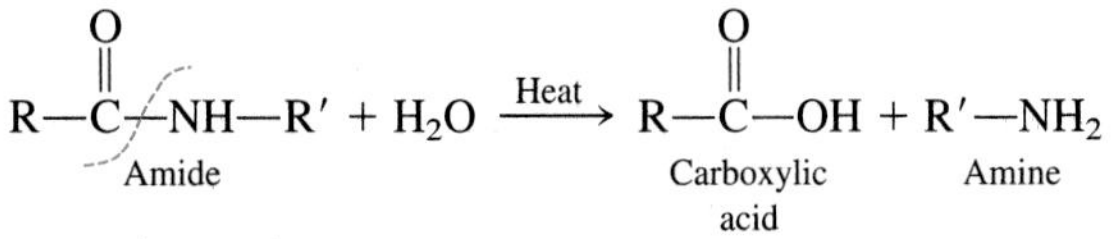

$$\underset{\text{Amide}}{R-\overset{\overset{\displaystyle O}{\|}}{C}-NH-R'} + H_2O \xrightarrow{\text{Heat}} \underset{\substack{\text{Carboxylic}\\\text{acid}}}{R-\overset{\overset{\displaystyle O}{\|}}{C}-OH} + \underset{\text{Amine}}{R'-NH_2}$$

Acidic or basic hydrolysis conditions have an effect on the products. *Acidic* conditions convert the product amine to an amine salt (Section 6.7). *Basic* conditions convert the product carboxylic acid to a carboxylic acid salt (Section 5.9).

Amide hydrolysis under basic conditions is also called amide saponification, just as ester hydrolysis under basic conditions is called ester saponification (Section 5.16).

$$
\underset{\text{Acidic hydrolysis of an amide}}{R-\overset{\overset{\displaystyle O}{\|}}{C}-NH-R' + H_2O + \boxed{HCl}} \xrightarrow{\text{Heat}} \underset{\text{Carboxylic acid}}{R-\overset{\overset{\displaystyle O}{\|}}{C}-OH} + \underset{\text{Amine salt}}{R'-\overset{+}{N}H_3\ Cl^-}
$$

$$
\underset{\text{Basic hydrolysis of an amide}}{R-\overset{\overset{\displaystyle O}{\|}}{C}-NH-R' + \boxed{NaOH}} \xrightarrow{\text{Heat}} \underset{\text{Carboxylic acid salt}}{R-\overset{\overset{\displaystyle O}{\|}}{C}-O^-\ Na^+} + \underset{\text{Amine}}{R'-NH_2}
$$

EXAMPLE 6.7

Predicting the Products of Amide Hydrolysis Reactions

■ Draw structural formulas for the organic products of each of the following amide hydrolysis reactions. Be sure to take into account whether the hydrolysis occurs under neutral, acidic, or basic conditions.

a.
$$
CH_3-CH_2-\overset{\overset{\displaystyle O}{\|}}{C}-NH-CH_3 + H_2O \xrightarrow{\text{Heat}}
$$

b.
$$
CH_3-CH_2-\overset{\overset{\displaystyle O}{\|}}{C}-NH-CH_2-CH_3 + H_2O \xrightarrow[\text{HCl}]{\text{Heat}}
$$

c.
$$
CH_3-\overset{\overset{\displaystyle O}{\|}}{\underset{\underset{\displaystyle CH_3}{|}}{C}}-N-CH_3 + H_2O \xrightarrow[\text{NaOH}]{\text{Heat}}
$$

d.
$$
CH_3-\underset{\underset{\displaystyle CH_3}{|}}{CH}-\overset{\overset{\displaystyle O}{\|}}{C}-NH_2 + H_2O \xrightarrow{\text{Heat}}
$$

Solution

a. This reaction is hydrolysis under neutral conditions. The products will be the "parent" acid and amine for the amide. These "parents" are

$$
CH_3-CH_2-\overset{\overset{\displaystyle O}{\|}}{C}-OH \qquad \text{and} \qquad CH_3-NH_2
$$

b. This reaction is hydrolysis under acidic conditions. The acid is hydrochloric acid (HCl). The products will be the "parent" carboxylic acid and the chloride salt of the amine. The HCl converts the amine to its chloride salt.

$$
CH_3-CH_2-\overset{\overset{\displaystyle O}{\|}}{C}-OH \qquad \text{and} \qquad CH_3-CH_2-\overset{+}{N}H_3\ Cl^-
$$

c. This reaction is hydrolysis under basic conditions. The base is sodium hydroxide (NaOH). The products will be the "parent" amine and the salt of the carboxylic acid. The NaOH converts the carboxylic acid to its sodium salt.

$$
CH_3-\overset{\overset{\displaystyle O}{\|}}{C}-O^-\ Na^+ \qquad \text{and} \qquad CH_3-NH-CH_3
$$

d. This reaction is hydrolysis under neutral conditions. The products will be the "parent" acid and amine of the amide. Because the amide is unsubstituted, the parent amine is actually ammonia.

$$\underset{\underset{CH_3}{|}}{CH_3-CH-\overset{\overset{O}{||}}{C}-OH} \quad \text{and} \quad NH_3$$

Practice Exercise 6.7

Draw structural formulas for the organic products of each of the following amide hydrolysis reactions. Be sure to take into account whether the hydrolysis occurs under neutral, acidic, or basic conditions.

a.

$$CH_3-\overset{\overset{O}{||}}{C}-NH-CH_3 + H_2O \xrightarrow[\text{NaOH}]{\text{Heat}}$$

b.

$$CH_3-\overset{\overset{O}{||}}{C}-NH-CH_3 + H_2O \xrightarrow[\text{HCl}]{\text{Heat}}$$

c.

$$CH_3-\overset{\overset{O}{||}}{C}-NH-CH_3 + H_2O \xrightarrow{\text{Heat}}$$

d.

$$\bigcirc\!\!\!\!\bigcirc-\overset{\overset{O}{||}}{C}-NH_2 + H_2O \xrightarrow{\text{Heat}}$$

The Chemistry at a Glance feature on page 208 summarizes the reactions that involve amines and amides.

6.18 Polyamides and Polyurethanes

Amide polymers—polyamides—are synthesized by combining diamines and dicarboxylic acids in a condensation polymerization reaction (Section 5.18). A **polyamide** *is a condensation polymer in which the monomers are joined through amide linkages.*

The most important synthetic polyamide is *nylon.* Nylon is used in clothing and hosiery, as well as in carpets, tire cord, rope, and parachutes. It also has nonfiber uses; for example, it is used in paint brushes, electrical parts, valves, and fasteners. It is a tough, strong, nontoxic, nonflammable material that is resistant to chemicals. Surgical suture is made of nylon because it is such a strong fiber.

There are actually many different types of nylon, all of which are based on diamine and diacid monomers. The most important nylon is nylon 66, which is made by using 1,6-hexanediamine and hexanedioic acid as monomers (Figure 6.16).

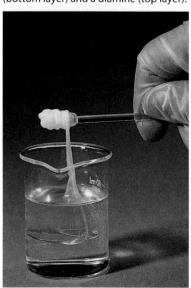

FIGURE 6.16 A white strand of a nylon polymer forms between the two layers of a solution containing a diacid (bottom layer) and a diamine (top layer).

$$\underset{\text{1,6-Hexanediamine}}{H-\overset{\overset{H}{|}}{N}-(CH_2)_6-\overset{\overset{H}{|}}{N}-H}$$

$$\underset{\text{Hexanedioic acid}}{HO-\overset{\overset{O}{||}}{C}-(CH_2)_4-\overset{\overset{O}{||}}{C}-OH}$$

The reaction of one acid group of the diacid with one amine group of the diamine initially produces an amide molecule; an acid group is left over on one end, and an amine group is left over on the other end.

$$HO-\overset{\overset{O}{||}}{C}-(CH_2)_4-\overset{\overset{O}{||}}{C}-OH + H-\overset{\overset{H}{|}}{N}-(CH_2)_6-\overset{\overset{H}{|}}{N}-H \longrightarrow$$

$$HO-\overset{\overset{O}{||}}{C}-(CH_2)_4-\overset{\overset{O}{||}}{C}-\overset{\overset{H}{|}}{N}-(CH_2)_6-\overset{\overset{H}{|}}{N}-H + H_2O$$

Leftover acid group that can react further   Amide linkage   Leftover amine group that can react further

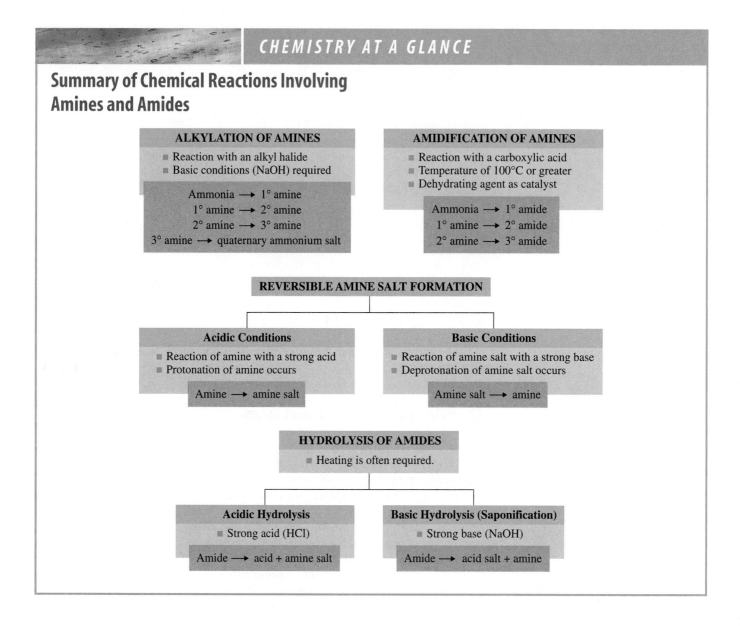

CHEMISTRY AT A GLANCE

Summary of Chemical Reactions Involving Amines and Amides

ALKYLATION OF AMINES
- Reaction with an alkyl halide
- Basic conditions (NaOH) required

Ammonia ⟶ 1° amine
1° amine ⟶ 2° amine
2° amine ⟶ 3° amine
3° amine ⟶ quaternary ammonium salt

AMIDIFICATION OF AMINES
- Reaction with a carboxylic acid
- Temperature of 100°C or greater
- Dehydrating agent as catalyst

Ammonia ⟶ 1° amide
1° amine ⟶ 2° amide
2° amine ⟶ 3° amide

REVERSIBLE AMINE SALT FORMATION

Acidic Conditions
- Reaction of amine with a strong acid
- Protonation of amine occurs

Amine ⟶ amine salt

Basic Conditions
- Reaction of amine salt with a strong base
- Deprotonation of amine salt occurs

Amine salt ⟶ amine

HYDROLYSIS OF AMIDES
- Heating is often required.

Acidic Hydrolysis
- Strong acid (HCl)

Amide ⟶ acid + amine salt

Basic Hydrolysis (Saponification)
- Strong base (NaOH)

Amide ⟶ acid salt + amine

The name *nylon 66* comes from the fact that each of the monomers has six carbon atoms.

This species then reacts further, and the process continues until a long polymeric molecule, nylon, has been produced.

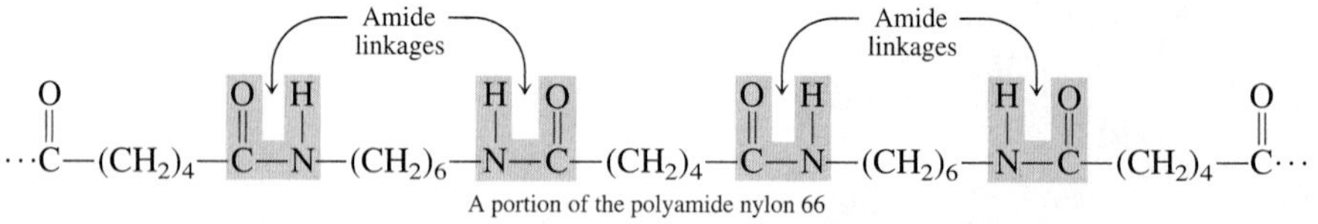

A portion of the polyamide nylon 66

Additional stiffness and toughness are imparted to polyamides if aromatic rings are present in the polymer "backbone." The polyamide Kevlar is now used in place of steel in bullet-resistant vests. The polymeric repeating unit in Kevlar is

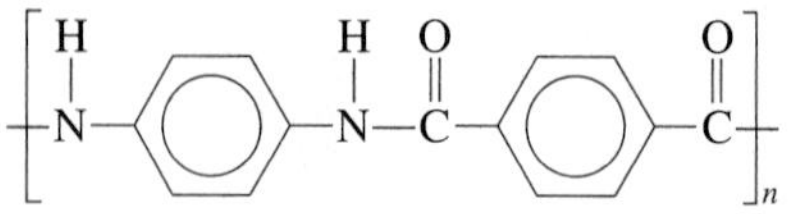

Kevlar

A uniform system of hydrogen bonds that holds polymer chains together accounts for the "amazing" strength of Kevlar (see Figure 6.17).

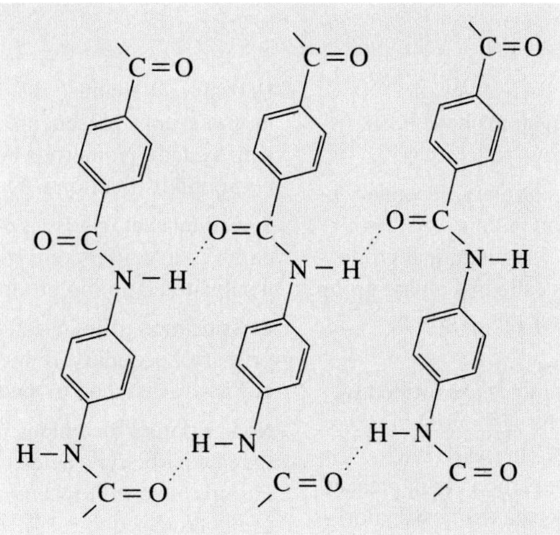

FIGURE 6.17 A regular hydrogen-bonding pattern among Kevlar polymer strands contributes to the great strength of this polymer.

Nomex is a polymer whose structure is a variation of that of Kevlar. With Nomex, the monomers are *meta* isomers rather than *para* isomers. Nomex is used in flame-resistant clothing for fire fighters and race car drivers (Figure 6.18).

Silk and wool are examples of *naturally occurring* polyamide polymers. Silk and wool are proteins, and proteins are polyamide polymers. Because much of the human body is protein material, much of the human body is polyamide polymer. The monomers for proteins are amino acids, difunctional molecules containing both amino and carboxyl groups. Here are some representative structures for amino acids, of which there are many (Section 9.2).

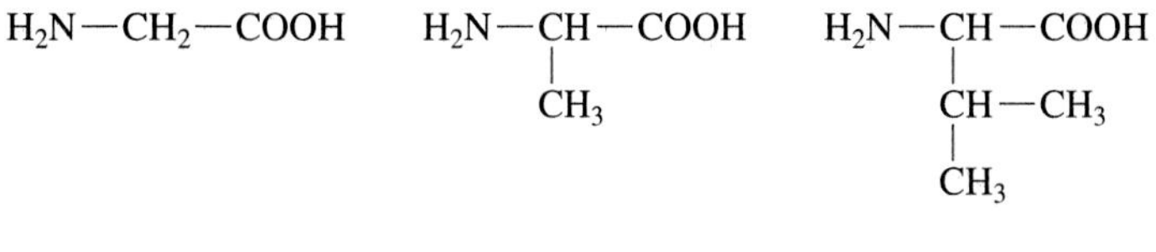

$$H_2N-CH_2-COOH \qquad H_2N-CH-COOH \qquad H_2N-CH-COOH$$
$$\qquad\qquad\qquad\qquad\qquad \mid \qquad\qquad\qquad\qquad \mid$$
$$\qquad\qquad\qquad\qquad\qquad CH_3 \qquad\qquad\qquad\qquad CH-CH_3$$
$$\qquad\qquad\qquad\qquad\qquad\qquad\qquad\qquad\qquad\qquad\qquad\quad \mid$$
$$\qquad\qquad\qquad\qquad\qquad\qquad\qquad\qquad\qquad\qquad\qquad\quad CH_3$$

Polyurethanes are polymers related to polyesters and polyamides. The backbone of a polyurethane polymer contains aspects of both ester and amide functional groups. The following is a portion of the structure of a typical polyurethane polymer.

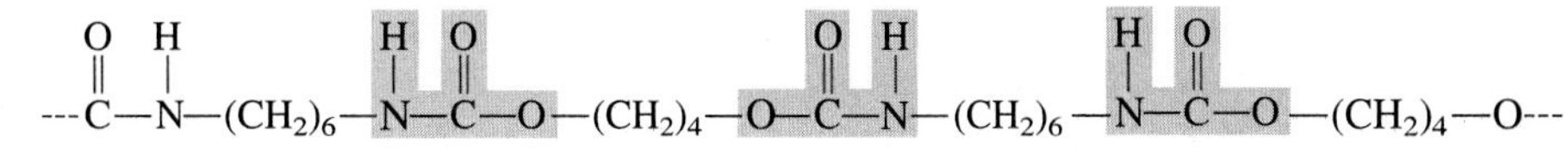

Foam rubber in furniture upholstery, packaging materials, life preservers, elastic fibers, and many other products contain polyurethane polymers (Figure 6.19).

FIGURE 6.18 Fire fighters with flame-resistant clothing containing Nomex.

FIGURE 6.19 Polyurethanes have medical applications. For example, polyurethane membranes are used as skin substitutes for severe burn victims. Because they pass only oxygen and water, these membranes help patients recover more rapidly.

CONCEPTS TO REMEMBER

Structural characteristics of amines. Amines are derivatives of ammonia (NH_3) in which one or more hydrogen atoms have been replaced by an alkyl, a cycloalkyl, or an aryl group (Section 6.2).

Classification of amines. Amines are classified as primary, secondary or tertiary, depending on the number of hydrocarbon groups (one, two, or three) directly attached to the nitrogen atom. The functional group present in a primary amine, the —NH_2 group, is called an *amino* group. Secondary and tertiary amines contain substituted amino groups (Section 6.2).

Nomenclature for amines. Common names for amines are formed by listing the hydrocarbon groups attached to the nitrogen atom in alphabetical order, followed by the suffix -*amine*. In the IUPAC system, the -*e* ending of the name of the longest carbon chain present is changed to -*amine,* and a number is used to locate the position of the amino group. Carbon-chain substituents are given numbers to designate their locations (Section 6.3).

Properties of amines. The methylamines and ethylamine are gases at room temperature; amines of higher molecular mass are usually liquids and smell like raw fish. Primary and secondary, but not tertiary, amines can participate in hydrogen bonding to other amine molecules (Section 6.5).

Basicity of amines. Amines are weak bases because of the ability of the unshared electron pair on the amine nitrogen atom to accept a proton in acidic solution (Section 6.6).

Amine salts. The reaction of a strong acid with an amine produces an amine salt. Such salts are more soluble in water than are the parent amines (Section 6.7).

Alkylation of ammonia and amines. Alkylation of ammonia, primary amines, secondary amines, and tertiary amines produces primary amines, secondary amines, tertiary amines, and quaternary ammonium salts, respectively (Section 6.8).

Heterocyclic amines. In a heterocyclic amine, the nitrogen atoms of amino groups present are part of either an aromatic or a nonaromatic ring system. Numerous heterocyclic amines are important biochemical compounds (Section 6.9).

Structural characteristics of amides. An amide is derived from a carboxylic acid by replacing the hydroxyl group with an amino or a substituted amino group (Section 6.12).

Classification of amides. Amides, like amines, can be classified as primary, secondary, or tertiary, depending on how many nonhydrogen atoms are attached to the nitrogen atom (Section 6.12).

Nomenclature for amides. The nomenclature for amides is derived from that for carboxylic acids by changing the -*oic acid* ending to -*amide*. Groups attached to the nitrogen atom of the amide are located using the prefix *N*- (Section 6.13).

Properties of amides. Amides do not exhibit basic properties in solution. Most unbranched amides are solids at room temperature and have correspondingly high boiling points because of strong hydrogen bonds between molecules (Section 6.15).

Preparation of amides. Reaction, at elevated temperature, of carboxylic acids with ammonia, primary amines, and secondary amines produces primary, secondary, and tertiary amides, respectively (Section 6.16).

Hydrolysis of amides. In amide hydrolysis, the bond between the carbonyl carbon atom and the nitrogen is broken, and free acid and free amine are produced. Acidic hydrolysis conditions convert the product amine to an amine salt. Basic hydrolysis conditions convert the product acid to an acid salt (Section 6.17).

Polyamides. Polyamides are condensation polymers with monomers joined together by amide linkages. The monomers for polyamides are diacids and diamines (Section 6.18).

KEY REACTIONS AND EQUATIONS

1. Reaction of amines with water to give a basic solution (Section 6.6)

$$R{-}NH_2 + H_2O \rightleftharpoons R{-}\overset{+}{N}H_3 + OH^-$$

2. Reaction of amines with acids to produce amine salts (Section 6.7)

$$R{-}NH_2 + HCl \longrightarrow R{-}\overset{+}{N}H_3\ Cl^-$$

3. Conversion of an amine salt to an amine (Section 6.7)

$$R{-}\overset{+}{N}H_3\ Cl^- + NaOH \longrightarrow R{-}NH_2 + NaCl + H_2O$$

4. Alkylation of ammonia to produce a primary amine (Section 6.8)

$$NH_3 + R{-}X + NaOH \longrightarrow R{-}NH_2 + NaX + H_2O$$

5. Alkylation of primary and secondary amines to produce, respectively, secondary and tertiary amines (Section 6.8)

$$RNH_2 + R{-}X + NaOH \longrightarrow R_2NH + NaX + H_2O$$

$$R_2NH + R{-}X + NaOH \longrightarrow R_3N + NaX + H_2O$$

6. Alkylation of a tertiary amine to produce a quaternary ammonium salt (Section 6.8)

$$R_3N + R{-}X \longrightarrow R_4\overset{+}{N}\ X^-$$

7. Reaction of amines with carboxylic acids to form amides (Section 6.16)

$$\underset{\text{O}}{\overset{\text{O}}{\underset{\|}{R{-}C{-}OH}}} + R{-}NH_2 \xrightarrow[\text{Catalyst}]{100°C} \underset{\overset{\|}{\text{O}}}{R{-}C{-}NH{-}R} + H_2O$$

8. Acid hydrolysis of amides to produce a carboxylic acid and an amine salt (Section 6.17)

$$\overset{\overset{\text{O}}{\|}}{R{-}C{-}NH{-}R} + H_2O + HCl \xrightarrow{\text{Heat}} \overset{\overset{\text{O}}{\|}}{R{-}C{-}OH} + R{-}\overset{+}{N}H_3\ Cl^-$$

9. Basic hydrolysis of amides to produce a carboxylic acid salt and an amine (Section 6.17)

$$\overset{\overset{\text{O}}{\|}}{R{-}C{-}NH{-}R} + NaOH \xrightarrow{\text{Heat}} \overset{\overset{\text{O}}{\|}}{R{-}C{-}O^-}\ Na^+ + R{-}NH_2$$

EXERCISES AND PROBLEMS

The members of each pair of problems in this section test similar material.

■ The Amine Functional Group (Section 6.2)

6.1 In which of the following compounds is an amine functional group present?

a. 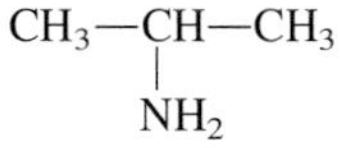 CH₃—CH—CH₃
       |
       NH₂

b. CH₃—NH—CH₃

c. 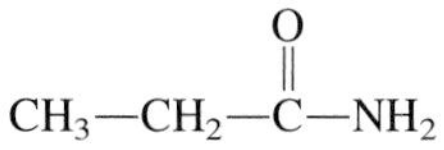
           O
           ‖
   CH₃—CH₂—C—NH₂

d. 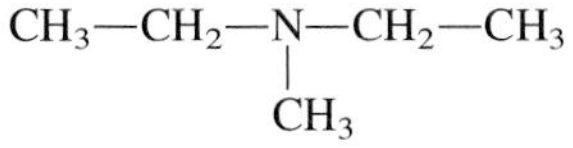 CH₃—CH₂—N—CH₂—CH₃
              |
              CH₃

e. 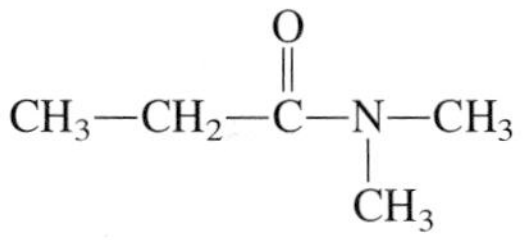
           O
           ‖
   CH₃—CH₂—C—N—CH₃
             |
             CH₃

f. 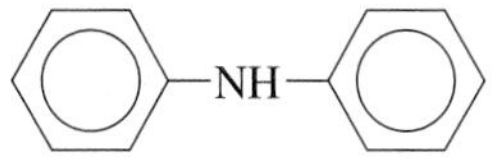 ⬡—NH—⬡

6.2 In which of the following compounds is an amine functional group present?

a. CH₃—CH₂—CH₂—NH₂

b. 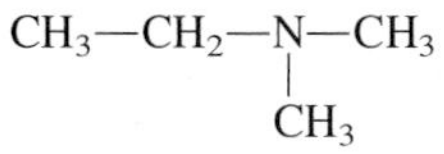 CH₃—CH₂—N—CH₃
            |
            CH₃

c. 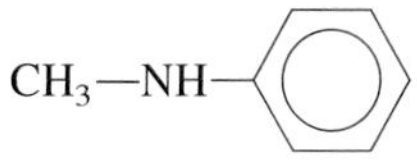 CH₃—NH—⬡

d. 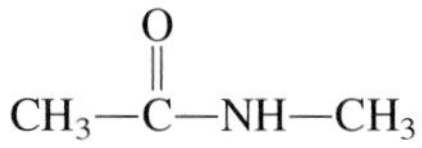
       O
       ‖
   CH₃—C—NH—CH₃

e. 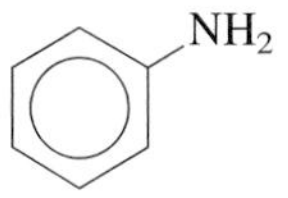 ⬡—NH₂

f. 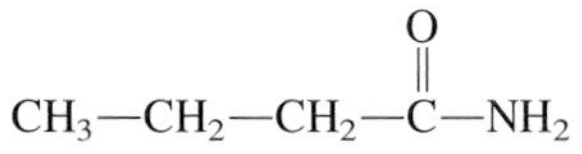
                       O
                       ‖
   CH₃—CH₂—CH₂—C—NH₂

■ Classification of Amines (Section 6.2)

6.3 Classify each of the following amines as a primary, secondary, or tertiary amine.

a. CH₃—NH₂

b. 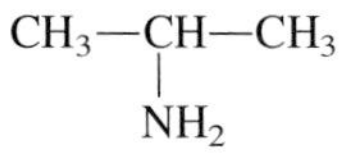 CH₃—CH—CH₃
       |
       NH₂

c. CH₃—NH—CH₃

d. 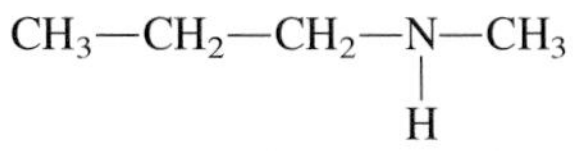 CH₃—CH₂—CH₂—N—CH₃
                |
                H

e. 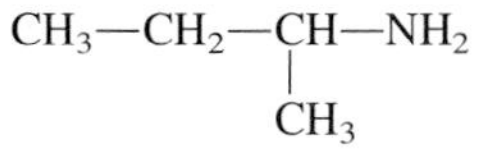 CH₃—CH₂—CH—NH₂
              |
              CH₃

f. 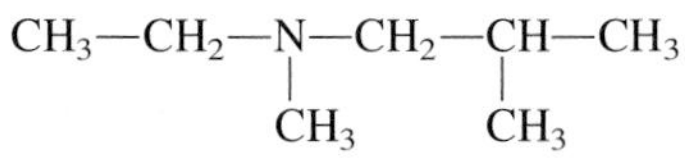 CH₃—CH₂—N—CH₂—CH—CH₃
            | |
            CH₃   CH₃

6.4 Classify each of the following amines as a primary, secondary, or tertiary amine.

a. 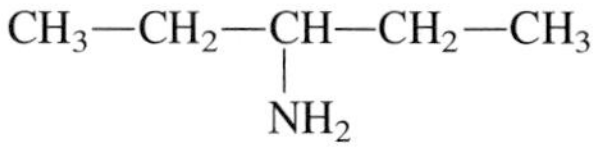 CH₃—CH₂—CH—CH₂—CH₃
              |
              NH₂

b.
       CH₃
       |
   CH₃—C—CH₃
       |
       NH₂

c. CH₃—N—CH₃
        |
        CH₃

d. CH₃—CH—NH—CH—CH₂
        | |
        CH₃   CH₃

e. CH₃—CH₂—CH₂—CH₂—NH₂

f. 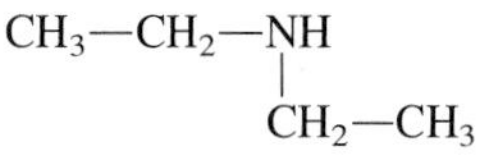 CH₃—CH₂—NH
              |
              CH₂—CH₃

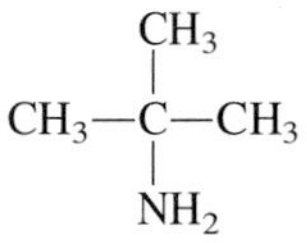

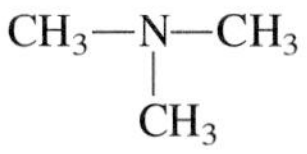

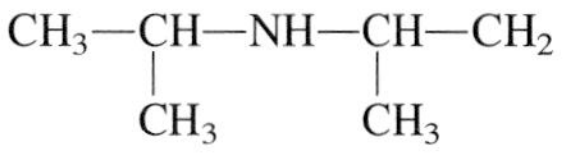

6.5 Classify each of the following amines as a primary, secondary, or tertiary amine.

a. 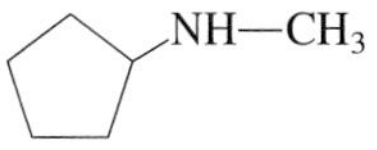 ⬠—NH—CH₃   b. 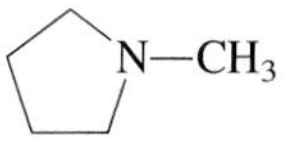 ⬠—N—CH₃

c. 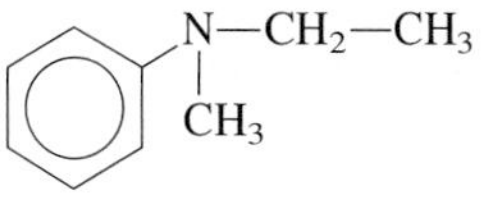 ⬡—N—CH₂—CH₃   d. 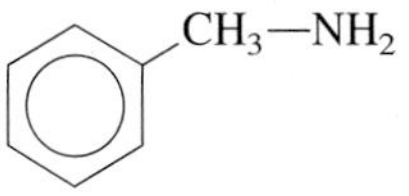 ⬡—CH₃—NH₂
      |
      CH₃

e. 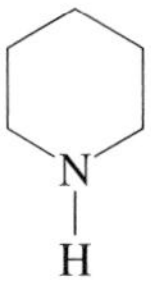 ⬡—N—H (piperidine)

f. 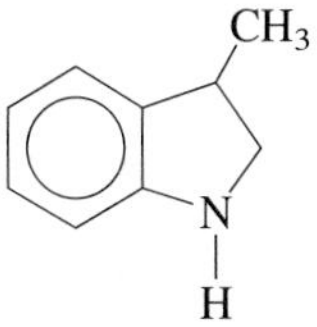 (indole with CH₃)

6.6 Classify each of the following amines as a primary, secondary, or tertiary amine.

a. 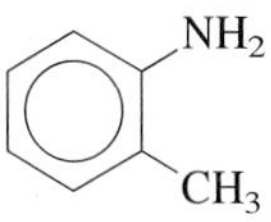 ⬡—NH₂ with CH₃   b. 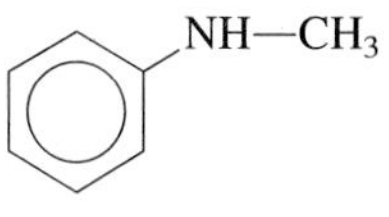 ⬡—NH—CH₃

c. 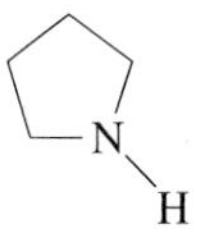 ⬠—N—H (pyrrolidine)   d. 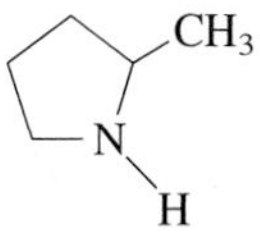 ⬠—N—H with CH₃

e. 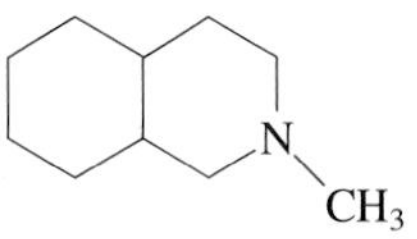 (decahydroquinoline with N—CH₃)   f. 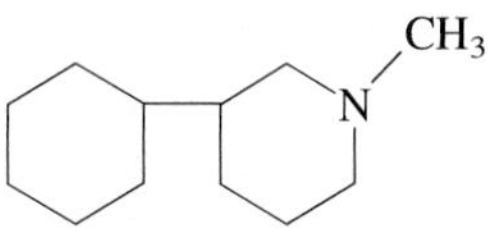 (cyclohexyl piperidine with N—CH₃)

■ Nomenclature for Amines (Section 6.3)

6.7 Assign a common name to each of the following amines.

a. CH₃—NH—CH₂—CH₃

b. CH₃—CH₂—CH₂—NH₂

c. 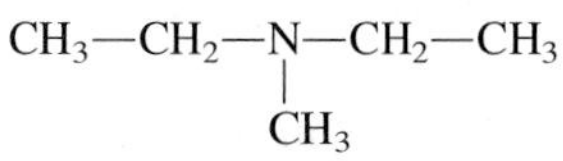 CH₃—CH₂—N—CH₂—CH₃
            |
            CH₃

d.

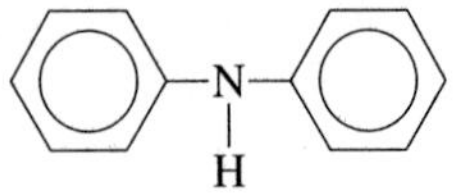

e. $CH_3—CH—NH—CH_3$
        $|$
        CH_3

f. $CH_3—CH—N—CH—CH_3$
        $|$ $|$ $|$
        CH_3 H CH_3

6.8 Assign a common name to each of the following amines.

a.

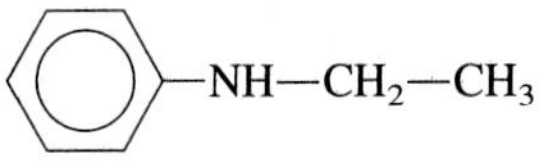

b. $CH_3—CH—CH_3$
        $|$
        NH_2

c. $H_2N—CH_2—CH_2—CH_2—CH_3$

d. $CH_3—CH_2—N—CH_2—CH_3$
              $|$
              $CH_2—CH_3$

e.

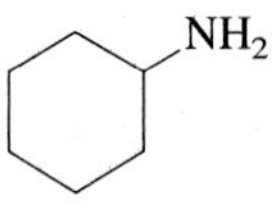

f. $CH_3—CH_2—CH_2—NH—CH—CH_3$
                        $|$
                        CH_3

6.9 Assign an IUPAC name to each of the following amines.

a. $CH_3—CH_2—CH—CH_2—CH_3$
                $|$
                NH_2

b. $CH_3—CH—CH—CH_2—CH_3$
        $|$ $|$
        CH_3 NH_2

c. $CH_3—CH_2—CH—CH_2—CH_3$
                $|$
                $NH—CH_3$

d. $H_2N—CH_2—CH_2—CH_2—CH_2—CH_2—NH_2$

e. $CH_3—CH—CH—CH_3$
        $|$ $|$
        NH_2 NH_2

f. $CH_3—CH_2—CH_2—CH_2—N—CH_3$
                              $|$
                              CH_3

6.10 Assign an IUPAC name to each of the following amines.

a. $CH_3—CH_2—CH_2—NH_2$

b. $CH_3—CH—NH_2$
        $|$
        CH_3

c. $CH_3—CH—CH—CH—CH_3$
        $|$ $|$ $|$
        NH_2 CH_3 NH_2

d. $CH_3—CH_2—CH_3—NH—CH_3$

e. $CH_3—CH_2—CH_2—N—CH_2—CH_3$
                      $|$
                      $CH_2—CH_3$

f. $CH_3—CH_2—CH_2—NH—CH_2—CH_3$

6.11 Assign an IUPAC name to each of the following amines.

a.

b.

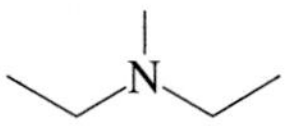

c.

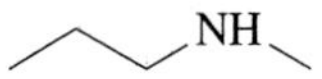

d.

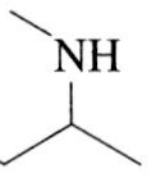

6.12 Assign an IUPAC name to each of the following amines.

a.

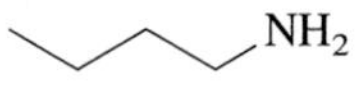

b.

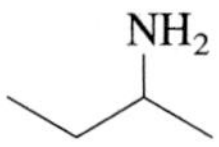

c.

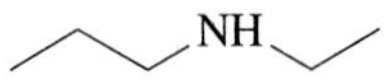

d.

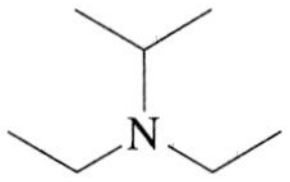

6.13 Name each of the following aromatic amines as a derivative of aniline.

a.

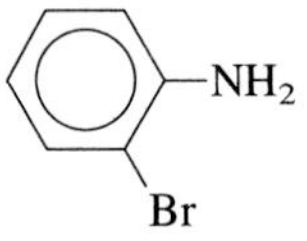

b.

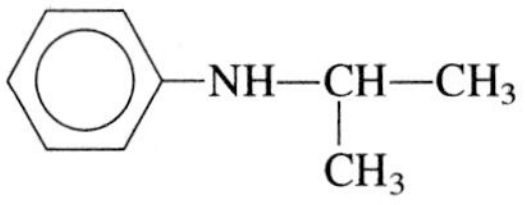

c.

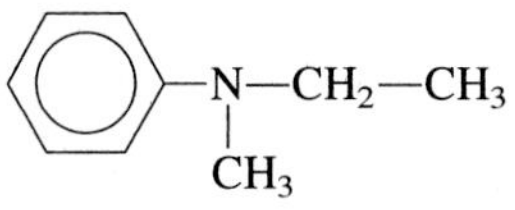

d.

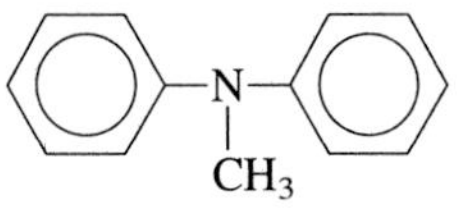

e. 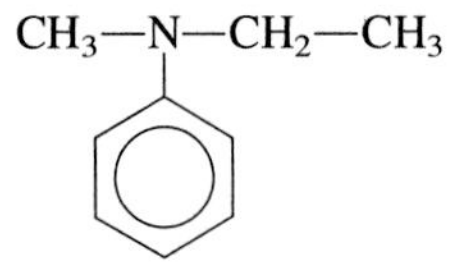 $CH_3—N—CH_2—CH_3$

f. 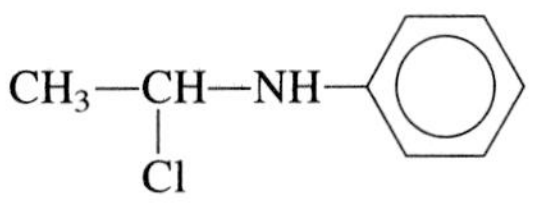 $CH_3—CH—NH—$
        $|$
        Cl

6.14 Name each of the following aromatic amines as a derivative of aniline.

a.

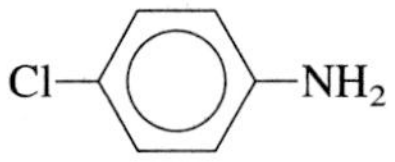

b.

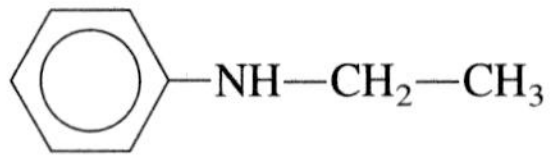

c.

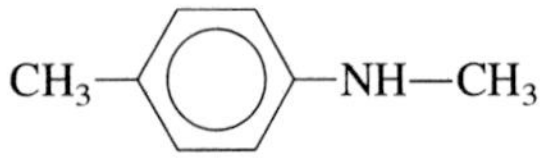

d.

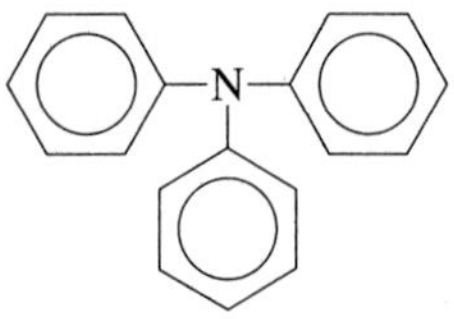

e.

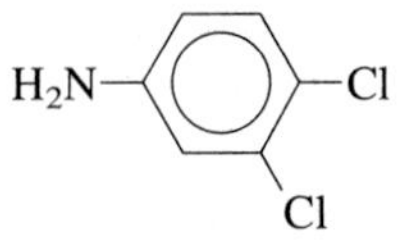

f.

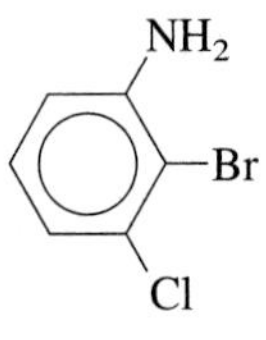

6.15 Draw a structural formula for each of the following amines.
   a. Ethylamine
   b. Triisopropylamine
   c. *o*-Methylaniline
   d. *N*-methylaniline
   e. 2-Methyl-2-butanamine
   f. 1,6-Hexanediamine
   g. 2-Amino-3-pentanone
   h. 2-Aminopropanoic acid

6.16 Draw a structural formula for each of the following amines.
   a. Ethylmethylamine
   b. Diethylpropylamine
   c. *p*-Nitroaniline
   d. *N,N*-dimethylaniline
   e. 2-Methyl-3-ethyl-1-hexanamine
   f. 1,3-Pentanediamine
   g. 3-Amino-2-pentanol
   h. *N,N*-dimethyl-1-butanamine

■ Isomerism for Amines (Section 6.4)

6.17 Draw condensed structural formulas for the eight isomeric primary amines that have the molecular formula $C_5H_{11}N$.

6.18 Draw condensed structural formulas for the six isomeric secondary amines that have the molecular formula $C_5H_{11}N$.

6.19 Give common names for the three isomeric tertiary amines that have the molecular formula $C_5H_{11}N$.

6.20 Give common names for the seven isomeric tertiary amines that have the molecular formula $C_6H_{13}N$.

6.21 Assign an IUPAC name to each of the four isomeric amines that have the molecular formula C_3H_9N.

6.22 Assign an IUPAC name to each of the eight isomeric amines that have the molecular formula C_4H_9N.

■ Physical Properties of Amines (Section 6.5)

6.23 Indicate whether each of the following amines is a liquid or a gas at room temperature.
   a. Butylamine   b. Dimethylamine
   c. Ethylamine   d. Dibutylamine

6.24 Indicate whether each of the following amines is a liquid or a gas at room temperature.
   a. Methylamine   b. Propylamine
   c. Trimethylamine   d. Pentylamine

6.25 Determine the maximum number of hydrogen bonds that can form between a methylamine molecule and
   a. other methylamine molecules
   b. water molecules

6.26 Determine the maximum number of hydrogen bonds that can form between a dimethylamine molecule and
   a. other dimethylamine molecules
   b. water molecules

6.27 Although they have similar molecular masses (73 and 72 amu, respectively), the boiling point of butylamine is much higher (78°C) than that of pentane (36°C). Explain why.

6.28 Although they have similar molecular masses (73 and 74 amu, respectively), the boiling point of 1-butanamine is much lower (78°C) than that of 1-butanol (118°C). Explain why.

6.29 Which compound in each of the following pairs of amines would you expect to be more soluble in water? Justify each answer.

   a. $CH_3-CH_2-NH_2$ and
      $CH_3-CH_2-CH_2-CH_2-CH_2-NH_2$
   b. $CH_3-CH_2-CH_2-NH_2$ and
      $H_2N-CH_2-CH_2-CH_2-NH_2$

6.30 Which compound in each of the following pairs of amines would you expect to be more soluble in water? Justify each answer.
   a. $CH_3-CH_2-CH_2-NH_2$ and
      $CH_3-CH_2-CH_2-CH_2-NH_2$
   b. $CH_3-CH_2-NH-CH_3$ and $CH_3-\underset{\underset{\displaystyle CH_3}{|}}{N}-CH_3$

■ Basicity of Amines (Section 6.6)

6.31 Show the structures of the missing substance(s) in each of the following acid–base equilibria.
   a. $CH_3-CH_2-NH_2 + H_2O \rightleftharpoons ? + OH^-$
   b.

   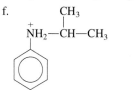

   c. $? + H_2O \rightleftharpoons CH_3-\underset{\underset{\displaystyle CH_3}{|}}{\overset{+}{C}H}-NH_2-CH_3 + OH^-$
   d. Diethylamine $+ H_2O \rightleftharpoons ? + ?$

6.32 Show the structures of the missing substance(s) in each of the following acid–base equilibria.
   a. $CH_3-CH_2-CH_2-NH_2 + H_2O \rightleftharpoons$
      $CH_3-CH_2-CH_2-\overset{+}{N}H_3 + ?$
   b. $? + H_2O \rightleftharpoons$ ⬡$-CH_2-\overset{+}{N}H_3 + OH^-$
   c. $CH_3-\underset{\underset{\displaystyle CH_3}{|}}{C}H-CH_2-NH-CH_3 + H_2O \rightleftharpoons ? + OH^-$
   d. Trimethylamine $+ H_2O \rightleftharpoons ? + ?$

6.33 Name each of the following substituted ammonium and substituted anilinium ions.
   a. $CH_3-\overset{+}{N}H_2-CH_3$
   b. $CH_3-CH_2-\underset{\underset{\displaystyle CH_2-CH_3}{|}}{\overset{+}{N}H}-CH_2-CH_3$
   c. $CH_3-CH_2-\overset{+}{N}H-CH_2-CH_3$
      ⬡
   d. $CH_3-CH_2-CH_2-\underset{\underset{\displaystyle CH_3}{|}}{\overset{+}{N}H}-CH_3$
   e. $CH_3-CH_2-CH_2-\overset{+}{N}H_3$
   f. $\underset{\displaystyle ⬡}{\overset{\displaystyle CH_3}{\underset{\displaystyle |}{}}}\overset{+}{N}H_2-CH-CH_3$

6.34 Name each of the following substituted ammonium and substituted anilinium ions.

a. $CH_3—\overset{+}{N}H_3$

b. $CH_3—CH_2—CH_2—\overset{+}{N}H_2—CH_3$

c. 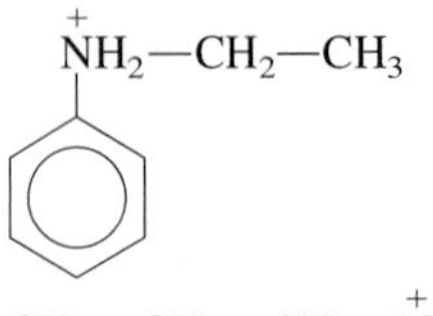 $\overset{+}{N}H_2—CH_2—CH_3$

d. $CH_3—CH_2—CH_2—\overset{+}{N}H—CH_2—CH_3$
$\qquad\qquad\qquad\quad |$
$\qquad\qquad\qquad\ CH_3$

e. $CH_3—CH_2—CH_2—\overset{+}{N}H_2—CH_2—CH_2—CH_3$

f. 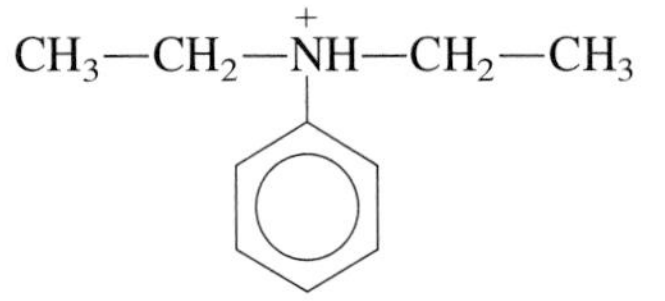 $CH_3—CH_2—\overset{+}{N}H—CH_2—CH_3$

6.35 Draw a structural formula for the "parent" amine of each of the substituted ammonium and substituted anilinium ions in Problem 6.33.

6.36 Draw a structural formula for the "parent" amine of each of the substituted ammonium and substituted anilinium ions in Problem 6.34.

■ Amine Salts (Section 6.7)

6.37 Draw the structure of the missing substance in each of the following reactions involving amine salts.

a. $CH_3—CH_2—NH_2 + HCl \longrightarrow ?$

b.

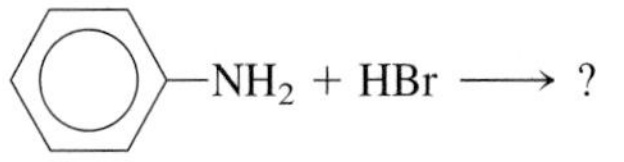

$—NH_2 + HBr \longrightarrow ?$

c.
$\qquad\qquad\qquad CH_3$
$\qquad\qquad\qquad\ |$
$? + HBr \longrightarrow CH_3—C—\overset{+}{N}H_3\ Br^-$
$\qquad\qquad\qquad\ |$
$\qquad\qquad\qquad CH_3$

d. $CH_3—CH_2—NH—CH_3 + ? \longrightarrow$
$\qquad\qquad\qquad CH_3—CH_2—\overset{+}{N}H_2—CH_3\ Cl^-$

6.38 Draw the structure of the missing substance in each of the following reactions involving amine salts.

a. $CH_3—CH_2—NH—CH_2—CH_3 + HBr \longrightarrow ?$

b. $CH_3—NH_2 + ? \longrightarrow CH_3—\overset{+}{N}H_3\ Cl^-$

c. $? + HBr \longrightarrow CH_3—CH—\overset{+}{N}H—CH_3\ Br^-$
$\qquad\qquad\qquad\qquad |\qquad\ |$
$\qquad\qquad\qquad\quad CH_3\ CH_3$

d.

$—NH—CH_3 + HCl \longrightarrow ?$

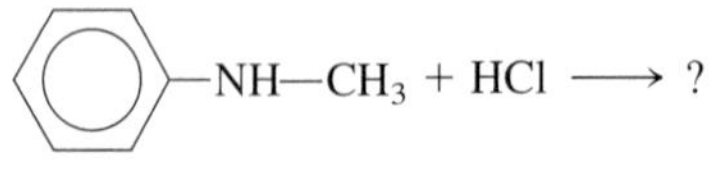

6.39 Draw the structures of the missing substance(s) in each of the following reactions involving amine salts.

a. $CH_3—CH—\overset{+}{N}H_3\ Cl^- + NaOH \longrightarrow ? + NaCl + H_2O$
$\qquad\quad\ |$
$\qquad\quad CH_3$

b. $? + NaOH \longrightarrow CH_3—NH—CH_3 + NaCl + H_2O$

c.

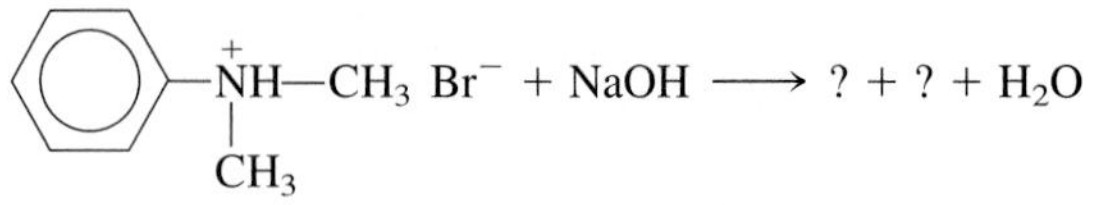

$—\overset{+}{N}H—CH_3\ Br^- + NaOH \longrightarrow ? + ? + H_2O$
$\ |$
CH_3

d. $CH_3—\overset{+}{N}H_2—CH_3\ Cl^- + NaOH \longrightarrow ? + NaCl + H_2O$

6.40 Draw the structures of the missing substance(s) in each of the following reactions involving amine salts.

a. $CH_3—CH_2—CH_2—\overset{+}{N}H_3\ Br^- + NaOH \longrightarrow$
$\qquad\qquad\qquad\qquad\qquad\qquad ? + NaBr + H_2O$

b. $? + NaOH \longrightarrow CH_3—N—CH_3 + NaBr + H_2O$
$\qquad\qquad\qquad\qquad\qquad |$
$\qquad\qquad\qquad\qquad\ CH_3$

c. $CH_3—CH—\overset{+}{N}H_2—CH_3\ Cl^- + NaOH \longrightarrow$
$\qquad\quad\ |$
$\qquad\quad CH_3 \qquad\qquad\qquad\quad ? + NaCl + H_2O$

d.

$—\overset{+}{N}H_3\ Cl^- + NaOH \longrightarrow ? + NaCl + ?$

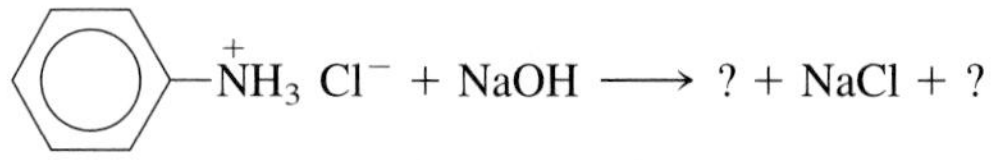

6.41 Name each of the following amine salts.

a. $CH_3—CH_2—CH_2—\overset{+}{N}H_3\ Cl^-$

b. $CH_3—CH_2—CH_2—\overset{+}{N}H_2\ Cl^-$
$\qquad\qquad\qquad\qquad |$
$\qquad\qquad\qquad\ CH_3$

c. $CH_3—CH_2—\overset{+}{N}H—CH_3\ Br^-$
$\qquad\qquad\qquad |$
$\qquad\qquad\quad CH_3$

d.

$—\overset{+}{N}H—CH_3\ Br^-$
$\ |$
CH_3

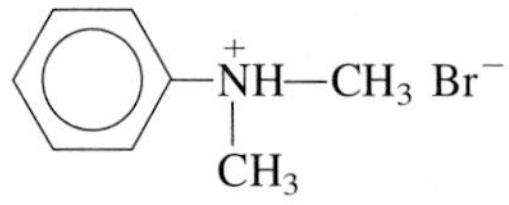

6.42 Name each of the following amine salts.

a. $CH_3—CH_2—\overset{+}{N}H_2\ Cl^-$
$\qquad\qquad\ |$
$\qquad\qquad CH_3$

b. $CH_3—CH_2—CH_2—CH_2—\overset{+}{N}H_3\ Cl^-$

c. $CH_3—CH—\overset{+}{N}H—CH_3\ Br^-$
$\qquad\quad\ |\qquad\ |$
$\qquad\ CH_3\ CH_3$

d.

$—\overset{+}{N}H_2\ Cl^-$
$\ |$
CH_3

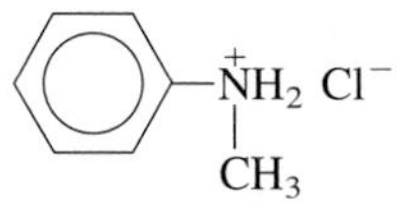

6.43 Why are drugs that contain the amine functional group most often administered to patients in the form of amine chloride or hydrogen sulfate salts?

6.44 Both heptylamine and heptyl alcohol are insoluble in water. If you were given a mixture of these two liquids, how could you separate them without heating (distilling) them?

6.45 How would the structure and name of the amine salt ethylmethylammonium chloride probably be written by someone in the pharmaceutical industry?

6.46 A student looking in an old chemistry book found the following name and structure for a compound.

$$CH_3—CH_2—NH_2 \cdot HBr$$
Ethylamine hydrobromide

What are the modern name and structural representation for this compound?

Alkylation of Ammonia and Amines (Section 6.8)

6.47 Identify the three products in each of the following reactions.

a. NH_3 + $CH_3—CH_2—CH_2—Cl$ + NaOH $\longrightarrow$

b. $CH_3—Br$ + $CH_3—CH—NH—CH_3$ + NaOH $\longrightarrow$
  $\quad\quad\quad\quad\quad\quad\quad\quad\quad |$
  $\quad\quad\quad\quad\quad\quad\quad\quad\quad CH_3$

c. $CH_3—CH_2$ + $NH_2—CH_3—CH_2—Cl$ + NaOH $\longrightarrow$

d. $\quad\quad\quad CH_3$
  $\quad\quad\quad\quad |$
  $CH_3—C—Br$ + NH_3 + NaOH $\longrightarrow$
  $\quad\quad\quad\quad |$
  $\quad\quad\quad CH_3$

6.48 Identify the three products in each of the following reactions.

a. $CH_3—CH—Cl$ + NH_3 + NaOH $\longrightarrow$
  $\quad\quad\quad |$
  $\quad\quad\quad CH_3$

b. $CH_3—NH—CH_3$ + $CH_3—Br$ + NaOH $\longrightarrow$

c. $CH_3—CH_2—CH_2—NH_2$ +
  $\quad\quad\quad\quad\quad CH_3—CH_2—Br$ + NaOH $\longrightarrow$

d. $CH_3—CH_2—CH—Cl$ +
  $\quad\quad\quad\quad\quad |$
  $\quad\quad\quad\quad\quad CH_3$
  $\quad CH_3—CH_2—NH—CH—CH_3$ + NaOH $\longrightarrow$
  $\quad\quad\quad\quad\quad\quad\quad\quad |$
  $\quad\quad\quad\quad\quad\quad\quad\quad CH_3$

6.49 List three different sets of alkyl chloride – secondary amine reactants that could be used to prepare the tertiary amine ethylmethylpropylamine.

6.50 List three different sets of alkyl chloride – secondary amine reactants that could be used to prepare the tertiary amine butylethylpropylamine.

6.51 Draw the structure of the amine or quaternary ammonium salt produced when each of the following pairs of compounds reacts in the presence of a strong base.
a. Trimethylamine and ethyl bromide
b. Diisopropylamine and methyl bromide
c. Ethylmethylpropylamine and methyl chloride
d. Ethylamine and ethyl chloride

6.52 Draw the structure of the amine or quaternary ammonium salt produced when each of the following pairs of compounds reacts in the presence of a strong base.
a. Dimethylamine and propyl bromide
b. Diethylmethylamine and isopropyl chloride
c. Methylpropylamine and ethyl chloride
d. Tripropylamine and propyl chloride

6.53 Classify each of the following salts as an amine salt or a quaternary ammonium salt.

a. $CH_3—\overset{+}{N}H—CH_3$ Br^-
  $\quad\quad\quad |$
  $\quad\quad\quad CH_3$

b. $\quad\quad\quad CH_3$
  $\quad\quad\quad\quad |$
  $CH_3—\overset{+}{N}—CH_3$ Cl^-
  $\quad\quad\quad\quad |$
  $\quad\quad\quad CH_3$

c. $CH_3—CH_2—\overset{+}{N}H_2—CH_3$ Br^-

d. $\quad\quad\quad\quad CH_3$
  $\quad\quad\quad\quad\quad |$
  $CH_3—CH_2—\overset{+}{N}—CH_2—CH_3$ Cl^-
  $\quad\quad\quad\quad\quad |$
  $\quad\quad\quad\quad CH_3$

6.54 Classify each of the following salts as an amine salt or a quaternary ammonium salt.

a. $\quad\quad\quad CH_3$
  $\quad\quad\quad\quad |$
  $CH_3—\overset{+}{N}—CH_2—CH_3$ Cl^-
  $\quad\quad\quad\quad |$
  $\quad\quad\quad CH_3$

b. $\quad\quad\quad H$
  $\quad\quad\quad\quad |$
  $CH_3—\overset{+}{N}—CH_2—CH_3$ Cl^-
  $\quad\quad\quad\quad |$
  $\quad\quad\quad CH_3$

c. $\quad\quad\quad H$
  $\quad\quad\quad\quad |$
  $CH_3—\overset{+}{N}—CH_3$ Br^-
  $\quad\quad\quad\quad |$
  $\quad\quad\quad H$

d. $\quad\quad\quad\quad\quad\quad CH_3$
  $\quad\quad\quad\quad\quad\quad\quad |$
  $CH_3—CH_2—CH_2—\overset{+}{N}—CH_3$ Br^-
  $\quad\quad\quad\quad\quad\quad\quad |$
  $\quad\quad\quad\quad\quad\quad CH_3$

6.55 Name each of the salts in Problem 6.53.

6.56 Name each of the salts in Problem 6.54.

Selected Important Amines (Sections 6.9 through 6.11)

6.57 With the help of Figure 6.8, identify the heterocyclic amine ring system or systems present in each of the following substances.
a. Caffeine   b. Heme
c. Histamine   d. Serotonin

6.58 With the help of Figure 6.8, identify the heterocyclic amine ring system or systems present in each of the following.
a. Nicotine
b. Quinine
c. "Odor of popcorn"
d. Porphyrin ring

6.59 Indicate whether each of the following statements about biochemically important amines is true or false.
a. Both caffeine and nicotine are alkaloids.
b. The alkaloid quinine is used medically to dilate the pupil of the eye.
c. Structurally, morphine and codeine differ by a methyl group.
d. Heroin is a naturally occurring substance.
e. Serotonin deficiency is associated with Parkinson's disease.
f. Adrenaline is another name for norepinephrine.

6.60 Indicate whether each of the following statements about biochemically important amines is true or false.
a. Epinephrine and norepinephrine are hormones produced by the adrenal glands.
b. Both serotonin and dopamine are found in the brain and are heterocyclic amines.
c. The alkaloid atropine is used in the treatment of malaria.
d. "Free" histamine in tissues and the blood causes the symptoms associated with hay fever.
e. Ephedrin and epinephrine are two names for the same compound.
f. Structurally, epinephrine and norepinephrine differ by a methyl group.

■ Structure of and Classification of Amides (Section 6.12)

6.61 Which of the following compounds contain an amide functional group?

a.
$$CH_3-CH_2-\overset{\overset{\displaystyle O}{\|}}{C}-NH_2$$

b.
$$\bigcirc-\overset{\overset{\displaystyle O}{\|}}{C}-\underset{\underset{\displaystyle CH_3}{|}}{N}-CH_2-CH_3$$

c.
$$CH_3-\overset{\overset{\displaystyle O}{\|}}{C}-CH_2-CH_2-NH_2$$

d.
$$CH_3-\overset{\overset{\displaystyle O}{\|}}{C}-NH-\bigcirc$$

e.
$$CH_3-\underset{\underset{\displaystyle NH_2}{|}}{CH}-\overset{\overset{\displaystyle O}{\|}}{C}-OH$$

f.
$$\bigcirc\!\!-N-H \text{ (with } =O\text{)}$$

6.62 Which of the following compounds contain an amide functional group?

a.
$$CH_3-\overset{\overset{\displaystyle O}{\|}}{C}-NH-CH_3$$

b.
$$CH_3-\overset{\overset{\displaystyle O}{\|}}{C}-\underset{\underset{\displaystyle CH_2-CH_3}{|}}{N}-CH_2-CH_2-CH_3$$

c.
$$\underset{\underset{\displaystyle NH_2}{|}}{\overset{\overset{\displaystyle O}{\|}}{C}}-CH_2-CH_3$$

d.
$$H_2N-\overset{\overset{\displaystyle O}{\|}}{C}-CH_2-CH_3$$

e.
$$CH_3-CH_2-\underset{\underset{\displaystyle CH_3}{|}}{CH}-\overset{\overset{\displaystyle O}{\|}}{C}-NH-CH_3$$

f.
$$\bigcirc\!\!-N-H \text{ (with } =O \text{ and } CH_3\text{)}$$

6.63 Classify each of the following amides as unsubstituted, monosubstituted, or disubstituted.

a.
$$CH_3-\overset{\overset{\displaystyle O}{\|}}{C}-NH-CH_3$$

b.
$$CH_3-\overset{\overset{\displaystyle O}{\|}}{C}-\underset{\underset{\displaystyle CH_3}{|}}{N}-CH_2-CH_3$$

c.
$$CH_3-\overset{\overset{\displaystyle O}{\|}}{C}-NH_2$$

d.
$$\bigcirc\!\!-N-H \text{ (with } =O\text{)}$$

6.64 Classify each of the following amides as unsubstituted, monosubstituted, or disubstituted.

a.
$$CH_3-\overset{\overset{\displaystyle O}{\|}}{C}-NH_2$$

b.
$$CH_3-CH_2-\underset{\underset{\displaystyle CH_3}{|}}{CH}-\overset{\overset{\displaystyle O}{\|}}{C}-NH-CH_3$$

c.
$$CH_3-\overset{\overset{\displaystyle O}{\|}}{C}-\underset{\underset{\displaystyle CH_3-CH_2-CH_3}{|}}{N}-CH_2-CH_2-CH_3$$

d.
$$\bigcirc\!\!-N-CH_3 \text{ (with } =O\text{)}$$

6.65 Classify each of the amides in Problem 6.63 as a primary, secondary, or tertiary amide.

6.66 Classify each of the amides in Problem 6.64 as a primary, secondary, or tertiary amide.

■ Nomenclature for Amides (Section 6.13)

6.67 Assign an IUPAC name to each of the following amides.

a.
$$CH_3-\overset{\overset{\displaystyle O}{\|}}{C}-NH-CH_2-CH_3$$

b.
$$CH_3-CH_2-\overset{\overset{\displaystyle O}{\|}}{C}-\underset{\underset{\displaystyle CH_3}{|}}{N}-CH_3$$

c.
$$H_2N-\overset{\overset{\displaystyle O}{\|}}{C}-CH_2-CH_2-CH_3$$

d.
$$H-\overset{\overset{\displaystyle O}{\|}}{C}-\underset{\underset{\displaystyle CH_3}{|}}{N}-H$$

e.
$$Cl-\underset{\underset{\displaystyle CH_3}{|}}{CH}-\overset{\overset{\displaystyle O}{\|}}{C}-NH_2$$

f.
$$CH_3-\underset{\underset{\displaystyle CH_3}{|}}{CH}-\overset{\overset{\displaystyle O}{\|}}{C}-NH-CH_3$$

6.68 Assign an IUPAC name to each of the following amides.

a.
$$CH_3-CH_2-\overset{\overset{\displaystyle O}{\|}}{C}-NH-CH_2-CH_3$$

b.
$$CH_3-CH_2-CH_2-CH_2-\overset{\overset{\displaystyle O}{\|}}{C}-NH_2$$

c.
$$CH_3-CH_2-CH_2-\overset{\overset{\displaystyle O}{\|}}{C}-\underset{\underset{\displaystyle CH_3}{|}}{N}-CH_3$$

d.
$$H_2N-\underset{\underset{\displaystyle CH_3}{|}}{C}=O$$

e.
$$CH_3-\underset{\underset{\displaystyle CH_3}{|}}{CH}-\underset{\underset{\displaystyle CH_3}{|}}{CH}-\overset{\overset{\displaystyle O}{\|}}{C}-NH_2$$

f.
$$CH_3-\underset{\underset{\displaystyle CH_3}{|}}{CH}-\underset{\underset{\displaystyle CH_3}{|}}{CH}-\overset{\overset{\displaystyle O}{\|}}{C}-NH-CH_3$$

6.69 Assign a common name to each of the amides in Problem 6.67.

6.70 Assign a common name to each of the amides in Problem 6.68.

6.71 Assign an IUPAC name to each of the following amides.

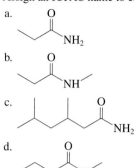

a.

b.

c.

d.

6.72 Assign an IUPAC name to each of the following amides.

a.

b.

c.

d.

6.73 Write a structural formula for each of the following amides.
a. *N,N*-dimethylacetamide
b. 2-Methylbutyramide
c. 3,*N*-dimethylbutanamide
d. Methanamide
e. *N*-phenylbenzamide
f. Formamide

6.74 Write a structural formula for each of the following amides.
a. *N,N*-diethylpropanamide
b. Propionamide
c. 3-Methylbutyramide
d. *N*-methylbenzamide
e. 3,3,*N*-trimethylbutyramide
f. *N*-methyl-*N*-phenylpentanamide

■ Properties of Amides (Section 6.15)

6.75 Although amides contain a nitrogen atom, they are not bases as amines are. Explain why.

6.76 Would you expect *N*-ethylacetamide or *N,N*-diethylacetamide to have the higher boiling point? Explain.

6.77 Determine the maximum number of hydrogen bonds that can form between an acetamide molecule and
a. other acetamide molecules
b. water molecules

6.78 Determine the maximum number of hydrogen bonds that can form between a propanamide molecule and
a. other propanamide molecules
b. water molecules

■ Preparation of Amides (Section 6.16)

6.79 Draw the structures of the missing substances in each of the following reactions involving amides.

a.

b.

c.

d.

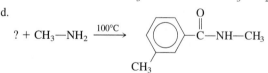

6.80 Draw the structures of the missing substances in each of the following reactions involving amides.

a.

b.

c.

d.

6.81 Draw the structures of the carboxylic acid and the amine from which each of the following amides could be formed.

a.

b. *N*-methylpentanamide

c.

d. 2,3,*N*-trimethylbutanamide

6.82 Draw the structures of the carboxylic acid and the amine from which each of the following amides could be formed.

a.

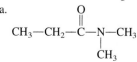

b. 2-Methylpentanamide

c.

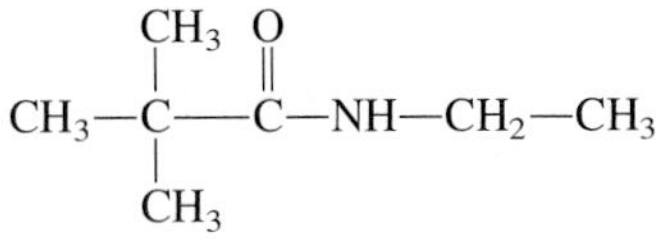

$$CH_3-\underset{\underset{CH_3}{|}}{\overset{\overset{CH_3}{|}}{C}}-\overset{\overset{O}{||}}{C}-NH-CH_2-CH_3$$

d. *N,N*-diethylacetamide

Hydrolysis of Amides (Section 6.17)

6.83 Draw the structures of the organic products in each of the following hydrolysis reactions.

a.

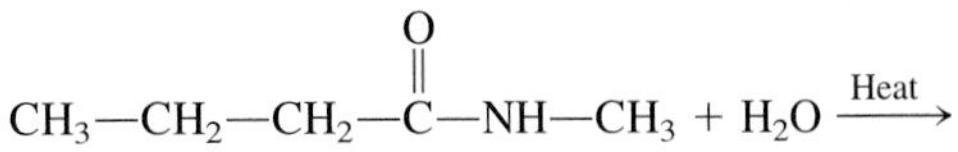

$$CH_3-CH_2-CH_2-\overset{\overset{O}{||}}{C}-NH-CH_3 + H_2O \xrightarrow{Heat}$$

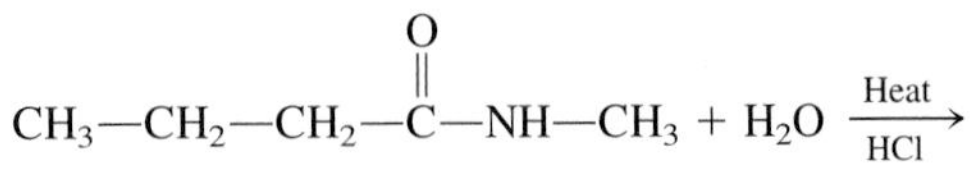

b.

$$CH_3-CH_2-CH_2-\overset{\overset{O}{||}}{C}-NH-CH_3 + H_2O \xrightarrow[HCl]{Heat}$$

c.

$$CH_3-CH_2-CH_2-\overset{\overset{O}{||}}{C}-NH-CH_3 + H_2O \xrightarrow[NaOH]{Heat}$$

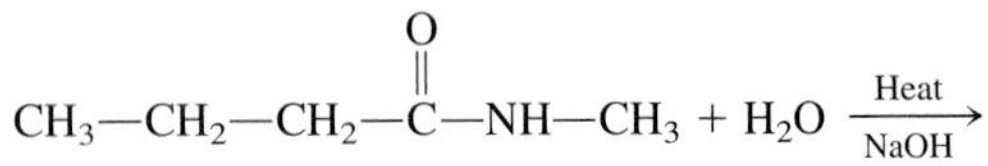

d.

(Ph)—$\overset{\overset{O}{||}}{C}$—$\underset{\underset{CH_3}{|}}{N}$—(Ph) + H$_2$O $\xrightarrow{Heat}$

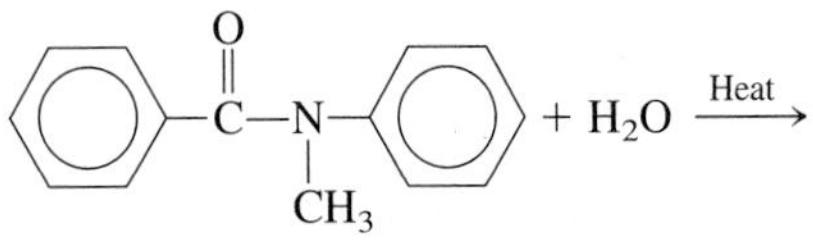

6.84 Draw the structures of the organic products in each of the following hydrolysis reactions.

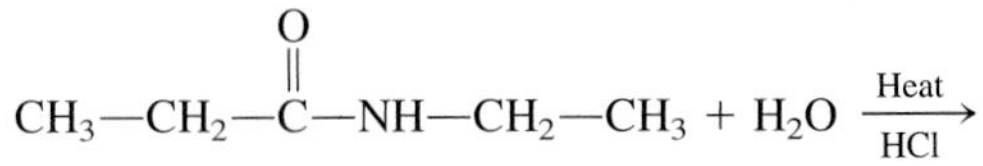

a.

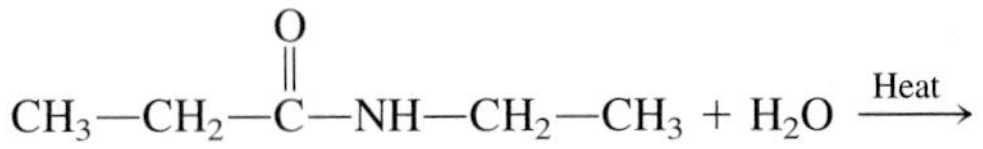

$$CH_3-CH_2-\overset{\overset{O}{||}}{C}-NH-CH_2-CH_3 + H_2O \xrightarrow{Heat}$$

b.

$$CH_3-CH_2-\overset{\overset{O}{||}}{C}-NH-CH_2-CH_3 + H_2O \xrightarrow[HCl]{Heat}$$

c.

$$CH_3-CH_2-\overset{\overset{O}{||}}{C}-NH-CH_2-CH_3 + H_2O \xrightarrow[NaOH]{Heat}$$

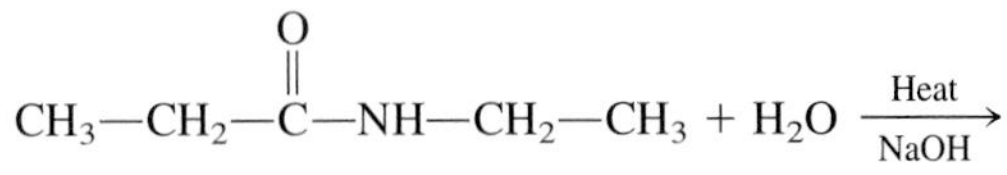

d.

$$CH_3-\underset{\underset{(Ph)}{|}}{CH}-\overset{\overset{O}{||}}{C}-NH_2 + H_2O \xrightarrow{Heat}$$

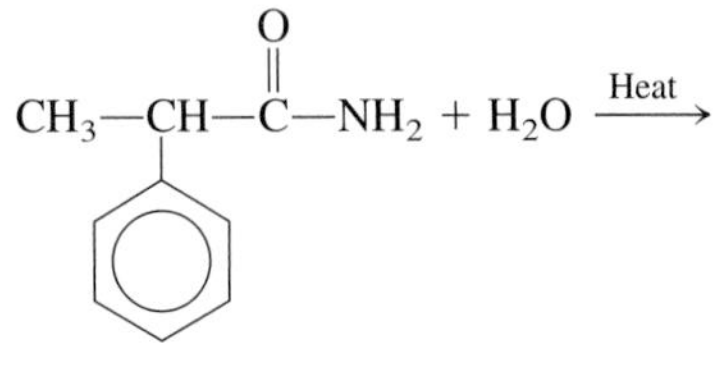

Polyamides and Polyurethanes (Section 6.18)

6.85 List the general characteristics of the monomers needed to produce a polyamide.

6.86 Contrast the monomers needed to produce a polyamide with those needed to produce a polyester.

6.87 Draw a structural representation for the polyamide formed from the reaction of succinic acid and 1,4-butanediamine.

6.88 Draw a structural representation for the polyamide formed from the reaction of adipic acid and 1,2-ethanediamine.

ADDITIONAL PROBLEMS

6.89 Draw structural formulas for the following compounds.
  a. Formamide
  b. 2-Pentamine
  c. 2-Methylpentanamide
  d. *N*-Isopropylethanamide
  e. Diethylammonium chloride
  f. Trimethylanilinium chloride

6.90 What is the structure of the organic product (or products) in each of the following reactions?

a. $CH_3-CH_2-\overset{+}{N}H_2-CH_3\ Br^- + NaOH \longrightarrow$

b.

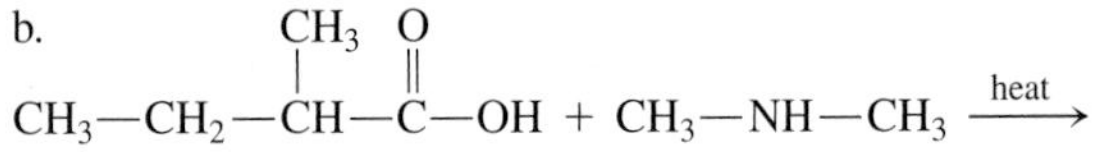

$$CH_3-CH_2-\underset{\underset{CH_3}{|}}{CH}-\overset{\overset{O}{||}}{C}-OH + CH_3-NH-CH_3 \xrightarrow{heat}$$

c.

$$CH_3-\underset{\underset{CH_3}{|}}{\overset{\overset{CH_3}{|}}{N}}-CH_3 + CH_3-Cl \xrightarrow{NaOH}$$

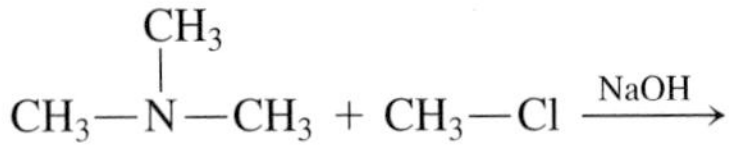

d.

$$CH_3-CH_2-\underset{\underset{CH_3}{|}}{CH}-\overset{\overset{O}{||}}{C}-NH_2 + H_2O \xrightarrow[heat]{NaOH}$$

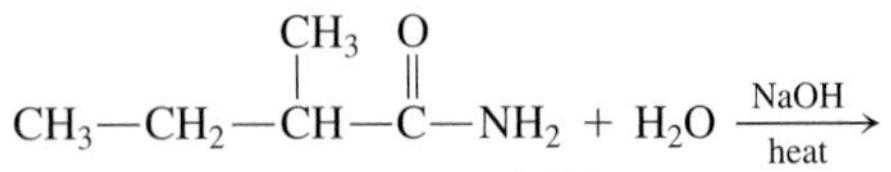

e. $CH_3-Br + CH_3-NH_2 \xrightarrow{NaOH}$

f. $CH_3-CH_2-NH_2 + H_2O \longrightarrow$

6.91 Draw structural formulas and assign IUPAC names to the four amide constitutional isomers with the formula C_3H_7ON.

6.92 Draw the structural formula of the quaternary ammonium salt with the formula $C_5H_{14}NCl$.

6.93 Classify each of the following amines or amides as unsubstituted, monosubstituted, or disubstituted.

  a. *o*-Methylbenzamide
  b. *N*-Methylbenzamide
  c. Cyclopentylmethylamine
  d. *N,N,*-dimethylhexanamide
  e. Isopropylamine
  f. 4-Methylheptanamine

6.94 Indicate whether each of the following compounds is an amine, an amide, both, or neither.

a. 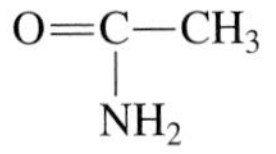 $O=\underset{\underset{NH_2}{|}}{C}-CH_3$

b. $CH_3-CH_2-\underset{\underset{NH_2}{|}}{CH}-\underset{\underset{NH_2}{|}}{CH}-CH_3$

c. (structure: methyl-substituted pyridinone ring)

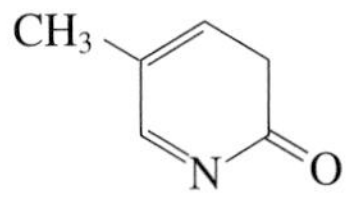

d. $CH_3-NH-CH_3$

e. 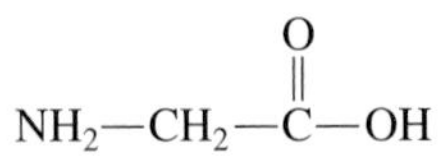 $NH_2-CH_2-\overset{\overset{O}{||}}{C}-OH$

f. 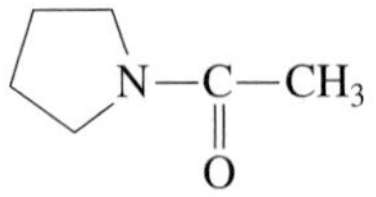 (pyrrolidine ring)$N-\overset{\overset{}{}}{\underset{\underset{O}{||}}{C}}-CH_3$

6.95 Assign IUPAC names to each of the following compounds.

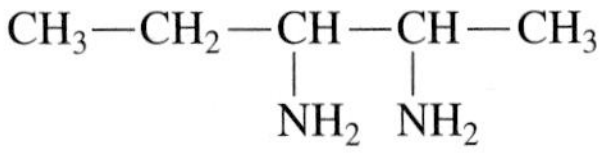

a. 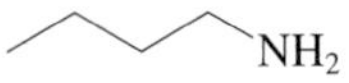 (structure with NH$_2$)

b. 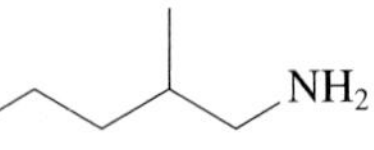 (structure with NH$_2$)

c. 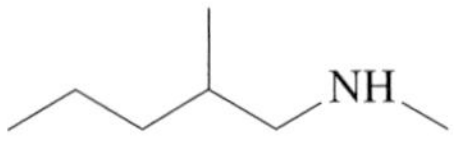 (structure with NH)

d.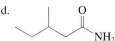

e. 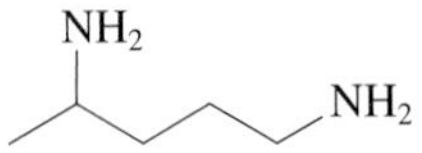 (structure with NH$_2$ and NH$_2$)

f.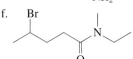

MULTIPLE-CHOICE PRACTICE TEST

6.96 Which of the following elements is not present in an unsubstituted amine?
a. Carbon   b. Hydrogen   c. Oxygen   d. Nitrogen

6.97 Which of the following amines is classified as a *secondary* amine?
a. 1-Butanamine   b. 2-Butanamine
c. *N*-methyl-2-butanamine   d. 3-Methyl-2-butanamine

6.98 Why are the boiling points of amines lower than those of alcohols of similar molecular mass?
a. Amines do not contain an oxygen atom as do alcohols.
b. Amine–amine hydrogen bonding is not possible.
c. N$\cdots$H hydrogen bonds are weaker than O$\cdots$H hydrogen bonds.
d. Amines are insoluble in water.

6.99 What is the molecular formula for the compound aniline?
a. C_6H_6N b. C_6H_7N c. C_6H_8N d. $C_6H_8N_2$

6.100 Which of the following sets of reactants, under appropriate conditions, produces a secondary amine?
a. Ammonia + alkyl halide
b. Ammonia + carboxylic acid
c. Primary amine + alkyl halide
d. Primary amine + carboxylic acid

6.101 Which of the following statements concerning amines and amides is correct?
a. Both amines and amides exhibit basic properties in aqueous solution.
b. Amines but not amides exhibit basic properties in aqueous solution.
c. Amides but not amines exhibit basic properties in aqueous solution.
d. Neither amines nor amides exhibit basic properties in aqueous solution.

6.102 Which of the following statements concerning amines and amides is correct?
a. Both amines and amides undergo hydrolysis reactions.
b. Amines but not amides undergo hydrolysis reactions.
c. Amides but not amines undergo hydrolysis reactions.
d. Neither amines nor amides undergo hydrolysis reactions.

6.103 What is the name of the amide produced by the reaction of butanoic acid and methyl amine?
a. *N*-methylbutanamide
b. 2-Methylbutanamide
c. Butyl amide
d. Methyl butyl amide

6.104 What are the organic products when an amide undergoes hydrolysis under basic conditions?
a. Carboxylic acid and amine salt
b. Carboxylic acid salt and amine
c. Carboxylic acid salt and amine salt
d. Carboxylic acid and amine

6.105 Which of the following sets of monomers would produce a polyamide?
a. Dicarboxylic acid and dialcohol
b. Dicarboxylic acid and diamine
c. Diamide and dialcohol
d. Diamide and diamine

7 Carbohydrates

Carbohydrates in the form of cotton and linen may be woven into clothing materials.

Beginning with this chapter on carbohydrates, we will focus almost exclusively on biochemistry, the chemistry of living systems. Like organic chemistry, biochemistry is a vast subject, and we can discuss only a few of its facets. Our approach to biochemistry will be similar to our approach to organic chemistry. We will devote individual chapters to each of the major classes of biochemical compounds, which are carbohydrates, lipids, proteins, and nucleic acids. Then we will examine the major types of chemical reactions in living organisms. In this first "biochapter," carbohydrates are considered.

The same functional groups found in organic compounds are also present in biochemical compounds. Usually, however, there is greater structural complexity associated with biochemical compounds as a result of polyfunctionality; several different functional groups are present. Often biochemical compounds interact with each other, within cells, to form larger structures. But the same chemical principles and chemical reactions associated with the various organic functional groups that we have studied apply to these larger biochemical structures as well.

7.1 Biochemistry — An Overview

Biochemistry *is the study of the chemical substances found in living organisms and the chemical interactions of these substances with each other.* Biochemistry is a field in which new discoveries are made almost daily about how cells manufacture the molecules needed

FIGURE 7.1 Mass composition data for the human body in terms of major types of biochemical substances.

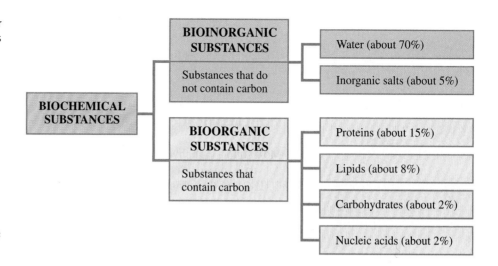

As isolated compounds, bioinorganic and bioorganic substances have no life in and of themselves. Yet when these substances are gathered together in a cell, their chemical interactions are able to sustain life.

It is estimated that more than half of all organic carbon atoms are found in the carbohydrate materials of plants.

Human uses for carbohydrates of the plant kingdom extend beyond food. Carbohydrates in the form of cotton and linen are used as clothing. Carbohydrates in the form of wood are used for shelter and heating and in making paper.

FIGURE 7.2 Most of the matter in plants, except water, is carbohydrate material. Photosynthesis, the process by which carbohydrates are made, requires sunlight.

for life and how the chemical reactions by which life is maintained occur. The knowledge explosion that has occurred in the field of biochemistry during the last decades of the twentieth century and the beginning of the twenty-first is truly phenomenal.

A **biochemical substance** *is a chemical substance found within a living organism.* Biochemical substances are divided into two groups: bioinorganic substances and bioorganic substances. *Bioinorganic substances* include water and inorganic salts. *Bioorganic substances* include carbohydrates, lipids, proteins, and nucleic acids. Figure 7.1 gives an approximate mass composition for the human body in terms of types of biochemical substances present.

Although we tend to think of the human body as made up of organic substances, bioorganic molecules make up only about one-fourth of body mass. The bioinorganic substance water constitutes over two-thirds of the mass of the human body, and another 4%–5% of body mass comes from inorganic salts.

7.2 Occurrence and Functions of Carbohydrates

Carbohydrates are the most abundant class of bioorganic molecules on planet Earth. Although their abundance in the human body is relatively low (Section 7.1), carbohydrates constitute about 75% by mass of dry plant materials (see Figure 7.2).

Green (chlorophyll-containing) plants produce carbohydrates via *photosynthesis.* In this process, carbon dioxide from the air and water from the soil are the reactants, and sunlight absorbed by chlorophyll is the energy source.

$$CO_2 + H_2O + \text{solar energy} \xrightarrow[\text{Plant enzymes}]{\text{Chlorophyll}} \text{carbohydrates} + O_2$$

Plants have two main uses for the carbohydrates they produce. In the form of *cellulose,* carbohydrates serve as structural elements, and in the form of *starch,* they provide energy reserves for the plants.

Dietary intake of plant materials is the major carbohydrate source for humans and animals. The average human diet should ideally be about two-thirds carbohydrate by mass. Carbohydrates have the following functions in humans:

1. Carbohydrate oxidation provides energy.
2. Carbohydrate storage, in the form of glycogen, provides a short-term energy reserve.
3. Carbohydrates supply carbon atoms for the synthesis of other biochemical substances (proteins, lipids, and nucleic acids).
4. Carbohydrates form part of the structural framework of DNA and RNA molecules.

5. Carbohydrates linked to lipids (Chapter 19) are structural components of cell membranes.
6. Carbohydrates linked to proteins (Chapter 20) function in a variety of cell–cell and cell–molecule recognition processes.

7.3 Classification of Carbohydrates

Most simple carbohydrates have empirical formulas that fit the general formula $C_nH_{2n}O_n$. An early observation by scientists that this general formula can also be written as $C_n(H_2O)_n$ is the basis for the term *carbohydrate*—that is, "hydrate of carbon." It is now known that this hydrate viewpoint is not correct, but the term *carbohydrate* still persists. Today the term is used to refer to an entire family of compounds, only some of which have the formula $C_nH_{2n}O_n$.

A **carbohydrate** *is a polyhydroxy aldehyde, a polyhydroxy ketone, or a compound that yields polyhydroxy aldehydes or polyhydroxy ketones upon hydrolysis.* The carbohydrate glucose is a polyhydroxy aldehyde, and the carbohydrate fructose is a polyhydroxy ketone.

Glucose
(a polyhydroxy aldehyde)

Fructose
(a polyhydroxy ketone)

A striking structural feature of carbohydrates is the large number of functional groups present. In glucose and fructose there is a functional group attached to each carbon atom.

Carbohydrates are classified on the basis of molecular size as monosaccharides, oligosaccharides, and polysaccharides.

A **monosaccharide** *is a carbohydrate that contains a single polyhydroxy aldehyde or polyhydroxy ketone unit.* Monosaccharides cannot be broken down into simpler units by hydrolysis reactions. Both glucose and fructose are monosaccharides. Naturally occurring monosaccharides have from three to seven carbon atoms; five- and six-carbon species are especially common. Pure monosaccharides are water-soluble, white, crystalline solids.

An **oligosaccharide** *is a carbohydrate that contains two to ten monosaccharide units covalently bonded to each other.* Disaccharides are the most common type of oligosaccharide. A **disaccharide** *is a carbohydrate that contains two monosaccharide units covalently bonded to each other.* Like monosaccharides, disaccharides are crystalline, water-soluble substances. Sucrose (table sugar) and lactose (milk sugar) are disaccharides.

Within the human body, oligosaccharides are often found associated with proteins and lipids in complexes that have both structural and regulatory functions. Free oligosaccharides, other than disaccharides, are seldom encountered in biochemical systems.

Complete hydrolysis of an oligosaccharide produces monosaccharides. Upon hydrolysis, a disaccharide produces two monosaccharides, a trisaccharide three monosaccharides, a hexasaccharide six monosaccharides, and so on.

A **polysaccharide** *is a polymeric carbohydrate that contains many monosaccharide units covalently bonded to each other.* Polysaccharides often contain several thousand monosaccharide units. Both cellulose and starch are polysaccharides. We encounter these two substances everywhere. The paper on which this book is printed is mainly cellulose, as are the cotton in our clothes and the wood in our houses. Starch is a component of many types of foods, including bread, pasta, potatoes, rice, corn, beans, and peas.

The term *monosaccharide* is pronounced "mon-oh-SACK-uh-ride."

The *oligo* in the term *oligosaccharides* comes from the Greek *oligos,* which means "small" or "few." The term *oligosaccharide* is pronounced "OL-ee-go-SACK-uh-ride."

Types of carbohydrates are related to each other through hydrolysis.

Polysaccharides
↓ Hydrolysis
Oligosaccharides
↓ Hydrolysis
Monosaccharides

Left   **Right**

Mirror image of left hand
is in the back of the mirror

FIGURE 7.3 The mirror image of the right hand is the left hand. Conversely, the mirror image of the left hand is the right hand.

7.4 Chirality: Handedness in Molecules

Monosaccharides are the simplest type of carbohydrate. Before considering specific structures for and specific reactions of monosaccharides, we will consider an important general structural property called *handedness,* which most monosaccharides exhibit. Most monosaccharides exist in two forms: a "left-handed" form and a "right-handed" form. These two forms are related to each other in the same way your left and right hands are related to each other. That relationship is that of *mirror images.* Figure 7.3 shows this mirror-image relationship for human hands.

The property of handedness is not restricted to carbohydrates. It is a general phenomenon found in all classes of organic compounds.

Mirror Images

The concept of *mirror images* is the key to understanding molecular handedness. All objects, including all molecules, have mirror images. A **mirror image** *is the reflection of an object in a mirror.* Objects can be divided into two classes on the basis of their mirror images: objects with *superimposable* mirror images and objects with *nonsuperimposable* mirror images. **Superimposable mirror images** *are images that coincide at all points when the images are laid upon each other.* A dinner plate with no design features has superimposable mirror images. **Nonsuperimposable mirror images** *are images where not all points coincide when the images are laid upon each other.* Human hands are nonsuperimposable mirror images, as Figure 7.4 shows; note in this figure that the two thumbs point in opposite directions and that the fingers do not align correctly. Like human hands, all objects with nonsuperimposable mirror images exist in "left-handed" and "right-handed" forms.

Every object has a mirror image. The question is, "Is the mirror image the same (superimposable) or different (nonsuperimposable)?"

The term *chiral* (rhymes with *spiral*) comes from the Greek word *cheir,* which means "hand." Chiral objects are said to possess "handedness."

Chirality

Of particular concern to us is the "handedness concept" as it applies to molecules. Not all molecules possess handedness. What, then, is the molecular structural feature that generates "handedness?" Any organic molecule that contains a carbon atom with four *different* groups attached to it in a tetrahedral orientation possesses handedness. Such a carbon atom is called a *chiral center.* A **chiral center** *is an atom in a molecule that has four different groups tetrahedrally bonded to it.*

A molecule that contains a chiral center is said to be *chiral.* A **chiral molecule** *is a molecule whose mirror images are not superimposable.* Chiral molecules have handedness. An **achiral molecule** *is a molecule whose mirror images are superimposable.* Achiral molecules do not possess handedness.

A trisubstituted methane molecule, such as bromochloroiodomethane, is the simplest example of a chiral organic molecule.

FIGURE 7.4 A person's left and right hands are not superimposable upon each other.

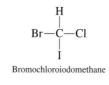

Bromochloroiodomethane

FIGURE 7.5 Examples of simple molecules that are chiral. (a) The mirror image forms of the molecule bromochloroiodomethane are nonsuperimposable. (b) The mirror-image forms of the molecule glyceraldehyde are nonsuperimposable.

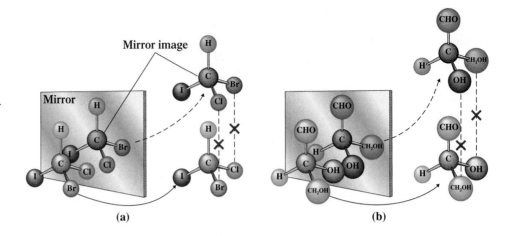

(a)   **(b)**

Note the four different groups attached to the carbon atom present: —H, —Br, —Cl, and —I. Figure 7.5a shows the nonsuperimposability of the two mirror-image forms of this molecule.

The simplest example of a chiral monosaccharide molecule is the three-carbon monosaccharide called glyceraldehyde.

$$
\begin{array}{c}
CHO \\
| \\
H-C-OH \\
| \\
CH_2OH
\end{array}
$$

Glyceraldehyde

There are a few chiral molecules known that do not have a chiral center. Such exceptions are not important for the applications of the chirality concept that we will make in this text.

The four different groups attached to the carbon atom at the chiral center in this molecule are —H, —OH, —CHO, and —CH$_2$OH. The nonsuperimposability of the two mirror-image forms of glyceraldehyde is shown in Figure 7.5b.

Chiral centers within molecules are often denoted by a small asterisk. Note the chiral centers in the following molecules.

$$
CH_3-CH_2-\overset{H}{\underset{OH}{*C}}-CH_3
\qquad
H-\overset{Cl}{\underset{I}{*C}}-CH_3
\qquad
CH_3-CH_2-CH_2-\overset{CH_3}{\underset{H}{*C}}-CH_2-CH_3
$$

2-Butanol   1-Chloro-1-iodoethane   3-Methylhexane

EXAMPLE 7.1

Identifying Chiral Centers in Molecules

■ Indicate whether the circled carbon atom in each of the following molecules is a chiral center.

a. CH$_3$—ⒸH—CH$_2$—CH$_3$
         |
         Cl

b. CH$_3$—CH$_2$—Ⓒ—CH$_3$
             ‖
             O

c. CH$_3$—CH$_2$—ⒸH—OH
              |
              CH$_2$
              |
              CH$_3$

d.
$$
\begin{array}{c}
Br \\
| \\
ⒸH \\
H_2C \quad\quad CH_2 \\
| \quad\quad\quad | \\
HC \quad\quad CH_2 \\
Br \quad CH_2
\end{array}
$$

Solution

a. This is a chiral center. The four different groups attached to the carbon atom are —CH$_3$, —Cl, —CH$_2$—CH$_3$, and —H.

b. No chiral center is present. The carbon atom is attached to only *three* groups.

c. No chiral center is present. Two of the groups attached to the carbon atom are identical.

d. The chirality rules for ring carbon atoms are the same as those for acyclic carbon atoms. A chiral center is present. Two of the groups are —H and —Br. The third group, obtained by proceeding clockwise around the ring, is —CH_2—CH_2—CH_2. The fourth group, obtained by proceeding counterclockwise around the ring, is —CH_2—CHBr—CH_2.

Practice Exercise 7.1

Indicate whether the circled carbon atom in each of the following molecules is a chiral center.

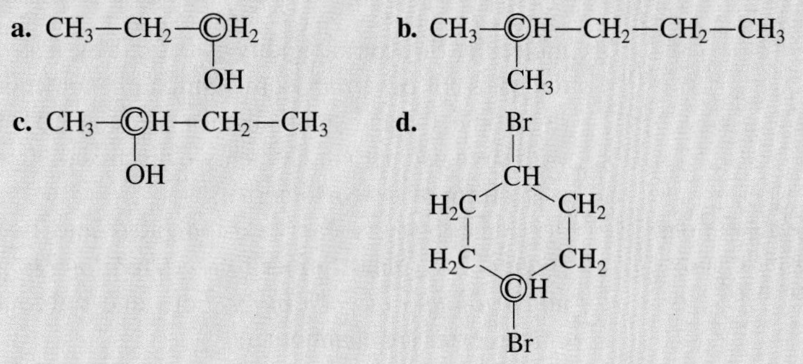

a. CH_3—CH_2—ⒸH_2
　　　　　　|
　　　　　　OH

b. CH_3—ⒸH—CH_2—CH_2—CH_3
　　　　　|
　　　　　CH_3

c. CH_3—ⒸH—CH_2—CH_3
　　　　|
　　　　OH

d.
　　　　　　Br
　　　　　　|
　　　　　　CH
　　　H_2C　　　CH_2
　　　H_2C　　　CH_2
　　　　　ⒸH
　　　　　|
　　　　　Br

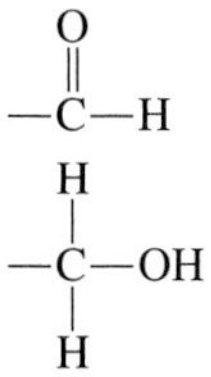

Organic molecules, especially monosaccharides, may contain more than one chiral center. For example, the following monosaccharide has two chiral centers.

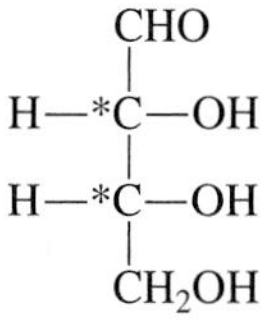

$$\begin{array}{c} CHO \\ | \\ H{-}{*}C{-}OH \\ | \\ H{-}{*}C{-}OH \\ | \\ CH_2OH \end{array}$$

What is the importance of the handedness that we have been discussing? In human body chemistry, right-handed and left-handed forms of a molecule often elicit different responses within the body. Sometimes both forms are biologically active, each form giving a different response; sometimes both elicit the same response, but one form's response is many times greater than that of the other; and sometimes only one of the two forms is biochemically active. For example, studies show that the body's response to the right-handed form of the hormone epinephrine (Section 6.10) is 20 times greater than its response to the left-handed form.

Naturally occurring monosaccharides are almost always "right-handed." Plants, our dietary source for carbohydrates, produce only right-handed monosaccharides. Interestingly, when we consider protein chemistry (Chapter 9), we will find that amino acids, the building blocks for proteins, are always left-handed molecules.

Remember the meaning of the structural notations —CHO and —CH_2OH.

$$-CHO \quad means \quad \begin{array}{c} O \\ \parallel \\ -C-H \end{array}$$

$$-CH_2OH \quad means \quad \begin{array}{c} H \\ | \\ -C-OH \\ | \\ H \end{array}$$

7.5 Stereoisomerism: Enantiomers and Diastereomers

The left- and right-handed forms of a chiral molecule are isomers. They are not *constitutional isomers,* the type of isomerism that we encountered repeatedly in the organic chemistry chapters of the text, but rather are *stereoisomers.* **Stereoisomers** *are isomers that have the same molecular and structural formulas but differ in the orientation of atoms in space.* By contrast, atoms are connected to each other in different ways in constitutional isomers (Section 1.6).

There are two major structural features that generate *stereoisomerism:* (1) the presence of a chiral center in a molecule and (2) the presence of "structural rigidity" in a

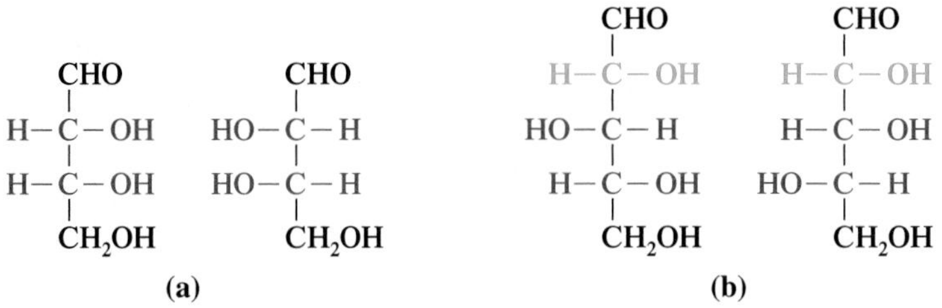

FIGURE 7.6 (a) Enantiomers are stereoisomers whose molecules are nonsuperimposable mirror images of each other, as in left-handed and right-handed forms of a molecule. (b) Diastereomers are stereoisomers whose molecules are not mirror images of each other.

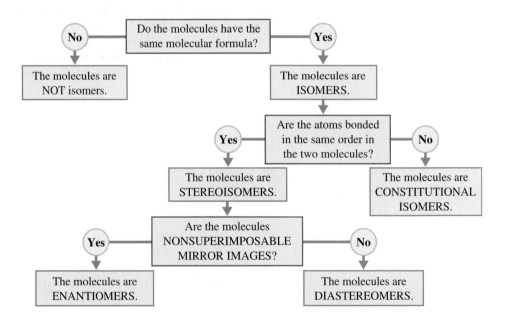

molecule. Structural rigidity is caused by restricted rotation about chemical bonds. It is the basis for *cis–trans* isomerism, a phenomenon found in some substituted cycloalkanes (Section 1.14) and some alkenes (Section 2.5). Thus handedness is our second encounter with stereoisomerism. (When we discussed *cis–trans* isomerism, we did not mention that it is a form of stereoisomerism.)

Stereoisomers can be subdivided into two types: *enantiomers* and *diastereomers* (Figure 7.6). **Enantiomers** *are stereoisomers whose molecules are nonsuperimposable mirror images of each other.* Left- and right-handed forms of a molecule with a single chiral center are enantiomers.

Diastereomers *are stereoisomers whose molecules are not mirror images of each other. Cis–trans* isomers (of both the alkene and the cycloalkane types) are diastereomers. Molecules that contain more than one chiral center can also exist in diastereomeric as well as enantiomeric forms, as is shown in Figure 7.6.

Figure 7.7 shows the "thinking pattern" involved in using the terms *stereoisomers, enantiomers,* and *diastereomers.*

> The term *enantiomer* comes from the Greek *enantios,* which means "opposite." It is pronounced "en-AN-tee-o-mer."

> Some textbooks use the term *diastereoisomers* instead of *diastereomers.* The pronunciation for *diastereomer* is "dye-a-STEER-ee-o-mer."

7.6 Designating Handedness Using Fischer Projections

Drawing *three*-dimensional representations of chiral molecules can be both time-consuming and awkward. Fischer projections represent a method for giving molecular chirality specifications in *two* dimensions. A **Fischer projection** *is a two-dimensional structural notation for showing the spatial arrangement of groups about chiral centers in molecules.*

> Fischer projections carry the name of their originator, the German chemist Hermann Emil Fischer (see Figure 7.8).

FIGURE 7.7 A summary of the "thought process" used in classifying molecules as enantiomers or diastereomers.

In a Fischer projection, a chiral center is represented as the intersection of vertical and horizontal lines. The atom at the chiral center, which is almost always carbon, is not explicitly shown.

The tetrahedral arrangement of the four groups attached to the atom at the chiral center is governed by the following conventions: (1) Vertical lines from the chiral center represent bonds to groups directed into the printed page. (2) Horizontal lines from the chiral center represent bonds to groups directed out of the printed page.

In Fischer projections for monosaccharides (the simplest type of carbohydrate; Section 7.2), the monosaccharide *carbon chain is* positioned vertically with the carbonyl group (aldehyde or ketone) at or near the top. The smallest monosaccharide that has a chiral center is the compound glyceraldehyde (2,3-dihydroxypropanal; Section 7.4).

The Fischer projections for the two enantiomers of glyceraldehyde are

L-Glyceraldehyde   D-Glyceraldehyde

The handedness (right and left) of these two enantiomers is specified by using the designations D and L. The enantiomer with the chiral center —OH group on the right in the Fischer projection is by definition the right-handed isomer (D-glyceraldehyde), and the enantiomer with the chiral center —OH group on the left in the Fischer projection is by definition the left-handed isomer (L-glyceraldehyde).

We now consider Fischer projections for the compound 2,3,4-trihydroxybutanal, a monosaccharide with four carbons and *two* chiral centers.

There are four stereoisomers for this compound—two pairs of enantiomers.

First enantiomeric pair   Second enantiomeric pair

FIGURE 7.8 The German chemist Hermann Emil Fischer (1852–1919), the developer of the two-dimensional system for specifying chirality, was one of the early greats in organic chemistry. He made many fundamental discoveries about carbohydrates, proteins, and other natural products. In 1902 he was awarded the second Nobel Prize in chemistry.

The D and L designations for the handedness of the two members of an enantiomeric pair come from the Latin words *dextro,* which means "right," and *levo,* which means "left."

To draw the mirror image of a Fischer projection structure, keep up-and-down and front-and-back aspects of the structure the same and reverse the left-and-right aspects.

> Any given molecular structure can have only one mirror image. Hence enantiomers always come in pairs; there can never be more than two.

In the first enantiomeric pair, both chiral-center —OH groups are on the same side of the Fischer projection, and in the second enantiomeric pair, the chiral-center —OH groups are on opposite sides of the Fischer projection. These are the only —OH group arrangements possible.

The D,L system used to designate the handedness of glyceraldehyde enantiomers is extended to monosaccharides with more than one chiral center in the following manner. The carbon chain is numbered starting at the carbonyl group end of the molecule, and the highest-numbered chiral center is used to determine D or L configuration.

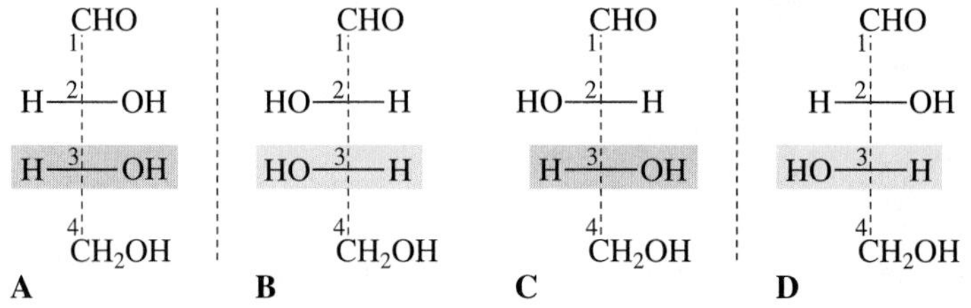

A D isomer | B L isomer | C D isomer | D L isomer

The D,L nomenclature gives the configuration (handedness) only at the highest-numbered chiral center. The configuration at other chiral centers in a molecule is accounted for by assigning a different common name to each pair of D,L enantiomers. In our present example, compounds A and B (the first enantiomeric pair) are D-erythrose and L-erythrose; compounds C and D (the second enantiomeric pair) are D-threose and L-threose.

> Diastereomers that have two chiral centers must have the same handedness (both left or both right) at one chiral center and opposite handedness (one left and one right) at the other chiral center.

What is the relationship between compounds A and C in our present example? They are diastereomers (Section 7.5), stereoisomers that are not mirror images of each other. Other diastereomeric pairs in our example are A and D, B and C, and B and D. The members of each of these four pairs are epimers. **Epimers** *are diastereomers whose molecules differ only in the configuration at one chiral center.*

EXAMPLE 7.2
Drawing Fischer Projections for Monosaccharides

■ Draw a Fischer projection for the enantiomer of each of the following monosaccharides.

a.

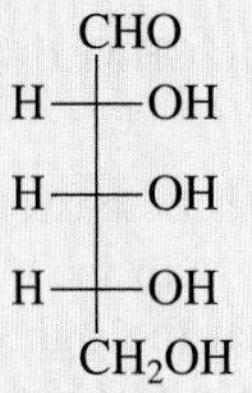

b.

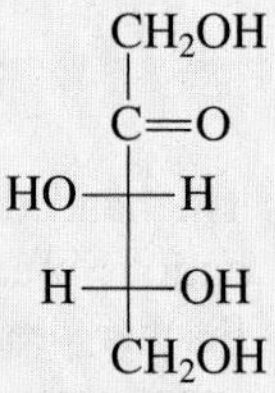

Solution

Given the Fischer projection of one member of an enantiomeric pair, we draw the other enantiomer's Fischer projection by reversing the substituents that are in horizontal positions *at each chiral center.*

a. Three chiral centers are present in this polyhydroxy aldehyde. Reversing the positions of the —H and —OH groups at each chiral center produces the Fischer projection of the other enantiomer.

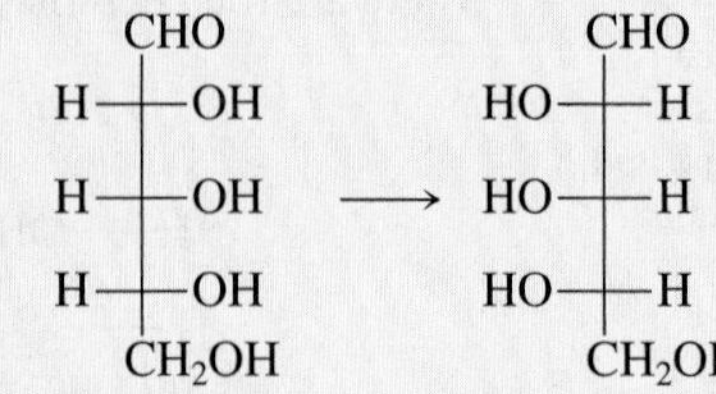

The given enantiomer — The other enantiomer

b. This monosaccharide is a polyhydroxy ketone with two chiral centers. Reversing the positions of the —H and —OH groups at both chiral centers generates the Fischer projection of the other enantiomer.

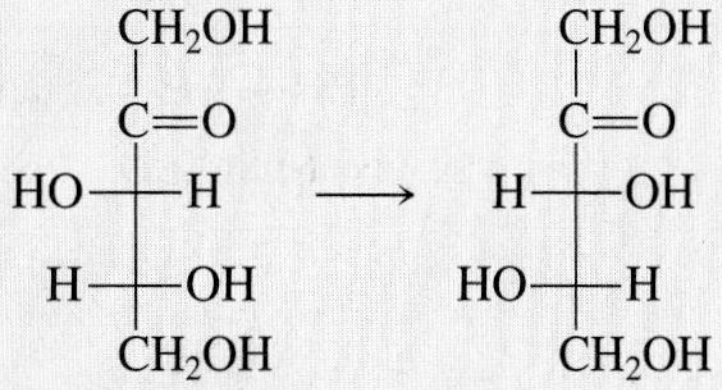

$$
\begin{array}{ccc}
\text{CH}_2\text{OH} & & \text{CH}_2\text{OH} \\
| & & | \\
\text{C}=\text{O} & & \text{C}=\text{O} \\
\text{HO}-\!\!\!\!-\text{H} & \longrightarrow & \text{H}-\!\!\!\!-\text{OH} \\
\text{H}-\!\!\!\!-\text{OH} & & \text{HO}-\!\!\!\!-\text{H} \\
\text{CH}_2\text{OH} & & \text{CH}_2\text{OH} \\
\text{The given enantiomer} & & \text{The other enantiomer}
\end{array}
$$

Practice Exercise 7.2

Draw a Fischer projection for the enantiomer of each of the following monosaccharides.

a.
$$
\begin{array}{c}
\text{CHO} \\
\text{H}-\!\!\!\!-\text{OH} \\
\text{HO}-\!\!\!\!-\text{H} \\
\text{HO}-\!\!\!\!-\text{H} \\
\text{CH}_2\text{OH}
\end{array}
$$

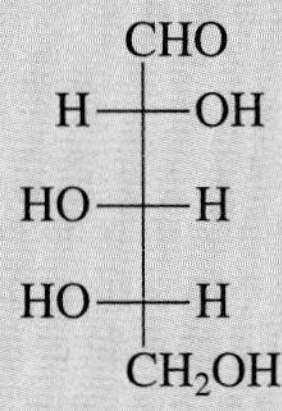

b.
$$
\begin{array}{c}
\text{CH}_2\text{OH} \\
| \\
\text{C}=\text{O} \\
\text{H}-\!\!\!\!-\text{OH} \\
\text{H}-\!\!\!\!-\text{OH} \\
\text{CH}_2\text{OH}
\end{array}
$$

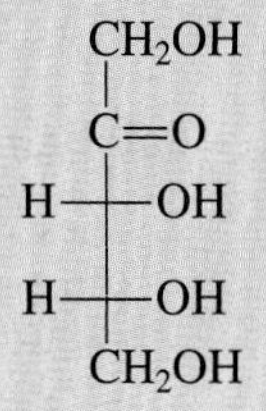

EXAMPLE 7.3

Classifying Monosaccharides as D or L Enantiomers

■ Classify each of the following monosaccharides as a D enantiomer or an L enantiomer.

a.
$$
\begin{array}{c}
^1\text{CHO} \\
\text{HO}-\!\!^2\!\!-\text{H} \\
\text{H}-\!\!^3\!\!-\text{OH} \\
\text{H}-\!\!^4\!\!-\text{OH} \\
^5\text{CH}_2\text{OH}
\end{array}
$$

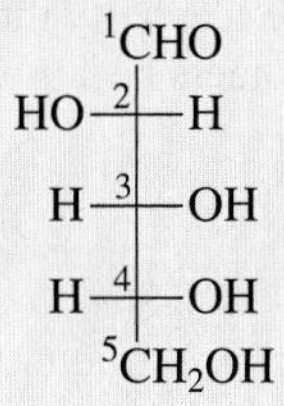

b.
$$
\begin{array}{c}
^1\text{CH}_2\text{OH} \\
| \\
^2\text{C}=\text{O} \\
\text{H}-\!\!^3\!\!-\text{OH} \\
\text{HO}-\!\!^4\!\!-\text{H} \\
\text{HO}-\!\!^5\!\!-\text{H} \\
^6\text{CH}_2\text{OH}
\end{array}
$$

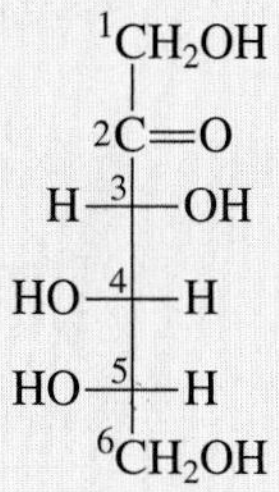

Solution

D or L configuration for a monosaccharide is determined by the highest-numbered chiral center, the one farthest from the carbonyl carbon atom.

a. The highest-numbered chiral center, which involves carbon 4, has the —OH group on the right. Thus this monosaccharide is a D enantiomer.

b. The highest-numbered chiral center, which involves carbon 5, has the —OH group on the left. Thus this monosaccharide is an L enantiomer.

Practice Exercise 7.3

Classify each of the following monosaccharides as a D enantiomer or an L enantiomer.

a.
$$
\begin{array}{c}
\text{CH}_2\text{OH} \\
| \\
\text{C}=\text{O} \\
\text{HO}-\!\!\!\!-\text{H} \\
\text{H}-\!\!\!\!-\text{OH} \\
\text{H}-\!\!\!\!-\text{OH} \\
\text{CH}_2\text{OH}
\end{array}
$$

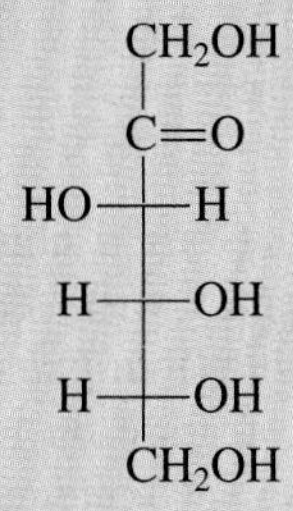

b.
$$
\begin{array}{c}
\text{CHO} \\
\text{H}-\!\!\!\!-\text{OH} \\
\text{HO}-\!\!\!\!-\text{H} \\
\text{H}-\!\!\!\!-\text{OH} \\
\text{HO}-\!\!\!\!-\text{H} \\
\text{CH}_2\text{OH}
\end{array}
$$

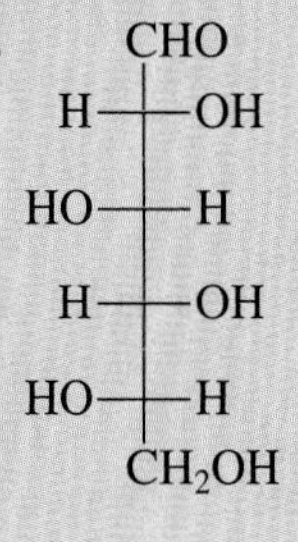

EXAMPLE 7.4

**Recognizing Enantiomers
and Diastereomers**

■ Characterize each of the following pairs of structures as enantiomers, diastereomers, or neither enantiomers nor diastereomers.

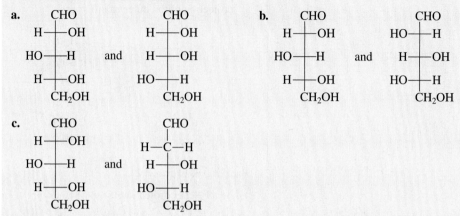

Solution

a. These two structures represent *diastereomers*—the arrangement of —H and —OH substituents is identical for at least one chiral center, whereas the arrangement of —H and —OH substituents at remaining chiral centers is that of mirror images. The —H and —OH substituent arrangement is the same at the first chiral center and is that of mirror images at the second and third chiral centers.

b. These two structures represent *enantiomers*—a mirror-image substituent relationship exists between the two isomers at *every* chiral center.

c. These two structures are *neither enantiomers nor diastereomers*. The connectivity of atoms differs in the two structures at carbon 2. Stereoisomers (enantiomers and diastereomers) must have the same connectivity throughout both structures. (The two structures are not even constitutional isomers because the first structure contains one more oxygen atom than the second.)

Practice Exercise 7.4

Characterize the following pairs of structures as enantiomers, diastereomers, or neither enantiomers nor diastereomers.

a.
```
    CHO              CHO
H ──┬── OH      HO ──┬── H
H ──┼── OH  and HO ──┼── H
H ──┼── OH       H ──┼── OH
    CH₂OH           CH₂OH
```

b.
```
    CHO              CHO
H ──┬── OH      HO ──┬── H
H ──┼── OH  and HO ──┼── H
H ──┼── OH      HO ──┼── H
    CH₂OH           CH₂OH
```

c.
```
    CHO              CHO
H ──┬── OH      HO ──┬── H
HO ──┼── H  and  H ──┼── OH
HO ──┼── H      HO ──┼── H
    CH₂OH           CH₂OH
```

We calculate 2^n to predict the maximum possible number of stereoisomers for a molecule with n chiral atoms. In a few cases, the actual number of stereoisomers is less than the maximum because of symmetry considerations that make some mirror images superimposable.

In general, a compound that has n chiral centers may exist in a *maximum* of 2^n stereoisomeric forms. For example, when three chiral centers are present, at most eight stereoisomers ($2^3 = 8$) are possible (four pairs of enantiomers).

The Chemistry at a Glance feature on page 231 summarizes information about the various types of isomers we have encountered so far in the text—the various subtypes of constitutional isomers and the various subtypes of stereoisomers.

CHEMISTRY AT A GLANCE

Constitutional Isomers and Stereoisomers

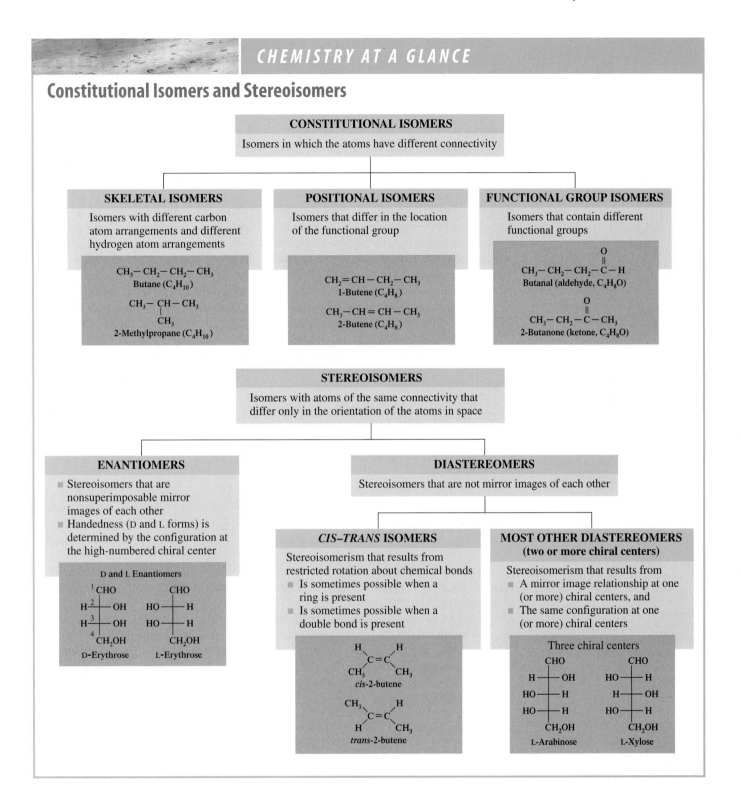

CONSTITUTIONAL ISOMERS

Isomers in which the atoms have different connectivity

SKELETAL ISOMERS

Isomers with different carbon atom arrangements and different hydrogen atom arrangements

$CH_3-CH_2-CH_2-CH_3$
Butane (C_4H_{10})

$CH_3-CH-CH_3$
          |
         CH_3
2-Methylpropane (C_4H_{10})

POSITIONAL ISOMERS

Isomers that differ in the location of the functional group

$CH_2=CH-CH_2-CH_3$
1-Butene (C_4H_8)

$CH_3-CH=CH-CH_3$
2-Butene (C_4H_8)

FUNCTIONAL GROUP ISOMERS

Isomers that contain different functional groups

$CH_3-CH_2-CH_2-C-H$
Butanal (aldehyde, C_4H_8O)

$CH_3-CH_2-C-CH_3$
2-Butanone (ketone, C_4H_8O)

STEREOISOMERS

Isomers with atoms of the same connectivity that differ only in the orientation of the atoms in space

ENANTIOMERS

- Stereoisomers that are nonsuperimposable mirror images of each other
- Handedness (D and L forms) is determined by the configuration at the high-numbered chiral center

D and L Enantiomers

D-Erythrose   L-Erythrose

DIASTEREOMERS

Stereoisomers that are not mirror images of each other

CIS–TRANS ISOMERS

Stereoisomerism that results from restricted rotation about chemical bonds
- Is sometimes possible when a ring is present
- Is sometimes possible when a double bond is present

cis-2-butene

trans-2-butene

MOST OTHER DIASTEREOMERS
(two or more chiral centers)

Stereoisomerism that results from
- A mirror image relationship at one (or more) chiral centers, and
- The same configuration at one (or more) chiral centers

Three chiral centers

L-Arabinose   L-Xylose

7.7 Properties of Enantiomers

Constitutional isomers differ in most chemical and physical properties. For example, constitutional isomers have different boiling points and melting points. Diastereomers also differ in most chemical and physical properties. They also have different boiling points and freezing points. In contrast, nearly all the properties of a pair of enantiomers

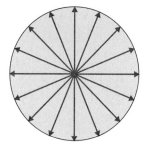

(a) Ordinary (unpolarized) light

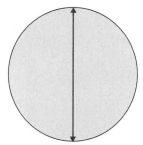

(b) Plane-polarized light

FIGURE 7.9 Vibrational characteristics of ordinary (unpolarized) light (a), and polarized light (b). The direction of travel of the light is toward the reader.

Achiral molecules are optically *inactive.* Chiral molecules are optically *active.*

Because of their ability to rotate the plane of polarized light, enantiomers are sometimes referred to as *optical isomers.*

In any pair of enantiomers, one, the (+)-enantiomer, always rotates the plane of polarized light to the right, and the other, the (−)-enantiomer, to the left.

are the same; for example, they have identical boiling points and freezing points. Enantiomers exhibit different properties in only two areas: (1) their interaction with plane-polarized light and (2) their interaction with other chiral substances.

■ Interaction of Enantiomers with Plane-Polarized Light

All light moves through space with a wave motion. Ordinary light waves—that is, unpolarized light waves—vibrate in *all* planes at right angles to their direction of travel. Plane-polarized light waves, by contrast, vibrate in *only one* plane at right angles to their direction of travel. Figure 7.9 contrasts the vibrational behavior of ordinary light with that of plane-polarized light.

Ordinary light can be converted to plane-polarized light by passing it through a *polarizer,* an instrument with lenses or filters containing special types of crystals. When plane-polarized light is passed through a solution containing a *single* enantiomer, the plane of the polarized light is rotated counterclockwise (to the left) or clockwise (to the right), depending on the enantiomer. The extent of rotation depends on the concentration of the enantiomer as well as on its identity. Furthermore, the two enantiomers of a pair rotate the plane-polarized light the same number of degrees, but in opposite directions. If a 0.50 M solution of one enantiomer rotates the light 30° to the right, then a 0.50 M solution of the other enantiomer rotates the light 30° to the left.

Instruments used to measure the degree of rotation of plane-polarized light by enantiomeric compounds are called *polarimeters.* The schematic diagram in Figure 7.10 shows the basis for these instruments.

■ Dextrorotatory and Levorotatory Compounds

Enantiomers are said to be optically active because of the way they interact with plane-polarized light. An **optically active compound** *is a compound that rotates the plane of polarized light.*

An enantiomer that rotates plane-polarized light in a clockwise direction (to the *right*) is said to be dextrorotatory (the Latin *dextro* means "right"). A **dextrorotatory compound** *is a chiral compound that rotates the plane of polarized light in a clockwise direction.* An enantiomer that rotates plane-polarized light in a counterclockwise direction (to the *left*) is said to be levorotatory (the Latin *levo* means "left"). A **levorotatory compound** *is a chiral compound that rotates the plane of polarized light in a counterclockwise direction.* If one member of an enantiomeric pair is dextrorotatory, then the other member must be levorotatory.

A plus or minus sign inside parentheses is used to denote the direction of rotation of plane-polarized light by a chiral compound. The notation (+) means rotation to the right (clockwise), and (−) means rotation to the left (counterclockwise). Thus the dextrorotatory enantiomer of glucose is (+)-glucose.

The handedness of enantiomers (D or L, Section 7.6) and the direction of rotation of plane-polarized light by enantiomers [(+) or (−)] are not connected entities. There is

FIGURE 7.10 Schematic depiction of how a polarimeter works.

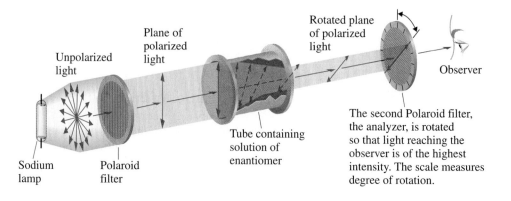

Unpolarized light

Plane of polarized light

Rotated plane of polarized light

Observer

Sodium lamp

Polaroid filter

Tube containing solution of enantiomer

The second Polaroid filter, the analyzer, is rotated so that light reaching the observer is of the highest intensity. The scale measures degree of rotation.

Both handedness and direction of rotation of plane-polarized light can be incorporated into the name of an enantiomer. For example, the notation D-(+)-mannose specifies that the right-handed isomer of the monosaccharide mannose rotates plane-polarized light in a clockwise direction (to the right).

no way of knowing which way an enantiomer will rotate light until it is examined with a polarimeter. Not all D enantiomers rotate plane-polarized light in the same direction, nor do all L enantiomers rotate plane-polarized light in the same direction. Some D enantiomers are dextrorotatory; others are levorotatory.

Interactions Between Chiral Compounds

A left-handed baseball player (chiral) and a right-handed baseball player (chiral) can use the same baseball bat (achiral) or wear the same baseball hat (achiral). However, left- and right-handed baseball players (chiral) cannot use the same baseball glove (chiral). This nonchemical example illustrates that the chirality of an object becomes important when the object interacts with another chiral object.

Applying this generalization to molecules, we find that the two members of an enantiomeric pair have the same interaction with achiral molecules and different interactions with chiral molecules. We find that

1. Enantiomers have identical boiling points, freezing points, and densities because such properties depend on the strength of intermolecular forces, and intermolecular force strength does not depend on chirality. Intermolecular force strength is the same for both forms of a chiral molecule because both forms have identical sets of functional groups.
2. A pair of enantiomers have the same solubility in an achiral solvent, such as ethanol, but differing solubilities in a chiral solvent, such as D-2-butanol.
3. The rate and extent of reaction of enantiomers with another reactant are the same if the reactant is achiral but differ if the reactant is chiral.
4. Receptor sites for molecules within the body have chirality associated with them. Thus enantiomers always generate different responses within the human body as they interact at such sites. Sometimes the responses are only slightly different, and at other times they are very different.

Let us consider two specific examples of differing chiral—chiral interactions involving enantiomers that occur within the human body. The first example involves taste perceptions. The distinctly different natural flavors "spearmint" and "caraway" are generated by molecules that are enantiomers interacting with chiral "taste receptors" (see Figure 7.11).

The second example involves the body's response to the enantiomer forms of the hormone epinephrine (adrenaline). The response of the body to the D isomer of the hormone is 20 times greater than its response to the L isomer of the hormone. Epinephrine binds to its cellular receptor site by means of a three-point contact, as is shown in Figure 7.12. D-Epinephrine makes a perfect three-point contact with the receptor surface, but the biochemically weaker L-epinephrine can make only a two-point contact. Because of the poorer fit, the binding of the L isomer is weaker, and less physiological response is observed.

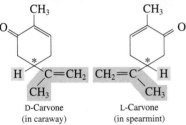

FIGURE 7.11 The distinctly different natural flavors of spearmint and caraway are caused by enantiomeric molecules. Spearmint leaves contain L-carvone, and caraway seeds contain D-carvone.

FIGURE 7.12 D-Epinephrine binds to the receptor at three points, whereas the biochemically weaker L-epinephrine binds at only two sites.

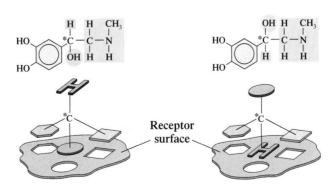

7.8 Classification of Monosaccharides

Now that we have considered molecular chirality and its consequences (Sections 7.4 through 7.7), we return to the subject of carbohydrates by considering further details about monosaccharides, the simplest carbohydrates (Section 7.3).

Although there is no limit to the number of carbon atoms that can be present in a monosaccharide, only monosaccharides with three to seven carbon atoms are commonly found in nature. A three-carbon monosaccharide is called a *triose,* and those that contain four, five, and six carbon atoms are called *tetroses, pentoses,* and *hexoses,* respectively.

Monosaccharides are classified as *aldoses* or *ketoses* on the basis of type of carbonyl group (Section 4.1) present. An **aldose** *is a monosaccharide that contains an aldehyde functional group.* Aldoses are polyhydroxy aldehydes. A **ketose** *is a monosaccharide that contains a ketone functional group.* Ketoses are polyhydroxy ketones.

Monosaccharides are often classified by both their number of carbon atoms and their functional group. A six-carbon monosaccharide with an aldehyde functional group is an *aldohexose;* a five-carbon monosaccharide with a ketone functional group is a *ketopentose.*

Monosaccharides are also often called sugars. Hexoses are six-carbon sugars, pentoses five-carbon sugars, and so on. The word *sugar* is associated with "sweetness," and most (but not all) monosaccharides have a sweet taste. The designation *sugar* is also applied to disaccharides, many of which also have a sweet taste. Thus **sugar** *is a general designation for either a monosaccharide or a disaccharide.*

The term *saccharide* comes from the Latin word for "sugar," which is *saccharum.*

EXAMPLE 7.5

Classifying Monosaccharides on the Basis of Structural Characteristics

■ Classify each of the following monosaccharides according to both the number of carbon atoms and the type of carbonyl group present.

a.
```
     CHO
HO──┼──H
 H──┼──OH
 H──┼──OH
   CH₂OH
```

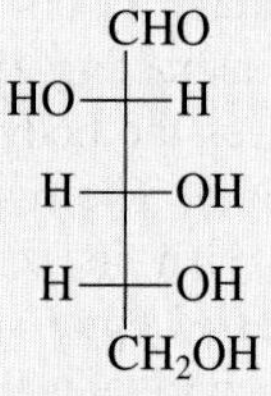

b.
```
   CH₂OH
    │
    C=O
 H──┼──OH
HO──┼──H
HO──┼──H
   CH₂OH
```

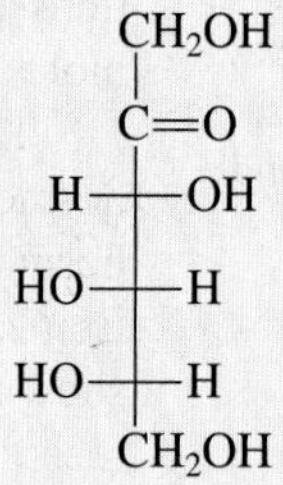

c.
```
     CHO
HO──┼──H
 H──┼──OH
HO──┼──H
HO──┼──H
   CH₂OH
```

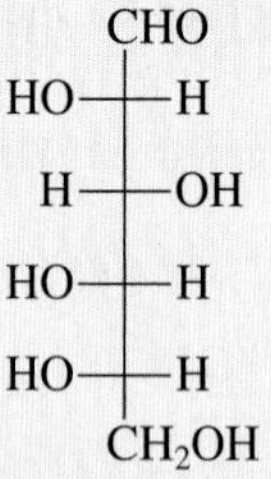

d.
```
   CH₂OH
    │
    C=O
 H──┼──OH
 H──┼──OH
   CH₂OH
```

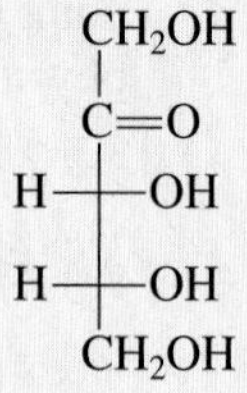

Solution

a. An aldehyde functional group is present as well as five carbon atoms. This monosaccharide is thus an *aldopentose.*

b. This monosaccharide contains a ketone group and six carbon atoms, so it is a *ketohexose.*

c. Six carbon atoms and an aldehyde group in a monosaccharide are characteristic of an *aldohexose.*

d. This monosaccharide is a *ketopentose.*

Practice Exercise 7.5

Classify each of the following monosaccharides according to both the number of carbon atoms and the type of carbonyl group present.

a.
$$
\begin{array}{c}
\text{CH}_2\text{OH} \\
|\\
\text{C}=\text{O} \\
\text{HO}-\!\!\!-\text{H} \\
\text{H}-\!\!\!-\text{OH} \\
\text{H}-\!\!\!-\text{OH} \\
\text{CH}_2\text{OH}
\end{array}
$$

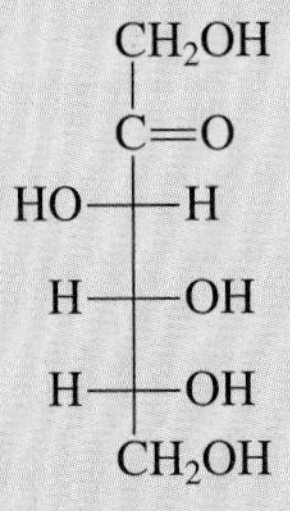

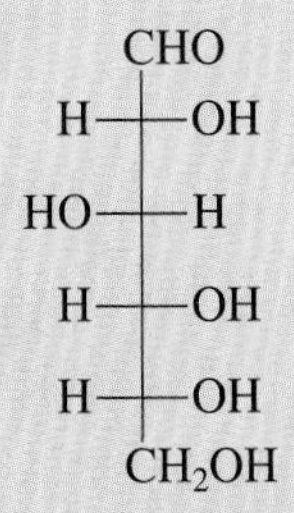

b.
$$
\begin{array}{c}
\text{CHO} \\
\text{H}-\!\!\!-\text{OH} \\
\text{HO}-\!\!\!-\text{H} \\
\text{H}-\!\!\!-\text{OH} \\
\text{H}-\!\!\!-\text{OH} \\
\text{CH}_2\text{OH}
\end{array}
$$

c.
$$
\begin{array}{c}
\text{CHO} \\
\text{H}-\!\!\!-\text{OH} \\
\text{H}-\!\!\!-\text{OH} \\
\text{CH}_2\text{OH}
\end{array}
$$

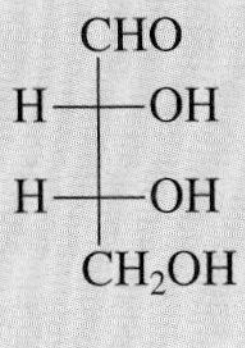

d.
$$
\begin{array}{c}
\text{CH}_2\text{OH} \\
|\\
\text{C}=\text{O} \\
\text{H}-\!\!\!-\text{OH} \\
\text{HO}-\!\!\!-\text{H} \\
\text{CH}_2\text{OH}
\end{array}
$$

Nearly all naturally occurring monosaccharides are D isomers. These D monosaccharides are important energy sources for the human body. L Monosaccharides, which can be produced in the laboratory, cannot be used by the body as energy sources. Body enzymes are specific for D isomers.

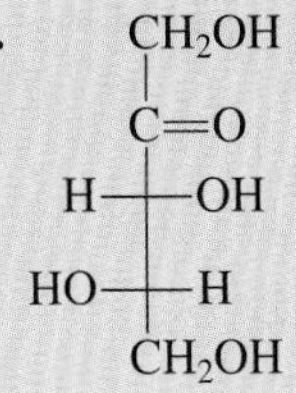

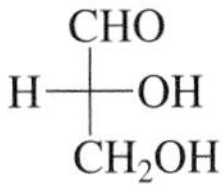

In terms of carbon atoms, trioses are the smallest monosaccharides that can exist. There are two such compounds, one an aldose (glyceraldehyde) and the other a ketose (dihydroxyacetone).

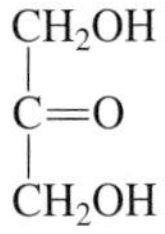

$$
\begin{array}{cc}
\text{CHO} & \text{CH}_2\text{OH} \\
\text{H}-\!\!\!-\text{OH} & \text{C}=\text{O} \\
\text{CH}_2\text{OH} & \text{CH}_2\text{OH} \\
\text{D-Glyceraldehyde} & \text{Dihydroxyacetone}
\end{array}
$$

These two triose structures serve as the reference points for consideration of the structures of aldoses and ketoses that contain more carbon atoms.

The structures of all D aldoses containing three, four, five, and six carbon atoms are given in Figure 7.13. Figure 7.13 starts with the triose glyceraldehyde at the top and proceeds downward through the tetroses, pentoses, and hexoses. The number of possible aldoses doubles each time an additional carbon atom is added because the new carbon atom is a chiral center. Glyceraldehyde has one chiral center, the tetroses two chiral centers, the pentoses three chiral centers, and the hexoses four chiral centers.

In aldose structures such as those shown in Figure 7.13, the chiral center farthest from the aldehyde group determines the D or L designation for the aldose. The configurations about the other chiral centers present are accounted for by assigning a different common name to each set of D and L enantiomers. (Only the D isomer is shown in Figure 7.13; the L isomer is the mirror image of the structure shown.)

A major difference between glyceraldehyde and dihydroxyacetone is that the latter does not possess a chiral carbon atom. Thus D and L forms are not possible for dihydroxyacetone. This reduces by half (compared with aldoses) the number of stereoisomers possible for ketotetroses, ketopentoses, and ketohexoses. An aldohexose has four chiral carbon atoms, but a ketohexose has only three. Figure 7.14 gives the projection formulas and common names for the D forms of ketoses containing three, four, five, and six carbon atoms.

All monosaccharides have names that end in -ose except the trioses glyceraldehyde and dihydroxyacetone.

You should memorize the structures of the six monosaccharides considered in this section.

7.9 Biochemically Important Monosaccharides

Of the many monosaccharides, six that are particularly important in the functioning of the human body are the trioses D-glyceraldehyde and dihydroxyacetone and the D forms of glucose, galactose, fructose, and ribose. Glucose and galactose are aldohexoses, fructose

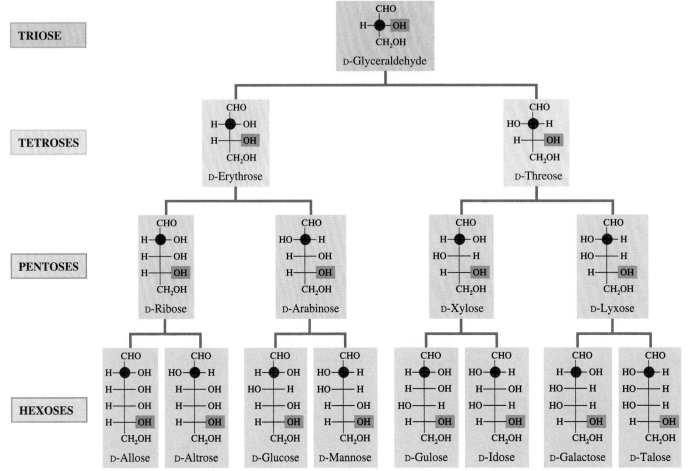

TRIOSE

TETROSES

PENTOSES

HEXOSES

FIGURE 7.13 Fischer projections and common names for D aldoses containing three, four, five, and six carbon atoms. The new chiral-center carbon atom added in going from triose to tetrose to pentose to hexose is marked in color. This new chiral center can have the hydroxyl group at the right or left in the Fischer projection, which doubles the number of stereoisomers. The hydroxyl group that specifies the D configuration is highlighted in red.

is a ketohexose, and ribose is an aldopentose. All six of these monosaccharides are water-soluble, white, crystalline solids.

■ D-Glyceraldehyde and Dihydroxyacetone

The simplest of the monosaccharides, these two trioses are important intermediates in the process of glycolysis (Section 13.2), a series of reactions whereby glucose is converted into two molecules of pyruvate. D-Glyceraldehyde is a chiral molecule but dihydroxyacetone is not.

$$
\begin{array}{cc}
\text{CHO} & \text{CH}_2\text{OH} \\
\text{H}\!-\!\!-\!\text{OH} & \text{C}\!=\!\text{O} \\
\text{CH}_2\text{OH} & \text{CH}_2\text{OH} \\
\text{D-Glyceraldehyde} & \text{Dihydroxyacetone}
\end{array}
$$

■ D-Glucose

D-Glucose tastes sweet, is nutritious, and is an important component of the human diet. L-Glucose, on the other hand, is tasteless, and the body cannot use it.

Of all monosaccharides, D-glucose is the most abundant in nature and the most important from a human nutritional standpoint. Its Fischer projection is

$$
\begin{array}{c}
\text{CHO} \\
\text{H}\!-\!\!-\!\text{OH} \\
\text{HO}\!-\!\!-\!\text{H} \\
\text{H}\!-\!\!-\!\text{OH} \\
\text{H}\!-\!\!-\!\text{OH} \\
\text{CH}_2\text{OH} \\
\text{D-Glucose}
\end{array}
$$

FIGURE 7.14 Fischer projections and common names for D ketoses containing three, four, five, and six carbon atoms. The new chiral-center carbon atom added in going from triose to tetrose to pentose to hexose is marked in color. This new chiral center can have the hydroxyl group at the right or left in the Fischer projection, which doubles the number of stereoisomers. The hydroxyl group that specifies the D configuration is highlighted in red.

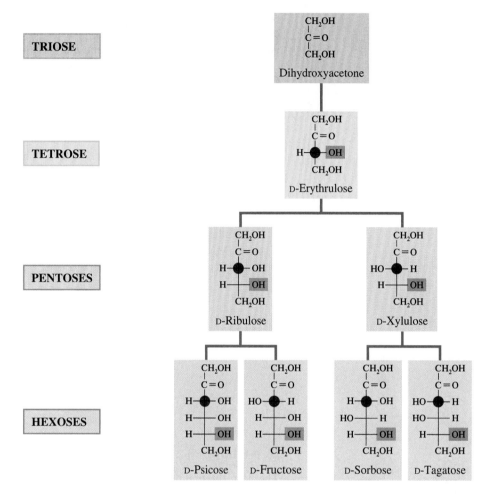

FIGURE 7.15 A 5% (m/v) glucose solution is often used in hospitals as an intravenous source of nourishment for patients who cannot take food by mouth. The body can use it as an energy source without digesting it.

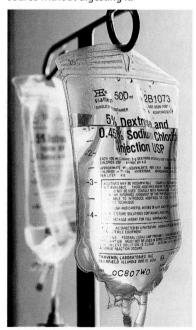

Ripe fruits, particularly ripe grapes (20%–30% glucose by mass), are a good source of glucose, which is often referred to as *grape sugar*. Two other names for D-glucose are *dextrose* and *blood sugar*. The name *dextrose* draws attention to the fact that the optically active D-glucose, in aqueous solution, rotates plane-polarized light to the right. The term *blood sugar* draws attention to the fact that blood contains dissolved glucose. The concentration of glucose in human blood is fairly constant; it is in the range of 70–100 mg per 100 mL of blood. Cells use this glucose as a primary energy source (see Figure 7.15).

▇ D-Galactose

A comparison of the Fischer projections for D-galactose and D-glucose shows that these two compounds differ only in the configuration of the —OH group and —H group on carbon 4.

D-Galactose and D-glucose are epimers (diastereomers that differ only in the configuration at one chiral center; Section 7.6).

D-Galactose is seldom encountered as a free monosaccharide. It is, however, a component of numerous important biochemical substances. In the human body, galactose is synthesized from glucose in the mammary glands for use in lactose (milk sugar), a disaccharide consisting of a glucose unit and a galactose unit (Section 7.13). D-Galactose is sometimes called *brain sugar* because it is a component of glycoproteins (protein–carbohydrate compounds; Section 7.18) found in brain and nerve tissue. D-Galactose is also present in the chemical markers that distinguish various types of blood—A, B, AB, and O (see the Chemical Connections feature on page 244).

■ D-Fructose

D-Fructose is biochemically the most important ketohexose. It is also known as *levulose* and *fruit sugar.* Aqueous solutions of naturally occurring D-fructose rotate plane-polarized light to the left; hence the name *levulose*. The sweetest-tasting of all sugars, D-fructose is found in many fruits and is present in honey in equal amounts with glucose. It is sometimes used as a dietary sugar, not because it has fewer calories per gram than other sugars but because less is needed for the same amount of sweetness.

From the third to the sixth carbon, the structure of D-fructose is identical to that of D-glucose. Differences at carbons 1 and 2 are related to the presence of a ketone group in fructose and of an aldehyde group in glucose.

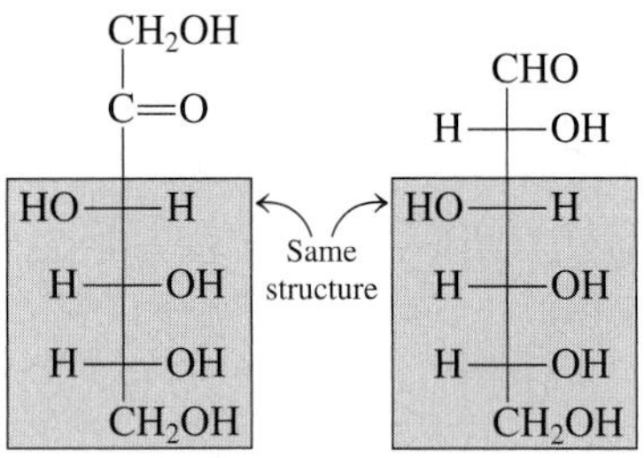

D-Fructose   D-Glucose

■ D-Ribose

The last three monosaccharides discussed in this section have all been hexoses. D-Ribose is a pentose. If carbon 3 and its accompanying —H and —OH groups were eliminated from the structure of D-glucose, the remaining structure would be that of D-ribose.

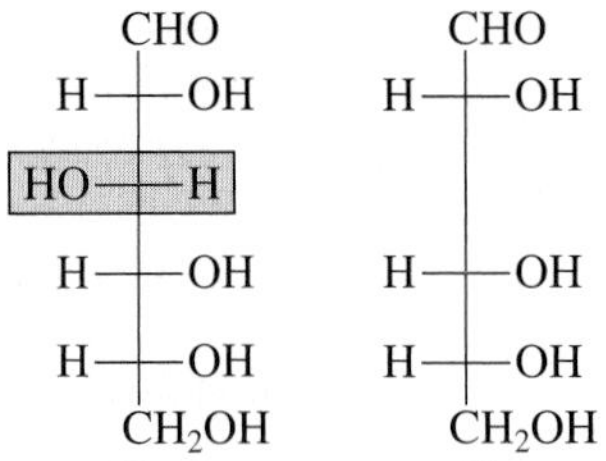

D-Glucose   D-Ribose

D-Ribose is a component of a variety of complex molecules, including ribonucleic acids (RNAs) and energy-rich compounds such as adenosine triphosphate (ATP). The compound 2-deoxy-D-ribose is also important in nucleic acid chemistry. This monosaccharide is a component of DNA molecules. The prefix *deoxy-* means "minus an oxygen"; the structures of ribose and 2-deoxyribose differ in that the latter compound lacks an oxygen atom at carbon 2.

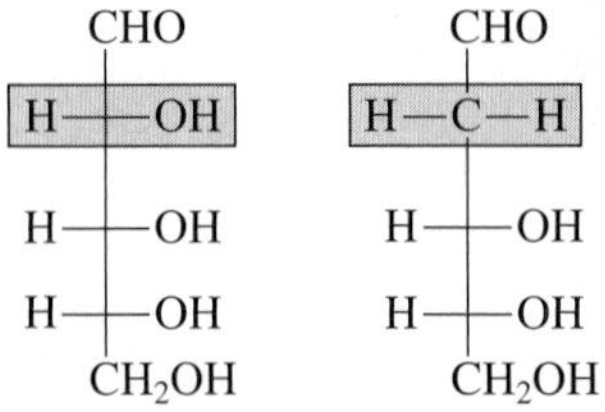

D-Ribose   2-Deoxy-D-ribose

7.10 Cyclic Forms of Monosaccharides

So far in this chapter, the structures of monosaccharides have been depicted as open-chain polyhydroxy aldehydes or ketones. However, experimental evidence indicates that for monosaccharides containing five or more carbon atoms, such open-chain structures are actually in equilibrium with two cyclic structures, and the cyclic structures are the dominant forms at equilibrium.

The cyclic forms of monosaccharides result from the ability of their carbonyl group to react intramolecularly with a hydroxyl group. The result is a cyclic hemiacetal (Section 4.10). Such an intramolecular cyclization reaction for D-glucose is shown in Figure 7.16.

In Figure 7.16, structure 2 is a rearrangement of the projection formula for D-glucose in which the carbon atoms have locations similar to those found for carbon atoms in a six-membered ring. All hydroxyl groups drawn to the right in the original Fischer projection formula appear below the ring. Those to the left in the Fischer projection formula appear above the ring.

Structure 3 in Figure 7.16 is obtained by rotating the groups attached to carbon 5 in a counterclockwise direction so that they are in the positions where it is easiest to visualize intramolecular hemiacetal formation. The intramolecular reaction occurs between the hydroxyl group on carbon 5 and the carbonyl group (carbon 1). The —OH group adds across the carbon–oxygen double bond, producing a heterocyclic ring that contains five carbon atoms and one oxygen atom.

Addition across the carbon–oxygen double bond with its accompanying ring formation produces a chiral center at carbon 1, so two stereoisomers are possible (see Figure 7.16, structures 4–6). These two forms differ in the orientation of the —OH group on the hemiacetal carbon atom (carbon 1). In α-D-glucose, the —OH group is on the opposite

Recall from Section 4.10 that hemiacetals have both an —OH group and an —OR group attached to the same carbon atom. In the cyclic hemiacetals that monosaccharides form, it is the carbonyl carbon atom that bears the —OH and —OR groups.

Cyclization of glucose (hemiacetal formation) creates a new chiral center at carbon 1, and the presence of this new chiral center produces two stereoisomers, called α and β isomers.

FIGURE 7.16 The cyclic hemiacetal forms of D-glucose result from the intramolecular reaction between the carbonyl group and the hydroxyl group on carbon 5.

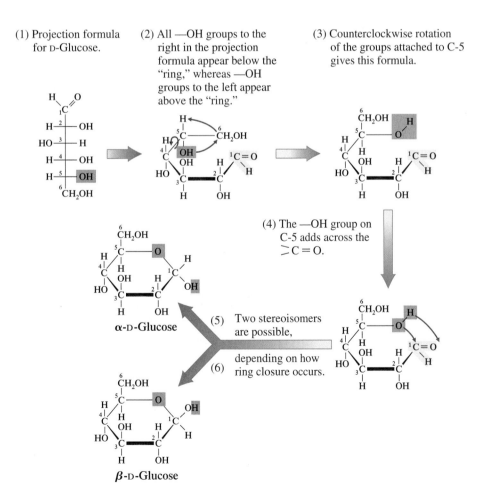

(1) Projection formula for D-Glucose.

(2) All —OH groups to the right in the projection formula appear below the "ring," whereas —OH groups to the left appear above the "ring."

(3) Counterclockwise rotation of the groups attached to C-5 gives this formula.

(4) The —OH group on C-5 adds across the ⊇C = O.

(5) Two stereoisomers are possible,

(6) depending on how ring closure occurs.

α-D-Glucose

β-D-Glucose

side of the ring from the CH_2OH group attached to carbon 5. In β-D-glucose, the CH_2OH group on carbon 5 and the —OH group on carbon 1 are on the same side of the ring.

In an aqueous solution of D-glucose, a dynamic equilibrium exists among the α, β, and open-chain forms, and there is continual interconversion among them. For example, a freshly mixed solution of pure α-D-glucose slowly converts to a mixture of both α- and β-D-glucose by an opening and a closing of the cyclic structure. When equilibrium is established, 63% of the molecules are β-D-glucose, 37% are α-D-glucose, and less than 0.01% are in the open-chain form.

α-D-Glucose $\rightleftharpoons$ Open-chain D-Glucose $\rightleftharpoons$ β-D-Glucose
(37%)   (less than 0.01%)   (63%)

Intramolecular cyclic hemiacetal formation and the equilibrium between forms associated with it are not restricted to glucose. All aldoses with five or more carbon atoms establish similar equilibria, but with different percentages of the alpha, beta, and open-chain forms. Fructose and other ketoses with a sufficient number of carbon atoms also cyclize.

Galactose, like glucose, forms a six-membered ring, but both D-fructose and D-ribose form a five-membered ring.

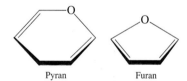

α-D-Fructose   α-D-Ribose

D-Fructose cyclization involves carbon 2 (the keto group) and carbon 5, which results in two CH_2OH groups being outside the ring (carbons 1 and 6). D-Ribose cyclization involves carbon 1 (the aldehyde group) and carbon 4.

A cyclic monosaccharide containing a six-atom ring is called a *pyranose,* and one containing a five-atom ring is called *furanose* because their ring structures resemble the ring structures in the cyclic ethers *pyran* and *furan* (Section 3.19), respectively.

Pyran   Furan

Such nomenclature leads to more specific names for the cyclic forms of monosaccharides—names that specify ring size. The more specific name for α-D-glucose is α-D-glucopyranose, and the more specific name for α-D-fructose is α-D-fructofuranose. The last part of each of these names specifies ring size.

7.11 Haworth Projection Formulas

The structural representations of the cyclic forms of monosaccharides found in the previous section are examples of Haworth projection formulas. A **Haworth projection** *is a two-dimensional structural notation that specifies the three-dimensional structure of a cyclic form of a monosaccharide.* Such projections carry the name of their originator, the British chemist Walter Norman Haworth (see Figure 7.17).

In a Haworth projection, the hemiacetal ring system is viewed "edge on" with the oxygen ring atom at the upper right (six-membered ring) or at the top (five-membered ring).

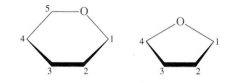

FIGURE 7.17 Walter Norman Haworth (1883–1950), the developer of Haworth projection formulas, was a British carbohydrate chemist. He helped determine the structures of the cyclic forms of glucose, was the first to synthesize vitamin C, and was a corecipient of the 1937 Nobel Prize in chemistry.

The D or L form of a monosaccharide is determined by the position of the terminal CH_2OH group on the highest-numbered ring carbon atom. In the D form, this group is positioned above the ring. In the L form, which is not usually encountered in biochemical systems, the terminal CH_2OH group is positioned below the ring.

α or β configuration is determined by the position of the —OH group on carbon 1 relative to the CH_2OH group that determines D or L series. In a β configuration, both of these groups point in the same direction; in an α configuration, the two groups point in opposite directions.

β-D-Monosaccharide α-D-Monosaccharide β-L-Monosaccharide

In situations where α or β configuration does not matter, the —OH group on carbon 1 is placed in a horizontal position, and a wavy line is used as the bond that connects it to the ring.

The specific identity of a monosaccharide is determined by the positioning of the other —OH groups in the Haworth projection. Any —OH group at a chiral center that is to the right in a Fischer projection formula points down in the Haworth projection. Any group to the left in a Fischer projection points up in the Haworth projection. The following is a matchup between the Haworth projection and a Fischer projection.

α Form ≡ β Form

Comparison of this Fischer projection with those given in Figure 7.13 reveals that the monosaccharide is D-mannose.

7.12 Reactions of Monosaccharides

Five important reactions of monosaccharides are oxidation to acidic sugars, reduction to sugar alcohols, glycoside formation, phosphate ester formation, and amino sugar formation. In considering these reactions, we will use glucose as the monosaccharide reactant. Remember, however, that other aldoses, as well as ketoses, undergo similar reactions.

Oxidation to Produce Acidic Sugars

The redox chemistry of monosaccharides is closely linked to that of the alcohol and aldehyde functional groups. This latter redox chemistry, which we considered in Chapters 3 and 4, is summarized in the following diagram.

Monosaccharide oxidation can yield three different types of *acidic sugars.* The oxidizing agent used determines the product.

Weak oxidizing agents, such as Tollens and Benedict's solutions (Section 4.9), oxidize the aldehyde end of an aldose to give an *aldonic acid.* Oxidation of the aldehyde end of glucose produces gluconic acid, and oxidation of the aldehyde end of galactose produces galactonic acid. The structures involved in the glucose reaction are

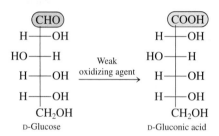

Because aldoses act as reducing agents in such reactions, they are called *reducing sugars.* With Tollens solution, glucose reduces Ag^+ ion to Ag, and with Benedict's solution, glucose reduces Cu^{2+} ion to Cu^+ ion (see Section 4.9). A **reducing sugar** *is a carbohydrate that gives a positive test with Tollens and Benedict's solutions.*

Under the basic conditions associated with Tollens and Benedict's solutions, ketoses are also reducing sugars. In this situation the ketose undergoes a structural rearrangement that produces an aldose, and the aldose then reacts. Thus all monosaccharides, both aldoses and ketoses, are reducing sugars.

Tollens and Benedict's solutions can be used to test for glucose in urine, a symptom of diabetes. For example, using Benedict's solution, we observe that if no glucose is present in the urine (a normal condition), the Benedict's solution remains blue. The presence of glucose is indicated by the formation of a red precipitate. Testing for the presence of glucose in urine is such a common laboratory procedure that much effort has been put into the development of easy-to-use test methods (Figure 7.18).

Strong oxidizing agents can oxidize both ends of a monosaccharide at the same time (the carbonyl group and the terminal primary alcohol group) to produce a dicarboxylic acid. Such polyhydroxy dicarboxylic acids are known as *aldaric acids.* For glucose, this oxidation produces glucaric acid.

FIGURE 7.18 The glucose content of urine can be determined by dipping a plastic strip treated with oxidizing agents into the urine sample and comparing the color change of the strip to a color chart that indicates glucose concentration.

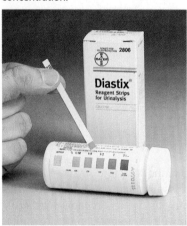

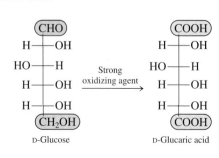

Although it is difficult to do in the laboratory, in biochemical systems enzymes can oxidize the primary alcohol end of an aldose such as glucose, without oxidation of the aldehyde group, to produce an *alduronic acid*. For glucose, such an oxidation produces D-glucuronic acid.

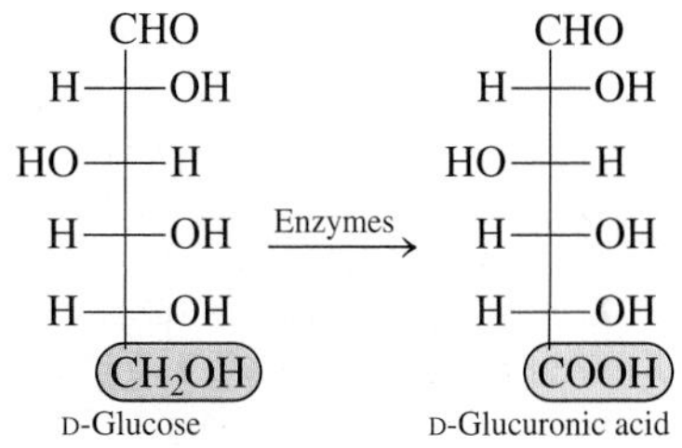

■ Reduction to Produce Sugar Alcohols

The carbonyl group present in a monosaccharide (either an aldose or a ketose) can be reduced to a hydroxyl group, using hydrogen as the reducing agent. For aldoses and ketoses, the product of the reduction is the corresponding polyhydroxy alcohol, which is sometimes called a *sugar alcohol*. For example, the reduction of D-glucose gives D-glucitol.

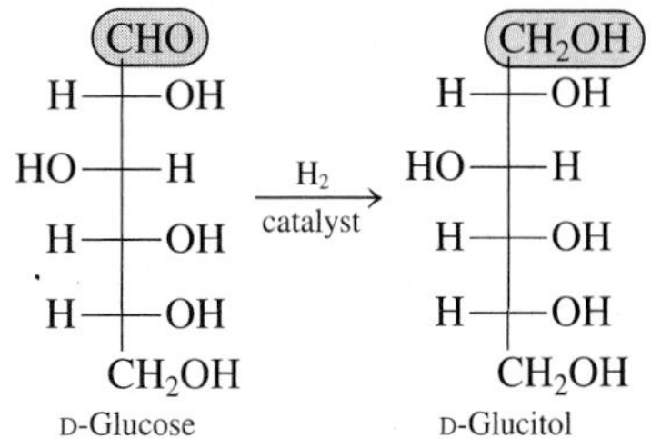

> D-Sorbitol accumulation in the eye is a major factor in the formation of cataracts due to diabetes.

D-Glucitol is also known by the common name D-sorbitol. Hexahydroxy alcohols such as D-sorbitol have properties similar to those of the trihydroxy alcohol *glycerol* (Section 3.4). These alcohols are used as moisturizing agents in foods and cosmetics because of their affinity for water. D-Sorbitol is also used as a sweetening agent in chewing gum; bacteria that cause tooth decay cannot use polyalcohols as food sources, as they can glucose and many other monosaccharides.

■ Glycoside Formation

> Remember, from Section 4.10, that acetals have two —OR groups attached to the same carbon atom.

In Section 4.10 we learned that hemiacetals can react with alcohols in acid solution to produce acetals. Because the cyclic forms of monosaccharides are hemiacetals, they react with alcohols to form acetals, as is illustrated here for the reaction of β-D-glucose with methyl alcohol.

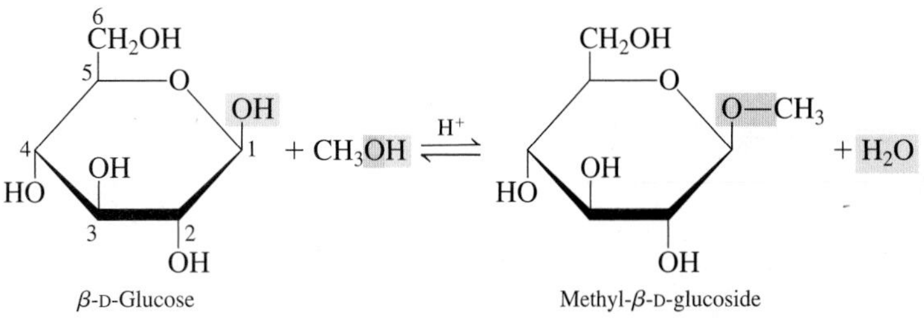

The general name for monosaccharide acetals is *glycoside*. A **glycoside** *is an acetal formed from a cyclic monosaccharide by replacement of the hemiacetal carbon —OH group with an —OR group*. More specifically, a glycoside produced from glucose is called a glucoside, that from galactose is called a galactoside, and so on. Glycosides, like

CHEMICAL CONNECTIONS   Blood Types and Monosaccharides

Human blood is classified into four types: A, B, AB, and O. If a blood transfusion is necessary and the patient's own blood is not available, the donor's blood must be matched to that of the patient. Blood of one type cannot be given to a recipient with blood of another type unless the two types are compatible. A transfusion of the wrong blood type can cause the blood cells to form clumps, a potentially fatal reaction. The following table shows compatibility relationships. People with type O blood are universal donors, and those with type AB blood are universal recipients.

Human Blood Group Compatibilities

Donor blood type	Recipient Blood Type			
	A	**B**	**AB**	**O**
A	+	−	+	−
B	−	+	+	−
AB	−	−	+	−
O	+	+	+	+

+ = compatible; − = incompatible

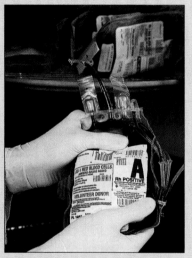

A unit of blood obtained from a blood bank.

In the United States, sampling studies show that 41% of the population has type A blood, 10% type B, 4% type AB, and 45% type O.

The biochemical basis for the various blood types involves monosaccharides. The plasma membranes of red blood cells carry biochemical markers made up of monosaccharides. Four monosaccharides are involved in the "marking system." One is the simple monosaccharide D-galactose and the other three are monosaccharide derivatives. Two of these are N-acetyl amino derivatives (Section 7.12), those of D-glucose and D-galactose. The third is L-fucose (6-deoxy-L-galactose), an L-galactose derivative in which the oxygen atom at carbon 6 has been removed (converting the —CH₂OH group to a —CH₃ group). The L configuration of this derivative is unusual in that L-monosaccharides are seldom found in the human body. The arrangement of these monosaccharides in the biochemical marker determines blood type.

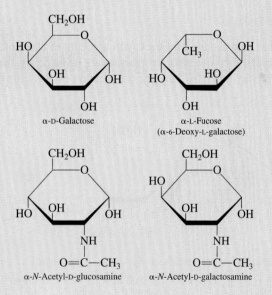

α-D-Galactose

α-L-Fucose
(α-6-Deoxy-L-galactose)

α-N-Acetyl-D-glucosamine

α-N-Acetyl-D-galactosamine

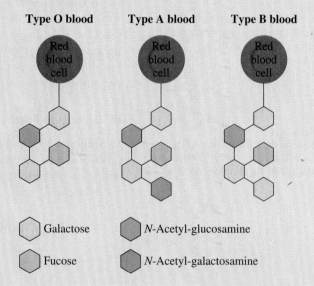

Note that all of the biochemical markers have a common structural portion that involves four monosaccharide units. Type A markers differ from type O markers in that an N-acetyl galactosamine unit is also present. In type B markers, a second galactose unit is present. Type AB blood contains both type A and type B markers.

the hemiacetals from which they are formed, can exist in both α and β forms. Glycosides are named by listing the alkyl or aryl group attached to the oxygen, followed by the name of the monosaccharide involved, with the suffix -*ide* appended to it.

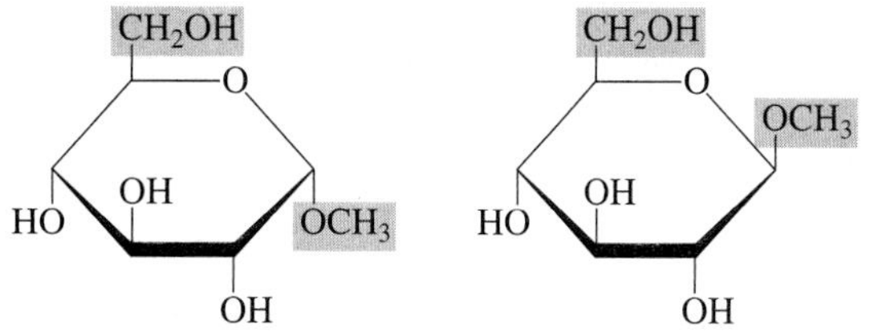

Methyl-α-D-glucoside   Methyl-β-D-glucoside

■ Phosphate Ester Formation

The hydroxyl groups of a monosaccharide can react with inorganic oxyacids to form inorganic esters (Section 5.20). Phosphate esters, formed from phosphoric acid and various monosaccharides, are commonly encountered in biochemical systems. For example, specific enzymes in the human body catalyze the esterification of the carbonyl group (carbon 1) and the primary alcohol group (carbon 6) in glucose to produce the compounds glucose 1-phosphate and glucose 6-phosphate, respectively.

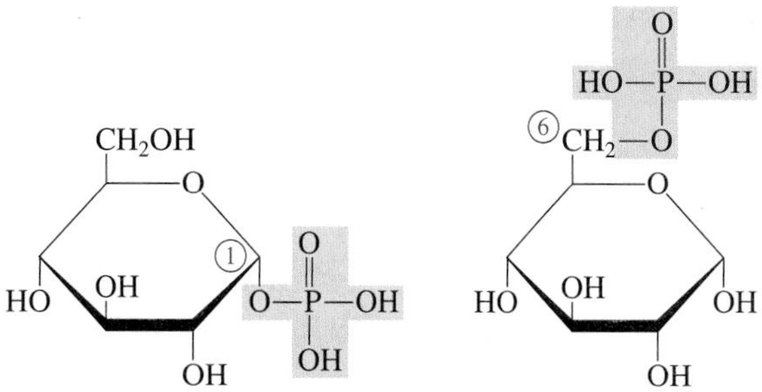

α-D-Glucose 1-phosphate α-D-Glucose 6-phosphate

These phosphate esters of glucose are stable in aqueous solution and play important roles in the metabolism of carbohydrates.

■ Amino Sugar Formation

If one of the hydroxyl groups of a monosaccharide is replaced with an amino group, an amino sugar is produced. In naturally occurring amino sugars, of which there are three common ones, the amino group replaces the carbon 2 hydroxyl group. The three common natural amino sugars are

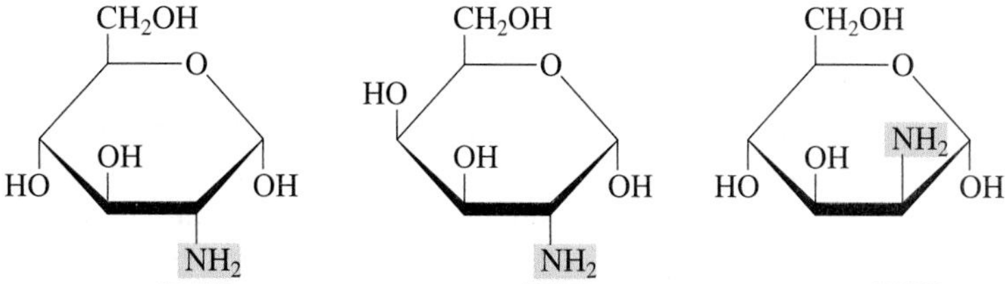

D-Glucosamine   D-Galactosamine   D-Mannosamine

Amino sugars and their *N*-acetyl derivatives are important building blocks of polysaccharides found in cartilage (Section 7.17). The *N*-acetyl derivatives of D-glucosamine and D-galactosamine are present in the biochemical markers on red blood cells, which distinguish the various blood types. (See the Chemical Connections feature on page 244.)

An *acetyl group* has the structure

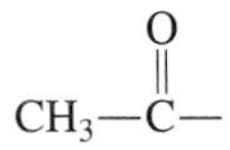

$$CH_3-C-$$

with an O double-bonded above the C.

It can be considered to be derived from acetic acid by removal of the —OH portion of that structure.

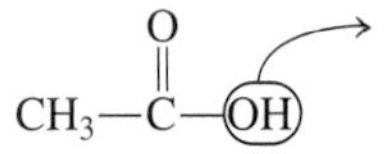

$$CH_3-C-(OH) \longrightarrow$$

Acetic acid

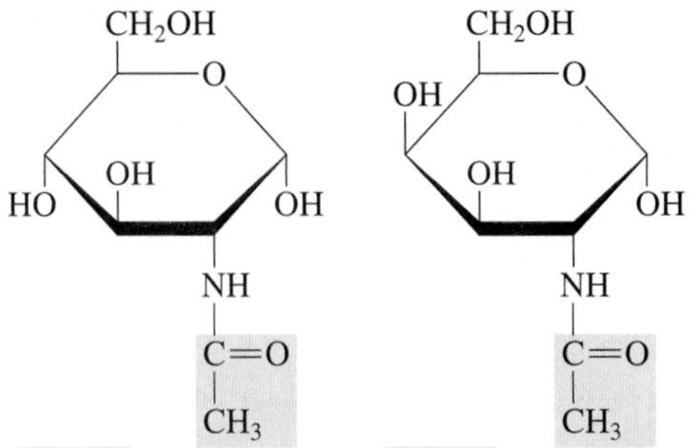

N-Acetyl-α-D-glucosamine

N-Acetyl-α-D-galactosamine

The Chemistry at a Glance feature on page 247 summarizes the "sugar terminology" associated with the common types of monosaccharides and monosaccharide derivatives that we have considered so far in this chapter.

7.13 Disaccharides

A monosaccharide that has cyclic forms (hemiacetal forms) can react with an alcohol to form a glycoside (acetal), as we noted in Section 7.12. This same type of reaction can be used to produce a *disaccharide,* a carbohydrate in which two monosaccharides are bonded together (Section 7.3). In disaccharide formation, one of the monosaccharide reactants functions as a hemiacetal, and the other functions as an alcohol.

Monosaccharide + monosaccharide $\longrightarrow$ disaccharide + H_2O

(Functioning as a hemiacetal) (Functioning as an alcohol) (Glycoside)

Glycosidic linkage

$$+ H_2O$$

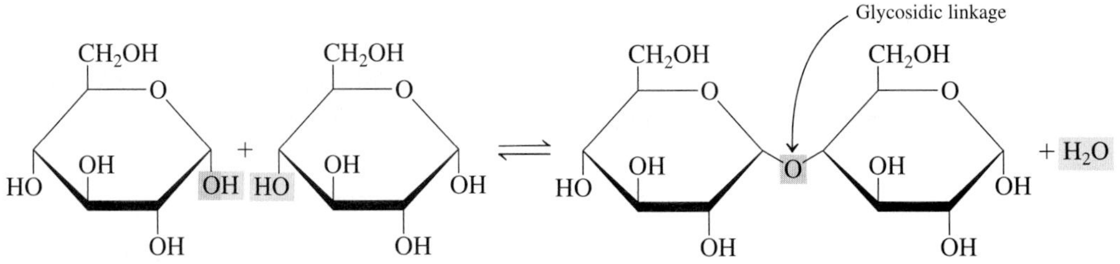

The bond that links the two monosaccharides of a disaccharide (glycoside) together is called a glycosidic linkage. A **glycosidic linkage** *is the bond in a disaccharide resulting from the reaction between the hemiacetal carbon atom —OH group of one monosaccharide and a —OH group on the other monosaccharide.* It is always a carbon–oxygen–carbon bond in a disaccharide.

We now examine the structures and properties of four important disaccharides: maltose, cellobiose, lactose, and sucrose. As we consider details of the structures of these compounds, we will find that the configuration (α or β) at carbon 1 of the reacting monosaccharides that functions as a hemiacetal is of prime importance.

■ Maltose

Maltose, often called *malt sugar,* is produced whenever the polysaccharide starch (Section 7.15) breaks down, as happens in plants when seeds germinate and in human beings during starch digestion. It is a common ingredient in baby foods and is found in malted milk. Malt (germinated barley that has been baked and ground) contains maltose; hence the name *malt sugar.*

CHEMISTRY AT A GLANCE

"Sugar Terminology" Associated with Monosaccharides and Their Derivatives

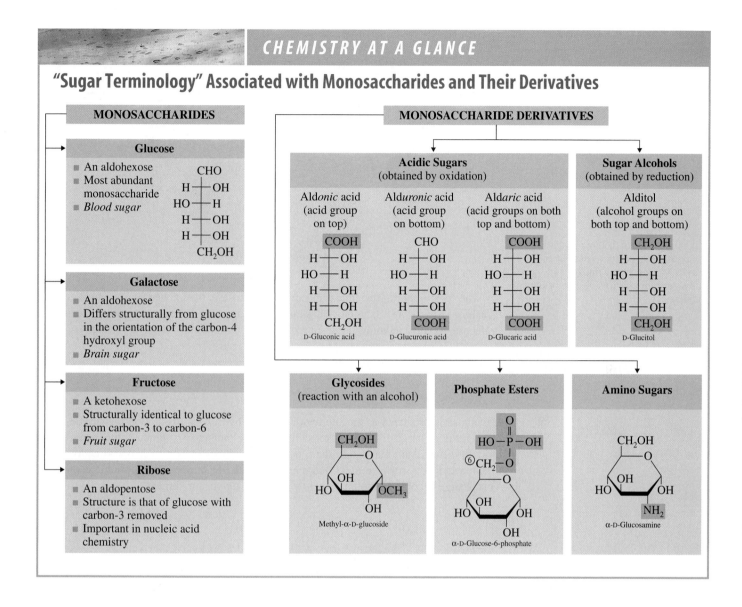

MONOSACCHARIDES

Glucose
- An aldohexose
- Most abundant monosaccharide
- *Blood sugar*

Galactose
- An aldohexose
- Differs structurally from glucose in the orientation of the carbon-4 hydroxyl group
- *Brain sugar*

Fructose
- A ketohexose
- Structurally identical to glucose from carbon-3 to carbon-6
- *Fruit sugar*

Ribose
- An aldopentose
- Structure is that of glucose with carbon-3 removed
- Important in nucleic acid chemistry

MONOSACCHARIDE DERIVATIVES

Acidic Sugars (obtained by oxidation)

Ald*onic* acid (acid group on top) — D-Gluconic acid

Ald*uronic* acid (acid group on bottom) — D-Glucuronic acid

Ald*aric* acid (acid groups on both top and bottom) — D-Glucaric acid

Sugar Alcohols (obtained by reduction)

Alditol (alcohol groups on both top and bottom) — D-Glucitol

Glycosides (reaction with an alcohol) — Methyl-α-D-glucoside

Phosphate Esters — α-D-Glucose-6-phosphate

Amino Sugars — α-D-Glucosamine

Structurally, maltose is made up of two D-glucose units, one of which must be α-D-glucose. The formation of maltose from two glucose molecules is as follows:

α-D-Glucose + D-Glucose ⟶ α(1 → 4) Linkage + H_2O

The glycosidic linkage between the two glucose units is called an $\alpha(1 \rightarrow 4)$ linkage. The two —OH groups that form the linkage are attached, respectively, to carbon 1 of the first glucose unit (in an α configuration) and to carbon 4 of the second.

Maltose is a reducing sugar (Section 7.12) because the glucose unit on the right has a hemiacetal carbon atom (C-1). Thus this glucose unit can open and close; it is in equilibrium with its open-chain aldehyde form (Section 7.10). This means there are actually three forms of the maltose molecule: α-maltose, β-maltose, and the open-chain form. Structures for these three maltose forms are shown in Figure 7.19. In the solid state, the β form is dominant.

FIGURE 7.19 The three forms of maltose present in aqueous solution.

α-Maltose

β-Maltose

Open-chain aldehyde form

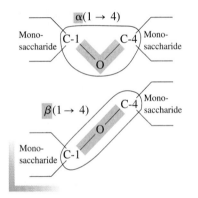

It is important to distinguish between the structural notation used for an α(1 → 4) glycosidic linkage and that used for a β(1 → 4) glycosidic linkage.

The most important chemical reaction of maltose is that of hydrolysis. Hydrolysis of D-maltose, whether in a laboratory flask or in a living organism, produces two molecules of D-glucose. Acidic conditions or the enzyme *maltase* is needed for the hydrolysis to occur.

$$\text{D-Maltose} + H_2O \xrightarrow{H^+ \text{ or maltase}} 2 \text{ D-glucose}$$

■ Cellobiose

Cellobiose is produced as an intermediate in the hydrolysis of the polysaccharide cellulose (Section 7.16). Like maltose, cellobiose contains two D-glucose monosaccharide units. It differs from maltose in that one of the D-glucose units—the one functioning as a hemiacetal—must have a β configuration instead of the α configuration for maltose. This change in configuration results in a β(1 → 4) glycosidic linkage.

β-D-Glucose   **D-Glucose**   **β(1→ 4) Linkage**

Like maltose, cellobiose is a reducing sugar, has three isomeric forms in aqueous solution, and upon hydrolysis produces two D-glucose molecules.

$$\text{D-Cellobiose} + H_2O \xrightarrow{H^+ \text{ or cellobiase}} 2 \text{ D-glucose}$$

Despite these similarities, maltose and cellobiose have different biochemical behaviors. These differences are related to the stereochemistry of their glycosidic linkages. Maltase, the enzyme that breaks the glucose–glucose α(1 → 4) linkage present in maltose, is found both in the human body and in yeast. Consequently, maltose is digested easily by humans and is readily fermented by yeast. Both the human body and yeast lack the enzyme cellobiase needed to break the glucose–glucose β(1 → 4) linkage of cellobiose. Thus cellobiose cannot be digested by humans or fermented by yeast.

■ Lactose

In both maltose and cellobiose, the monosaccharide units present are identical—two glucose units in each case. However, the two monosaccharide units in a disaccharide need

CHEMICAL CONNECTIONS — Lactose Intolerance and Galactosemia

Lactose is the principal carbohydrate in milk. Human mother's milk obtained by nursing infants contains 7%–8% lactose, almost double the 4%–5% lactose found in cow's milk.

For many people, the digestion and absorption of lactose are a problem. This problem, called *lactose intolerance,* is a condition in which people lack the enzyme *lactase,* which is needed to hydrolyze lactose to galactose and glucose.

$$\text{Lactose} + \text{H}_2\text{O} \xrightarrow{\text{Lactase}} \text{glucose} + \text{galactose}$$

Deficiency of lactase can be caused by a genetic defect, by physiological decline with age, or by injuries to the mucosa lining the intestines. When lactose molecules remain in the intestine undigested, they attract water to themselves, causing fullness, discomfort, cramping, nausea, and diarrhea. Bacterial fermentation of the lactose further along the intestinal tract produces acid (lactic acid) and gas, adding to the discomfort.

The level of the enzyme lactase in humans varies with age. Most children have sufficient lactase during the early years of their life when milk is a much-needed source of calcium in their

diet. In adulthood, the enzyme level decreases, and lactose intolerance develops. This explains the change in milk-drinking habits of many adults. Some researchers estimate that as many as one of three adult Americans exhibits a degree of lactose intolerance.

The level of the enzyme lactase in humans varies widely among ethnic groups, indicating that the trait is genetically determined (inherited). The occurrence of lactose intolerance is lowest among Scandinavians and other northern Europeans and highest among native North Americans, Southeast Asians, Africans, and Greeks. The estimated prevalence of lactose intolerance is as follows:

80% Asian Americans	60% Inuits
80% Native Americans	50% Hispanics
75% African Americans	20% Caucasians
70% Mediterranean peoples	10% Northern Europeans

After lactose has been degraded into glucose and galactose, the galactose has to be converted into glucose before it can be used by cells. In humans, the genetic condition called *galactosemia* is caused by the absence of one or more of the enzymes needed for this conversion. In people with this condition, galactose and its toxic metabolic derivative galactitol (dulcitol) accumulate in the blood.

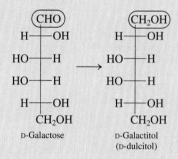

If not treated, galactosemia can cause mental retardation in infants and even death. Treatment involves exclusion of milk and milk products from the diet.

---

not be identical. *Lactose* is made up of a β-D-galactose unit and a D-glucose unit joined by a $\beta(1 \rightarrow 4)$ glycosidic linkage.

The α form of lactose is sweeter to the taste and more soluble in water than the β form. The β form can be found in ice cream that has been stored for a long time; it crystallizes and gives the ice cream a gritty texture.

The glucose hemiacetal center is unaffected when galactose bonds to glucose in the formation of lactose, so lactose is a reducing sugar (the glucose ring can open to give an aldehyde).

Lactose is the major sugar found in milk. This accounts for its common name, *milk sugar*. Enzymes in mammalian mammary glands take glucose from the bloodstream and synthesize lactose in a four-step process. Epimerization (Section 7.9) of glucose yields galactose, and then the $\beta(1 \rightarrow 4)$ linkage forms between a galactose and a glucose unit. Lactose is an important ingredient in commercially produced infant formulas that are designed to simulate mother's milk. Souring of milk is caused by the conversion of lactose to lactic acid by bacteria in the milk. Pasteurization of milk is a quick-heating process that kills most of the bacteria and retards the souring process.

Lactose can be hydrolyzed by acid or by the enzyme *lactase,* forming an equimolar mixture of galactose and glucose.

$$\text{D-Lactose} + H_2O \xrightarrow{\text{H}^+ \text{ or lactase}} \text{D-galactose} + \text{D-glucose}$$

In the human body, the galactose so produced is then converted to glucose by other enzymes. The genetic condition *lactose intolerance,* an inability of the human digestive system to hydrolyze lactose, is considered in the Chemical Connections feature on page 249.

Sucrose

Sucrose, common *table sugar,* is the most abundant of all disaccharides and occurs throughout the plant kingdom. It is produced commercially from the juice of sugar cane and sugar beets. Sugar cane contains up to 20% by mass sucrose, and sugar beets contain up to 17% by mass sucrose. Figure 7.20 shows a molecular model for sucrose.

The two monosaccharide units present in a D-sucrose molecule are α-D-glucose and β-D-fructose. The glycosidic linkage is not a $(1 \rightarrow 4)$ linkage, as was the case for maltose, cellobiose, and lactose. It is instead an $\alpha,\beta(1 \rightarrow 2)$ glycosidic linkage. The —OH group on carbon 2 of D-fructose (the hemiacetal carbon) reacts with the —OH group on carbon 1 of D-glucose (the hemiacetal carbon).

The enzyme needed to break the $\beta(1 \rightarrow 4)$ linkage in lactose is different from the one needed to break the $\beta(1 \rightarrow 4)$ linkage in cellobiose. Because the two disaccharides have slightly different structures, different enzymes are required—*lactase* for lactose and *cellobiase* for cellobiose.

The glycosidic linkage in sucrose is very different from that in maltose, cellobiose, and lactose. The linkages in the latter three compounds can be characterized as "head-to-tail" linkages—that is, the front end (carbon 1) of one monosaccharide is linked to the back end (carbon 4) of the other monosaccharide. Sucrose has a "head-to-head" glycosidic linkage; the front ends of the two monosaccharides (carbon 1 for glucose and carbon 2 for fructose) are linked.

The term *invert sugar* comes from the observation that the direction of rotation of plane-polarized light (Section 7.7) changes from positive (clockwise) to negative (counterclockwise) when sucrose is hydrolyzed to invert sugar. The rotation is +66° for sucrose. The *net* rotation for the invert sugar mixture of fructose (−92°) and glucose (+52°) is −40°.

FIGURE 7.20 Space-filling model of the disaccharide sucrose. Average per capita consumption of sucrose in the United States is approximately 100 pounds per year. Two-thirds of this is sucrose that is added to food for extra sweetening.

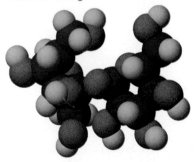

Sucrose, unlike maltose, cellobiose, and lactose, is a *nonreducing sugar.* No hemiacetal is present in the molecule, because the glycosidic linkage involves the reducing ends of both monosaccharides. Sucrose, in the solid state and in solution, exists in only one form—there are no α and β isomers, and an open-chain form is not possible.

Sucrase, the enzyme needed to break the $\alpha,\beta(1 \rightarrow 2)$ linkage in sucrose, is present in the human body. Hence sucrose is an easily digested substance. Sucrose hydrolysis (digestion) produces an equimolar mixture of glucose and fructose called *invert sugar* (see Figure 7.21).

$$\text{D-Sucrose} + H_2O \xrightarrow{\text{H}^+ \text{ or sucrase}} \underbrace{\text{D-glucose} + \text{D-fructose}}_{\text{Invert sugar}}$$

When sucrose is cooked with acid-containing foods such as fruits or berries, partial hydrolysis takes place, forming some invert sugar. Jams and jellies prepared in this manner are actually sweeter than the pure sucrose added to the original mixture because one-to-one mixtures of glucose and fructose taste sweeter than sucrose. The Chemical Connections feature on page 252 discusses alternatives to sucrose use.

7.14 General Characteristics of Polysaccharides

A **polysaccharide** *is a polymer that contains many monosaccharide units bonded to each other by glycosidic linkages.* Polysaccharides are often also called *glycans.* **Glycan** *is an alternate name for a polysaccharide.*

Important parameters that distinguish various polysaccharides (or glycans) from each other are:

1. *The identity of the monosaccharide repeating unit(s) in the polymer chain.* The more abundant polysaccharides in nature contain only one type of monosaccharide repeating unit. Such polysaccharides, including starch, glycogen, cellulose, and chitin, are examples of *homopolysaccharides.* A **homopolysaccharide** *is a polysaccharide in which only one type of monosaccharide monomer is present.* Polysaccharides whose structures contain two or more types of monosaccharide monomers, including hyaluronic acid and heparin, are called *heteropolysaccharides.* A **heteropolysaccharide** *is a polysaccharide in which more than one (usually two) type of monosaccharide monomer is present.*

2. *The length of the polymer chain.* Polysaccharide chain length can vary from less than a hundred monomer units to up to a million monomer units.

3. *The type of glycosidic linkage between monomer units.* As with disaccharides (Section 7.13), several different types of glycosidic linkages are encountered in polysaccharide structures.

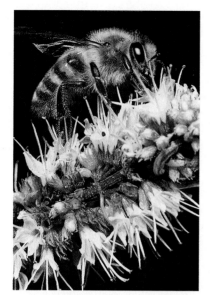

FIGURE 7.21 Honeybees and many other insects possess an enzyme called *invertase* that hydrolyzes sucrose to invert sugar. Thus honey is predominantly a mixture of D-glucose and D-fructose with some unhydrolyzed sucrose. Honey also contains flavoring agents obtained from the particular flowers whose nectars are collected. Whether a person eats monosaccharides individually, as in honey, or linked together, as in sucrose, they end up the same way in the human body: as glucose and fructose.

FIGURE 7.22 The polymer chain of a polysaccharide may be unbranched or branched. The monosaccharide monomers in the polymer chain may all be identical or two or more kinds of monomers may be present.

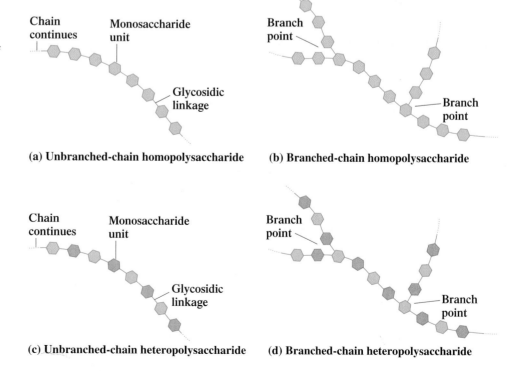

(a) **Unbranched-chain homopolysaccharide**

(b) **Branched-chain homopolysaccharide**

(c) **Unbranched-chain heteropolysaccharide**

(d) **Branched-chain heteropolysaccharide**

Artificial Sweeteners

Because of the high caloric value of sucrose, it is often difficult to satisfy a demanding "sweet tooth" with sucrose without adding pounds to the body frame or inches to the waistline. Artificial sweeteners, which provide virtually no calories, are now used extensively as a solution to the "sucrose problem."

Three artificial sweeteners that have been widely used are saccharin, sodium cyclamate, and aspartame.

Saccharin

Sodium cyclamate

Aspartame

All three of these artificial sweeteners have received much publicity because of concern about their safety.

Saccharin is the oldest of the artificial sweeteners, having been in use for over 100 years. Questions about its safety arose in 1977 from a study that suggested that large doses of saccharin caused bladder tumors in rats. As a result, the FDA proposed banning saccharin, but public support for its use caused Congress to impose a moratorium on the ban. In 1991, on the basis of many further studies, the FDA withdrew its proposal to ban saccharin.

Sodium cyclamate, approved by the FDA in 1949, dominated the artificial sweetener market for 20 years. In 1969, principally on the basis of one study suggesting that it caused cancer in laboratory animals, the FDA banned its use. Further studies have shown that neither sodium cyclamate nor its metabolites cause cancer in animals. Reapproval of sodium cyclamate has been suggested, but there has been little action by the FDA. Interestingly, Canada has approved sodium cyclamate use but banned the use of saccharin.

Aspartame (Nutra-Sweet), approved by the FDA in 1981, is used in both the United States and Canada and accounts for three-fourths of current artificial sweetener use. It tastes like sucrose but is 180 times sweeter. It provides 4 kcal/g, as does sucrose, but because so little is used, its calorie contribution is negligible. Aspartame has quickly found its way into almost every diet food on the market today.

The safety of aspartame lies with its hydrolysis products: the amino acids aspartic acid and phenylalanine. These amino acids are identical to those obtained from digestion of proteins. The only danger aspartame poses is that it contains phenylalanine, an amino acid that can lead to mental retardation among young children suffering from PKU (phenylketonuria). Labels on all products containing aspartame warn phenylketonurics of this potential danger.

Sucralose, a derivative of sucrose, is a new low-calorie entry into the artificial sweetener market. It is synthesized from sucrose by substitution of three chlorine atoms for hydroxyl groups.

An advantage of sucralose over aspartame is that it is heat-stable and can therefore be used in cooked food. Aspartame loses its sweetness when heated. Sucralose is 600 times sweeter than sucrose and has a similar taste. It is calorie-free because it cannot be hydrolyzed as it passes through the digestive tract.

Name	Type	Sweetness*
lactose	disaccharide	16
glucose	monosaccharide	74
sucrose	disaccharide	100
fructose	monosaccharide	173
sodium cyclamate	noncarbohydrate	3,000
aspartame	noncarbohydrate	15,000
saccharin	noncarbohydrate	35,000
sucralose	disaccharide derivative	60,000

*Sweetness is compared to table sugar (sucrose), which is 100 on the scale.

In nutrition discussions, monosaccharides and disaccharides are called *simple carbohydrates,* and polysaccharides are called *complex carbohydrates.*

4. *The degree of branching of the polymer chain.* The ability to form *branched-chain* structures distinguishes polysaccharides from the other two major types of biochemical polymers: proteins (Chapter 9) and nucleic acids (Chapter 11), which occur only as linear (unbranched) polymers.

Figure 7.22 illustrates important general structural considerations relative to polysaccharides.

Unlike monosaccharides and most disaccharides, polysaccharides are not sweet and do not test positive in Tollens and Benedict's solutions. They have limited water solubility because of their size. However, the —OH groups present can individually become hydrated by water molecules. The result is usually a thick colloidal suspension of the polysaccharide in water. Polysaccharides, such as flour and cornstarch, are often used as thickening agents in sauces, desserts, and gravy.

Although there are many naturally occurring polysaccharides of biochemical importance, we will focus on only six of them: starch, glycogen, cellulose, chitin, hyaluronic acid, and heparin. Starch and glycogen are examples of *storage polysaccharides,* cellulose and chitin are *structural polysaccharides,* and hyaluronic acid and heparin are *acidic polysaccharides.*

7.15 Storage Polysaccharides

A **storage polysaccharide** *is a polysaccharide that is a storage form for monosaccharides and is used as an energy source in cells.* In cells, monosaccharides are stored in the form of polysaccharides rather than as individual monosaccharides in order to lower the osmotic pressure within cells. Osmotic pressure depends on the number of *individual* molecules present. Incorporating many monosaccharide molecules into a single polysaccharide molecule results in a dramatic reduction in molecular numbers. The most important storage polysaccharides are starch (in plant cells) and glycogen (in animal and human cells).

■ Starch

Starch is a homopolysaccharide containing only glucose monosaccharide units. It is the energy-storage polysaccharide in plants. If excess glucose enters a plant cell, it is converted to starch and stored for later use. When the cell cannot get enough glucose from outside the cell, it hydrolyzes starch to release glucose.

Two different polyglucose polysaccharides can be isolated from most starches: amylose and amylopectin. *Amylose,* a straight-chain glucose polymer, usually accounts for 15%–20% of the starch; *amylopectin,* a branched glucose polymer, accounts for the remaining 80%–85% of the starch.

In amylose's non-branched structure, the glucose units are connected by $\alpha(1 \rightarrow 4)$ glycosidic linkages.

> Amylose and cellulose are both linear chains of D-glucose molecules. They are stereoisomers that differ in the configuration at carbon 1 of each D-glucose unit. In amylose, α-D-glucose is present; in cellulose, β-D-glucose.

> The glucose polymers amylose, amylopectin, and glycogen compare as follows in molecular size and degree of branching.
>
> Amylose: Up to 1000 glucose units; no branching
> Amylopectin: Up to 100,000 glucose units; branch points every 25–30 glucose units
> Glycogen: Up to 1,000,000 glucose units; branch points every 8–12 glucose units

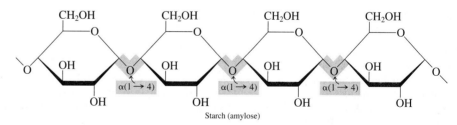

Starch (amylose)

The number of glucose units present in an amylose chain depends on the source of the starch; 300–500 monomer units are usually present.

Amylopectin, the other polysaccharide in starch, has a high degree of branching in its polyglucose structure. A branch occurs about once every 25–30 glucose units. The branch points involve $\alpha(1 \rightarrow 6)$ linkages (Figure 7.23). Because of the branching, amylopectin has a larger average molecular mass than the linear amylose. Up to 100,000 glucose units may be present in an amylopectin polymer chain.

All of the glycosidic linkages in starch (both amylose and amylopectin) are of the α type. In amylose, they are all $(1 \rightarrow 4)$; in amylopectin, both $(1 \rightarrow 4)$ and $(1 \rightarrow 6)$ linkages are present. Because both types of α linkages can be broken through hydrolysis within the human digestive tract (with the help of the enzyme *amylase*), starch has nutritional value for humans. The starches present in potatoes and cereal grains (wheat, rice, corn, etc.) account for approximately two-thirds of the world's food consumption.

FIGURE 7.23 Two perspectives on the structure of the polysaccharide amylopectin. (a) Molecular structure of amylopectin. (b) An overview of the branching that occurs in the amylopectin structure. Each dot is a glucose unit.

An $\alpha (1 \rightarrow 6)$ linkage is present in the amylopectin structure at each branch point.

(a) **(b)**

Iodine is often used to test for the presence of starch in solution. Starch-containing solutions turn a dark blue-black when iodine is added (see Figure 7.24). As starch is broken down through acid or enzymatic hydrolysis to glucose monomers, the blue-black color disappears.

Glycogen

Glycogen, like starch, is a polysaccharide containing only glucose units. It is the glucose storage polysaccharide in humans and animals. Its function is thus similar to that of starch in plants, and it is sometimes referred to as *animal starch.* Liver cells and muscle cells are the storage sites for glycogen in humans.

Glycogen has a structure similar to that of amylopectin; all glycosidic linkages are of the α type, and both $(1 \rightarrow 4)$ and $(1 \rightarrow 6)$ linkages are present. Glycogen and amylopectin differ in the number of glucose units between branches and in the total number of glucose units present in a molecule. Glycogen is about three times more highly branched than amylopectin, and it is much larger, with up to 1,000,000 glucose units present.

When excess glucose is present in the blood (normally from eating too much starch), the liver and muscle tissue convert the excess glucose to glycogen, which is then stored in these tissues. Whenever the glucose blood level drops (from exercise, fasting, or normal activities), some stored glycogen is hydrolyzed back to glucose. These two opposing processes are called *glycogenesis* and *glycogenolysis,* the formation and decomposition of glycogen, respectively.

$$\text{Glucose} \underset{\text{Glycogenolysis}}{\overset{\text{Glycogenesis}}{\rightleftharpoons}} \text{glycogen}$$

The amount of stored glycogen in the human body is relatively small. Muscle tissue is approximately 1% glycogen, liver tissue 2%–3%. However, this amount is sufficient to take care of normal-activity glucose demands for about 15 hours. During strenuous exercise, glycogen supplies can be exhausted rapidly. At this point, the body begins to oxidize fat as a source of energy.

Many marathon runners eat large quantities of starch foods the day before a race. This practice, called *carbohydrate loading,* maximizes body glycogen reserves.

FIGURE 7.24 Use of iodine to test for starch. Starch-containing solutions turn dark blue-black when iodine is added.

FIGURE 7.25 The small, dense particles within this electron micrograph of a liver cell are glycogen granules.

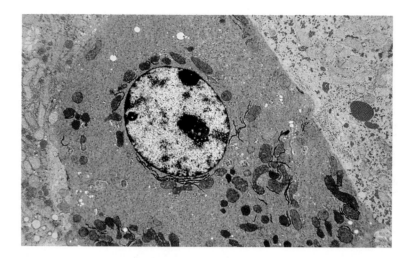

Glycogen is an ideal storage form for glucose. The large size of these macromolecules prevents them from diffusing out of cells. Also, conversion of glucose to glycogen reduces osmotic pressure. Cells would burst because of increased osmotic pressure if all of the glucose in glycogen were present in cells in free form. High concentrations of glycogen in a cell sometimes precipitate or crystallize into *glycogen granules*. These granules are discernible in photographs of cells under electron microscope magnification (Figure 7.25).

7.16 Structural Polysaccharides

A **structural polysaccharide** *is a polysaccharide that serves as a structural element in plant cell walls and animal exoskeletons.* Two of the most important structural polysaccharides are cellulose and chitin. Both are homopolysaccharides.

■ Cellulose

Cellulose, the structural component of plant cell walls, is the most abundant naturally-occurring polysaccharide. The "woody" portions of plants—stems, stalks, and trunks—have particularly high concentrations of this fibrous, water-insoluble substance.

Like amylose, cellulose is an unbranched glucose polymer. The structural difference between cellulose and amylose, which gives them completely different properties, is that the glucose residues present in cellulose have a beta-configuration whereas the glucose residues in amylose have an alpha-configuration. The glycosidic linkages in cellulose are therefore $\beta(1 \rightarrow 4)$ linkages rather than $\alpha(1 \rightarrow 4)$ linkages.

This difference in glycosidic linkage type causes cellulose and amylose to have different molecular shapes. Amylose molecules tend to have spiral-like structures whereas cellulose molecules tend to have linear structures. The linear (straight-chain) cellulose

FIGURE 7.26 A sandwich such as this is high in dietary fiber; that is, it is a cellulose-rich "meal."

The word *chitin* is pronounced "kye-ten"; it rhymes with *Titan.*

FIGURE 7.27 Chitin, a linear $\beta(1 \rightarrow 4)$ polysaccharide, produces the rigidity in the exoskeletons of crabs and other arthropods.

molecules, when aligned side by side, become water-insoluble fibers because of interchain hydrogen bonding involving the numerous hydroxy groups present.

Typically, cellulose chains contain about 5000 glucose units, which gives macromolecules with molecular masses of about 900,000 amu. Cotton is almost pure cellulose (95%), and wood is about 50% cellulose.

Even though it is a glucose polymer, cellulose is not a source of nutrition for human beings. Humans lack the enzymes capable of catalyzing the hydrolysis of $\beta(1 \rightarrow 4)$ linkages in cellulose. Even grazing animals lack the enzymes necessary for cellulose digestion. However, the intestinal tracts of animals such as horses, cows, and sheep contain bacteria that produce *cellulase,* an enzyme that can hydrolyze cellulose $\beta(1 \rightarrow 4)$ linkages and produce free glucose from cellulose. Thus grasses and other plant materials are a source of nutrition for grazing animals. The intestinal tracts of termites contain the same microorganisms, which enable termites to use wood as their source of food. Microorganisms in the soil can also metabolize cellulose, which makes possible the biodegradation of dead plants.

Despite its nondigestibility, cellulose is still an important component of a balanced diet. It serves as dietary fiber. Dietary fiber provides the digestive tract with "bulk" that helps move food through the intestinal tract and facilitates the excretion of solid wastes. Cellulose readily absorbs water, leading to softer stools and frequent bowel action. Links have been found between the length of time stools spend in the colon and possible colon problems.

High-fiber food may also play a role in weight control. Obesity is not seen in parts of the world where people eat large amounts of fiber-rich foods (see Figure 7.26). Many of the weight-loss products on the market are composed of bulk-inducing fibers such as methylcellulose.

Some dietary fibers bind lipids such as cholesterol (Section 8.9) and carry them out of the body with the feces. This lowers blood lipid concentrations and, possibly, the risk of heart and artery disease.

About 25–35 grams of dietary fiber daily is a desirable intake. This is two to three times higher than the average intake of Americans.

■ Chitin

Chitin is a polysaccharide that is similar to cellulose in both function and structure. Its function is to give rigidity to the exoskeletons of crabs, lobsters, shrimp, insects, and other arthropods (see Figure 7.27). It also occurs in the cell walls of fungi.

Structurally, chitin is a linear polymer (no branching) with all $\beta(1 \rightarrow 4)$ glycosidic linkages, as is cellulose. Chitin differs from cellulose in that the monosaccharide present is an *N*-acetyl amino derivative of D-glucose (Section 7.12). Figure 7.28 contrasts the structures of chitin and cellulose.

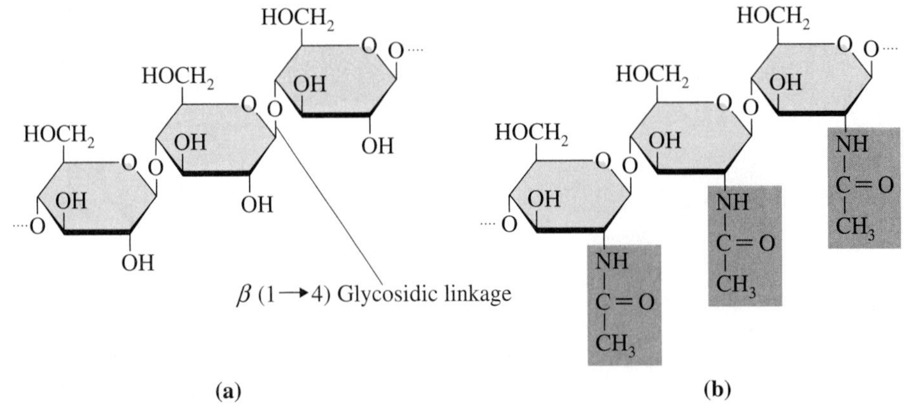

FIGURE 7.28 The structures of cellulose (a) and chitin (b). In both substances, all glycosidic linkages are of the $\beta(1 \rightarrow 4)$ type.

$\beta (1 \rightarrow 4)$ Glycosidic linkage

(a)

(b)

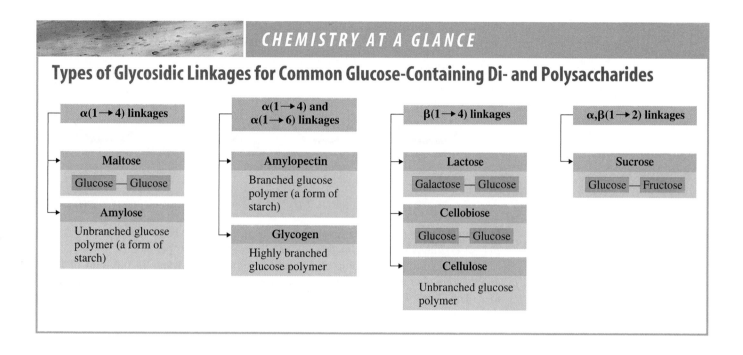

CHEMISTRY AT A GLANCE

Types of Glycosidic Linkages for Common Glucose-Containing Di- and Polysaccharides

The Chemistry at a Glance feature summarizes the types of glycosidic linkages present in commonly encountered glucose-containing di- and polysaccharides.

7.17 Acidic Polysaccharides

An **acidic polysaccharide** *is a polysaccharide with a disaccharide repeating unit in which one of the disaccharide components is an amino sugar and one or both disaccharide components has a negative charge due to a sulfate group or a carboxyl group.* Unlike the polysaccharides discussed in the previous two sections, acidic polysaccharides are *heteropolysaccharides;* two different monosaccharides are present in an alternating pattern. Acidic polysaccharides are involved in a variety of cellular functions and tissues (Figure 7.29). Two of the most well-known acidic polysaccharides are hyaluronic acid and heparin, both of which have unbranched-chain structures.

■ Hyaluronic Acid

The structure of hyaluronic acid contains alternating residues of *N*-acetyl-*β*-D-glucosamine and D-glucuronic acid.

FIGURE 7.29 Acidic polysaccharides associated with the connective tissue of joints give hurdlers such as these the flexibility needed to accomplish their task.

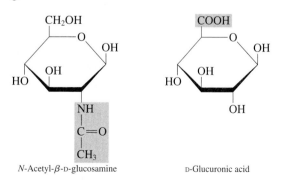

N-Acetyl-*β*-D-glucosamine   D-Glucuronic acid

Both of these monosaccharide derivatives have been encountered previously. *N*-acetyl-*β*-D-glucosamine is the repeating unit in chitin (see Section 7.16). Glucuronic acid is derived from glucose by oxidation of the —OH group at carbon 6 to an acid group (see Section 7.12). A section of the polymeric structure for hyaluronic acid is

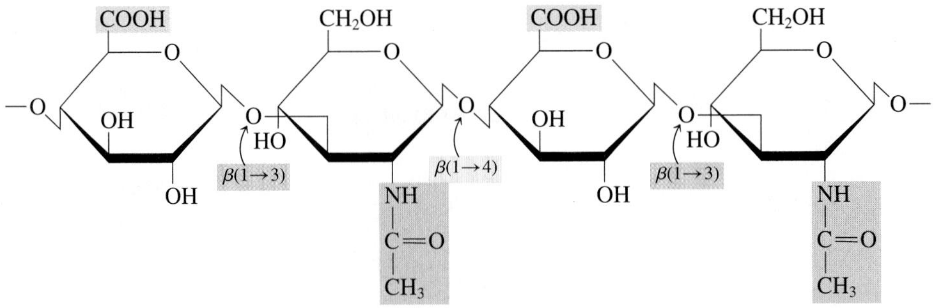

In this structure, note the alternating pattern of glycosidic bond types, $\beta(1 \rightarrow 3)$ and $\beta(1 \rightarrow 4)$. There are approximately 50,000 disaccharide units per chain.

Highly viscous hyaluronic acid solutions serve as lubricants in the fluid of joints and they are also associated with the jelly-like consistency of the vitreous humor of the eye. (The Greek word *hyalos* means "glass"; hyaluronic acid solutions have a glass-like appearance.)

■ Heparin

The best known of heparin's biochemical functions is that of an anticoagulant; it helps prevent blood clots. It binds strongly to a protein involved in terminating the process of blood clotting, thus inhibiting blood clotting.

The monosaccharides present in the disaccharide repeating unit for heparin are D-glucuronate-2-sulfate and *N*-sulfo-D-glucosamine-6-sulfate, both of which contain two negatively charged acidic groups.

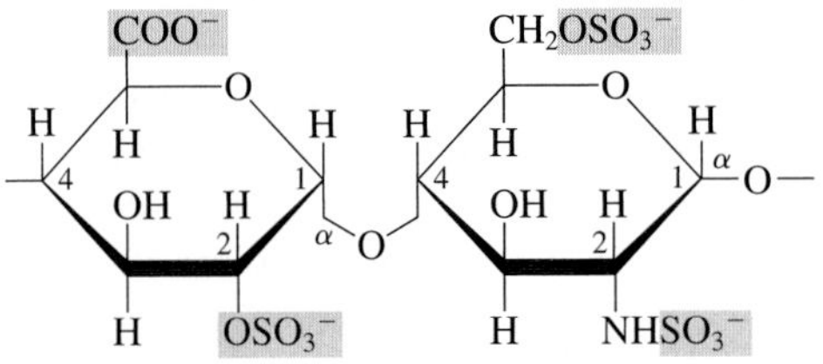

D-Glucuronate-2-sulfate

N-Sulfo-D-glucosamine-6-sulfate

Heparin is a small polysaccharide with only 15–90 disaccharide residues per chain.

7.18 Glycolipids and Glycoproteins: Cell Recognition

Prior to 1960, the biochemistry of carbohydrates was thought to be rather simple. These compounds served (1) as energy sources for plants, humans, and animals and (2) as structural materials for plants and arthropods.

Research since then has shown that oligosaccharides (Section 7.2) attached through glycosidic linkages to lipid molecules or to protein molecules have a wide variety of cellular functions including the process of cell recognition. Such molecules, called *glycolipids* and *glycoproteins*, respectively, often govern how individual cells of differing function within a biochemical system recognize each other and how cells interact with invading bacteria and viruses.

The lipid or protein part of the glycolipid or glycoprotein is incorporated into the cell membrane structure and the carbohydrate (oligosaccharide) part functions as a marker on the outer cell membrane surface. Cell recognition generally involves the interaction between the carbohydrate marker of one cell and a protein imbedded into the cell membrane of another cell.

The prefix *glyco-*, used in the terms *glycolipid* and *glycoprotein*, is derived from the Greek word *glykys*, which means "sweet." Most monosaccharides and disaccharides have a sweet taste.

"Good and Bad Carbs": The Glycemic Index

The glycemic index (GI) is a dietary carbohydrate rating system that indicates how fast a particular carbohydrate is broken down into glucose (through hydrolysis) and the level of blood glucose that results. Its focal point is thus blood glucose levels.

Slow generation of glucose, a modest rise in blood glucose, and a smooth return to normal blood glucose levels are desirable. A rapid increase (surge) in blood glucose levels, with a resulting overcorrection (from excess insulin production) that drops glucose levels below normal, is undesirable. Low-GI foods promote the first of these two effects, and high-GI foods promote the latter.

Selected examples of GI ratings for foods are given in the accompanying tables.

Considerations relative to use of GI values such as these include the following:

1. At least one low-GI food should be part of each meal.
2. Fruits, vegetables, and legumes tend to have low-GI ratings.
3. Whole-grain foods, substances high in fiber, tend to have slower digestion rates.
4. High-GI rated foods should still be consumed, but as part of meals that also contain low-GI foods.

Low-glycemic foods (less than 55)		Intermediate-glycemic foods (55 to 70)		High-glycemic foods (over 70)	
Low-fat yogurt, artificially sweetened	14	Brown rice	55	Corn chips	72
Grapefruit	25	Popcorn, sweet corn	55	Watermelon†	72
Kidney beans	27	Long-grain white rice	56	Cheerios†	74
Low-fat yogurt, sugar sweetened	33	Mini shredded wheats	58	French fries	76
Apple, pear	38	Cheese pizza	60	Rice cakes	82
Spaghetti	41	Coca-Cola*	63	Pretzels	83
Orange	44	Raisins	64	Cornflakes	84
Low-fat ice cream*	50	Table sugar*	65	Baked potato†	85
Potato chips*	54	White bread	70	Dried dates†	103

*High in a lot of "empty calories," so eat sparingly. Otherwise, they'll crowd out more nutritious foods.

†Don't avoid these healthful foods. Instead, combine them with low-glycemic choices.

In the human reproductive process, fertilization involves a binding interaction between oligosaccharide markers on the outer membrane surface of an ovulated egg and protein receptor sites on a sperm cell membrane. This binding process is followed by release of enzymes by the sperm cell which dissolve the egg cell membrane allowing for entry of the sperm and the ensuing egg–sperm fertilization process.

The oligosaccharide markers on cell surfaces that are the basis for blood types were considered in the Chemical Connection feature on page 244.

7.19 Dietary Considerations and Carbohydrates

Foods high in carbohydrate content constitute over 50% of the diet of most people of the world—rice in Asia, corn in South America, cassava (a starchy root vegetable) in parts of Africa, the potato and wheat in North America, and so on. Current nutritional recommendations support such a situation; a balanced diet should ideally be about 60% carbohydrate.

Nutritionists usually subdivide dietary carbohydrates into the categories *simple* and *complex*. A **simple carbohydrate** *is a dietary monosaccharide or disaccharide.* Simple carbohydrates are usually sweet to the taste and are commonly referred to as sugars (Section 7.8). A **complex carbohydrate** *is a dietary polysaccharide.* The main complex carbohydrates are starch and cellulose, substances not generally sweet to the taste.

Simple carbohydrates provide 20% of the energy in the U.S. diet. Half of this energy content comes from *natural sugars* and the other half from *refined sugars* added to foods. A **natural sugar** *is a sugar naturally present in whole foods.* Milk and fresh fruit are two important sources of natural sugars. A **refined sugar** *is a sugar that has been separated from its plant source.* Sugar beets and sugar cane are major sources for refined sugars.

Despite claims to the contrary, refined sugars are chemically and structurally no different from the sugars naturally present in foods. The only difference is that the refined sugar is in a pure form, whereas natural sugars are part of mixtures of substances obtained from a plant source.

Refined sugars are often said to provide *empty Calories* because they provide energy but few other nutrients. Natural sugars, on the other hand, are accompanied by nutrients. A tablespoon of sucrose (table sugar) provides 50 Calories of energy just as a small orange does. The small orange, however, also supplies vitamin C, potassium, calcium, and fiber; table sugar provides no other nutrients.

The major dietary source for complex carbohydrates in the U.S. diet is grains, a source of both starch and fiber as well as of protein, vitamins, and minerals. The pulp of a potato provides starch, and the skin provides fiber. Vegetables such as broccoli and green beans are low in starch but high in fiber.

A developing concern about dietary intake of carbohydrates involves *how fast* a given dietary carbohydrate is broken down to generate glucose within the human body. The term *glycemic effect* refers to how quickly carbohydrates are digested (broken down into glucose), how high blood glucose levels rise, and how quickly blood glucose levels return to normal. A measurement system called the *glycemic index (GI)* has been developed for rating foods in terms of their glycemic effect. The Chemical Connections feature on page 259 discusses this topic further.

CONCEPTS TO REMEMBER

Biochemistry. Biochemistry is the study of the chemical substances found in living systems and the chemical interactions of these substances with each other (Section 7.1).

Carbohydrates. Carbohydrates are polyhydroxy aldehydes, polyhydroxy ketones, or compounds that yield such substances upon hydrolysis. Plants contain large quantities of carbohydrates produced via photosynthesis (Section 7.2).

Carbohydrate classification. Carbohydrates are classified into three groups: monosaccharides, oligosaccharides, and polysaccharides (Section 7.3).

Chirality and achirality. A chiral object is not identical to its mirror image. An achiral object is identical to its mirror image (Section 7.4).

Chiral center. A chiral center is an atom in a molecule that has four different groups tetrahedrally bonded to it. Molecules that contain a single chiral center exist in a left-handed and a right-handed form (Section 7.4).

Stereoisomerism. The atoms of stereoisomers are connected in the same way but are arranged differently in space. The major causes of stereoisomerism in molecules are structural rigidity and the presence of a chiral center (Section 7.5).

Enantiomers and diastereomers. Two types of stereoisomers exist: enantiomers and diastereomers. Enantiomers have structures that are nonsuperimposable mirror images of each other. Enantiomers have identical achiral properties but different chiral properties. Diastereomers have structures that are not mirror images of each other (Section 7.5).

Fischer projections. Fischer projections are two-dimensional structural formulas used to depict the three-dimensional shapes of molecules with chiral centers (Section 7.6).

Chirality of monosaccharides. Monosaccharides are classified as D or L stereoisomers on the basis of the configuration of the chiral center farthest from the carbonyl group (Section 7.6).

Optical activity. Chiral compounds are optically active—that is, they rotate the plane of polarized light. Enantiomers rotate the plane of polarized light in opposite directions. The prefix (+) indicates that

the compound rotates the plane of polarized light in a clockwise direction, whereas compounds that rotate the plane of polarized light in a counterclockwise direction have the prefix (−) (Section 7.7).

Classification of monosaccharides. Monosaccharides are classified as aldoses or ketoses on the basis of the type of carbonyl group present. They are further classified as trioses, tetroses, pentoses, etc. on the basis of the number of carbon atoms present (Section 7.8).

Important monosaccharides. Important monosaccharides include glucose, galactose, fructose, and ribose. Glucose and galactose are aldohexoses, fructose is a ketohexose, and ribose is an aldopentose (Section 7.9).

Cyclic monosaccharides. Cyclic monosaccharides form through an intramolecular reaction between the carbonyl group and an alcohol group of an open-chain monosaccharide. These cyclic forms predominate in solution (Section 7.10).

Reactions of monosaccharides. Five important reactions of monosaccharides are (1) oxidation to an acidic sugar, (2) reduction to a sugar alcohol, (3) glycoside formation, (4) phosphate ester formation, and (5) amino sugar formation (Section 7.12).

Disaccharides. Disaccharides are glycosides formed from the linkage of two monosaccharides. The most important disaccharides are maltose, cellobiose, lactose, and sucrose. Each of these has at least one glucose unit in its structure (Section 7.13).

Polysaccharides. Polysaccharides are polymers in which monosaccharides are the monomers. In homopolysaccharides only one type of monomer is present. Two or more monosaccharide monomers are present in heteropolysaccharides. Storage polysaccharides (starch, glycogen) are storage molecules for monosaccharides. Structural polysaccharides (cellulose, chitin) serve as structural elements in plant cell walls and animal exoskeletons (Sections 7.14 to 7.17).

Glycolipids and glycoproteins. Glycolipids and glycoproteins are molecules in which oligosaccharides are attached through glycosidic linkages to lipids and proteins, respectively. Such molecules often govern how cells of differing function interact with each other (Section 7.18).

KEY REACTIONS AND EQUATIONS

1. Monosaccharide oxidation (Section 7.12)

   Aldose or ketose + weak oxidizing agent $\longrightarrow$ acidic sugar

2. Monosaccharide reduction (Section 7.12)

   Aldose or ketose + H_2 $\xrightarrow{\text{Catalyst}}$ sugar alcohol

3. Glycoside (acetal) formation (Section 7.12)

   Cyclic monosaccharide + alcohol $\longrightarrow$ glycoside (acetal) + H_2O

4. Monosaccharide ester formation (Section 7.12)

   Monosaccharide + oxyacid $\longrightarrow$ ester + H_2O

5. Hydrolysis of disaccharide (Section 7.13)

   Disaccharide + H_2O $\xrightarrow{\text{Catalyst}}$ two monosaccharides

6. Hydrolysis of maltose (Section 7.13)

   D-Maltose + H_2O $\xrightarrow{\text{H}^+ \text{ or maltase}}$ 2 D-glucose

7. Hydrolysis of cellobiose (Section 7.13)

   D-Cellobiose + H_2O $\xrightarrow{\text{H}^+ \text{ or cellobiase}}$ 2 D-glucose

8. Hydrolysis of lactose (Section 7.13)

   D-Lactose + H_2O $\xrightarrow{\text{H}^+ \text{ or lactase}}$ D-galactose + D-glucose

9. Hydrolysis of sucrose (Section 7.13)

   D-Sucrose + H_2O $\xrightarrow{\text{H}^+ \text{ or sucrase}}$ D-fructose + D-glucose

10. Complete hydrolysis of starch (Section 7.15)

    Starch + H_2O $\xrightarrow{\text{H}^+ \text{ or enzymes}}$ many D-glucose

11. Complete hydrolysis of glycogen (Section 7.15)

    Glycogen + H_2O $\xrightarrow{\text{H}^+ \text{ or enzymes}}$ many D-glucose

EXERCISES AND PROBLEMS

The members of each pair of problems in this section test similar material.

▍ Biochemical Substances (Section 7.1)

7.1 Define the term *biochemistry*.

7.2 What are the two general groups of biochemical substances?

7.3 What are the four major types of bioorganic substances?

7.4 For each of the following pairs of bioorganic substances, indicate which member of the pair is more abundant in the human body.
   a. Proteins and nucleic acids
   b. Proteins and carbohydrates
   c. Lipids and carbohydrates
   d. Lipids and nucleic acids

▍ Occurrence of Carbohydrates (Section 7.2)

7.5 Write a general chemical equation for photosynthesis.

7.6 What role does chlorophyll play in photosynthesis?

7.7 What are the two major functions of carbohydrates in the plant kingdom?

7.8 What are the six major functions of carbohydrates in the human body?

▍ Structural Characteristics of Carbohydrates (Section 7.3)

7.9 Define the term *carbohydrate*.

7.10 What functional group is present in all carbohydrates?

7.11 Explain the difference between
   a. a monosaccharide and an oligosaccharide.
   b. a disaccharide and a tetrasaccharide.

7.12 Explain the difference between
   a. an oligosaccharide and a polysaccharide.
   b. a trisaccharide and an oligosaccharide.

▍ Chirality (Section 7.4)

7.13 Explain what the term *superimposable* means.

7.14 Explain what the term *nonsuperimposable* means.

7.15 In each of the following lists of objects, identify those objects that are chiral.
   a. Nail, hammer, screwdriver, drill bit
   b. Your hand, your foot, your ear, your nose
   c. The words TOT, TOOT, POP, PEEP

7.16 In each of the following lists of objects, identify those objects that are chiral.
   a. Baseball cap, glove, shoe, scarf
   b. Pliers, scissors, spoon, fork
   c. The words MOM, DAD, AHA, WAX

7.17 Indicate whether the circled carbon atom in each of the following molecules is a chiral center.
   a. CH_3—ⒸH_2—OH
   b. CH_3—ⒸH—OH
        |
        CH_3
   c. CH_3—ⒸH—OH
        |
        Cl
   d. CH_3—CH_2—ⒸH—OH
                    |
                    CH_3

7.18 Indicate whether the circled carbon atom in each of the following molecules is a chiral center.
   a. CH_3—ⒸH_2—NH_2
   b. CH_3—ⒸH—CH_3
        |
        NH_2
   c. CH_3—ⒸH—NH_2
        |
        CH_3
   d. CH_3—ⒸH—NH_2
        |
        Cl

7.19 Use asterisks to show the chiral center(s) in the following structures.

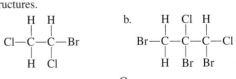

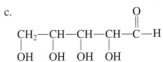

7.20 Use asterisks to show the chiral center(s) in the following structures.

a.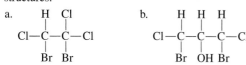

b.

Cl—C—C—C—Cl (H H H / Br OH Br)

c.

CH₃—CH—CH—CH—C—H (OH OH OH, O double bond)

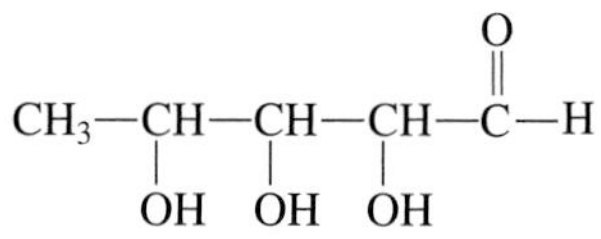

d. 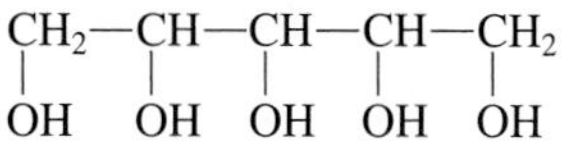 CH₂—CH—CH—CH—CH₂ (OH OH OH OH OH)

7.21 How many chiral centers are present in each of the following molecular structures?

a. 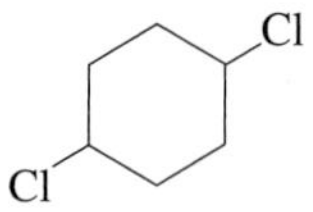 (cyclohexane with Cl and Cl)

b. 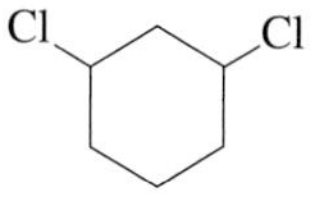 (cyclohexane with Cl, Cl)

c. 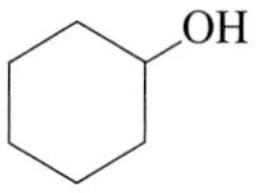 (cyclohexane with OH)

d. 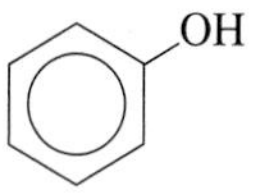 (benzene with OH)

7.22 How many chiral centers are present in each of the following molecular structures?

a. 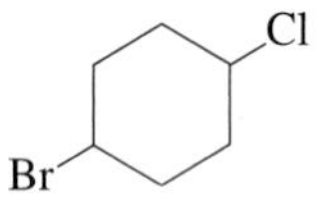 (cyclohexane with Cl and Br)

b. 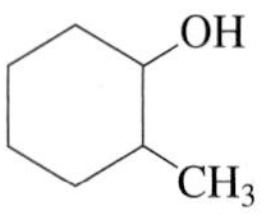 (cyclohexane with OH and CH₃)

c. 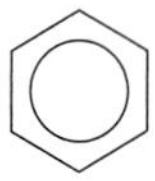 (benzene ring)

d. 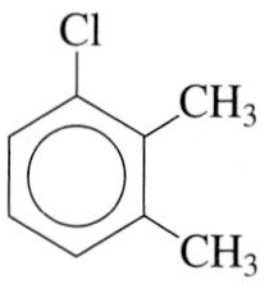 (benzene with Cl, CH₃, CH₃)

■ Stereoisomerism: Enantiomers and Diastereomers (Section 7.5)

7.23 What is the difference between constitutional isomers and stereoisomers?

7.24 Both enantiomers and diastereomers are stereoisomers. How do they differ?

■ Fischer Projections (Section 7.6)

7.25 Draw the Fischer projection for each of the following molecules.

a. 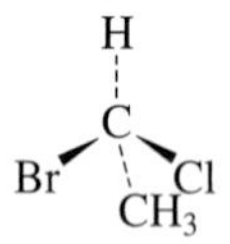 H—C with Br, Cl, CH₃

b. CH₃—C with Br, Cl, H

c. CH₃—C with Br, Cl, H

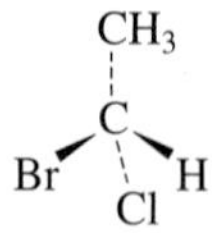

d. 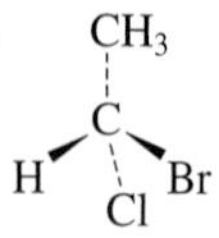 CH₃—C with H, Br, Cl

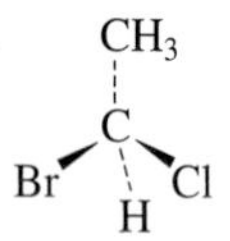

7.26 Draw the Fischer projection for each of the following molecules.

a. 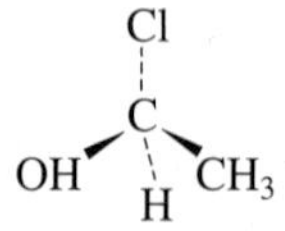 Cl—C with OH, CH₃, H

b. OH—C with Cl, CH₃, H

c. CH₃ / C with HO, Cl, H

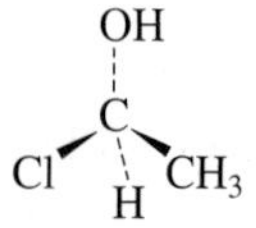

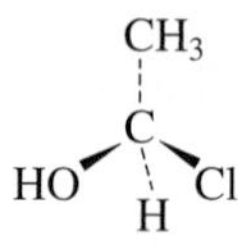

d. 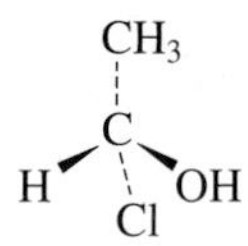 CH₃ / C with H, OH, Cl

7.27 Draw a Fischer projection for the enantiomer of each of the following monosaccharides.

a. 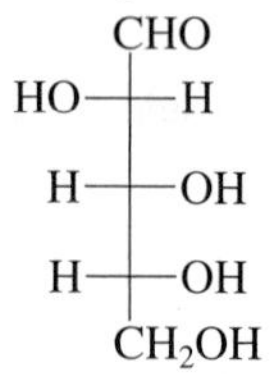 CHO
HO——H
H——OH
H——OH
CH₂OH

b. 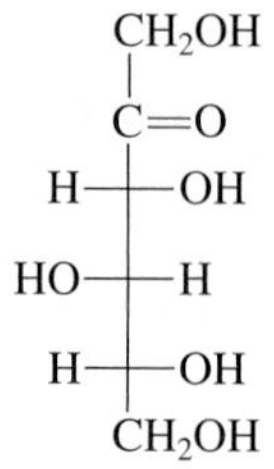 CH₂OH
C=O
H——OH
HO——H
H——OH
CH₂OH

c. CHO
H——OH
H——OH
H——OH
HO——H
CH₂OH

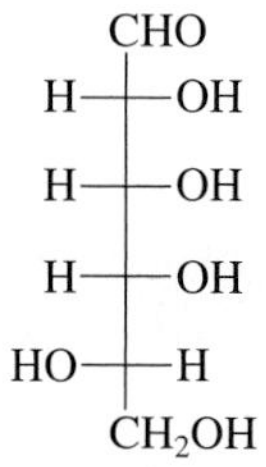

d. 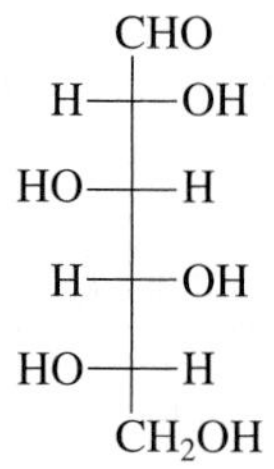 CHO
H——OH
HO——H
H——OH
HO——H
CH₂OH

7.28 Draw a Fischer projection for the enantiomer of each of the following monosaccharides.

a. 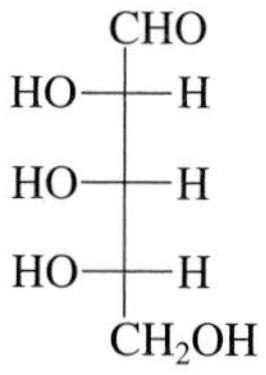 CHO
HO——H
HO——H
HO——H
CH₂OH

b. 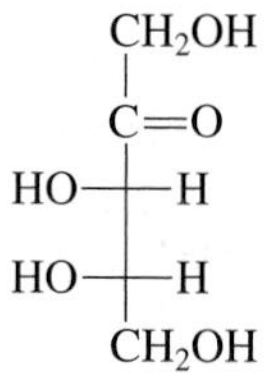 CH₂OH
C=O
HO——H
HO——H
CH₂OH

c. CHO
HO——H
HO——H
H——OH
H——OH
CH₂OH

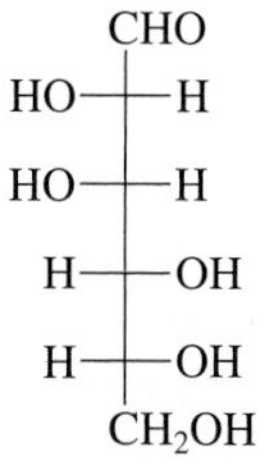

d. 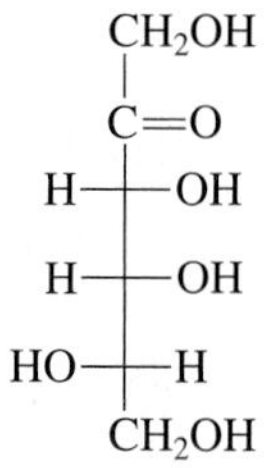 CH₂OH
C=O
H——OH
H——OH
HO——H
CH₂OH

7.29 Classify each of the molecules in Problem 7.27 as a D enantiomer or an L enantiomer.

7.30 Classify each of the molecules in Problem 7.28 as a D enantiomer or an L enantiomer.

7.31 Characterize the members of each of the following pairs of structures as (1) enantiomers, (2) diastereomers, or (3) neither enantiomers nor disastereomers.

a. 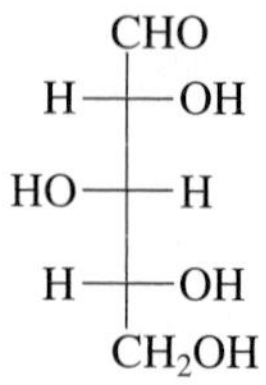 CHO
H——OH
HO——H   and
H——OH
CH₂OH

CHO
H——OH
H——OH
H——OH
CH₂OH

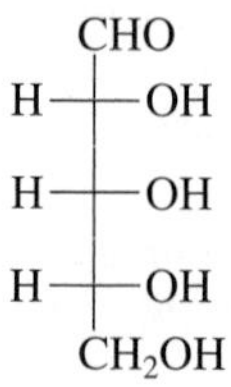

b. 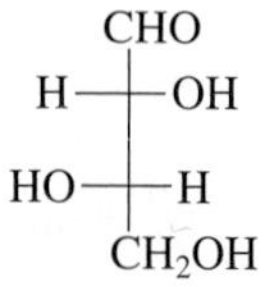 CHO
H——OH
HO——H   and
CH₂OH

CHO
H——H
HO——H
CH₂OH

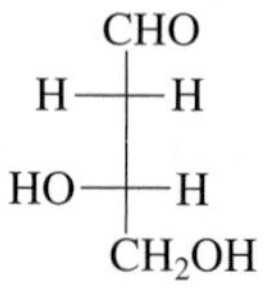

c.

```
      CHO                CHO
  H——OH            HO——H
 HO——H       and    H——OH
  H——OH            HO——H
 HO——H              H——OH
     CH₂OH              CH₂OH
```

d.

```
     CH₂OH             CH₂OH
      C=O               C=O
 HO——H        and  HO——H
 HO——H              H——OH
     CH₂OH             CH₂OH
```

7.32 Characterize the members of each of the following pairs of structures as (1) enantiomers, (2) diastereomers, or (3) neither enantiomers nor disastereomers.

a.

```
      CHO               CHO
  H——OH           HO——H
  H——OH     and  HO——H
 HO——H            H——OH
     CH₂OH            CH₂OH
```

b.

```
      CHO               CHO
  H——OH            H——OH
  H——OH     and  HO——H
     CH₂OH            CH₂OH
```

c.

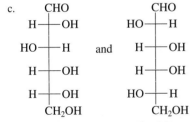

d.

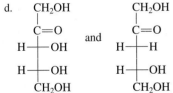

■ Properties of Enantiomers (Section 7.7)

7.33 D-glucose and L-glucose would be expected to show differences in which of the following properties?
  a. Solubility in an achiral solvent
  b. Density
  c. Melting point
  d. Effect on plane-polarized light

7.34 D-glucose and L-glucose would be expected to show differences in which of the following properties?
  a. Solubility in a chiral solvent
  b. Freezing point
  c. Reaction with ethanol
  d. Reaction with (+)-lactic acid

7.35 Compare (+)-lactic acid and (−)-lactic acid with respect to each of the following properties.
  a. Boiling point
  b. Optical activity
  c. Solubility in water
  d. Reaction with (+)-2,3-butanediol

7.36 Compare (+)-glyceraldehyde and (−)-glyceraldehyde with respect to each of the following properties.
  a. Freezing point
  b. Rotation of plane-polarized light
  c. Reaction with ethanol
  d. Reaction with (−)-2,3-butanediol

■ Classification of Monosaccharides (Section 7.8)

7.37 Classify each of the following monosaccharides as an aldose or a ketose.

a.
```
      CHO
  H—C—OH
 HO—C—H
 HO—C—H
  H—C—OH
     CH₂OH
```
b.
```
     CH₂OH
      C=O
  H—C—OH
  H—C—OH
  H—C—OH
     CH₂OH
```
c.
```
   CH₂OH
    C=O
   CH₂OH
```
d.
```
     CH₂OH
      C=O
  HO—C—H
     CH₂OH
```

7.38 Classify each of the following monosaccharides as an aldose or a ketose.

a.
```
      CHO
 HO—C—H
 HO—C—H
 HO—C—H
  H—C—OH
     CH₂OH
```
b.
```
     CH₂OH
      C=O
 HO—C—H
 HO—C—H
     CH₂OH
```
c.
```
      CHO
 HO—C—H
  H—C—OH
     CH₂OH
```
d.
```
     CH₂OH
      C=O
 HO—C—H
 HO—C—H
  H—C—OH
     CH₂OH
```

7.39 Classify each monosaccharide in Problem 7.37 by its number of carbon atoms and its type of carbonyl group.

7.40 Classify each monosaccharide in Problem 7.38 by its number of carbon atoms and its type of carbonyl group.

7.41 Using the information in Figures 7.13 and 7.14, assign a name to each of the monosaccharides in Problem 7.37.

7.42 Using the information in Figures 7.13 and 7.14, assign a name to each of the monosaccharides in Problem 7.38.

■ **Biologically Important Monosaccharides (Section 7.9)**

7.43 Indicate at what carbon atom(s) the structures of each of the following pairs of monosaccharides differ.
a. D-Glucose and D-galactose
b. D-Glucose and D-fructose
c. D-Glyceraldehyde and dihydroxyacetone
d. D-Ribose and 2-deoxy-D-ribose

7.44 Indicate whether the members of each of the following pairs of monosaccharides have the same molecular formula.
a. D-Glucose and D-galactose
b. D-Glucose and D-fructose
c. D-Glyceraldehyde and dihydroxyacetone
d. D-Ribose and 2-deoxy-D-ribose

7.45 Indicate which of the terms *aldoses, ketoses, hexoses,* and *aldohexoses* apply to both members of each of the following pairs of monosaccharides. More than one term may apply in a given situation.
a. D-Glucose and D-galactose
b. D-Glucose and D-fructose
c. D-Galactose and D-fructose
d. D-Glyceraldehyde and D-ribose

7.46 Indicate which of the terms *aldoses, ketoses, trioses,* and *aldohexoses* apply to both members of each of the following pairs of monosaccharides. More than one term may apply in a given situation.
a. D-Glucose and D-ribose
b. D-Fructose and dihydroxyacetone
c. D-Glyceraldehyde and dihydroxyacetone
d. D-Galactose and D-ribose

7.47 Draw the Fischer projection for each of the following monosaccharides.
a. D-Glucose   b. D-Glyceraldehyde
c. D-Fructose   d. L-Galactose

7.48 Draw the Fischer projection for each of the following monosaccharides.
a. D-Galactose   b. D-Ribose
c. Dihydroxyacetone   d. L-Glucose

7.49 To which of the common monosaccharides does each of the following terms apply?
a. Levulose   b. Grape sugar   c. Brain sugar

7.50 To which of the common monosaccharides does each of the following terms apply?
a. Dextrose   b. Fruit sugar   c. Blood sugar

■ **Cyclic Forms of Monosaccharides (Section 7.10)**

7.51 The intermolecular reaction that produces the cyclic forms of monosaccharides involves functional groups on which two carbon atoms in the case of each of the following?
a. D-Glucose   b. D-Galactose
c. D-Fructose   d. D-Ribose

7.52 How many carbon atoms and how many oxygen atoms are present in the ring portion of the cyclic forms of each of the following monosaccharides?
a. D-Glucose   b. D-Galactose
c. D-Fructose   d. D-Ribose

7.53 What is the structural difference between the alpha and beta forms of D-glucose?

7.54 What is the structural difference between the alpha forms of D-glucose and D-galactose?

7.55 Fructose contains six carbon atoms, and ribose has only five carbon atoms. Why do both of these monosaccharides have cyclic forms that involve a five-membered ring?

7.56 Fructose and glucose both contain six carbon atoms. Why do the cyclic forms of fructose have a five-membered ring instead of the six-membered ring found in the cyclic forms of glucose?

7.57 The structure of glucose is sometimes written in an open-chain form and sometimes as a cyclic hemiacetal structure. Explain why either form is acceptable.

7.58 When pure α-D-glucose is dissolved in water, β-D-glucose and α-D-glucose are both soon present. Explain how this is possible.

■ **Haworth Projection Formulas (Section 7.11)**

7.59 Identify each of the following structures as an α-D-monosaccharide or a β-D-monosaccharide.

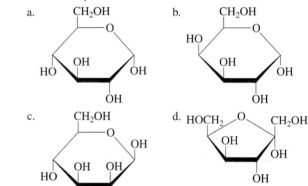

7.60 Identify each of the following structures as an α-D-monosaccharide or a β-D-monosaccharide.

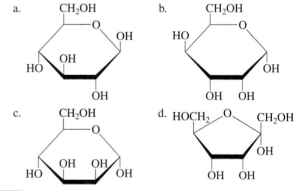

7.61 Identify whether each of the structures in Problem 7.59 is that of a hemiacetal.

7.62 Identify whether each of the structures in Problem 7.60 is that of a hemiacetal.

7.63 Draw the open-chain form for each of the monosaccharides in Problem 7.59.

7.64 Draw the open-chain form for each of the monosaccharides in Problem 7.60.

7.65 Using the information in Figures 7.13 and 7.14, assign a name to each of the monosaccharides in Problem 7.59.

7.66 Using the information in Figures 7.13 and 7.14, assign a name to each of the monosaccharides in Problem 7.60.

7.67   Draw the Haworth projection for each of the following monosaccharides.
a. α-D-Galactose   b. β-D-Galactose
c. α-L-Galactose   d. β-L-Galactose

7.68   Draw the Haworth projection for each of the following monosaccharides.
a. α-D-Mannose   b. β-D-Mannose
c. α-L-Mannose   d. β-L-Mannose

■ **Reactions of Monosaccharides (Section 7.12)**

7.69   Which of the following monosaccharides is a *reducing
sugar*?
a. D-Glucose   b. D-Galactose
c. D-Fructose   d. D-Ribose

7.70   Which of the following monosaccharides will give a positive
test with Benedict's solution?
a. D-Glucose   b. D-Galactose
c. D-Fructose   d. D-Ribose

7.71   In terms of oxidation and reduction, explain what occurs to
both D-glucose and Tollens solution when they react with
each other.

7.72   Describe the chemical reaction used to detect glucose in urine
that involves Benedict's solution.

7.73   Draw structures for the following compounds.
a. Galactonic acid
b. Galactaric acid
c. Galacturonic acid
d. Galactitol

7.74   Draw structures for the following compounds.
a. Mannonic acid
b. Mannaric acid
c. Mannuronic acid
d. Mannitol

7.75   Indicate whether each of the following structures is that of
a glycoside.

a. 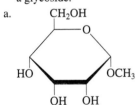   b.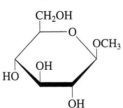

c. 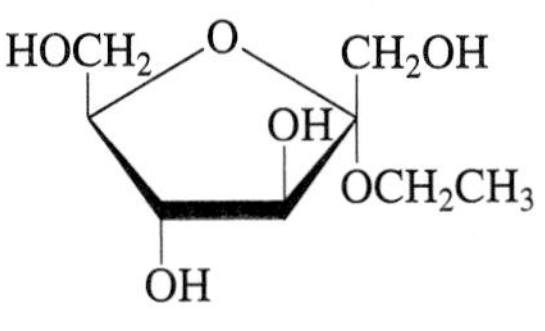 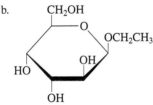   d.

7.76   Indicate whether each of the following structures is that of a
glycoside.
a.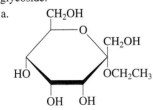

b.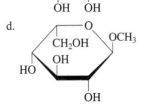

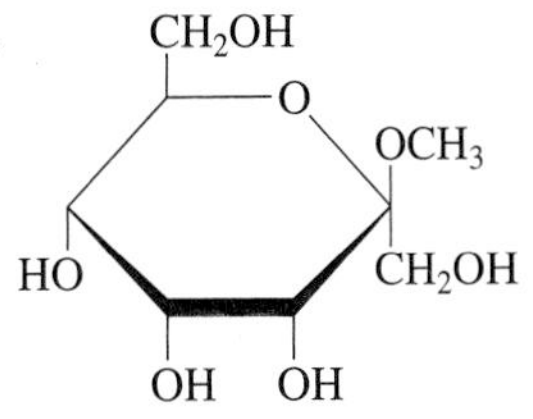

c.

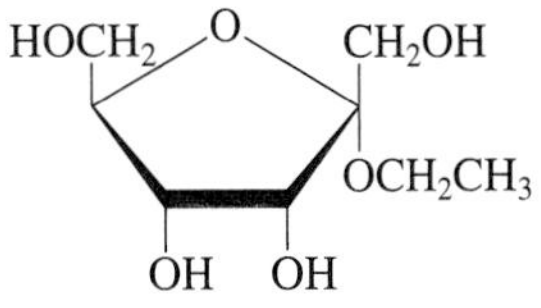

d.

7.77   For each structure in Problem 7.75, identify the configuration
at the acetal carbon atom as α or β.

7.78   For each structure in Problem 7.76, identify the configuration
at the acetal carbon atom as α or β.

7.79   Identify the alcohol needed to produce each of the compounds
in Problem 7.75 by reaction of the alcohol with the appropriate monosaccharide.

7.80   Identify the alcohol needed to produce each of the compounds
in Problem 7.76 by reaction of the alcohol with the appropriate monosaccharide.

7.81   What is the difference in meaning between the terms *glycoside*
and *glucoside*?

7.82   What is the difference in meaning between the terms *glycoside*
and *galactoside*?

7.83   Draw structures for the following compounds.
a. Ethyl-β-D-glucoside
b. Methyl-α-D-galactoside

7.84   Draw structures for the following compounds.
a. Ethyl-α-D-galactoside
b. Methyl-β-D-glucoside

7.85   Draw structures for the following compounds.
a. α-D-Galactose 6-phosphate
b. *N*-Acetyl-α-D-galactosamine

7.86   Draw structures for the following compounds.
a. α-D-Mannose 6-phosphate
b. *N*-Acetyl-α-D-mannosamine

■ **Disaccharides (Section 7.13)**

7.87   What monosaccharides are produced from the hydrolysis of
the following disaccharides?
a. Sucrose   b. Maltose
c. Lactose   d. Cellobiose

7.88   What type of glycosidic linkage [α(1 → 4), etc.] is present
in each of the following disaccharides?
a. Sucrose   b. Maltose
c. Lactose   d. Cellobiose

7.89   Explain why lactose is a reducing sugar.

7.90   Explain why sucrose is not a reducing sugar.

7.91 Indicate whether each of the following disaccharides gives a positive or a negative Benedict's test.
  a. Sucrose
  b. Maltose
  c. Lactose
  d. Cellobiose

7.92 Indicate whether each of the following disaccharides gives a positive or a negative Tollens test.
  a. Maltose
  b. Lactose
  c. Cellobiose
  d. Sucrose

7.93 What type of glycosidic linkage [$\alpha(1 \rightarrow 4)$, etc.] is present in each of the following disaccharides?
  a.
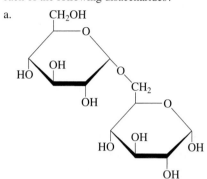
  b.
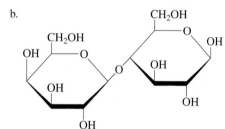
  c.
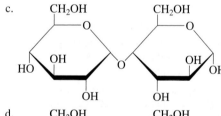
  d.
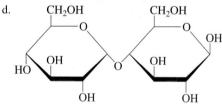

7.94 What type of glycosidic linkage [$\alpha(1 \rightarrow 4)$, etc.] is present in each of the following disaccharides?
  a.
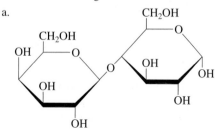

b.
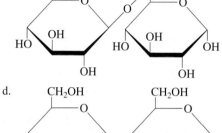

c.
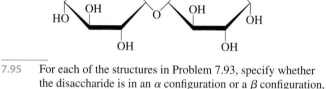

d.

7.95 For each of the structures in Problem 7.93, specify whether the disaccharide is in an α configuration or a β configuration, or neither.

7.96 For each of the structures in Problem 7.94, specify whether the disaccharide is in an α configuration or a β configuration, or neither.

7.97 Identify each of the structures in Problem 7.93 as a reducing sugar or a nonreducing sugar.

7.98 Identify each of the structures in Problem 7.94 as a reducing sugar or a nonreducing sugar.

7.99 Using the information in Figures 7.13 and 7.14, assign a name to each monosaccharide present in each of the structures in Problem 7.93.

7.100 Using the information in Figures 7.13 and 7.14, assign a name to each monosaccharide present in each of the structures in Problem 7.94.

■ Polysaccharides (Sections 7.14 through 7.17)

7.101 Indicate whether or not both members of the following pairs of polysaccharides are *homopolysaccharides*.
  a. Glycogen and starch
  b. Amylose and amylopectin
  c. Cellulose and chitin
  d. Heparin and hyaluronic acid

7.102 Indicate whether or not both members of the following pairs of polysaccharides are *heteropolysaccharides*.
  a. Glycogen and cellulose
  b. Starch and chitin
  c. Amylose and heparin
  d. Amylopectin and hyaluronic acid

7.103 Indicate whether or not both members of each pair of polysaccharides in Problem 7.101 are *storage polysaccharides*.

7.104 Indicate whether or not both members of each pair of polysaccharides in Problem 7.102 are *structural polysaccharides*.

7.105 Indicate whether or not both members of each pair of polysaccharides in Problem 7.101 contain *branched-chain polymers*.

7.106 Indicate whether or not both members of each pair of polysaccharides in Problem 7.102 contain *unbranched-chain polymers*.

7.107 Describe the structural differences and similarities between the following pairs of polysaccharides.
   a. Glycogen and amylopectin   b. Amylose and cellulose

7.108 Describe the structural differences and similarities between the following pairs of polysaccharides.
   a. Amylose and glycogen   b. Amylose and amylopectin

7.109 Match each of the following structural characteristics to the polysaccharides amylopectin, amylose, glycogen, cellulose, and chitin. A specific characteristic may apply to more than one polysaccharide.
   a. Contains both $\alpha(1 \rightarrow 4)$ and $\alpha(1 \rightarrow 6)$ glycosidic linkages
   b. Composed of glucose monosaccharide units
   c. Composed of unbranched molecular chains
   d. Contains only $\beta(1 \rightarrow 4)$ glycosidic linkages

7.110 Match each of the following structural characteristics to the polysaccharides amylopectin, amylose, glycogen, cellulose, and chitin. A specific characteristic may apply to more than one polysaccharide.
   a. Contains acetal linkages between monosaccharide units
   b. Contains only $\alpha(1 \rightarrow 4)$ glycosidic linkages
   c. Monosaccharide units are derivatives of glucose
   d. Composed of highly branched molecular chains

7.111 Why, when both contain D-glucose, can humans digest starch but not cellulose?

7.112 What is the difference between plant starch and animal starch?

■ Dietary Considerations and Carbohydrates (Section 7.19)

7.113 In a dietary context, what is the difference between a *simple* carbohydrate and a *complex* carbohydrate?

7.114 In a dietary context, what is the difference between a *natural* sugar and a *refined* sugar?

7.115 In a dietary context, what are *empty* Calories?

7.116 In a dietary context, what is the *glycemic effect?*

ADDITIONAL PROBLEMS

7.117 Indicate whether each of the following compounds is chiral or achiral.
   a. 1-Chloro-2-methylpentane   b. 2-Chloro-2-methylpentane
   c. 2-Chloro-3-methylpentane   d. 3-Chloro-2-methylpentane

7.118 Indicate whether each of the following compounds is optically active or optically inactive.

a.
```
        H
        |
  Cl————Cl
        |
        Cl
```

b.
```
        H
        |
  Cl————Cl
        |
        H
```

c.
```
       COOH
        |
  H————OH
        |
       COOH
```

d.
```
       CH3
        |
   F————Cl
        |
        H
```

7.119 In which of the following pairs of monosaccharides do both members of the pair contain the same number of carbon atoms?
   a. Glyceraldehyde and glucose
   b. Dihydroxyketone and ribose
   c. Ribose and deoxyribose
   d. Glyceraldehyde and dihydroxyketone

7.120 Draw Fischer projections for the four stereoisomers of the molecule

```
                    O
                    ‖
  CH2—CH—CH—C—CH2
   |    |   |        |
   OH  OH  OH      OH
```

7.121 What is the alkane of lowest molecular mass that is a chiral compound?

7.122 What monosaccharide(s) is (are) obtained from the hydrolysis of each of the following?
   a. Sucrose   b. Glycogen   c. Starch   d. Amylose

7.123 Classify each of the following carbohydrates as a glucose polymer or a glucose-derivative polymer.

   a. Chitin   b. Amylopectin
   c. Hyaluronic acid   d. Glycogen

7.124 List the reactant(s) necessary to effect the following chemical changes.

a.
```
      CHO                      COOH
       |                        |
  H—C—OH          ?        H—C—OH
       |          ⟶              |
  H—C—OH                   H—C—OH
       |                        |
     CH2OH                    COOH
```

b.
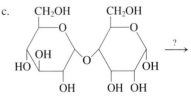

c.
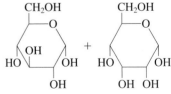

d.
```
      CHO              CHO
       |                |
  H—C—OH     ?     H—C—OH
       |        ⟶         |
  H—C—OH           H—C—OH
       |                |
     CH2OH            COOH
```

MULTIPLE-CHOICE PRACTICE TEST

7.125 Which of the following statements is *incorrect*?
 a. A chiral center is an atom in a molecule that has four different groups tetrahedrally bonded to it.
 b. A chiral molecule is a molecule whose mirror images are superimposable.
 c. Naturally occurring monosaccharides are almost always "right-handed."
 d. The simplest example of a chiral monosaccharide is glyceraldehyde.

7.126 Which of the following statements concerning the D and L forms of a monosaccharide is *incorrect*?
 a. Structurally they are nonsuperimposable mirror images of each other.
 b. They must contain the same number of chiral centers.
 c. They are enantiomers.
 d. They are diastereomers.

7.127 Which of the following is a correct characterization for the monosaccharide glucose?
 a. Aldopentose   b. Aldohexose
 c. Ketopentose   d. Ketohexose

7.128 The structures of D-glucose and D-fructose differ at which carbon atom(s)?
 a. Carbon 1 only   b. Carbon 2 only
 c. Carbon 1 and carbon 2   d. Carbon 1 and carbon 6

7.129 How many different forms of a D-monosaccharide are present, at equilibrium, in an aqueous solution of the monosaccharide?

 a. One   b. Two
 c. Three   d. Four

7.130 Which of the following disaccharides produces both D-glucose and D-fructose upon hydrolysis?
 a. Sucrose   b. Lactose
 c. Maltose   d. Cellobiose

7.131 In which of the following pairs of disaccharides do both members of the pair have the same type of glycosidic linkage?
 a. Sucrose and lactose   b. Cellobiose and maltose
 c. Lactose and cellobiose   d. Sucrose and maltose

7.132 In which of the following pairs of carbohydrates are both members of the pair heteropolysaccharides?
 a. Cellulose and amylose
 b. Starch and chitin
 c. Hyaluronic acid and heparin
 d. Glycogen and amylopectin

7.133 In which of the following pairs of polysaccharides are both members of the pair structural polysaccharides?
 a. Glycogen and cellulose   b. Starch and chitin
 c. Glycogen and starch   d. Cellulose and chitin

7.134 The carbohydrate portion of glycolipids and glycoproteins that are involved in cell recognition processes is which of the following?
 a. Monosaccharide   b. Glucose molecule
 c. Oligosaccharide   d. Polysaccharide

8 Lipids

Fats and oils are the most widely occurring types of lipids. Thick layers of fat help insulate polar bears against the effects of low temperatures.

There are four major classes of bioorganic substances: carbohydrates, lipids, proteins, and nucleic acids (Section 7.1). In the previous chapter we considered the first of these classes, carbohydrates. We now turn our attention to the second of the bioorganic classes, the compounds we call lipids.

Lipids known as fats provide a major way of storing chemical energy and carbon atoms in the body. Fats also surround and insulate vital body organs, providing protection from mechanical shock and preventing excessive loss of heat energy. Phospholipids, glycolipids, and cholesterol (a lipid) are the basic components of cell membranes. Several cholesterol derivatives function as chemical messengers (hormones) within the body.

8.1 Structure and Classification of Lipids

Unlike carbohydrates and most other classes of compounds, lipids do not have a common structural feature that serves as the basis for defining such compounds. Instead, their characterization is based on solubility characteristics. A **lipid** *is an organic compound found in living organisms that is insoluble (or only sparingly soluble) in water but soluble in nonpolar organic solvents.* When a biochemical material (human, animal, or plant tissue) is homogenized in a blender and mixed with a nonpolar organic solvent, the substances that dissolve in the solvent are the lipids.

Figure 8.1 shows the structural diversity that is associated with lipid molecules. Some are esters, some are amides, and some are alcohols; some are acyclic, some are cyclic, and

FIGURE 8.1 The structural formulas of these types of lipids illustrate the great structural diversity among lipids. The defining parameter for lipids is solubility rather than structure.

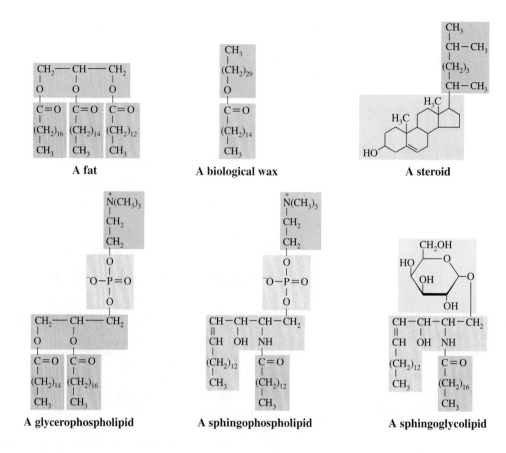

some are polycyclic. The common thread that ties all of the compounds of Figure 8.1 together is solubility rather than structure. All are insoluble in water.

For purposes of study, we will divide lipids into five categories on the basis of lipid function:

1. **Energy-storage lipids** (triacylglycerols)
2. **Membrane lipids** (phospholipids, sphingoglycolipids, and cholesterol)
3. **Emulsification lipids** (bile acids)
4. **Messenger lipids** (steroid hormones and eicosanoids)
5. **Protective-coating lipids** (biological waxes)

Our entry point into a discussion of these five general types of lipids is a consideration of molecules called *fatty acids*. Fatty acids are structural components of all the lipids that we consider in this chapter except cholesterol, bile acids, and steroid hormones. Familiarity with the structural characteristics and physical properties of fatty acids makes it easier to understand the behavior of the many fatty-acid-containing lipids found in the human body.

8.2 Fatty Acids: Lipid Building Blocks

A **fatty acid** *is a naturally occurring monocarboxylic acid.* Because of the pathway by which they are biosynthesized (Section 14.7), fatty acids nearly always contain an even number of carbon atoms and have a carbon chain that is unbranched. In terms of carbon chain length, fatty acids are characterized as *long-chain fatty acids* (C_{12} to C_{26}), *medium-chain fatty acids* (C_8 and C_{10}), or *short-chain fatty acids* (C_4 and C_6). Fatty acids are rarely found free in nature but rather occur as part of the structure of more complex lipid molecules.

■ Saturated and Unsaturated Fatty Acids

The carbon chain of a fatty acid may or may not contain carbon–carbon double bonds. On the basis of this consideration, fatty acids are classified as saturated fatty acids (SFAs), monounsaturated fatty acids (MUFAs), or polyunsaturated fatty acids (PUFAs).

A **saturated fatty acid** *is a fatty acid with a carbon chain in which all carbon–carbon bonds are single bonds.* The structural formula for the 16-carbon SFA is

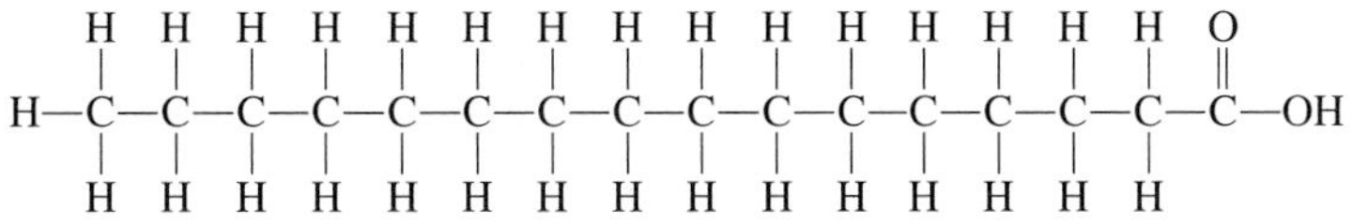

IUPAC name: hexadecanoic acid
Common name: palmitic acid

The structural formula for a fatty acid is usually written in a more condensed form than the preceding structural formula. Two alternative structural notations for palmitic acid are

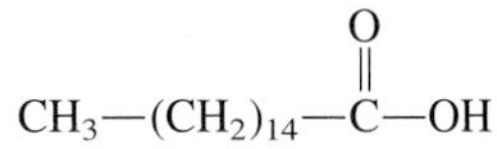

$$CH_3-(CH_2)_{14}-\overset{\displaystyle O}{\overset{\displaystyle \|}{C}}-OH$$

and

(We first encountered line-angle formulas in Section 1.9.)

A **monounsaturated fatty acid** *is a fatty acid with a carbon chain in which one carbon–carbon double bond is present.* In biochemically important MUFAs, the configuration about the double bond is nearly always *cis* (Section 2.5). Different ways of depicting the structure of a MUFA follow.

More than 500 different fatty acids have been isolated from the lipids of microorganisms, plants, animals, and humans. These fatty acids differ from one another in the length of their carbon chains, their degree of unsaturation (number of double bonds), and the positions of the double bonds in the chains.

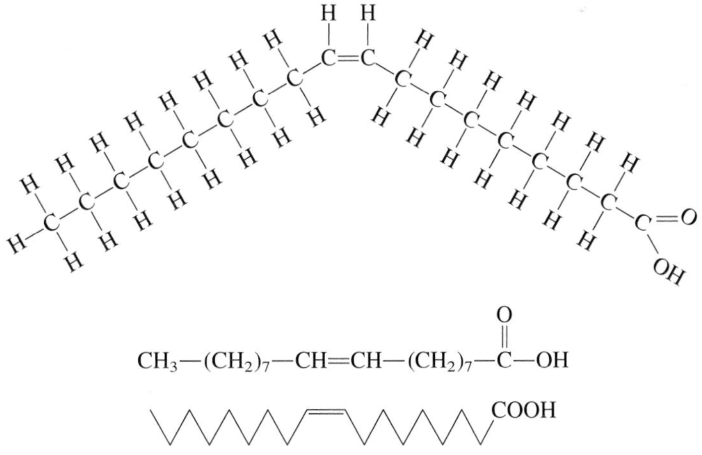

$$CH_3-(CH_2)_7-CH=CH-(CH_2)_7-\overset{\displaystyle O}{\overset{\displaystyle \|}{C}}-OH$$

IUPAC name: *cis*-9-octadecenoic acid
Common name: oleic acid

The first of these structures correctly emphasizes that the presence of a *cis* double bond in the carbon chain puts a rigid 30° bend in the chain. Such a bend affects the physical properties of a fatty acid, as we will see in Section 8.3.

A **polyunsaturated fatty acid** *is a fatty acid with a carbon chain in which two or more carbon–carbon double bonds are present.* Up to six double bonds are found in biochemically important PUFAs.

Fatty acids are nearly always referred to using their common names. IUPAC names for fatty acids, although easily constructed, are usually quite long. These two types of names for an 18-carbon PUFA containing *cis* double bonds in the 9 and 12 positions are as follows:

IUPAC name: *cis,cis*-9,12-octadecadienoic acid
Common name: linoleic acid

Unsaturated Fatty Acids and Double-Bond Position

A numerically based shorthand system exists for specifying key structural parameters for fatty acids. In this system, two numbers separated by a colon are used to specify the number of carbon atoms and the number of carbon–carbon double bonds present. The notation 18:0 denotes a C_{18} fatty acid with no double bonds, whereas the notation 18:2 signifies a C_{18} fatty acid in which two double bonds are present.

To specify double-bond positioning within the carbon chain of an unsaturated fatty acid, the preceding notation is expanded by adding the Greek capital letter delta (Δ) followed by one or more superscript numbers. The notation $18:3(\Delta^{9,12,15})$ denotes a C_{18} PUFA with three double bonds at locations between carbons 9 and 10, 12 and 13, and 15 and 16.

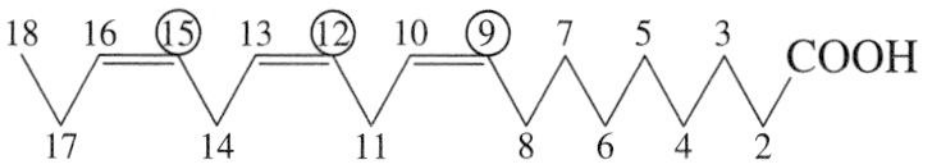

MUFAs are usually Δ^9 acids, and the first two additional double bonds in PUFAs are generally at the Δ^{12} and Δ^{15} locations. [A notable exception to this generalization is the biochemically important arachidonic acid, a PUFA with the structural parameters $20:4(\Delta^{5,8,11,14})$]. Denoting double-bond locations using this "delta notation" always assumes a numbering system in which the carboxyl carbon atom is C-1.

Several different "families" of unsaturated fatty acids exist. These family relationships become apparent when double-bond position is specified relative to the methyl (noncarboxyl) end of the fatty acid carbon chain. Double-bond positioning determined in this manner is denoted by using the Greek lower-case letter omega (ω). An **omega-3 fatty acid** *is an unsaturated fatty acid with its endmost double bond three carbon atoms away from its methyl end.* An example of an omega-3 fatty acid is

An **omega-6 fatty acid** *is an unsaturated fatty acid with its endmost double bond six carbon atoms away from its methyl end.*

The following three acids all belong to the omega-6 fatty acid family.

The structural feature common to these omega-6 fatty acids is highlighted with color in the preceding structural formulas. All the members of an omega family of fatty acids have structures in which the same "methyl end" is present.

Table 8.1 gives the names and structures of the fatty acids most commonly encountered as building blocks in biochemically important lipid structures, as well as the "delta" and "omega" notations for the acids.

TABLE 8.1
Selected Fatty Acids of Biological Importance

Structure Notation		Common Name	Structure
Saturated Fatty Acids			
12:0		lauric acid	
14:0		myristic acid	
16:0		palmitic acid	
18:0		stearic acid	
20:0		arachidic acid	
Monounsaturated Fatty Acids			
16:1 Δ^9	ω-7	palmitoleic acid	
18:1 Δ^9	ω-9	oleic acid	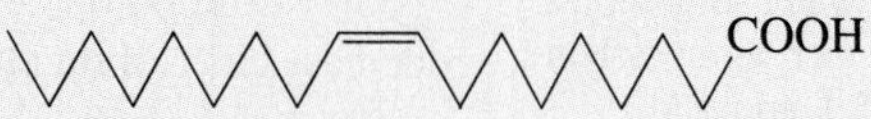
Polyunsaturated Fatty Acids			
18:2 $\Delta^{9,12}$	ω-6	linoleic acid	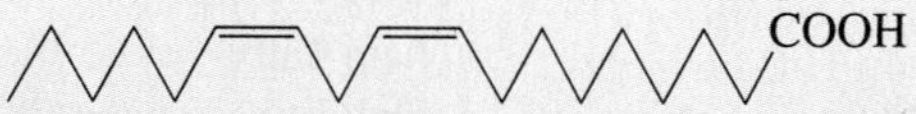
18:3 $\Delta^{9,12,15}$	ω-3	linolenic acid	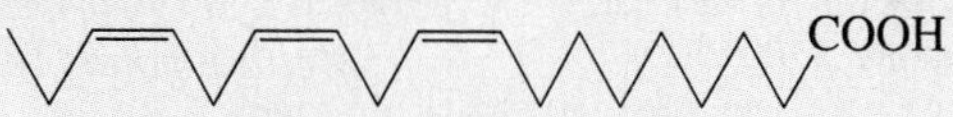
20:4 $\Delta^{5,8,11,14}$	ω-6	arachidonic acid	
20:5 $\Delta^{5,8,11,14,17}$	ω-3	EPA (eicosapentaenoic acid)	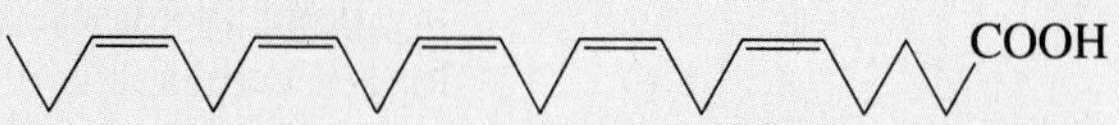
22:6 $\Delta^{4,7,10,13,16,19}$	ω-3	DHA (docosahexaenoic acid)	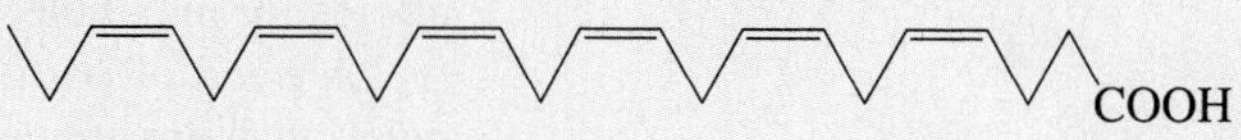

EXAMPLE 8.1

Classifying Fatty Acids on the Basis of Structural Characteristics

■ Classify the fatty acid with the following structural formula in the ways indicated.

a. What is the type designation (SFA, MUFA, or PUFA) for this fatty acid?
b. On the basis of carbon chain length and degree of unsaturation, what is the numerical shorthand designation for this fatty acid?
c. To which "omega" family of fatty acids does this fatty acid belong?
d. What is the "delta" designation for the carbon chain double-bond locations for this fatty acid?

Solution

a. Two carbon–carbon double bonds are present in this molecule, which makes it a *polyunsaturated fatty acid (PUFA)*.
b. Eighteen carbon atoms and two carbon–carbon double bonds are present. The short-hand numerical designation for this fatty acid is thus *18:2*.
c. Counting from the methyl end of the carbon chain, the first double bond encountered involves carbons 6 and 7. This fatty acid belongs to the *omega-6* family of fatty acids.
d. Counting from the carboxyl end of the carbon chain, with C-1 being the carboxyl group, the double-bond locations are 9 and 12. This is a $\Delta^{9,12}$ fatty acid.

(continued)

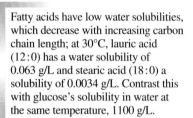
8.3 Physical Properties of Fatty Acids

The physical properties of fatty acids, and of lipids that contain them, are largely determined by the length and degree of unsaturation of the fatty acid carbon chain.

Water solubility for fatty acids is a direct function of carbon chain length; solubility decreases as carbon chain length increases. Short-chain fatty acids have a slight solubility in water. Long-chain fatty acids are essentially insoluble in water. The slight solubility of short-chain fatty acids is related to the polarity of the carboxyl group present. In longer-chain fatty acids, the nonpolar nature of the hydrocarbon chain completely dominates solubility considerations.

Melting points for fatty acids are strongly influenced by both carbon chain length and degree of unsaturation (number of double bonds present). Figure 8.2 shows melting-point variation as a function of both of these variables. As carbon chain length increases, melting point increases. This trend is related to the greater surface area associated with a longer carbon chain and to the increased opportunities that this greater surface area affords for intermolecular attractions between fatty acid molecules.

A trend of particular significance is that saturated fatty acids have higher melting points than unsaturated fatty acids with the same number of carbon atoms. The greater the degree of unsaturation, the greater the reduction in melting points. Figure 8.2 shows this effect for the 18-carbon acids with zero, one, two, and three double bonds. Long-chain saturated fatty acids tend to be solids at room temperature, whereas long-chain unsaturated fatty acids tend to be liquids at room temperature.

Fatty acids have low water solubilities, which decrease with increasing carbon chain length; at 30°C, lauric acid (12:0) has a water solubility of 0.063 g/L and stearic acid (18:0) a solubility of 0.0034 g/L. Contrast this with glucose's solubility in water at the same temperature, 1100 g/L.

FIGURE 8.2 The melting point of a fatty acid depends on the length of the carbon chain and on the number of double bonds present in the carbon chain.

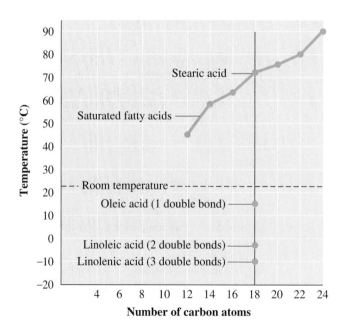

FIGURE 8.3 Space-filling models of four 18-carbon fatty acids, which differ in the number of double bonds present. Note how the presence of double bonds changes the shape of the molecule.

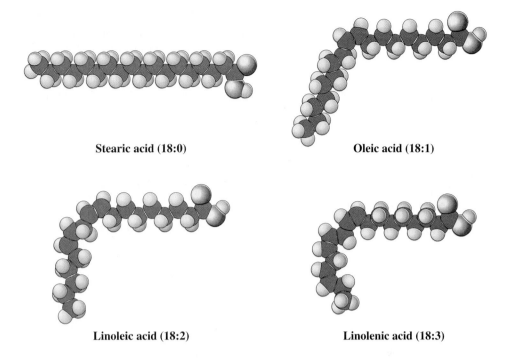

Stearic acid (18:0)

Oleic acid (18:1)

Linoleic acid (18:2)

Linolenic acid (18:3)

The decreasing melting point associated with increasing degree of unsaturation in fatty acids is explained by decreased molecular attractions between carbon chains. The double bonds in unsaturated fatty acids, which generally have the *cis* configuration, produce "bends" in the carbon chains of these molecules (see Figure 8.3). These "bends" prevent unsaturated fatty acids from packing together as tightly as saturated fatty acids. The greater the number of double bonds, the less efficient the packing. As a result, unsaturated fatty acids always have fewer intermolecular attractions, and therefore lower melting points, than their saturated counterparts.

FIGURE 8.4 An electron micrograph of adipocytes, the body's triacylglycerol-storing cells. Note the bulging spherical shape.

8.4 Energy-Storage Lipids: Triacylglycerols

With the notable exception of nerve cells, human cells store small amounts of energy-providing materials for use when energy demand is high. The most widespread energy-storage material within cells is the carbohydrate glycogen (Section 7.15); it is present in small amounts in most cells.

Lipids known as triacylglycerols also function within the body as energy-storage materials. Rather than being widespread, triacylglycerols are concentrated primarily in special cells (adipocytes) that are nearly filled with the material. Adipose tissue containing these cells is found in various parts of the body: under the skin, in the abdominal cavity, in the mammary glands, and around various organs (see Figure 8.4). Triacylglycerols are much more efficient at storing energy than is glycogen because large quantities of them can be packed into a very small volume. These energy-storage lipids are the most abundant type of lipid present in the human body.

In terms of functional groups present, triacylglycerols are triesters; three ester functional groups are present. Recall from Section 5.11 that an ester is a compound produced from the reaction of an alcohol with a carboxylic acid. The alcohol involved in triacylglycerol formation is always glycerol, a three-carbon alcohol with three hydroxyl groups.

$$\text{CH}_2\text{—OH}$$
$$|$$
$$\text{CH—OH}$$
$$|$$
$$\text{CH}_2\text{—OH}$$
Glycerol

Fatty acids are the carboxylic acids involved in triacylglycerol formation. In the esterification reaction producing a triacylglycerol, a single molecule of glycerol reacts with three fatty acid molecules; each of the three hydroxyl groups present is esterified. Figure 8.5 shows the triple esterification reaction that occurs between glycerol and three molecules of stearic acid (18:0); note the production of three molecules of water as a by-product of the reaction.

Two general ways to represent the structure of a triacylglycerol are

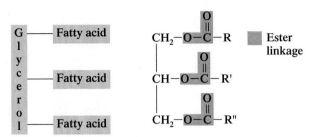

The first representation, a block diagram, shows the four subunits present in the structure: glycerol and three fatty acids. The second representation, a general structural formula, shows the three ester linkages present in a triacylglycerol. Each of the fatty acids is attached to glycerol through an ester linkage.

Formally defined, a **triacylglycerol** *is a lipid formed by esterification of three fatty acids to a glycerol molecule.* Within the name *triacylglycerol* is the term *acyl*. An **acyl group** *is the portion of a carboxylic acid that remains after the —OH group is removed from the carbonyl carbon atom.* The structural representation for an acyl group is

$$
\begin{array}{c}
O \\
\parallel \\
R\!-\!C\!- \\
\end{array}
$$

An acyl group

Thus, as the name implies, triacylglycerol molecules contain three fatty acid residues (three acyl groups) attached to a glycerol residue. An older name that is still frequently used for a triacylglycerol is *triglyceride*.

The triacylglycerol produced from glycerol and three molecules of stearic acid (as in Figure 8.5) is an example of a simple triacylglycerol. A **simple triacylglycerol** *is a triester formed from the esterification of glycerol with three identical fatty acid molecules.* If the reacting fatty acid molecules are not all identical, then the result is a mixed triacylglycerol. A **mixed triacylglycerol** *is a triester formed from the esterification of glycerol with more than one kind of fatty acid molecule.* Figure 8.6 shows the structure of a mixed triacylglycerol in which one fatty acid is saturated, another monounsaturated, and the third polyunsaturated. Naturally occurring *simple* triacylglycerols are rare. Most biochemically important triacylglycerols are *mixed* triacylglycerols.

> Triacylglycerols do not actually contain glycerol and three fatty acids, as the block diagram for a triacylglycerol implies. They actually contain a glycerol *residue* and three fatty acid *residues*. In the formation of the triacylglycerol, three molecules of water have been removed from the structural components of the triacylglycerol, leaving residues of the reacting molecules.

FIGURE 8.5 Structure of the simple triacylglycerol produced from the triple esterification reaction between glycerol and three molecules of stearic acid (18:0 acid). Three molecules of water are a by-product of this reaction.

Glycerol   Three fatty acids   Triester of glycerol   Three water molecules

FIGURE 8.6 Structure of a mixed triacylglycerol in which three different fatty acid residues are present.

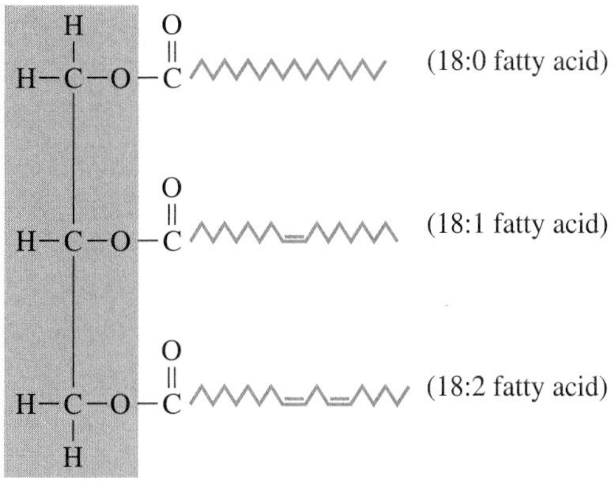

(18:0 fatty acid)

(18:1 fatty acid)

(18:2 fatty acid)

■ Fats and Oils

Fats are naturally occurring complex mixtures of triacylglycerol molecules in which many different kinds of triacylglycerol molecules are present. *Oils* are also naturally occurring complex mixtures of triacylglycerol molecules in which there are many different kinds of triacylglycerol molecules present. Given that both are triacylglycerol mixtures, what distinguishes a fat from an oil? The answer is physical state at room temperature. A **fat** *is a triacylglycerol mixture that is a solid or a semi-solid at room temperature (25°C).* Generally, fats are obtained from animal sources. An **oil** *is a triacylglycerol mixture that is a liquid at room temperature (25°C).* Generally, oils are obtained from plant sources. Because they are mixtures, no fat or oil can be represented by a single specific chemical formula. Many different fatty acids are represented in the triacylglycerol molecules present in the mixture. The actual composition of a fat or oil varies even for the species from which it is obtained. Composition depends on both dietary and climatic factors. For example, fat obtained from corn-fed hogs has a different overall composition than fat obtained from peanut-fed hogs. Flax seed grown in warm climates gives oil with a different composition from that obtained from flax seed grown in colder climates.

Additional generalizations and comparisons between fats and oils follow.

Petroleum oils (Section 12.15) are structurally different from *lipid oils*. The former are mixtures of alkanes and cycloalkanes. The latter are mixtures of triesters of glycerol.

1. Fats are composed largely of triacylglycerols in which saturated fatty acids predominate, although some unsaturated fatty acids are present. Such triacylglycerols can pack closely together because of the "linearity" of their fatty acid chains (Figure 8.7a), thus causing the higher melting points associated with fats. Oils contain triacylglycerols with larger amounts of mono- and polyunsaturated fatty acids than those in fats. Such triacylglycerols cannot pack as tightly together because of "bends" in their fatty acid chains (Figure 8.7b). The result is lower melting points.
2. Fats are generally obtained from animals; hence the term *animal fat*. Although fats are solids at room temperature, the warmer body temperature of the living animal keeps the fat somewhat liquid (semi-solid) and thus allows for movement. Oils

FIGURE 8.7 Representative triacylglycerols from (a) a fat and (b) an oil.

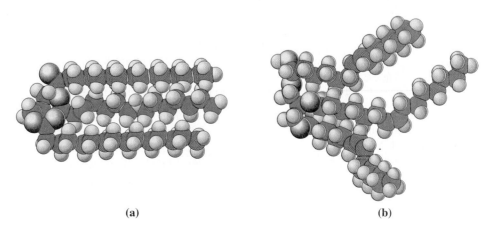

(a)   (b)

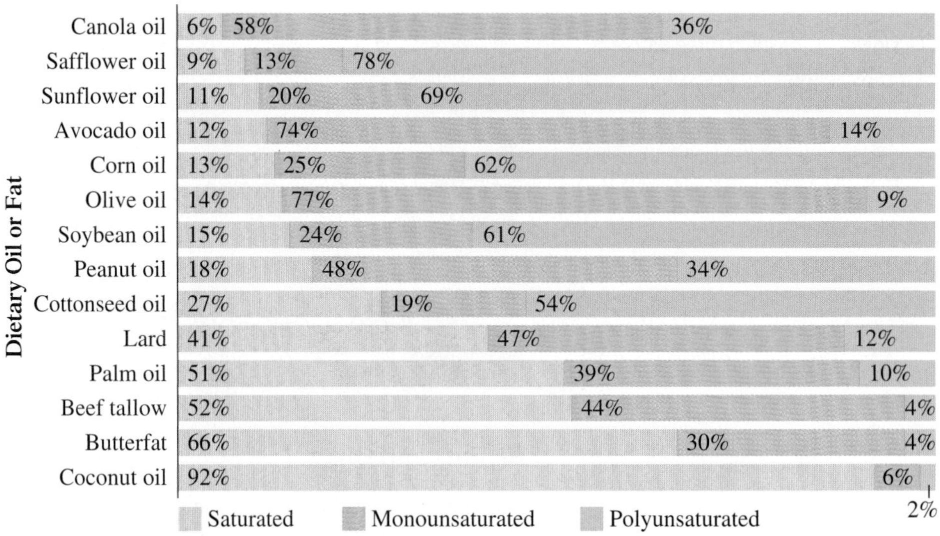

FIGURE 8.8 Percentages of saturated, monounsaturated, and polyunsaturated fatty acids in the triacylglycerols of various dietary fats and oils.

All oils, even polyunsaturated oils, contain some *saturated* fatty acids. All fats, even highly saturated fats, contain some *unsaturated* fatty acids.

typically come from plants, although there are also fish oils. A fish would have some serious problems if its triacylglycerols "solidified" when it encountered cold water.

3. Pure fats and pure oils are colorless, odorless, and tasteless. The tastes, odors, and colors associated with dietary plant oils are caused by small amounts of other naturally occurring substances present in the plant that have been carried along during processing. The presence of these "other" compounds is usually considered desirable.

Figure 8.8 gives the percentages of saturated, monounsaturated, and polyunsaturated fatty acids found in common dietary oils and fats. In general, a higher degree of fatty acid unsaturation is associated with oils than with fats. A notable exception to this generalization is coconut oil, which is highly saturated. This oil is a liquid not because it contains many double bonds within the fatty acids but because it is rich in *shorter-chain* fatty acids, particularly lauric acid (12:0).

8.5 Dietary Considerations and Triacylglycerols

In the past two decades, considerable research has been carried out concerning the role of dietary factors as a cause of disease (obesity, diabetes, cancer, hypertension, and atherosclerosis). Numerous studies have shown that, *in general,* nations whose citizens have high dietary intakes of triacylglycerols (fats and oils) tend to have higher incidences of heart disease and certain types of cancers. This is the reason for concern that the typical American diet contains too much fat and the call for Americans to reduce their total dietary fat intake.

Contrary to the general trend, however, there are several areas of the world where high dietary fat intake does not translate into high risks for cardiovascular disease, obesity, and certain types of cancers. These exceptions, which include some Mediterranean countries and the Inuit people of Greenland, suggest that relationships between dietary triacylglycerol intake and risk factors for disease involve more than simply the *total amount* of triacylglycerols consumed.

A grain- and vegetable-rich diet that contains small amounts of extra-virgin olive oil (three to four teaspoons daily) has been found to help people with high blood pressure reduce the amount of blood pressure medication they require, on average, by 48%. Substitution of sunflower oil for the olive oil resulted in only a 4% reduction in medication dosage.

The blood-pressure-reduction benefits of olive oil do not relate to the triacylglycerols present but rather come from "other" compounds naturally present, namely from antioxidant polyphenols olive oil contains. These antioxidants help promote the relaxation of blood vessels.

■ "Good Fats" Versus "Bad Fats"

In dietary discussions, the term *fat* is used as a substitute for the term *triacylglycerol.* Thus a dietary fat can be either a "fat" or an "oil." Ongoing studies indicate that both the *type of dietary fat* consumed and the *amount of dietary fat* consumed are important factors in determining human body responses to dietary fat. Current dietary fat recommendations are that people limit their total fat intake to 30% of total calories—with up to 15% coming from monounsaturated fat, up to 10% from polyunsaturated fat, and less than 10% from saturated fats.

FIGURE 8.9 Fish that live in deep, cold water — mackerel, herring, tuna, and salmon — are better sources of omega-3 fatty acids than other fish.

These recommendations imply correctly that different types of dietary fat have different effects. In simplified terms, research studies indicate that saturated fats are "bad fat," monounsaturated fats are "good fat," and polyunsaturated fats can be both "good fat" and "bad fat." In the latter case, fatty acid omega type (Section 8.2) becomes important, a situation addressed later in this section. Studies indicate that saturated fat can increase heart disease risk, that monounsaturated fat can decrease both heart disease and breast cancer risk, and that polyunsaturated fat can reduce heart disease risk but promote the risk of certain types of cancers.

Referring to Figure 8.8, note the wide variance in the three general types of fatty acids (SFAs, MUFAs, and PUFAs) present in various kinds of dietary fats. Dietary fats high in "good" monounsaturated fatty acids include olive, avocado, and canola oils. Monounsaturated fatty acids help reduce the stickiness of blood platelets. This helps prevent the formation of blood clots and may also dissolve clots once they form.

Many people do not realize that most tree nuts and peanuts are good sources of MUFAs. The Chemical Connections feature on page 280 looks at recent research on the fat content of nuts.

Omega-3 and Omega-6 Fatty Acids

In the 1980s, researchers found that the Inuit people of Greenland exhibit a low incidence of heart disease despite having a diet very high in fat. This contrasts markedly with studies on the U.S. population, which show a correlation between a high-fat diet and a high incidence of heart disease. What accounts for the difference between the two peoples? The Inuit diet is high in omega-3 fatty acids (from fish), and the U.S. diet is high in omega-6 fatty acids (from plant oils). An American consumes about double the amount of omega-6 fatty acids and half the amount of omega-3 fatty acids that an Inuit consumes.

Several large studies now confirm that benefits can be derived from eating several servings of fish each week. The choice of fish is important, however. Not all fish are equal in omega-3 fatty acid content. Cold-water fish, also called fatty fish because of the extra amounts of fat they have for insulation against the cold, contain more omega-3 acids than leaner, warm-water fish. Fatty fish include albacore tuna, salmon, and mackerel (see Figure 8.9). Leaner, warm-water fish, which include cod, catfish, halibut, sole, and snapper, do not appear to offer as great a positive effect on heart health as do their "fatter" counterparts. (Note that most of the fish used in fish and chips (e.g., cod, halibut) is on the low end of the omega-3 scale.) Table 8.2 gives the actual omega-3 fatty acid concentrations associated with various kinds of cold-water fish.

TABLE 8.2
Omega-3 Fatty Acid Amounts Associated with Various Kinds of Cold-Water Fish

Per 3.5-oz. Serving (raw)	Omega-3s (grams)*
mackerel	2.3
albacore tuna	2.1
herring, Atlantic	1.6
anchovy	1.5
salmon, wild king (Chinook)	1.4
salmon, wild sockeye (red)	1.2
tuna, bluefin	1.2
salmon, wild pink	1.0
salmon, wild Coho (silver)	0.8
oysters, Pacific	0.7
salmon, farm-raised Atlantic	0.6
swordfish	0.6
trout, rainbow	0.6

*Omega-3 content of fish can vary depending on harvest location and time of year.

CHEMICAL CONNECTIONS   The Fat Content of Tree Nuts and Peanuts

People who bypass the nut tray at holiday parties usually believe in a myth—that nuts are *unhealthful* high-fat foods. Indeed, nuts are high-fat food. However, the fat is "good fat" rather than "bad fat" (Section 8.5); that is, the fatty acids present are MUFAs and PUFAs rather than SFAs. In most cases, a handful of nuts is better for you than a cookie or bagel.

Numerous studies now indicate that eating nuts can have a strong protective effect against coronary heart disease. The most improvement comes from adding small amounts of nuts—an ounce (3–4 teaspoons)—to the diet five or more times a week. Raw, dry-roasted, or lightly salted varieties are best.

The recommendation of only one ounce of nuts per day relates to the high calorie content of nuts, which is 160 to 200 calories per ounce. The number of nuts and number of calories per ounce for common types of nuts is as follows:

Nuts	Calories
18 cashews	160
20 peanuts	160
47 pistachios	160
24 almonds	166
14 walnut halves	180
8 Brazil nuts	186
12 hazelnuts	188
15 pecan halves	190
12 macadamias	200

The amount of fat present in nuts ranges from 74% in the macadamia nut, 68% in pecans, and 63% in hazelnuts to around 50% in nuts such as the almond, cashew, peanut, and pistachio, as is shown in the accompanying table.

Fat and Fatty Acid Composition of Selected Nuts

	Total Fat (percentage of weight)	SFA	MUFA	PUFA	UFA/SFA Ratio
		(percentage of total fat)			
almonds	52	10	68	22	9.0
cashews	46	20	62	18	3.9
hazelnuts	63	8	82	10	11.9
macadamias	74	16	82	2	5.4
peanuts	49	15	51	34	5.7
pecans	68	8	66	26	10.9
pistachios	48	13	72	15	6.6
walnuts	62	10	24	66	9.0

The different fatty acid fractions (SFAs, MUFAs, and PUFAs) present in nuts also vary, but with definite trends. Unsaturated fatty acids always significantly dominate saturated fatty acids. The unsaturation/saturation ratio is highest for hazelnuts (11.9), pecans (10.9), walnuts (9.0), and almonds (9.0) and is lowest for cashews (3.9).

Their low amounts of saturated fatty acids are not the only reason why nuts help reduce the risk of coronary heart disease. Nuts also offer valuable antioxidant vitamins, minerals, and plant fiber protein. The protein content is highest (18%–26%) in the cashew, pistachio, almond, and peanut; here the amount of protein is about the same as in meat, fish, and cheese. The carbohydrate content of nuts is relatively low, less than 10% in most cases.

An unexpected discovery involving the anticancer drug Taxol and hazelnuts was made in the year 2000. The active chemical component in this drug, paclitaxel, was found in hazelnuts. It was the first report of this potent chemical being found in a plant other than in the bark of the Pacific yew tree, a slow-growing plant found in limited quantities in the Pacific Northwest. Although the amount of the chemical found in a hazelnut tree is about one-tenth that found in yew bark, the effort required to extract paclitaxel from these sources is comparable. Because hazelnut trees are more common, this finding could reduce the cost of the commercial drug and make it more readily available.

■ Essential Fatty Acids

An **essential fatty acid** *is a fatty acid needed in the human body that must be obtained from dietary sources because it cannot be synthesized within the body, in adequate amounts, from other substances.* There are two essential fatty acids: *linoleic acid* and *linolenic acid.* Linoleic acid (18:2) is the primary member of the omega-6 acid family, and linolenic acid (18:3) is the primary member of the omega-3 acid family. Their structures were given in Table 8.1.

These two acids (1) are needed for proper membrane structure and (2) serve as starting materials for the production of several nutritionally important longer-chain omega-6 and

TABLE 8.3
Biochemically Important Omega-3 and Omega-6 Fatty Acids

Omega-3 Acids	Omega-6-Acids
linolenic acid (18:3) (lin-oh-LEN-ic) eicosapentaenoic acid (20:5) (EYE-cossa-PENTA-ee-NO-ic) docosahexaenoic acid (20:6) (DOE-cossa-HEXA-ee-NO-ic)	linoleic acid (18:2) (lin-oh-LAY-ic) arachidonic acid (20:4) (a-RACK-ih-DON-ic)

In 2001 the FDA gave approval for manufacturers of baby formula to add the fatty acids DHA (docosahexaenoic acid) and AA (arachidonic acid) to infant formulas. Human breast milk naturally contains these acids, which are important in brain and vision development. Because not all mothers can breast-feed, health officials regulate the ingredients in infant formula so that formula-fed babies get the next best thing to mother's milk.

omega-3 acids. When these two acids are missing from the diet, the skin reddens and becomes irritated, infections and dehydration are likely to occur, and the liver may develop abnormalities. If the fatty acids are restored, then the conditions reverse themselves. Infants are especially in need of these acids for their growth. Human breast milk has a much higher percentage of the essential fatty acids than cow's milk.

Linoleic acid is the starting material for the biosynthesis of arachidonic acid.

$$\text{Linoleic acid (18:2)} \longrightarrow \text{arachidonic acid (20:4)}$$
Omega-6 fatty acids

Arachidonic acid is the major starting material for eicosanoids (Section 8.12), substances that help regulate blood pressure, clotting, and several other important body functions.

Linolenic acid is the starting material for the biosynthesis of two additional omega-3 fatty acids.

$$\text{Linolenic acid (18:3)} \longrightarrow \text{EPA (20:5)} \longrightarrow \text{DHA (22:6)}$$
Omega-3 fatty acids

EPA (eicosapentaenoic acid) and DHA (docosahexaenoic acid) are important constituents of the communication membranes of the brain and are necessary for normal brain development. EPA and DHA are also active in the retina of the eye.

Table 8.3 gives pronunciation guidelines for the names of the two essential fatty acids and of the other acids mentioned that are biosynthesized from them.

■ Fat Substitutes (Artificial Fats)

In response to consumer demand for low-fat, low-calorie foods, food scientists have developed several types of "artificial fats." Such substances replicate the taste, texture, and cooking properties of fats but are themselves not lipids. See the Chemical Connections feature on page 283 for further discussion of this topic.

8.6 Chemical Reactions of Triacylglycerols

The chemical properties of triacylglycerols (fats and oils) are typical of esters and alkenes because these are the two functional groups present in triacylglycerols. Four important triacylglycerol reactions are hydrolysis, saponification, hydrogenation, and oxidation.

■ Hydrolysis

Naturally occurring mono- and diacylglycerols are seldom encountered. *Synthetic* mono- and diacylglycerols are used as emulsifiers in many food products. Emulsifiers prevent suspended particles in colloidal solutions from coalescing and settling. Emulsifiers are usually present in so-called fat-free cakes and other fat-free products.

Hydrolysis of a triacylglycerol is the reverse of the esterification reaction by which it was formed (see Figure 8.5). *Complete* hydrolysis of a triacylglycerol molecule always gives one glycerol molecule and three fatty acid molecules as products (Figure 8.10a).

Triacylglycerol hydrolysis within the human body requires the help of enzymes (protein catalysts; Section 10.1) produced by the pancreas. These enzymes cause the triacylglycerol to be hydrolyzed in a *stepwise* fashion. First one of the outer fatty acids is removed, then the other outer one, leaving a monoacylglycerol. In most cases this is the end product of the initial hydrolysis (digestion) of the triacylglycerol (Figure 8.10b). Sometimes, enzymes remove all three fatty acids, leaving a free molecule of glycerol.

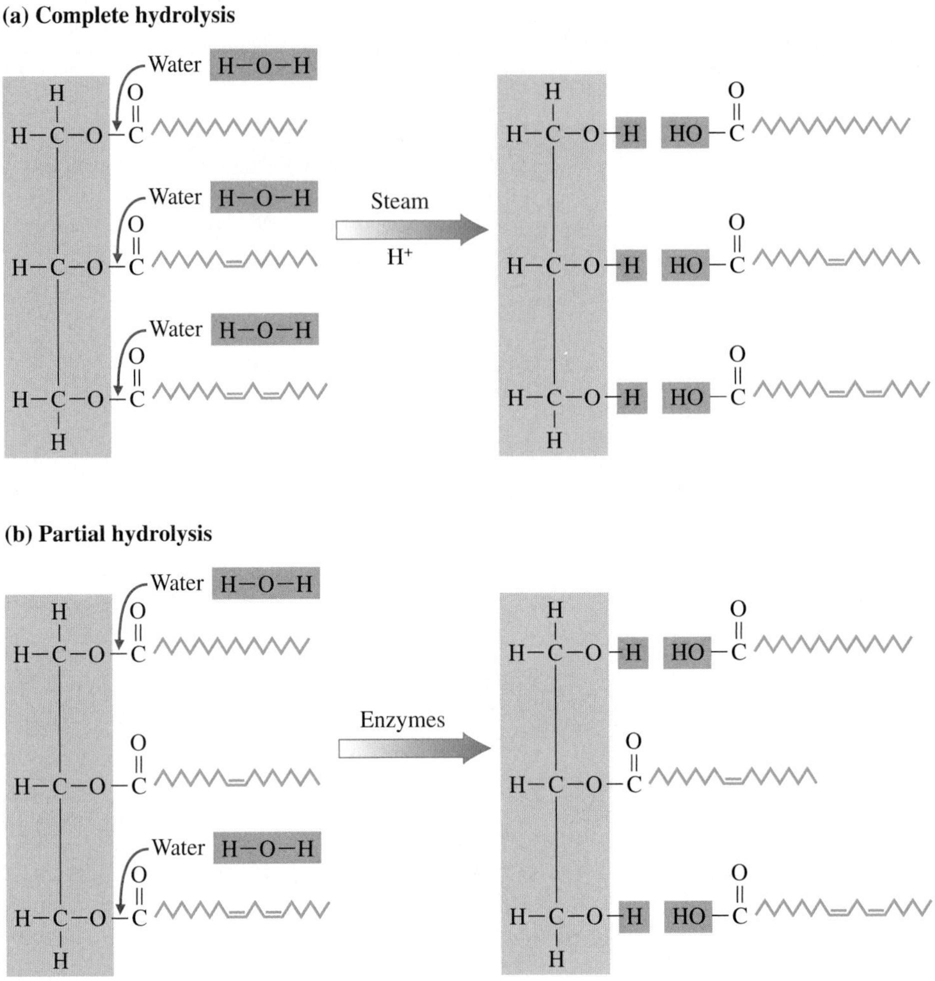

FIGURE 8.10 Complete and partial hydrolysis of a triacylglycerol. (a) Complete hydrolysis of a triacylglycerol produces glycerol and three fatty acid molecules. (b) Partial hydrolysis (during digestion) of a triacylglycerol produces a monoacylglycerol and two fatty acid molecules.

(a) Complete hydrolysis

(b) Partial hydrolysis

Recall from Section 5.9 the structural difference between a carboxylic acid and a carboxylic acid salt.

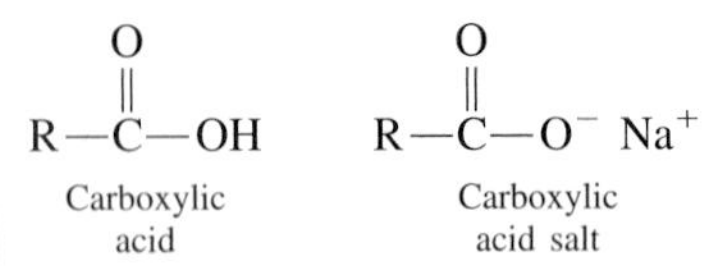

■ Saponification

Saponification (Section 5.16) is a hydrolysis reaction carried out in an alkaline (basic) solution. For fats and oils, the products of saponification are glycerol and fatty acid *salts*.

The overall reaction of triacylglycerol saponification can be thought of as occurring in two steps. The first step is the hydrolysis of the ester linkages to produce glycerol and three fatty acid molecules:

$$\text{Fat or oil} + 3H_2O \longrightarrow 3 \text{ fatty acids} + \text{glycerol}$$

The second step involves a reaction between the fatty acid molecules and the base (usually NaOH) in the alkaline solution. This is an acid–base reaction that produces water plus salts:

$$3 \text{ Fatty acids} + 3NaOH \longrightarrow 3 \text{ fatty acid salts} + 3H_2O$$

Saponification of animal fat is the process by which soap was made in pioneer times. Soap making involved heating lard (fat) with lye (ashes of wood, an impure form of KOH). Today most soap is prepared by hydrolyzing fats and oils (animal fat and coconut oil) under high pressure and high temperature. Sodium carbonate is used as the base.

The cleansing action of soap is related to the structure of the carboxylate ions present in the fatty acid salts of soap and the fact that these ions readily participate in micelle formation. A **micelle** *is a spherical cluster of molecules in which the polar portions of the molecules are on the surface, and the nonpolar portions are located in the interior.* The Chemical Connections feature on page 284 discusses micelle formation further as it relates to the cleansing action of soap.

CHEMICAL CONNECTIONS   Artificial Fat Substitutes

Artificial sweeteners (sugar substitutes) have been an accepted part of the diet of most people for many years. New since the 1990s are artificial fats—substances that create the sensations of "richness" of taste and "creaminess" of texture in food without the negative effects associated with dietary fats (heart disease and obesity).

Food scientists have been trying to develop fat substitutes since the 1960s. Now available for consumer use are two types of fat substitutes: *calorie-reduced* fat substitutes and *calorie-free* substitutes. They differ in their chemical structures and therefore in how the body handles them.

Simplesse, the best-known calorie-reduced fat substitute, received FDA marketing approval in 1990. It is made from the protein of fresh egg whites and milk by a procedure called microparticulation. This procedure produces tiny, round protein particles so fine that the tongue perceives them as a fluid rather than as the solid they are. Their fineness creates a sensation of smoothness, richness, and creaminess on the tongue.

In the body, Simplesse is digested and absorbed, contributing to energy intake. But 1 g of Simplesse provides 1.3 cal, compared with the 9 cal provided by 1 g of fat. Simplesse is used only to replace fats in *formulated* foods such as salad dressings, cheeses, sour creams, and other dairy products. Simplesse is unsuitable for frying or baking because it turns rubbery or rigid (gels) when heated. Consequently, it is not available for home use.

Olestra, the best-known calorie-free fat substitute, received FDA marketing approval in 1996. It is produced by heating cottonseed and/or soybean oil with sucrose in the presence of methyl alcohol. Chemically, olestra has a structure somewhat similar to that of a triacylglycerol; sucrose takes the place of the glycerol molecule, and six to eight fatty acids are attached by ester linkages to it rather than the three fatty acids in a triacylglycerol. Unlike triacylglycerols, however, olestra cannot be

Olestra

hydrolyzed by the body's digestive enzymes and therefore passes through the digestive tract undigested.

Olestra looks, feels, and tastes like dietary fat and can substitute for fats and oils in foods such as shortenings, oils, margarines, snacks, ice creams, and other desserts. It has the same cooking properties as fats and oils.

In the digestive tract, Olestra interferes with the absorption of both dietary and body-produced cholesterol; thus it may lower total cholesterol levels. A problem with its use is that it also reduces the absorption of the fat-soluble vitamins A, D, E, and K. To avoid such depletion, Olestra is fortified with these vitamins. Another problem with Olestra use is that in some individuals it can cause gastrointestinal irritation and/or diarrhea. All products containing Olestra must carry the following label: "Olestra may cause abdominal cramping and loose stools. Olestra inhibits the absorption of some vitamins and other nutrients. Vitamins A, D, E, and K have been added."

■ Hydrogenation

Hydrogenation is a chemical reaction we first encountered in Section 2.8. It involves hydrogen addition across carbon–carbon multiple bonds, which increases the degree of saturation as some double bonds are converted to single bonds. With this change, there is a corresponding increase in the melting point of the substance.

Hydrogenation involving just one carbon–carbon bond within a fatty acid residue of a triacylglycerol can be diagrammed as follows:

$$---CH_2-CH_2-\boxed{CH=CH}-CH_2-CH_2--- + \boxed{H_2} \longrightarrow ---CH_2-CH_2-\boxed{CH_2-CH_2}-CH_2-CH_2---$$

Portion of an unsaturated fatty acid residue in a triacylglycerol containing one double bond

The double bond has been converted to a single bond; the degree of saturation has increased

The structural equation for the complete hydrogenation of a triacylglycerol in which all three fatty acid residues are oleic acid (18:1) is shown in Figure 8.11.

The Cleansing Action of Soap

The cleansing action of soap is directly related to the structure of the carboxylate ions present in soap within fatty acid salts. Their structure is such that they exhibit a "dual polarity." The hydrocarbon portion of the carboxylate ion is nonpolar, and the carboxyl portion is polar. This dual polarity for the fatty acid salt *sodium stearate,* which is representative of all fatty acid salts present in soap, is as follows:

$$\text{COO}^-\,\text{Na}^+$$

Nonpolar portion   Polar portion

Soap solubilizes oily and greasy materials in the following manner: The nonpolar portion of the carboxylate ion dissolves in the nonpolar oil or grease, and the polar carboxyl portion maintains its solubility in the polar water.

The penetration of the oil or grease by the nonpolar end of the carboxylate ion is followed by the formation of micelles (see the accompanying diagram). The carboxyl groups (the micelle exterior) and water molecules are attracted to each other, causing the solubilizing of the micelle.

The micelles do not combine into larger drops because their surfaces are all negatively charged, and like charges repel each

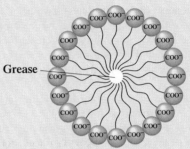

Fatty acid micelle

other. The water-soluble micelles are subsequently rinsed away, leaving a material devoid of oil and grease.

For most cleansing purposes, synthetic detergents have largely replaced soaps. The basis of the cleansing action of synthetic detergents is very similar to that of soaps because their structures are very similar. The structure of the sodium salt of a benzene sulfonic acid is typical of the types of molecules used in detergents.

$$\text{S}-\text{O}^-\,\text{Na}^+$$

Many food products are produced via partial hydrogenation. In partial hydrogenation some, but not all, of the double bonds present are converted into single bonds. In this manner, liquids (usually plant oils) are converted into semi-solid materials.

Peanut butter is produced from peanut oil through partial hydrogenation. Solid cooking shortenings and stick margarine are produced from liquid plant oils through partial hydrogenation. Soft-spread margarines are also partial-hydrogenation products. Here, the extent of hydrogenation is carefully controlled to make the margarine soft at refrigerator temperatures (4°C). Concern has arisen about food products obtained from hydrogenation processes because the hydrogenation process itself converts some *cis* double bonds within fatty acid residues into *trans* double bonds. The Chemical Connections feature on page 285 explores this issue further.

FIGURE 8.11 Structural equation for the complete hydrogenation of a triacylglycerol with oleic acid (18:1) fatty acid residues.

Hydrogenation

3H₂

CHEMICAL CONNECTIONS — *Trans* Fatty Acids and Blood Cholesterol Levels

All current dietary recommendations stress reducing saturated fat intake. In accordance with such recommendations, many people have switched from butter to margarine and now use partially hydrogenated vegetable oils rather than animal fat for cooking. However, recent studies suggest that partially hydrogenated products also play a role in raising blood cholesterol levels. Why would this be so?

It is now known that when triacylglycerols are subjected to partial hydrogenation (Section 8.6) two types of changes occur in the fatty acid residues present: (1) some of the *cis* double bonds present are converted to single bonds (the objective of the process), and (2) some of the remaining *cis* double bonds are converted to *trans* double bonds (an unanticipated result of the process). These latter *cis–trans* conversions affect the general shape of the fatty acid residues present in triacylglycerols, which in turn affects the biochemical behavior of the triacylglycerols. In

the preceding diagram, note how conversion of a *cis,cis*-18:2 fatty acid to a *trans,trans*-18:2 fatty acid affects molecular shape. The *trans,trans*-18:2 fatty acid has a shape very much like that of an 18:0 saturated fatty acid (the structure on the right).

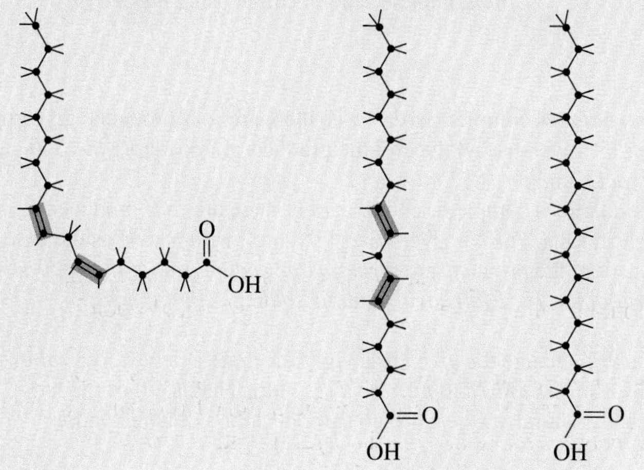

Studies show that fatty acids with *trans* double bonds affect blood cholesterol levels in a manner similar to saturated fatty acids.

Trans fatty acids (*trans* fat) make up approximately 5% of the fat intake in the typical diet in the United States, and the amount of *trans* fat a person consumes depends on the amount of fat eaten and on the types of foods selected. The best example of a *trans* fat food may be stick margarine, but it is also found in crackers, cookies, pastries, and deep-fried fast foods. Spreadable margarine in tubs, though, contains little if any *trans* fat.

Beginning in 2006, the U.S. Food and Drug Administration (FDA) requires that the *trans* fat content of a food be included in the nutrition facts panel found on all food products. Prior to this rule change, the only way consumers could determine whether a food included *trans* fat was to look for the word *hydrogenated* on the list of ingredients. A food that lists partially hydrogenated oils among its first three ingredients usually contains substantial amounts of *trans* fatty acids as well as some saturated fat.

The health implications of *trans* fatty acids is an area of active research; many answers are yet to be found. Preliminary studies indicate that *trans* fat raises bad (LDL) cholesterol, but it does not raise good (HDL) cholesterol. Saturated fat, on the other hand, raises both bad and good cholesterol. Thus, just as too much saturated fat isn't healthy, too much *trans* fat is also not healthy. Recommendations are that total fat intake be limited to 30% of daily calories and that combined saturated fat and *trans* fat intake should be limited to 10% or less of daily calories.

18:2 *(cis, cis)*　　**18:2** *(trans, trans)*　　**18:0**

■ Oxidation

The carbon–carbon double bonds present in the fatty acid residues of a triacylglycerol are subject to oxidation with molecular oxygen (from air) as the oxidizing agent. Such oxidation breaks the carbon–carbon bonds, producing both aldehyde and carboxylic acid products.

Antioxidants are compounds that are easily oxidized. When added to foods, they are more easily oxidized than the food. Thus they prevent the food from being oxidized (see Section 3.1).

$$-CH{=}CH- \xrightarrow{\text{Oxidation}} \underset{\text{Short-chain aldehydes}}{-\overset{O}{\overset{\|}{C}}-H + H-\overset{O}{\overset{\|}{C}}-} \xrightarrow{\text{Oxidation}} \underset{\text{Short-chain carboxylic acids}}{-\overset{O}{\overset{\|}{C}}-OH + HO-\overset{O}{\overset{\|}{C}}-}$$

Unsaturated fatty acids

The short-chain aldehydes and carboxylic acids so produced often have objectionable odors, and fats and oils containing them are said to have become *rancid*. To avoid this unwanted oxidation process, commercially prepared foods containing fats and oils nearly always contain *antioxidants*—substances that are more easily oxidized than the food. Two naturally occurring antioxidants are vitamin C (Section 10.13) and vitamin E (Section 10.14). Two synthetic oxidation inhibitors are BHA and BHT (Section 3.14). In the presence of air, antioxidants, rather than food, are oxidized.

EXAMPLE 8.2

Determining the Products for Reactions That Triacylglycerols Undergo

■ Using words rather than structural formulas, characterize the products formed when the following triacylglycerol undergoes the reactions listed.

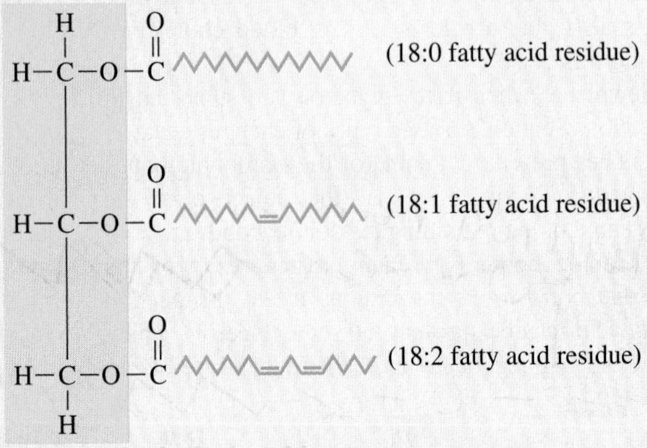

(18:0 fatty acid residue)

(18:1 fatty acid residue)

(18:2 fatty acid residue)

a. Complete hydrolysis **b.** Complete saponification using NaOH
c. Complete hydrogenation

Solution

a. When a triacylglycerol undergoes complete hydrolysis, there are four products: glycerol and three fatty acids. For the given triacylglycerol the products are *glycerol, an 18:0 fatty acid, an 18:1 fatty acid,* and *an 18:2 fatty acid.*
b. When a triacylglycerol undergoes complete saponification, there are four products: glycerol and three fatty acid salts. For the given triacylglycerol, with NaOH as the base involved in the saponification, the products are *glycerol, the sodium salt of the 18:0 fatty acid, the sodium salt of the 18:1 fatty acid,* and *the sodium salt of the 18:2 fatty acid.*
c. Complete hydrogenation will change the given triacylglycerol into a *triacylglycerol in which all three fatty acid residues are 18:0 fatty acid residues.* That is, all of the fatty acid residues are completely saturated (there are no carbon–carbon double bonds).

Practice Exercise 8.2

Using words rather than structural formulas, characterize the products formed when the following triacylglycerol undergoes the reactions listed.

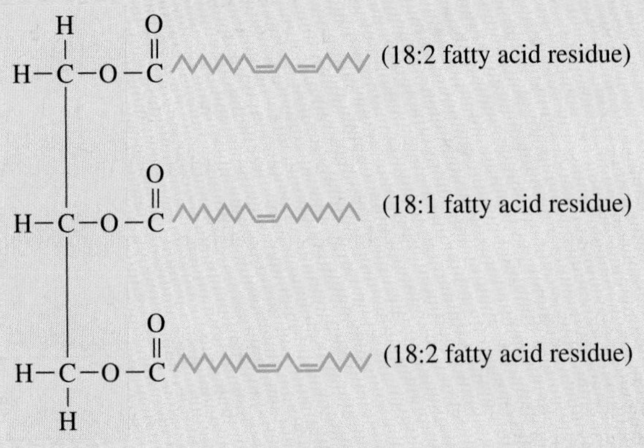

(18:2 fatty acid residue)

(18:1 fatty acid residue)

(18:2 fatty acid residue)

a. Complete hydrolysis **b.** Complete saponification using NaOH
c. Complete hydrogenation

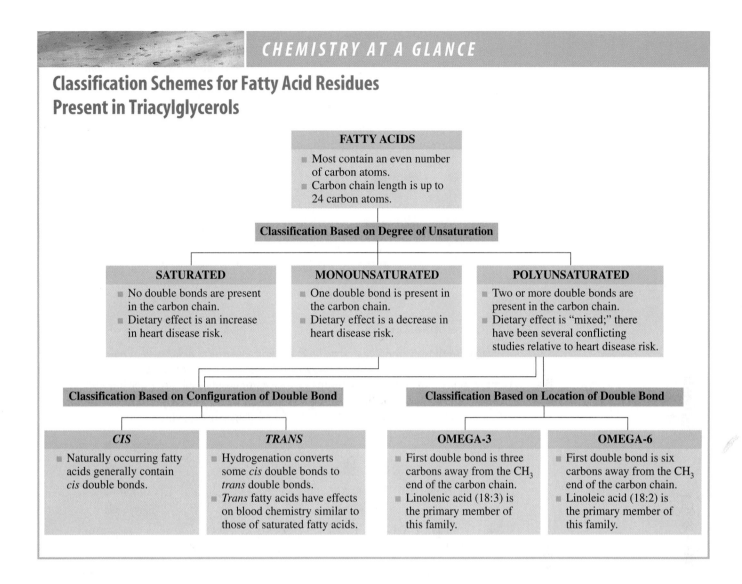

Perspiration generated by strenuous exercise or by "hot and muggy" climatic conditions contains numerous triacylglycerols (oils). Rapid oxidation of these oils, promoted by microorganisms on the skin, generates the body odor that accompanies most "sweaty" people (see Figure 8.12).

The Chemistry at a Glance on this page contains a summary of the terminology used in characterizing the properties of the fatty acid residues that are part of the structure of triacylglycerols (fats and oils).

8.7 Membrane Lipids: Phospholipids

All cells are surrounded by a membrane that confines their contents. Up to 80% of the mass of a cell membrane can be lipid materials; the rest is primarily protein. It is membranes that give cells their individuality by separating them from their environment.

There are three common types of membrane lipids: phospholipids, sphingoglycolipids, and cholesterol. We consider phospholipids in this section and the other two types of membrane lipids in the next two sections.

FIGURE 8.12 The oils (triacylglycerols) present in skin perspiration rapidly undergo oxidation. The oxidation products, short-chain aldehydes and short-chain carboxylic acids, often have strong odors.

Phospholipids are the most abundant type of membrane lipid. A **phospholipid** *is a lipid that contains one or more fatty acids, a phosphate group, a platform molecule to which the fatty acid(s) and the phosphate group are attached, and an alcohol that is attached to the phosphate group.* The platform molecule on which a phospholipid is built may be the 3-carbon alcohol *glycerol* or a more complex C_{18} aminodialcohol called *sphingosine.* Glycerol-based phospholipids are called *glycerophospholipids,* and those based on sphingosine are called *sphingophospholipids.* The general block diagrams for a glycerophospholipid and a sphingophospholipid are as follows:

> An aminodialcohol contains two hydroxyl groups, —OH, and an amino group, —NH₂.

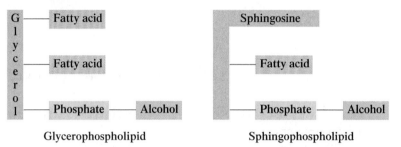

Glycerophospholipid   Sphingophospholipid

▇ Glycerophospholipids

A **glycerophospholipid** *is a lipid that contains two fatty acids and a phosphate group esterified to a glycerol molecule and an alcohol esterified to the phosphate group.* All attachments (bonds) between groups in a glycerophospholipid are ester linkages, a situation similar to that in triacylglycerols (Section 8.4). However, glycerophospholipids have four ester linkages as contrasted to three ester linkages in triacylglycerols.

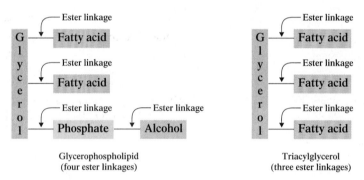

Glycerophospholipid   Triacylglycerol
(four ester linkages)   (three ester linkages)

Because of the ester linkages present, glycerophospholipids undergo hydrolysis and saponification reactions in a manner similar to that for triacylglycerols (Section 8.6). There will be five reaction products, however, instead of the four for triacylglycerols.

The alcohol attached to the phosphate group in a glycophospholipid is usually one of three amino alcohols: choline, ethanolamine, or serine. The structures of these three amino alcohols, given in terms of the charged forms (Sections 6.8 and 9.43) that they adopt in neutral solution, are

$$HO-CH_2-CH_2-\overset{+}{N}(CH_3)_3 \qquad HO-CH_2-CH_2-\overset{+}{N}H_3 \qquad HO-CH_2-\underset{\underset{COO^-}{|}}{CH}-\overset{+}{N}H_3$$

Choline   Ethanolamine   Serine
(a quaternary ammonium ion)   (positive-ion form)   (two ionic groups present)

Glycerophospholipids containing these three amino alcohols are respectively known as phosphatidylcholines, phosphatidylethanolamines, and phosphatidylserines. The fatty acid, glycerol, and phosphate portions of a glycerophospholipid structure constitute a *phosphatidyl* group.

Although the general structural features of glycerophospholipids are similar in many respects to those of triacylglycerols, these two types of lipids have quite different biochemical functions. Triacylglycerols serve as storage molecules for metabolic fuel.

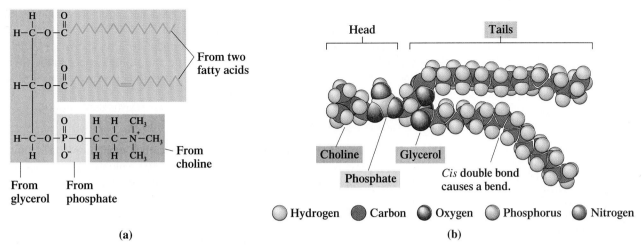

FIGURE 8.13 (a) Structural formula and (b) molecular model showing the "head and two tails" structure of a phosphatidylcholine molecule containing stearic acid (18:0) and oleic acid (18:1).

Glycerophospholipids have a *hydrophobic* ("water-hating") portion, the nonpolar fatty acid groups, and a *hydrophilic* ("water-loving") portion, the polar head group.

The amino alcohol in phosphatidylcholines (pronounced fahs-fuh-TIDE-ul-KOH-leen) is choline.

Glycerophospholipids function almost exclusively as components of cell membranes (Section 8.10) and are not stored. A major structural difference between the two types of lipids, that of polarity, is related to their differing biochemical functions. Triacylglycerols are a nonpolar class of lipids, whereas glycerophospholipids are polar. In general, membrane lipids have polarity associated with their structures.

Further consideration of general phosphoacylglycerol structure reveals an additional structural characteristic of most membrane lipids. Let us consider a phosphatidylcholine containing stearic and oleic acids to illustrate this additional feature. The chemical structure of this molecule is shown in Figure 8.13a.

A molecular model for this compound, which gives the orientation of groups in space, is illustrated in Figure 8.13b. There are two important things to notice about this model: (1) There is a "head" part, the choline and phosphate, and (2) there are two "tails," the two fatty acid carbon chains. The head part is polar. The two tails, the carbon chains, are nonpolar.

All glycerophospholipids have structures similar to that shown in Figure 8.13. All have a "head" and two "tails." A simplified representation for this structure uses a circle to represent the polar head and two wavy lines to represent the nonpolar tails.

The polar head group of a glycerophospholipid is soluble in water. The nonpolar tail chains are insoluble in water but soluble in nonpolar substances. This dual polarity, which we previously encountered when we discussed soaps (see the Chemical Connections feature on page 284), is a structural characteristic of most membrane lipids.

Phosphatidylcholines are also known as *lecithins*. There are a number of different phosphatidylcholines because different fatty acids may be bonded to the glycerol portion of the phosphatidylcholine structure. In general, phosphatidylcholines are waxy solids that form colloidal suspensions in water. Egg yolks and soybeans are good dietary sources of these lipids. Within the body, phosphatidylcholines are prevalent in cell membranes.

Periodically, claims arise that phosphatidylcholine should be taken as a nutritive supplement; some even maintain it will improve memory. There is no evidence that these supplements are useful. The enzyme *lecithinase* in the intestine hydrolyzes most of the phosphatidylcholine taken orally before it passes into body fluids, so it does not reach body tissues. The phosphatidylcholine present in cell membranes is made by the liver; thus phosphatidylcholines are not essential nutrients.

The food industry uses phosphatidylcholines as emulsifiers to promote the mixing of otherwise immiscible materials. Mayonnaise, ice cream, and custards are some of the products they are found in. It is the polar–nonpolar (head–tail) structure of phosphatidyl-cholines that enables them to function as emulsifiers.

Phosphatidylethanolamines and phosphatidylserines are also known as *cephalins.* These compounds are found in heart and liver tissue and in high concentrations in the brain. They are important in blood clotting. Much is yet to be learned about how these compounds function within the human body.

■ Sphingophospholipids

> When sphingolipids were discovered over a century ago by the physician–chemist Johann Thudichum (1829–1901), their biochemical role seemed as enigmatic as the Sphinx, for which he named them.

Sphingophospholipids have structures based on the 18-carbon monounsaturated aminodi-alcohol *sphingosine.* A **sphingophospholipid** *is a lipid that contains one fatty acid and one phosphate group attached to a sphingosine molecule and an alcohol attached to the phosphate group.*

The structure of sphingosine, the platform molecule for a sphingophospholipid, is

$$CH_3-(CH_2)_{12}-CH\text{=}CH-\underset{\underset{OH}{|}}{CH}-\underset{\underset{NH_2}{|}}{CH}-\underset{\underset{OH}{|}}{CH_2}$$

Sphingosine

All phospholipids derived from sphingosine have (1) the fatty acid attached to the sphingosine —NH$_2$ group via an *amide linkage,* (2) the phosphate group attached to the sphingosine terminal —OH group via an *ester linkage,* and (3) an additional alcohol es-terified to the phosphate group. The general block diagram for a sphingophospholipid is

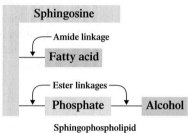

Sphingophospholipid
(two ester linkages and one amide linkage)

> In sphingophospholipids, the first three carbon atoms at the polar end of sphingosine are analogous to the three carbon atoms of glycerol in glycerophospholipids.

Molecular models showing orientation of atoms in space for sphingosine itself and for a sphingophospholipid are given in Figure 8.14. Note that, as in glycerophospholipids, the "head and two tails" structure is present in sphingophospholipids. For sphingophospho-lipids, the fatty acid is one of the tails, and the long carbon chain of sphingosine itself is the other tail. The polar head is the phosphate group with its esterified alcohol.

Like glycerophospholipids, sphingophospholipids participate in saponification reac-tions. Amide linkages behave much as ester linkages do in this type of reaction.

Sphingophospholipids in which the alcohol esterified to the phosphate group is *choline* are called *sphingomyelins.* Sphingomyelins are found in all cell membranes and are important structural components of the myelin sheath, the protective and insulating

FIGURE 8.14 Molecular models for (a) sphingosine and (b) a sphingophospholipid. The particular sphingophospholipid shown has choline as the alcohol esterified to the phosphate group. Note the "head and two tails" structure for the sphingophospholipid.

(a) Sphingosine　　　　　　　　**(b) A sphingomyelin**

coating that surrounds nerves. The molecule depicted in Figure 8.14b is a sphingomyelin. The structural formula for a sphingomyelin in which stearic acid (18:0) is the fatty acid is

Sphingosine

$$HO-CH-CH=CH-(CH_2)_{12}-CH_3$$

$$CH-NH-\overset{\overset{\displaystyle O}{\|}}{C}-(CH_2)_{16}-CH_3$$

Stearic acid (18:0)

$$CH_2-O-\overset{\overset{\displaystyle O}{\|}}{\underset{\underset{\displaystyle O^-}{|}}{P}}-O-CH_2-CH_2-\overset{\overset{\displaystyle CH_3}{|}}{\underset{\underset{\displaystyle CH_3}{|}}{\overset{+}{N}}}-CH_3$$

Phosphate   Choline

8.8 Membrane Lipids: Sphingoglycolipids

The second of the three major types of membrane lipids is *sphingoglycolipids*. A **sphingoglycolipid** *is a lipid that contains both a fatty acid and a carbohydrate component attached to a sphingosine molecule.* A fatty acid is attached to the sphingosine through an amide linkage, and a monosaccharide or oligosaccharide (Section 7.3) is attached to the sphingosine at the terminal —OH carbon atom through a glycosidic linkage (Section 7.13). The generalized block diagram for a sphingoglycolipid is

Sphingosine

— Amide linkage
Fatty acid

— Glycosidic linkage
Monosaccharide or Oligosaccharide

Sphingoglycolipids undergo saponification reactions; both the amide and the glycosidic linkages can be hydrolyzed under saponification conditions.

The simplest sphingoglycolipids, which are called *cerebrosides,* contain a single monosaccharide unit—either glucose or galactose. As the name suggests, cerebrosides occur primarily in the brain (7% of dry mass). They are also present in the myelin sheath of nerves. The specific structure for a cerebroside in which stearic acid (18:0) is the fatty acid and galactose is the monosaccharide is

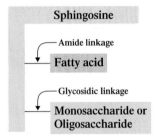

Sphingosine

$$HO-CH-CH=CH-(CH_2)_{12}-CH_3$$

$$CH-NH-\overset{\overset{\displaystyle O}{\|}}{C}-(CH_2)_{16}-CH_3 \quad \text{Stearic acid (18:0)}$$

Galactose

CHEMISTRY AT A GLANCE

Terminology for and Structural Relationships Among Various Types of Fatty-Acid-Containing Lipids

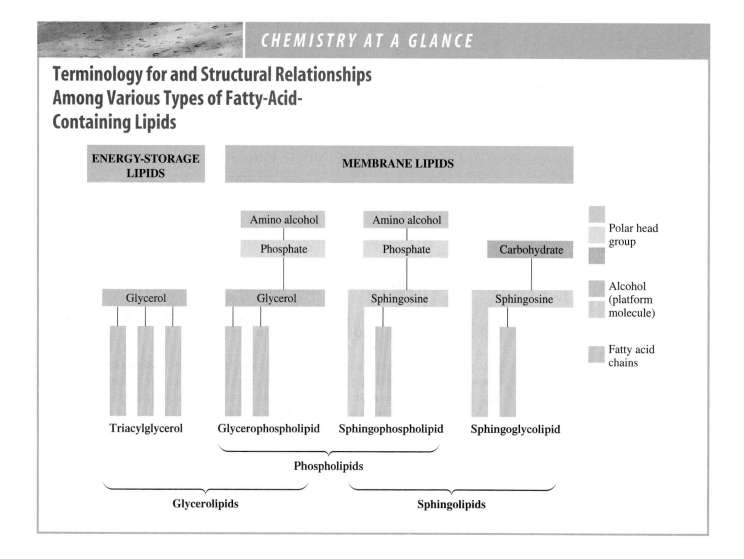

More complex sphingoglycolipids, called *gangliosides,* contain a branched chain of up to seven monosaccharide residues. These substances occur in the gray matter of the brain as well as in the myelin sheath.

The Chemistry at a Glance feature summarizes terminology and structural relationships among the types of lipids that we have considered up to this point. The common thread among all of the structures is the presence of at least one fatty acid residue.

8.9 Membrane Lipids: Cholesterol

Cholesterol, the third of the three major types of membrane lipids, is a specific compound rather than a family of compounds like the phospholipids (Section 8.7) and sphingoglycolipids (Section 8.8). Cholesterol's structure differs markedly from that of other membrane lipids in that (1) there are no fatty acid residues present and (2) neither glycerol nor sphingosine is present as the platform molecule.

Cholesterol is a *steroid.* A **steroid** *is a lipid whose structure is based on a fused-ring system that involves three 6-membered rings and one 5-membered ring.* This steroid fused-ring system, which is called the *steroid nucleus,* has the following structure:

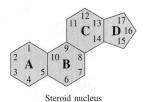

Steroid nucleus

Note that each of the rings of the steroid nucleus carries a letter designation and that a "consecutive" numbering system is used to denote individual carbon atoms.

Numerous steroids have been isolated from plants, animals, and human beings. Location of double bonds within the fused-ring system and the nature and location of substituents distinguish one steroid from another. Most steroids have an oxygen functional group ($=O$ or $-OH$) at carbon 3 and some kind of side chain at carbon 17. Many also have a double bond from carbon 5 to either carbon 4 or carbon 6.

Cholesterol *is a C_{27} steroid molecule that is a component of cell membranes and a precursor for other steroid-based lipids.* It is the most abundant steroid in the human body. The *-ol* ending in the name cholester*ol* conveys the information that an alcohol functional group is present in this molecule; it is located on carbon 3 of the steroid nucleus. In addition, cholesterol has methyl group attachments at carbons 10 and 13, a carbon–carbon double bond between carbons 5 and 6, and an eight-carbon branched side chain at carbon 17. Figure 8.15 gives both the structural formula and a molecular model for cholesterol. The molecular model shows the rather compact nature of the cholesterol molecule. The "head and two tails" arrangement found in other membrane lipids is not present. The lack of a *large* polar head group causes cholesterol to have limited water solubility. The $-OH$ group on carbon 3 is considered the head of the molecule.

Within the human body, cholesterol is found in cell membranes (up to 25% by mass), in nerve tissue, in brain tissue (about 10% by dry mass), and in virtually all fluids. Every 100 mL of human blood plasma contains about 50 mg of free cholesterol and about 170 mg of cholesterol esterified with various fatty acids.

Although a portion of the body's cholesterol is obtained from dietary intake, most of it is biosynthesized by the liver and (to a lesser extent) the intestine. Typically, 800–1000 mg are biosynthesized each day. Ingested cholesterol decreases biosynthetic cholesterol production. However, the reduction is less than the amount ingested. Therefore, total body cholesterol levels increase with increased dietary intake of cholesterol.

Biosynthetic cholesterol is distributed to cells throughout the body for various uses via the bloodstream. Because cholesterol is only sparingly soluble in water (blood), a protein carrier system is used for its distribution. These cholesterol–protein combinations are called *lipoproteins*.

The lipoproteins that carry cholesterol *from the liver* to various tissues are called LDLs (low-density lipoproteins), and those that carry excess cholesterol from tissues *back to the liver* are called HDLs (high-density lipoproteins). If too much cholesterol is being

Besides being an important molecule in and of itself, cholesterol serves as a precursor for several other important steroid molecules including bile acids (Section 8.11), steroid hormones (Section 8.12), and vitamin D (Section 10.14).

FIGURE 8.15 Structural formula and molecular model for the cholesterol molecule.

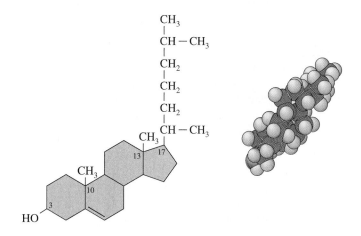

FIGURE 8.16 A severely occluded artery—the result of the buildup of cholesterol-containing plaque deposits.

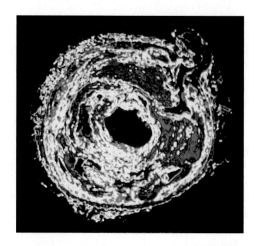

transported by LDLs or too little by HDLs, the imbalance results in an increase in blood cholesterol levels. High blood cholesterol levels contribute to atherosclerosis, a form of cardiovascular disease characterized by the buildup of plaque along the inner walls of arteries. Plaque is a mound of lipid material mixed with smooth muscle cells and calcium. Much of the lipid material in plaque is cholesterol. Plaque deposits in the arteries that serve the heart reduce blood flow to the heart muscle and can lead to a heart attack. Figure 8.16 shows the occlusion that can occur in an artery as a result of plaque buildup.

The cholesterol associated with LDLs is often called "bad cholesterol" because it contributes to increased blood cholesterol levels, and the cholesterol associated with HDLs is often called "good cholesterol" because it contributes to reduced blood cholesterol levels. The Chemical Connections feature titled "Lipoproteins and Heart Attack Risk" on page 342 in the next chapter considers this topic in further detail.

Much still needs to be learned concerning the actual role played by serum cholesterol in plaque buildup within arteries. Current knowledge suggests that it makes good sense to reduce the amount of cholesterol (as well as saturated fats) taken into the body through dietary intake. People who want to reduce dietary cholesterol intake should reduce the amount of animal products they eat (meat, dairy products, etc.) and eat more fruit and vegetables. Plant foods contain negligible amounts of cholesterol; cholesterol is found primarily in foods of animal origin. Table 8.4 gives cholesterol amounts associated with selected foods.

8.10 Cell Membranes

Cell membranes are also commonly called *plasma membranes* because they separate the cytoplasm (aqueous contents) of a cell from its surroundings.

Prior to discussing additional types of lipid molecules—emulsification lipids (Section 8.11), messenger lipids (Sections 8.12 and 8.13), and protective-coating lipids (Section 8.14)—we will extend our discussion of membrane lipids to include how these types of lipids interact with each other to form cell membranes.

TABLE 8.4
The Amount of Cholesterol Found in Various Foods

Food	Cholesterol (mg)
liver (3 oz)	410
egg (1 large)	213
shrimp (3 oz)	166
pork chop (3 oz)	83
chicken (3 oz)	75
beef steak (3 oz)	70
fish fillet (3 oz)	54
whole milk (1 cup)	33
cheddar cheese (1 oz)	30
Swiss cheese (1 oz)	26
low-fat milk (1 cup)	22

FIGURE 8.17 Cross section of a lipid bilayer. The circles represent the polar heads of the lipid components, and the wavy lines represent the nonpolar tails of the lipid components. The heads occupy "surface" positions, and the tails occupy "internal" positions.

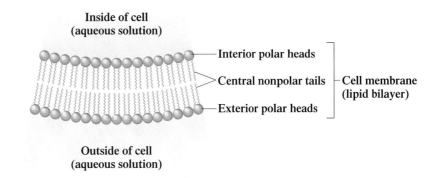

Inside of cell
(aqueous solution)

Interior polar heads

Central nonpolar tails — Cell membrane (lipid bilayer)

Exterior polar heads

Outside of cell
(aqueous solution)

The percentage of lipid and protein components in a cell membrane is related to the function of the cell. The lipid/protein ratio ranges from about 80% lipid/20% protein by mass in the myelin sheath of nerve cells to the unique 20% lipid/80% protein ratio for the inner mitochondrial membrane (Section 12.2). Red blood cell membranes contain approximately equal amounts of lipid and protein. A typical membrane also has a carbohydrate content that varies between 2% and 10% by mass.

The glycerophospholipids and sphingoglycolipids found in a lipid bilayer are chiral molecules with the chiral center (Section 7.4) at carbon 2 of the glycerol or sphingosine components of the molecules. The stereoconfiguration of these chiral molecules is always "left-handed," that is, L isomers. The fact that they all have the same configuration enhances their ability to aggregate together in the lipid bilayer.

Living cells contain an estimated 10,000 different kinds of molecules in an aqueous environment confined by a *cell membrane*. A **cell membrane** *is a lipid-based structure that separates a cell's aqueous-based interior from the aqueous environment surrounding the cell.* Besides its "separation" function, a cell membrane also controls the movement of substances into and out of the cell. Up to 80% of the mass of a cell membrane is lipid material consisting primarily of the three types of membrane lipids we have just discussed: phospholipids, glycolipids, and cholesterol.

The keys to understanding the structural basis for a cell membrane are (1) the virtually insoluble nature of membrane lipids in water and (2) the "head and two tails" structure (Section 8.7) of phospholipids and sphingoglycolipids. When these lipids are placed in water, the polar heads of phospholipids and sphingoglycolipids favor contact with water, whereas their nonpolar tails interact with one another rather than with water. The result is a remarkable bit of molecular architecture called a *lipid bilayer*. A **lipid bilayer** *is a two-layer-thick structure of phospholipids and glycolipids in which the nonpolar tails of the lipids are in the middle of the structure and the polar heads are on the outside surfaces of the structure.* Such a bilayer is six-billionths to nine-billionths of a meter thick—that is, 6 to 9 nanometers thick. There are three distinct parts to the bilayer: the exterior polar "heads," the interior polar "heads," and the central nonpolar "tails," as shown in Figure 8.17.

Figure 8.18, which is based on space-filling models for phospholipids, gives a "close-up" view of the arrangement of lipid molecules in a section of a lipid bilayer. Note the "exterior" nature of the polar heads of these membrane lipids.

A lipid bilayer is held together by intermolecular interactions, not by covalent bonds. This means each phospholipid or sphingolipid is free to diffuse laterally within the lipid bilayer. Most lipid molecules in the bilayer contain at least one unsaturated fatty acid. The presence of such acids, with the kinks in their carbon chains (Section 8.3), prevents tight packing of fatty acid chains (Figure 8.19). The open packing imparts a liquidlike character to the membrane—a necessity because numerous types of biochemicals must pass into and out of a cell.

FIGURE 8.18 Space-filling model of a section of a lipid bilayer. The key to the structure is the "head and two tails" structure of the membrane lipids that constitute the bilayer.

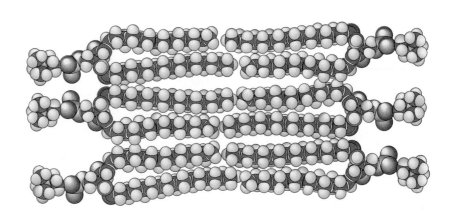

FIGURE 8.19 The kinks associated with *cis* double bonds in fatty acid chains prevent tight packing of the lipid molecules in a lipid bilayer.

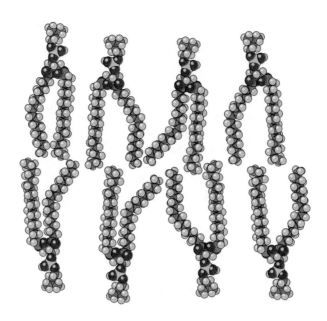

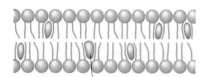

Cholesterol

FIGURE 8.20 Cholesterol molecules fit between fatty acid chains in a lipid bilayer.

Cholesterol molecules are also components of cell membranes. They regulate membrane fluidity. Because of their compact shape (Section 8.9; Figure 8.15), cholesterol molecules fit between the fatty acid chains of the lipid bilayer (Figure 8.20), restricting movement of the fatty acid chains. Within the membrane, the cholesterol molecule orientation is "head" to the outside (the hydroxyl group) and "tail" to the inside (the steroid ring structure with its attached alkyl groups).

Proteins are also components of lipid bilayers. The proteins are responsible for moving substances such as nutrients and electrolytes across the membrane, and they also act as receptors that bind hormones and neurotransmitters.

There are two general types of membrane proteins: *integral* and *peripheral*. An **integral protein** *is a membrane protein that penetrates the cell membrane.* Some membrane proteins penetrate only partially through the lipid bilayer while others go completely from one side to the other side of the lipid bilayer. A **peripheral protein** *is a nonpenetrating membrane protein located on the surface of the cell membrane.* Intermolecular forces rather than chemical bonds govern the interactions between membrane proteins and the lipid bilayer. Figure 8.21 shows diagrammatically the relationship between membrane proteins and the overall structure of a cell membrane.

FIGURE 8.21 Proteins are important structural components of cell membranes.

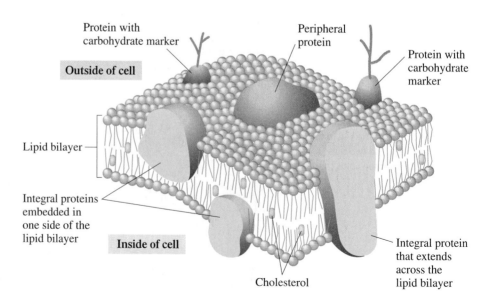

Small carbohydrate molecules are also components of cell membranes. They are found on the *outer* membrane surface covalently bonded to protein molecules (a glycoprotein) or lipid molecules (a glycolipid). The carbohydrate portions of glycoproteins and glycolipids function as *markers,* substances that play key roles in the process by which different cells recognize each other.

■ Transport Across Cell Membranes

In order for cellular processes to be maintained, molecules of various types must be able to cross cell membranes. Three common transport mechanisms exist by which molecules can enter and leave cells. They are *passive* transport, *facilitated* transport, and *active* transport.

Passive transport *is the transport process in which a substance moves across a cell membrane by diffusion from a region of higher concentration to a region of lower concentration without the expenditure of any cellular energy.* Only a few types of molecules, including O_2, N_2, H_2O, urea, and ethanol, can cross membranes in this manner. Passive transport is closely related to the process of osmosis.

Facilitated transport *is the transport process in which a substance moves across a cell membrane, with the aid of membrane proteins, from a region of higher concentration to a region of lower concentration without the expenditure of cellular energy.* The specific protein molecules involved in the process are called *carriers* or *transporters.* A carrier protein forms a complex with a specific molecule at one surface of the membrane. Formation of the complex induces a conformational change in the protein that allows the molecule to move through a "gate" to the other side of the membrane. Once the molecule is released, the protein returns to its original conformation. Glucose, chloride ion, and bicarbonate ion cross membranes in this manner.

Active transport *is the transport process in which a substance moves across a cell membrane, with the aid of membrane proteins, against a concentration gradient with the expenditure of cellular energy.* Proteins involved in active transport are called "pumps," because they require energy much as a water pump requires energy in order to function. The needed energy is supplied by molecules such as ATP (Section 12.3). The need for energy expenditure is related to the molecules moving against a concentration gradient— from lower to higher concentration. It is essential to life processes to have some solutes "permanently" at different concentrations on the two sides of a membrane, a situation contrary to the natural tendency (osmosis) to establish equal concentrations on both sides of a membrane. Hence the need for active transport. Sodium, potassium, and hydronium ions cross membranes through active transport.

Figure 8.22 contrasts the processes of passive transport, facilitated transport, and active transport.

FIGURE 8.22 Three processes by which substances can cross plasma membranes: (a) passive transport, (b) facilitated transport, and (c) active transport.

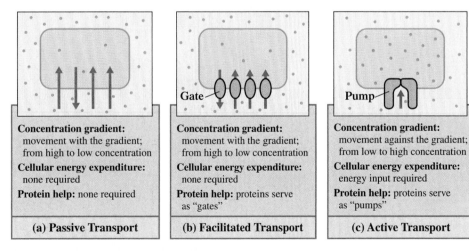

(a) Passive Transport	(b) Facilitated Transport	(c) Active Transport
Concentration gradient: movement with the gradient; from high to low concentration	**Concentration gradient:** movement with the gradient; from high to low concentration	**Concentration gradient:** movement against the gradient; from low to high concentration
Cellular energy expenditure: none required	**Cellular energy expenditure:** none required	**Cellular energy expenditure:** energy input required
Protein help: none required	**Protein help:** proteins serve as "gates"	**Protein help:** proteins serve as "pumps"

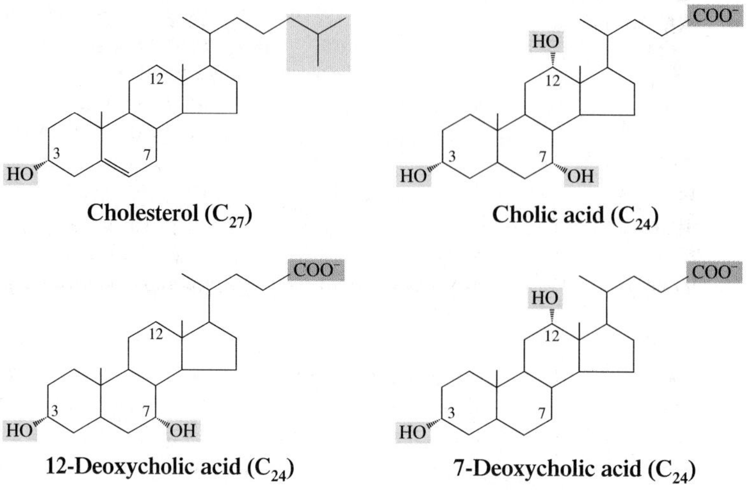

FIGURE 8.23 Structural formulas for cholesterol, cholic acid, and two deoxycholic acids.

Cholesterol (C_{27})

Cholic acid (C_{24})

12-Deoxycholic acid (C_{24})

7-Deoxycholic acid (C_{24})

8.11 Emulsification Lipids: Bile Acids

An **emulsifier** *is a substance that can disperse and stabilize water-insoluble substances as colloidal particles in an aqueous solution.* Cholesterol derivatives called *bile acids* function as emulsifying agents that facilitate the absorption of dietary lipids in the intestine. Their mode of action is much like that of soap during washing (see the Chemical Connections feature on page 284).

A **bile acid** *is a cholesterol derivative that functions as a lipid-emulsifying agent in the aqueous environment of the digestive tract.* Approximately one-third of the daily production of cholesterol by the liver is converted to bile acids. Obtained by oxidation of cholesterol, bile acids differ structurally from cholesterol in three respects:

1. They are tri- or dihydroxy cholesterol derivatives.
2. The carbon 17 side chain of cholesterol has been oxidized to a carboxylic acid.
3. The oxidized acid side chain is bonded to an amino acid (either glycine or taurine) through an amide linkage.

Figure 8.23 gives structural formulas for the three major types of bile acids produced from cholesterol by biochemical oxidation: cholic acid, 7-deoxycholic acid, and 12-de-

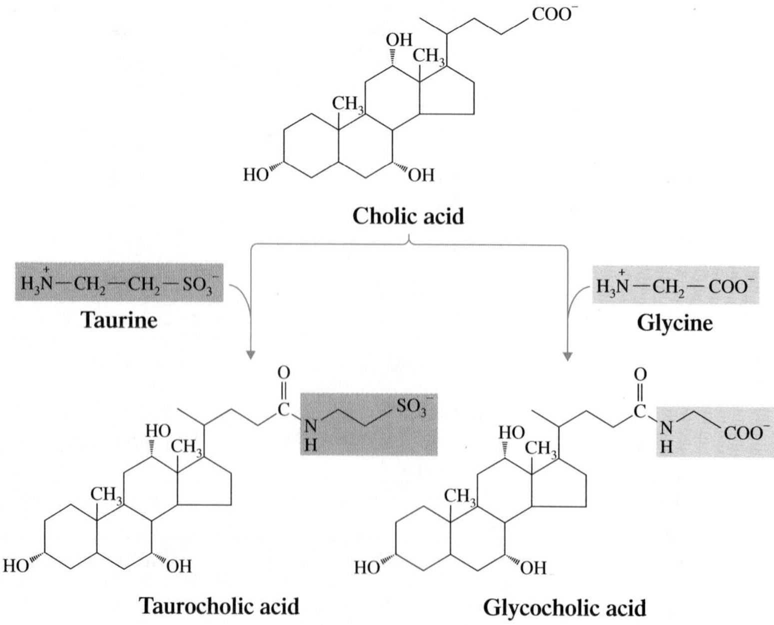

FIGURE 8.24 The structures of glycocholic acid and taurocholic acid.

Cholic acid

Taurine

Glycine

Taurocholic acid

Glycocholic acid

FIGURE 8.25 A large percentage of gallstones, the causative agent for many "gallbladder attacks," are almost pure crystallized cholesterol that has precipitated from bile solution.

The average bile acid composition in normal human adult bile is 38% cholic acid derivatives, 34% 7-deoxycholic acid derivatives, and 28% 12-deoxycholic acid derivatives. Glycine-containing derivatives predominate over taurine-containing derivatives by a 3:1 to 4:1 ratio. Uncomplexed (free) bile acids are not present in bile.

oxycholic acid. The structural formulas are those for these bile acids prior to the attachment of the amino acid to the carbon 17 side chain.

Bile acids always carry an amino acid (either glycine or taurine) attached to the side-chain carboxyl group via an amide linkage. The presence of this amino acid attachment increases both the polarity of the bile acid and its water solubility. Figure 8.24 shows the structures of glycocholic acid (glycine is the amino acid) and taurocholic acid (taurine is the amino acid).

The medium through which bile acids are supplied to the small intestine is *bile*. **Bile** *is a fluid containing emulsifying agents that is secreted by the liver, stored in the gallbladder, and released into the small intestine during digestion.* Besides bile acids, bile also contains bile pigments (breakdown products of hemoglobin; Section 15.7), cholesterol itself, and electrolytes such as bicarbonate ion. The bile acids that are present increase the solubility of the cholesterol in the bile fluid.

A number of factors, including increased secretion of cholesterol and a decrease in the size of the bile pool, can upset the balance between the cholesterol present in bile and the bile acid derivatives needed to maintain cholesterol's solubility in the bile. The result is the precipitation of crystallized cholesterol from the bile and the resulting formation of gallstones in the gallbladder. In Western countries, approximately 80% of gallstones are almost pure cholesterol (see Figure 8.25).

8.12 Messenger Lipids: Steroid Hormones

We have previously considered lipids that function as energy-storage molecules (triacylglycerols; Section 8.4), as components of cell membranes (phospholipids, sphingoglycolipids, and cholesterol; Sections 8.7 through 8.9), and as emulsifying agents (bile acids; Section 8.11). An additional role played by lipids is that of "chemical messenger." *Steroid hormones* and *eicosanoids* are two large families of lipids that have messenger functions. In this section we consider steroid hormones, which are cholesterol derivatives. In Section 8.13 we consider eicosanoids, which are fatty acid derivatives.

A **hormone** *is a biochemical substance, produced by a ductless gland, that has a messenger function.* Hormones serve as a means of communication *between* various tissues. Some hormones, though not all, are lipids.

A **steroid hormone** *is a hormone that is a cholesterol derivative.* There are two major classes of steroid hormones: (1) sex hormones, which control reproduction and secondary sex characteristics, and (2) adrenocorticoid hormones, which regulate numerous biochemical processes in the body.

■ Sex Hormones

The sex hormones can be classified into three major groups:

1. Estrogens—the female sex hormones
2. Androgens—the male sex hormones
3. Progestins—the pregnancy hormones

Estrogens are synthesized in the ovaries and adrenal cortex and are responsible for the development of female secondary sex characteristics at the onset of puberty and for regulation of the menstrual cycle. They also stimulate the development of the mammary glands during pregnancy and induce estrus (heat) in animals.

Androgens are synthesized in the testes and adrenal cortex and promote the development of secondary male characteristics. They also promote muscle growth.

Progestins are synthesized in the ovaries and the placenta and prepare the lining of the uterus for implantation of the fertilized ovum. They also suppress ovulation.

Figure 8.26a, gives the structure of the primary hormone in each of the three subclasses of sex hormones. Other members of these hormone families are metabolized forms of the primary hormone.

Estrogens are a class of molecules rather than a single molecule. Statements like "the estrogen level is high" should be rephrased as "there is a high level of estrogens."

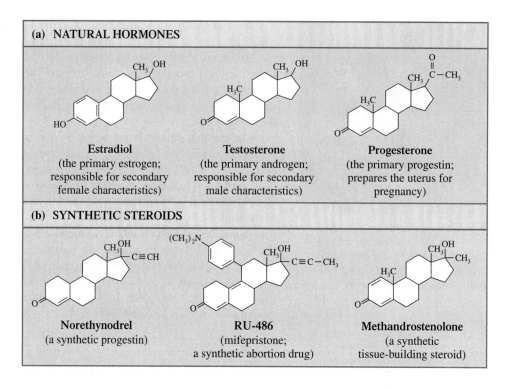

(a) NATURAL HORMONES

Estradiol
(the primary estrogen; responsible for secondary female characteristics)

Testosterone
(the primary androgen; responsible for secondary male characteristics)

Progesterone
(the primary progestin; prepares the uterus for pregnancy)

(b) SYNTHETIC STEROIDS

Norethynodrel
(a synthetic progestin)

RU-486
(mifepristone; a synthetic abortion drug)

Methandrostenolone
(a synthetic tissue-building steroid)

Note, in Figure 8.26a, how similar the structures are for these principal hormones, and yet how different their functions. The fact that seemingly minor changes in structure effect great changes in biofunction points out, again, the extreme specificity (Section 10.5) of the enzymes that control biochemical reactions.

Increased knowledge of the structures and functions of sex hormones has led to the development of a number of *synthetic* steroids whose actions often mimic those of the natural steroid hormones. The best known types of synthetic steroids are oral contraceptives and anabolic agents.

Oral contraceptives are used to suppress ovulation as a method of birth control. Generally, a mixture of a synthetic estrogen and a synthetic progestin is used. The synthetic estrogen regulates the menstrual cycle, and the synthetic progestin prevents ovulation, thus creating a false state of pregnancy. The structure of norethynodrel (Enovid), a synthetic progestin, is given in Figure 8.26b. Compare its structure to that of progesterone (the real hormone); the structures are very similar.

Interestingly, the controversial "morning after" pill developed in France and known as RU-486, is also similar in structure to progesterone. RU-486 interferes with gestation of a fertilized egg and terminates a pregnancy within the first 9 weeks of gestation more effectively and safely than surgical methods. The structure of RU-486 appears next to that of norethynodrel in Figure 8.26b.

Anabolic agents include the illegal steroid drugs used by some athletes to build up muscle strength and enhance endurance. Anabolic agents are now known to have serious side effects on the user. The Chemical Connections feature on page 302 focuses on the use of anabolic steroids. The structure of one of the more commonly used anabolic agents, methandrostenolone, is given in Figure 8.26b. Note the similarities between its structure and that of the naturally occurring testosterone.

■ **Adrenocorticoid Hormones**

The second major group of steroid hormones consists of the adrenocorticoid hormones. Produced by the adrenal glands, small organs located on top of each kidney, at least 28 different hormones have been isolated from the adrenal cortex (the outer part of the glands).

The C≡C functional group, which occurs in both norethynodrel (Enovid) and RU-486, is rarely found in biomolecules.

FIGURE 8.27 Structures of selected adrenocorticoid hormones and related synthetic compounds.

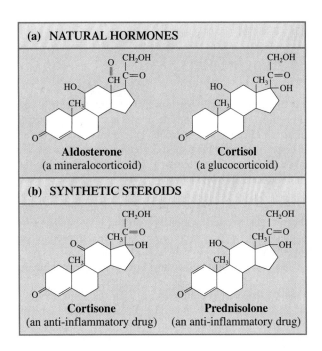

There are two types of adrenocorticoid hormones.

1. *Mineralocorticoids* control the balance of Na⁺ and K⁺ ions in cells.
2. *Glucocorticoids* control glucose metabolism and counteract inflammation.

The major mineralocorticoid is aldosterone, and the major glucocorticoid is cortisol (hydrocortisone). Cortisol is the hormone synthesized in the largest amount by the adrenal glands. Cortisol and its synthetic ketone derivative cortisone exert powerful anti-inflammatory effects in the body. Both cortisone and prednisolone, a similar synthetic derivative, are used as prescription drugs to control inflammatory diseases such as rheumatoid arthritis. Figure 8.27 gives the structures of these adrenocorticoid hormones.

8.13 Messenger Lipids: Eicosanoids

An **eicosanoid** *is an oxygenated C₂₀ fatty acid derivative that functions as a messenger lipid.* The term *eicosanoid* is derived from the Greek word *eikos,* which means "twenty." The metabolic precursor for most eicosanoids is arachidonic acid, the 20:4 fatty acid.

Almost all cells, except red blood cells, produce eicosanoids. These substances, like hormones, have profound physiological effects at extremely low concentrations. Eicosanoids are hormonelike molecules rather than true hormones because they are not transported in the bloodstream to their site of action as true hormones are. Instead, they exert their effects in the tissues where they are synthesized. Eicosanoids usually have a very short "life," being broken down, often within seconds of their synthesis, to inactive residues (which are eliminated in urine). For this reason, they are difficult to study and monitor within cells.

The physiological effects of eicosanoids include mediation of

Eicosanoids exert their effects at very low concentrations, sometimes less than one part in a billion (10^9).

1. The inflammatory response, a normal response to tissue damage
2. The production of pain and fever
3. The regulation of blood pressure
4. The induction of blood clotting
5. The control of reproductive functions, such as induction of labor
6. The regulation of the sleep/wake cycle

There are three principal types of eicosanoids: prostaglandins, thromboxanes, and leukotrienes.

CHEMICAL CONNECTIONS — Steroid Drugs in Sports

The steroid hormone testosterone is the principal male sex hormone. It has masculinizing (androgenic) effects and muscle-building (anabolic) effects. Masculinizing effects of testosterone include the growth of facial and body hair, deepening of the voice, and maturation of the male sex organs. Testosterone's anabolic effects are responsible for the muscle development that boys experience at puberty.

Some of the many synthetic testosterone derivatives exert primarily androgenic effects, whereas others exert primarily anabolic effects. Androgenic compounds can be used to correct hormonal imbalances in the body. Anabolic steroids can be used to prevent the withering of muscle in persons recovering from major surgery or serious injuries.

Anabolic steroids have also been "discovered" by athletes, who have found that these compounds can be used to help build muscle mass and reduce the healing time for muscle injuries. Often the net result of anabolic hormone use by an athlete is enhanced athletic performance.

The International Olympic Committee, as well as the NBA and NFL prohibit steroid use and test for it. Recently major league baseball adopted a policy regarding the use of steroids. There are two reasons for steroid prohibition in athletics: (1) Their use is considered a form of cheating because it confers an unfair advantage, and (2) their "beneficial effects" are far outweighed by serious negative side effects.

Current medical evidence indicates that using anabolic steroids is dangerous. Steroid abuse is associated with a wide range of adverse side effects ranging from some that are physically unattractive, such as acne and breast development in men, to others that are life-threatening, such as heart attacks and liver problems. Most of these alarming effects are reversible if the abuser stops taking the drugs, but some are permanent.

Steroid abuse disrupts the normal production of hormones in the body, causing both reversible and irreversible change. The male reproductive system is altered, causing testicular shrinkage and decreased sperm production. Both of these effects are reversible, but breast development is an irreversible change.

In the female body, steroid abuse causes masculinization. Breast size and body fat decrease, the skin becomes coarse, and the voice deepens. Some women may experience excessive growth of body hair but lose hair from the scalp.

A definite link exists between steroid abuse and cardiovascular diseases, including heart attacks and strokes, even in persons younger than 30. Steroids, particularly the oral types, increase the level of low-density lipoprotein (LDL) and decrease the level of high-density lipoprotein (HDL). High LDL and low HDL levels increase the risk of atherosclerosis. Steroids also increase the risk that blood clots will form in blood vessels.

■ Prostaglandins

A **prostaglandin** *is a messenger lipid that is a C_{20}-fatty-acid derivative that contains a cyclopentane ring and oxygen-containing functional groups.* Twenty-carbon fatty acids are converted into a prostaglandin structure when the eighth and twelfth carbon atoms of the fatty acid become connected to form a five-membered ring (Figure 8.28b).

Prostaglandins are named after the prostate gland, which was first thought to be their only source. Today, more than 20 prostaglandins have been discovered in a variety of tissues in both males and females.

FIGURE 8.28 Relationship of the structures of various eicosanoids to their precursor, arachidonic acid.

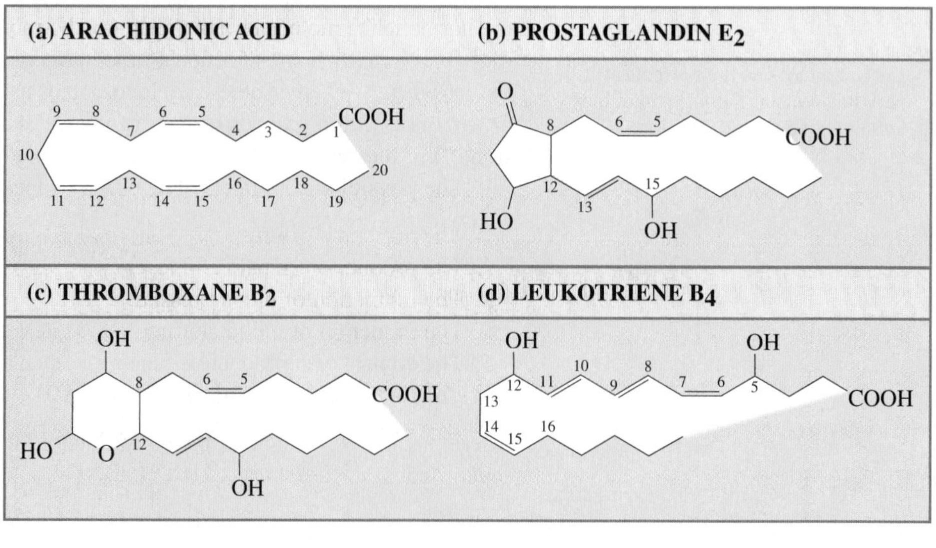

The capital letter–numerical subscript designations for individual eicosanoids is based on selected structural characteristics of the molecules. The numerical subscript indicates the number of carbon–carbon double bonds present. The letters denote subgroups of molecules. The prostaglandin E group, for example, has a carbonyl group on carbon 9.

CHEMICAL CONNECTIONS — The Mode of Action for Anti-Inflammatory Drugs

Injury or damage to bodily tissue is associated with the process of inflammation. This inflammation response is mediated by prostaglandin molecules (Section 8.13). The mode of action for most anti-inflammatory drugs now in use involves decreasing prostaglandin synthesis within the body by inhibiting the action of one or more of the enzymes (biochemical catalysts; Chapter 10) needed for prostaglandin synthesis.

Prostaglandin molecules are derivatives of arachidonic acid, a $20:4$ fatty acid (Section 8.13). Anti-inflammatory steroid drugs such as cortisone (Section 8.12) inhibit the action of the enzyme *phospholipase A$_2$*, the enzyme that facilitates the breakdown of complex arachidonic acid-containing lipids to produce free arachidonic acid. Inhibiting arachidonic acid release stops the prostaglandin synthesis process, which in turn prevents (or diminishes) inflammation.

Besides anti-inflammatory steroid drugs, many nonsteroidal anti-inflammatory drugs (NSAIDs) are also available for inflammation control. The most frequently used NSAIDs are the over-the-counter pain relievers aspirin, ibuprofen (Advil), and naproxen (Aleve). These substances, which have anti-pain, anti-fever, and anti-inflammatory properties, prevent prostaglandin synthesis by inhibiting the enzyme needed for the ring closure reaction at carbons 8 and 12 in arachidonic acid, a necessary step in prostaglandin synthesis (Section 8.13). The enzyme that NSAIDs inhibit is called *cyclooxygenase,* an enzyme known by the acronym *COX.*

There are actually two forms of the COX enzyme: COX-1 and COX-2. The COX-1 enzyme is involved in the normal physiological production of prostaglandin molecules (a desirable situation) and the COX-2 enzyme is responsible for the prostaglandin production associated with the inflammation response (a situation that is desirable to control). NSAIDs such

as aspirin, ibuprofen, and naproxen inhibit both the COX-1 and COX-2 enzymes (the good and the bad).

A new generation of prescription anti-inflammatory agents are now available that are COX-2 inhibitors but not COX-1 inhibitors. They have been touted by some as "super aspirins." The best known of the COX-2 inhibitors are Vioxx and Celebrex, whose chemical structures are

Vioxx   **Celebrex**

Like almost all anti-inflammatory drugs, these drugs have ulcer-causing side effects.

In 2004, Vioxx was withdrawn from the market because of concerns relative to heart attacks and strokes. A study indicated that patients taking Vioxx were twice as likely to suffer a heart attack or stroke as a control group involved in the study who were taking a placebo. The actual risk was 3.5% in the Vioxx group, compared with 1.9% in the control group, according to the FDA. The difference became apparent after 18 months of Vioxx use. The use of COX-2 inhibitors is now a topic under intense scrunity.

Within the human body, prostaglandins are involved in many regulatory functions, including raising body temperature, inhibiting the secretion of gastric juices, relaxing and contracting smooth muscle, directing water and electrolyte balance, intensifying pain, and enhancing inflammation responses. Aspirin reduces inflammation and fever because it inactivates the enzyme needed for prostaglandin synthesis.

■ Thromboxanes

A **thromboxane** *is a messenger lipid that is a C$_{20}$-fatty-acid derivative that contains a cyclic ether ring and oxygen-containing functional groups.* As with prostaglandins, the cyclic structure involves a bond between carbons 8 and 12 (Figure 8.28c). An important function of thromboxanes is to promote the formation of blood clots. Thromboxanes are produced by blood platelets and promote platelet aggregation.

■ Leukotrienes

A **leukotriene** *is a messenger lipid that is a C$_{20}$-fatty-acid derivative that contains three conjugated double bonds and hydroxy groups.* Fatty acids and their derivatives do not normally contain *conjugated* double bonds (see the Chemical Connections feature "Carotenoids: A Source of Color" on page 48 in Chapter 2), as is the case in leukotrienes (Figure 8.28d). Leukotrienes are found in leukocytes (white blood cells). Their source and the presence of the

FIGURE 8.29 A biological wax has a structure with a small, weakly polar "head" and two long, nonpolar "tails." The polarity of the small "head" is not sufficient to impart any degree of water solubility to the molecule.

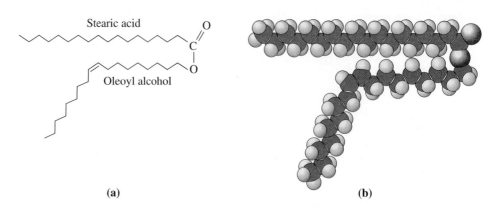

(a)　　　　(b)

three conjugated double bonds account for their name. Various inflammatory and hypersensitivity (allergy) responses are associated with elevated levels of leukotrienes. The development of drugs that inhibit leukotriene synthesis has been an active area of research.

8.14 Protective-Coating Lipids: Biological Waxes

A **biological wax** *is a lipid that is a monoester of a long-chain fatty acid and a long-chain alcohol.* Biological waxes are *monoesters,* unlike fats and oils (Section 8.4), which are *triesters.* The fatty acids found in biological waxes generally are saturated and contain from 14 to 36 carbon atoms. The alcohols found in biological waxes may be saturated or unsaturated and may contain from 16 to 30 carbon atoms.

The block diagram for a biological wax is

| Long-chain fatty acid |———| Long-chain alcohol |

with the fatty acid and alcohol linked through an ester linkage. An actual structural formula for a biological wax that bees secrete and use as a structural material is

The term *wax* derives from the old English word *weax,* which means "the material of the honeycomb."

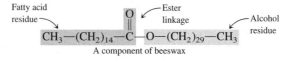

$$CH_3\!-\!(CH_2)_{14}\!-\!C\!-\!O\!-\!(CH_2)_{29}\!-\!CH_3$$

Fatty acid residue — Ester linkage — Alcohol residue

A component of beeswax

Note that the general structural formula for a biological wax is the same as that for a simple ester (Section 5.10).

$$R\!-\!\overset{O}{\overset{\|}{C}}\!-\!O\!-\!R'$$

FIGURE 8.30 Plant leaves often have a biological wax coating to prevent excessive loss of water.

However, for waxes both R and R′ must be long carbon chains (usually 20–30 carbon atoms).

The water-insoluble, water-repellent properties of biological waxes result from the complete dominance of the nonpolar nature of the long hydrocarbon chains present (from the alcohol and the fatty acid) over the weakly polar nature of the ester functional group that links the two carbon chains together (see Figure 8.29).

In living organisms biological waxes have numerous functions, all of which are related directly or indirectly to their water-repellent properties. Both humans and animals possess skin glands that secrete biological waxes to protect hair and skin and to keep it pliable and lubricated. With animal fur, waxes impart water repellency to the fur. Birds, particularly aquatic birds, rely on waxes secreted from preen glands to keep their feathers water repellent. Such wax coatings also help minimize loss of body heat when the bird is in cold water. Many plants, particularly those that grow in arid regions, have leaves that are coated with a thin layer of biological waxes, which serve to prevent excessive evaporation of water and to protect against parasite attack (see Figure 8.30). Similarly, insects with a high surface-area-to-volume ratio are often coated with a protective biological wax.

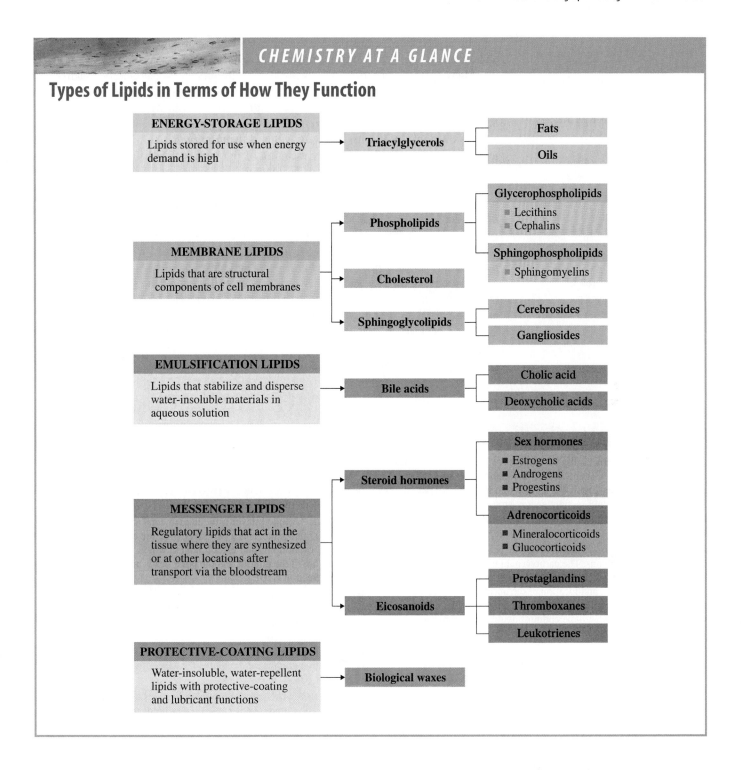

CHEMISTRY AT A GLANCE

Types of Lipids in Terms of How They Function

ENERGY-STORAGE LIPIDS

Lipids stored for use when energy demand is high

→ **Triacylglycerols**
— Fats
— Oils

MEMBRANE LIPIDS

Lipids that are structural components of cell membranes

→ **Phospholipids**
— **Glycerophospholipids**
  ▪ Lecithins
  ▪ Cephalins
— **Sphingophospholipids**
  ▪ Sphingomyelins

→ **Cholesterol**

→ **Sphingoglycolipids**
— **Cerebrosides**
— **Gangliosides**

EMULSIFICATION LIPIDS

Lipids that stabilize and disperse water-insoluble materials in aqueous solution

→ **Bile acids**
— Cholic acid
— Deoxycholic acids

MESSENGER LIPIDS

Regulatory lipids that act in the tissue where they are synthesized or at other locations after transport via the bloodstream

→ **Steroid hormones**
— **Sex hormones**
  ▪ Estrogens
  ▪ Androgens
  ▪ Progestins
— **Adrenocorticoids**
  ▪ Mineralocorticoids
  ▪ Glucocorticoids

→ **Eicosanoids**
— **Prostaglandins**
— **Thromboxanes**
— **Leukotrienes**

PROTECTIVE-COATING LIPIDS

Water-insoluble, water-repellent lipids with protective-coating and lubricant functions

→ **Biological waxes**

When aquatic birds are caught in an oil spill, the oil dissolves the wax coating on their feathers. This causes the birds to lose their buoyancy (they cannot swim properly) and compromises their protection against the effects of *cold* water.

Biological waxes find use in the pharmaceutical, cosmetics, and "polishing" industries. Carnauba wax (obtained from a species of Brazilian palm tree) is a particularly hard wax whose uses involve high-gloss finishes: automobile wax, boat wax, floor wax, and shoe wax.

$$CH_3-(CH_2)_{28}-\overset{\overset{\displaystyle O}{\|}}{C}-O-(CH_2)_{31}-CH_3$$

A component of carnauba wax

Naturally occuring waxes, such as beeswax, are usually mixtures of several monoesters rather than being a single monoester. This parallels the situation for fats and oils, which are mixtures of numerous triesters (triacylglycerols).

Human ear wax, which acts as a protective barrier against infection by capturing airborne particles, is not a true biological wax—that is, it is not a mixture of simple esters. Human ear wax is a yellow waxy secretion that is a mixture of triacylglycerols, phospholipids, and esters of cholesterol. Its medical name is *cerumen*.

Lanolin, a mixture of waxes obtained from sheep wool, is used as a base for skin creams and ointments intended to enhance retention of water (which softens the skin).

Many synthetic materials are now available with properties that closely match—and even improve on—the properties of biological waxes. Such synthetic materials, which are generally polymers, have now replaced biological waxes in many cosmetics, ointments, and the like. The synthetic *carbowax,* for example, is a polyether.

Throughout this discussion, we have used the term *biological wax* rather than just *wax.* This is because the everyday meaning of the term *wax* is broader in scope than the chemical definition of the term *biological wax.* In general discussions, a **wax** *is a pliable, water-repelling substance used particularly in protecting surfaces and producing polished surfaces.* This broadened definition for waxes includes not only *biological waxes* but also *mineral waxes.* A **mineral wax** *is a mixture of long-chain alkanes obtained from the processing of petroleum.* (How mineral waxes are obtained from petroleum was considered in Section 1.15.) Mineral waxes, which are also called *paraffin waxes,* resist moisture and chemicals and have no odor or taste. They serve as a waterproof coating for such paper products as milk cartons and waxed paper. Most candles are made from mineral waxes. Some "wax products" are a blend of biological and mineral waxes. For example, beeswax is sometimes a component of candle wax.

The Chemistry at a Glance feature on page 305 summarizes the function-based lipid classifications we have considered in this chapter. Subclassifications within each function classification are also given in this summary.

 ## CONCEPTS TO REMEMBER

Lipids. Lipids are a structurally heterogeneous group of compounds of biochemical origin that are soluble in nonpolar organic solvents and insoluble in water. Lipids are divided into five major types on the basis of biochemical function: energy-storage lipids, membrane lipids, emulsification lipids, messenger lipids, and protective-coating lipids (Section 8.1).

Fatty acids. Fatty acids are monocarboxylic acids that contain long, unbranched carbon chains. The carbon chain may be saturated, monounsaturated, or polyunsaturated. Length of carbon chain, degree of unsaturation, and location of the unsaturation influence the properties of fatty acids. Omega-3 and omega-6 fatty acids are unsaturated fatty acids with the endmost double bond three and six carbons, respectively, away from the methyl end of the carbon chain (Section 8.2).

Triacylglycerols. Triacylglycerols are energy-storage lipids formed by esterification of three fatty acids to a glycerol molecule. Fats are triacylglycerol mixtures that are solids or semi-solids at room temperature; they contain a relatively high percentage of saturated fatty acid residues. Oils are triacylglycerol mixtures that are liquids at room temperature; they contain a relatively high percentage of unsaturated fatty acid residues (Section 8.4).

Phospholipids. Phospholipids are membrane lipids that contain one or more fatty acids, a phosphate group, a platform molecule to which the fatty acid(s) and phosphate group are attached, and an alcohol attached to the phosphate group. The platform molecule is either glycerol (glycerophospholipids) or sphingosine (sphingophospholipids). Phospholipids have a "head and two tails" structure. Lecithins, cephalins, and sphingomyelins are types of phospholipids (Section 8.7).

Sphingoglycolipids. Sphingoglycolipids are membrane lipids in which a fatty acid and a mono- or oligosaccharide are attached to the platform molecule sphingosine. Cerebrosides and gangliosides are types of sphingoglycolipids (Section 8.8).

Cholesterol. Cholesterol is a membrane lipid whose structure contains a steroid nucleus. It is the most abundant type of steroid. Besides its membrane functions, it also serves as a precursor for several other types of lipids (Section 8.9).

Lipid bilayer. A lipid bilayer is the fundamental structure associated with a cell membrane. It is a two-layer structure of lipid molecules (mostly phospholipids and glycolipids) in which the nonpolar tails of the lipids are in the interior and the polar heads are on the outside surfaces (Section 8.10).

Membrane transport mechanisms. The transport mechanisms by which molecules enter and leave cells include *passive* transport, *facilitated* transport, and *active* transport. Passive and facilitated transport follow a concentration gradient and do not involve cellular energy expenditure. Active transport involves movement against a concentration gradient and requires the expenditure of cellular energy (Section 8.10).

Bile acids. Bile acids are cholesterol derivatives that function as emsulsification lipids. They cause dietary lipids to be soluble in the aqueous environment of the digestive tract. Cholic acid and deoxycholic acids are the major types of bile acids (Section 8.11).

Steroid hormones. Steroid hormones are cholesterol derivatives that function as messenger lipids. The two major types of steroid hormones are sex hormones and adrenocorticoid hormones (Section 8.12).

Eicosanoids. Eicosanoids are fatty acid derivatives that function as messenger lipids. The major classes of eicosanoids are prostaglandins, thromboxanes, and leukotrienes (Section 8.13).

Biological waxes. Biological waxes are protective-coating lipids formed through the esterification of a long-chain fatty acid to a long-chain alcohol (Section 8.14).

KEY REACTIONS AND EQUATIONS

1. Formation of a triacylglycerol (Section 8.6)

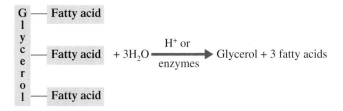

2. Hydrolysis of a triacylglycerol to produce glycerol and fatty acids (Section 8.6)

$$\text{Glycerol—Fatty acid} + 3H_2O \xrightarrow{H^+ \text{ or enzymes}} \text{Glycerol} + 3 \text{ fatty acids}$$

3. Saponification of a triacylglycerol to produce glycerol and fatty acid salts (Section 8.6)

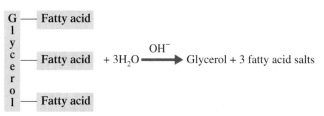

$$+ 3H_2O \xrightarrow{OH^-} \text{Glycerol} + 3 \text{ fatty acid salts}$$

4. Hydrogenation of a triacylglycerol to reduce the unsaturation of its fatty acid components (Section 8.6)

EXERCISES AND PROBLEMS

The members of each pair of problems in this section test similar material.

Structure and Classification of Lipids (Section 8.1)

8.1 What characteristic do all lipids have in common?

8.2 What structural feature, if any, do all lipid molecules have in common? Explain your answer.

8.3 Would you expect lipids to be soluble or insoluble in each of the following solvents?
a. H_2O (polar)
b. CH_3—CH_2—O—CH_2—CH_3 (nonpolar)
c. CH_3—OH (polar)
d. CH_3—CH_2—CH_2—CH_2—CH_3 (nonpolar)

8.4 Would you expect lipids to be soluble or insoluble in each of the following solvents?
a. CH_3—$(CH_2)_7$—CH_3 (nonpolar)
b. CH_3—Cl (polar)
c. CCl_4 (nonpolar)
d. CH_3—CH_2—OH (polar)

8.5 In terms of biochemical function, what are the five major categories of lipids?

8.6 What is the biochemical function of each of the following types of lipids?
a. Triacylglycerols
b. Bile acids
c. Sphingoglycolipids
d. Eicosanoids

Fatty Acids (Sections 8.2 and 8.3)

8.7 Classify each of the following fatty acids as long-chain, medium-chain, or short-chain.
a. Myristic (14:0)
b. Caproic (6:0)
c. Arachidic (20:0)
d. Capric (10:0)

8.8 Classify each of the following fatty acids as long-chain, medium-chain, or short-chain.
a. Lauric (12:0)
b. Oleic (18:1)
c. Butyric (4:0)
d. Stearic (18:0)

8.9 Classify each of the following fatty acids as saturated, monounsaturated, or polyunsaturated.
a. Stearic (18:0)
b. Linolenic (18:3)
c. Docosahexaenoic (22:6)
d. Oleic (18:1)

8.10 Classify each of the following fatty acids as saturated, monounsaturated, or polyunsaturated.
a. Palmitic (16:0)
b. Linoleic (18:2)
c. Arachidonic (20:4)
d. Palmitoleic (16:1)

8.11 Structurally, what is the difference between a SFA and a MUFA?

8.12 Structurally, what is the difference between a MUFA and a PUFA?

8.13 With the help of Table 8.1, classify each of the acids in Problem 8.9 as an omega-3 acid, an omega-6 acid, or neither an omega-3 nor an omega-6 acid.

8.14 With the help of Table 8.1, classify each of the acids in Problem 8.10 as an omega-3 acid, an omega-6 acid, or neither an omega-3 nor an omega-6 acid.

8.15 Draw the condensed structural formula for the fatty acid whose numerical shorthand designation is 18:2 ($\Delta^{9,12}$).

8.16 Draw the condensed structural formula for the fatty acid whose numerical shorthand designation is 20:4 ($\Delta^{5,8,11,14}$).

8.17 Why does the introduction of double bonds into a fatty acid molecule lower its melting point?

8.18   What effect does a *cis* double bond have on the shape of a fatty acid molecule?

8.19   In each of the following pairs of fatty acids, select the fatty acid that has the lower melting point.
a. 18:0 acid and 18:1 acid   b. 18:2 acid and 18:3 acid
c. 14:0 acid and 16:0 acid   d. 18:1 acid and 20:0 acid

8.20   In each of the following pairs of fatty acids, select the fatty acid that has the higher melting point.
a. 14:0 acid and 18:0 acid   b. 20:4 acid and 20:5 acid
c. 18:3 acid and 20:3 acid   d. 16:0 acid and 16:1 acid

8.21   Using the structural information given in Table 8.1, assign an IUPAC name to each of the following fatty acids.
a. Myristic acid   b. Palmitoleic acid

8.22   Using the structural information given in Table 8.1, assign an IUPAC name to each of the following fatty acids.
a. Stearic acid   b. Linolenic acid

■ Triacylglycerols (Section 8.4)

8.23   What are the four structural subunits that contribute to the structure of a triacylglycerol?

8.24   Draw the general block diagram for a triacylglycerol.

8.25   How many different kinds of functional groups are present in a triacylglycerol in which all three fatty acid residues come from saturated fatty acids?

8.26   How many different kinds of functional groups are present in a triacylglycerol in which all three fatty acid residues come from unsaturated fatty acids?

8.27   Draw the condensed structural formula of a triacylglycerol formed from glycerol and three molecules of palmitic acid.

8.28   Draw the condensed structural formula of a triacylglycerol formed from glycerol and three molecules of stearic acid.

8.29   Draw block diagram structures for the four different triacylglycerols that can be produced from glycerol, stearic acid, and linolenic acid.

8.30   Draw block diagram structures for the three different triacylglycerols that can be produced from glycerol, palmitic acid, stearic acid, and linolenic acid.

8.31   Identify the fatty acids present in each of the following triacylglycerols.

a.

$$CH_2-O-\overset{\overset{\displaystyle O}{\|}}{C}-(CH_2)_{14}-CH_3$$
$$CH-O-\overset{\overset{\displaystyle O}{\|}}{C}-(CH_2)_{12}-CH_3$$
$$CH_2-O-\overset{\overset{\displaystyle O}{\|}}{C}-(CH_2)_7-CH=CH-(CH_2)_7-CH_3$$

b.

8.32   Identify the fatty acids present in each of the following triacylglycerols.

a.

$$CH_2-O-\overset{\overset{\displaystyle O}{\|}}{C}-(CH_2)_{16}-CH_3$$
$$CH-O-\overset{\overset{\displaystyle O}{\|}}{C}-(CH_2)_7-CH=CH-(CH_2)_7-CH_3$$
$$CH_2-O-\overset{\overset{\displaystyle O}{\|}}{C}-(CH_2)_{12}-CH_3$$

b.

8.33   For each of the acyl groups present in the triacylglycerol of Problem 8.31a, indicate how many carbon atoms are present and how many oxygen atoms are present.

8.34   For each of the acyl groups present in the triacylglycerol of Problem 8.32a, indicate how many carbon atoms are present and how many oxygen atoms are present.

8.35   What is the difference in meaning, if any, between the members of each of the following pairs of terms?
a. Triacylglycerol and triglyceride
b. Triacylglycerol and fat
c. Triacylglycerol and mixed triacylglycerol
d. Fat and oil

8.36   What is the difference in meaning, if any, between the members of each of the following pairs of terms?
a. Triacylglycerol and oil
b. Triacylglycerol and simple triacylglycerol
c. Simple triacylglycerol and mixed triacylglycerol
d. Triglyceride and fat

■ Dietary Considerations and Triacylglycerols (Section 8.5)

8.37   In a dietary context, indicate whether each of the following pairings of concepts is correct.
a. "Saturated fat" and "good fat"
b. "Polyunsaturated fat" and "bad fat"

8.38   In a dietary context, indicate whether each of the following pairings of concepts is correct.
a. "Monounsaturated fat" and "good fat"
b. "Saturated fat" and "good and bad fat"

8.39   In a dietary context, which of the following pairings of concepts is correct?
a. "Cold-water fish" and "high in omega-3 fatty acids"
b. "Fatty fish" and "low in omega-3 fatty acids"

8.40   In a dietary context, which of the following pairings of concepts is correct?
a. "Warm-water fish" and "low in omega-3 fatty acids"
b. "Fish and chips" and "high in omega-3 fatty acids"

8.41   In a dietary context, classify each of the following fatty acids as an essential fatty acid or as a nonessential fatty acid.
a. Lauric acid (12:0)   b. Linoleic acid (18:2)
c. Myristic acid (14:0)   d. Palmitoleic (16:1)

8.42   In a dietary context, classify each of the following fatty acids as an essential fatty acid or as a nonessential fatty acid.
a. Stearic acid (18:0)   b. Linolenic acid (18:3)
c. Oleic acid (18:1)   d. Arachidic acid (20:0)

■ **Chemical Reactions of Triacylglycerols (Section 8.6)**

8.43 Name, in general terms, the products of the complete
a. hydrolysis of a fat b. saponification of an oil

8.44 Name, in general terms, the products of the complete
a. saponification of a fat b. hydrolysis of an oil

8.45 Draw condensed structural formulas for all products you would obtain from the complete hydrolysis of the following triacylglycerol.

$$CH_2-O-\overset{\overset{\textstyle O}{\|}}{C}-(CH_2)_{14}-CH_3$$
$$CH-O-\overset{\overset{\textstyle O}{\|}}{C}-(CH_2)_{12}-CH_3$$
$$CH_2-O-\overset{\overset{\textstyle O}{\|}}{C}-(CH_2)_7-CH=CH-(CH_2)_7-CH_3$$

8.46 Draw condensed structural formulas for all products you would obtain from the complete hydrolysis of the following triacylglycerol.

$$CH_2-O-\overset{\overset{\textstyle O}{\|}}{C}-(CH_2)_{16}-CH_3$$
$$CH-O-\overset{\overset{\textstyle O}{\|}}{C}-(CH_2)_{12}-CH_3$$
$$CH_2-O-\overset{\overset{\textstyle O}{\|}}{C}-(CH_2)_6-(CH_2-CH=CH)_3-CH_2-CH_3$$

8.47 With the help of Table 8.1, determine the names of each of the products obtained in Problem 8.45.

8.48 With the help of Table 8.1, determine the names of each of the products obtained in Problem 8.46.

8.49 Draw condensed structural formulas for all products you would obtain from the saponification with NaOH of the triacylglycerol in Problem 8.45.

8.50 Draw condensed structural formulas for all products you would obtain from the saponification with KOH of the triacylglycerol in Problem 8.46.

8.51 With the help of Table 8.1, name each of the products obtained in Problem 8.49.

8.52 With the help of Table 8.1, name each of the products obtained in Problem 8.50.

8.53 Why can only unsaturated triacylglycerols undergo hydrogenation?

8.54 A food package label lists an oil as "partially hydrogenated." What does this mean?

8.55 How many molecules of H_2 will react with one molecule of the following triacylglycerol?

$$CH_2-O-\overset{\overset{\textstyle O}{\|}}{C}-(CH_2)_6-(CH_2-CH=CH)_2-(CH_2)_4-CH_3$$
$$CH-O-\overset{\overset{\textstyle O}{\|}}{C}-(CH_2)_7-CH=CH-(CH_2)_7-CH_3$$
$$CH_2-O-\overset{\overset{\textstyle O}{\|}}{C}-(CH_2)_6-(CH_2-CH=CH)_3-CH_2-CH_3$$

8.56 How many molecules of H_2 will react with one molecule of the following triacylglycerol?

$$CH_2-O-\overset{\overset{\textstyle O}{\|}}{C}-(CH_2)_7-CH=CH-(CH_2)_7-CH_3$$
$$CH-O-\overset{\overset{\textstyle O}{\|}}{C}-(CH_2)_{16}-CH_3$$
$$CH_2-O-\overset{\overset{\textstyle O}{\|}}{C}-(CH_2)_6-(CH_2-CH=CH)_3-CH_2-CH_3$$

8.57 Draw block diagram structures for all possible products of the partial hydrogenation, with two molecules of H_2, of the following molecules.

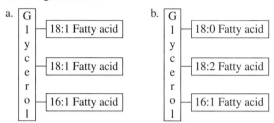

8.58 Draw block diagram structures for all possible products of the partial hydrogenation, with two molecules of H_2, of the following molecules.

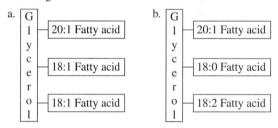

8.59 Why do animal fats and vegetable oils become rancid when exposed to moist, warm air?

8.60 Why are the compounds BHA and BHT often added to foods that contain fats and oils?

■ **Phospholipids (Section 8.7)**

8.61 What are the two common types of platform molecules for a phospholipid?

8.62 How many fatty acid residues are present in a phospholipid?

8.63 Draw the general block diagram for a glycerophospholipid.

8.64 Draw the general block diagram for a sphingophospholipid.

8.65 Draw the structures of the three amino alcohols commonly esterified to the phosphate group in a glycerophospholipid.

8.66 What structural subunits are present in a phosphatidyl group?

8.67 Sphingophospholipids have a "head and two tails" structure. Give the chemical identity of the head and of each of the two tails.

8.68 Glycerophospholipids have a "head and two tails" structure. Give the chemical identity of the head and of each of the two tails.

8.69 Which portion of the structure of a phospholipid has hydrophobic characteristics?

8.70 Which portion of the structure of a phospholipid has hydrophilic characteristics?

8.71 Indicate how many ester linkages are present in the structure of a
a. glycerophospholipid b. sphingophospholipid

8.72 Indicate how many amide linkages are present in the structure of a
 a. glycerophospholipid   b. sphingophospholipid

8.73 Structurally, what is the difference between a lecithin and a phosphatidylserine?

8.74 Structurally, what is the difference between a lecithin and a sphingomyelin?

■ Sphingoglycolipids (Section 8.8)

8.75 Draw the general block diagram for a sphingoglycolipid.

8.76 How many of each of the following types of linkages are present in a sphingoglycolipid?
 a. Ester linkages   b. Amide linkages
 c. Glycosidic linkages

8.77 How does the general structure of a sphingoglycolipid differ from that of a sphingophospholipid?

8.78 Structurally, what is the difference between a cerebroside and a ganglioside?

■ Cholesterol (Section 8.9)

8.79 Draw and number the fused hydrocarbon ring system characteristic of all steroids.

8.80 What positions in the steroid nucleus are particularly likely to bear substituents?

8.81 Describe the structure of cholesterol in terms of substituents attached to the steroid nucleus.

8.82 Structurally, what is considered the "head" of a cholesterol molecule?

8.83 In a dietary context, what is the difference between "good cholesterol" and "bad cholesterol"?

8.84 In a dietary context, how do HDL and LDL differ in function?

■ Cell Membranes (Section 8.10)

8.85 What are the three major types of lipids present in cell membranes?

8.86 What is the structural characteristic common to all the nonsteroid lipids present in cell membranes?

8.87 What is a lipid bilayer?

8.88 What is the basic structure of a cell membrane?

8.89 What is the function of unsaturation in the hydrocarbon tails of membrane lipids?

8.90 What function does cholesterol serve when it is present in cell membranes?

8.91 What is the difference between *passive transport* and *facilitated transport*?

8.92 What is the difference between *facilitated transport* and *active transport*?

8.93 Match each of the following statements related to membrane transport processes to the appropriate term: *passive transport, facilitated transport, active transport*. More than one term may apply in a given situation.
 a. Movement across the membrane is against the concentration gradient.
 b. Proteins serve as "gates."
 c. Expenditure of cellular energy is required.
 d. Movement across the membrane is from a high to a low concentration.

8.94 Match each of the following statements related to membrane transport processes to the appropriate term: *passive transport, facilitated transport, active transport*. More than one term may apply in a given situation.
 a. Movement across the membrane is with the concentration gradient.
 b. Proteins serve as "pumps."
 c. Expenditure of cellular energy is not required.
 d. Movement across the membrane is from a low to a high concentration.

■ Bile Acids (Section 8.11)

8.95 Describe the structural differences between a bile acid and cholesterol.

8.96 Describe the structural differences between cholic acid and a deoxycholic acid.

8.97 Describe the structural differences between glycocholic acid and taurocholic acid.

8.98 Describe the structural differences between glycocholic acid and glyco-7-deoxycholic acid.

8.99 What is the medium through which bile acids are supplied to the small intestine?

8.100 What is the chemical composition of bile?

8.101 At what location in the body are bile acids stored until needed?

8.102 What is the chemical composition of the majority of gallstones?

■ Steroid Hormones (Section 8.12)

8.103 What are the two major classes of steroid hormones?

8.104 Describe the general function of each of the following types of steroid hormones.
 a. Estrogens   b. Androgens
 c. Progestins   d. Mineralocorticoids

8.105 How do the sex hormones estradiol and testosterone differ in structure?

8.106 What functional groups are present in each of the following steroid hormones?
 a. Estradiol   b. Testosterone
 c. Progesterone   d. Cortisone

■ Eicosanoids (Section 8.13)

8.107 What is the major structural difference between a prostaglandin and its parent fatty acid?

8.108 What is the major structural difference between a leukotriene and its parent fatty acid?

8.109 List six physiological processes that are regulated by eicosanoids.

8.110 What is the biochemical basis for the effectiveness of aspirin in decreasing inflammation?

■ Biological Waxes (Section 8.14)

8.111 Draw the general block diagram for a biological wax.

8.112 Draw the condensed structural formula of a wax formed from palmitic acid (Table 8.1) and cetyl alcohol, $CH_3—(CH_2)_{14}—CH_2—OH$.

8.113 What is the difference between a biological wax and a mineral wax?

8.114 Biological waxes have a "head and two tails" structure. Give the chemical identity of the head and of the two tails.

ADDITIONAL PROBLEMS

8.115 Classify each of the following types of lipids as (1) glycerol-based, (2) sphingosine-based, or (3) neither glycerol-based nor sphingosine-based.
  a. Bile acids   b. Fats   c. Thromboxanes
  d. Gangliosides   e. Waxes   f. Leukotrienes

8.116 Indicate whether each of the lipid types in Problem 8.115 has a "head and two tails" structure.

8.117 Identify the type of lipid that fits each of the following "structural component" characterizations.
  a. Sphingosine + fatty acid + phosphoric acid + choline
  b. Glycerol + three fatty acids
  c. Fused-ring system with three 6-membered rings and one 5-membered ring
  d. 20-carbon fatty acid + three conjugated double bonds
  e. 20-carbon fatty acid + cyclopentane ring
  f. Sphingosine + fatty acid + monosaccharide

8.118 Classify each of the following types of lipids as (1) an energy-storage lipid, (2) a membrane lipid, (3) an emulsification lipid, (4) a messenger lipid, or (5) a protective-coating lipid.
  a. Fats   b. Cholic acid
  c. Cholesterol   d. Estrogens
  e. Sphingomyelins   f. Prostaglandins

8.119 Which of the terms *glycerolipid, sphingolipid,* and *phospholipid* apply to each of the following lipids? More than one term may apply in a given situation.
  a. Triacylglycerol   b. Sphingoglycolipid
  c. Glycerophospholipid   d. Sphingophospholipid

8.120 Specify the numbers of ester linkages, amide linkages, and glycosidic linkages present in each of the following types of lipids.
  a. Oils   b. Lecithins
  c. Sphinogomyelins   d. Biological waxes
  e. Cerebrosides   f. Phosphatidylcholines

8.121 Indicate whether each of the following types of lipids contain a "steroid nucleus" as part of its structure.
  a. Prostaglandins   b. Cortisone
  c. Cholesterol   d. Bile acids
  e. Estrogens   f. Leukotrienes

MULTIPLE-CHOICE PRACTICE TEST

8.122 Which of the following statements concerning fatty acids is *correct*?
  a. They are naturally occurring dicarboxylic acids.
  b. They are rarely found in the free state in nature.
  c. Their carbon chains always contain at least two double bonds.
  d. They almost always contain an odd number of carbon atoms.

8.123 Which of the following is a distinguishing characteristic between fats and oils?
  a. Physical state at room temperature
  b. Identity of the alcohol component present
  c. Number of structural subunits present
  d. Number of fatty acid residues present

8.124 *Partial* hydrogenation of a fat or an oil does which of the following?
  a. Produces fatty acid salts
  b. Increases the degree of fatty acid unsaturation
  c. Increases the melting point
  d. Decreases the number of fatty acid residues present

8.125 In the oxidation of fats and oils, which part of the molecule is attacked by the oxidizing agent?
  a. Carbon–carbon double bonds
  b. Ester linkages
  c. Hydroxyl groups
  d. Carboxyl groups

8.126 In which of the following pairs of lipids are both members of the pair *membrane* lipids?
  a. Triacylglycerols and cholesterol
  b. Triacylglycerols and sphingophospholipids
  c. Sphingophospholipids and sphingoglycolipids
  d. Eicosanoids and bile salts

8.127 Which of the following types of lipids does *not* have a "head and two tails" structure?
  a. Glycerophospholipids
  b. Sphingophospholipids
  c. Sphingoglycolipids
  d. Triacylglycerols

8.128 The "steroid nucleus" of steroid lipids involves a fused-ring system that has how many rings?
  a. Two
  b. Three
  c. Four
  d. Five

8.129 Which of the following polarity-based descriptions is correct for a lipid bilayer?
  a. Both the outer and inner surfaces contain polar "heads."
  b. Both the outer and inner surfaces contain nonpolar "heads."
  c. Both the outer and the inner surfaces contain polar "tails."
  d. Both the outer and the inner surfaces contain nonpolar "tails."

8.130 Based on function, eicosanoids are classified as which of the following?
  a. Membrane lipids
  b. Emulsification lipids
  c. Messenger lipids
  d. Protective-coating lipids

8.131 How many structural subunits are present in the "block diagram" for a biological wax?
  a. Two
  b. Three
  c. Four
  d. Five

9 Proteins

The protein made by spiders to produce a web is a form of silk that can be exceptionally strong.

In this chapter we consider the third of the bioorganic classes of molecules (Section 7.1), the compounds called proteins. An extraordinary number of different proteins, each with a different function, exist in the human body. A typical human cell contains about 9000 different kinds of proteins, and the human body contains about 100,000 different proteins. Proteins are needed for the synthesis of enzymes, certain hormones, and some blood components; for the maintenance and repair of existing tissues; for the synthesis of new tissue; and sometimes for energy.

9.1 Characteristics of Proteins

Next to water, proteins are the most abundant substances in nearly all cells—they account for about 15% of a cell's overall mass (Section 7.1) and for almost half of a cell's dry mass. All proteins contain the elements carbon, hydrogen, oxygen, and nitrogen; most also contain sulfur. The presence of nitrogen in proteins sets them apart from carbohydrates and lipids, which generally do not contain nitrogen. The average nitrogen content of proteins is 15.4% by mass. Other elements, such as phosphorus and iron, are essential constituents of certain specialized proteins. Casein, the main protein of milk, contains phosphorus, an element very important in the diet of infants and children. Hemoglobin, the oxygen-transporting protein of blood, contains iron.

A **protein** *is a naturally-occurring, unbranched polymer in which the monomer units are amino acids.* Thus the starting point for a discussion of proteins is an understanding of the structures and chemical properties of amino acids.

The word protein comes from the Greek *proteios,* which means "of first importance." This reflects the key role that proteins play in life processes.

9.2 Amino Acids: The Building Blocks for Proteins

An **amino acid** *is an organic compound that contains both an amino (*—NH_2*) group and a carboxyl (*—COOH*) group.* The amino acids found in proteins are always α-amino acids. An **α-amino acid** *is an amino acid in which the amino group and the carboxyl group are attached to the α-carbon atom.* The general structural formula for an α-amino acid is

The R group present in an α-amino acid is called the amino acid *side chain.* The nature of this side chain distinguishes α-amino acids from each other. Side chains vary in size, shape, charge, acidity, functional groups present, hydrogen-bonding ability, and chemical reactivity.

Over 700 different naturally occurring amino acids are known, but only 20 of them, called standard amino acids, are normally present in proteins. A **standard amino acid** *is one of the 20 α-amino acids normally found in proteins.* The structures of the 20 standard amino acids are given in Table 9.1. Within Table 9.1, amino acids are grouped according to side-chain polarity. In this system there are four categories: (1) nonpolar amino acids, (2) polar neutral amino acids, (3) polar acidic amino acids, and (4) polar basic amino acids. This classification system gives insights into how various types of amino acid side chains help determine the properties of proteins (Section 9.11).

A **nonpolar amino acid** *is an amino acid that contains one amino group, one carboxyl group, and a nonpolar side chain.* When incorporated into a protein, such amino acids are *hydrophobic* ("water-fearing"); that is, they are not attracted to water molecules. They are generally found in the interior of proteins, where there is limited contact with water. There are nine nonpolar amino acids. Tryptophan is a borderline member of this group because water can weakly interact through hydrogen bonding with the NH ring location on tryptophan's side-chain ring structure. Thus, some textbooks list tryptophan as a polar neutral amino acid.

The three types of polar amino acids have varying degrees of affinity for water. Within a protein, such amino acids are said to be *hydrophilic* ("water-loving"). Hydrophilic amino acids are often found on the surfaces of proteins.

A **polar neutral amino acid** *is an amino acid that contains one amino group, one carboxyl group, and a side chain that is polar but neutral.* In solution at physiological pH, the side chain of a polar neutral amino acid is neither acidic nor basic. There are six polar neutral amino acids. These amino acids are more soluble in water than the nonpolar amino acids as, in each case, the R group present can hydrogen bond to water.

A **polar acidic amino acid** *is an amino acid that contains one amino group and two carboxyl groups, the second carboxyl group being part of the side chain.* In solution at physiological pH, the side chain of a polar acidic amino acid bears a negative charge; the side-chain carboxyl group has lost its acidic hydrogen atom. There are two polar acidic amino acids: aspartic acid and glutamic acid.

A **polar basic amino acid** *is an amino acid that contains two amino groups and one carboxyl group, the second amino group being part of the side chain.* In solution at physiological pH, the side chain of a polar basic amino acid bears a positive charge; the nitrogen atom of the amino group has accepted a proton (basic behavior; Section 6.6). There are three polar basic amino acids: lysine, arginine, and histidine.

The names of the standard amino acids are often abbreviated using three-letter codes. Except in four cases, these abbreviations are the first three letters of the amino acid's name. In addition, a new one-letter code for amino acid names is currently gaining popularity (particularly in computer applications). Both sets of abbreviations are used extensively in describing peptides and proteins, which contain tens and hundreds of amino acid units. Both types of abbreviations are given in Table 9.1.

In an α-amino acid, the carboxyl group and the amino group are attached to the same carbon atom.

The nature of the side chain (R group) distinguishes α-amino acids from each other, both physically and chemically.

The nonpolar amino acid *proline* has a structural feature not found in any other standard amino acid. Its side chain, a propyl group, is bonded to both the α-carbon atom and the amino nitrogen atom, giving a cyclic side chain.

Proline

A variety of functional groups are present in the side chains of the 20 standard amino acids: six have alkyl groups (Section 1.8), three have aromatic groups (Section 2.11), two have sulfur-containing groups (Section 3.20), two have hydroxyl (alcohol) groups (Section 3.2), three have amino groups (Section 6.2), two have carboxyl groups (Section 5.1), and two have amide groups (Section 6.12).

TABLE 9.1
The 20 Standard Amino Acids, Grouped According to Side-Chain Polarity
Below each amino acid's structure are its name (with pronunciation), its three-letter abbreviation, and its one-letter abbreviation.

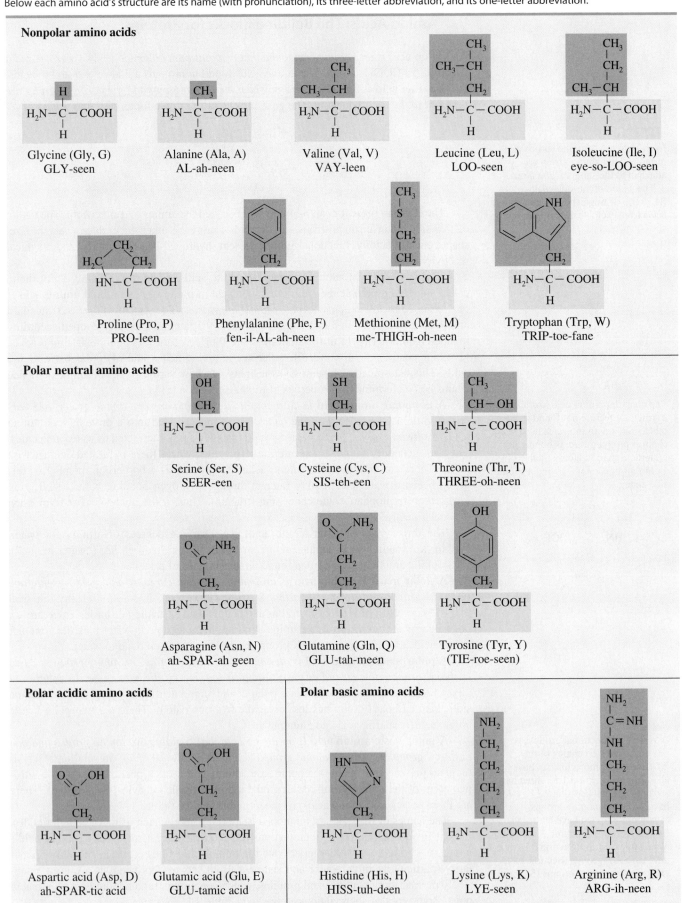

Nonpolar amino acids

Glycine (Gly, G)
GLY-seen

Alanine (Ala, A)
AL-ah-neen

Valine (Val, V)
VAY-leen

Leucine (Leu, L)
LOO-seen

Isoleucine (Ile, I)
eye-so-LOO-seen

Proline (Pro, P)
PRO-leen

Phenylalanine (Phe, F)
fen-il-AL-ah-neen

Methionine (Met, M)
me-THIGH-oh-neen

Tryptophan (Trp, W)
TRIP-toe-fane

Polar neutral amino acids

Serine (Ser, S)
SEER-een

Cysteine (Cys, C)
SIS-teh-een

Threonine (Thr, T)
THREE-oh-neen

Asparagine (Asn, N)
ah-SPAR-ah geen

Glutamine (Gln, Q)
GLU-tah-meen

Tyrosine (Tyr, Y)
(TIE-roe-seen)

Polar acidic amino acids

Aspartic acid (Asp, D)
ah-SPAR-tic acid

Glutamic acid (Glu, E)
GLU-tamic acid

Polar basic amino acids

Histidine (His, H)
HISS-tuh-deen

Lysine (Lys, K)
LYE-seen

Arginine (Arg, R)
ARG-ih-neen

CHEMICAL CONNECTIONS

The Essential Amino Acids

All of the amino acids in Table 9.1 are necessary constituents of human protein. Adequate amounts of 11 of the 20 amino acids can be synthesized from carbohydrates and lipids in the body if a source of nitrogen is also available. Because the human body is incapable of producing 9 of these 20 acids fast enough or in sufficient quantities to sustain normal growth, these 9 amino acids, called essential amino acids, must be obtained from food. An **essential amino acid** *is an amino acid needed in the human body that must be obtained from dietary sources because it cannot be synthesized within the body from other substances in adequate amounts.* The following table lists the *essential* amino acids for humans.

The Essential Amino Acids for Humans

arginine[a]	methionine
histidine	phenylalanine
isoleucine	threonine
leucine	tryptophan
lysine	valine

[a]Arginine is required for growth in children but is not required by adults.

The human body can synthesize small amounts of some of the essential amino acids, but not enough to meet its needs, especially in the case of growing children.

A **complete dietary protein** *is a protein that contains all the essential amino acids in approximately the same relative amounts in which the human body needs them.* A complete dietary protein may or may not contain all the nonessential amino acids. Most animal proteins, including casein from milk and proteins found in meat, fish, and eggs, are complete proteins, although gelatin is an exception (it lacks tryptophan). Proteins from plants (vegetables, grains, and legumes) have quite diverse amino acid patterns and some tend to be limited in one or more essential amino acids. Some plant proteins (for example, corn protein) are far from complete. Others (for example, soy protein) are complete. Thus vegetarians must eat a variety of plant foods to obtain all of the essential amino acids in appropriate quantities.

The following table lists the essential amino acid deficiencies associated with selected vegetables and grains.

Amino Acids Missing in Selected Vegetables and Grains

Food Source	Amino Acid Deficiency
soy	none
wheat, rice, oats	lysine
corn	lysine, tryptophan
beans	methionine, tryptophan
peas	methionine
almonds, walnuts	lysine, tryptophan

9.3 Chirality and Amino Acids

Glycine, the simplest of the standard amino acids, is achiral. All of the other standard amino acids are chiral.

Four different groups are attached to the α-carbon atom in all of the standard amino acids except glycine, where the R group is a hydrogen atom.

$$H_2N - \overset{\overset{\textstyle R}{|}}{\underset{\underset{\textstyle H}{|}}{C}} - COOH$$

This means that the structures of 19 of the 20 standard amino acids possess a chiral center (Section 7.4) at this location, so enantiomeric forms (left- and right-handed forms; Section 7.5) exist for each of these amino acids.

With few exceptions (in some bacteria), the amino acids found in nature and in proteins are L isomers. Thus, as is the case with monosaccharides (Section 7.8), nature favors one mirror-image form over the other. Interestingly, for amino acids the L isomer is the preferred form, whereas for monosaccharides the D isomer is preferred.

The rules for drawing Fischer projections (Section 7.6) for amino acid structures follow.

Because only L amino acids are constituents of proteins, the enantiomer designation of L or D will be omitted in subsequent amino acid and protein discussions. It is understood that it is the L isomer that is always present.

1. The —COOH group is put at the top of the projection, the R group at the bottom. This positions the carbon chain vertically.
2. The —NH$_2$ group is in a horizontal position. Positioning it on the left denotes the L isomer, and positioning it on the right denotes the D isomer.

FIGURE 9.1 Designation of handedness in standard amino acid structures involves aligning the carbon chain vertically and looking at the position of the horizontally aligned —NH$_2$ group. The L form has the —NH$_2$ group on the left, and the D form has the —NH$_2$ group on the right.

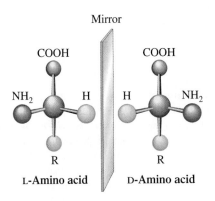

Mirror

L-Amino acid   D-Amino acid

Figure 9.1 shows molecular models that illustrate the use of these rules. Fischer projections for both enantiomers of the amino acids alanine and serine follow.

L-Alanine   D-Alanine   L-Serine   D-Serine

Two of the 19 chiral standard amino acids, isoleucine and threonine, possess two chiral centers (see Table 9.1). With two chiral centers present, four stereoisomers are possible for these amino acids. However, only one of the L isomers is found in proteins.

9.4 Acid–Base Properties of Amino Acids

In pure form, amino acids are white crystalline solids with relatively high decomposition points. (Most amino acids decompose before they melt.) Also, most amino acids are *not* very soluble in water because of strong intermolecular forces within their crystal structures. Such properties are those often exhibited by compounds in which charged species are present. Studies of amino acids confirm that they are charged species both in the solid state and in solution. Why is this so?

Both an acidic group (—COOH) and a basic group (—NH$_2$) are present on the same carbon in an α-amino acid.

In drawing amino acid structures, where handedness designation is not required, the placement of the four groups about the α-carbon atom is arbitrary. From this point on in the text, we will draw amino acid structures such that the —COOH group is on the left, the —NH$_2$ group is on the right, the R group points down, and the H atom points up. Drawing amino acids in this "arrangement" makes it easier to draw structures where amino acids are linked together to form longer amino acid chains.

Basic group → H$_2$N—C—COOH ← Acidic group

In Section 5.8, we learned that in neutral solution, carboxyl groups have a tendency to lose protons (H$^+$), producing a negatively charged species:

$$—COOH \longrightarrow —COO^- + H^+$$

In Section 6.6, we learned that in neutral solution, amino groups have a tendency to accept protons (H$^+$), producing a positively charged species:

$$—NH_2 + H^+ \longrightarrow —\overset{+}{N}H_3$$

Consistent with the behavior of these groups, in neutral solution, the —COOH group of an amino acid donates a proton to the —NH$_2$ of the same amino acid. We can characterize this behavior as an *internal* acid–base reaction. The net result is that in neutral solution, amino acid molecules have the structure

$$H_3\overset{+}{N}—\overset{\overset{\displaystyle H}{|}}{\underset{\underset{\displaystyle R}{|}}{C}}—COO^-$$

Strong intermolecular forces between the positive and negative centers of zwitterions are the cause of the high melting points of amino acids.

Such a molecule is known as a zwitterion, from the German term meaning "double ion." A **zwitterion** *is a molecule that has a positive charge on one atom and a negative charge on another atom, but which has no net charge.* Note that the net charge on a zwitterion is zero even though parts of the molecule carry charges. In solution and also in the solid state, α-amino acids exist as zwitterions.

Zwitterion structure changes when the pH of a solution containing an amino acid is changed from neutral either to acidic (low pH) by adding an acid such as HCl or to basic (high pH) by adding a base such as NaOH. In an acidic solution, the zwitterion accepts a proton (H^+) to form a positively charged ion.

$$\underset{\substack{\text{Zwitterion (no net charge)}}}{H_3\overset{+}{N}-\underset{\underset{R}{|}}{\overset{\overset{H}{|}}{C}}- COO^-} + H_3O^+ \longrightarrow \underset{\substack{\text{Positively charged ion}}}{H_3\overset{+}{N}-\underset{\underset{R}{|}}{\overset{\overset{H}{|}}{C}}- COOH} + H_2O$$

From this point on in the text, the structures of amino acids will be drawn in their zwitterion form unless information given about the pH of the solution indicates otherwise.

In basic solution, the $-\overset{+}{N}H_3$ of the zwitterion loses a proton, and a negatively charged species is formed.

$$\underset{\substack{\text{Zwitterion} \\ \text{(no net charge)}}}{H_3\overset{+}{N}-\underset{\underset{R}{|}}{\overset{\overset{H}{|}}{C}}-COO^-} + OH^- \longrightarrow \underset{\substack{\text{Negatively charged ion}}}{H_2N-\underset{\underset{R}{|}}{\overset{\overset{H}{|}}{C}}-COO^-} + H_2O$$

The ability of amino acids to react with both H_3O^+ and OH^- ions means that amino acid solutions can function as buffers. The same is true for proteins, which are amino acid polymers (Section 9.1). The buffering action of proteins present in blood is a major function of such proteins.

Thus, in solution, three different amino acid forms can exist (zwitterion, negative ion, and positive ion). The three species are actually in equilibrium with each other, and the equilibrium shifts with pH change. The overall equilibrium process can be represented as follows:

$$\underset{\substack{\text{Acidic solution} \\ \text{(low pH)}}}{H_3\overset{+}{N}-\underset{\underset{R}{|}}{\overset{\overset{H}{|}}{C}}-COOH} \underset{H_3O^+}{\overset{OH^-}{\rightleftharpoons}} \underset{\substack{\text{Neutral solution} \\ \text{(pH = 7.0)}}}{H_3\overset{+}{N}-\underset{\underset{R}{|}}{\overset{\overset{H}{|}}{C}}-COO^-} \underset{H_3O^+}{\overset{OH^-}{\rightleftharpoons}} \underset{\substack{\text{Basic solution} \\ \text{(high pH)}}}{H_2N-\underset{\underset{R}{|}}{\overset{\overset{H}{|}}{C}}-COO^-}$$

In acidic solution, the positively charged species on the left predominates; nearly neutral solutions have the middle species (the zwitterion) as the dominant species; in basic solution, the negatively charged species on the right predominates.

EXAMPLE 9.1

Determining Amino Acid Form in Solutions of Various pH

■ Draw the structural form of the amino acid alanine that predominates in solution at each of the following pH values.

a. pH = 1.0
b. pH = 7.0
c. pH = 11.0

Solution

At low pH, both amino and carboxyl groups are protonated. At high pH, both groups have lost their protons. At neutral pH, the zwitterion is present.

a.
$$H_3\overset{+}{N}-\underset{\underset{CH_3}{|}}{\overset{\overset{H}{|}}{C}}-COOH$$
pH = 1.0
(net charge of +1)

b.
$$H_3\overset{+}{N}-\underset{\underset{CH_3}{|}}{\overset{\overset{H}{|}}{C}}-COO^-$$
pH = 7.0
(no net charge)

c.
$$H_2N-\underset{\underset{CH_3}{|}}{\overset{\overset{H}{|}}{C}}-COO^-$$
pH = 11.0
(net charge of −1)

(continued)

Guidelines for amino acid form as a function of solution pH follow.

Low pH: All acid groups are protonated (—COOH). All amino groups are protonated (—$\overset{+}{N}H_3$).

High pH: All acid groups are deprotonated (—COO⁻). All amino groups are deprotonated (—NH_2).

Neutral pH: All acid groups are deprotonated (—COO⁻). All amino groups are protonated (—$\overset{+}{N}H_3$).

The term *protonated* denotes gain of a H⁺ ion, and the term *deprotonated* denotes loss of a H⁺ ion.

Side-chain carboxyl groups are weaker acids than α-carbon carboxyl groups.

Practice Exercise 9.1

Draw the structural form of the amino acid valine that predominates in solution at each of the following pH values.

a. pH = 7.0
b. pH = 12.0
c. pH = 2.0

The previous discussion assumed that the side chain (R group) of an amino acid remains unchanged in solution as the pH is varied. This is the case for neutral amino acids but not for acidic or basic ones. For these latter compounds, the side chain can also acquire a charge because it contains an amino or a carboxyl group that can, respectively, gain or lose a proton.

Because of the extra site that can be protonated or deprotonated, acidic and basic amino acids have four charged forms in solution. These four forms for aspartic acid, one of the acidic amino acids, are

Low-pH form (+1 charge) Moderately-low-pH form (no net charge) (zwitterion) Intermediate-pH form (−1 net charge) High-pH form (−2 net charge)

The existence of two low-pH forms for aspartic acid results from the two carboxyl groups being deprotonated at different pH values. For basic amino acids, two high-pH forms exist because deprotonation of the amino groups does not occur simultaneously. The side-chain amino group deprotonates before the α-amino group.

■ Isoelectric Points and Electrophoresis

The amounts of the various forms of an amino acid—zwitterion, negative ion(s), and positive ion(s)—that are present in an aqueous solution of the amino acid vary with solution pH. There is no pH at which ionic amino acid forms are absent, but there is a pH at which there is an equal number of positive and negative charges present, which produces a "no net charge" situation. The "no net charge" pH value for an amino acid solution is called its *isoelectric point*. An **isoelectric point** *is the pH at which an amino acid solution has no net charge because an equal number of positive and negative charges are present.* At the isoelectric point, almost all amino acid molecules in a solution (more than 99%) are present in their zwitterion form.

Every amino acid has a different isoelectric point. Fifteen of the 20 amino acids, those with nonpolar or polar neutral side chains (Table 9.1), have isoelectric points in the range of 4.8–6.3. The three basic amino acids have higher isoelectric points (His = 7.59, Lys = 9.74, Arg = 10.76), and the two acidic amino acids have lower ones (Asp = 2.77, Glu = 3.22).

The isoelectric point of an amino acid is measured by observing its behavior in an electric field. In an electric field, a charged molecule is attracted to (migrates toward) the electrode of opposite charge. At a high pH, an amino acid has a net negative charge and migrates toward the positive electrode. At a low pH, the opposite is true; with a net positive charge, the amino acid migrates toward the negative electrode. At the isoelectric point, migration does not occur because the zwitterions present have no net charge.

Mixtures of amino acids in solution can be separated by using their different migration patterns at various pH values. This type of analytical separation is called electrophoresis. **Electrophoresis** *is the process of separating charged molecules on the basis of their migration toward charged electrodes associated with an electric field.*

FIGURE 9.2 Separation, at a pH of 5.5, of the three amino acids Lys, Phe, and Glu using electrophoresis.

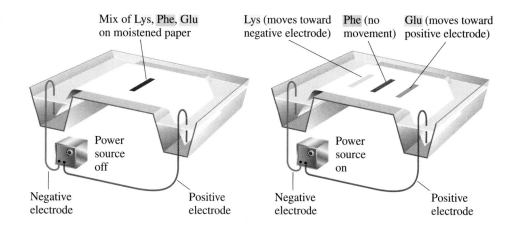

Figure 9.2 schematically shows the separation of the amino acids Lys, Phe, and Glu by electrophoresis. At a pH of 5.5, these amino acids exist in the following forms:

$$Lys \ (+1 \ charge) \qquad Phe \ (no \ net \ charge) \qquad Glu \ (-1 \ charge)$$

When a current is applied, Phe does not move (it has no net charge); Lys, because of its positive charge, migrates toward the negative electrode; and Glu, with a negative charge, moves toward the positive electrode.

Proteins, which are amino acid polymers (Section 9.8), also have isoelectric points and also can be separated via electrophoresis techniques.

EXAMPLE 9.2

Migration Patterns of Amino Acids at Various pH Values in an Electric Field

■ Predict the direction of migration (if any) toward the positively or negatively charged electrode for the following amino acids in solutions of the specified pH. Write "isoelectric" if no migration occurs.

a. Lysine at pH = 7.0
b. Glutamic acid at pH = 7.0
c. Serine at pH = 1.0

Solution

a. Lysine at pH = 7.0

$$H_3\overset{+}{N}-\overset{\overset{\displaystyle H}{|}}{\underset{\underset{\displaystyle CH_2}{|}}{\underset{\underset{\displaystyle CH_2}{|}}{\underset{\underset{\displaystyle CH_2}{|}}{\underset{\underset{\displaystyle CH_2}{|}}{\underset{\displaystyle {}^+NH_3}{}}}}} -COO^-$$

Net positive charge (2 "+" and 1 "−")

Migrates toward negatively charged electrode

(continued)

b. Glutamic acid at pH = 7.0

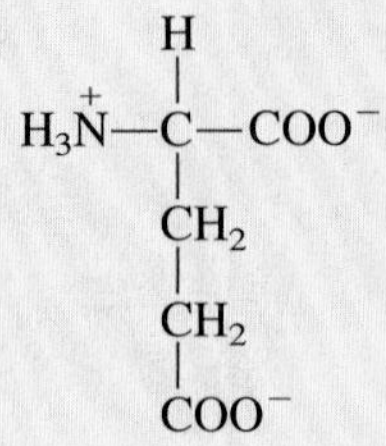

$$H_3\overset{+}{N}-\underset{\underset{\underset{COO^-}{\overset{|}{CH_2}}}{\overset{|}{CH_2}}}{\overset{|}{C}}-COO^-$$

Net negative charge (1 "+" and 2 "−")

Migrates toward positively charged electrode

c. Serine at pH = 1.0

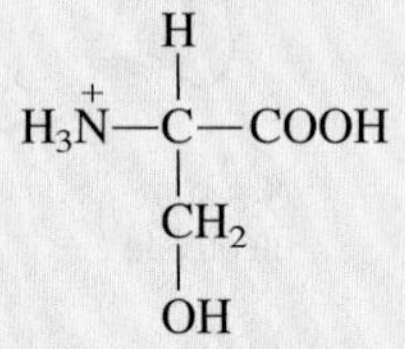

$$H_3\overset{+}{N}-\underset{\underset{OH}{\overset{|}{CH_2}}}{\overset{|}{C}}-COOH$$

One positive charge

Migrates toward negatively charged electrode

Practice Exercise 9.2

Predict the direction of migration (if any) toward the positively or negatively charged electrode for the following amino acids in solutions of the specified pH. Write "isoelectric" if no migration occurs.

a. Lysine at pH = 12.0   **b.** Glutamic acid at pH = 2.0   **c.** Serine at pH = 7.0

9.5 Cysteine: A Chemically Unique Amino Acid

Cysteine is the only standard amino acid (Table 9.1) that has a side chain that contains a sulfhydryl group (—SH group; Section 3.20). The presence of this sulfhydryl group imparts to cysteine a chemical property that is unique among the standard amino acids. Cysteine, in the presence of mild oxidizing agents, readily *dimerizes,* that is, reacts with another cysteine molecule to form a cystine molecule. (A *dimer* is a molecule that is made up of two like subunits.) In cystine, the two cysteine residues are linked via a covalent disulfide bond.

Cystine contains two *cysteine* residues linked by a disulfide bond.

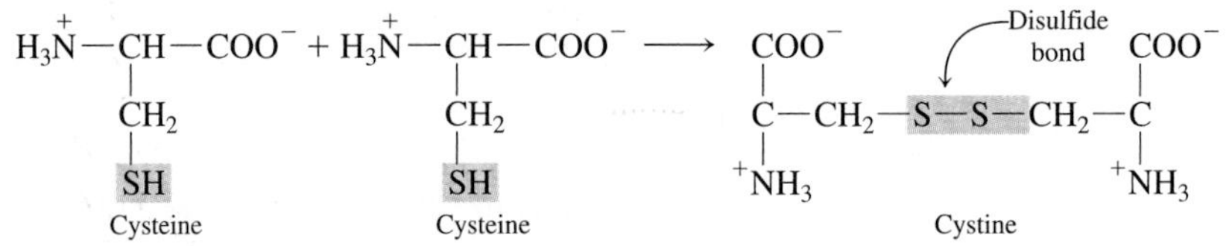

The covalent disulfide bond of cystine is readily broken, using reducing agents, to regenerate two cysteine molecules. This oxidation–reduction behavior involving sulfhydryl groups and disulfide bonds was previously encountered in Section 3.20 when the reactions of thioalcohols were considered.

$$-SH + HS- \underset{\text{Reduction}}{\overset{\text{Oxidation}}{\rightleftharpoons}} -S-S- + 2H$$

As we shall see in Section 9.10, the formation of disulfide bonds between cysteine residues present in protein molecules has important consequences relative to protein structure and protein shape.

9.6 Peptide Formation

In Section 6.15, we learned that a carboxylic acid and an amine can react to produce an amide. The general equation for this reaction is

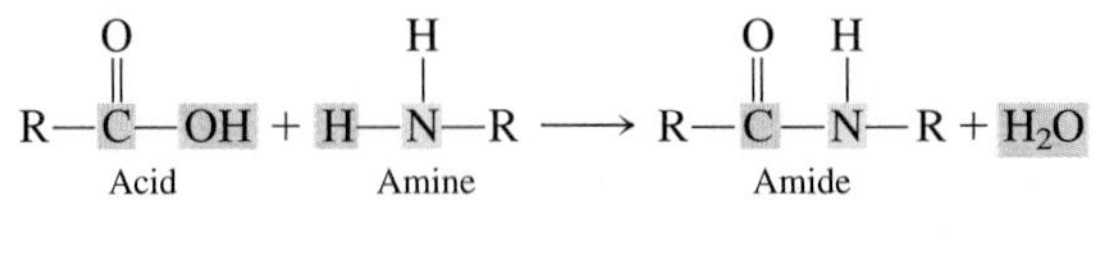

$$\underset{\text{Acid}}{R-\overset{\overset{\displaystyle O}{\|}}{C}-OH} + \underset{\text{Amine}}{H-\overset{\overset{\displaystyle H}{|}}{N}-R} \longrightarrow \underset{\text{Amide}}{R-\overset{\overset{\displaystyle O}{\|}}{C}-\overset{\overset{\displaystyle H}{|}}{N}-R} + H_2O$$

Two amino acids can combine in a similar way—the carboxyl group of one amino acid interacts with the amino group of the other amino acid. The products are a molecule of water and a molecule containing the two amino acids linked by an amide bond.

$$\underset{R_1}{\overset{H}{H_3\overset{+}{N}-C-COO^-}} + \underset{R_2}{\overset{H}{H_3\overset{+}{N}-C-COO^-}} \longrightarrow \underset{R_1 \qquad R_2}{\overset{H \quad O \quad H \quad H}{H_3\overset{+}{N}-C-C-N-C-COO^-}} + H_2O$$

Amide bond

Removal of the elements of water from the reacting carboxyl and amino groups and the ensuing formation of the amide bond are better visualized when expanded structural formulas for the reacting groups are used.

$$\overset{O}{\underset{}{-C}}-O + H-\overset{H}{\underset{H}{N}}- \longrightarrow \overset{O \quad H}{\underset{}{-C}}-N- + H_2O$$

Carboxyl   Amino   Amide
group   group   bond
$(-COO^-)$ $(H_3\overset{+}{N}-)$

In amino acid chemistry, amide bonds that link amino acids together are given the specific name of peptide bond. A **peptide bond** *is a covalent bond between the carboxyl group of one amino acid and the amino group of another amino acid.*

Under proper conditions, many amino acids can bond together to give an unbranched chain of amino acids containing numerous peptide bonds. For example, four peptide bonds are present in a chain of five amino acids.

> Peptide bond formation is an example of a condensation reaction.

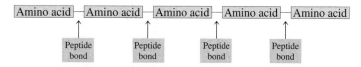

An unbranched chain of amino acids, such as the preceding one, is called a *peptide.* A **peptide** *is a molecule containing two or more amino acids in which the amino acids are joined together through peptide bonds.* Peptides are further classified by the number of amino acid units present in the chain. A compound containing two amino acids joined by a peptide bond is specifically called a *dipeptide;* three amino acids in a chain constitute a *tripeptide;* and so on. The name *oligopeptide* is loosely used to refer to peptides with 10 to 20 amino acid residues, and the name *polypeptide* is used to refer to longer peptides. A **polypeptide** *is a long chain of amino acids, each joined to the next by a peptide bond.*

In all peptides, long or short, the amino acid at one end of the amino acid sequence has a free $H_3\overset{+}{N}$ group, and the amino acid at the other end of the sequence has a free COO^- group. The end with the free $H_3\overset{+}{N}$ group is called the *N-terminal end,* and the end with the free COO^- group is called the *C-terminal end.* By convention, the sequence of amino acids in a peptide is written with the N-terminal end amino acid on the left. The individual amino acids within a peptide chain are called *amino acid residues.* An **amino acid residue** *is the portion of an amino acid structure that remains, after the release of H_2O, when an amino acid participates in peptide bond formation as it becomes part of a peptide chain.*

The structural formula for a peptide may be written out in full, or the sequence of amino acids present may be indicated by using the standard three-letter amino acid abbreviations. The abbreviated formula for the tripeptide.

> A peptide chain has *directionality* because its two ends are different. There is an N-terminal end and a C-terminal end. By convention, the direction of the peptide chain is always
>
> N-terminal end $\longrightarrow$ C-terminal end
>
> The N-terminal end is always on the left, and the C-terminal end is always on the right.

N-terminal $\longrightarrow$ $\underset{H}{\overset{H \quad O}{H_3\overset{+}{N}-C-C-}}\underset{CH_3}{\overset{H \quad H \quad O}{N-C-C-}}\underset{\underset{OH}{CH_2}}{\overset{H \quad H}{N-C-COO^-}}$ $\longleftarrow$ C-terminal
end   end

Glycine   Alanine   Serine

which contains the amino acids glycine, alanine, and serine, is Gly–Ala–Ser. When we use this abbreviated notation, by convention, the amino acid at the N-terminal end of the peptide is always written on the left.

The repeating sequence of peptide bonds and α-carbon —CH groups in a peptide is referred to as the *backbone* of the peptide.

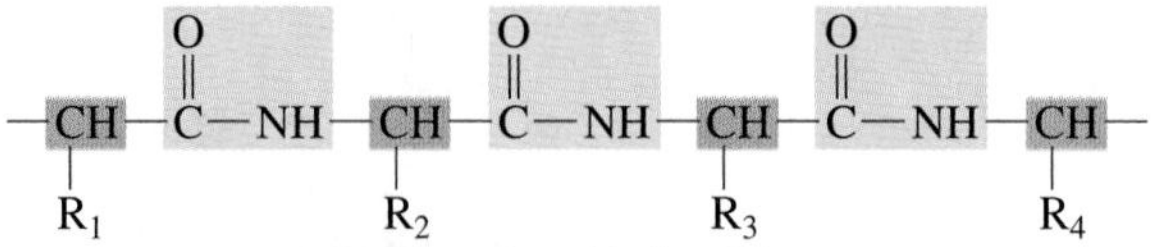

$$-CH-\overset{O}{\overset{\|}{C}}-NH-CH-\overset{O}{\overset{\|}{C}}-NH-CH-\overset{O}{\overset{\|}{C}}-NH-CH-$$

R₁   R₂   R₃   R₄

Backbone of peptide (in color)

The R group side chains are considered substituents on the backbone.

Thus, structurally, a peptide has a regularly repeating part (the backbone) and a variable part (the sequence of R groups). It is the variable R group sequence that distinguishes one peptide from another.

■ Draw the structural formula for the tripeptide Ala–Gly–Val.

Solution

Step 1: The N-terminal end of the peptide involves alanine. Its structure is written first.

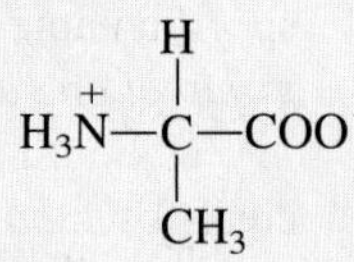

$$H_3\overset{+}{N}-\overset{\overset{\displaystyle H}{|}}{\underset{\underset{\displaystyle CH_3}{|}}{C}}-COO^-$$

Step 2: The structure of glycine is written to the right of the alanine structure, and a peptide bond is formed between the two amino acids by removing the elements of H_2O and bonding the N of glycine to the carboxyl C of alanine.

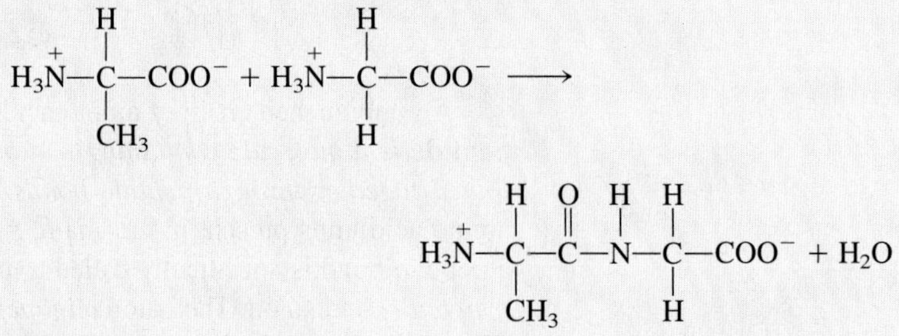

$$H_3\overset{+}{N}-\overset{\overset{\displaystyle H}{|}}{\underset{\underset{\displaystyle CH_3}{|}}{C}}-COO^- + H_3\overset{+}{N}-\overset{\overset{\displaystyle H}{|}}{\underset{\underset{\displaystyle H}{|}}{C}}-COO^- \longrightarrow$$

$$H_3\overset{+}{N}-\overset{\overset{\displaystyle H}{|}}{\underset{\underset{\displaystyle CH_3}{|}}{C}}-\overset{O}{\overset{\|}{C}}-\overset{\overset{\displaystyle H}{|}}{\underset{\underset{\displaystyle H}{|}}{N}}-\overset{\overset{\displaystyle H}{|}}{C}-COO^- + H_2O$$

Step 3: To the right of the just-formed dipeptide, draw the structure of valine. Then repeat Step 2 to form the desired tripeptide.

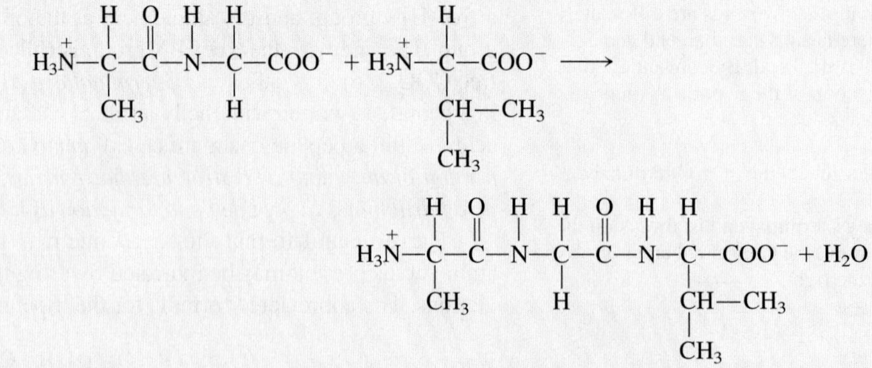

$$H_3\overset{+}{N}-\overset{\overset{\displaystyle H}{|}}{\underset{\underset{\displaystyle CH_3}{|}}{C}}-\overset{O}{\overset{\|}{C}}-\overset{\overset{\displaystyle H}{|}}{\underset{\underset{\displaystyle H}{|}}{N}}-\overset{\overset{\displaystyle H}{|}}{C}-COO^- + H_3\overset{+}{N}-\overset{\overset{\displaystyle H}{|}}{\underset{\underset{\displaystyle CH-CH_3}{|}}{\underset{\underset{\displaystyle CH_3}{|}}{C}}}-COO^- \longrightarrow$$

$$H_3\overset{+}{N}-\overset{\overset{\displaystyle H}{|}}{\underset{\underset{\displaystyle CH_3}{|}}{C}}-\overset{O}{\overset{\|}{C}}-\overset{\overset{\displaystyle H}{|}}{\underset{\underset{\displaystyle H}{|}}{N}}-\overset{\overset{\displaystyle H}{|}}{C}-\overset{O}{\overset{\|}{C}}-\overset{\overset{\displaystyle H}{|}}{N}-\overset{\overset{\displaystyle H}{|}}{\underset{\underset{\underset{\displaystyle CH_3}{\displaystyle |}}{\underset{\displaystyle CH-CH_3}{|}}}{C}}-COO^- + H_2O$$

Practice Exercise 9.3

Draw the structural formula for the tripeptide Val–Ala–Gly.

Peptides that contain the same amino acids but in different order are different molecules (constitutional isomers) with different properties. For example, two different dipeptides can be formed from one molecule of alanine and one molecule of glycine.

Amino acid sequence in a peptide has biochemical importance. Isomeric peptides give different biochemical responses; that is, they have different biochemical specificities.

$$\overset{H}{\underset{CH_3}{\overset{|}{H_3\overset{+}{N}-C}}}\overset{O}{\underset{}{\overset{||}{-C}}}\overset{H}{\underset{}{\overset{|}{-N}}}\overset{H}{\underset{H}{\overset{|}{-C}}}-COO^-$$

Ala–Gly

$$\overset{H}{\underset{H}{\overset{|}{H_3\overset{+}{N}-C}}}\overset{O}{\underset{}{\overset{||}{-C}}}\overset{H}{\underset{}{\overset{|}{-N}}}\overset{H}{\underset{CH_3}{\overset{|}{-C}}}-COO^-$$

Gly–Ala

In the first dipeptide, the alanine is the N-terminal residue, and in the second molecule, it is the C-terminal residue. These two compounds are isomers with different chemical and physical properties.

IUPAC rules for naming small peptides are as follows:

1. The C-terminal amino acid residue (located at the far right of the structure) keeps its full amino acid name.
2. All of the other amino acid residues have names that end in -*yl*. The -*yl* suffix replaces the -*ine* or -*ic acid* ending of the amino acid name, except for tryptophan, for which -*yl* is added to the name.
3. The amino acid naming sequence begins at the N-terminal amino acid residue.

For a peptide containing one each of *n* different kinds of amino acids, the number of constitutional isomers is given by *n!* (*n* factorial).

$$5! = 5 \times 4 \times 3 \times 2 \times 1 = 120$$

The two preceding dipeptides are alanylglycine (Ala-Gly) and glycylalanine (Gly-Ala). The tripeptide Val-Ser-Asp has the name valylserylaspartic acid.

The number of isomeric peptides possible increases rapidly as the length of the peptide chain increases. Let us consider the tripeptide Ala–Ser–Cys as another example. In addition to this sequence, five other arrangements of these three components are possible, each representing another isomeric tripeptide: Ala–Cys–Ser, Ser–Ala–Cys, Ser–Cys–Ala, Cys–Ala–Ser, and Cys–Ser–Ala. For a pentapeptide containing 5 different amino acids, 120 isomers are possible.

9.7 Biochemically Important Small Peptides

Many relatively small peptides have been shown to be biochemically active. Functions for them include hormonal action, neurotransmission, and antioxidant activity.

■ Small Peptide Hormones

The two best-known peptide hormones, both produced by the pituitary gland, are *oxytocin* and *vasopressin*. Each hormone is a nonapeptide (nine amino acid residues) with six of the residues held in the form of a loop by a disulfide bond formed from the interaction of two cysteine residues (Section 9.5). Structurally, these nonapeptides differ in the amino acid present in positions 3 and 8 of the peptide chain. In both structures an amide group replaces the C terminal oxygen atom.

Oxytocin plays a role in stimulating the flow of milk in a nursing mother. The baby's suckling action sends nerve signals to the mother's brain, triggering the release of oxytocin, via the blood, to the mammary glands. The oxytocin causes muscle contraction in the mammary gland, forcing out milk. As suckling continues, more oxytocin is released and more milk is available for the baby.

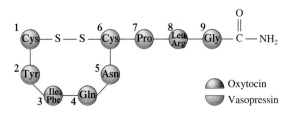

Oxytocin regulates uterine contractions and lactation. Vasopressin regulates the excretion of water by the kidneys; it also affects blood pressure.

■ **Small Peptide Neurotransmitters**

Enkephalins are pentapeptide neurotransmitters produced by the brain itself that bind at receptor sites in the brain to reduce pain. The two best-known enkephalins are Met-enkephalin and Leu-enkephalin, whose structures are

Met-enkephalin: Tyr–Gly–Gly–Phe–Met

Leu-enkephalin: Tyr–Gly–Gly–Phe–Leu

The two enkephalins differ structurally only in the amino acid at the end of the chain.

The pain-reducing effects of enkephalin action play a role in the "high" reported by long-distance runners, in the competitive athlete's managing to finish the game despite being injured, and in the pain-relieving effects of acupuncture.

The action of the prescription painkillers morphine and codeine is based on their binding at the same receptor sites in the brain as the naturally occurring enkephalins.

■ **Small Peptide Antioxidants**

The tripeptide *glutathione* (Glu–Cys–Gly) is present in significant concentrations in most cells and is of considerable physiological importance as a regulator of oxidation–reduction reactions. Specifically, glutathione functions as an antioxidant (Section 3.13), protecting cellular contents from oxidizing agents such as peroxides and superoxides (highly reactive forms of oxygen often generated within the cell in response to bacterial invasion) (Section 12.11).

The tripeptide structure of glutathione has an unusual feature. The amino acid Glu, an acidic amino acid, is bonded to Cys through the side-chain carboxyl group rather than through its α-carbon carboxyl group.

Other antioxidants previously considered are BHA and BHT (Section 3.13) and β-carotene (Section 2.6).

$$\overset{+}{H_3N}-CH-CH_2-CH_2-\overset{\overset{\displaystyle O}{\|}}{C}-NH-CH-\overset{\overset{\displaystyle O}{\|}}{C}-NH-CH-COO^-$$

Glu | Cys | Gly

9.8 General Structural Characteristics of Proteins

In Section 9.1, we defined a protein simply as a naturally-occurring, unbranched polymer in which the monomer units are amino acids. A more specific protein definition is now in order. A **protein** *is a peptide in which at least 50 amino acid residues are present.* The defining line governing the use of the term *protein*—50 amino acid residues—is an arbitrary line. The terms *polypeptide* and *protein* are often used interchangeably; a protein is a relatively long polypeptide. The key point is that the term *protein* is reserved for peptides with a large number of amino acids; it is not correct to call a tripeptide a protein. Over 10,000 amino acid residues are present in several proteins; 400–500 amino acid residues are common in proteins; small proteins contain 50–100 amino acid residues.

More than one peptide chain may be present in a protein. On this basis, proteins are classified as *monomeric* or *multimeric*. A **monomeric protein** *is a protein in which only one peptide chain is present.* Large proteins, those with many amino acid residues, usually are multimeric. A **multimeric protein** *is a protein in which more than one peptide chain is present.* The peptide chains present in multimeric proteins are called *protein subunits*. The protein subunits within a multimeric protein may all be identical to each other or different kinds of subunits may be present. Proteins with up to 12 subunits are known. The small protein insulin, which functions as a hormone in the human body, is a multimeric protein with two protein subunits; one subunit contains 21 amino acid

Proteins are the second type of biochemical polymer we have encountered; the other was polysaccharides (Section 7.14). Protein monomers are amino acids, whereas polysaccharide monomers are monosaccharides.

TABLE 9.2
Types of Conjugated Proteins

Class	Prosthetic Group	Specific Example	Function of Example
hemoproteins	heme unit	hemoglobin	carrier of O_2 in blood
		myoglobin	oxygen binder in muscles
lipoproteins	lipid	low-density lipoprotein (LDL)	lipid carrier
		high-density lipoprotein (HDL)	lipid carrier
glycoproteins	carbohydrate	gamma globulin	antibody
		mucin	lubricant in mucous secretions
		interferon	antiviral protection
phosphoproteins	phosphate group	glycogen phosphorylase	enzyme in glycogen phosphorylation
nucleoproteins	nucleic acid	ribosomes	site for protein synthesis in cells
		viruses	self-replicating, infectious complex
metalloproteins	metal ion	iron–ferritin	storage complex for iron
		zinc–alcohol dehydrogenase	enzyme in alcohol oxidation

FIGURE 9.3 The British biochemist Frederick Sanger (1918–) determined the primary structure of the protein hormone *insulin* in 1953. His work is a landmark in biochemistry because it showed for the first time that a protein has a precisely defined amino acid sequence. Sanger was awarded the Nobel Prize in chemistry in 1958 for this work. Later, in 1980, he was awarded a second Nobel Prize in chemistry, this time for work that involved the sequencing of units in nucleic acids (Chapter 11).

residues and the other 30 amino acid residues. The structure of insulin is considered in more detail in Section 9.11.

Proteins, on the basis of chemical composition, are classified as *simple* or *complex*. A **simple protein** *is a protein in which only amino acid residues are present.* More than one protein subunit may be present in a simple protein, but all subunits contain only amino acids. A **conjugated protein** *is a protein that has one or more non-amino acid entities present in its structure in addition to one or more peptide chains.* These non-amino acid components, which may be organic or inorganic, are called *prosthetic groups.* A **prosthetic group** *is a non-amino acid group present in a conjugated protein.*

Conjugated proteins may be further classified according to the nature of the prosthetic group(s) present. *Lipoproteins* contain lipid prosthetic groups, *glycoproteins* contain carbohydrate groups, *metalloproteins* contain a specific metal, and so on (see Table 9.2). Some proteins contain more than one type of prosthetic group. In general, prosthetic groups have important roles in the biochemical functions for conjugated proteins. Several examples of glycoproteins and lipoproteins are discussed in Sections 9.16 and 9.17, respectively.

In general, the three-dimensional structures of proteins, even those with just a single peptide chain, are more complex than those of carbohydrates and lipids—the biomolecules discussed in the two previous chapters. Our approach to describing and understanding this complexity in protein structure involves considering this structure at four levels. These four protein structural levels, listed in order of increasing complexity, are *primary* structure, *secondary* structure, *tertiary* structure, and *quaternary* structure. They are the subject matter for the next four sections of this chapter.

9.9 Primary Structure of Proteins

Primary protein structure *is the order in which amino acids are linked together in a protein.* Every protein has its own unique amino acid sequence. Primary protein structure always involves more than just the numbers and kinds of amino acids present; it also involves the *order of attachment* of the amino acids.

Insulin, the hormone that regulates blood-glucose levels, was the first protein for which primary structure was determined; the "sequencing" of its 51 amino acids was completed in 1953, after 8 years of work by the British biochemist Frederick Sanger (see Figure 9.3). Today, primary structures are known for thousands of proteins, and the sequencing procedures involve automated methods that require relatively short periods of

FIGURE 9.4 The primary structure of human myoglobin. This diagram gives only the sequence of the amino acids present and conveys no information about the actual three-dimensional shape of the protein. The "wavy" pattern for the 153 amino acid sequence was chosen to minimize the space used to present the needed information. The actual shape of the protein is determined by secondary and tertiary levels of protein structure, levels yet to be discussed.

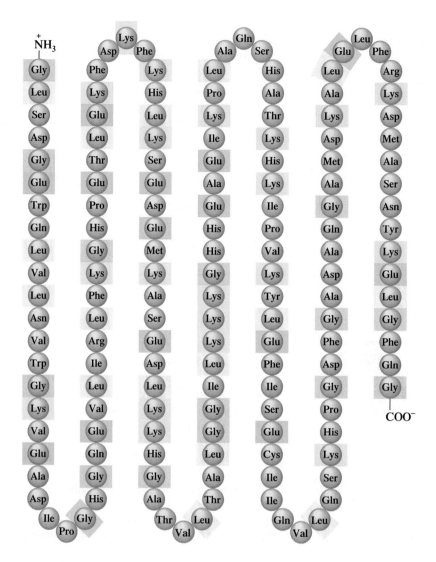

The primary structure of a protein is the *sequence* of amino acids in a protein chain—that is, the order in which the amino acids are connected to each other.

time (days). Figure 9.4 shows the primary structure of myoglobin, a protein involved in oxygen transport in muscles; it contains 153 amino acids assembled in the particular, definite order shown in this diagram.

The primary structure of a specific protein is always the same regardless of where the protein is found within an organism. The structures of certain proteins are even similar among different species of animals. For example, the primary structures of insulin in cows, pigs, sheep, and horses are very similar both to each other and to human insulin. Until recently, this similarity was particularly important for diabetics who required supplemental injections of insulin. (See the Chemical Connections feature on page 327.)

An analogy is often drawn between the primary structure of proteins and words. Words, which convey information, are formed when the 26 letters of the English alphabet are properly sequenced. Proteins are formed from proper sequences of the 20 standard amino acids. Just as the proper sequence of letters in a word is necessary for it to make sense, the proper sequence of amino acids is necessary to make biochemically active protein. Furthermore, the letters that form a word are written from left to right, as are amino acids in protein formulas. As any dictionary of the English language will document, a tremendous variety of words can be formed by different letter sequences. Imagine the number of amino acid sequences possible for a large protein. There are 1.55×10^{66} sequences possible for the 51 amino acids found in insulin! From these possibilities, the body reliably produces only *one*, illustrating the

CHEMICAL CONNECTIONS   Substitutes for Human Insulin

In humans, an insufficient production of insulin results in the disease *diabetes mellitus*. Treatment of this disease involves giving the patient extra insulin via subcutaneous injection. For many years, because of the limited availability of human insulin, most insulin used by diabetics was obtained from the pancreases of slaughter-house animals. Such animal insulin, primarily from cows and pigs, was used by most diabetics without serious side effects because it is structurally very similar to human insulin. Immunological reactions gradually do increase over time, however, because the animal insulin is foreign to the human body.

A comparison of the primary structure of human insulin with pig and cow insulins shows differences at only 4 of the 51 amino acid positions: positions 8, 9, and 10 on chain A and position 30 on chain B (see Figure 9.11 and the following table).

Species	Chain A			Chain B
	#8	**#9**	**#10**	**#30**
human	Thr	Ser	Ile	Thr
pig (porcine)	Thr	Ser	Ile	Ala
cow (bovine)	Ala	Ser	Val	Ala

The dependence of diabetics on animal insulin has declined because of the availability of human insulin produced by genetically engineered bacteria (Section 11.14). These bacteria carry a gene that directs the synthesis of human insulin. Such bacteria-produced insulin is fully functional. All diabetics now have the choice of using human insulin or using animal insulin. Many still continue to use the animal insulin because it is cheaper.

remarkable precision of life processes. From the simplest bacterium to the human brain cell, only those amino acid sequences needed by the cell are produced. The fascinating process of protein biosynthesis and the way in which genes in DNA direct this process will be discussed in Chapter 11.

9.10 Secondary Structure of Proteins

Secondary protein structure *is the arrangement in space adopted by the backbone portion of a protein.* The two most common types of secondary structure are the *alpha helix* (α helix) and the *beta pleated sheet* (β pleated sheet). The type of interaction responsible for both of these types of secondary structure is hydrogen bonding between a carbonyl oxygen atom of a peptide linkage and the hydrogen atom of an amino group of another peptide linkage farther along the backbone. Information about the geometry associated with these peptide linkages is helpful in understanding how hydrogen bonding interactions occur between peptide linkages of a protein backbone. Important geometrical considerations are:

1. The peptide linkages are essentially planar. This means that for two amino acids linked through a peptide linkage, six atoms lie in the same plane: the α-carbon atom and the C=O group from the first amino acid and the N–H group and the α-carbon atom from the second amino acid.

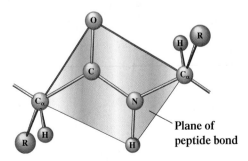

Plane of peptide bond

2. The planar peptide linkage structure has considerable rigidity, which means that rotation of groups about the C–N bond is hindered, and *cis–trans* isomerism is possible

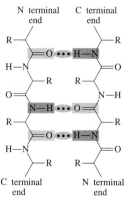

FIGURE 9.5 The hydrogen bonding between the carbonyl oxygen atom of one peptide linkage and the amide hydrogen atom of another peptide linkage.

The hydrogen bonding present in an α helix is *intra*molecular. In a β pleated sheet, the hydrogen bonding can be *inter*molecular (between two different chains) or *intra*molecular (a single chain folding back on itself).

about this bond. The *trans* isomer orientation is the preferred orientation, as shown in the preceding diagram. The O atom of the C=O group and the H atom of the N–H group are positioned *trans* to each other.

Figure 9.5 shows the hydrogen bonding possibilities that exist between carbonyl oxygen atoms and amide hydrogen atoms associated with different peptide linkages in a protein backbone. The protein backbone segments shown can be two segments of the same backbone or two segments from different backbones. We consider both of these situations in further detail in this section.

■ The Alpha Helix

An **alpha helix structure** *is a protein secondary structure in which a single protein chain adopts a shape that resembles a coiled spring (helix), with the coil configuration maintained by hydrogen bonds.* The hydrogen bonds are between $\diagdown$N—H and $\diagdown$C=O groups of every fourth amino acid, as is shown diagrammatically in Figure 9.6.

Proteins have varying amounts of α-helical secondary structure, ranging from a few percent to nearly 100%. In an α helix, all of the amino acid side chains (R groups) lie outside the helix; there is not enough room for them in the interior. Figure 9.6d illustrates this situation.

■ The Beta Pleated Sheet

A **beta pleated sheet structure** *is a protein secondary structure in which two fully extended protein chain segments in the same or different molecules are held together by hydrogen bonds.* Hydrogen bonds form between oxygen and hydrogen peptide linkage atoms that are either in different parts of a single chain that folds back on itself (intrachain bonds) or between atoms in different peptide chains in those proteins that contain more than one chain (interchain bonds). In molecules where the β pleated sheet involves a single molecule, several U-turns in the protein chain arrangement are needed in order to form the structure.

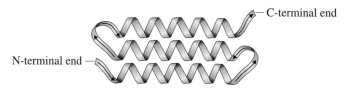

This "U-turn structure" is the most frequently encountered type of β pleated sheet structure.

FIGURE 9.6 Four representations of the α helix protein secondary structure. (a) Arrangement of protein backbone with no detail shown. (b) Backbone arrangement with hydrogen-bonding interactions shown. (c) Backbone atomic detail shown, as well as hydrogen-bonding interactions. (d) Top view of an α helix showing that amino acid side chains (R groups) point away from the long axis of the helix.

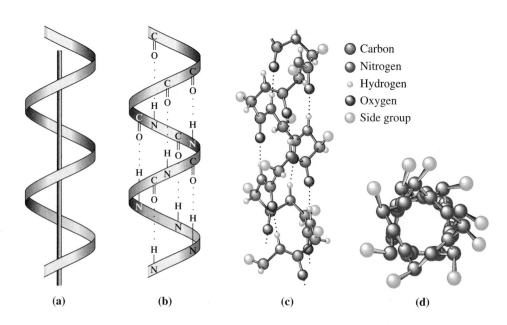

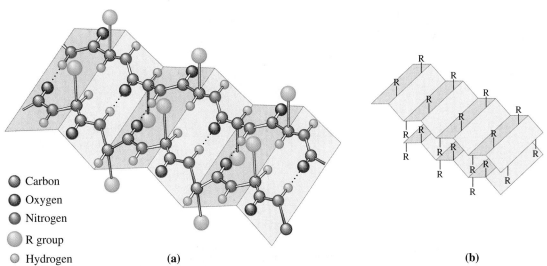

Carbon
Oxygen
Nitrogen
R group
Hydrogen

(a)   **(b)**

FIGURE 9.7 Two representations of the β pleated sheet protein structure. (a) A representation emphasizing the hydrogen bonds between protein chains. (b) A representation emphasizing the pleats and the location of the R groups.

Figure 9.7a shows a representation of the β pleated sheet structure that occurs when portions of two different peptide chains are aligned parallel to each other (interchain bonds). The term *pleated sheet* arises from the repeated zigzag pattern in the structure (Figure 9.7b). Note how in a pleated sheet structure the amino acid side chains are positioned above and below the plane of the sheet.

Very few proteins have entirely α helix or β pleated sheet structures. Instead, most proteins have only certain portions of their molecules in these conformations. The rest of the molecule assumes an "irregular structure." It is possible to have both α helix and β pleated sheet structures within the same protein. Figure 9.8 is a diagram of a protein chain where both helical and pleated sheet segments, as well as irregular structure, are present within a single peptide chain. The β pleated sheet segment involves a single peptide chain folding back on itself (intrachain bonds). Helical structure and pleated sheet structure are found only in portions of a protein where the amino acid R groups present are relatively small; large R groups tend to disrupt both of these types of secondary structure. The term *irregular structure* used in describing portions of a protein structure is somewhat of a misnomer, because all molecules of a given protein exhibit *identical* irregular structure.

The β pleated sheet is found extensively in the protein of silk. Because such proteins are already fully extended, silk fibers cannot be stretched. When wool, which has an α helix structure, becomes wet, it stretches as hydrogen bonds of the helix are broken. The wool returns to its original shape as it dries. Wet stretched wool, dried under tension, maintains its stretched length because it has assumed a β pleated sheet configuration.

FIGURE 9.8 The secondary structure of a single protein often shows areas of α helix and β pleated sheet configurations, as well as areas of random coiling.

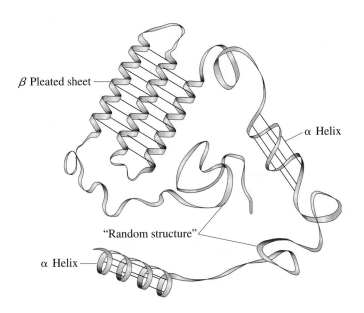

β Pleated sheet

α Helix

"Random structure"

α Helix

FIGURE 9.9 A telephone cord has three levels of structure. These structural levels are a good analogy for the first three levels of protein structure.

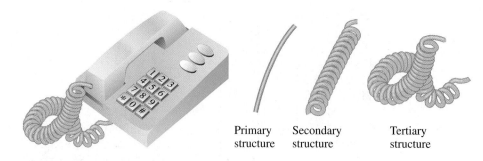

Primary structure   Secondary structure   Tertiary structure

FIGURE 9.9 A telephone cord has three levels of structure. These structural levels are a good analogy for the first three levels of protein structure.

9.11 Tertiary Structure of Proteins

Tertiary protein structure *is the overall three-dimensional shape of a protein that results from the interactions between amino acid side chains (R groups) that are widely separated from each other within a peptide chain.*

A good analogy for the relationships among the primary, secondary, and tertiary structures of a protein is that of a telephone cord (Figure 9.9). The primary structure is the long, straight cord. The coiling of the cord into a helical arrangement gives the secondary structure. The supercoiling arrangement the cord adopts after you hang up the receiver is the tertiary structure.

■ Interactions Responsible for Tertiary Structure

Four types of attractive interactions contribute to the tertiary structure of a protein: (1) covalent disulfide bonds, (2) electrostatic attractions (salt bridges), (3) hydrogen bonds, and (4) hydrophobic attractions. All four of these interactions are interactions between amino acid R groups. This is a major distinction between tertiary-structure interactions and secondary-structure interactions. Tertiary-structure interactions involve the R groups of amino acids; secondary-structure interactions involve the peptide linkages between amino acid residues.

Disulfide bonds, the strongest of the tertiary-structure interactions, result from the —SH groups of two cysteine residues reacting with each other to form a *covalent* disulfide bond (Section 9.5). This type of interaction is the only one of the four tertiary-structure interactions that involves a covalent bond. Disulfide bond formation may involve two cysteine units in the same peptide chain (an intramolecular disulfide bond; see Figure 9.10a) or two cysteine units in different chains (an intermolecular disulfide bond; see Figure 9.10b). Figure 9.11 gives the structure of the protein hormone insulin, a protein that has two peptide chains and a total of 51 amino acid residues; both inter- and intramolecular disulfide bonds are present in its structure.

Electrostatic interactions, also called *salt bridges,* always involve amino acids with charged side chains. These amino acids are the acidic and basic amino acids. The two

Cysteine is the only α-amino acid that contains a sulfhydryl group (—SH).

FIGURE 9.10 Disulfide bonds involving cysteine residues can form in two different ways: (a) between two SH groups on the same chain or (b) between two SH groups on different chains.

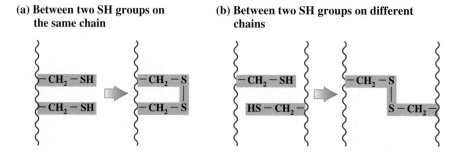

(a) Between two SH groups on the same chain

(b) Between two SH groups on different chains

FIGURE 9.11 Human insulin, a small two-chain protein, has both intrachain and interchain disulfide linkages as part of its tertiary structure.

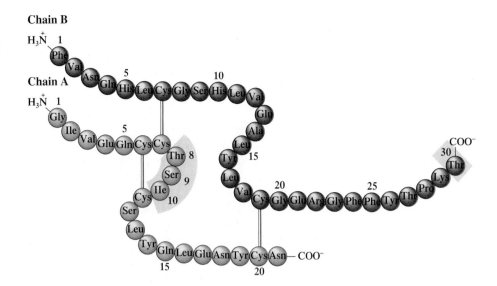

R groups, one acidic and one basic, interact through ion–ion attractions. Figure 9.12b shows an electrostatic interaction.

Hydrogen bonds can occur between amino acids with polar R groups. A variety of polar side chains can be involved, especially those that possess the following functional groups:

$$-\text{OH} \qquad -\text{NH}_2 \qquad -\overset{\displaystyle \overset{O}{\|}}{\text{C}}-\text{OH} \qquad -\overset{\displaystyle \overset{O}{\|}}{\text{C}}-\text{NH}_2$$

Hydrogen bonds are relatively weak and are easily disrupted by changes in pH and temperature. Figure 9.12c shows the hydrogen-bonding interactions between the R groups of glutamine and serine.

Hydrophobic interactions result when two nonpolar side chains are close to each other. In aqueous solution, many proteins have their polar R groups outward, toward the aqueous solvent (which is also polar), and their nonpolar R groups inward (away from the polar water molecules). The nonpolar R groups then interact with each other. Hydrophobic interactions are common between phenyl rings and alkyl side chains. Although hydrophobic interactions are weaker than hydrogen bonds or electrostatic interactions, they are a significant force in some proteins because there are so many of them; their cumulative effect can be greater in magnitude than the effects of hydrogen bonding. Figure 9.12d shows the hydrophobic interactions between the R groups of phenylalanine and leucine.

In 1959, a protein tertiary structure was determined for the first time. The determination involved myoglobin, a conjugated protein (Section 9.8) whose function is oxygen storage in muscle tissue. Figure 9.13 shows myoglobin's tertiary structure. It involves a single peptide chain of 153 amino acids with numerous α helix segments within the chain. The structure also contains a prosthetic heme group, an iron-containing group with the ability to bind molecular oxygen.

FIGURE 9.12 Four types of interactions between amino acid R groups produce the tertiary structure of a protein. (a) Disulfide bonds. (b) Electrostatic interactions (salt bridges). (c) Hydrogen bonds. (d) Hydrophobic interactions.

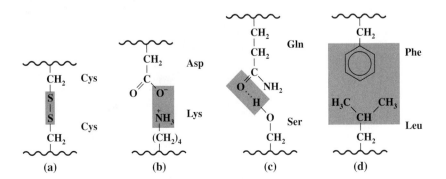

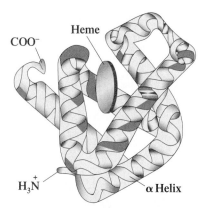

FIGURE 9.13 A schematic diagram showing the tertiary structure of the single-chain protein myoglobin.

A comparison of Figure 9.13 (myoglobin's tertiary structure) with Figure 9.4 (myoglobin's primary structure) shows how different the perspectives of primary and tertiary structure are for a protein.

9.12 Quaternary Structure of Proteins

Quaternary structure is the highest level of protein organization. It is found only in multimeric proteins (Section 9.8). Such proteins have structures involving two or more peptide chains that are independent of each other—that is, are not covalently bonded to each other. **Quaternary protein structure** *is the organization among the various peptide chains in a multimeric protein.*

Most multimeric proteins contain an even number of subunits (two subunits = a dimer, four subunits = a tetramer, and so on). The subunits are held together mainly by hydrophobic interactions between amino acid R groups.

The noncovalent interactions that contribute to tertiary structure (electrostatic interactions, hydrogen bonds, and hydrophobic interactions) are also responsible for the maintenance of quaternary structure. The noncovalent interactions that contribute to quaternary structure are, however, more easily disrupted. For example, only small changes in cellular conditions can cause a tetrameric protein to fall apart, dissociating into dimers or perhaps four separate subunits, with a resulting temporary loss of protein activity. As original cellular conditions are restored, the tertiary structure automatically re-forms, and normal protein function is restored.

An example of a protein with quaternary structure is hemoglobin, the oxygen-carrying protein in blood (Figure 9.14). It is a tetramer in which there are two identical α chains and two identical β chains. Each chain enfolds a heme group, the site where oxygen binds to the protein.

The Chemistry at a Glance feature on page 333 reviews what we have said about protein structural levels.

9.13 Fibrous and Globular Proteins

On the basis of secondary, tertiary, and quaternary structural features, proteins can be classified into two major types: *fibrous* proteins and *globular* proteins. A **fibrous protein** *is a protein in which peptide chains are arranged in long strands or sheets.* Such proteins have long rodlike molecules that can intertwine with one another and form strong fibers. A **globular protein** *is a protein in which peptide chains are folded into spherical or globular shapes.* Table 9.3 gives examples of selected fibrous and globular proteins.

FIGURE 9.14 A schematic diagram showing the tertiary and quaternary structure of the oxygen-carrying protein hemoglobin.

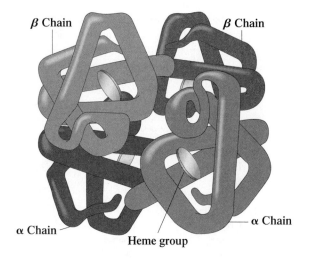

CHEMISTRY AT A GLANCE

Protein Structure

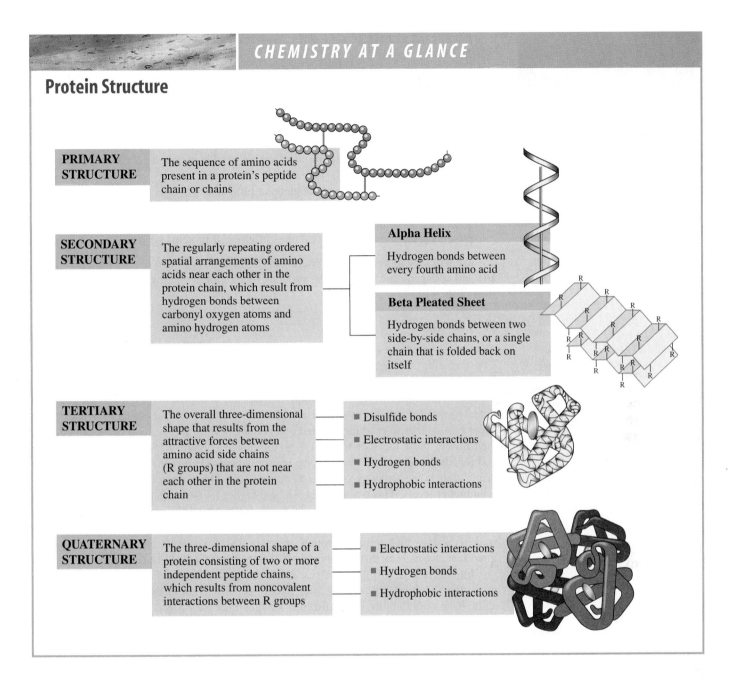

PRIMARY STRUCTURE
The sequence of amino acids present in a protein's peptide chain or chains

SECONDARY STRUCTURE
The regularly repeating ordered spatial arrangements of amino acids near each other in the protein chain, which result from hydrogen bonds between carbonyl oxygen atoms and amino hydrogen atoms

Alpha Helix
Hydrogen bonds between every fourth amino acid

Beta Pleated Sheet
Hydrogen bonds between two side-by-side chains, or a single chain that is folded back on itself

TERTIARY STRUCTURE
The overall three-dimensional shape that results from the attractive forces between amino acid side chains (R groups) that are not near each other in the protein chain

- Disulfide bonds
- Electrostatic interactions
- Hydrogen bonds
- Hydrophobic interactions

QUATERNARY STRUCTURE
The three-dimensional shape of a protein consisting of two or more independent peptide chains, which results from noncovalent interactions between R groups

- Electrostatic interactions
- Hydrogen bonds
- Hydrophobic interactions

TABLE 9.3
Some Common Fibrous and Globular Proteins

Name	Occurrence and Function
Fibrous proteins (insoluble)	
keratins	found in wool, feathers, hooves, silk, and fingernails
collagens	found in tendons, bone, and other connective tissue
elastins	found in blood vessels and ligaments
myosins	found in muscle tissue
fibrin	found in blood clots
Globular proteins (soluble)	
insulin	regulatory hormone for controlling glucose metabolism
myoglobin	involved in oxygen storage in muscles
hemoglobin	involved in oxygen transport in blood
transferrin	involved in iron transport in blood
immunoglobulins	involved in immune system responses

FIGURE 9.15 The tail feathers of a peacock contain the fibrous protein α *keratin.*

Natural silk (silkworm silk) and spider silk (spider webs) are made of *fibroin,* a fibrous protein that exists mainly in a beta pleated sheet form. The great strength and toughness of silk fibers, which exceed those of many synthetic fibers, is related to the *close* stacking of the beta sheets. A high percentage of the amino acid residues (primary structure) in silk are either glycine (R═H) or alanine (R═CH₃). It is the smallness of these two R groups that makes the close stacking possible.

Fibrous and globular proteins differ in several properties:

1. Fibrous proteins are generally water-insoluble, whereas globular proteins dissolve in water. This enables globular proteins to travel through the blood and other body fluids to sites where their activity is needed.
2. Fibrous proteins usually have a single type of secondary structure, whereas globular proteins often contain several types of secondary structure.
3. Fibrous proteins generally have structural functions that provide support and external protection, whereas globular proteins are involved in metabolic chemistry, performing functions such as catalysis, transport, and regulation.
4. The number of different kinds of globular protein far exceeds the number of different kinds of fibrous protein. However, because the most abundant proteins in the human body are fibrous proteins rather than globular proteins, the total mass of fibrous proteins present exceeds the total mass of globular proteins present.

We will now more closely examine the characteristics of two fibrous proteins (α keratin and collagen) and two globular proteins (hemoglobin and myoglobin) as representatives of their types.

■ α Keratin

The fibrous protein α *keratin* is particularly abundant in nature, where it is found in protective coatings for organisms. It is the major protein constituent of hair, feathers (Figure 9.15), wool, fingernails and toenails, claws, scales, horns, turtle shells, quills, and hooves.

The structure of a typical α keratin, that of hair, is depicted in Figure 9.16. The individual molecules are almost wholly α-helical (Figure 9.16a). Pairs of these helices twine about one another to produce a coiled coil (Figure 9.16b). In hair, two of the coiled coils then further twist together to form a protofilament (Figure 9.16c). Protofilaments then coil together in groups of four to form microfilaments (Figure 9.16d), which become the "core" unit in the structure of the α-keratin of hair. These microfilaments in turn coil at even higher levels. This coiling at higher and higher levels is what produces the strength associated with α-keratin-containing proteins. All levels of coiling organization are stabilized by attractive forces of the types previously considered in the discussion of generalized secondary and tertiary protein structure (Sections 9.10 and 9.11). Particularly important are *inter*coil disulfide bridges that form between cysteine residues.

Introduction of disulfide bridges within the several levels of coiling structure determines the "hardness" of an α keratin. "Hard" keratins, such as those found in horns and nails, have considerably more disulfide bridges than their softer counterparts found in hair, wool, and feathers.

FIGURE 9.16 The coiled-coil structure of the fibrous protein α keratin.

(a)
α **Helix**

(b)
Coiled coil of
two α helices

(c)
Protofilament
(pair of coiled
coils)

(d)
Microfilament
(four coiled
protofilaments)

Collagen, pronounced "kahl-uh-jen," is the most abundant protein in the human body.

The Collagen Content of Selected Body Tissues

Tissue	Collagen (% dry mass)
Achilles tendon	86
aorta	12–24
bone (mineral-free)	88
cartilage	46–63
cornea	68
ligament	17
skin	72

The hemoglobin of a fetus is slightly different in structure from adult hemoglobin. Called *fetal hemoglobin,* this hemoglobin has a greater affinity for oxygen than the mother's hemoglobin. This ensures a steady flow of oxygen to the fetus. Shortly after birth, a baby's body ceases to produce fetal hemoglobin, and its production of "adult" hemoglobin begins.

FIGURE 9.17 A schematic diagram emphasizing how three helical polypeptide chains intertwine to form a triple helix. The chains are partially unwound and cut away to show their structure.

Collagen

Collagen, the most abundant of all proteins in humans (30% of total body protein), is a major structural material in tendons, ligaments, blood vessels, and skin; it is also the organic component of bones and teeth. Table 9.4 gives the collagen content of selected body tissues. The predominant structural feature within collagen molecules is a *triple helix* formed when three chains of amino acids wrap around each other to give a ropelike arrangement of polypeptide chains (see Figure 9.17).

The rich content of the amino acid proline (up to 20%) in collagen is one reason why it has a triple-helix conformation rather than the simpler α helix structure (Section 9.10). Proline amino acid residues do not fit into regular α helices because of the cyclic nature of the side chain present and its accompanying different "geometry."

Portion of a collagen chain

Collagen molecules (triple helices) are very long, thin, and rigid. Many such molecules, lined up alongside each other, combine to make collagen fibrils. Cross-linking between helices gives the fibrils extra strength. The greater the number of cross links, the more rigid the fibril is. The stiffening of skin and other tissues associated with aging is thought to result, at least in part, from an increasing amount of cross-linking between collagen molecules. The process of tanning, which converts animal hides to leather, involves increasing the degree of cross-linking.

Figure 9.18 shows an electron micrograph of collagen fibers.

Hemoglobin

The globular protein *hemoglobin* transports oxygen from the lungs to tissue. Its tertiary structure was shown in Figure 9.14. It is a tetramer (four peptide chains) with each subunit also containing a *heme group,* the entity that binds oxygen. With four heme groups present, a hemoglobin molecule can transport four oxygen molecules at the same time.

The structure of a heme group is

Heme

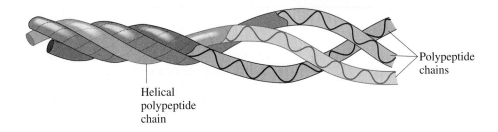

Protein Structure and the Color of Meat

The meat that humans eat is composed primarily of muscle tissue. The major proteins present in such muscle tissue are *myosin* and *actin,* which lie in alternating layers and which slide past each other during muscle contraction. Contraction is temporarily maintained through interactions between these two types of proteins.

Structurally, myosin consists of a rodlike coil of two alpha helices (fibrous protein) with two globular protein heads. It is the "head portions" of myosin that interact with the actin.

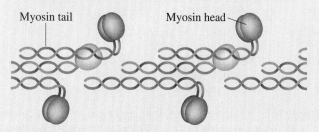

Myosin tail — Myosin head —

Structurally, actin has the appearance of two filaments spiraling about one another (see diagram below). Each circle in this structural diagram represents a monomeric unit of actin (called globular actin). The monomeric actin units associate to form a long polymer (called fibrous actin). Each identical monomeric actin unit is a globular protein containing many amino acid residues.

The chemical process associated with muscle contraction (interaction between myosin and actin) requires molecular oxygen. The oxygen storage protein *myoglobin* (Section 9.13) is the oxygen source. The amount of myoglobin present in a muscle is determined by how the muscle is used. Heavily used muscles require larger amounts of myoglobin than infrequently used muscles require.

The amount of myoglobin present in muscle tissue is a major determiner of the color of the muscle tissue. Myoglobin molecules have a red color when oxygenated and a purple color when deoxygenated. Thus, heavily worked muscles have a darker color than infrequently used muscles.

The different colors of meat reflect the concentration of myoglobin in the muscle tissue. In turkeys and chickens, which walk around a lot but rarely fly, the leg meat is dark, the breast meat is white. On the other hand, game birds that do fly a lot have dark breast meat. In general, game animals (which use all of their muscles regularly) tend to have darker meat than domesticated animals.

All land animals and birds need to support their own weight. Fish, on the other hand, are supported by water as they swim, which reduces the need for myoglobin oxygen support. Hence fish tend to have lighter flesh. Fish that spend most of their time lying at the bottom of a body of water have the lightest (whitest) flesh of all. Salmon flesh contains additional pigments that give it its characteristic "orange-pink" color.

Meat, when cooked, turns brown as the result of changes in myoglobin structure caused by the heat; the iron atom in the heme unit of myoglobin (Section 9.13) becomes oxidized. When meat is heavily salted with preservatives (NaCl, $NaNO_2$, or the like), as in the preparation of ham, the myoglobin picks up nitrite ions, and its color changes to pink.

FIGURE 9.18 Electron micrograph of collagen fibers.

The function of hemoglobin is oxygen *transfer,* and the function of myoglobin is oxygen *storage.*

It is the iron atom at the center of the heme molecule that actually interacts with the O_2.

■ Myoglobin

The globular protein *myoglobin* functions as an oxygen storage molecule in muscles. Its tertiary structure was shown in Figure 9.13. Myoglobin is a monomer, whereas hemoglobin is a tetramer. That is, myoglobin consists of a single peptide chain and a heme unit, and hemoglobin has four peptide chains and four heme units. Thus only one O_2 molecule can be carried by a myoglobin molecule. The tertiary structure of the single peptide chain of myoglobin is almost identical to the tertiary structure of each of the subunits of hemoglobin.

Myoglobin has a higher affinity for oxygen than does hemoglobin. Thus the transfer of oxygen from hemoglobin to myoglobin occurs readily. Oxygen stored in myoglobin molecules serves as a reserve oxygen source for working muscles when their demand for oxygen exceeds that which can be supplied by hemoglobin.

The Chemical Connections feature considers how the amount of myoglobin present in muscle tissue is related to the color of the meats that humans eat.

9.14 Protein Hydrolysis

When a protein or smaller peptide in a solution of strong acid or strong base is heated, the peptide bonds of the amino acid chain are hydrolyzed and free amino acids are produced. The hydrolysis reaction is the reverse of the formation reaction for a peptide bond. Amine and carboxylic acid functional groups are regenerated.

Let us consider the hydrolysis of the tripeptide Ala–Gly–Cys under acidic conditions. Complete hydrolysis produces one unit each of the amino acids alanine, glycine, and cysteine. The equation for the hydrolysis is

$$\overset{+}{H_3N}-\underset{\underset{CH_3}{|}}{\overset{\overset{H}{|}}{C}}-\overset{\overset{O}{\|}}{C}-\underset{\underset{H}{|}}{\overset{\overset{H}{|}}{N}}-\underset{\underset{H}{|}}{\overset{\overset{H}{|}}{C}}-\overset{\overset{O}{\|}}{C}-\underset{\underset{\underset{SH}{|}}{CH_2}}{\overset{\overset{H}{|}}{N}}-\overset{\overset{H}{|}}{C}-COOH \xrightarrow[\text{heat}]{H_2O,\ H^+} \overset{+}{H_3N}-\underset{\underset{CH_3}{|}}{\overset{\overset{H}{|}}{C}}-COOH + \overset{+}{H_3N}-\underset{\underset{H}{|}}{\overset{\overset{H}{|}}{C}}-COOH + \overset{+}{H_3N}-\underset{\underset{\underset{SH}{|}}{CH_2}}{\overset{\overset{H}{|}}{C}}-COOH$$

Ala–Gly–Cys   Ala   Gly   Cys

Note that the product amino acids in this reaction are written in positive-ion form because of the acidic reaction conditions.

Protein digestion (Section 15.1) is simply enzyme-catalyzed hydrolysis of ingested protein. The free amino acids produced from this process are absorbed through the intestinal wall into the bloodstream and transported to the liver. Here they become the raw materials for the synthesis of new protein tissue. Also, the hydrolysis of cellular proteins to amino acids is an ongoing process, as the body resynthesizes needed molecules and tissue.

> Protein hydrolysis produces free amino acids. This process is the reverse of protein synthesis, where free amino acids are combined.

9.15 Protein Denaturation

Protein denaturation *is the partial or complete disorganization of a protein's characteristic three-dimensional shape as a result of disruption of its secondary, tertiary, and quaternary structural interactions.* Because the biochemical function of a protein depends on its three-dimensional shape, the result of denaturation is loss of biochemical activity. Protein denaturation does not affect the primary structure of a protein.

Although some proteins lose all of their three-dimensional structural characteristics upon denaturation (Figure 9.19), most proteins maintain some three-dimensional structure. Often, for limited denaturation changes, it is possible to find conditions under which the effects of denaturation can be reversed; this restoration process, in which the protein is "refolded," is called *renaturation*. However, for extensive denaturation changes, the process is usually irreversible.

> A consequence of protein denaturation, the partial or complete loss of a protein's three-dimensional structure, is loss of biochemical activity for the protein.

FIGURE 9.19 Protein denaturation involves loss of the protein's three-dimensional structure. Complete loss of such structure produces a random-coil protein strand.

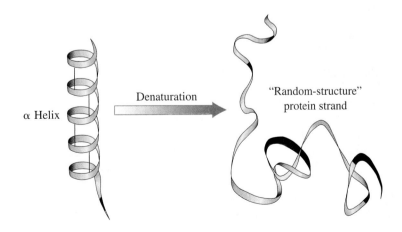

α Helix   Denaturation →   "Random-structure" protein strand

FIGURE 9.20 Heat denatures the protein in egg white, producing a white jellylike solid. The primary structure of the protein remains intact, but all higher levels of protein structure are disrupted.

Loss of water solubility is a frequent physical consequence of protein denaturation. The precipitation out of biochemical solution of denatured protein is called *coagulation.*

A most dramatic example of protein denaturation occurs when egg white (a concentrated solution of the protein albumin) is poured onto a hot surface. The clear albumin solution immediately changes into a white solid with a jelly-like consistency (see Figure 9.20). A similar process occurs when hamburger juices encounter a hot surface. A brown jelly-like solid forms.

When protein-containing foods are cooked, protein denaturation occurs. Such "cooked" protein is more easily digested because it is easier for digestive enzymes to "work on" denatured (unraveled) protein. Cooking foods also kills microorganisms through protein denaturation. For example, ham and bacon can harbor parasites that cause trichinosis. Cooking the ham or bacon denatures parasite protein.

In surgery, heat is often used to seal small blood vessels. This process is called *cauterization.* Small wounds can also be sealed by cauterization. Heat-induced denaturation is used in sterilizing surgical instruments and in canning foods; bacteria are destroyed when the heat denatures their protein.

The body temperature of a patient with fever may rise to 102°F, 103°F, or even 104° without serious consequences. A temperature above 106°F (41°C) is extremely dangerous, for at this level the enzymes of the body begin to be inactivated. Enzymes, which function as catalysts for almost all body reactions, are protein. Inactivation of enzymes, through denaturation, can have lethal effects on body chemistry.

The effect of ultraviolet radiation from the sun, a nonionizing radiation, is similar to that of heat. Denatured skin proteins cause most of the problems associated with sunburn.

A curdy precipitate of casein, the principal protein in milk, is formed in the stomach when the hydrochloric acid of gastric juice denatures the casein. The curdling of milk that takes place when milk sours or cheese is made (see Figure 9.21) results from the presence of lactic acid, a by-product of bacterial growth. Yogurt is prepared by growing lactic-acid-producing bacteria in skim milk. The coagulated denatured protein gives yogurt its semi-solid consistency.

Serious eye damage can result from eye tissue contact with acids or bases, when irreversibly denatured and coagulated protein causes a clouded cornea. This reaction is part of the basis for the rule that students wear protective eyewear in the chemistry laboratory.

Alcohols are an important type of denaturing agent. Denaturation of bacterial protein takes place when isopropyl or ethyl alcohol is used as a disinfectant—hence the common practice of swabbing the skin with alcohol before giving an injection. Interestingly, pure isopropyl or ethyl alcohol is less effective than the commonly used 70% alcohol solution. Pure alcohol quickly denatures and coagulates the bacterial surface, thereby forming an effective barrier to further penetration by the alcohol. The 70% solution denatures more slowly and allows complete penetration to be achieved before coagulation of the surface proteins takes place.

The process of giving a person a "hair permanent" involves protein denaturation through the use of reducing agents and oxidizing agents (see the Chemical Connections feature on page 339).

Table 9.5 is a listing of selected physical and chemical agents that cause protein denaturation. The effectiveness of a given denaturing agent depends on the type of protein upon which it is acting.

Globular proteins denature more readily than fibrous proteins because of weaker secondary and tertiary attractive forces.

FIGURE 9.21 Storage room for cheese; during storage cheese "matures" as bacteria and enzymes ferment the cheese, giving it a stronger flavor.

9.16 Glycoproteins

A **glycoprotein** *is a conjugated protein that contains carbohydrates or carbohydrate derivatives in addition to amino acids.* The carbohydrate content of glycoproteins is variable (from a few percent up to 85%), but it is fixed for any specific glycoprotein.

Glycoproteins include a number of very important substances; two of these, collagen and immunoglobulins, are considered in this section. Many of the proteins in cell membranes

CHEMICAL CONNECTIONS

Denaturation and Human Hair

The process used in waving hair—that is, in a hair permanent—involves reversible denaturation. Hair is protein in which many disulfide (—S—S—) linkages occur as part of its tertiary structure; 16%–18% of hair is the amino acid cysteine. It is these disulfide linkages that give hair protein its overall shape. When a permanent is administered, hair is first treated with a reducing agent (ammonium thioglycolate) that breaks the disulfide linkages in the hair, producing two sulfhydryl (—SH) groups:

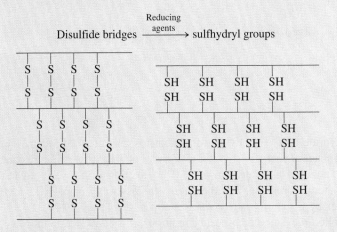

Disulfide bridges $\xrightarrow{\text{Reducing agents}}$ sulfhydryl groups

The "reduced" hair, whose tertiary structure has been disrupted, is then wound on curlers to give it a new configuration.

Finally, the reduced and rearranged hair is treated with an oxidizing agent (potassium bromate) to form disulfide linkages at new locations within the hair:

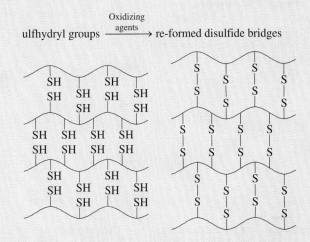

ulfhydryl groups $\xrightarrow{\text{Oxidizing agents}}$ re-formed disulfide bridges

The new shape and curl of the hair are maintained by the newly formed disulfide bonds and the resulting new tertiary structure accompanying their formation. Of course, as new hair grows in, the "permanent" process has to be repeated.

TABLE 9.5
Selected Physical and Chemical Denaturing Agents

Disulfide bridges, which involve covalent bonds, impart considerable resistance to denaturation because they are much stronger than the noncovalent interactions otherwise present.

Denaturing Agent	Mode of Action
heat	disrupts hydrogen bonds by making molecules vibrate too violently; produces coagulation, as in the frying of an egg
microwave radiation	causes violent vibrations of molecules that disrupt hydrogen bonds
ultraviolet radiation	operates very similarly to the action of heat (e.g., sunburning)
violent whipping or shaking	causes molecules in globular shapes to extend to longer lengths, which then entangle (e.g., beating egg white into meringue)
detergent	affects R-group interactions
organic solvents (e.g., ethanol, 2-propanol, acetone)	interfere with R-group interactions because these solvents also can form hydrogen bonds; quickly denature proteins in bacteria, killing them (e.g., the disinfectant action of 70% ethanol)
strong acids and bases	disrupt hydrogen bonds and salt bridges; prolonged action leads to actual hydrolysis of peptide bonds
salts of heavy metals (e.g., salts of Hg^{2+}, Ag^+, Pb^{2+})	metal ions combine with —SH groups and form poisonous salts
reducing agents	reduce disulfide linkages to produce —SH groups

(lipid bilayers; see Section 8.10) are actually glycoproteins. The blood group markers of the ABO system (see the Chemical Connections feature on page 244 in Chapter 7) are also glycoproteins in which the carbohydrate content can reach 85%.

■ Collagen

The fibrous protein collagen (Section 9.13) qualifies as a *glycoprotein* because carbohydrate units are present in its structure. A structural feature of collagen not considered in Section 9.13 is the presence of the *nonstandard* amino acids 4-hydroxyproline (5%) and 5-hydroxylysine (1%)—derivatives of the standard amino acids proline and lysine (Table 9.1).

4-Hydroxyproline

5-Hydroxylysine

> Nonstandard amino acids consist of amino acid residues that have been chemically modified after their incorporation into a protein (as is the case with 4-hydroxyproline and 5-hydroxylysine) and amino acids that occur in living organisms but are not found in proteins.

The presence of carbohydrate units (mostly glucose, galactose, and their disaccharides) attached by glycosidic linkages (Section 7.13) to collagen at its 5-hydroxylysine residues causes collagen to be classified as a *glycoprotein*. The function of the carbohydrate groups in collagen is related to cross-linking; they direct the assembly of collagen triple helices into more complex aggregations called *collagen fibrils*.

When collagen is boiled in water, under basic conditions, it is converted to the water-soluble protein gelatin. This process involves both denaturation (Section 9.15) and hydrolysis (Section 9.14). Heat acts as a denaturant, causing rupture of the hydrogen bonds supporting collagen's triple-helix structure. Regions in the amino acid chains where proline and hydroxyproline concentrations are high are particularly susceptible to hydrolysis, which breaks up the polypeptide chains. Meats become more tender when cooked because of the conversion of some collagen to gelatin. Tougher cuts of meat (more cross-linking), such as stew meat, need longer cooking times.

> The primary biochemical function of vitamin C involves the hydroxylation of proline and lysine during collagen formation. These hydroxylation processes require the enzymes *proline hydroxylase* and *lysine hydroxylase.* These enzymes can function only in the presence of vitamin C.

■ Immunoglobulins

Immunoglobulins are among the most important and interesting of the soluble proteins in the human body. An **immunoglobulin** *is a glycoprotein produced by an organism as a protective response to the invasion of microorganisms or foreign molecules.* Different classes of immunoglobulins, identified by differing carbohydrate content and molecular mass, exist.

Immunoglobulins serve as *antibodies* to combat invasion of the body by *antigens.* An **antigen** *is a foreign substance, such as a bacterium or virus, that invades the human body.* An **antibody** *is a biochemical molecule that counteracts a specific antigen.* The immune system of the human body has the capability to produce immunoglobulins that respond to several thousand different antigens.

All types of immunoglobulin molecules have much the same basic structure, which includes the following features:

1. Four polypeptide chains are present: two identical heavy (H) chains and two identical light (L) chains.
2. The H chains, which usually contain 400–500 amino acid residues, are approximately twice as long as the L chains.
3. Both the H and L chains have constant and variable regions. The constant regions have the same amino acid sequence from immunoglobulin to immunoglobulin, and the variable regions have a different amino acid sequence in each immunoglobulin.

CHEMICAL CONNECTIONS   Cyclosporine: An Antirejection Drug

The survival rate for patients undergoing human organ transplant operations such as heart, liver, or kidney replacement has risen dramatically since the late 1980s. This increased success coincides with the introduction of a new drug for controlling transplant rejection by a patient's own immune system. This new immunosuppressive agent (antirejection drug) is cyclosporine, a substance obtained from a particular type of soil fungus.

The primary structure of cyclosporine is that of a cyclic peptide containing 11 amino acid units. Ten of these are amino acids with simple side chains (four or fewer carbon atoms). The eleventh amino acid, which is the key to cyclosporine's pharmacological activity, had not been previously reported. This novel amino acid has a 7-carbon branched, unsaturated, hydroxylated side chain with the structure

$$-CH-CH-CH_2-CH=CH-CH_3$$
$$\quad\,| \qquad |$$
$$\quad OH \quad CH_3$$

The following diagram shows the amino acid sequence within the cyclosporine ring. Seven of the amino acid units,

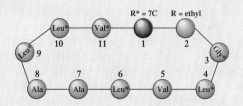

denoted by asterisks, have their nitrogen atom methylated; that is, a methyl group has replaced the hydrogen atom. This unique structural feature makes cyclosporine water-insoluble but fat-soluble.

The fat solubility of cyclosporine allows it to cross cell membranes readily and to be widely distributed in the body. It is administered either intravenously or orally. Because of its low water solubility, the drug is supplied in olive oil for oral administration. Cyclosporine has a narrow therapeutic index. When the blood concentration is too low, inadequate immunosuppression occurs. On the other hand, a high cyclosporine concentration can lead to kidney problems.

---

4. The carbohydrate content of various immunoglobulins varies from 1% to 12% by mass.
5. The secondary and tertiary structures are similar for all immunoglobulins. They involve a Y-shaped conformation (Figure 9.22) with disulfide linkages between H and L chains stabilizing the structure.

The interaction of an immunoglobulin molecule with an antigen occurs at the "tips" (upper-most part) of the Y structure. These tips are the variable-composition region of the immunoglobulin structure. It is here that the antigen binds specifically, and it is here that the amino acid sequence differs from one immunoglobulin to another.

Each immunoglobulin has two identical active sites and can thus bind to two molecules of the antigen it is "designed for." The action of many such immunoglobulins of a

FIGURE 9.22 This schematic diagram shows the structure of an immunoglobulin. Two heavy (H) polypeptide chains and two light (L) polypeptide chains are cross-linked by disulfide bridges. The purple areas are the constant amino acid regions, and the areas shown in red are the variable amino acid regions of each chain. Carbohydrate molecules attached to the heavy chains aid in determining the destinations of immunoglobulins in the tissues.

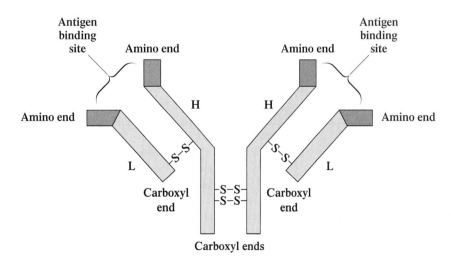

CHEMICAL CONNECTIONS
Lipoproteins and Heart Attack Risk

The lipoproteins present in blood serum are classified according to their density, which is related to the fractions of protein and lipid present. The more protein in the lipoprotein, the higher its density. On a density basis, there are three general categories of blood serum lipoproteins: (1) very-low-density lipoprotein (VLDL), (2) low-density lipoprotein (LDL), and (3) high-density lipoprotein (HDL). Characteristics of these three types of lipoproteins are given in the following table.

Type of Lipoprotein	Density Range (g/mL)	Approximate Percent-by-mass Protein
VLDL	1.006–1.019	5
LDL	1.019–1.063	25
HDL	1.063–1.21	50

The various types of lipoproteins have different functions. VLDLs are the principal carriers of triacylglycerols in the blood. (As cells remove triacylglycerols as needed from VLDLs, the VLDLs become LDLs.) Both LDLs and HDLs are involved in cholesterol transport. LDLs carry approximately 80% of this substance, and HDLs the remainder.

Of significance, LDLs and HDLs carry cholesterol for different purposes. LDLs carry cholesterol to cells for their use, whereas HDLs carry excess cholesterol away from cells to the liver for processing and excretion from the body.

Studies show that LDL levels correlate *directly* with heart disease, whereas HDL levels correlate *inversely* with heart disease risk. Thus HDL is sometimes referred to as "good" cholesterol (*H*DL = *H*ealthy) and LDL as "bad" cholesterol (*L*DL = *L*ess healthy).

The goal of dietary measures to slow the advance of atherosclerosis is to reduce LDL cholesterol levels. Reduction in the dietary intake of saturated fat appears to be a key action

(see the Chemical Connections feature on page 285 in Chapter 8).

High HDL levels are desirable because they give the body an efficient means of removing excess cholesterol. Low HDL levels can result in excess cholesterol depositing within the circulatory system.

In general, women have higher HDL levels than men—an average of 55 mg per 100 mL of blood serum versus 45 mg per 100 mL. This may explain in part why proportionately fewer women have heart attacks than men. Nonsmokers have uniformly higher HDL levels than smokers. Exercise on a regular basis tends to increase HDL levels. This discovery has increased the popularity of walking and running exercise. Genetics also plays a role in establishing HDL as well as other lipoprotein concentrations in the blood.

A person's *total* blood cholesterol level does not necessarily correlate with that individual's real risk for heart and blood vessel disease. A better measure is the *cholesterol ratio,* which is defined as

$$\text{Cholesterol ratio} = \frac{\text{total cholesterol}}{\text{HDL cholesterol}}$$

For example, if a person's total cholesterol is 200 and his or her HDL is 45, then the cholesterol ratio would be 4.4. According to the accompanying guidelines for interpreting cholesterol ratio values, this indicates an average risk for heart disease.

What Your Cholesterol Ratio Means	
Ratio	**Heart Disease Risk**
6.0	high
5.0	above average
4.5	average
4.0	below average
3.0	low

given type in concert with each other creates an *antigen–antibody complex* that precipitates from solution (Figure 9.23). Eventually, an invading antigen can be eliminated from the body through such precipitation. The bonding of an antigen to the variable region of an immunoglobulin occurs through hydrophobic interactions, dipole–dipole interactions, and hydrogen bonds rather than covalent bonds.

The importance of immunoglobulins is amply and tragically demonstrated by the effects of AIDS (acquired immunodeficiency syndrome). The AIDS virus upsets the body's normal production of immunoglobulins and leaves the body susceptible to what would otherwise not be debilitating and deadly infections.

Individuals who receive organ transplants must be given drugs to suppress the production of immunoglobulins against foreign proteins in the new organ, thus preventing rejection of the organ. The major reason for the increasing importance of organ transplants is the successful development of drugs that can properly manipulate the body's immune system (see the Chemical Connections feature on cyclosporine on page 341).

FIGURE 9.23 In this immunoglobulin – antigen comple: note that more than one immunoglobulin molecule can atta itself to a given antigen. Also, any g immunoglobulin has only two sites where antigen can bind.

ntigen

-feed a newborn infant. One of the most two or three days of lactation, the breasts ing immunoglobulins from the mother's infant from those infections to which ases are the ones in her environment— n. Breast milk, once it is produced, is a ort time. (After the first week of nursing, rease rapidly.) Infant formula used as a nally equivalent, but it does not contain

9.17 Lipoproteins

A **lipoprotein** *is a conjugated protein that contains lipids in addition to amino acids.* The major function of such proteins is to help suspend lipids and transport them through the bloodstream. Lipids, in general, are insoluble in blood (an aqueous medium) because of their nonpolar nature (Section 8.1).

The presence or absence of various types of lipoproteins in the blood appears to have implications for the health of the heart and blood vessels. Lipoprotein levels in the blood are now used as an indicator of heart attack risk (see the Chemical Connections feature on page 342).

CONCEPTS TO REMEMBER

Protein. A protein is a polymer in which the monomer units are amino acids (Section 9.1).

α-Amino acid. An α-amino acid is an amino acid in which the amino group and the carboxyl group are both attached to the α-carbon atom (Section 9.2).

Standard amino acid. A standard amino acid is one of the 20 α-amino acids that are normally present in protein (Section 9.2).

Amino acid classifications. Amino acids are classified as nonpolar, polar neutral, polar basic, or polar acidic depending on the nature of the side chain (R group) present (Section 9.2).

Chirality of amino acids. Amino acids found in proteins are always left-handed (L isomer) (Section 9.3).

Zwitterion. A zwitterion is a molecule that has a positive charge on one atom and a negative charge on another atom. In neutral solution

and in the solid state, amino acids exist as zwitterions. For amino acids in solution, the isoelectric point is the pH at which the solution has no net charge because an equal number of positive and negative charges are present (Section 9.4).

Disulfide bond formation. The amino acid cysteine readily dimerizes; the —SH groups of two cysteine molecules interact to form a covalent disulfide bond (Section 9.5).

Peptide bond. A peptide bond is an amide bond involving the carboxyl group of one amino acid and the amino group of another amino acid. In a protein, the amino acids are linked to each other through peptide bonds (Section 9.6).

Biochemically important peptides. Numerous small peptides are biochemically active. Their functions include hormonal action, neuro-transmission functions, and antioxidant activity (Section 9.7).

General characteristics of proteins. Proteins are peptides with at least 50 amino acid residues. A single peptide chain is present in a monomeric protein and two or more peptide chains are present in a multimeric protein. A simple protein contains only one or more peptide chains. A conjugated protein contains one or more additional chemical components, called prosthetic groups, in addition to peptide chains (Section 9.8).

Primary protein structure. The primary structure of a protein is the sequence of amino acids present in the peptide chain or chains of the protein (Section 9.9).

Secondary protein structure. The secondary structure of a protein is the arrangement in space of the backbone portion of the protein. The two major types of protein secondary structure are the α helix and the β pleated sheet (Section 9.10).

Tertiary protein structure. The tertiary structure of a protein is the overall three-dimensional shape that results from the attractive forces among amino acid side chains (R groups) (Section 9.11).

Quaternary protein structure. The quaternary structure of a protein involves the associations among the peptide chains present in a multimeric protein (Section 9.12).

Fibrous and globular proteins. Fibrous proteins are generally insoluble in water and have a long, thin, fibrous shape. α Keratin and collagen are important fibrous proteins. Globular proteins are generally soluble in water and have a roughly spherical or globular overall shape. Hemoglobin and myoglobin are important globular proteins (Section 9.13).

Protein hydrolysis. Protein hydrolysis is a chemical reaction in which peptide bonds within a protein are broken through reaction with water. Complete hydrolysis produces free amino acids (Section 9.14).

Protein denaturation. Protein denaturation is the partial or complete disorganization of a protein's characteristic three-dimensional shape as a result of disruption of its secondary, tertiary, and quaternary structural interactions (Section 9.15).

Glycoproteins. Glycoproteins are conjugated proteins that contain carbohydrates or carbohydrate derivatives in addition to amino acids. Collagen and immunoglobulins are important glycoproteins (Section 9.16).

Lipoproteins. Lipoproteins are conjugated proteins that are composed of both lipids and amino acids. Lipoproteins are classified on the basis of their density (Section 9.17).

KEY REACTIONS AND EQUATIONS

1. Formation of a zwitterion at pH 7 (Section 9.4)

$$H_2N-\underset{\underset{R}{|}}{CH}-COOH \longrightarrow \overset{+}{H_3N}-\underset{\underset{R}{|}}{CH}-COO^-$$

2. Conversion of a zwitterion to a positive ion in acidic solution (Section 9.4)

$$\overset{+}{H_3N}-\underset{\underset{R}{|}}{CH}-COO^- + H_3O^+ \longrightarrow \overset{+}{H_3N}-\underset{\underset{R}{|}}{CH}-COOH + H_2O$$

3. Conversion of a zwitterion to a negative ion in basic solution (Section 9.4)

$$\overset{+}{H_3N}-\underset{\underset{R}{|}}{CH}-COO^- + OH^- \longrightarrow H_2N-\underset{\underset{R}{|}}{CH}-COO^- + H_2O$$

4. Formation of a peptide bond (Section 9.6)

$$\overset{+}{H_3N}-\underset{\underset{R}{|}}{CH}-COO^- + \overset{+}{H_3N}-\underset{\underset{R'}{|}}{CH}-COO^- \longrightarrow$$

$$\overset{+}{H_3N}-\underset{\underset{R}{|}}{CH}-\overset{\overset{O}{\|}}{C}-\underset{\underset{}{|}}{\overset{\overset{H}{|}}{N}}-\underset{\underset{R'}{|}}{CH}-COO^- + H_2O$$

5. Hydrolysis of a protein in acidic solution (Section 9.14)

$$\text{Protein} + H_2O \xrightarrow{H^+} \text{smaller peptides} \xrightarrow{H^+} \text{amino acids}$$

6. Denaturation of a protein (Section 9.15)

$$\begin{matrix} \text{Protein with} \\ 1°, 2°, \text{and} \\ 3° \text{ structure} \end{matrix} \xrightarrow[\text{agent}]{\text{Denaturing}} \begin{matrix} \text{Protein with } 1° \\ \text{structure only} \end{matrix}$$

EXERCISES AND PROBLEMS

The members of each pair of problems in this section test similar material.

■ Amino Acid Structural Characteristics (Section 9.2)

9.1 Which of the following structures represent α-amino acids?

a.
$$H_2N-\underset{\underset{CH_3}{|}}{\overset{\overset{H}{|}}{C}}-COOH$$

b.
$$H_2N-\underset{\underset{CH_3}{|}}{\overset{\overset{H}{|}\ \ \ \overset{COOH}{|}}{C}}-CH_2$$

c.
$$H_2N-CH_2-\underset{\underset{H}{|}}{\overset{\overset{H}{|}}{C}}-COOH$$

d.
$$H_2N-\underset{\underset{CH_2}{|}}{\underset{\underset{CH_3}{|}}{\overset{\overset{H}{|}}{C}}}-COOH$$

9.2 What is the significance of the prefix α in the designation α amino acid?

9.3 What is the major structural difference among the various standard amino acids?

9.4 On the basis of polarity, what are the four types of side chains found in the standard amino acids?

9.5 With the help of Table 9.1, determine which of the standard amino acids have a side chain with the following characteristics.
a. Contains an aromatic group
b. Contains the element sulfur

c. Contains a carboxyl group
d. Contains a hydroxyl group

9.6   With the help of Table 9.1, determine which of the standard amino acids have a side chain with the following characteristics.
   a. Contains only carbon and hydrogen
   b. Contains an amino group
   c. Contains an amide group
   d. Contains more than four carbon atoms

9.7   What is the distinguishing characteristic of a polar basic amino acid?

9.8   What is the distinguishing characteristic of a polar acidic amino acid?

9.9   In what way is the structure of the amino acid proline different from that of the other 19 standard amino acids?

9.10 Which two of the standard amino acids are constitutional isomers?

Amino Acid Nomenclature (Section 9.2)

9.11 What amino acids do these abbreviations stand for?
   a. Ala   b. Leu   c. Met   d. Trp

9.12 What amino acids do these abbreviations stand for?
   a. Asp   b. Cys   c. Phe   d. Val

9.13 Which four standard amino acids have three-letter abbreviations that are not the first three letters of their common names?

9.14 What are the three-letter abbreviations for the three polar basic amino acids?

9.15 Classify each of the following amino acids as nonpolar, polar neutral, polar acidic, or polar basic.
   a. Asn   b. Glu   c. Pro   d. Ser

9.16 Classify each of the following amino acids as nonpolar, polar neutral, polar acidic, or polar basic.
   a. Gly   b. Thr   c. Tyr   d. His

Chirality and Amino Acids (Section 9.3)

9.17 To which family of mirror-image isomers do nearly all naturally occurring amino acids belong?

9.18 In what way is the structure of glycine different from that of the other 19 common amino acids?

9.19 Draw Fischer projection formulas for the following amino acids.
   a. L-Serine   b. D-Serine
   c. D-Alanine   d. L-Leucine

9.20 Draw Fischer projection formulas for the following amino acids.
   a. L-Cysteine   b. D-Cysteine
   c. L-Alanine   d. L-Valine

Acid–Base Properties of Amino Acids (Section 9.4)

9.21 At room temperature, amino acids are solids with relatively high decomposition points. Explain why.

9.22 Amino acids exist as zwitterions in the solid state. Explain why.

9.23 Draw the zwitterion structure for each of the following amino acids.
   a. Leucine   b. Isoleucine
   c. Cysteine   d. Glycine

9.24 Draw the zwitterion structure for each of the following amino acids.
   a. Serine   b. Methionine
   c. Threonine   d. Phenylalanine

9.25 Draw the structure of serine at each of the following pH values.
   a. 7.0   b. 1.0   c. 12.0   d. 3.0

9.26 Draw the structure of glycine at each of the following pH values.
   a. 7.0   b. 13.0   c. 2.0   d. 11.0

9.27 Explain what is meant by the term *isoelectric point.*

9.28 Most amino acids have isoelectric points between 5.0 and 6.0, but the isoelectric point of lysine is 9.7. Explain why lysine has such a high value for its isoelectric point.

9.29 Glutamic acid exists in two low-pH forms instead of the usual one. Explain why.

9.30 Arginine exists in two high-pH forms instead of the usual one. Explain why.

9.31 Predict the direction of movement of each of the following amino acids in a solution at the pH value specified under the influence of an electric field. Indicate the direction as toward the positive electrode or toward the negative electrode. Write "isoelectric" if no net movement occurs.
   a. Alanine at pH = 12.0   b. Valine at pH = 7.0
   c. Aspartic acid at pH = 1.0   d. Arginine at pH = 13.0

9.32 Predict the direction of movement of each of the following amino acids in a solution at the pH value specified under the influence of an electric field. Indicate the direction as toward the positive electrode or toward the negative electrode. Write "isoelectric" if no net movement occurs.
   a. Alanine at pH = 2.0   b. Valine at pH = 12.0
   c. Aspartic acid at pH = 13.0   d. Arginine at pH = 1.0

9.33 A direct current was passed through a solution containing valine, histidine, and aspartic acid at a pH of 6.0. One amino acid migrated to the positive electrode, one migrated to the negative electrode, and one did not migrate to either electrode. Which amino acids went where?

9.34 A direct current was passed through a solution containing alanine, arginine, and glutamic acid at a pH of 6.0. One amino acid migrated to the positive electrode, one migrated to the negative electrode, and one did not migrate to either electrode. Which amino acids went where?

Cysteine and Disulfide Bonds (Section 9.5)

9.35 When two cysteine molecules dimerize, what happens to the R groups present?

9.36 What chemical reaction involving the cysteine molecule produces a disulfide bond?

Peptide Formation (Section 9.6)

9.37 What two functional groups are involved in the formation of a peptide bond?

9.38 What is meant by the N-terminal end and the C-terminal end of a peptide?

9.39 Write out the full structure of the tripeptide Val–Phe–Cys.

9.40 Write out the full structure of the tripeptide Glu–Ala–Leu.

9.41 Explain why the notations Ser–Cys and Cys–Ser represent two different molecules rather than the same molecule.

9.42 Explain why the notations Ala–Gly–Val–Ala and Ala–Val–Gly–Ala represent two different molecules rather than the same molecule.

9.43 There are a total of six different amino acid sequences for a tripeptide containing one molecule each of serine, valine, and glycine. Using three-letter abbreviations for the amino acids, draw the six possible sequences of amino acids.

9.44 There are a total of six different amino acid sequences for a tetrapeptide containing two molecules each of serine and valine. Using three-letter abbreviations for the amino acids, draw the six possible sequences of amino acids.

9.45 Identify the amino acids contained in each of the following tripeptides.

a.

$$H_3\overset{+}{N}-CH-\overset{O}{\overset{||}{C}}-\overset{H}{\overset{|}{N}}-CH-\overset{O}{\overset{||}{C}}-\overset{H}{\overset{|}{N}}-CH-COO^-$$

with side chains: CH_2-OH ; CH_3 ; CH_2-SH

b.

$$H_3\overset{+}{N}-CH-\overset{O}{\overset{||}{C}}-\overset{H}{\overset{|}{N}}-CH-\overset{O}{\overset{||}{C}}-\overset{H}{\overset{|}{N}}-CH-COO^-$$

with side chains: CH_2-COO^- ; $CH-OH$ with CH_3 ; CH_2-C-NH_2 with O

9.46 Identify the amino acids contained in each of the following tripeptides.

a.

$$H_3\overset{+}{N}-CH_2-\overset{O}{\overset{||}{C}}-\overset{H}{\overset{|}{N}}-CH-\overset{O}{\overset{||}{C}}-\overset{H}{\overset{|}{N}}-CH_2-COO^-$$

with side chain: $CH-CH_3$ with CH_3

b.

$$H_3\overset{+}{N}-CH-\overset{O}{\overset{||}{C}}-\overset{H}{\overset{|}{N}}-CH-\overset{O}{\overset{||}{C}}-\overset{H}{\overset{|}{N}}-CH-COO^-$$

with side chains: CH_2-OH ; $CH_2-CH-CH_3$ with CH_3 ; $CH_2-CH_2-COO^-$

9.47 How many peptide bonds are present in each of the molecules in Problem 9.45?

9.48 How many peptide bonds are present in each of the molecules in Problem 9.46?

9.49 With the help of Table 9.1, assign an IUPAC name to each of the following small peptides.
a. Ser–Cys   b. Gly–Ala–Val
c. Tyr–Asp–Gln   d. Leu–Lys–Trp–Met

9.50 With the help of Table 9.1, assign an IUPAC name to each of the following small peptides.
a. Cys–Ser   b. Val–Ala–Gly
c. Tyr–Gln–Asp   d. Phe–Met–Try–Asn

9.51 What are the two repeating units present in the "backbone" of a peptide?

9.52 For a peptide, describe
a. the regularly repeating part of its structure.
b. the variable part of its structure.

■ **Biochemically Important Small Peptides (Section 9.7)**

9.53 Contrast the structures of the protein hormones oxytocin and vasopressin in terms of
a. what they have in common.
b. how they differ.

9.54 Contrast the protein hormones oxytocin and vasopressin in terms of their biochemical functions.

9.55 Contrast the binding-site locations in the brain for enkephalins and the prescription painkillers morphine and codeine.

9.56 Contrast the structures of the peptide neurotransmitters Met-enkephalin and Leu-enkephalin in terms of
a. what they have in common.
b. how they differ.

9.57 What is the unusual structural feature present in the molecule glutathione?

9.58 What is the major biochemical function of glutathione?

■ **General Structural Characteristics of Proteins (Section 9.8)**

9.59 What is the major difference between a monomeric protein and a multimeric protein?

9.60 What is the major difference between a simple protein and a conjugated protein?

9.61 Indicate whether each of the following statements about proteins is true or false.
a. Two or more peptide chains are always present in a multimeric protein.
b. A simple protein contains only one type of amino acid.
c. A conjugated protein can also be a monomeric protein.
d. The prosthetic group(s) present in a glycoprotein are carbohydrate groups.

9.62 Indicate whether each of the following statements about proteins is true or false.
a. Conjugated proteins always have only one peptide chain.
b. All peptide chains in a multimeric protein must be identical to each other.
c. A simple protein can also be a multimeric protein.
d. Both monomeric proteins and multimeric proteins can contain prosthetic groups.

■ **Levels of Protein Structure (Sections 9.9–9.12)**

9.63 What is primary protein structure?

9.64 Two proteins with the same amino acid composition do not have to have the same primary structure. Explain why.

9.65 What are the two common types of secondary protein structure?

9.66 Hydrogen bonding between which functional groups stabilizes protein secondary structure arrangements?

9.67 The β pleated sheet secondary structure can be formed through either intramolecular hydrogen bonding or intermolecular hydrogen bonding. Explain why.

9.68 The α helix secondary structure always involves intramolecular hydrogen bonding and never involves intermolecular hydrogen bonding. Explain why.

9.69 Can more than one type of secondary structure be present in the same protein molecule? Explain your answer.

9.70 What is meant by the statement that a section of a protein has a "random structure" arrangement?

9.71 What is the difference between the types of hydrogen bonding that occur in secondary and tertiary protein structures?

9.72 State the four types of attractive interactions that give rise to tertiary protein structure.

9.73 Specify the nature of each of the following tertiary-structure interactions, using the choices hydrophobic, electrostatic, hydrogen bonding, and disulfide bond.
 a. Phenylalanine and leucine
 b. Arginine and glutamic acid
 c. Two cysteines
 d. Serine and tyrosine

9.74 Specify the nature of each of the following tertiary-structure interactions using the choices hydrophobic, electrostatic, hydrogen bonding, and disulfide bond.
 a. Lysine and aspartic acid b. Threonine and tyrosine
 c. Alanine and valine d. Leucine and isoleucine

■ **Fibrous and Globular Proteins (Section 9.13)**

9.75 Contrast fibrous and globular proteins in terms of
 a. solubility characteristics in water.
 b. general biochemical function.

9.76 Contrast fibrous and globular proteins in terms of
 a. general secondary structure.
 b. relative abundance within the human body.

9.77 Classify each of the following proteins as a globular protein or a fibrous protein.
 a. α Keratin b. Collagen
 c. Hemoglobin d. Myoglobin

9.78 What is the major biochemical function for each of the following proteins?
 a. α Keratin b. Collagen
 c. Hemoglobin d. Myoglobin

9.79 Contrast the structures of the proteins α keratin and collagen.

9.80 Contrast the structures of the proteins myoglobin and hemoglobin.

■ **Protein Hydrolysis (Section 9.14)**

9.81 Will hydrolysis of the dipeptides Ala–Val and Val–Ala yield the same products? Explain your answer.

9.82 A shampoo bottle lists "partially hydrolyzed protein" as one of its ingredients. What is the difference between partially hydrolyzed protein and completely hydrolyzed protein?

9.83 Drugs that are proteins, such as insulin, must always be injected rather than taken by mouth. Explain why.

9.84 Which structural levels of a protein are affected by hydrolysis?

9.85 Identify the primary structure of a hexapeptide containing six different amino acids if the following smaller peptides are among the partial-hydrolysis products:
Ala–Gly, His–Val–Arg, Ala–Gly–Met, and Gly–Met–His.

9.86 Identify the primary structure of a hexapeptide containing five different amino acids if the following smaller peptides are among the partial-hydrolysis products:
Gly–Cys, Ala–Ser, Ala–Gly, and Cys–Val–Ala.

9.87 How many different di- and tripeptides could be present in a solution of partially hydrolyzed Ala–Gly–Ser–Tyr?

9.88 How many different di- and tripeptides could be present in a solution of partially hydrolyzed Ala–Gly–Ala–Gly?

■ **Protein Denaturation (Section 9.15)**

9.89 Which structural levels of a protein are affected by denaturation?

9.90 Suppose a sample of protein is completely hydrolyzed and another sample of the same protein is denatured. Compare the final products of these processes.

9.91 In what way is the protein in a cooked egg the same as that in a raw egg?

9.92 Why is 70% ethanol rather than pure ethanol preferred for use as an antiseptic agent?

■ **Glycoproteins (Section 9.16)**

9.93 What two nonstandard amino acids are present in collagen?

9.94 Where are the carbohydrate units located in collagen?

9.95 What is the function of the carbohydrate groups present in collagen?

9.96 What is the role of vitamin C in the biosynthesis of collagen?

9.97 What is the difference between an antigen and an antibody?

9.98 What is an immunoglobulin?

9.99 Describe the structural features of a typical immunoglobulin molecule.

9.100 Describe the process by which blood immunoglobulins help protect the body from invading bacteria and viruses.

■ **Lipoproteins (Section 9.17)**

9.101 What is the major biochemical function of lipoproteins?

9.102 What is the basis for the classification of blood serum lipoproteins into groups?

ADDITIONAL PROBLEMS

9.103 State whether each of the following statements applies to primary, secondary, tertiary, or quaternary protein structure.
 a. A disulfide bond forms between two cysteine residues in different protein chains.
 b. A salt bridge forms between amino acids with acidic and basic side chains.
 c. Hydrogen bonding between carbonyl oxygen atoms and nitrogen atoms of amino groups causes a peptide to coil into a helix.
 d. Peptide linkages hold amino acids together in a polypeptide chain.

9.104 What is the common name for each of the following IUPAC-named standard amino acids?

 a. 2-Aminopropanoic acid
 b. 2-Amino-4-methylbutanoic acid
 c. 2-Amino-3-hydroxybutanoic acid
 d. 2-Aminobutanedioic acid

9.105 What is the net charge at a pH of 1.0 for each of the following peptides?
 a. Val–Ala–Leu b. Tyr–Trp–Thr
 c. Asp–Asp–Glu–Gly d. His–Arg–Ser–Ser

9.106 What is the net charge at a pH of 13.0 for each of the peptides in Problem 9.105?

9.107 The amino acid isoleucine possesses two chiral centers. Draw Fischer projection formulas for the four stereoisomers that are possible for this amino acid.

9.108   Indicate how many structurally isomeric tetrapeptides are possible for a tetrapeptide in which
   a. four different amino acids are present.
   b. three different amino acids are present.
   c. two different amino acids are present.

9.109   Draw the structures of the three hydrolysis products obtained when the tripeptide in part a of Problem 9.45 undergoes hydrolysis under
   a. low-pH (acidic) conditions.
   b. high-pH (basic) conditions.

9.110   Classify each of the following proteins as a simple protein, a conjugated protein, a glycoprotein, a lipoprotein, a fibrous protein, or a globular protein. More than one classification may apply to a given protein.
   a. α Keratin   b. Hemoglobin
   c. Myoglobin   d. Collagen

MULTIPLE-CHOICE PRACTICE TEST

9.111   Which of the following sets of four elements are found in all amino acids?
   a. C, H, O, S   b. C, H, S, N
   c. C, H, O, N   d. C, H, S, N

9.112   Which of the following statements concerning the structure of α-amino acids is *correct*?
   a. The amino group is attached to the carbon atom of the carboxyl group.
   b. The amino group and the carboxyl group are directly bonded to the same carbon atom.
   c. The amino acid contains only two carbon atoms.
   d. The amino acid contains only one carbon atom.

9.113   Which of the following is an *incorrect* statement about glycine, the amino acid with the simplest structure?
   a. It does not contain a chiral center.
   b. It has a side chain that does not contain the element carbon.
   c. It is one of the 20 standard amino acids.
   d. Its amino group and carboxyl group are directly bonded to each other.

9.114   In a solution of high pH, all of the acidic and basic sites in an amino acid are which of the following?
   a. Protonated   b. Deprotonated
   c. Positively charged   d. Negatively charged

9.115   Which of the standard amino acids exist as zwitterions in the solid state?
   a. All of them.
   b. Only those that have nonpolar side chains.
   c. Only those that are polar neutral.
   d. Only those that are acidic or basic.

9.116   Which of the following statements concerning the tripeptide Val-Ala-Gly is *correct*?
   a. The C-terminal amino acid residue is Val.
   b. The N-terminal amino acid residue is Gly.
   c. Three peptide linkages are present.
   d. It is constitutionally isomeric with five other tripeptides.

9.117   Which of the following types of bonding is responsible for protein secondary structure?
   a. Peptide linkages
   b. Amide linkages
   c. Hydrogen bonds
   d. Bonds involving R groups

9.118   R-group interaction between which of the following pairs of amino acids produces a covalent bond?
   a. Cysteine–cysteine
   b. Proline–proline
   c. Alanine–glycine
   d. Valine–lysine

9.119   Which of the following levels of protein structure is not disrupted when protein denaturation occurs?
   a. Primary structure
   b. Secondary structure
   c. Tertiary structure
   d. Quaternary structure

9.120   In which of the following pairs of proteins are both members of the pair *fibrous* proteins?
   a. α-Keratin and collagen
   b. Collagen and hemoglobin
   c. Hemoglobin and myoglobin
   d. α-Keratin and hemoglobin

Enzymes and Vitamins

Yellow- and orange-colored vegetables such as pumpkins and squash have significant vitamin A activity due to the presence of the molecule beta-carotene.

In this chapter we consider two topics: enzymes and vitamins. Enzymes govern all chemical reactions in living organisms. They are specialized proteins that, with fascinating precision and selectivity, catalyze biochemical reactions that store and release energy, make pigments in our hair and eyes, digest the food we eat, synthesize cellular building materials, and protect us by repairing cellular damage and clotting our blood. Enzymes are sensitive to their environment, responding quickly to changes in the cell. The deficiency or excess of particular enzymes can cause certain diseases or signal problems such as heart attacks and other organ damage. Our knowledge of protein structure (Chapter 9) can help us appreciate and better understand how enzymes function in living cells.

Vitamins, which are necessary components of a healthful diet, play important roles in cellular metabolism. In most cases, they function as enzyme cofactors or carriers of functional groups during biosynthesis.

10.1 General Characteristics of Enzymes

An **enzyme** *is an organic compound that acts as a catalyst for a biochemical reaction.* Each cell in the human body contains thousands of different enzymes because almost every reaction in a cell requires its own specific enzyme. Enzymes cause cellular reactions to occur millions of times faster than corresponding uncatalyzed reactions. As catalysts, enzymes are not consumed during the reaction but merely help the reaction occur more rapidly.

FIGURE 10.1 Bread dough rises as a result of the action of yeast enzymes.

The word *enzyme* comes from the Greek words *en,* which means "in," and *zyme,* which means "yeast." Long before their chemical nature was understood, yeast enzymes were used in the production of bread and alcoholic beverages. The action of yeast on sugars produces the carbon dioxide gas that causes bread to rise (see Figure 10.1) Fermentation of sugars in fruit juices with the same yeast enzymes produces alcoholic beverages.

Most enzymes are globular proteins (Section 9.14). Some are simple proteins, consisting entirely of amino acid chains. Others are conjugated proteins, containing additional chemical components (Section 10.3). Until the 1980s, it was thought that *all* enzymes were proteins. A few enzymes are now known that are made of ribonucleic acids (RNA; Section 11.7) and that catalyze cellular reactions involving nucleic acids. In this chapter, we will consider only enzymes that are proteins.

Enzymes undergo all the reactions of proteins, including *denaturation* (Section 9.16). Slight alterations in pH, temperature, or other protein denaturants affect enzyme activity dramatically. Good cooks realize that overheating yeast kills the action of the yeast. A person suffering from a high fever (greater than 106°F) runs the risk of denaturing certain enzymes. The biochemist must exercise extreme caution in handling enzymes to avoid the loss of their activity. Even vigorous shaking of an enzyme solution can destroy enzyme activity.

Enzymes differ from nonbiochemical (laboratory) catalysts in that their activity is usually regulated by other substances present in the cell in which they are found. Most laboratory catalysts need to be removed from a reaction mixture to stop their catalytic action; this is not so with enzymes. In some cases, if a certain chemical is needed in the cell, the enzyme responsible for its production is activated by other cellular components. When a sufficient quantity has been produced, the enzyme is then deactivated. In other situations, the cell may produce more or less enzyme as required. Because different enzymes are required for nearly all cellular reactions, certain necessary reactions can be accelerated or decelerated without affecting the rest of the cellular chemistry.

10.2 Nomenclature and Classification of Enzymes

> Enzymes, the most efficient catalysts known, increase the rates of biochemical reactions by factors of up to 10^{20} over uncatalyzed reactions. Nonenzymatic catalysts, on the other hand, typically enhance the rate of a reaction by factors of 10^2 to 10^4.

Enzymes are most commonly named by using a system that attempts to provide information about the *function* (rather than the structure) of the enzyme. Type of reaction catalyzed and *substrate* identity are focal points for the nomenclature. A **substrate** *is the reactant in an enzyme-catalyzed reaction.* The substrate is the substance upon which the enzyme "acts."

Three important aspects of the enzyme-naming process are the following:

1. The suffix *-ase* identifies a substance as an enzyme. Thus ure*ase,* sucr*ase,* and lip*ase* are all enzyme designations. The suffix *-in* is still found in the names of some of the first enzymes studied, many of which are digestive enzymes. Such names include *trypsin, chymotrypsin,* and *pepsin.*
2. The type of reaction catalyzed by an enzyme is often noted with a prefix. An *oxidase* enzyme catalyzes an oxidation reaction, and a *hydrolase* enzyme catalyzes a hydrolysis reaction.
3. The identity of the substrate is often noted in addition to the type of reaction. Enzyme names of this type include *glucose oxidase, pyruvate carboxylase,* and *succinate dehydrogenase.* Infrequently, the substrate but not the reaction type is given, as in the names *urease* and *lactase.* In such names, the reaction involved is hydrolysis; *urease* catalyzes the hydrolysis of urea, *lactase* the hydrolysis of lactose.

EXAMPLE 10.1

Predicting Enzyme Function from an Enzyme's Name

■ Predict the function of the following enzymes.

a. Cellulase   **b.** Sucrase
c. L-Amino acid oxidase   **d.** Aspartate aminotransferase

Solution

a. Cellulase catalyzes the hydrolysis of cellulose.
b. Sucrase catalyzes the hydrolysis of the disaccharide sucrose.

c. L-Amino acid oxidase catalyzes the oxidation of L-amino acids.
d. Aspartate aminotransferase catalyzes the transfer of an amino group from aspartate to a different molecule.

Practice Exercise 10.1

Predict the function of the following enzymes.

a. Maltase   **b.** Lactate dehydrogenase
c. Fructose oxidase   **d.** Maleate isomerase

Enzymes are grouped into six major classes on the basis of the types of reactions they catalyze.

1. An **oxidoreductase** *is an enzyme that catalyzes oxidation–reduction reactions.*
2. A **transferase** *is an enzyme that catalyzes the transfer of a functional group from one molecule to another.*
3. A **hydrolase** *is an enzyme that catalyzes hydrolysis reactions in which the addition of a water molecule to a bond causes the bond to break.*
4. A **lyase** *is an enzyme that catalyzes the addition of a group to a double bond or the removal of a group to form a double bond in a manner that does not involve hydrolysis or oxidation.*

TABLE 10.1
Main Classes and Subclasses of Enzymes

Main Classes	Selected Subclasses	Type of Reaction Catalyzed
oxidoreductases	oxidases	oxidation of a substrate
	reductases	reduction of a substrate
	dehydrogenases	introduction of double bond (oxidation) by formal removal of two H atoms from substrate, the H being accepted by a coenzyme
transferases	transaminases	transfer of an amino group between substrates
	kinases	transfer of a phosphate group between substrates
hydrolases	lipases	hydrolysis of ester linkages in lipids
	proteases	hydrolysis of amide linkages in proteins
	nucleases	hydrolysis of sugar–phosphate ester bonds in nucleic acids
	carbohydrases	hydrolysis of glycosidic bonds in carbohydrates
	phosphatases	hydrolysis of phosphate–ester bonds
lyases	dehydratases	removal of H_2O from substrate
	decarboxylases	removal of CO_2 from substrate
	deaminases	removal of NH_3 from substrate
	hydratases	addition of H_2O to a substrate
isomerases	racemases	conversion of D to L isomer, or vice versa
	mutases	conversion of one constitutional isomer into another
ligases	synthetases	formation of new bond between two substrates, with participation of ATP
	carboxylases	formation of new bond between a substrate and CO_2, with participation of ATP

5. An **isomerase** *is an enzyme that catalyzes the rearrangement of functional groups within a molecule, converting the molecule into another molecule isomeric with it.*

6. A **ligase** *is an enzyme that catalyzes the bonding together of two molecules into one, with the participation of ATP.*

Within each of these six main classes of enzymes there are subclasses. Table 10.1 gives further information about enzyme subclass terminology.

10.3 Enzyme Structure

Enzymes can be divided into two general structural classes: simple enzymes and conjugated enzymes. A **simple enzyme** *is an enzyme composed only of protein (amino acid chains).* A **conjugated enzyme** *is an enzyme that has a nonprotein part in addition to a protein part.* By itself, neither the protein part nor the nonprotein portion of a conjugated enzyme has catalytic properties. An **apoenzyme** *is the protein part of a conjugated enzyme.* A **cofactor** *is the nonprotein part of a conjugated enzyme.* It is the combination of apoenzyme with cofactor that produces a biochemically active enzyme. A **holoenzyme** *is the biochemically active conjugated enzyme produced from an apoenzyme and a cofactor.*

$$\text{apoenzyme} + \text{cofactor} = \text{holoenzyme}$$

Why do apoenzymes need cofactors? Cofactors provide additional chemically reactive functional groups besides those present in the amino acid side chains of apoenzymes.

A cofactor is generally either a small organic molecule or an inorganic ion (usually a metal ion). A **coenzyme** *is a small organic molecule that serves as a cofactor in a conjugated enzyme.* Many vitamins (Section 10.12) have coenzyme functions in the human body.

Typical inorganic ion cofactors include Zn^{2+}, Mg^{2+}, Mn^{2+}, and $Fe^{10.}$ The nonmetallic Cl^- ion occasionally acts as a cofactor. Dietary minerals are an important source of inorganic ion cofactors.

10.4 Models of Enzyme Action

Explanations of *how* enzymes function as catalysts in biochemical systems are based on the concepts of an enzyme active site and enzyme–substrate complex formation.

▪ Enzyme Active Site

Studies show that only a small portion of an enzyme molecule called the active site participates in the interaction with a substrate or substrates during a reaction. The **active site** *is the relatively small part of an enzyme's structure that is actually involved in catalysis.*

The active site in an enzyme is a three-dimensional entity formed by groups that come from different parts of the protein chain(s); these groups are brought together by the folding and bending (secondary and tertiary structure; Sections 9.10 and 9.11) of the protein. The active site is usually a "crevicelike" location in the enzyme (see Figure 10.2).

▪ Enzyme–Substrate Complex

Catalysts offer an alternative pathway with lower activation energy through which a reaction can occur. In enzyme-controlled reactions, this alternative pathway involves the formation of an enzyme–substrate complex as an intermediate species in the reaction. An **enzyme–substrate complex** *is the intermediate reaction species that is formed when*

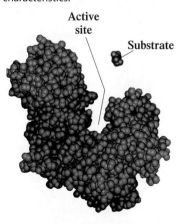

FIGURE 10.2 The active site of an enzyme is usually a crevicelike region formed as a result of the protein's secondary and tertiary structural characteristics.

Active site

Substrate

FIGURE 10.3 The lock-and-key model
for enzyme activity. Only a substrate
whose shape and chemical nature are
complementary to those of the active
site can interact with the enzyme.

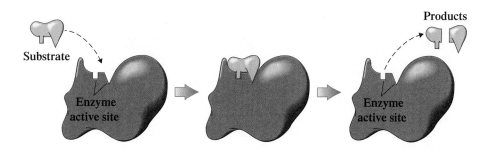

a substrate binds to the active site of an enzyme. Within the enzyme– substrate complex, the substrate encounters more favorable reaction conditions than if it were free. The result is faster formation of product.

Lock-and-Key Model

To account for the highly specific way an enzyme recognizes a substrate and binds it to the active site, researchers have proposed several models. The simplest of these models is the lock-and-key model.

In the lock-and-key model, the active site in the enzyme has a fixed, rigid geometrical conformation. Only substrates with a complementary geometry can be accommodated at such a site, much as a lock accepts only certain keys. Figure 10.3 illustrates the lock-and-key concept of substrate–enzyme interaction.

> The lock-and-key model is more than just a "shape fit." In addition, there are weak binding forces (R group interactions) between parts.

Induced-Fit Model

The lock-and-key model explains the action of numerous enzymes. It is, however, too restrictive for the action of many other enzymes. Experimental evidence indicates that many enzymes have flexibility in their shapes. They are not rigid and static; there is constant change in their shape. The induced-fit model is used for this type of situation.

The induced-fit model allows for small changes in the shape or geometry of the active site of an enzyme to accommodate a substrate. A good analogy is the changes that occur in the shape of a glove when a hand is inserted into it. The induced fit is a result of the enzyme's flexibility; it adapts to accept the incoming substrate. This model, illustrated in Figure 10.4, is a more thorough explanation for the active-site properties of an enzyme because it includes the specificity of the lock-and-key model coupled with the flexibility of the enzyme protein.

The forces that draw the substrate into the active site are many of the same forces that maintain tertiary structure in the folding of peptide chains. Electrostatic interactions, hydrogen bonds, and hydrophobic interactions all help attract and bind substrate molecules. For example, a protonated (positively charged) amino group in a substrate could be attracted and held at the active site by a negatively charged aspartate or glutamate residue. Alternatively, cofactors such as positively charged metal ions often help bind substrate molecules. Figure 10.5 is a schematic representation of the amino acid R group interactions that bind a substrate to an enzyme active site.

FIGURE 10.4 The induced-fit model
for enzyme activity. The enzyme
active site, although not exactly
complementary in shape to that of the
substrate, is flexible enough that it can
adapt to the shape of the substrate.

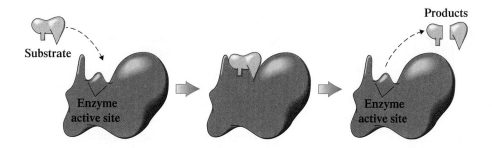

FIGURE 10.5 A schematic diagram representing amino acid R group interactions that bind a substrate to an enzyme active site. The R group interactions that maintain the three-dimensional structure of the enzyme (secondary and tertiary structure) are also shown.

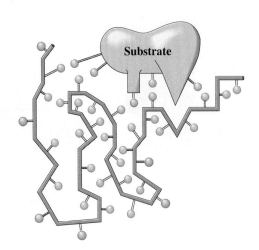

● R group interactions that bind the substrate to the enzyme active site

● R group interactions that maintain the three-dimensional structure of the enzyme

● Noninteracting R groups that help determine the solubility of the enzyme

10.5 Enzyme Specificity

Enzymes exhibit different levels of selectivity, or specificity, for substrates. The degree of enzyme specificity is determined by the active site. Some active sites accommodate only one particular compound, whereas others can accommodate a "family" of closely related compounds. Types of enzyme specificity include

1. *Absolute Specificity.* Such specificity means an enzyme will catalyze a particular reaction for *only one* substrate. This most restrictive of all specificities is not common. Urease is an enzyme with absolute specificity.
2. *Stereochemical Specificity.* Such specificity means an enzyme can distinguish between stereoisomers. Chirality is inherent in an active site, because amino acids are chiral compounds. L-Amino-acid oxidase will catalyze reactions of L-amino acids but not of D-amino acids.
3. *Group Specificity.* Such specificity involves structurally similar compounds that have the same functional groups. Carboxypeptidase is group-specific; it cleaves amino acids, one at a time, from the carboxyl end of the peptide chain.
4. *Linkage Specificity.* Such specificity involves a particular type of bond, irrespective of the structural features in the vicinity of the bond. Phosphatases hydrolyze phosphate–ester bonds in all types of phosphate esters. Linkage specificity is the most general of the specificities considered.

FIGURE 10.6 A graph showing the effect of temperature on the rate of an enzymatic reaction.

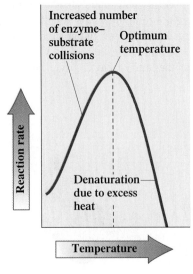

10.6 Factors That Affect Enzyme Activity

Enzyme activity *is a measure of the rate at which an enzyme converts substrate to products in a biochemical reaction.* Four factors affect enzyme activity: temperature, pH, substrate concentration, and enzyme concentration.

■ Temperature

Temperature is a measure of the kinetic energy (energy of motion) of molecules. Higher temperatures mean molecules are moving faster and colliding more frequently. This concept applies to collisions between substrate molecules and enzymes. As the temperature of an enzymatically catalyzed reaction increases, so does the rate (velocity) of the reaction.

However, when the temperature increases beyond a certain point, the increased energy begins to cause disruptions in the tertiary structure of the enzyme; denaturation is occurring. Change in tertiary structure at the active site impedes catalytic action, and the enzyme activity quickly decreases as the temperature climbs past this point (Figure 10.6). The temperature that produces maximum activity for an enzyme is known as the optimum

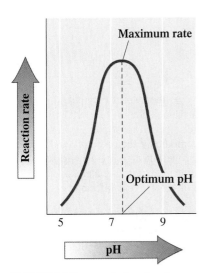

FIGURE 10.7 A graph showing the effect of pH on the rate of an enzymatic reaction.

The upper temperature limit for life now stands at 121°C as the result of the discovery, in 2004, of a new "heat-loving" microbe. The microbe was found in a water sample from a hydrothermal vent deep in the Northeast Pacific Ocean. Its method of respiration involves reduction of Fe(III) to Fe(II) to produce energy.

FIGURE 10.8 A graph showing the change in enzyme activity with a change in substrate concentration at constant temperature, pH, and enzyme concentration. Enzyme activity remains constant after a certain substrate concentration is reached.

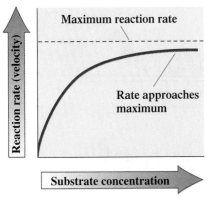

temperature for that enzyme. **Optimum temperature** *is the temperature at which an enzyme exhibits maximum activity.*

For human enzymes, the optimum temperature is around 37°C, normal body temperature. A person who has a fever where body core temperature exceeds 40°C can be in a life-threatening situation because such a temperature is sufficient to initiate enzyme denaturation. The loss of function of critical enzymes, particularly those of the central nervous system, can result in dysfunction sufficient to cause death.

The "destroying" effect of temperature on bacterial enzymes is used in a hospital setting to sterilize medical instruments and laundry. In high-temperature high-pressure vessels called *autoclaves* super-heated steam is used to produce a temperature sufficient to denature bacterial enzymes.

Not all enzymes have optimal temperatures around the physiological temperature of 37°C. This is particularly true for enzymes found in microbes associated with hydrothermal areas such as those in Yellowstone National Park and hydrothermal vents on the ocean floor, where temperature and pressure can be extremely high. The ability of microbial enzymes to survive under such harsh conditions is related to the amino acid sequences in their protein structures, sequences that are stable under such extraordinary conditions. Microbial enzymes that survive in such extreme environments are collectively called *extremozymes.*

The study of extremozymes is an area of special interest for industrial chemists. Enzymes can function as catalysts for industrial processes, just as they do for biochemical reactions, provided they can survive the conditions associated with the process. Because industrial processes usually require higher temperature and pressure than physiological processes, extremozymes can be useful. The enzymes present in some detergent formulations, which must function in hot water, are the result of research associated with high-temperature microbial enzymes.

■ pH

The pH of an enzyme's environment can affect its activity. This is not surprising because the *charge* on acidic and basic amino acids (Section 9.2) located at the active site depends on pH. Small changes in pH (less than one unit) can result in enzyme denaturation (Section 9.16) and subsequent loss of catalytic activity.

Most enzymes exhibit maximum activity over a very narrow pH range. Only within this narrow pH range do the enzyme's amino acids exist in properly charged forms (Section 9.4). **Optimum pH** *is the pH at which an enzyme exhibits maximum activity.* Figure 10.7 shows the effect of pH on an enzyme's activity. Biochemical buffers help maintain the optimum pH for an enzyme.

Each enzyme has a characteristic optimum pH, which usually falls within the physiological pH range of 7.0–7.5. Notable exceptions to this generalization are the digestive enzymes pepsin and trypsin. Pepsin, which is active in the stomach, functions best at a pH of 2.0. On the other hand, trypsin, which operates in the small intestine, functions best at a pH of 8.0. The amino acid sequences present in pepsin and trypsin are those needed such that the R groups present can maintain protein tertiary structure (Section 9.11) at low (2.0) and high (8.0) pH values, respectively.

A variation from normal pH can also affect substrates, causing either protonation or deprotonation of groups on the substrate. The interaction between the altered substrate and the enzyme active site may be less efficient than normal—or even impossible.

■ Substrate Concentration

When the concentration of an enzyme is kept constant and the concentration of substrate is increased, the enzyme activity pattern shown in Figure 10.8 is obtained. This activity pattern is called a *saturation curve.* Enzyme activity increases up to a certain substrate concentration and thereafter remains constant.

What limits enzymatic activity to a certain maximum value? As substrate concentration increases, the point is eventually reached where enzyme capabilities are used to their maximum extent. The rate remains constant from this point on (Figure 10.8). Each

CHEMICAL CONNECTIONS *H. pylori* and Stomach Ulcers

Helicobacter pylori, commonly called *H. pylori,* is a bacterium that can function in the highly acidic environment of the stomach. The discovery in 1982 of the existence of this bacterium in the stomach was startling to the medical profession because conventional thought at the time was that bacteria could not survive at the stomach's pH of about 1.4.

It is now known that *H. pylori* causes more than 90% of duodenal ulcers and up to 80% of gastric ulcers. Before this discovery, it was thought that most ulcers were caused by excess stomach acid eating the stomach lining. Contributory causes were thought to be spicy food and stress. Conventional treatment involved acid-suppression or acid-neutralization medications. Now, treatment regimens involve antibiotics. The medical profession was slow to accept the concept of a bacterial cause for most ulcers, and it was not until the mid-1990s that antibiotic treatment became common.

How the enzymes present in the *H. pylori* bacterium can function in the acidic environment of the stomach (where they should be denatured) is now known. Present on the surface of the bacterium is the enzyme *urease,* an enzyme that converts urea to the basic substance ammonia. The ammonia then neutralizes acid present in its immediate vicinity; a protective barrier is thus created. The *urease* itself is protected from denaturation by its complex quaternary structure.

H. pylori causes ulcers by weakening the protective mucous coating of the stomach and duodenum, which allows acid to get through to the sensitive lining beneath. Both the acid and the bacteria irritate the lining and cause a sore—the ulcer. Ultimately the *H. pylori* themselves burrow into the lining to an acid-safe area within the lining.

Approximately two-thirds of the world's population is infected with *H. pylori.* In the United States 30% of the adult population is infected, with the infection most prevalent among older adults. About 20% of people under the age of 40 and half of those over 60 have it. Only one out of every six people infected with *H. pylori* ever suffer symptoms related to ulcers. Why *H. pylori* does not cause ulcers in every infected person is not known.

H. pylori bacteria are most likely spread from person to person through fecal–oral or oral–oral routes. Possible environmental sources include contaminated water sources. The infection is more common in crowded living conditions with poor sanitation. In countries with poor sanitation, 90% of the adult population can be infected.

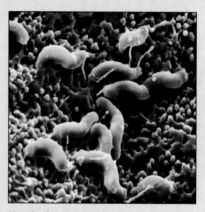

H. pylori bacteria.

substrate must occupy an enzyme active site for a finite amount of time, and the products must leave the site before the cycle can be repeated. When each enzyme molecule is working at full capacity, the incoming substrate molecules must "wait their turn" for an empty active site. At this point, the enzyme is said to be under saturating conditions.

The rate at which an enzyme accepts and releases substrate molecules at substrate saturation is given by its turnover number. An enzyme's **turnover number** *is the number of substrate molecules transformed per minute by one molecule of enzyme under optimum conditions of temperature, pH, and saturation.* Table 10.2 gives

TABLE 10.2
Turnover Numbers for Selected Enzymes

Enzyme	Turnover Number (per minute)	Reaction Catalyzed
carbonic anhydrase	36,000,000	$CO_2 + H_2O \rightleftharpoons H_2CO_3$
catalase	5,600,000	$2H_2O_2 \rightleftharpoons 2H_2O + O_2$
cholinesterase	1,500,000	hydrolysis of acetylcholine
penicillinase	120,000	hydrolysis of penicillin
lactate dehydrogenase	60,000	conversion of pyruvate to lactate
DNA polymerase I	900	addition of nucleotides to DNA chains

CHEMICAL CONNECTIONS
Enzymatic Browning: Discoloration of Fruits and Vegetables

Everyone is familiar with the way fruits such as apples, pears, peaches, apricots, and bananas, and vegetables such as potatoes quickly turn brown when their tissue is exposed to oxygen. Such oxygen exposure occurs when the food is sliced or bitten into or when it has sustained bruises, cuts or other injury to the peel. This "browning reaction" is related to the work of an enzyme called phenolase (or polyphenoloxidase), a conjugated enzyme in which copper is present.

Phenolase is classified as an oxidoreductase. The substrates for phenolase are phenolic compounds present in the tissues of the fruits and vegetables. Phenolase hydroxylates monophenols to *o*-diphenols and oxidizes *o*-diphenols to *o*-quinones (see chemical equations below). The *o*-quinones then enter into a number of other reactions, which produce the "undesirable" brown discolorations. Quinone formation is enzyme- and oxygen-dependent. Once the quinones have formed, the subsequent reactions occur spontaneously and no longer depend on the presence of phenolase or oxygen.

Enzymatic browning can be prevented or slowed in several ways. Immersing the "injured" food (for example, apple slices) in cold water slows the browning process. The lower temperature decreases enzyme activity, and the water limits the enzyme's access to oxygen. Refrigeration slows enzyme activity even more, and boiling temperatures destroy (denature) the

At left, a freshly cut apple. Brownish oxidation products form in a few minutes (at right).

enzyme. A long-used method for preventing browning involves lemon juice. Phenolase works very slowly in the acidic environment created by the lemon juice's presence. In addition, the vitamin C (ascorbic acid) present in lemon juice functions as an antioxidant. It is more easily oxidized than the phenolic-derived compounds, and its oxidation products are colorless.

Monophenol derivatives → (O₂, Phenolase) → *o*-Diphenol derivatives → (O₂, Phenolase) → *o*-Quinone derivatives → Brownish oxidation products

---

FIGURE 10.9 A graph showing the change in reaction rate with a change in enzyme concentration for an enzymatic reaction. Temperature, pH, and substrate concentration are constant. The substrate concentration is high relative to enzyme concentration.

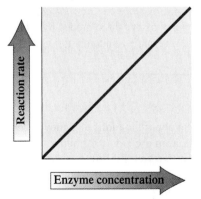

Reaction rate — Enzyme concentration

turnover numbers for selected enzymes. Some enzymes have a much faster mode of operation than others.

Enzyme Concentration

Because enzymes are not consumed in the reactions they catalyze, the cell usually keeps the number of enzymes low compared with the number of substrate molecules. This is efficient; the cell avoids paying the energy costs of synthesizing and maintaining a large work force of enzyme molecules. Thus, in general, the concentration of substrate in a reaction is much higher than that of the enzyme.

If the amount of substrate present is kept constant and the enzyme concentration is increased, the reaction rate increases because more substrate molecules can be accommodated in a given amount of time. A plot of enzyme activity versus enzyme concentration, at a constant substrate concentration that is high relative to enzyme concentration, is shown in Figure 10.9. The greater the enzyme concentration, the greater the reaction rate.

The Chemistry at a Glance feature on page 358 reviews what we have said about enzyme activity.

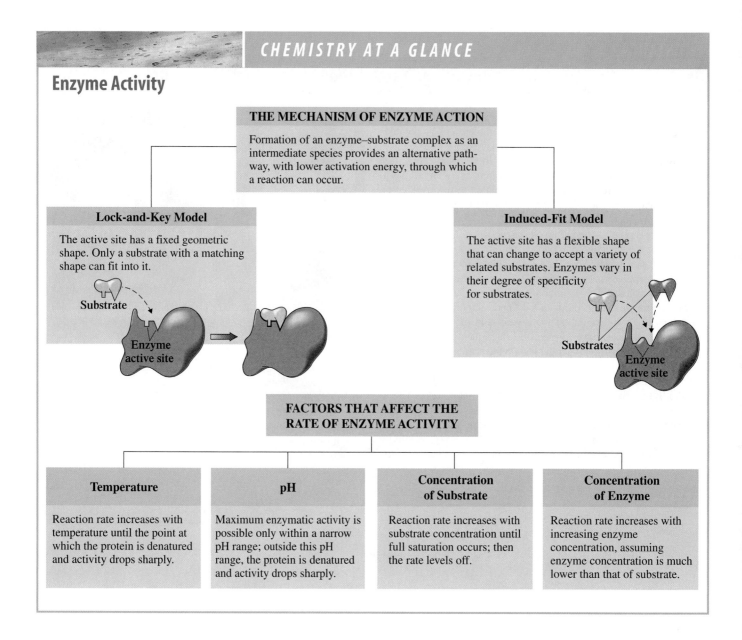

CHEMISTRY AT A GLANCE

Enzyme Activity

THE MECHANISM OF ENZYME ACTION

Formation of an enzyme–substrate complex as an intermediate species provides an alternative pathway, with lower activation energy, through which a reaction can occur.

Lock-and-Key Model

The active site has a fixed geometric shape. Only a substrate with a matching shape can fit into it.

Substrate

Enzyme active site

Induced-Fit Model

The active site has a flexible shape that can change to accept a variety of related substrates. Enzymes vary in their degree of specificity for substrates.

Substrates

Enzyme active site

FACTORS THAT AFFECT THE RATE OF ENZYME ACTIVITY

Temperature	**pH**	**Concentration of Substrate**	**Concentration of Enzyme**
Reaction rate increases with temperature until the point at which the protein is denatured and activity drops sharply.	Maximum enzymatic activity is possible only within a narrow pH range; outside this pH range, the protein is denatured and activity drops sharply.	Reaction rate increases with substrate concentration until full saturation occurs; then the rate levels off.	Reaction rate increases with increasing enzyme concentration, assuming enzyme concentration is much lower than that of substrate.

10.7 Enzyme Inhibition

The treatment for methanol poisoning involves giving a patient intravenous ethanol (Section 3.5). This action is based on the principle of competitive enzyme inhibition. The same enzyme, *alcohol dehydrogenase,* detoxifies both methanol and ethanol. Ethanol has 10 times the affinity for the enzyme that methanol has. Keeping the enzyme busy with ethanol as the substrate gives the body time to excrete the methanol before it is oxidized to the potentially deadly formaldehyde (Section 3.4).

The rates of enzyme-catalyzed reactions can be *decreased* by a group of substances called inhibitors. An **enzyme inhibitor** *is a substance that slows or stops the normal catalytic function of an enzyme by binding to it.* In this section, we consider three modes by which inhibition takes place: reversible competitive inhibition, reversible noncompetitive inhibition, and irreversible inhibition.

■ Reversible Competitive Inhibition

In Section 10.5 we noted that enzymes are quite specific about the molecules they accept at their active sites. Molecular shape and charge distribution are key determining factors in whether an enzyme accepts a molecule. A **competitive enzyme inhibitor** *is a molecule that sufficiently resembles an enzyme substrate in shape and charge distribution that it can compete with the substrate for occupancy of the enzyme's active site.*

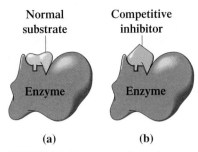

FIGURE 10.10 A comparison of an enzyme with a substrate at its active site (a) and an enzyme with a competitive inhibitor at its active site (b).

When a competitive inhibitor binds to an enzyme active site, the inhibitor remains unchanged (no reaction occurs), but its physical presence at the site prevents a normal substrate molecule from occupying the site. The result is a decrease in enzyme activity.

The formation of an enzyme–competitive inhibitor complex is a reversible process because it is maintained by weak interactions (hydrogen bonds, etc.). With time (fractions of a second), the complex breaks up. The empty active site is then available for a new occupant. Substrate and inhibitor again compete for the empty active site. Thus the active site of an enzyme binds either inhibitor or normal substrate on a random basis. If inhibitor concentration is greater than substrate concentration, the inhibitor dominates the occupancy process. The reverse is also true. Competitive inhibition can be reduced by simply increasing the concentration of the substrate.

Figure 10.10 compares the binding of a normal substrate and that of a competitive inhibitor at an enzyme's active site. Note that the portions of these two molecules that bind to the active site have the same shape but that the two molecules differ in *overall* shape. It is because of this overall difference in shape that the substrate reacts at the active site but the inhibitor does not.

Numerous drugs act by means of competitive inhibition. For example, antihistamines are competitive inhibitors of histidine decarboxylation, the enzymatic reaction that converts histidine to histamine. Histamine causes the usual allergy and cold symptoms: watery eyes and runny nose.

■ Reversible Noncompetitive Inhibition

A **noncompetitive enzyme inhibitor** *is a molecule that decreases enzyme activity by binding to a site on an enzyme other than the active site.* The substrate can still occupy the active site, but the presence of the inhibitor causes a change in the structure of the enzyme sufficient to prevent the catalytic groups at the active site from properly effecting their catalyzing action. Figure 10.11 contrasts the processes of reversible competitive inhibition and reversible noncompetitive inhibition.

Unlike the situation in competitive inhibition, increasing the concentration of substrate does not completely overcome the inhibitory effect in this case. However, lowering the concentration of a noncompetitive inhibitor sufficiently does free up many enzymes, which then return to normal activity.

FIGURE 10.11 The difference between a reversible competitive inhibitor and a reversible noncompetitive inhibitor.

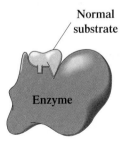

(a)
An enzyme-substrate complex in absence of an inhibitor

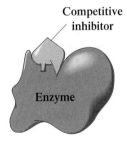

(b)
A competitive inhibitor binds to the active site; the normal substrate cannot bind.

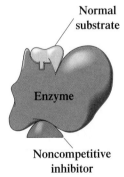

(c)
A noncompetitive inhibitor binds to a site other than the active site; the normal substrate still binds but the enzyme cannot catalyze the reaction due to the presence of the inhibitor.

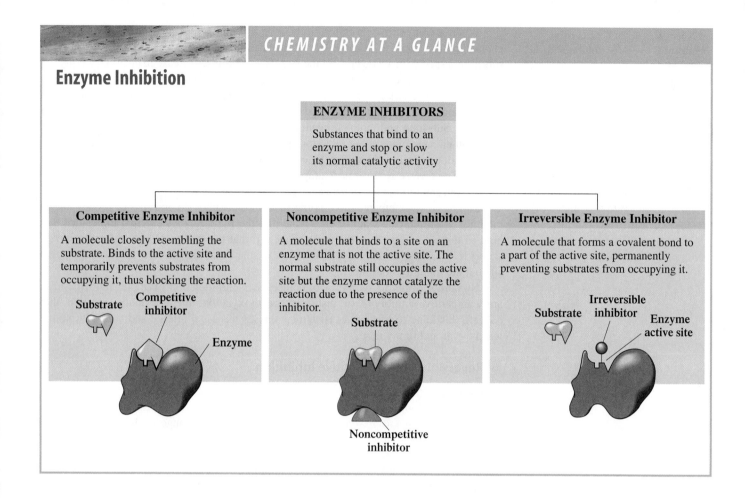

CHEMISTRY AT A GLANCE

Enzyme Inhibition

ENZYME INHIBITORS

Substances that bind to an enzyme and stop or slow its normal catalytic activity

Competitive Enzyme Inhibitor

A molecule closely resembling the substrate. Binds to the active site and temporarily prevents substrates from occupying it, thus blocking the reaction.

Substrate   Competitive inhibitor

Enzyme

Noncompetitive Enzyme Inhibitor

A molecule that binds to a site on an enzyme that is not the active site. The normal substrate still occupies the active site but the enzyme cannot catalyze the reaction due to the presence of the inhibitor.

Substrate

Noncompetitive inhibitor

Irreversible Enzyme Inhibitor

A molecule that forms a covalent bond to a part of the active site, permanently preventing substrates from occupying it.

Irreversible Substrate   inhibitor   Enzyme active site

Examples of noncompetitive inhibitors include the heavy metal ions Pb^{2+}, Ag^+, and $Hg^{10.}$ The binding sites for these ions are sulfhydryl (—SH) groups located away from the active site. Metal disulfide linkages are formed, an effect that disrupts secondary and tertiary structure.

■ Irreversible Inhibition

An **irreversible enzyme inhibitor** *is a molecule that inactivates enzymes by forming a strong covalent bond to an amino acid side-chain group at the enzyme's active site.* In general, such inhibitors do *not* have structures similar to that of the enzyme's normal substrate. The inhibitor–active site bond is sufficiently strong that addition of excess substrate does not reverse the inhibition process. Thus the enzyme is permanently deactivated. The actions of chemical warfare agents (nerve gases) and organophosphate insecticides are based on irreversible inhibition.

The Chemistry at a Glance feature summarizes what we have considered concerning enzyme inhibition.

10.8 Regulation of Enzyme Activity: Allosteric Enzymes

In the previous section, we looked at the decrease in enzyme activity caused by inhibiting agents that were "foreign" to normal cells. In this section, we consider the regulation of enzyme activity by substances produced within a cell—that is, regulation by "normal" cell components. The concept of noncompetitive inhibition that was developed in the previous section will be part of our discussion.

■ Allosteric Enzymes

Many, but not all, of the molecules responsible for regulating cellular processes are a special group of enzymes called *allosteric enzymes*. Characteristics of such enzymes are as follows:

1. All allosteric enzymes have quaternary structure; that is, they are composed of two or more protein chains.
2. All allosteric enzymes have two kinds of binding sites: those for substrate and those for regulators.
3. Active and regulatory binding sites are distinct from each other in both location and shape. Often the regulatory site is on one protein chain and the active site is on another.
4. Binding of a molecule at the regulatory site causes changes in the overall three-dimensional structure of the enzyme, including structural changes at the active site.

The term *allosteric* comes from the Greek *allo,* which means "other," and *stereos,* which means "site or space."

Thus an **allosteric enzyme** *is an enzyme with two or more protein chains (quaternary structure) and two kinds of binding sites (substrate and regulator).*

Substances that bind at regulatory sites of allosteric enzymes are called *regulators*. The binding of a *positive regulator* increases enzyme activity; the shape of the active site is changed such that it can more readily accept substrate. The binding of a *negative regulator* (a noncompetitive inhibitor) decreases enzyme activity; changes to the active site are such that substrate is less readily accepted.

Some regulators of allosteric enzyme function are inhibitors (negative regulators), and some increase enzyme activity (positive regulators).

■ Feedback Control

One of the mechanisms by which allosteric enzyme activity is regulated is feedback control. **Feedback control** *is a process in which activation or inhibition of the first reaction in a reaction sequence is controlled by a product of the reaction sequence.*

To illustrate the feedback control mechanism, let us consider a biochemical process within a cell that occurs in several steps, each step catalyzed by a different enzyme.

Most biochemical processes within cells take place in several steps rather than in a single step. A different enzyme is required for each step of the process.

$$A \xrightarrow{\text{Enzyme 1}} B \xrightarrow{\text{Enzyme 2}} C \xrightarrow{\text{Enzyme 3}} D$$

The product of each step is the substrate for the next enzyme.

What will happen in this reaction series if the final product (D) is a negative regulator of the first enzyme (enzyme 1)? At low concentrations of D, the reaction sequence proceeds rapidly. At higher concentrations of D, the activity of enzyme 1 becomes inhibited (by feedback), and eventually the activity stops. At the stopping point, there is sufficient D present in the cell to meet its needs. Later, when the concentration of D decreases through use in other cell reactions, the activity of enzyme 1 increases and more D is produced.

$$\underbrace{\overbrace{A}^{\text{Feedback control}}_{\text{Inhibition of enzyme 1 by product D}} \xrightarrow{\text{Enzyme 1}} B \xrightarrow{\text{Enzyme 2}} C \xrightarrow{\text{Enzyme 3}} D}$$

The general term *allosteric control* is often used to describe a process in which a regulatory molecule that binds at one site in an enzyme influences substrate binding at the active site in the enzyme.

Feedback control is not the only mechanism by which an allosteric enzyme can be regulated; it is just one of the more common ways. Regulators of a particular allosteric enzyme may be products of entirely different pathways of reaction within the cell, or they may even be compounds produced outside the cell (hormones).

10.9 Regulation of Enzyme Activity: Zymogens

A **proteolytic enzyme** *is an enzyme that catalyzes the breaking of peptide bonds that maintain the primary structure of a protein.* Because they would otherwise destroy the tissues that produce them, proteolytic enzymes are generated in an inactive form and then

FIGURE 10.12 Conversion of a zymogen (the inactive form of a proteolytic enzyme) to a proteolytic enzyme (the active form of the enzyme) often involves removal of a peptide chain segment from the zymogen structure.

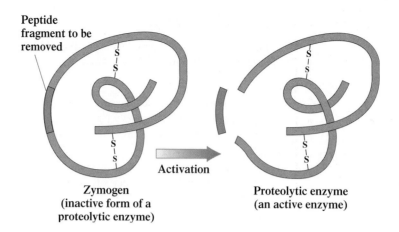

Peptide fragment to be removed

Activation

Zymogen
(inactive form of a
proteolytic enzyme)

Proteolytic enzyme
(an active enzyme)

The names of zymogens can be recognized by the suffix *-ogen* or the prefix *pre-* or *pro-*.

later, when they are needed, are converted to their active form. Most digestive and blood-clotting enzymes are proteolytic enzymes. The inactive forms of proteolytic enzymes are called *zymogens*. A **zymogen** *is the inactive precursor of a proteolytic enzyme.* (An alternative, but less often used, name for a zymogen is *proenzyme.*)

Activation of a zymogen requires an enzyme-controlled reaction that either adds to the zymogen structure or removes some part of it. Such modification changes the three-dimensional structure (secondary and tertiary structure) of the zymogen, which affects active site conformation. For example, the zymogen pepsinogen is converted to the active enzyme pepsin in the stomach, where it then functions as a digestive enzyme. Pepsin would digest the tissues of the stomach wall if it were prematurely generated in active form. Pepsinogen activation involves removal of a peptide fragment from its structure (Figure 10.12).

10.10 Antibiotics That Inhibit Enzyme Activity

An **antibiotic** *is a substance that kills bacteria or inhibits their growth.* Antibiotics exert their action selectively on bacteria and do not affect the normal metabolism of the host organism. Antibiotics usually inhibit specific enzymes essential to the life processes of bacteria. In this section we consider the actions of two families of antibiotics, sulfa drugs and penicillins, as well as the specific antibiotic *Cipro.*

■ Sulfa Drugs

The antibiotic activity of a compound called sulfanilamide was first discovered in 1932 by a German chemist who was synthesizing sulfur-containing dyes, some of which were sulfanilamide derivatives. Since that time, scientists have synthesized many biochemically active derivatives of sulfanilamide, which are collectively called sulfa drugs.

Sulfanilamide inhibits bacterial growth because it is structurally similar to PABA (*p*-aminobenzoic acid).

Sulfanilamide

p-Aminobenzoic acid
(PABA)

Many bacteria need PABA in order to produce an important coenzyme, folic acid. Sulfanilamide acts as a competitive inhibitor to enzymes in the biosynthetic pathway for converting PABA into folic acid in these bacteria. Folic acid deficiency retards growth of the bacteria and can eventually kill them.

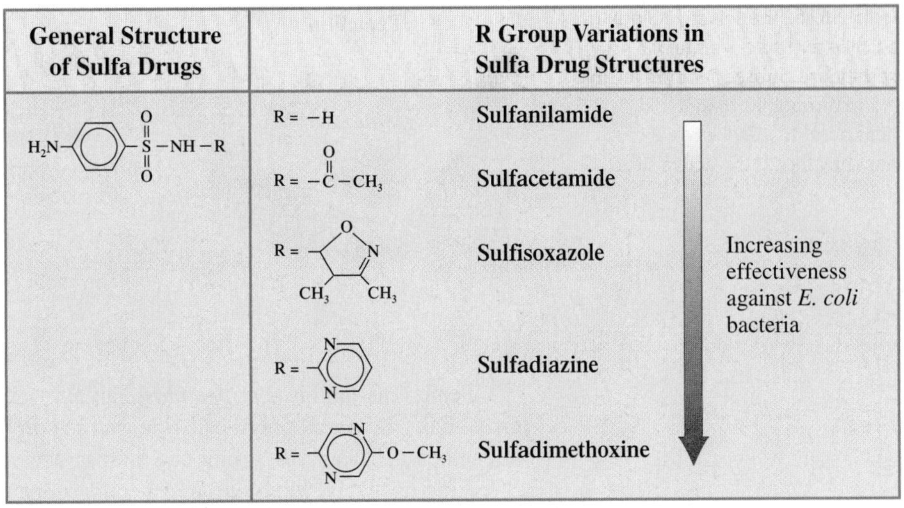

FIGURE 10.13 Structures of selected sulfa drugs in use today as antibiotics.

General Structure of Sulfa Drugs	R Group Variations in Sulfa Drug Structures	
H_2N—⟨benzene⟩—$\overset{O}{\underset{O}{\overset{\|}{\underset{\|}{S}}}}$—NH—R	R = —H	Sulfanilamide
	R = $-\overset{O}{\overset{\|}{C}}-CH_3$	Sulfacetamide
	R = ⟨isoxazole ring with CH_3, CH_3⟩	Sulfisoxazole
	R = ⟨pyrimidine ring⟩	Sulfadiazine
	R = ⟨pyrimidine ring⟩—O—CH_3	Sulfadimethoxine

Increasing effectiveness against *E. coli* bacteria

Sulfa drugs selectively inhibit only bacteria metabolism and growth because humans absorb folic acid from their diet and thus do not use PABA for its synthesis. A few of the most common sulfa drugs and their structures are shown in Figure 10.13.

■ Penicillins

Penicillin, one of the most widely used antibiotics, was accidentally discovered by Alexander Fleming in 1928 while he was working with cultures of an infectious staphylococcus bacterium. A decade later, the scientists Howard Flory and Ernst Chain isolated penicillin in pure form and proved its effectiveness as an antibiotic.

Several naturally occurring penicillins have now been isolated, and numerous derivatives of these substances have been synthetically produced. All have structures containing a four-membered β-lactam ring (Section 6.12) fused with a five-membered thiazolidine ring (Figure 10.14). As with sulfa drugs, derivatives of the basic structure differ from each other in the identity of a particular R group.

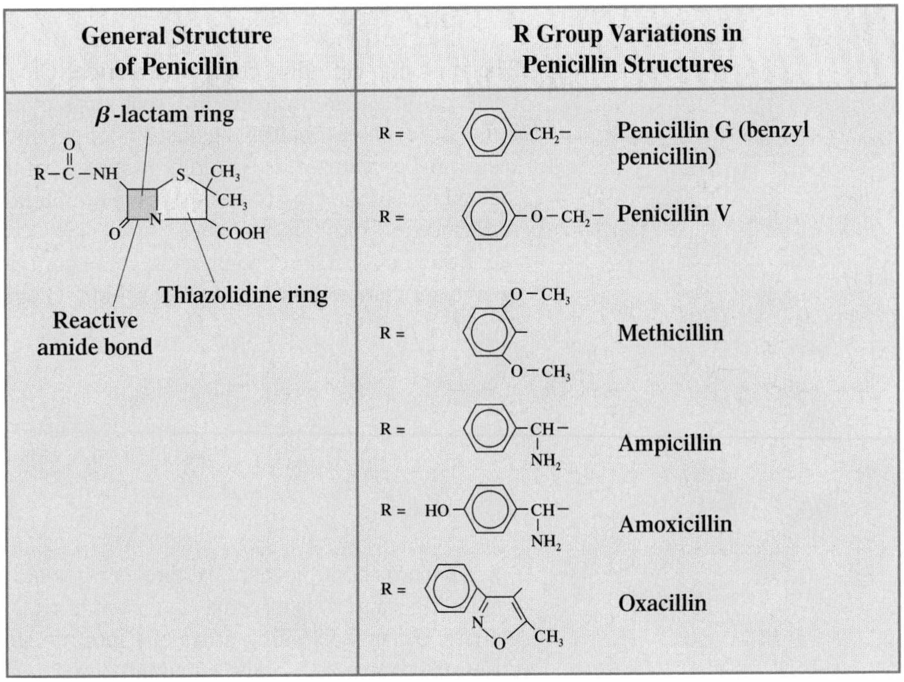

FIGURE 10.14 Structures of selected penicillins in use today as antibiotics.

General Structure of Penicillin	R Group Variations in Penicillin Structures	
	R = ⟨benzene⟩—CH_2—	Penicillin G (benzyl penicillin)
	R = ⟨benzene⟩—O—CH_2—	Penicillin V
	R = ⟨benzene with O—CH_3, O—CH_3⟩	Methicillin
	R = ⟨benzene⟩—$\overset{}{\underset{NH_2}{CH}}$—	Ampicillin
	R = HO—⟨benzene⟩—$\overset{}{\underset{NH_2}{CH}}$—	Amoxicillin
	R = ⟨isoxazole ring with CH_3⟩	Oxacillin

β-lactam ring · Reactive amide bond · Thiazolidine ring

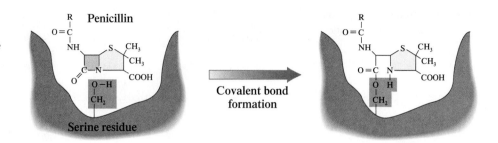

FIGURE 10.15 The selective binding of penicillin to the active site of transpeptidase. Subsequent irreversible inhibition through formation of a covalent bond to a serine residue permanently blocks the active site.

Penicillins inhibit *transpeptidase,* an enzyme that catalyzes the formation of peptide cross links between polysaccharide strands in bacterial cell walls. These cross links strengthen cell walls. A strong cell wall is necessary to protect the bacterium from lysis (breaking open). By inhibiting transpeptidase, penicillin prevents the formation of a strong cell wall. Any osmotic or mechanical shock then causes lysis, killing the bacterium.

Penicillin's unique action depends on two aspects of enzyme deactivation that we have discussed before: structural similarity to the enzyme's natural substrate and irreversible inhibition. Penicillin is *highly specific* in binding to the active site of transpeptidase. In this sense, it acts as a very selective competitive inhibitor. However, unlike a normal competitive inhibitor, once bound to the active site, the β-lactam ring opens as the highly reactive amide bond forms a covalent linkage bond to a critical serine residue required for normal catalytic action. The result is an irreversibly inhibited transpeptidase enzyme (Figure 10.15).

Some bacteria produce the enzyme *penicillinase,* which protects them from penicillin. Penicillinase selectively binds penicillin and catalyzes the opening of the β-lactam ring before penicillin can form a covalent bond to the enzyme. Once the ring is opened, the penicillin is no longer capable of inactivating transpeptidase.

Certain semi-synthetic penicillins such as methicillin and amoxicillin have been produced that are resistant to penicillinase activity and are thus clinically important.

Penicillin does not usually interfere with normal metabolism in humans because of its highly selective binding to bacterial transpeptidase. This selectivity makes penicillin an extremely useful antibiotic.

Cipro

The antibiotic ciprofloxacin hydrochloride (Cipro for short) is an effective agent against bacterial infections in many different parts of the body. It is effective against skin and bone infections as well as against infections involving the urinary, gastrointestinal, and respiratory systems. It is the drug of choice for treatment of traveler's diarrhea. It is considered one of the best broad-spectrum antibiotics available. Bacteria are slow to acquire resistance to Cipro.

Structurally, Cipro contains several common functional groups (carboxylic acid, ketone, and amine), as well as two seldom-encountered groups (fluoro and cyclopropyl).

Concern about biochemical threats associated with terrorism has thrust Cipro into the spotlight because it is effective against anthrax. As early as 1990, the U.S. Department of Defense began stockpiling doses of Cipro; the government's 2001 order was 100 million doses. Cipro is not the only antianthrax drug available. The FDA recommends the use of

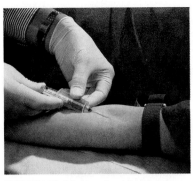

FIGURE 10.16 Drawing of a blood sample. Determination of enzyme concentrations in blood provides important information about the "state" of various organs within the human body.

doxycycline as the first-line treatment for anthrax because this compound can deal with all the strains of anthrax that are currently encountered. Authorities would rather keep Cipro in reserve; widespread current use of the drug could speed up the evolution of drug-resistant organisms.

Cipro is believed to attack the enzyme *DNA gyrase,* which controls how DNA in a bacterial chromosome coils into its tertiary structure. When tertiary structure is disrupted, replication and transcription of the DNA cannot occur.

10.11 Medical Uses of Enzymes

Enzymes can be used to diagnose certain diseases. Although blood serum contains many enzymes, some enzymes are not normally found in the blood but are produced only inside cells of certain organs and tissues. The appearance of these enzymes in the blood often indicates that there is tissue damage in an organ and that cellular contents are spilling out (leaking) into the bloodstream (see Figure 10.16). Assays of abnormal enzyme activity in blood serum can be used to diagnose many disease states, some of which are listed in Table 10.3. The Chemical Connections feature on page 367 examines the use of enzymes to diagnose heart attacks (myocardial infarctions).

Enzymes can also be used in the treatment of diseases. A recent advance in treating heart attacks is the use of tissue plasminogen activator (TPA), which activates the enzyme plasminogen. When so activated, this enzyme dissolves blood clots in the heart and often provides immediate relief.

Another medical use for enzymes is in clinical laboratory chemical analysis. For example, no simple direct test for the measurement of urea in the blood is available. However, if the urea in the blood is converted to ammonia via the enzyme *urease,* the ammonia produced, which is easily measured, becomes an indicator of urea. This *blood urea nitrogen (BUN) test* is a common clinical laboratory procedure. High urea levels in the blood indicate kidney malfunction.

10.12 Vitamins

This section and the two that follow deal with vitamins. Vitamins are considered in conjunction with enzymes because many enzymes contain vitamins as part of their structure. Recall from Section 10.3 that conjugated enzymes have a protein part (apoenzyme) and a nonprotein part (cofactor). Vitamins, in many cases, are cofactors in conjugated enzymes.

A **vitamin** *is an organic compound, essential in some amounts for the proper functioning of the human body, that must be obtained from dietary sources because the body cannot synthesize it.*

Vitamins differ from the major classes of foods (carbohydrates, lipids, and proteins) in the amount required; for vitamins it is *micro*gram or *milli*gram quantities

TABLE 10.3
Selected Blood Enzyme Assays Used in Diagnostic Medicine

Enzyme	Condition Indicated by Abnormal Level
lactate dehydrogenase (LDH)	heart disease, liver disease
creatine phosphokinase (CPK)	heart disease
aspartate transaminase (AST)	heart disease, liver disease, muscle damage
alanine transaminase (ALT)	heart disease, liver disease, muscle damage
gamma-glutamyl transpeptidase (GGTP)	heart disease, liver disease
alkaline phosphatase (ALP)	bone disease, liver disease

TABLE 10.4
The General Properties of Water-Soluble Vitamins and Fat-Soluble Vitamins

	Water-Soluble Vitamins (B vitamins and vitamin C)	Fat-Soluble Vitamins (vitamins A, D, E, and K)
Absorption	directly into the blood	first enter into the lymph system
Transport	travel without carriers	many require protein carriers
Storage	circulate in the water-filled parts of the body	found in the cells associated with fat
Excretion	kidneys remove excess in urine	tend to remain in fat-storage sites
Toxicity	not likely to reach toxic levels when consumed from supplements	likely to reach toxic levels when consumed from supplements
Requirements	needed in frequent doses	needed in periodic doses

The spelling of the term *vitamin* was originally *vitamine,* a word derived from the Latin *vita,* meaning "life," and from the fact that these substances were all thought to contain the *amine* functional group. When this supposition was found to be false, the final *e* was dropped from *vitamine,* and the term *vitamin* came into use. Some vitamins contain amine functional groups, but others do not.

per day compared with 50–200 grams per day for the major food categories. To illustrate the small amount of vitamins needed by the human body, consider the recommended daily allowance (RDA) of vitamin B_{12}, which is 2.0 micrograms per day for an adult. Just 1.0 gram of this vitamin could theoretically supply the daily needs of 500,000 people.

A well-balanced diet usually meets all the body's vitamin requirements. However, supplemental vitamins are often required for women during pregnancy and for people recovering from certain illnesses.

One of the most common myths associated with the nutritional aspects of vitamins is that vitamins from natural sources are superior to synthetic vitamins. In truth, synthetic vitamins, manufactured in the laboratory, are identical to the vitamins found in foods. The body cannot tell the difference and gets the same benefits from either source.

There are 13 known vitamins, and scientists believe that the discovery of additional vitamins is unlikely. Despite searches for new vitamins, it has been over 50 years since the last of the known vitamins (B_{12}) was discovered. Strong evidence that the vitamin family is complete comes from the fact that many people have lived for years being fed, intravenously, solutions containing the known vitamins and nutrients, and they have not developed any known vitamin deficiency disease.

Solubility characteristics divide the vitamins into two major classes: the water-soluble vitamins and the fat (lipid)-soluble vitamins. Water-soluble vitamins must be constantly replenished in the body because they are rapidly eliminated from the body in the urine. They are carried in the bloodstream, are needed in frequent, small doses, and are unlikely to be toxic except when taken in unusually large doses. The fat-soluble vitamins are found dissolved in lipid materials. They are, in general, carried in the blood by protein carriers, are stored in fat tissues, are needed in periodic doses, and are more likely to be toxic when consumed in excess of need.

An important difference exists, in terms of function, between water-soluble and fat-soluble vitamins. Water-soluble vitamins function as coenzymes for a number of important biochemical reactions in humans, animals, and microorganisms. Fat-soluble vitamins generally do not function as coenzymes in humans and animals and are rarely utilized in any manner by microorganisms. Other differences between the two categories of vitamins are summarized in Table 10.4. A few exceptions occur, but the differences shown in this table are generally valid.

10.13 Water-Soluble Vitamins

There are nine water-soluble vitamins: vitamin C and eight B vitamins. These vitamins got their names from the labels B and C on the test tubes in which they were first collected. Later, test tube B was found to contain more than one vitamin.

CHEMICAL CONNECTIONS   Heart Attacks and Enzyme Analysis

The symptoms of a heart attack—that is, a *myocardial infarction (MI)*—include irregular breathing and pain in the left chest that may radiate to the neck, left shoulder, and arm. An initial diagnosis of an MI is based on these and other physical symptoms, and treatment is initiated on this basis.

Physicians then use enzyme analysis to confirm the diagnosis and to monitor the course of treatment. The blood levels of three enzymes are commonly assayed in MI situations: creatine phosphokinase (CPK), aspartate transaminase (AST), and lactate dehydrogenase (LDH). The CPK level rises and falls relatively rapidly after a heart attack, reaching a maximum after about 30 hours at a level approximately six times normal. The AST level triples after about 40 hours. LDH, whose concentrations rise slowly, is used to monitor the later stages of the MI and to assess the extent of heart damage. The accompanying graph shows blood levels of these three enzymes as a function of time in an MI situation.

Further information about the seriousness of an MI is obtained by studying isoenzymes. **Isoenzymes** *are isomeric forms of the same enzyme with slightly different amino acid sequences.* Lactate dehydrogenase is a mixture of five isoenzymes denoted LDH_1 to LDH_5. Creatine phosphokinase is a mixture of three isoenzymes denoted CK-MM, CK-MB, and CK-BB.

Consideration of further details about the LDH isoenzymes shows how isoenzymes give information about whether a heart attack has occurred or not. Note from the following table that heart tissue is particularly high in LDH_1 and that liver and skeletal muscle are particularly high in LDH_5.

Tissue	LDH_1	LDH_2	LDH_3	LDH_4	LDH_5
brain	23	34	30	10	3
heart	50	36	9	3	2
kidney	28	34	21	11	6
liver	4	6	17	16	57
lung	10	20	30	25	15
serum	28	41	19	7	5
skeletal muscle	5	5	10	22	58

When a heart attack occurs, some heart muscle cells are damaged (destroyed), and their enzymes "leak" into the bloodstream.

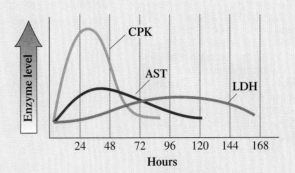

This changes the ratios among the various LDHs present in blood, because heart muscle is particularly high in LDH_1. The accompanying graph of an LDH isoenzyme assay for a heart attack victim shows that the LDH_1/LDH_2 ratio, which is normally less than one in blood serum, is now greater than one.

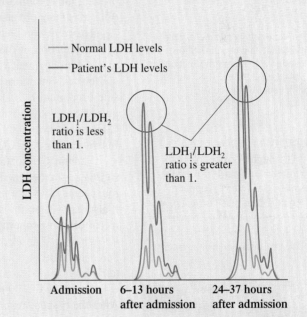

- Normal LDH levels
- Patient's LDH levels

LDH_1/LDH_2 ratio is less than 1.

LDH_1/LDH_2 ratio is greater than 1.

Admission   6–13 hours after admission   24–37 hours after admission

▪ Vitamin C

Vitamin C, which has the simplest structure of the 13 vitamins, exists in two active forms in the human body: an oxidized form and a reduced form.

$$CH_2\!-\!CH \overset{O}{\underset{HO \quad\; OH}{\diagup}} O \quad \underset{\text{Reduction}}{\overset{\text{Oxidation}}{\rightleftharpoons}} \quad CH_2\!-\!CH \overset{O}{\underset{O \quad\; O}{\diagup}} O \;+\; 2H$$

Ascorbic acid (reduced)   Dehydroascorbic acid (oxidized)

Vitamin C, the best known of all vitamins, was the first vitamin to be discovered (1928), the first to be structurally characterized (1933), and the first to be synthesized in the laboratory (1933). Laboratory production of vitamin C, which exceeds 80 million pounds per year, is greater than the combined production of all the other vitamins. In addition to its use as a vitamin supplement, synthetic vitamin C is used as a food additive (preservative), a flour additive, and an animal feed additive.

Why is vitamin C called ascorbic *acid* when there is no carboxyl group (acid group) present in its structure? Vitamin C is a cyclic ester in which a carbon 1 carboxyl group has reacted with a carbon 4 hydroxyl group, forming the ring structure.

Other naturally occurring dietary antioxidants include glutathione (Section 9.7), vitamin E (Section 10.14), beta-carotene (Section 10.14), and flavonoids (Section 12.11).

FIGURE 10.17 Rows of cabbage plants. Although many people think citrus fruits (50 mg per 100 g) are the best source of vitamin C, peppers (128 mg per 100 g), cauliflower (70 mg per 100 g), strawberries (60 mg per 100 g), and spinach or cabbage (60 mg per 100 g) are all richer in vitamin C.

Humans, monkeys, apes, and guinea pigs are among the relatively few species that require dietary sources of vitamin C. Other species synthesize vitamin C from carbohydrates. Vitamin C's biosynthesis involves L-gulonic acid, an acid derivative of the monosaccharide L-gulose (see Figure 7.13). L-Gulonic acid is reduced by the enzyme *lactonase* to give a cyclic ester (lactone, Section 5.11); ring closure involves carbons 1 and 4. An *oxidase* then introduces a double bond into the ring, producing L-ascorbic acid.

L-Gulonic acid → (Lactonase) → γ-L-Gulonolactone → (Oxidase) → L-Ascorbic acid

The four —OH groups present in vitamin C's reduced form are suggestive of its biosynthetic monosaccharide (*polyhydroxy* aldehyde) origins. Its chemical name, L-ascorbic acid, correctly indicates that vitamin C is a weak acid. Although no carboxyl group is present, the carbon 3 hydroxyl group hydrogen atom exhibits acidic behavior as a result of its attachment to an unsaturated carbon atom.

The most completely characterized role of vitamin C is its function as a cosubstrate in the formation of the structural protein collagen (Section 9.17), which makes up much of the skin, ligaments, and tendons and also serves as the matrix on which bone and teeth are formed. Specifically, biosynthesis of the amino acids hydroxyproline and hydroxylysine (important in binding collagen fibers together) from proline and lysine requires the presence of both vitamin C and iron. Iron serves as a cofactor in the reaction, and vitamin C maintains iron in the oxidation state that allows it to function. In this role, vitamin C is functioning as a *specific* antioxidant.

Vitamin C also functions as a *general* antioxidant (Section 3.14) for water-soluble substances in the blood and other body fluids. Its antioxidant properties are also beneficial for several other vitamins. The active form of vitamin E is regenerated by vitamin C, and it also helps keep the active form of folate (a B vitamin) in its reduced state. Because of its antioxidant properties, vitamin C is often added to foods as a preservative.

Vitamin C is also involved in the metabolism of several amino acids that end up being converted to the hormones norepinephrine and thyroxine. The adrenal glands contain a higher concentration of vitamin C than any other organ in the body.

An intake of 100 mg/day of vitamin C saturates all body tissues with the compound. After the tissues are saturated, all additional vitamin C is excreted. The RDA for vitamin C varies from country to country. It is 30 mg/day in Great Britain, 60 mg/day in the United States and Canada, and 75 mg/day in Germany. A variety of fruits and vegetables have a relatively high vitamin C content (see Figure 10.17).

■ Vitamin B

There are eight B vitamins. Our discussion of them involves four topics: nomenclature, function, structural characteristics, and dietary sources.

Much confusion exists about the B vitamins' names. Many have "number" names as well as "word" names (often several). The *preferred* names for the B vitamins (alternative names in parentheses) are

1. **Thiamin** (vitamin B_1)
2. **Riboflavin** (vitamin B_2)
3. **Niacin** (nicotinic acid, nicotinamide, vitamin B_3)
4. **Vitamin B_6** (pyridoxine, pyridoxal, pyridoxamine)

Pronunciation guidelines for the standard names of the B vitamins:

THIGH-ah-min
RYE-boh-flay-vin
NIGH-a-sin
FOLL-ate
PAN-toe-THEN-ick acid
BY-oh-tin

5. **Folate** (folic acid)
6. **Vitamin B$_{12}$** (cobalamin)
7. **Pantothenic acid** (vitamin B$_5$)
8. **Biotin**

B vitamin structure is very diverse. The only common thread among structures is that all structures, except that of pantothenic acid, involve heterocyclic nitrogen ring systems. The element sulfur is present in two structures (thiamin and biotin), and vitamin B$_{12}$ contains a metal atom (cobalt). (Biotin does not contain a tin atom, as the name might imply.) Table 10.5 gives structural forms for the eight B vitamins. Note that for two B vitamins (niacin and vitamin B$_6$), more than one form of the vitamin exists.

The major function of B vitamins within the human body is as components of coenzymes. Unlike vitamin C, all of the B vitamins must be chemically modified before they become functional within the coenzymes. For example, thiamine is converted to thiamine pyrophosphate (TPP), which then serves as the coenzyme in several reactions involving carbohydrate metabolism.

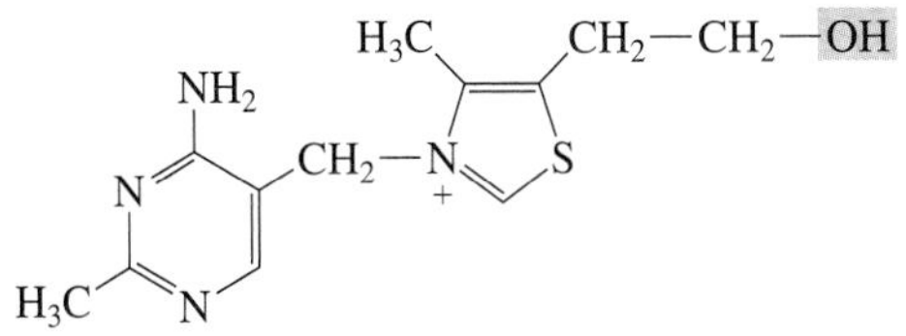

Thiamine (vitamin B$_1$) 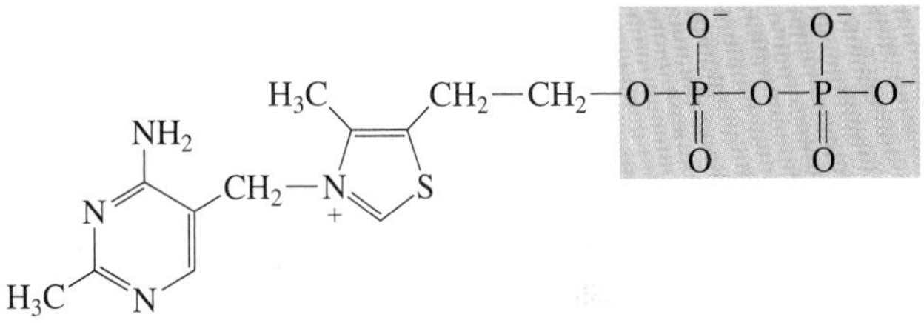   Thiamine pyrophosphate (TPP)

Another example of chemical modification for a B vitamin is the conversion of folate to tetrahydrofolate (THF).

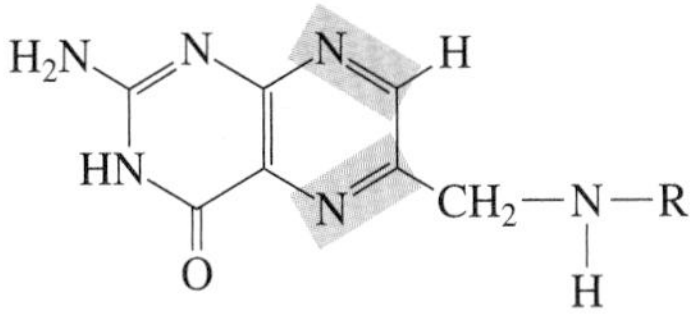

Folate 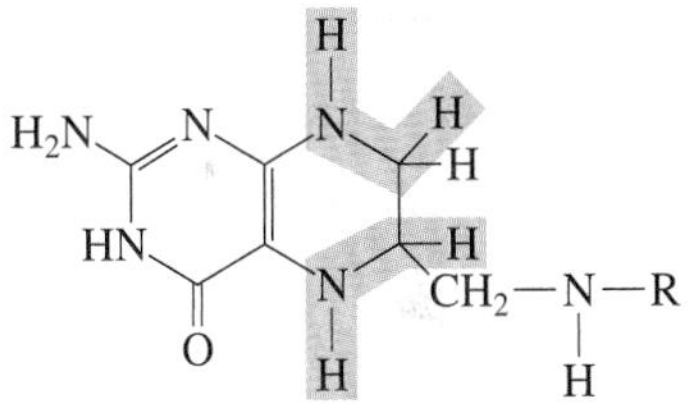   Tetrahydrofolate (THF)

Table 10.6 lists selected important coenzymes that involve B vitamins and indicates how these enzymes function. In general, coenzymes serve as temporary carriers of atoms or functional groups in redox and group transfer reactions.

An ample supply of the B vitamins can be obtained from normal dietary intake as long as a variety of foods are consumed. A certain food may be a better source of a particular B vitamin than others; however, there are multiple sources for each of the B vitamins as Table 10.7 shows. Note from Table 10.7 that fruits, in general, are very poor sources of B vitamins and that only certain vegetables are good B vitamin sources. Vitamin B$_{12}$ is unique among the vitamins in being found almost exclusively in food derived from animals. Legislation that dates back to the 1940s requires that all grain products that cross state lines be enriched in thiamin, riboflavin, and niacin. Folate was added to the legislated enrichment list in 1996 when research showed that folate was essential in the prevention of certain birth defects.

Both niacin and folate have been linked positively to improvement in cardiovascular health. Adding prescription-strength, extended-release niacin to cholesterol-lowering statin medications slows the progression of atherosclerosis among people with coronary heart disease and low HDL levels better than statin therapy alone. Additionally, prescription niacin is the most effective treatment currently available to increase low levels of HDL.

TABLE 10.5
Structures of the Eight B Vitamins

Thiamin	Riboflavin

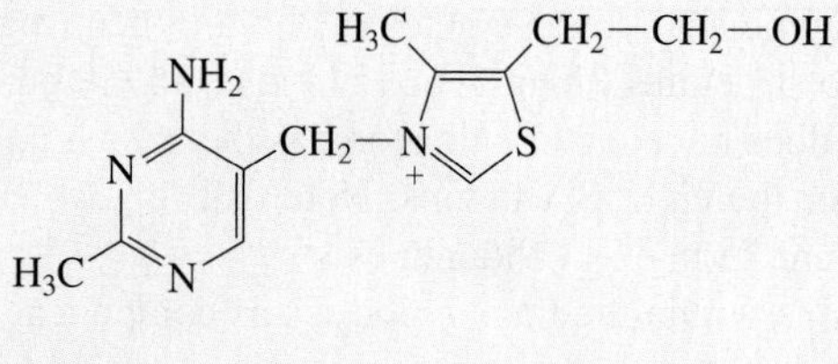

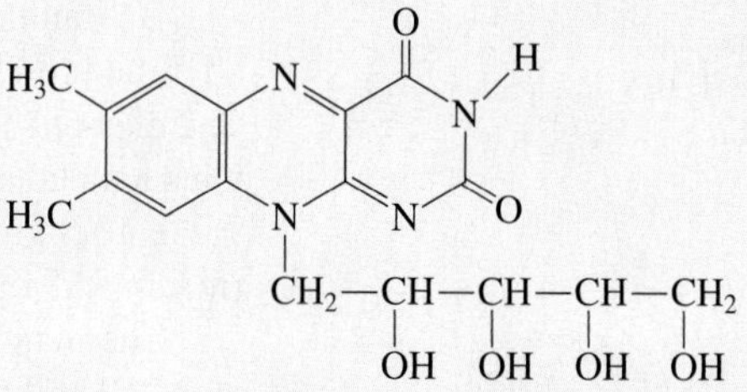

Niacin (two forms)	Pantothenic Acid
Nicotinic acid Nicotinamide	

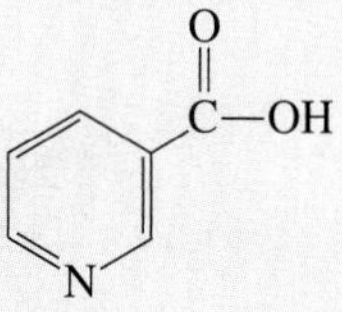

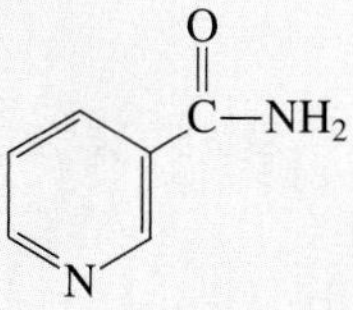

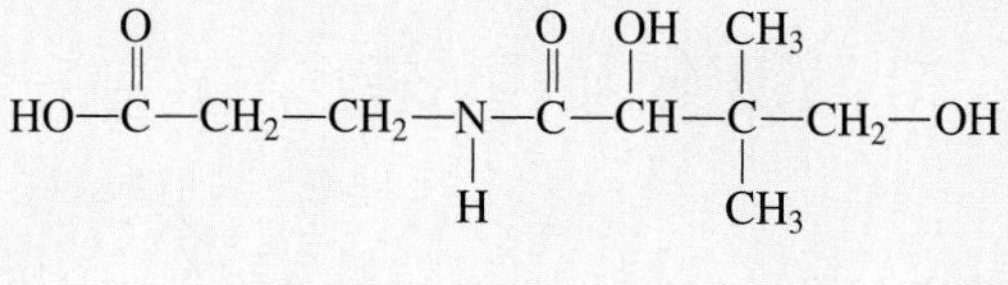

Vitamin B$_6$ (three forms)

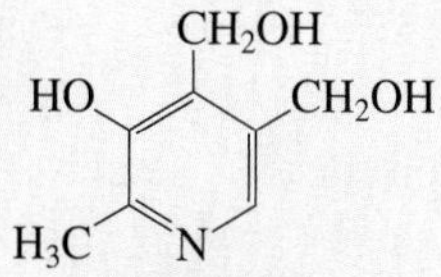

Pyridoxine 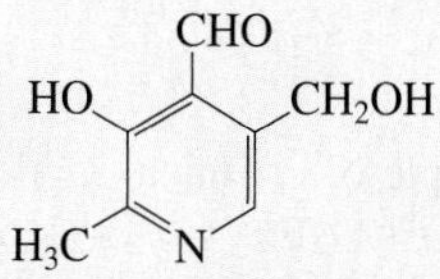   Pyridoxal 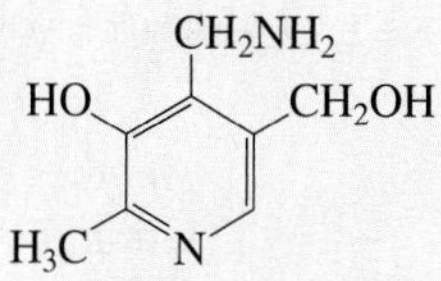   Pyridoxamine

Folate

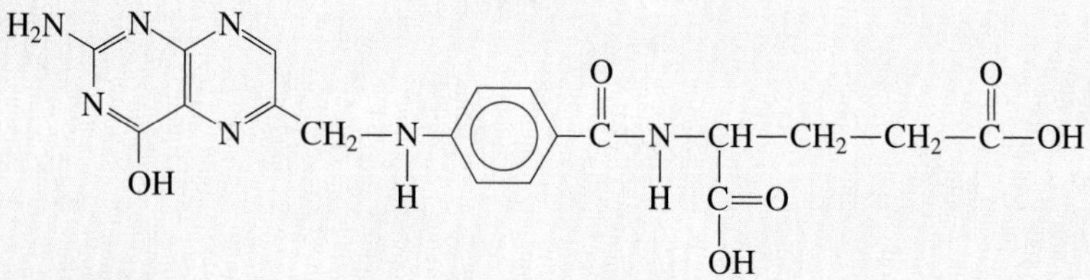

Vitamin B$_{12}$	Biotin

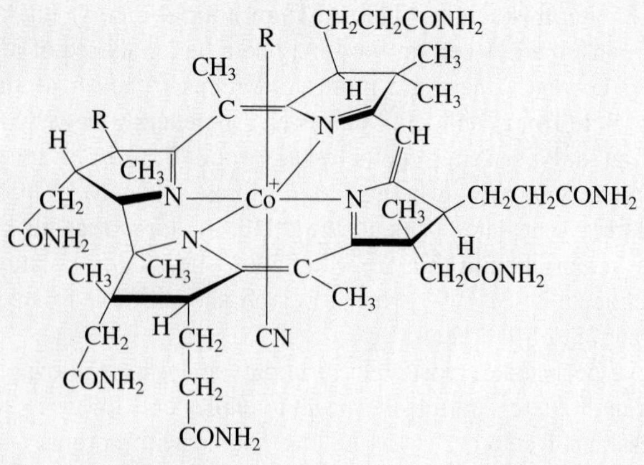

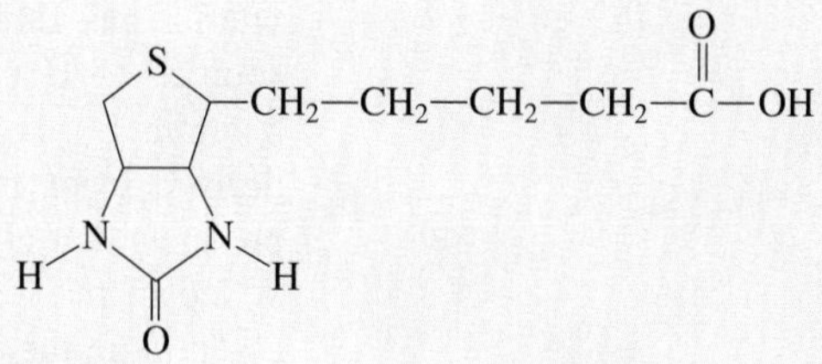

TABLE 10.6
Selected Important Coenzymes in Which B Vitamins Are Present

B Vitamin	Coenzymes	Groups Transferred
thiamine	thiamine pyrophosphate (TPP)	aldehydes
riboflavin	flavin mononucleotide (FMN) flavin adenine dinucleotide (FAD)	hydrogen atoms
niacin	nicotinamide adenine dinucleotide (NAD^+) nicotinamide adenine dinucleotide phosphate ($NADP^+$)	hydride ion (H^-)
vitamin B_6	pyridoxal-5-phosphate (PLP)	amino groups
folate	tetrahydrofolate (THF)	one-carbon groups other than CO_2
vitamin B_{12}	5′-deoxyadenosylcobalamine	alkyl groups, hydrogen atoms
pantothenic acid	coenzyme A (CoA) acyl carrier protein (ACP)	acyl groups
biotin	biocytin	carbon dioxide

TABLE 10.7
A Summary of Dietary Sources of B Vitamins

Vitamin	Vegetable Group	Fruit Group	Bread, Cereal, Rice, and Pasta Group	Milk, Yogurt, and Cheese Group	Meats	Dry Beans	Eggs	Nuts and Seeds
thiamin		watermelon	whole and enriched grains		pork, organ meats	legumes		sunflower seeds
riboflavin	mushrooms, asparagus, broccoli, leafy greens		whole and enriched grains	milk, cheeses	liver, red meat, poultry, fish	legumes	eggs	
niacin	mushrooms, asparagus, potato		whole and enriched grains, wheat bran		tuna, chicken, beef, turkey	legumes, peanuts		sunflower seeds
vitamin B_6	broccoli, spinach, potato, squash	bananas, watermelon	whole wheat, brown rice		chicken, fish, pork, organ meats	soybeans		sunflower seeds
folate	mushrooms, leafy greens, broccoli, asparagus, corn	oranges	fortified grains		organ meats (muscle meats are poor sources)	legumes		sunflower seeds, nuts
vitamin B_{12}				milk products	beef, poultry, fish, shellfish		egg yolk	
pantothenic acid	mushrooms, broccoli, avocados		whole grains		meat	legumes	egg yolk	
biotin			fortified cereals	yogurt	liver (muscle meats are poor sources)	soybeans	egg yolk	nuts

The "Meats," "Dry Beans," "Eggs," and "Nuts and Seeds" columns fall under the spanning header **Meat, Poultry, Fish, Dry Beans, Eggs, and Nuts Group**.

Another study shows that younger women (26–46 years old) who consume 800 μg of folate per day reduce the risk of developing high blood pressure by almost a third compared to those who consume less than 200 μg/day.

10.14 Fat-Soluble Vitamins

There are four fat-soluble vitamins, denoted by the letters A, D, E, and K. Many of the functions of the fat-soluble vitamins involve processes that occur in cell membranes. The structures of the fat-soluble vitamins are more hydrocarbon-like, with fewer functional groups than the water-soluble vitamins. Their structures as a whole are nonpolar, which enhances their solubility in cell membranes.

■ Vitamin A

Beta-carotene is a deep yellow (almost orange) compound. If a plant food is white or colorless, it possesses little or no vitamin A activity. Potatoes, pasta, and rice are foods in this category.

Normal dietary intake provides a person with both *preformed* and *precursor forms* (provitamin forms) of vitamin A. Preformed vitamin A forms are called *retinoids.* The retinoids include retin*al*, retin*ol*, and retin*oic acid.*

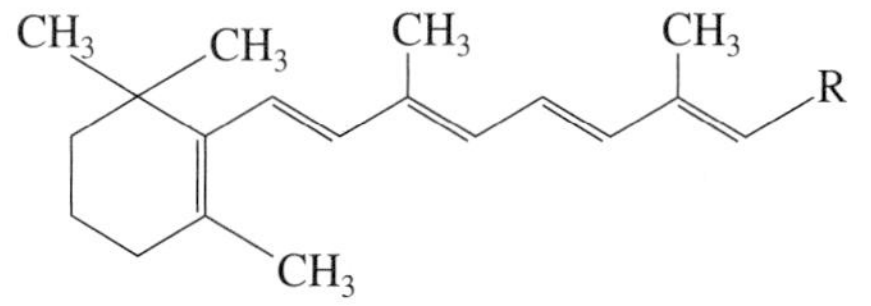

R = CH$_2$—OH (Retinol)
R = CHO (Retinal)
R = COOH (Retinoic acid)

Retinoids

Foods derived from animals, including egg yolks and dairy products, provide compounds (retinyl esters) that are easily hydrolyzed to retinoids in the intestine.

Foods derived from plants provide carotenoids (see the Chemical Connections feature "Carotenoids: A Source of Color" on page 48 in Chapter 2), which serve as precursor forms of vitamin A. The major carotenoid with vitamin A activity is beta-carotene (β-carotene), which can be cleaved to yield two molecules of vitamin A.

The retinoids are terpenes (Section 2.6) in which four isoprene units are present. Beta-carotene has an eight-unit terpene structure.

Beta-carotene cleavage does not always occur in the "middle" of the molecule, so only one molecule of vitamin A is produced. Furthermore, not all beta-carotene is converted to vitamin A, and its absorption is not as efficient as that of vitamin A itself. It is estimated that 6 mg of beta-carotene is needed to produce 1 mg of retinol. Unconverted beta-carotene serves as an antioxidant (see Section 2.6), a role independent of its conversion to vitamin A.

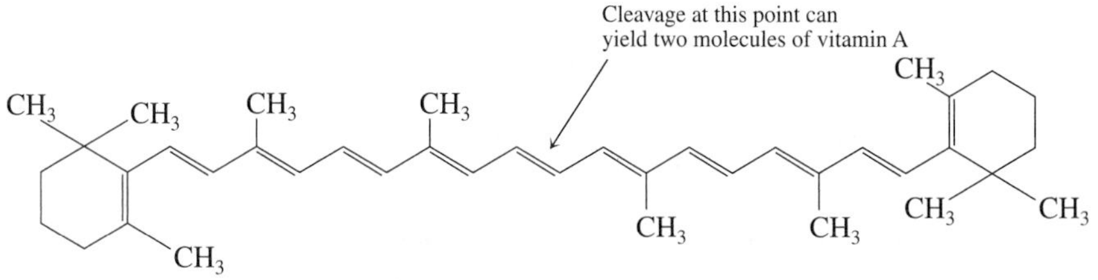

Cleavage at this point can yield two molecules of vitamin A

Beta-carotene, a precursor for vitamin A

Beta-carotene is a yellow to red-orange pigment plentiful in carrots, squash, cantaloupe, apricots, and other yellow vegetables and fruits, as well as in leafy green vegetables (where the yellow pigment is masked by green chlorophyll).

Vitamin A has four major functions in the body.

1. *Vision.* In the eye, vitamin A combines with the protein opsin to form the visual pigment rhodopsin (see the Chemical Connections feature "*Cis–Trans* Isomerism and Vision" on page 45 in Chapter 2). Rhodopsin participates in the conversion of light energy into nerve impulses that are sent to the brain. Although vitamin A's involvement in the process of vision is its best-known function ("Eat your carrots and you'll see better"), only 0.1% of the body's vitamin A is found in the eyes.
2. *Regulating Cell Differentiation.* Cell differentiation is the process whereby immature cells change in structure and function to become specialized cells.

FIGURE 10.18 The quantity of vitamin D synthesized by exposure of the skin to sunlight (ultraviolet radiation) varies with latitude, the length of exposure time, and skin pigmentation. (Darker-skinned people synthesize less vitamin D because the pigmentation filters out ultraviolet light.)

For example, some immature bone marrow cells differentiate into white blood cells and others into red blood cells. In the cellular differentiation process, vitamin A binds to protein receptors; these vitamin A–protein receptor complexes then bind to regulatory regions of DNA molecules.

3. *Maintenance of the Health of Epithelial Tissues.* Epithelial tissue covers outer body surfaces as well as lining internal cavities and tubes. It includes skin and the linings of the mouth, stomach, lungs, vagina, and bladder. Lack of vitamin A causes such surfaces to become drier and harder than normal. Vitamin A's role here is related to cellular differentiation involving mucus-secreting cells.

4. *Reproduction and Growth.* In men, vitamin A participates in sperm development. In women, normal fetal development during pregnancy requires vitamin A. Again vitamin A's role is related to cellular differentiation processes.

■ Vitamin D

The two most important members of the vitamin D family of molecules are vitamin D_3 (cholecalciferol) and vitamin D_2 (ergocalciferol). Vitamin D_3 is produced in the skin of humans and animals by the action of sunlight (ultraviolet light) on its precursor molecule, the cholesterol derivative 7-dehydrocholesterol (a normal metabolite of cholesterol found in the skin). Absorption of light energy induces breakage of the 9, 10 carbon bond; a spontaneous isomerization (shifting of double bonds) then occurs (see Figure 10.18).

7-Dehydrocholesterol → (UV) → Pre-vitamin D_3 → (Spontaneous conversion) → Vitamin D_3 (cholecalciferol)

Vitamin D_3 (cholecalciferol) is sometimes called the "sunshine vitamin" because of its synthesis in the skin by sunlight irradiation.

Vitamin D_2 (ergocalciferol) differs from vitamin D_3 only in the side-chain structure. It is produced from the plant sterol ergosterol through the action of light.

Vitamin D

D_2 series
R =

D_3 series
R =

Both the cholecalciferol and the ergocalciferol forms of vitamin D must undergo two further hydroxylation steps before the vitamin D becomes fully functional. The first step, which occurs in the liver, adds a —OH group to carbon 25. The second step, which occurs in the kidneys, adds a —OH group to carbon 1.

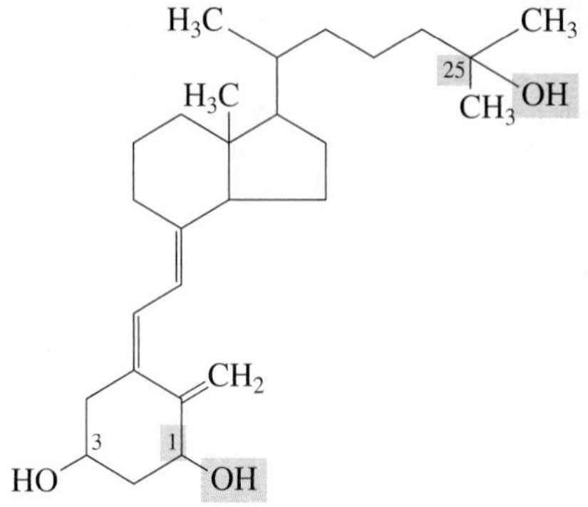

1,25-Dihydroxyvitamin D₃

Milk is enriched in vitamin D by exposure to ultraviolet light. Cholesterol in milk is converted to cholecalciferol (vitamin D) by ultraviolet light.

Only a few foods, including liver, fatty fish (such as salmon), and egg yolks, are good natural sources of vitamin D. Such vitamin D is vitamin D₃. Foods fortified with vitamin D include milk and margarine. The rest of the body's vitamin D supplies are made within the body (skin) with the help of sunlight.

The principal function of vitamin D is to maintain normal blood levels of calcium ion and phosphate ion so that bones can absorb these ions. Vitamin D stimulates absorption of these ions from the gastrointestinal tract and aids in their retention by the kidneys. Vitamin D triggers the deposition of calcium salts into the organic matrix of bones by activating the biosynthesis of calcium-binding proteins.

When it comes to strong bones, calcium won't do you a lot of good unless you are also getting enough vitamin D. In one study, women consuming 500 IU of vitamin D a day had a 37% lower risk of hip fracture than women consuming only 140 IU daily of vitamin D. Some researchers now recommend a standard of 800–1000 IU per day instead of the long-established standard recommendation for vitamin D of 400 IU per day.

Vitamin E

There are four forms of vitamin E: alpha-, beta-, delta-, and gamma-tocopherol. These forms differ from each other structurally in what substituents (—CH₃ or —H) are present at two positions on an aromatic ring.

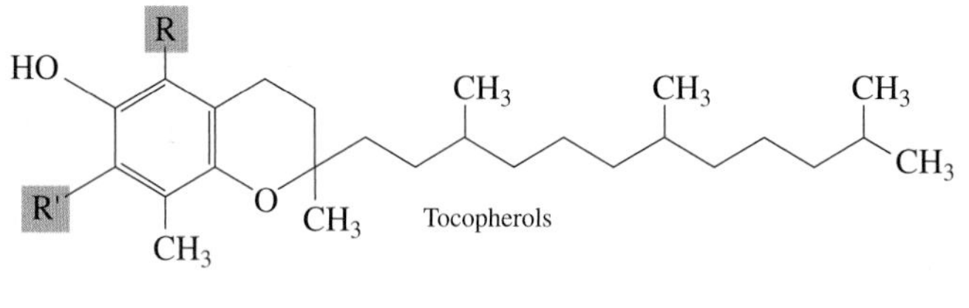

The word *tocopherol* is pronounced "tuh-KOFF-er-ol."

The tocopherol form with the greatest biochemical activity is alpha-tocopherol, the vitamin E form in which methyl groups are present at both the R and R′ positions on the aromatic ring. Gamma-tocopherol is the main form of vitamin E in vitamin-E rich foods.

Plant oils (margarine, salad dressings, and shortenings), green and leafy vegetables, and whole-grain products are sources of vitamin E.

The primary function of vitamin E in the body is as an antioxidant—a compound that protects other compounds from oxidation by being oxidized itself. Vitamin E is particularly important in preventing the oxidation of polyunsaturated fatty acids (Section 8.2) in membrane lipids. It also protects vitamin A from oxidation. After vitamin E is "spent" as an antioxidant, its antioxidant function can be restored by vitamin C.

Vitamin E is unique among the vitamins in that antioxidant activity is its *principal* biochemical role.

A most important location in the human body where vitamin E exerts its antioxidant effect is the lungs, where exposure of cells to oxygen (and air pollutants) is greatest. Both red and white blood cells that pass through the lungs, as well as the cells of the lung tissue itself, benefit from vitamin E's protective effect.

Infants, particularly premature infants, do not have a lot of vitamin E, which is passed from the mother to the infant only in the last weeks of pregnancy. Often, premature infants require oxygen supplementation for the purpose of controlling respiratory distress. In such situations, vitamin E is administered to the infant along with oxygen to give antioxidant protection.

Vitamin E has also been found to be involved in the conversion of arachidonic acid (20:4) to prostaglandins (Section 8.13.)

◼ Vitamin K

Like the other fat-soluble vitamins, vitamin K occurs in several forms. *Menaquinones* (vitamin K_1) are found in fish oils and meats and are synthesized by bacteria, including those in the human intestinal tract. *Phylloquinones* (vitamin K_2) are found in plants. Typically, about half of the body's vitamin K comes from the diet and half from synthesis by intestinal bacteria. Menaquinones are the form found in vitamin K supplements.

Vitamin K_1 and vitamin K_2 differ structurally in the length and degree of unsaturation of a side chain.

Vitamin K

K_1: R =

K_2: R =

All of the fat-soluble vitamins share a common structural feature; they all have terpenelike structures. That is, they are all made up of five-carbon isoprene units (Section 2.5) No common structural pattern exists for the water-soluble vitamins. On the other hand, the water-soluble vitamins have *functional* uniformity, whereas the fat-soluble vitamins have diverse functions.

The best dietary sources of vitamin K are dark green, leafy vegetables and liver. Milk, meat, eggs, and cereals contain smaller amounts.

Vitamin K is essential to the blood-clotting process. Over a dozen different proteins and the mineral calcium are involved in the formation of a blood clot. Vitamin K is essential for the formation of prothrombin and at least five other proteins involved in the regulation of blood clotting. Vitamin K is sometimes given to presurgical patients to ensure adequate prothrombin levels and prevent hemorrhaging.

Vitamin K is also required for the biosynthesis of several other proteins found in the plasma, bone, and kidney.

CONCEPTS TO REMEMBER

Enzymes. Enzymes are highly specialized protein molecules that act as biochemical catalysts. Enzymes have common names that provide information about their function rather than their structure. The suffix *-ase* is characteristic of most enzyme names (Section 10.1).

Enzyme classification. There are six classes of enzymes based on function: oxidoreductases, transferases, hydrolases, lyases, isomerases, and ligases (Section 10.2).

Enzyme structure. Simple enzymes are composed only of protein (amino acids). Conjugated enzymes have a nonprotein portion (cofactor) in addition to a protein portion (apoenzyme). Cofactors may be small organic molecules (coenzymes) or inorganic ions (Section 10.3).

Enzyme active site. An enzyme active site is the relatively small part of the enzyme that is actually involved in catalysis. It is where substrate binds to the enzyme (Section 10.4).

Lock-and-key model of enzyme activity. The active site in an enzyme has a fixed, rigid geometrical conformation. Only substrates with a complementary geometry can be accommodated at the active site (Section 10.4).

Induced-fit model of enzyme activity. The active site in an enzyme can undergo small changes in geometry in order to accommodate a series of related substrates (Section 10.4).

Enzyme activity. Enzyme activity is a measure of the rate at which an enzyme converts substrate to products. Four factors that affect enzyme activity are temperature, pH, substrate concentration, and enzyme concentration (Section 10.6).

Enzyme inhibition. An enzyme inhibitor slows or stops the normal catalytic function of an enzyme by binding to it. Three modes of inhibition are reversible competitive inhibition, reversible noncompetitive inhibition, and irreversible inhibition (Section 10.7).

Allosteric enzyme. An allosteric enzyme is an enzyme with two or more protein chains and two kinds of binding sites (for substrate and regulator) (Section 10.8).

Zymogen. A zymogen is an inactive precursor of a proteolytic enzyme; the zymogen is activated by a chemical reaction that removes or adds to part of its structure (Section 10.9).

Vitamins. A vitamin is an organic compound necessary in small amounts for the normal growth of humans and some animals. Vitamins must be obtained from dietary sources because they cannot be synthesized in the body (Section 10.12).

Water-soluble vitamins. Vitamin C and the eight B vitamins are the water-soluble vitamins. Vitamin C is essential for the proper formation of bones and teeth and is also an important antioxidant. All eight B vitamins function as coenzymes (Section 10.13).

Fat-soluble vitamins. The four fat-soluble vitamins are vitamins A, D, E, and K. The best-known function of vitamin A is its role in vision. Vitamin D is essential for the proper use of calcium and phosphorus to form bones and teeth. The primary function of vitamin E is as an antioxidant. Vitamin K is essential in the regulation of blood clotting (Section 10.14).

KEY REACTIONS AND EQUATIONS

1. Conversion of an apoenzyme to an active enzyme (Section 10.3)

   Apoenzyme + cofactor $\longrightarrow$ holoenzyme (active enzyme)

2. Mechanism of enzyme action (Section 10.4)

   Enzyme + substrate $\longrightarrow$ enzyme–substrate complex
   Enzyme–substrate complex $\longrightarrow$ enzyme + product

3. Enzyme inhibition (Section 10.7)

   Enzyme + inhibitor $\longrightarrow$ inactive enzyme

4. Conversion of a zymogen to an active enzyme (Section 10.9)

   Zymogen $\xrightarrow{\text{Enzyme}}$ active enzyme + peptide fragment

EXERCISES AND PROBLEMS

The members of each pair of problems in this section test similar material.

 Importance of Enzymes (Section 10.1)

10.1 What is the general role of enzymes in the human body?

10.2 Why does the body need so many different enzymes?

10.3 List two ways in which enzymes differ from inorganic laboratory catalysts.

10.4 Occasionally we refer to the "delicate" nature of enzymes. Explain why this adjective is appropriate.

 Enzyme Nomenclature (Section 10.2)

10.5 Which of the following substances are enzymes?
   a. Sucrase
   b. Galactose
   c. Trypsin
   d. Xylulose reductase

10.6 Which of the following substances are enzymes?
   a. Sucrose
   b. Pepsin
   c. Glutamine synthetase
   d. Cellulase

10.7 Predict the function of each of the following enzymes.
   a. Pyruvate carboxylase
   b. Alcohol dehydrogenase
   c. L-Amino acid reductase
   d. Maltase

10.8 Predict the function of each of the following enzymes.
   a. Cytochrome oxidase
   b. *Cis–trans* isomerase
   c. Succinate dehydrogenase
   d. Lactase

10.9 Suggest a name for an enzyme that catalyzes each of the following reactions.
   a. Hydrolysis of sucrose
   b. Decarboxylation of pyruvate
   c. Isomerization of glucose
   d. Removal of hydrogen from lactate

10.10 Suggest a name for an enzyme that catalyzes each of the following reactions.
   a. Hydrolysis of lactose
   b. Oxidation of nitrite
   c. Decarboxylation of citrate
   d. Reduction of oxalate

10.11 Give the name of the substrate on which each of the following enzymes acts.
   a. Pyruvate carboxylase
   b. Galactase
   c. Alcohol dehydrogenase
   d. L-Amino acid reductase

10.12 Give the name of the substrate on which each of the following enzymes acts.
   a. Cytochrome oxidase
   b. Lactase
   c. Succinate dehydrogenase
   d. Tyrosine kinase

10.13 To which of the six major classes of enzymes does each of the following belong?
   a. Mutase
   b. Dehydratase
   c. Carboxylase
   d. Kinase

10.14 To which of the six major classes of enzymes does each of the following belong?
   a. Protease
   b. Racemase
   c. Dehydrogenase
   d. Synthetase

10.15 To which of the six major classes of enzymes does the enzyme that catalyzes each of the following reactions belong?
   a. A *cis* double bond is converted to a *trans* double bond.
   b. An alcohol is dehydrated to form a compound with a double bond.
   c. An amino group is transferred from one substrate to another.
   d. An ester linkage is hydrolyzed.

10.16 To which of the six major classes of enzymes does the enzyme that catalyzes each of the following reactions belong?
   a. An L isomer is converted to a D isomer.
   b. A phosphate group is transferred from one substrate to another.
   c. An amide linkage is hydrolyzed.
   d. Hydrolysis of a carbohydrate to monosaccharides occurs.

10.17 Identify the enzyme needed in each of the following reactions as an isomerase, a decarboxylase, a dehydrogenase, a lipase, or a phosphatase.

a.
$$CH_3-\overset{O}{\overset{\|}{C}}-COOH \rightarrow CH_3-\overset{O}{\overset{\|}{C}}-H + CO_2$$

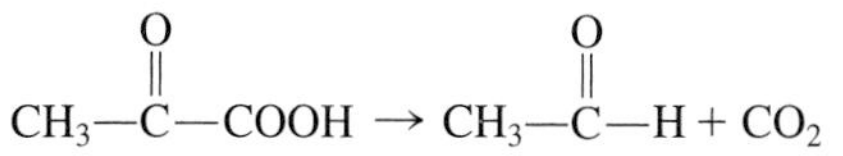

b.
$$\begin{array}{l} CH_2-O-\overset{O}{\overset{\|}{C}}-R \\ | \\ CH-O-\overset{O}{\overset{\|}{C}}-R \\ | \\ CH_2-O-\overset{O}{\overset{\|}{C}}-R \end{array} + 3H_2O \rightarrow \begin{array}{l} CH_2-OH \\ | \\ CH-OH \\ | \\ CH_2-OH \end{array} + 3R-COOH$$

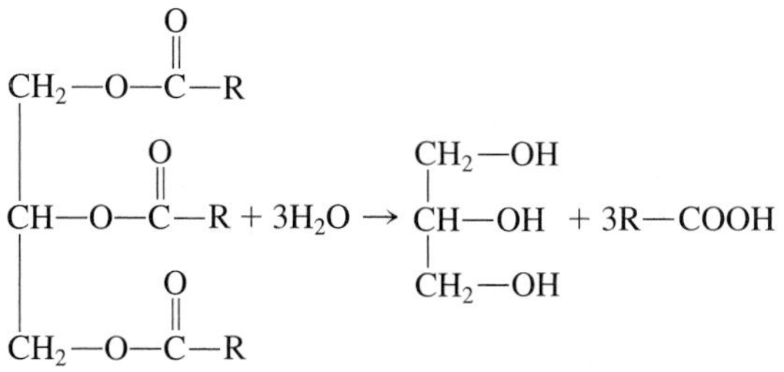

c.
$$\overset{+}{H_3N}-CH-COO^- + H_2O \rightarrow$$
$$\begin{array}{l} | \\ CH_2 \\ | \\ OPO_3^{2-} \end{array}$$
$$\overset{+}{H_3N}-CH-COO^- + HPO_4^{2-}$$
$$\begin{array}{l} | \\ CH_2 \\ | \\ OH \end{array}$$

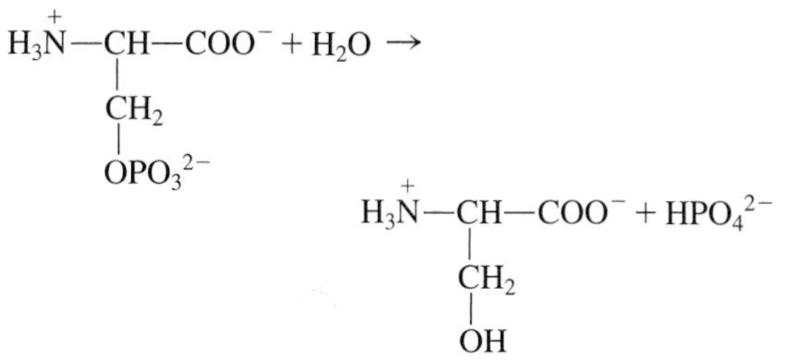

d.
$$\begin{array}{l} OH \\ | \\ CH_3-CH-COOH + NAD^+ \rightarrow \end{array}$$
$$CH_3-\overset{O}{\overset{\|}{C}}-COOH + NADH + H^+$$

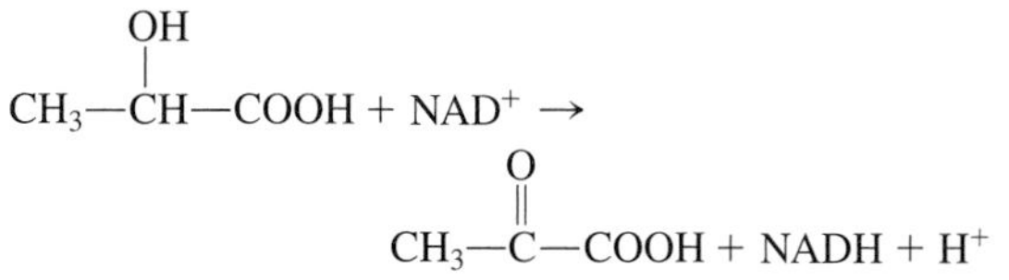

10.18 Identify the enzyme needed in each of the following reactions as an isomerase, a decarboxylase, a dehydrogenase, a protease, or a phosphatase.

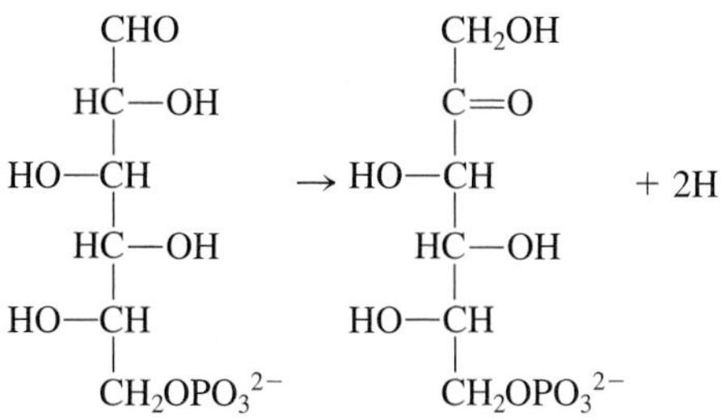

a.
$$CH_3-\overset{O}{\overset{\|}{C}}-\overset{O}{\overset{\|}{C}}-OH \rightarrow CH_3-\overset{O}{\overset{\|}{C}}-H + CO_2$$

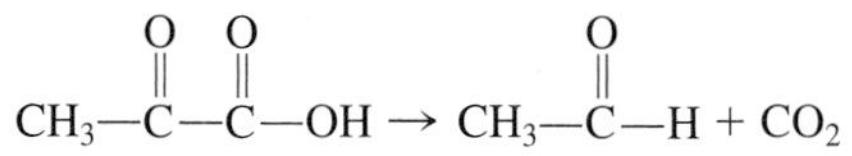

b.
$$\begin{array}{l} CHO \\ | \\ HC-OH \\ | \\ HO-CH \\ | \\ HC-OH \\ | \\ HO-CH \\ | \\ CH_2OPO_3^{2-} \end{array} \rightarrow \begin{array}{l} CH_2OH \\ | \\ C=O \\ | \\ HO-CH \\ | \\ HC-OH \\ | \\ HO-CH \\ | \\ CH_2OPO_3^{2-} \end{array} + 2H$$

c.
$$HO-\overset{O}{\overset{\|}{C}}-CH_2-CH_2-\overset{O}{\overset{\|}{C}}-OH \rightarrow$$
$$HO-\overset{O}{\overset{\|}{C}}-CH=CH-\overset{O}{\overset{\|}{C}}-OH + 2H$$

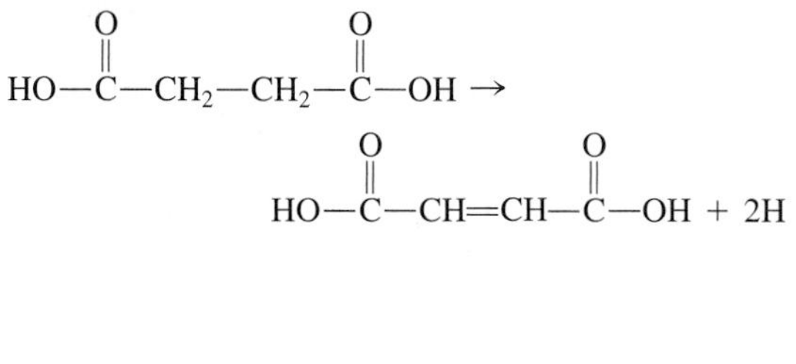

d.
$$\overset{+}{H_3N}-CH-\overset{O}{\overset{\|}{C}}-NH-CH-COO^- + H_2O \rightarrow$$
$$\begin{array}{cc} | & | \\ CH_3 & CH_3 \end{array}$$
$$2\ \overset{+}{H_3N}-CH-COO^-$$
$$\begin{array}{c} | \\ CH_3 \end{array}$$

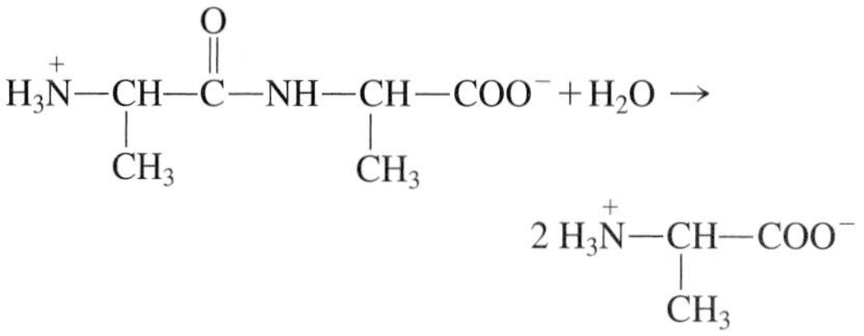

■ Enzyme Structure (Section 10.3)

10.19 Indicate whether each of the following phrases describes a simple or a conjugated enzyme.
   a. An enzyme that has both a protein and a nonprotein portion
   b. An enzyme that requires Mg^{2+} ion for activity
   c. An enzyme in which only amino acids are present
   d. An enzyme in which a cofactor is present

10.20 Indicate whether each of the following phrases describes a simple or a conjugated enzyme.
   a. An enzyme that contains a carbohydrate portion
   b. An enzyme that contains only protein
   c. A holoenzyme
   d. An enzyme that has a vitamin as part of its structure

10.21 What is the difference between a cofactor and a coenzyme?

10.22 All coenzymes are cofactors, but not all cofactors are coenzymes. Explain this statement.

10.23 Why are cofactors present in most enzymes?

10.24 What is the difference between an apoenzyme and a holoenzyme?

■ Models of Enzyme Action (Section 10.4)

10.25 What is an enzyme active site?

10.26 What is an enzyme–substrate complex?

10.27 How does the lock-and-key model of enzyme action explain the highly specific way some enzymes select a substrate?

10.28 How does the induced-fit model of enzyme action explain the broad specificities of some enzymes?

10.29 What types of forces hold a substrate at an enzyme active site?

10.30 The forces that hold a substrate at an enzyme active site are not covalent bonds. Explain why not.

■ Enzyme Specificity (Section 10.5)

10.31 Define the following terms dealing with enzyme specificity.
   a. Absolute specificity
   b. Linkage specificity

10.32 Define the following terms dealing with enzyme specificity.
   a. Group specificity
   b. Stereochemical specificity

10.33 Which type(s) of enzyme specificity are best accounted for by the lock-and-key model of enzyme action?

10.34 Which type(s) of enzyme specificity are best accounted for by the induced-fit model of enzyme action?

10.35 Which enzyme in each of the following pairs would be more limited in its catalytic scope?
   a. An enzyme that exhibits absolute specificity or an enzyme that exhibits group specificity
   b. An enzyme that exhibits stereochemical specificity or an enzyme that exhibits linkage specificity

10.36 Which enzyme in each of the following pairs would be more limited in its catalytic scope?

a. An enzyme that exhibits linkage specificity or an enzyme that exhibits absolute specificity

b. An enzyme that exhibits group specificity or an enzyme that exhibits stereochemical specificity

■ Factors That Affect Enzyme Activity (Section 10.6)

10.37 Temperature affects enzymatic reaction rates in two ways. An increase in temperature can accelerate the rate of a reaction or it can stop the reaction. Explain each of these effects.

10.38 Define the optimum temperature for an enzyme.

10.39 Explain why all enzymes do not possess the same optimum pH.

10.40 Why does an enzyme lose activity when the pH is drastically changed from the optimum pH?

10.41 Draw a graph that shows the effect of increasing substrate concentration on the rate of an enzyme-catalyzed reaction (at constant temperature, pH, and enzyme concentration).

10.42 Draw a graph that shows the effect of increasing enzyme concentration on the rate of an enzyme-catalyzed reaction (at constant temperature, pH, and substrate concentration).

10.43 In an enzyme-catalyzed reaction, all of the enzyme active sites are saturated by substrate molecules at a certain substrate concentration. What happens to the rate of the reaction when the substrate concentration is doubled?

10.44 What is an enzyme turnover number?

■ Enzyme Inhibition (Section 10.7)

10.45 In competitive inhibition, can both the inhibitor and the substrate bind to an enzyme at the same time? Explain your answer.

10.46 Compare the sites where competitive and noncompetitive inhibitors bind to enzymes.

10.47 Indicate whether each of the following statements describes a reversible competitive inhibitor, a reversible noncompetitive inhibitor, or an irreversible inhibitor. More than one answer may apply.
a. Both inhibitor and substrate bind at the active site on a random basis.
b. The inhibitor effect cannot be reversed by the addition of more substrate.
c. Inhibitor structure does not have to resemble substrate structure.
d. The inhibitor can bind to the enzyme at the same time as substrate.

10.48 Indicate whether each of the following statements describes a reversible competitive inhibitor, a reversible noncompetitive inhibitor, or an irreversible inhibitor. More than one answer may apply.
a. It bonds covalently to the enzyme active site.
b. The inhibitor effect can be reversed by the addition of more substrate.
c. Inhibitor structure must be somewhat similar to that of substrate.
d. The inhibitor cannot bind to the enzyme at the same time as substrate.

■ Regulation of Enzyme Activity (Sections 10.8 and 10.9)

10.49 What is an allosteric enzyme?

10.50 What is a regulator molecule?

10.51 What is feedback control?

10.52 What is the difference between positive and negative feedback to an allosteric enzyme?

10.53 What is the general relationship between zymogens and proteolytic enzymes?

10.54 What, if any, is the difference in meaning between the terms *zymogen* and *proenzyme*?

10.55 Why are proteolytic enzymes always produced in an inactive form?

10.56 What is the mechanism by which most zymogens are activated?

■ Antibiotics That Inhibit Enzyme Activity (Section 10.10)

10.57 By what mechanism do sulfa drugs kill bacteria?

10.58 By what mechanism do penicillins kill bacteria?

10.59 Why is penicillin toxic to bacteria but not to higher organisms?

10.60 What amino acid in transpeptidase forms a covalent bond to penicillin?

10.61 What situation has made Cipro a prominent antibiotic?

10.62 Describe the structure of Cipro in terms of common and "unusual" functional groups present.

■ Vitamins (Sections 10.12 – 10.14)

10.63 What is a vitamin?

10.64 List a way in which vitamins differ from carbohydrates, fats, and proteins (the major classes of food).

10.65 Indicate whether each of the following is a fat-soluble or a water-soluble vitamin.
a. Vitamin K   b. Vitamin B_{12}
c. Vitamin C   d. Thiamin

10.66 Indicate whether each of the following is a fat-soluble or a water-soluble vitamin.
a. Vitamin A   b. Vitamin B_6
c. Vitamin E   d. Riboflavin

10.67 Indicate whether each of the vitamins in Problem 10.65 would be likely or unlikely to be toxic when consumed in excess.

10.68 Indicate whether each of the vitamins in Problem 10.66 would be likely or unlikely to be toxic when consumed in excess.

10.69 Describe the two most completely characterized roles of vitamin C in the body.

10.70 Structurally, how do the oxidized and reduced forms of vitamin C differ?

10.71 What is the dominant function within the human body of the B vitamins as a group?

10.72 With the help of Table 10.6, identify the B vitamin or vitamins to which each of the following characterizations applies.
a. Is part of the coenzymes NAD and NADP
b. Is part of coenzyme A
c. Is part of THF
d. Is part of TPP

10.73 With the help of Table 10.6, indicate whether each of the following B vitamins exists in more than one structural form.
a. Folate   b. Niacin   c. Vitamin B_6 d. Biotin

10.74 With the help of Table 10.6, identify the B vitamin or vitamins to which each of the following characterizations applies.
a. Contains a heterocyclic nitrogen ring system
b. Has the most complex structure
c. Contains a metal atom as part of its structure
d. Contains sulfur as part of its structure

10.75 Describe the structural differences among the three retinoic forms of vitamin A.

10.76 What is the relationship between the plant pigment beta-carotene and vitamin A?

10.77 What is *cell differentiation* and how does vitamin A participate in this process?

10.78 List four major functions for vitamin A in the human body.

10.79 How do vitamin D_2 and vitamin D_3 differ in structure?

10.80 In terms of source, how do vitamin D_2 and vitamin D_3 differ?

10.81 What is the principal function of vitamin D in the human body?

10.82 Why is vitamin D often called the sunshine vitamin?

10.83 Which form of tocopherol (vitamin E) exhibits the greatest biochemical activity?

10.84 How do the various forms of tocopherol differ in structure?

10.85 What is the principal function of vitamin E in the human body?

10.86 Why is vitamin E often given to premature infants that are on oxygen therapy?

10.87 How do vitamin K_1 and vitamin K_2 differ in structure?

10.88 In terms of source, how do vitamin K_1 and vitamin K_2 differ?

10.89 How are *menaquinones, phylloquinones,* and vitamin K related?

10.90 What is the principal function of vitamin K in the human body?

 ADDITIONAL PROBLEMS

10.91 Explain the difference, if any, between the following types of enzymes.
  a. Apoenzyme and proenzyme
  b. Simple enzyme and allosteric enzyme
  c. Coenzyme and isoenzyme
  d. Conjugated enzyme and holoenzyme

10.92 Identify the functional groups present in a molecule of each of the following vitamins.
  a. Vitamin C   b. Vitamin A (retinol)
  c. Vitamin D   d. Vitamin K

10.93 Indicate whether each of the following vitamins functions as a coenzyme.
  a. Vitamin C   b. Vitamin A
  c. Vitamin D   d. Niacin
  e. Riboflavin   f. Biotin

10.94 Which vitamin has each of the following functions?
  a. Water-soluble antioxidant
  b. Fat-soluble antioxidant
  c. Involved in the process of calcium deposition in bone
  d. Involved in the blood-clotting process
  e. Involved in cell differentiation
  f. Involved in vision
  g. Involved in collagen formation
  h. Involved in prostaglandin formation

10.95 What general kinds of reactions do the following types of enzymes catalyze?
  a. Oxidoreductases   b. Lyases
  c. Isomerases   d. Ligases
  e. Hydrolases   f. Transferases

10.96 Explain what is meant by the equation
$$E + S \rightleftharpoons ES \rightarrow E + P$$
given that ES stands for enzyme–substrate complex.

10.97 Alcohol dehydrogenase catalyzes the conversion of ethanol to acetaldehyde. This enzyme, in its active state, consists of a protein molecule and a zinc ion. On the basis of this information, identify the following for this chemical system.
  a. Substrate   b. Cofactor
  c. Apoenzyme   d. Holoenzyme

10.98 Each of the following is an abbreviation for an enzyme used in the diagnosis and/or treatment of heart attacks. What does each abbreviation stand for?
  a. TPA   b. LDH
  c. CPK   d. AST

 MULTIPLE-CHOICE PRACTICE TEST

10.99 Which are the two most common endings for the name of an enzyme?
  a. -ase and -ose   b. -ase and -in
  c. -in and -ogen   d. -in and -ine

10.100 Which of the following pairings of enzyme type and enzyme function is *incorrect*?
  a. Kinase and transfer of a phosphate group between substrates
  b. Mutase and introduction of a double bond within a molecule
  c. Protease and hydrolysis of amide linkages in proteins
  d. Decarboxylase and removal of CO_2 from a substrate

10.101 Which of the following is true for a conjugated enzyme?
  a. It contains only protein.
  b. It does not contain protein.
  c. It has a nonprotein part.
  d. It always contains a metal ion.

10.102 Which of the following statements concerning *cofactors* is *incorrect*?
  a. All conjugated enzymes contain cofactors.
  b. Some cofactors are metal ions.
  c. A cofactor is the nonprotein portion of an enzyme.
  d. Vitamins cannot be cofactors.

10.103 What happens to substrate molecules at an enzyme active site?
   a. They always react with O_2.
   b. They become covalently bonded to the enzyme.
   c. They become catalysts.
   d. They undergo change to a desired product.

10.104 Which of the following statements about a reversible noncompetitive inhibitor is *correct*?
   a. It prevents substrate from occupying the enzyme active site.
   b. It must resemble the substrate in general shape.
   c. It and the substrate can simultaneously occupy the active site.
   d. It binds to the enzyme at a location other than the active site.

10.105 What is the shape of a plot of enzyme activity (*y*-axis) versus temperature (*x*-axis) with other variables constant?
   a. Straight line with an upward slope
   b. Line with an upward slope and a long flat top
   c. Line with an upward slope followed by a downward slope
   d. Straight horizontal line

10.106 Which of the following statements concerning the B vitamins is *correct*?
   a. Structurally, they are all very similar.
   b. All except two of them are water soluble.
   c. In chemically modified form they serve as cofactors in enzymes.
   d. Fruits, in general, are very good sources of these vitamins.

10.107 Cholesterol is a precursor for which of the following vitamins?
   a. Vitamin A   b. Vitamin C
   c. Vitamin D   d. Vitamin E

10.108 Beta-carotene is a precursor for which of the following vitamins?
   a. Vitamin A   b. Vitamin D
   c. Vitamin E   d. Vitamin K

11 Nucleic Acids

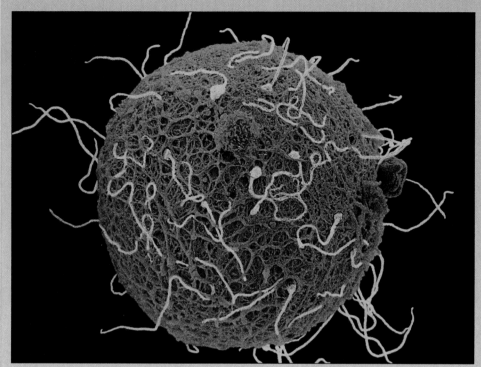

Human egg and sperm.

A most remarkable property of living cells is their ability to produce exact replicas of themselves. Furthermore, cells contain all the instructions needed for making the complete organism of which they are a part. The molecules within a cell that are responsible for these amazing capabilities are nucleic acids.

The Swiss physiologist Friedrich Miescher (1844–1895) discovered nucleic acids in 1869 while studying the nuclei of white blood cells. The fact that they were initially found in cell nuclei and are acidic accounts for the name *nucleic acid*. Although we now know that nucleic acids are found throughout a cell, not just in the nucleus, the name is still used for such materials.

11.1 Types of Nucleic Acids

Two types of nucleic acids are found within cells of higher organisms: *deoxyribonucleic acid* (DNA) and *ribonucleic acid* (RNA). Nearly all the DNA is found within the cell nucleus. Its primary function is the storage and transfer of genetic information. This information is used (indirectly) to control many functions of a living cell. In addition, DNA is passed from existing cells to new cells during cell division. RNA occurs in all parts of a cell. It functions primarily in synthesis of proteins, the molecules that carry out essential cellular functions. The structural distinctions between DNA and RNA molecules are considered in Section 11.3.

All nucleic acid molecules are polymers. A **nucleic acid** *is a polymer in which the monomer units are nucleotides*. Thus the starting point for a discussion of nucleic acids is an understanding of the structures and chemical properties of nucleotides.

It was not until 1944, 75 years after the discovery of nucleic acids, that scientists obtained the first evidence that these molecules are responsible for the storage and transfer of genetic information.

11.2 Nucleotides: Building Blocks of Nucleic Acids

Proteins are polypeptides, many carbohydrates are polysaccharides, and nucleic acids are polynucleotides.

A **nucleotide** *is a three-subunit molecule in which a pentose sugar is bonded to both a phosphate group and a nitrogen-containing heterocyclic base.* With a three-subunit structure, nucleotides are more complex monomers than the monosaccharides of polysaccharides (Section 7.8) and the amino acids of proteins (Section 9.2). A block structural diagram for a nucleotide is

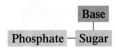

Pentose Sugars

The sugar unit of a nucleotide is either the pentose *ribose* or the pentose *2-deoxyribose.*

The systems for numbering the atoms in the pentose and nitrogen-containing base subunits of a nucleotide are important and will be used extensively in later sections of this chapter. The convention is that

1. Pentose ring atoms are designated with *primed* numbers.
2. Nitrogen-containing base ring atoms are designated with *unprimed* numbers.

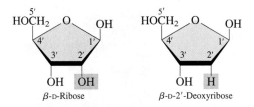

Structurally, the only difference between these two sugars occurs at carbon 2′. The —OH group present on this carbon in ribose becomes a —H atom in 2-deoxyribose. (The prefix *deoxy-* means "without oxygen.")

RNA and DNA differ in the identity of the sugar unit in their nucleotides. In RNA the sugar unit is *ribose*—hence the *R* in RNA. In DNA the sugar unit is *2-d*eoxyribose— hence the *D* in DNA.

Nitrogen-Containing Heterocyclic Bases

Five nitrogen-containing heterocyclic bases are nucleotide components. Three of them are derivatives of pyrimidine (Section 6.9), a monocyclic base with a six-membered ring, and two are derivatives of purine (Section 6.9), a bicyclic base with fused five- and six-membered rings.

A pyrimidine derivative that we have encountered previously is the B vitamin thiamin (see Section 10.13).

Caffeine, the most widely used nonprescription central nervous system stimulant, is the 3,7-dimethyl-2,6-dioxo derivative of purine (Section 6.9).

Pyrimidine   Purine

Both of these heterocyclic compounds are bases because they contain amine functional groups (secondary or tertiary), and amine functional groups exhibit basic behavior (proton acceptors; Section 6.6).

The three pyrimidine derivatives found in nucleotides are thymine (T), cytosine (C), and uracil (U).

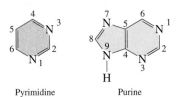

Thymine (T)   Cytosine (C)   Uracil (U)

FIGURE 11.1 Space-filling model of the molecule adenine, a nitrogen-containing heterocyclic base present in both DNA and RNA.

Thymine is the 5-methyl-2,4-dioxo derivative, cytosine the 4-amino-2-oxo derivative, and uracil the 2,4-dioxo derivative of pyrimidine.

The two purine derivatives found in nucleotides are adenine (A) and guanine (G).

Adenine (A)   Guanine (G)

Adenine is the 6-amino derivative of purine, and guanine is the 2-amino-6-oxo purine derivative. A space-filling model for adenine is shown in Figure 11.1.

Adenine, guanine, and cytosine are found in both DNA and RNA. Uracil is found only in RNA, and thymine usually occurs only in DNA. Figure 11.2 summarizes the occurrences of nitrogen-containing heterocyclic bases in nucleic acids.

Phosphate

Phosphate, the third component of a nucleotide, is derived from phosphoric acid (H_3PO_4). Under cellular pH conditions, the phosphoric acid loses two of its hydrogen atoms to give a hydrogen phosphate ion (HPO_4^{2-}).

Phosphoric acid   Hydrogen phosphate ion

Nucleotide Formation

The formation of a nucleotide from sugar, base, and phosphate can be visualized as occurring in the following manner:

Phosphate   Sugar   Nucleotide

Important characteristics of this combining of three molecules into one molecule (the nucleotide) are that

1. Condensation, with formation of a water molecule, occurs at two locations: between sugar and base and between sugar and phosphate.
2. The base is always attached at the C-1′ position of the sugar. For purine bases, attachment is through N-9; for pyrimidine bases, N-1 is involved. The C-1′ carbon atom of the ribose unit is always in a β configuration (Section 7.10), and the bond connecting the sugar and base is a β-N-glycosidic linkage (Section 7.13).

FIGURE 11.2 Two purine bases and three pyrimidine bases are found in the nucleotides present in nucleic acids.

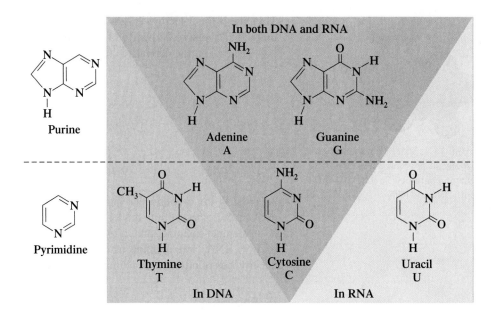

To remember which two of the five nucleotide bases are the purine derivatives (fused rings), use the phrase "pure silver" and the chemical symbol for silver, which is Ag.

pure Ag
purine A and G

3. The phosphate group is attached to the sugar at the C-5′ position through a phosphate–ester linkage.

There are four possible RNA nucleotides, differing in the base present (A, C, G, or U), and four possible DNA nucleotides, differing in the base present (A, C, G, or T).

■ Nucleotide Nomenclature

The common names and abbreviations for the eight nucleotides of DNA and RNA molecules are given in Table 11.1. It is important to be familiar with them because they are frequently encountered in biochemistry.

We can make several generalizations about the nomenclature given in Table 11.1.

1. All of the names end in 5′-monophosphate, which signifies the presence of a phosphate group attached to the 5′ carbon atom of ribose or deoxyribose. (In Chapter 12 we will encounter nucleotides that contain two or three phosphate groups—diphosphates and triphosphates.)
2. Preceding the monophosphate ending is the name of the base present in a modified form. The suffix -*osine* is used with purine bases, the suffix -*idine* with pyrimidine bases.
3. The prefix *deoxy-* at the start of the name signifies that the sugar present is deoxyribose. When no prefix is present, the sugar is ribose.

TABLE 11.1
The Names of the Eight Nucleotides Found in DNA and RNA

Base	Sugar	Nucleotide Name	Nucleotide Abbreviation
DNA Nucleotides			
adenine	deoxyribose	deoxyadenosine 5′-monophosphate	dAMP
guanine	deoxyribose	deoxyguanosine 5′-monophosphate	dGMP
cytosine	deoxyribose	deoxycytidine 5′-monophosphate	dCMP
thymine	deoxyribose	deoxythymidine 5′-monophosphate	dTMP
RNA Nucleotides			
adenine	ribose	adenosine 5′-monophosphate	AMP
guanine	ribose	guanosine 5′-monophosphate	GMP
cytosine	ribose	cytidine 5′-monophosphate	CMP
uracil	ribose	uridine 5′-monophosphate	UMP

CHEMICAL
CONNECTIONS

Use of Synthetic Nucleic Acid Bases in Medicine

Many hundreds of modified nucleic acid bases have been prepared in laboratories and their effects on nucleic acid synthesis investigated. Several of them are now in clinical use as drugs for controlling, at the cellular level, cancers and other related disorders.

The theory behind the use of these modified bases involves their masquerading as legitimate nucleic acid building blocks. The enzymes associated with the DNA replication process (Section 11.5) incorporate the modified bases into growing nucleic acid chains. The presence of these "pseudonucleotides" in the chain stops further growth of the chain, thus interfering with nucleic acid synthesis.

Examples of drugs now in use include 5-fluorouracil, which is employed against a variety of cancers, especially those of the breast and digestive tract, and 6-mercaptopurine, which is used in the treatment of leukemia.

6-Mercaptopurine
(a modified adenine)   and   Adenine

The rapidly dividing cells that are characteristic of cancer require large quantities of DNA. Anticancer drugs based on modified nucleic acid bases block DNA synthesis and therefore block the increase in the number of cancer cells. Cancer cells are generally affected to a greater extent than normal cells because of this rapid growth. Eventually, the normal cells are affected to such a degree that use of the drugs must be discontinued. 5-Fluorouracil inhibits the formation of thymine-containing nucleotides required for DNA synthesis. 6-Mercaptopurine, which substitutes for adenine, inhibits the synthesis of nucleotides that incorporate adenine and guanine.

5-Fluorouracil
(a modified thymine)   and   Thymine

4. The abbreviations in Table 11.1 for the nucleotides come from the one-letter symbols for the bases (A, C, G, T, and U), the use of MP for monophosphate, and a lower-case *d* at the start of the abbreviation whenever deoxyribose is the sugar.

The use of synthetic nucleic acid bases in medicine is considered in the Chemical Connections feature on this page.

11.3 Primary Nucleic Acid Structure

> Nucleotides are related to nucleic acids in the same way that amino acids are related to proteins.

Nucleic acids are polymers in which the repeating units, the monomers, are nucleotides (Section 11.2). The nucleotide units within a nucleic acid molecule are linked to each other through sugar–phosphate bonds. The resulting molecular structure (Figure 11.3) involves a chain of alternating sugar and phosphate groups with a base group protruding from the chain at regular intervals.

We can now define, in terms of structure, the two major types of nucleic acids: ribonucleic acids and deoxyribonucleic acids (Section 11.1). A **ribonucleic acid (RNA)** *is a nucleotide polymer in which each of the monomers contains ribose, a phosphate group, and one of the heterocyclic bases adenine, cytosine, guanine, or uracil.* Two changes to

FIGURE 11.3 The general structure of a nucleic acid in terms of nucleotide subunits.

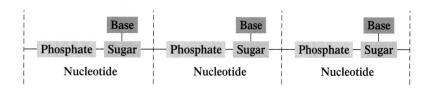

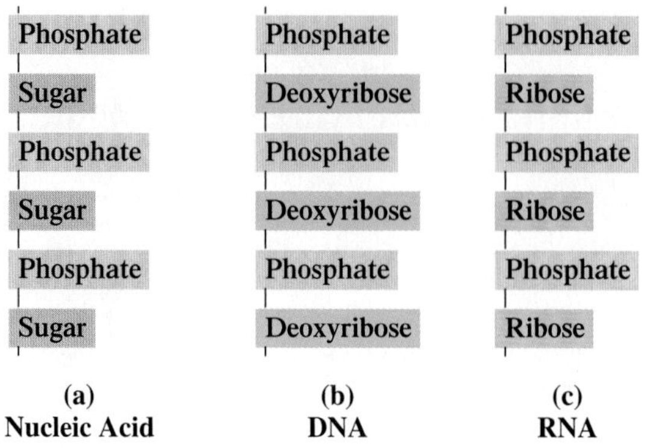

FIGURE 11.4 (a) The generalized backbone structure of a nucleic acid. (b) The specific backbone structure for a deoxyribonucleic acid (DNA). (c) The specific backbone structure for a ribonucleic acid (RNA).

Phosphate	Phosphate	Phosphate
Sugar	Deoxyribose	Ribose
Phosphate	Phosphate	Phosphate
Sugar	Deoxyribose	Ribose
Phosphate	Phosphate	Phosphate
Sugar	Deoxyribose	Ribose

(a)	(b)	(c)
Nucleic Acid	**DNA**	**RNA**

this definition generate the deoxyribonucleic acid definition; deoxyribose replaces ribose and thymine replaces uracil. A **deoxyribonucleic acid** *is a nucleotide polymer in which each of the monomers contains deoxyribose, a phosphate group, and one of the heterocyclic bases adenine, cytosine, guanine, or thymine.*

The alternating sugar–phosphate chain in a nucleic acid structure is often called the *nucleic acid backbone.* This backbone is constant throughout the entire nucleic acid structure. For DNA molecules, the backbone consists of alternating phosphate and *deoxyribose* sugar units; for RNA molecules, the backbone consists of alternating phosphate and *ribose* sugar units. Figure 11.4 contrasts the generalized backbone structure for a nucleic acid with the specific backbone structures of DNAs and RNAs.

The variable portion of nucleic acid structure is the sequence of bases attached to the sugar units of the backbone. The sequence of these base side chains distinguishes various DNAs from each other and various RNAs from each other. Only four types of bases are found in any given nucleic acid structure. This situation is much simpler than that for proteins, where 20 side-chain entities (amino acids) are available (Section 9.2). In both RNA and DNA, adenine, guanine, and cytosine are encountered as side-chain components; thymine is found mainly in DNA, and uracil is found only in RNA (Figure 11.2).

Primary nucleic acid structure *is the order in which nucleotides are linked together in a nucleic acid.* Because the sugar–phosphate backbone of a given nucleic acid does not vary, the primary structure of the nucleic acid depends only on the sequence of bases present. Further information about nucleic acid structure can be obtained by considering the detailed four-nucleotide segment of a DNA molecule shown in Figure 11.5.

The following list describes some important points about nucleic acid structure that are illustrated in Figure 11.5.

1. Each nonterminal phosphate group of the sugar–phosphate backbone is bonded to two sugar molecules through a *3′,5′-phosphodiester linkage.* There is a phosphoester bond to the 5′ carbon of one sugar unit and a phosphoester bond to the 3′ carbon of the other sugar.
2. A nucleotide chain has *directionality.* One end of the nucleotide chain, the *5′ end,* normally carries a free phosphate group attached to the 5′ carbon atom. The other end of the nucleotide chain, the *3′ end,* normally has a free hydroxyl group attached to the 3′ carbon atom. By convention, the sequence of bases of a nucleic acid strand is read from the 5′ end to the 3′ end.
3. Each nonterminal phosphate group in the backbone of a nucleic acid carries a -1 charge. The parent phosphoric acid molecule from which the phosphate was derived originally had three —OH groups (Section 11.2). Two of these become involved in the 3′,5′-phosphodiester linkage. The remaining —OH group is free to exhibit acidic behavior—that is, to produce a H^+ ion.

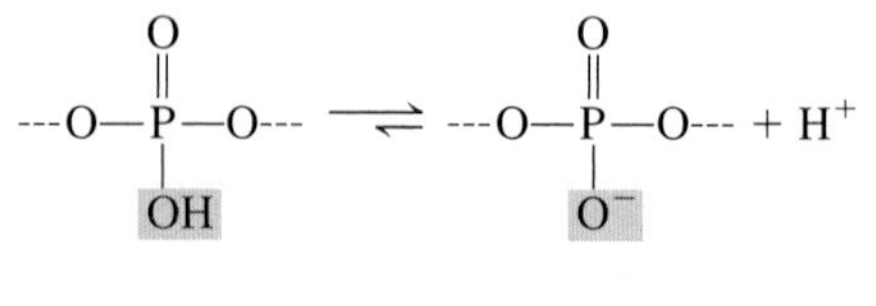

The backbone of a nucleic acid structure is always an alternating sequence of phosphate and sugar groups. The sugar is ribose in RNA and deoxyribose in DNA.

Just as the order of amino acid side chains determines the primary structure of a protein (Section 9.9), the order of nucleotide bases determines the primary structure of a nucleic acid.

For both nucleic acids and proteins, a distinction is made between the two ends of the polymer chain. For nucleic acids there is a 5′ end and a 3′ end; for proteins there is an N-terminal end and a C-terminal end (Section 9.6).

$$\text{---O---}\overset{\displaystyle\overset{O}{\|}}{\underset{\underset{\displaystyle OH}{|}}{P}}\text{---O---} \;\rightleftharpoons\; \text{---O---}\overset{\displaystyle\overset{O}{\|}}{\underset{\underset{\displaystyle O^-}{|}}{P}}\text{---O---} + H^+$$

FIGURE 11.5 A four-nucleotide-long segment of DNA. (The choice of bases was arbitrary.)

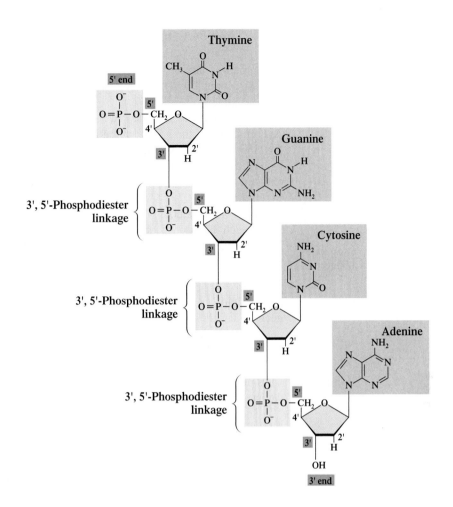

This behavior by the many phosphate groups in a nucleic acid backbone gives nucleic acids their acidic properties.

Three parallels between primary nucleic acid structure and primary protein structure (Section 9.9) are worth noting.

1. Both nucleic acids and proteins have backbones that do not vary in structure (see Figure 11.6).
2. The differences among various nucleic acids and among various proteins are related to the order in which groups are attached to the backbones (nitrogen bases in nucleic acids and amino acid R groups in proteins).
3. Both nucleic acid polymer chains and protein polymer chains have directionality; for nucleic acids there is a 5′ end and a 3′ end, and for proteins there is an N-terminal end and a C-terminal end.

FIGURE 11.6 A comparison of the general primary structures of nucleic acids and proteins.

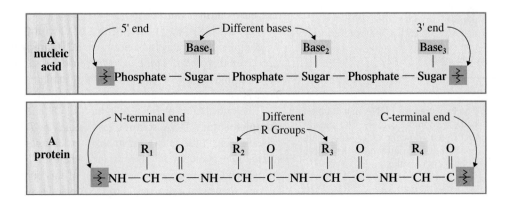

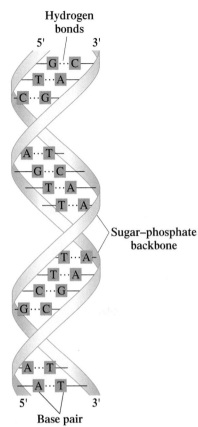

Hydrogen bonds

Sugar–phosphate backbone

Base pair

FIGURE 11.7 A schematic drawing of the DNA double helix that emphasizes the hydrogen bonding between bases on the two chains.

The α-helix secondary structure of proteins involves *one* polypeptide chain; the double-helix secondary structure of DNA involves *two* polynucleotide chains. In the α helix of proteins, the R groups are on the *outside* of the helix; in the double helix of DNA, the bases are on the *inside* of the double helix.

The *antiparallel* nature of the two polynucleotide chains in the DNA double helix means that there is a 5′ end and a 3′ end at both ends of the double helix.

The two strands of DNA in a double helix are complementary. This means that if you know the order of bases in one strand, you can predict the order of bases in the other strand.

11.4 The DNA Double Helix

Like proteins, nucleic acids have secondary, or three-dimensional, structure as well as primary structure. The secondary structures of DNAs and RNAs differ, and we will discuss them separately.

The amounts of the bases A, T, G, and C present in DNA molecules were the key to determination of the general three-dimensional structure of DNA molecules. Base composition data for DNA molecules from many different organisms revealed a definite pattern of base occurrence. The amounts of A and T were always equal, and the amounts of C and G were always equal, as were the amounts of total purines and total pyrimidines.

The relative amounts of these base pairs in DNA vary depending on the life form from which the DNA is obtained. (Each animal or plant has a unique base composition.) However, the relationships

$$\%A = \%T \qquad \text{and} \qquad \%C = \%G$$

always hold true. For example, human DNA contains 30% adenine, 30% thymine, 20% guanine, and 20% cytosine.

In 1953, an explanation for the base composition patterns associated with DNA molecules was proposed by the American microbiologist James Watson and the English biophysicist Francis Crick. Their model, which has now been validated in numerous ways, involves a double-helix structure that accounts for the equality of bases present, as well as for other known DNA structural data.

The DNA double helix involves two polynucleotide strands coiled around each other in a manner somewhat like a spiral staircase. The sugar–phosphate backbones of the two polynucleotide strands can be thought of as being the outside banisters of the spiral staircase (see Figure 11.7). The bases (side chains) of each backbone extend inward toward the bases of the other strand. The two strands are connected by *hydrogen bonds* between their bases. Additionally, the two strands of the double helix are *antiparallel*—that is, they run in opposite directions. One strand runs in the 5′-to-3′ direction, and the other is oriented in the 3′-to-5′ direction.

■ Base Pairing

A physical restriction, the size of the interior of the DNA double helix, limits the base pairs that can hydrogen-bond to one another. Only pairs involving one small base (a pyrimidine) and one large base (a purine) correctly "fit" within the helix interior. There is not enough room for two large purine bases to fit opposite each other (they overlap), and two small pyrimidine bases are too far apart to hydrogen-bond to one another effectively. Of the four possible purine–pyrimidine combinations (A–T, A–C, G–T, and G–C), hydrogen-bonding possibilities are *most favorable* for the A–T and G–C pairings, and these two combinations are the *only two* that normally occur in DNA. Figure 11.8 shows the specific hydrogen-bonding interactions for the four possible purine–pyrimidine base-pairing combinations.

The pairing of A with T and that of G with C are said to be *complementary*. A and T are complementary bases, as are G and C. **Complementary bases** *are pairs of bases in a nucleic acid structure that can hydrogen-bond to each other*. The fact that complementary base pairing occurs in DNA molecules explains, very simply, why the amounts of the bases A and T present are always equal, as are the amounts of G and C.

The two strands of DNA in a double helix are *not identical*—they are complementary. **Complementary DNA strands** *are strands of DNA in a double helix with base pairing such that each base is located opposite its complementary base*. Wherever G occurs in one strand, there is a C in the other strand; wherever T occurs in one strand, there is an A in the other strand. An important ramification of this complementary relationship is that knowing the base sequence of one strand of DNA enables us to predict the base sequence of the complementary strand.

FIGURE 11.8 Hydrogen-bonding possibilities are more favorable when A–T and G–C base pairing occurs than when A–C and G–T base pairing occurs. (a) Two and three hydrogen bonds can form, respectively, between A–T and G–C base pairs. These combinations are present in DNA molecules. (b) Only one hydrogen bond can form between G–T and A–C base pairs. These combinations are not present in DNA molecules.

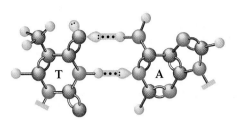
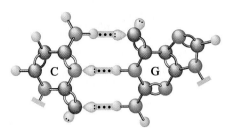

Thymine–Adenine Base Pairing
(two hydrogen bonds form)

Cytosine–Guanine Base Pairing
(three hydrogen bonds form)

(a)

A mnemonic device for recalling base-pairing combinations in DNA involves listing the base abbreviations in alphabetical order. Then the first and last bases pair, and so do the middle two bases.

DNA: A C G T

Another way to remember these base-pairing combinations is to note that AT spells a word and that C and G look very much alike.

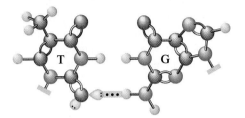
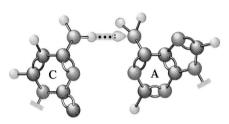

Thymine–Guanine Base Pairing
(only one hydrogen bond forms)

Cytosine–Adenine Base Pairing
(only one hydrogen bond forms)

(b)

● Carbon   ● Oxygen   Lone pair   ••• Hydrogen bond

● Nitrogen   ● Hydrogen   Attachment to backbone

In specifying the base sequence of a segment of a strand of DNA (or RNA), we list the bases in sequential order (using their one-letter abbreviations) in the direction from the 5′ end to the 3′ end of the segment.

5′ A–A–G–C–T–A–G–C–T–T–A–C–T 3′

EXAMPLE 11.1

Predicting Base Sequence in a Complementary DNA Strand

■ Predict the sequence of bases in the DNA strand that is complementary to the single DNA strand shown.

5′ C–G–A–A–T–C–C–T–A 3′

Solution

Because only A forms a complementary base pair with T, and only G with C, the complementary strand is as follows:

Given: 5′ C–G–A–A–T–C–C–T–A 3′
Complementary strand: 3′ G–C–T–T–A–G–G–A–T 5′

Note the reversal of the numbering of the ends of the complementary strand compared to the given strand. This is due to the antiparallel nature of the two strands in a DNA double helix.

Practice Exercise 11.1

Predict the sequence of bases in the DNA strand complementary to the single DNA strand shown.

5′ A–A–T–G–C–A–G–C–T 3′

Hydrogen bonding between base pairs is an important factor in stabilizing the DNA double helix structure. Although hydrogen bonds are relatively weak forces, each DNA

Hydrogen bonding is responsible for the secondary structure (double helix) of DNA. Hydrogen bonding is also responsible for secondary structure in proteins (Section 9.10).

molecule has so many base pairs that collectively these hydrogen bonds are a force of significant strength. In addition to hydrogen bonding, base-stacking interactions also contribute to DNA double-helix stabilization.

Base-Stacking Interactions

The bases in a DNA double helix are positioned with the planes of their rings parallel (like a stack of coins). Stacking interactions involving a given base and the parallel bases directly above it and below it also contribute to the stabilization of the DNA double helix. These stacking interactions are as important in their stabilization effects as is the hydrogen bonding associated with base pairing—perhaps even more important. Purine and pyrimidine bases are hydrophobic in nature, so their stacking interactions are those associated with hydrophobic molecules—mainly London forces. The concept of hydrophobic interactions has been encountered twice previously. Hydrophobic interactions involving the nonpolar tails of membrane lipids contribute to the structural stability of cell membranes (Section 8.10), and hydrophobic interactions involving nonpolar R groups of amino acids contribute to protein tertiary structure stability (Section 9.11).

11.5 Replication of DNA Molecules

DNA molecules are the carriers of genetic information within a cell; that is, they are the molecules of heredity. Each time a cell divides, an exact copy of the DNA of the parent cell is needed for the new daughter cell. The process by which new DNA molecules are generated is DNA replication. **DNA replication** *is the biochemical process by which DNA molecules produce exact duplicates of themselves.* The key concept in understanding DNA replication is the base pairing associated with the DNA double helix.

DNA Replication Overview

To understand DNA replication, we must regard the two strands of the DNA double helix as a pair of *templates,* or patterns. During replication, the strands separate. Each can then act as a template for the synthesis of a new, complementary strand. The result is two daughter DNA molecules with base sequences identical to those of the parent double helix. Let us consider details of this replication.

Under the influence of the enzyme *DNA helicase,* the DNA double helix unwinds, and the hydrogen bonds between complementary bases are broken. This unwinding process, as shown in Figure 11.9, is somewhat like opening a zipper.

The bases of the separated strands are no longer connected by hydrogen bonds. They can pair with *free* individual nucleotides present in the cell's nucleus. As shown in

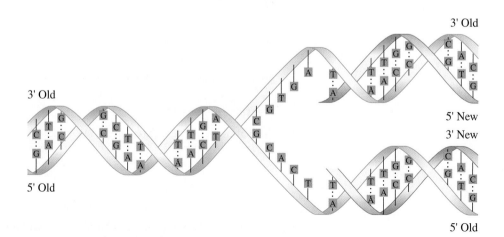

FIGURE 11.9 In DNA replication, the two strands of the DNA double helix unwind, the separated strands serving as templates for the formation of new DNA strands. Free nucleotides pair with the complementary bases on the separated strands of DNA. This process ultimately results in the complete replication of the DNA molecule.

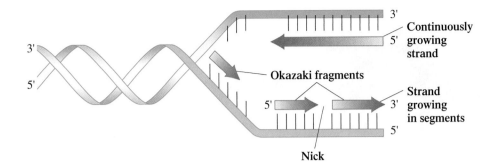

FIGURE 11.10 Because the enzyme *DNA polymerase* can act only in the 5'-to-3' direction, one strand (top) grows continuously in the direction of the unwinding, and the other strand grows in segments in the opposite direction. The segments in this latter chain are then connected by a different enzyme, *DNA ligase*.

Figure 11.9, the base pairing always involves C pairing with G and A pairing with T. The pairing process occurs one nucleotide at a time. After a free nucleotide has formed hydrogen bonds with a base of the old strand (the template), the enzyme *DNA polymerase* verifies that the base pairing is correct and then catalyzes the formation of a new phosphodiester linkage between the nucleotide and the growing stand (represented by the darker blue ribbons in Figure 11.9). The *DNA polymerase* then slides down the strand to the next unpaired base of the template, and the same process is repeated.

Each of the two daughter molecules of double-stranded DNA formed in the DNA replication process contains one strand from the original parent molecule and one newly formed strand.

■ The Replication Process in Finer Detail

Though simple in principle, the DNA replication process has many intricacies.

1. The enzyme *DNA polymerase* can operate on a forming DNA daughter strand only in the 5'-to-3' direction. Because the two strands of parent DNA run in opposite directions (one is 5' to 3' and the other 3' to 5'; Section 11.4), only one strand can grow continuously in the 5'-to-3' direction. The other strand must be formed in short segments, called *Okazaki fragments* (after their discoverer, Reiji Okazaki), as the DNA unwinds (see Figure 11.10). The breaks or gaps in this daughter strand are called *nicks*. To complete the formation of this strand, the Okazaki fragments are connected by action of the enzyme *DNA ligase*.

2. The process of DNA unwinding does not have to begin at an end of the DNA molecule. It may occur at any location within the molecule. Indeed, studies show that unwinding usually occurs at several interior locations simultaneously and that DNA replication is bidirectional for these locations; that is, it proceeds in both directions from the unwinding sites. As shown in Figure 11.11, the result of this multiple-site

FIGURE 11.11 DNA replication usually occurs at multiple sites within a molecule, and the replication is bidirectional from these sites.

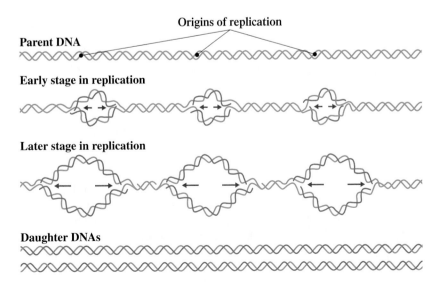

FIGURE 11.12 Identical twins share identical physical characteristics because they received identical DNA from their parents.

Chromosomes are *nucleoproteins*. They are a combination of nucleic acid (DNA) and various proteins.

replication process is formation of "bubbles" of newly synthesized DNA. The bubbles grow larger and eventually coalesce, giving rise to two complete daughter DNAs. Multiple-site replication enables large DNA molecules to be replicated rapidly.

■ Chromosomes

Once the DNA within a cell has been replicated, it interacts with specific proteins in the cell called *histones* to form structural units that provide the most stable arrangement for the long DNA molecules. These histone–DNA complexes are called *chromosomes*. A **chromosome** *is an individual DNA molecule bound to a group of proteins*. Typically, a chromosome is about 15% by mass DNA and 85% by mass protein.

Cells from different kinds of organisms have different numbers of chromosomes. A normal human has 46 chromosomes per cell, a mosquito 6, a frog 26, a dog 78, and a turkey 82.

Chromosomes occur in matched (*homologous*) pairs. The 46 chromosomes of a human cell constitute 23 homologous pairs. One member of each homologous pair is derived from a chromosome inherited from the father, and the other is a copy of one of the chromosomes inherited from the mother. Homologous chromosomes have similar, but not identical, DNA base sequences; both code for the same traits but for different forms of the trait (for example, blue eyes versus brown eyes). Offspring are like their parents, but they are different as well; part of their DNA came from one parent and part from the other parent. Occasionally, identical twins are born (see Figure 11.12). Such twins have received identical DNA from their parents.

The Chemistry at a Glance feature on page 393 summarizes the steps in DNA replication.

11.6 Overview of Protein Synthesis

We saw in the previous section how the replication of DNA makes it possible for a new cell to contain the same genetic information as its parent cell. We will now consider how the genetic information contained in a cell is expressed in cell operation. This brings us to the topic of protein synthesis. The synthesis of proteins (skin, hair, enzymes, hormones, and so on) is under the direction of DNA molecules. It is this role of DNA that establishes the similarities between parent and offspring that we regard as hereditary characteristics.

We can divide the overall process of protein synthesis into two steps. The first step is called transcription and the second translation. The following diagram summarizes the relationship between transcription and translation.

$$\boxed{\text{DNA}} \xrightarrow{\text{Transcription}} \boxed{\text{RNA}} \xrightarrow{\text{Translation}} \boxed{\text{protein}}$$

Before discussing the details of transcription and translation, we need to learn more about RNA molecules. We will be particularly concerned with differences between RNA and DNA and among various types of RNA molecules.

11.7 Ribonucleic Acids

Four major differences exist between RNA molecules and DNA molecules.

The bases thymine (T) and uracil (U) have similar structures. Thymine is a methyluracil (Section 11.2). The hydrogen-bonding patterns (Figure 11.7) for the A–U base pair (RNA) and the A–T base pair (DNA) are identical.

1. The sugar unit in the backbone of RNA is ribose; it is deoxyribose in DNA.
2. The base thymine found in DNA is replaced by uracil in RNA (Figure 11.2). Uracil, instead of thymine, pairs with (forms hydrogen bonds with) adenine in RNA.
3. RNA is a single-stranded molecule; DNA is double-stranded (double helix). Thus RNA, unlike DNA, does not contain equal amounts of specific bases.
4. RNA molecules are much smaller than DNA molecules, ranging from 75 nucleotides to a few thousand nucleotides.

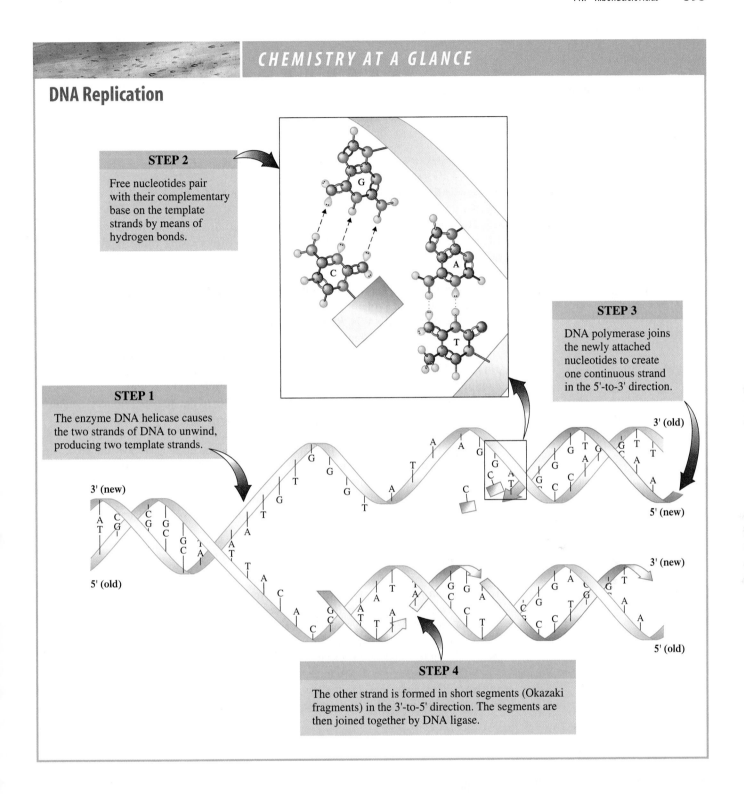

CHEMISTRY AT A GLANCE

DNA Replication

STEP 2

Free nucleotides pair with their complementary base on the template strands by means of hydrogen bonds.

STEP 3

DNA polymerase joins the newly attached nucleotides to create one continuous strand in the 5'-to-3' direction.

STEP 1

The enzyme DNA helicase causes the two strands of DNA to unwind, producing two template strands.

3' (new)

5' (old)

3' (old)

5' (new)

3' (new)

5' (old)

STEP 4

The other strand is formed in short segments (Okazaki fragments) in the 3'-to-5' direction. The segments are then joined together by DNA ligase.

We should note that the single-stranded nature of RNA does not prevent *portions* of an RNA molecule from folding back upon itself and forming double-helical regions. If the base sequences along two portions of an RNA strand are complementary, a structure with a hairpin loop results, as shown in Figure 11.13. The amount of double-helical structure present in an RNA varies with RNA type, but a value of 50% is not atypical.

Heterogeneous nuclear RNA (hnRNA) also goes by the name *primary transcript RNA* (ptRNA).

Types of RNA Molecules

RNA molecules found in human cells are categorized into five major types, distinguished by their function. These five RNA types are heterogeneous nuclear RNA (hnRNA),

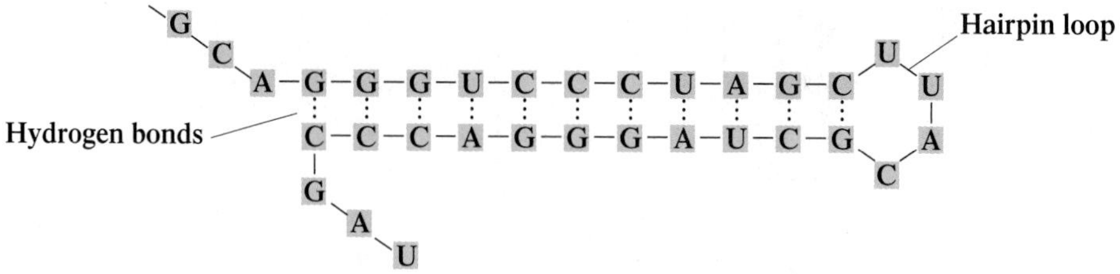

FIGURE 11.13 A hairpin loop is produced when single-stranded RNA doubles back on itself and complementary base pairing occurs.

Hydrogen bonds

Hairpin loop

messenger RNA (mRNA), small nuclear RNA (snRNA), ribosomal RNA (rRNA), and transfer RNA (tRNA).

Heterogeneous nuclear RNA *is RNA formed directly by DNA transcription from which messenger RNA is formed.* Post-transcription processing converts the hnRNA to mRNA.

Messenger RNA *is RNA that carries instructions for protein synthesis (genetic information) from DNA to the sites for protein synthesis.* The molecular mass of mRNA varies with the length of the protein whose synthesis it will direct.

Small nuclear RNA *is RNA that facilitates the conversion of hnRNA to mRNA.* It contains from 100 to 200 nucleotides.

Ribosomal RNA *is RNA that combines with specific proteins to form ribosomes, the physical sites for protein synthesis.* Ribosomes have molecular masses on the order of 3 million. The rRNA present in ribosomes has no informational function.

Transfer RNA *is RNA that delivers amino acids to the sites for protein synthesis.* Transfer RNAs are the smallest of the RNAs, possessing only 75–90 nucleotide units.

At a nondetail level, a cell consists of a nucleus and an extranuclear region called the cytoplasm. The process of DNA transcription occurs in the nucleus, as does the processing of hnRNA to mRNA. [DNA replication (Section 11.5) also occurs in the nucleus.] The mRNA formed in the nucleus travels to the cytoplasm where translation (protein synthesis) occurs. Figure 11.14 summarizes the transcription and translation processes in terms of the types of RNA involved and the cellular locations where the processes occur.

> The most abundant type of RNA in a cell is ribosomal RNA (75% to 80% by mass). Transfer RNA constitutes 10%–15% of cellular RNA; messenger RNA and its precursor, heterogeneous nuclear RNA, make up the 5%–10% of RNA material in the cell.

> A detailed look at cellular structure is found in Section 12.2.

11.8 Transcription: RNA Synthesis

Transcription *is the process by which DNA directs the synthesis of mRNA molecules that carry the coded information needed for protein synthesis.* Messenger RNA production via transcription is actually a "two-step" process in which an hnRNA molecule is initially produced and then is "edited" to yield the desired mRNA molecule. The mRNA molecule so produced then functions as the carrier of the information needed to direct protein synthesis.

Within a strand of a DNA molecule are instructions for the synthesis of numerous hnRNA/mRNA molecules. During transcription, a DNA molecule unwinds, under enzyme influence, at the particular location where the appropriate base sequence is found for the

FIGURE 11.14 An overview of types of RNA in terms of cellular locations where they are encountered and processes in which they are involved.

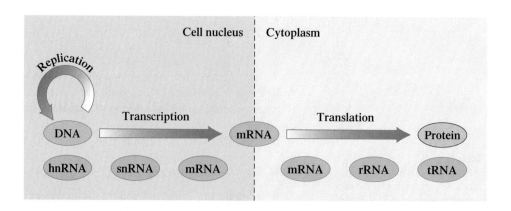

hnRNA/mRNA of concern, and the "exposed" base sequence is transcribed. A short segment of a DNA strand so transcribed, which contains instructions for the formation of a particular hnRNA/mRNA, is called a *gene*. A **gene** *is a segment of a DNA strand that contains the base sequence for the production of a specific hnRNA/mRNA molecule.*

In humans, most genes are composed of 1000–3500 nucleotide units. Hundreds of genes can exist along a DNA strand. Obtaining information concerning the total number of genes and the total number of nucleotide base pairs present in human DNA has been an area of intense research activity for the last two decades. The central activity in this research has been the *Human Genome Project,* a decade-long internationally-based research project to determine the location and base sequence of each of the genes in the human *genome.* A **genome** *is all of the genetic material (the total DNA) contained in the chromosomes of an organism.*

Before the Human Genome Project began, current biochemical thought predicted the presence of about 100,000 genes in the human genome. Initial results of the human genome project, announced in 2001, paired this number down to 30,000–40,000 genes and also indicated that the base pairs present in these genes constitute only a very small percentage (2%) of the 2.9 billion base pairs present in the chromosomes of the human genome. In 2004, based on reanalysis of human genome project information, the human gene count was pared down further to 20,000–25,000 genes. (Later in this section, the significance and ramifications of this dramatic decrease in estimates of the human gene count are considered.)

■ Steps in the Transcription Process

The mechanics of transcription are in many ways similar to those of DNA replication. Four steps are involved.

1. A *portion* of the DNA double helix unwinds, exposing some bases (a gene). The unwinding process is governed by the enzyme *RNA polymerase* rather than by *DNA helicase* (replication enzyme).
2. Free *ribo*nucleotides, one nucleotide at a time, align along *one* of the exposed strands of DNA bases, the *template* strand, forming new base pairs. In this process, U rather than T aligns with A in the base-pairing process. Because ribonucleotides rather than deoxyribonucleotides are involved in the base pairing, ribose, rather than deoxyribose, becomes incorporated into the new nucleic acid backbone.
3. *RNA polymerase* is involved in the linkage of ribonucleotides, one by one, to the growing RNA molecule.
4. Transcription ends when the *RNA polymerase* enzyme encounters a sequence of bases that is "read" as a stop signal. The newly formed RNA molecule and the *RNA polymerase* enzyme are released, and the DNA then rewinds to re-form the original double helix.

Figure 11.15 shows the overall process of transcription of DNA to form RNA.

In DNA–RNA base pairing, the complementary base pairs are

DNA   RNA
A — U
G — C
C — G
T — A

RNA molecules contain the base U instead of the base T.

EXAMPLE 11.2

Base Pairing Associated with the Transcription Process

■ From the base sequence 5′ A–T–G–C–C–A 3′ in a DNA template strand, determine the base sequence in the RNA synthesized from the DNA template strand.

Solution

An RNA molecule cannot contain the base T. The base U is present instead. Therefore, U–A base pairing will occur instead of T–A base pairing. The other base-pairing combination, G–C, remains the same. The RNA product of the transcription process will therefore be

DNA template: 5′ A–T–G–C–C–A 3′

⋮ ⋮ ⋮ ⋮ ⋮ ⋮

RNA molecule: 3′ U–A–C–G–G–U 5′

(continued)

FIGURE 11.15 The transcription of DNA to form RNA involves an unwinding of a portion of the DNA double helix. Only one strand of the DNA is copied during transcription.

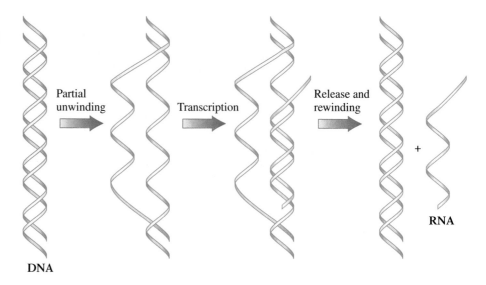

Note that the direction of the RNA strand is antiparallel to that of the DNA template. This will always be the case during transcription.

It is standard procedure, when writing and reading base sequences for nucleic acids (both DNAs and RNAs), always to specify base sequence in the $5' \longrightarrow 3'$ direction unless otherwise directed. Thus

$3'$ U–A–C–G–G–U $5'$ becomes $5'$ U–G–G–C–A–U $3'$

Practice Exercise 11.2

From the base sequence $5'$ T–A–A–C–C–T $3'$ in a DNA template strand, determine the base sequence in the RNA synthesized from the DNA template strand.

Post-Transcription Processing: Formation of mRNA

The RNA produced from a gene through transcription is hnRNA, the precursor for mRNA. The conversion of hnRNA to mRNA involves *post-transcription processing* of the hnRNA. In this processing, certain portions of the hnRNA are deleted and the retained parts are then spliced together. This process leads us to the concepts of *exons* and *introns*.

It is now known that not all bases in a gene convey genetic information. Instead, a gene is *segmented;* it has portions called *exons* that contain genetic information and portions called *introns* that do not convey genetic information.

An **exon** *is a gene segment that conveys (codes for) genetic information.* *Ex*ons are DNA segments that help *ex*press a genetic message. An **intron** *is a gene segment that does not convey (code for) genetic information.* *Int*rons are DNA segments that *int*errupt a genetic message. A gene consists of alternating exon and intron segments (Figure 11.16).

FIGURE 11.16 Heterogeneous nuclear RNA contains both exons and introns. Messenger RNA is heterogeneous nuclear RNA from which the introns have been excised.

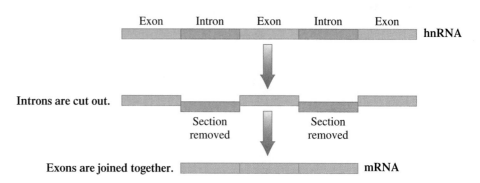

TABLE 11.2

The Universal Genetic Code

The code is composed of 64 three-nucleotide sequences (codons), which can be read from the table. The left-hand column indicates the nucleotide base found in the first (5′) position of the codon. The nucleotides in the second (middle) position of the codon are in the middle columns. The right-hand column indicates the nucleotide found in the third (3′) position. Thus the codon ACG encodes for the amino acid Thr, and the codon GGG encodes for the amino acid Gly.

First Position (5′ end)	Second Position				Third Position (3′ end)
	U	C	A	G	
U	Phe	Ser	Tyr	Cys	U
	Phe	Ser	Tyr	Cys	C
	Leu	Ser	Stop	Stop	A
	Leu	Ser	Stop	Trp	G
C	Leu	Pro	His	Arg	U
	Leu	Pro	His	Arg	C
	Leu	Pro	Gln	Arg	A
	Leu	Pro	Gln	Arg	G
A	Ile	Thr	Asn	Ser	U
	Ile	Thr	Asn	Ser	C
	Ile	Thr	Lys	Arg	A
	Met	Thr	Lys	Arg	G
G	Val	Ala	Asp	Gly	U
	Val	Ala	Asp	Gly	C
	Val	Ala	Glu	Gly	A
	Val	Ala	Glu	Gly	G

2. *There is a pattern to the arrangement of synonyms in the genetic code table.* All synonyms for an amino acid fall within a single box in Table 11.2, unless there are more than four synonyms, where two boxes are needed. The significance of the "single box" pattern is that with synonyms, the first two bases of the codon are the same—they differ only in the third base. For example, the four synonyms for the amino acid Pro are CCU, CCC, CCA, and CCG.

3. *The genetic code is almost universal.* Although Table 11.2 does not show this feature, studies of many organisms indicate that with minor exceptions, the code is the same in all of them. The same codon specifies the same amino acid whether the cell is a bacterial cell, a corn plant cell, or a human cell.

4. *An initiation codon exists.* The existence of "stop" codons (UAG, UAA, and UGA) suggests the existence of "start" codons. There is one initiation codon. Besides coding for the amino acid methionine, the codon AUG functions as an initiator of protein synthesis when it occurs as the first codon in an amino acid sequence.

EXAMPLE 11.3

Using the Genetic Code and mRNA Codons to Predict Amino Acid Sequences

■ Using the genetic code in Table 11.2, determine the sequence of amino acids encoded by the mRNA codon sequence

5′ GCC–AUG–GUA–AAA–UGC–GAC–CCA 3′

Solution

Matching the codons with the amino acids, using Table 11.2, yields

mRNA: 5′ GCC–AUG–GUA–AAA–UGC–GAC–CCA 3′
Peptide: Ala   Met   Val   Lys   Cys   Asp   Pro

Practice Exercise 11.3

Using the genetic code in Table 11.2, determine the sequence of amino acids encoded by the mRNA codon sequence

5′ CAU–CCU–CAC–ACU–GUU–UGU–UGG 3′

EXAMPLE 11.4

Relating Protein Amino Acid
Sequence to Base Sequence on
a DNA Template Strand

Introns and exons are actually never as
short as those given in this simplified
example.

■ Sections A, C, and E of the following base sequence section of a DNA template strand
are exons, and sections B and D are introns.

DNA   5′ ATT – CGT – TGT – TTT – CCC – AGT – GCC 3′
            A   B   C   D   E

a. What is the structure of the hnRNA transcribed from this template?
b. What is the structure of the mRNA obtained by splicing the hnRNA?
c. What polypeptide amino acid sequence will be synthesized using the mRNA?

Solution

a. The base sequence in the hnRNA will be complementary to that of the template DNA,
except that U is used in the RNA instead of T. The hnRNA will have a directionality
antiparallel to that of the DNA sequence.

hnRNA   3′ UAA – GCA – ACA – AAA – GGG – UCA – CGG 5′

b. In the splicing process, introns are removed and the exons combined to give the mRNA.

mRNA   3′ UAA – ACA – AAA – CGG 5′

c. The codons in an mRNA must be read in the 5′-to-3′ direction to use the genetic code
correctly to determine the sequence of amino acids in the peptide. Rewriting the mRNA
structure in the 5′-to-3′ direction gives

mRNA   5′ GGC – AAA – ACA – AAU 3′

The Universal Genetic Code (Table 11.2), reveals that this mRNA sequence codes for the
amino acid sequence

Gly–Lys–Thr–Ile

Practice Exercise 11.4

Sections A, C, and E of the following base sequence section of a DNA template strand are
exons, and sections B and D are introns.

DNA   5′ CGC – CGT – AGT – TGG – CCC – GGA – GGA 3′
            A   B   C   D   E

a. What is the structure of the hnRNA transcribed from this template?
b. What is the structure of the mRNA obtained by splicing the hnRNA?
c. What polypeptide amino acid sequence will be synthesized using the mRNA?

11.10 Anticodons and tRNA Molecules

The amino acids used in protein synthesis do not directly interact with the codons of an
mRNA molecule. Instead, tRNA molecules function as intermediaries that deliver amino
acids to the mRNA. At least one type of tRNA molecule exists for each of the 20 amino
acids found in proteins.

All tRNA molecules have the same general shape, and this shape is crucial to how they
function. Figure 11.18a shows the general *two-dimensional* "cloverleaf" shape of a tRNA
molecule, a shape produced by the molecule's folding and twisting into regions of parallel
strands and regions of hairpin loops. (The actual three-dimensional shape of a tRNA molecule involves considerable additional twisting of the "cloverleaf" shape—Figure 11.18b.)

Two features of the tRNA structure are of particular importance.

1. The 3′ end of the open part of the cloverleaf structure is where an amino acid
   becomes *covalently* bonded to the tRNA molecule through an ester bond. Each
   of the different tRNA molecules is specifically recognized by an *aminoacyl*

FIGURE 11.18 A tRNA molecule. The amino acid attachment site is at the open end of the cloverleaf (the 3′ end), and the anticodon is located in the hairpin loop opposite the open end.

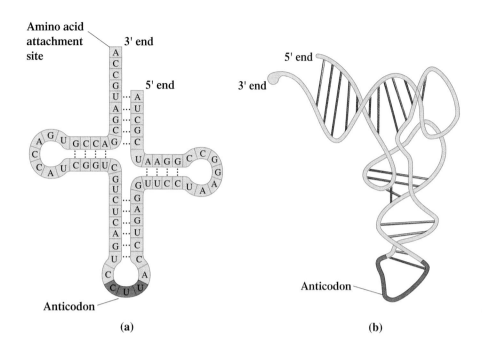

(a)

(b)

synthetase enzyme. These enzymes also recognize the one kind of amino acid that "belongs" with the particular tRNA and facilitates its bonding to the tRNA (see Figure 11.19).

2. The loop *opposite* the open end of the cloverleaf is the site for a sequence of three bases called an anticodon. An **anticodon** *is a three-nucleotide sequence on a tRNA molecule that is complementary to a codon on a mRNA molecule.*

The interaction between the anticodon of the tRNA and the codon of the mRNA leads to the proper placement of an amino acid into a growing peptide chain during protein synthesis. This interaction, which involves complementary base pairing, is shown in Figure 11.20.

11.11 Translation: Protein Synthesis

Translation *is the process by which mRNA codons are deciphered and a particular protein molecule is synthesized.*

The substances needed for the translation phase of protein synthesis are mRNA molecules, tRNA molecules, amino acids, ribosomes, and a number of different enzymes. A **ribosome** *is an rRNA–protein complex that serves as the site for the translation phase of protein synthesis.* Ribosomes have structures involving two subunits—a large subunit and a small subunit (see Figure 11.21). Each subunit is approximately 65% rRNA and 35%

FIGURE 11.19 An aminoacyl–tRNA synthetase has an active site for tRNA and a binding site for the particular amino acid that is to be attached to that tRNA.

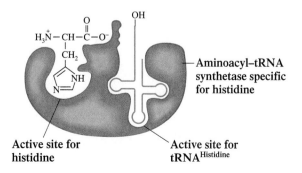

Active site for histidine

Aminoacyl–tRNA synthetase specific for histidine

Active site for tRNA^Histidine

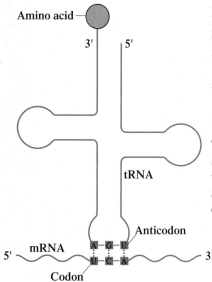

FIGURE 11.20 The interaction between anticodon (tRNA) and codon (mRNA), which involves complementary base pairing, governs the proper placement of amino acids in a protein.

protein. The rRNA provides binding sites for mRNA, and it provides the functional groups that promote all of the catalytic activity of the ribosome, including the peptide bond formation that occurs during protein synthesis. The ribosomal proteins help maintain the structure of the rRNA.

There are five general steps to the translation process: (1) activation of tRNA, (2) initiation, (3) elongation, (4) termination, and (5) post-translational processing.

■ Activation of tRNA

There are two steps involved in tRNA activation. First, an amino acid interacts with an activator molecule (ATP; Section 12.3) to form a highly energetic complex. This complex then reacts with the appropriate tRNA molecule to produce an *activated tRNA molecule,* a tRNA molecule that has an amino acid covalently bonded to it at its 3′ end through an ester linkage.

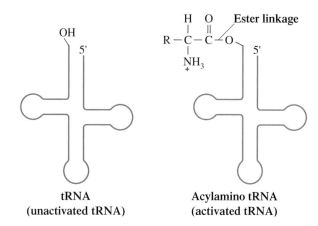

■ Initiation

The initiation of protein synthesis in human cells begins when mRNA attaches itself to the surface of a small ribosomal subunit such that its first codon, which is always the initiating codon AUG, occupies a site called the P site (peptidyl site.) (See Figure 11.22a.) An activated tRNA molecule with anticodon complementary to the codon AUG attaches itself, through complementary base pairing, to the AUG codon (Figure 11.22b). The resulting complex then interacts with a large ribosomal subunit to complete the formation of an initiation complex (Figure 11.22c). (Since the initiating codon AUG codes for the amino acid methionine, the first amino acid in a developing human protein chain will always be methionine.)

■ Elongation

Next to the P site in an mRNA–ribosome complex is a second binding site called the A site (aminoacyl site). (See Figure 11.23a.) At this second site the next mRNA codon is

FIGURE 11.21 Ribosomes, which contain both rRNA and protein, have structures that contain two subunits. One subunit is much larger than the other.

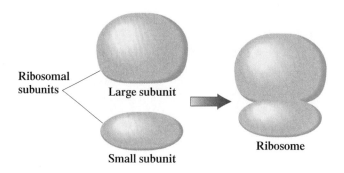

CHEMICAL CONNECTIONS   Antibiotics That Inhibit Bacterial Protein Synthesis

Some antibiotics work because they inhibit protein synthesis in bacteria but not in humans. They inhibit one specific enzyme or another in the bacterial ribosomes. These antibiotics are useful in treating disease and in studying protein synthesis mechanisms in bacteria. The accompanying table lists a few of the most commonly encountered antibiotics and their modes of action relative to protein synthesis.

Antibiotic	Biological action	Antibiotic	Biological action
chloramphenicol	inhibits an important enzyme (peptidyl transferase) in the large ribosomal subunit	streptomycin	inhibits initiation of protein synthesis and also causes the mRNA codons to be read incorrectly
erythromycin	binds to the large subunit and stops the ribosome from moving along the mRNA from one codon to the next	tetracycline	binds to the small ribosomal subunit and inhibits the binding of incoming tRNA molecules
puromycin	induces premature polypeptide chain termination		

exposed, and a tRNA with the appropriate anticodon binds to it (Figure 11.23b). With amino acids in place at both the P and the A sites, the enzyme *peptidyl transferase* effects the linking of the P site amino acid to the A site amino acid to form a dipeptide. Such peptide bond formation leaves the tRNA at the P site empty and the tRNA at the A site bearing the dipeptide (Figure 11.23c).

The empty tRNA at the P site now leaves that site and is free to pick up another molecule of its specific amino acid. Simultaneously with the release of tRNA from the P site, the ribosome shifts along the mRNA. This shift puts the newly formed dipeptide at the P site, and the third codon of mRNA is now available, at site A, to accept a tRNA molecule whose anticodon complements this codon (see Figure 11.23d). The movement of a ribosome along a mRNA molecule is called *translocation*. **Translocation** *is the part of translation in which a ribosome moves down a mRNA molecule three base positions (one codon) so that a new codon can occupy the ribosomal A site.*

Now a repetitious process begins. The third codon, now at the A site, accepts an incoming tRNA with its accompanying amino acid; and then the entire dipeptide at the P site is transferred and bonded to the A site amino acid to give a tripeptide (see Figure 11.23e). The empty tRNA at the P site is released, the ribosome shifts along the mRNA, and the process continues.

In elongation, the polypeptide chain grows one amino acid at a time.

FIGURE 11.22 Initiation of protein synthesis begins with the formation of an initiation complex.

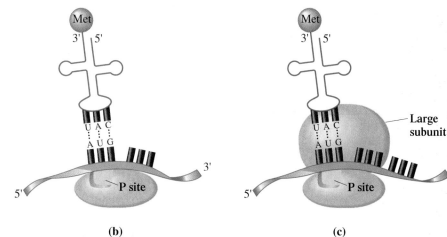

(a)   (b)   (c)

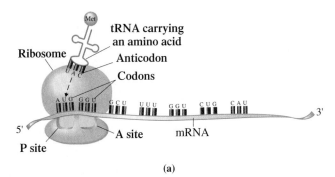

(a)

The initiation tRNA carrying the amino acid Met binds at the P site.

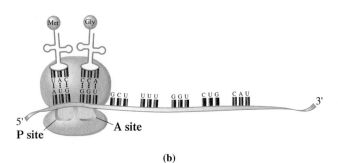

(b)

A tRNA with amino acid 2 binds at the A site.

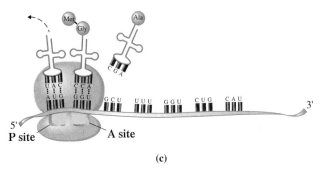

(c)

A peptide bond forms between amino acid 1 and amino acid 2 as amino acid 1 moves from the P site to the A site.

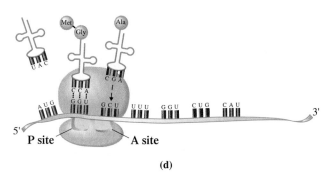

(d)

The first tRNA is released, the ribosome moves one codon to the right, translocating the dipeptide to the P site, and the tRNA with amino acid 3 occupies the A site.

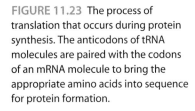

FIGURE 11.23 The process of translation that occurs during protein synthesis. The anticodons of tRNA molecules are paired with the codons of an mRNA molecule to bring the appropriate amino acids into sequence for protein formation.

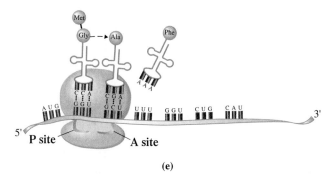

(e)

Elongation continues as the dipeptide at the P site is bonded to the amino acid at the A site to form a tripeptide.

■ Termination

The polypeptide continues to grow by way of translocation until all necessary amino acids are in place and bonded to each other. Appearance in the mRNA codon sequence of one of the three stop codons (UAA, UAG, or UGA) terminates the process. No tRNA has an anticodon that can base-pair with these stop codons. The polypeptide is then cleaved from the tRNA through hydrolysis.

■ Post-Translation Processing

Some modification of proteins usually occurs after translation. For example, most proteins do not have Met (the initiation codon) as their first amino acid. Cleavage of N-terminal Met is part of post-translation processing. Formation of S—S bonds between cysteine units is another example of post-translation processing.

FIGURE 11.24 Several ribosomes can simultaneously proceed along a single strand of mRNA one after another. Such a complex of mRNA and ribosomes is called a polysome.

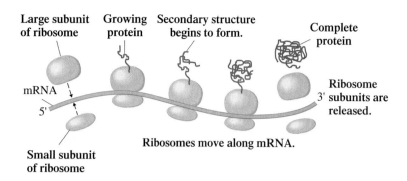

■ Efficiency of mRNA Utilization

Many ribosomes can move simultaneously along a single mRNA molecule (Figure 11.24). In this highly efficient arrangement, many identical protein chains can be synthesized almost at the same time from a single strand of mRNA. This multiple use of mRNA molecules reduces the amount of resources and energy that the cell expends to synthesize needed protein. Such complexes of several ribosomes and mRNA are called polyribosomes or polysomes. A **polysome** *is a complex of mRNA and several ribosomes.*

The Chemistry at a Glance feature on page 698 summarizes the steps in protein synthesis.

11.12 Mutations

A **mutation** *is an error in base sequence in a gene that is reproduced during DNA replication.* Such errors alter the genetic information that is passed on during transcription. The altered information can cause changes in amino acid sequence during protein synthesis. Sometimes, such changes have a profound effect on an organism.

A **mutagen** *is a substance or agent that causes a change in the structure of a gene.* Radiation and chemical agents are two important types of mutagens. Radiation, in the form of ultraviolet light, X rays, radioactivity, and cosmic rays, has the potential to be mutagenic. Ultraviolet light from the sun is the radiation that causes sunburn and can induce changes in the DNA of the skin cells. Sustained exposure to ultraviolet light can lead to serious problems such as skin cancer.

Chemical agents can also have mutagenic effects. Nitrous acid (HNO_2) is a mutagen that causes deamination of heterocyclic nitrogen bases. For example, HNO_2 can convert cytosine to uracil.

$$\text{Cytosine} \xrightarrow{\ HNO_2\ } \text{Uracil}$$

Deamination of a cytosine that was part of an mRNA codon would change the codon; for example, CGG would become UGG.

A variety of chemicals—including nitrites, nitrates, and nitrosamines—can form nitrous acid in the body. The use of nitrates and nitrites as preservatives in foods such as bologna and hot dogs is a cause of concern because of their conversion to nitrous acid in the body and possible damage to DNA.

Fortunately, the body has *repair enzymes* that recognize and replace altered bases. Normally, the vast majority of altered DNA bases are repaired, and mutations are avoided. Occasionally, however, the damage is not repaired, and the mutation persists.

Protein Synthesis: Transcription and Translation

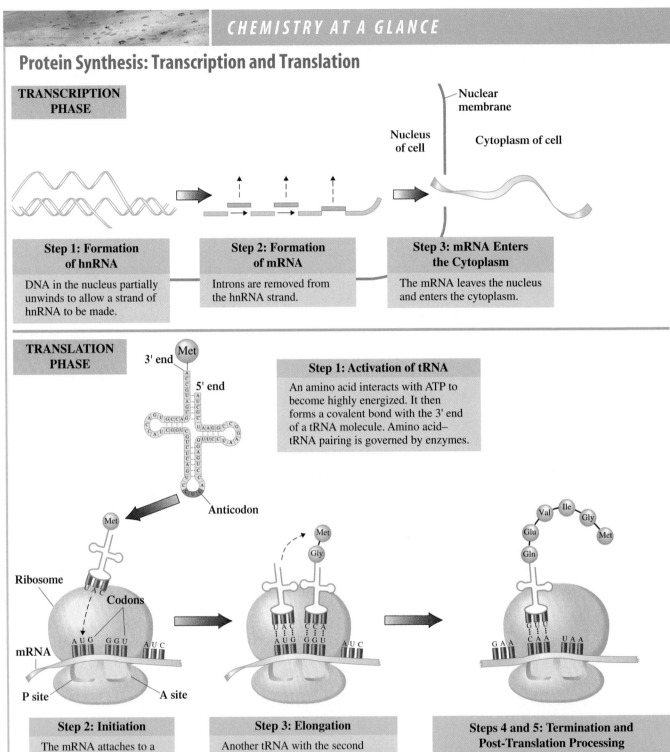

TRANSCRIPTION PHASE

Nuclear membrane

Nucleus of cell

Cytoplasm of cell

Step 1: Formation of hnRNA

DNA in the nucleus partially unwinds to allow a strand of hnRNA to be made.

Step 2: Formation of mRNA

Introns are removed from the hnRNA strand.

Step 3: mRNA Enters the Cytoplasm

The mRNA leaves the nucleus and enters the cytoplasm.

TRANSLATION PHASE

Met

3' end

5' end

Anticodon

Step 1: Activation of tRNA

An amino acid interacts with ATP to become highly energized. It then forms a covalent bond with the 3' end of a tRNA molecule. Amino acid–tRNA pairing is governed by enzymes.

Met

Ribosome

Codons

mRNA

P site

A site

Met

Gly

Val Ile Gly
Glu Met
Gln

Step 2: Initiation

The mRNA attaches to a ribosome so that the first codon (AUG) is at the P site. A tRNA carrying methionine attaches to the first codon.

Step 3: Elongation

Another tRNA with the second amino acid binds at the A site. The methionine transfers from the P site to the A site. The ribosome shifts to the next codon, making its A site available for the tRNA carrying the third amino acid.

Steps 4 and 5: Termination and Post-Translation Processing

The polypeptide chain continues to lengthen until a stop codon appears on the mRNA. The new protein is cleaved from the last tRNA.

During post-translation processing, cleavage of Met (the initiation codon) usually occurs. S—S bonds between Cys units also can form.

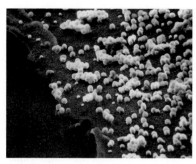

FIGURE 11.25 An electron microscope image of an influenza virus.

Viral infections are more difficult to treat than bacterial infections because viruses, unlike bacteria, replicate inside cells. It is difficult to design drugs that prevent the replication of the virus that do not also affect the normal activities of the host cells.

11.13 Nucleic Acids and Viruses

Viruses are very small disease-causing agents that are considered the lowest order of life. Indeed, their structure is so simple that some scientists do not consider them truly alive because they are unable to reproduce in the absence of other organisms. Figure 11.25 shows an electron microscope image of an influenza virus.

A **virus** *is a small particle that contains DNA or RNA (but not both) surrounded by a coat of protein and that cannot reproduce without the aid of a host cell.* Viruses do not possess the nucleotides, enzymes, amino acids, and other molecules necessary to replicate their nucleic acid or to synthesize proteins. To reproduce, viruses must invade the cells of another organism and cause these host cells to carry out the reproduction of the virus. Such an invasion disrupts the normal operation of cells, causing diseases within the host organism. The only function of a virus is reproduction; viruses do not generate energy.

There is no known form of life that is not subject to attack by viruses. Viruses attack bacteria, plants, animals, and humans. Many human diseases are of viral origin. Among them are the common cold, mumps, measles, smallpox, rabies, influenza, infectious mononucleosis, hepatitis, and AIDS.

Viruses most often attach themselves to the outside of specific cells in a host organism. An enzyme within the protein overcoat of the virus catalyzes the breakdown of the cell membrane, opening a hole in the membrane. The virus then injects its DNA or RNA into the cell. Once inside, this nucleic acid material is mistaken by the host cell for its own, whereupon that cell begins to translate and/or transcribe the viral nucleic acid. When all the virus components have been synthesized by the host cell, they assemble automatically to form many new virus particles. Within 20 to 30 minutes after a single molecule of viral nucleic acid enters the host cell, hundreds of new virus particles have formed. So many are formed that they eventually burst the host cell and are free to infect other cells.

If a virus contains DNA, the host cell replicates the viral DNA in a manner similar to the way it replicates its own DNA. The newly produced viral DNA then proceeds to make the proteins needed for the production of protein coats for additional viruses.

An RNA-containing virus is called a *retrovirus.* Once inside a host, such viruses first make viral DNA. This *reverse* synthesis is governed by the enzyme *reverse transcriptase.* The template is the viral RNA rather than DNA. The viral DNA so produced then produces additional viral DNA and the proteins necessary for the protein coats.

The AIDS (acquired immunodeficiency syndrome) virus is an example of a retrovirus. This virus has an affinity for a specific type of white blood cell called a *helper T cell,* which is an important part of the body's immune system. When helper T cells are unable to perform their normal functions as a result of such viral infection, the body becomes more susceptible to infection and disease.

A **vaccine** *is a preparation containing an inactive or weakened form of a virus or bacterium.* The antibodies produced by the body against these specially modified viruses or bacteria effectively act against the naturally occurring active forms as well. Thanks to vaccination programs, many diseases, such as polio and mumps (caused by RNA-containing viruses) and smallpox and yellow fever (caused by DNA-containing viruses), are now seldom encountered.

11.14 Recombinant DNA and Genetic Engineering

Increased knowledge about how DNA molecules function under various chemical conditions has opened the door to the field of technology called *genetic engineering* or *biotechnology.* Techniques now exist whereby a "foreign" gene can be added to an organism, and the organism will produce the protein associated with the added gene.

As an example of benefits that can come from genetic engineering, consider the case of human insulin. For many years, because of the very limited availability of human insulin, the insulin used by diabetics was obtained from the pancreases of slaughterhouse animals. Such insulin is structurally very similar to human insulin (see the Chemical Connections feature "Substitutes for Human Insulin" in Chapter 9) and can be substituted for it. Today, diabetics can also choose to use "real" human insulin produced by genetically altered bacteria. Such "genetically engineered" bacteria are grown in large numbers, and the insulin they produce is harvested in a manner similar to the way some antibiotics are obtained from cultured microorganisms. Human growth hormone is another substance that is now produced by genetically altered bacteria.

Genetic engineering procedures involve a type of DNA called *recombinant DNA.* **Recombinant DNA** *is DNA that contains genetic material from two different organisms.* Let us examine the theory and procedures used in obtaining recombinant DNA through genetic engineering.

The bacterium *E. coli,* which is found in the intestinal tract of humans and animals, is the organism most often used in recombinant DNA experiments. Yeast cells are also used, with increasing frequency, in this research.

In addition to their chromosomal DNA, *E. coli* (and other bacteria) contain DNA in the form of small, circular, double-stranded molecules called *plasmids.* These plasmids, which carry only a few genes, replicate independently of the chromosome. Also, they are transferred relatively easily from one cell to another. Plasmids from *E. coli* are used in recombinant DNA work.

The procedure used to obtain *E. coli* cells that contain recombinant DNA involves the following steps (see Figure 11.26).

Step 1: **Cell membrane dissolution.** *E. coli* cells of a specific strain are placed in a solution that dissolves cell membranes, thus releasing the contents of the cells.

Step 2: **Isolation of plasmid fraction.** The released cell components are separated into fractions, one fraction being the plasmids. The isolated plasmid fraction is the material used in further steps.

Step 3: **Cleavage of plasmid DNA.** A special enzyme, called a *restriction enzyme,* is used to cleave the double-stranded DNA of a circular plasmid. The result is a linear (noncircular) DNA molecule.

FIGURE 11.26 Recombinant DNA is made by inserting a gene obtained from DNA of one organism into the DNA from another kind of organism.

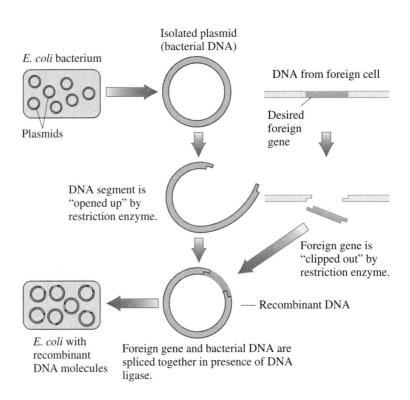

Step 4: **Gene removal from another organism.** The same restriction enzyme is then used to remove a desired gene from a chromosome of another organism.

Step 5: **Gene–plasmid splicing.** The gene (from Step 4) and the opened plasmid (from Step 3) are mixed in the presence of the enzyme *DNA ligase,* which splices the two together. This splicing, which attaches one end of the gene to one end of the opened plasmid and attaches the other end of the gene to the other end of the plasmid, results in an altered circular plasmid (the recombinant DNA).

Step 6: **Uptake of recombinant DNA.** The altered plasmids (recombinant DNA) are placed in a live *E. coli* culture, where they are taken up by the *E. coli* bacteria. The *E. coli* culture into which the plasmids are placed need not be identical to that from which the plasmids were originally obtained.

We noted in Step 3 that the conversion of a circular plasmid into a linear DNA molecule requires a restriction enzyme. A **restriction enzyme** *is an enzyme that recognizes specific base sequences in DNA and cleaves the DNA in a predictable manner at these sequences.* The discovery of restriction enzymes made genetic engineering possible.

Restriction enzymes occur naturally in numerous types of bacterial cells. Their function is to protect the bacteria from invasion by foreign DNA by catalyzing the cleavage of the invading DNA. The term *restriction* relates to such enzymes placing a "restriction" on the type of DNA allowed into the bacterial cells.

To understand how a restriction enzyme works, let us consider one that cleaves DNA between G and A bases in the 5′-to-3′ direction in the sequence G–A–A–T–T–C. This enzyme will cleave the double-helix structure of a DNA molecule in the manner shown in Figure 11.27.

Note that the double helix is not cut straight across; the individual strands are cut at different points, giving a staircase cut. (Both cuts must be between G and A in the 5′-to-3′ direction.) This staircase cut leaves unpaired bases on each cut strand. These ends with unpaired bases are called "sticky ends" because they are ready to "stick to" (pair up with) a complementary section of DNA if they can find one.

If the same restriction enzyme used to cut a plasmid is also used to cut a gene from another DNA molecule, the sticky ends of the gene will be complementary to those of the plasmid. This enables the plasmid and gene to combine readily, forming a new, modified plasmid molecule. This modified plasmid molecule is called recombinant DNA. In addition to the newly spliced gene, the recombinant DNA plasmid contains all of the genes and characteristics of the original plasmid. Figure 11.28 shows diagrammatically the match between sticky ends that occurs when plasmid and gene combine.

Step 6 involves inserting the recombinant DNA (modified plasmids) back into *E. coli* cells. The process is called transformation. **Transformation** *is the process of incorporating recombinant DNA into a host cell.*

The transformed cells then reproduce, resulting in large numbers of identical cells called clones. **Clones** *are cells with identical DNA that have descended from a single cell.* Within a few hours, a single genetically altered bacterial cell can give rise to thousands of clones. Each clone has the capacity to synthesize the protein directed by the foreign gene it carries.

FIGURE 11.27 Cleavage pattern resulting from the use of a restriction enzyme that cleaves DNA between G and A bases in the 5′-to-3′ direction in the sequence G–A–A–T–T–C. The double-helix structure is not cut straight across.

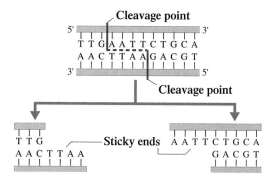

FIGURE 11.28 The "sticky ends" of the cut plasmid and the cut gene are complementary and combine to form recombinant DNA.

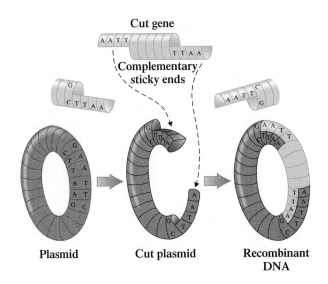

Plasmid   Cut plasmid   Recombinant DNA

Researchers are not limited to selection of naturally occurring genes for transforming bacteria. Chemists have developed nonenzymatic methods of linking nucleotides together such that they can construct artificial genes of any sequence they desire. In fact, benchtop instruments are now available that can be programmed by a microprocessor to synthesize any DNA base sequence *automatically*. The operator merely enters a sequence of desired bases, starts the instrument, and returns later to obtain the product. Such flexibility in manufacturing DNA has opened many doors, accelerated the pace of recombinant DNA research, and redefined the term *designer genes*!

11.15 The Polymerase Chain Reaction

The **polymerase chain reaction (PCR)** *is a method for rapidly producing multiple copies of a DNA nucleotide sequence.* Billions of copies of a specific DNA sequence (gene) can be produced in a few hours via this reaction. The PCR is easy to carry out, requiring only a few chemicals, a container, and a source of heat. (In actuality, the PCR process is now completely automated.)

By means of the PCR process, DNA that is available only in very small quantities can be amplified to quantities large enough to analyze. The PCR process, devised in 1983, has become a valuable tool for diagnosing diseases and detecting pathogens in the body. It is now used in the prenatal diagnosis of a number of genetic disorders, including muscular dystrophy and cystic fibrosis, and in the identification of bacterial pathogens. It is also the definitive way to detect the AIDS virus.

The PCR process has also proved useful in certain types of forensic investigations. A DNA sample may be obtained from a single drop of blood or semen or a single strand of hair at a crime scene and amplified by the PCR process. A forensic chemist can then compare the amplified samples with DNA samples taken from suspects. Work with DNA in the forensic area is often referred to as *DNA fingerprinting.*

DNA polymerase, an enzyme present in all living organisms, is a key substance in the PCR process. It can attach additional nucleotides to a short starter nucleotide chain, called a *primer,* when the primer is bound to a complementary strand of DNA that functions as a template. The original DNA is heated to separate its strands, and then primers, DNA polymerase, and deoxyribonucleotides are added so that the *DNA polymerase* can replicate the original strand. The process is repeated until, in a short time, millions of copies of the original DNA have been made.

Figure 11.29 shows diagrammatically, in very simplified terms, the basic steps in the PCR process.

PCR temperature conditions are higher than those in the human body. This is possible because the DNA polymerase used was isolated from an organism that lives in the "hot pots" of Yellowstone National Park at temperatures of 70°C–75°C.

After *n* cycles of the PCR process, the amount of DNA will have increased 2^n times.

2^{10} is approximately 1000.

2^{20} is approximately 1,000,000.

Twenty-five cycles of the PCR can be carried out in an hour in a process that is fully automated.

FIGURE 11.29 The basic steps in simplified terms, of the polymerase chain reaction process. Each cycle of the polymerase chain reaction doubles the number of copies of the target DNA sequence.

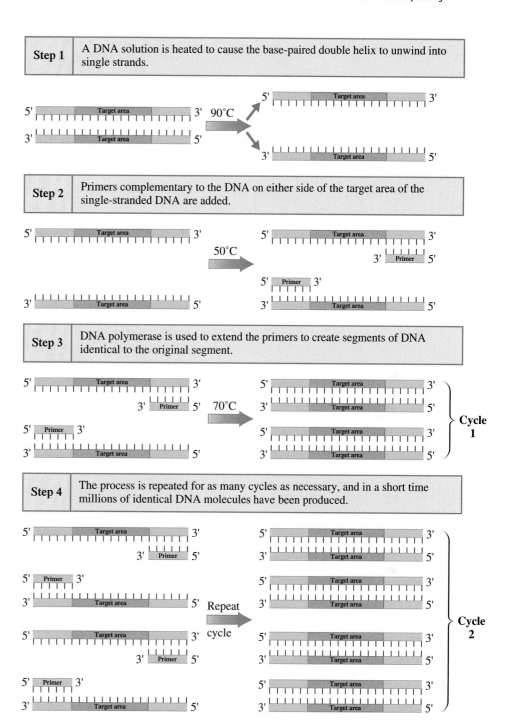

| Step 1 | A DNA solution is heated to cause the base-paired double helix to unwind into single strands. |

| Step 2 | Primers complementary to the DNA on either side of the target area of the single-stranded DNA are added. |

| Step 3 | DNA polymerase is used to extend the primers to create segments of DNA identical to the original segment. |

| Step 4 | The process is repeated for as many cycles as necessary, and in a short time millions of identical DNA molecules have been produced. |

11.16 DNA Sequencing

DNA sequencing *is a method by which the base sequence in a DNA molecule (or a portion of it) is determined.* Discovered in 1977, this is the process that made the Human Genome Project (Section 11.8) possible. Today, thanks to computer technology, sequencing a nucleic acid is a fairly routine, fully automated process.

The key concept in DNA sequencing is the *selective interruption* of polynucleotide synthesis. This interruption of synthesis, which is caused to occur at every possible nucleotide site, depends on the presence of 2′,3′-*dideoxy*ribonucleotide triphosphates (ddNTPs) in the synthesis mixture. Such compounds are synthetic analogs of the standard deoxyribonucleotide triphosphates in which both the 2′ and the 3′ hydroxy groups of deoxyribose have been replaced by hydrogen substituents.

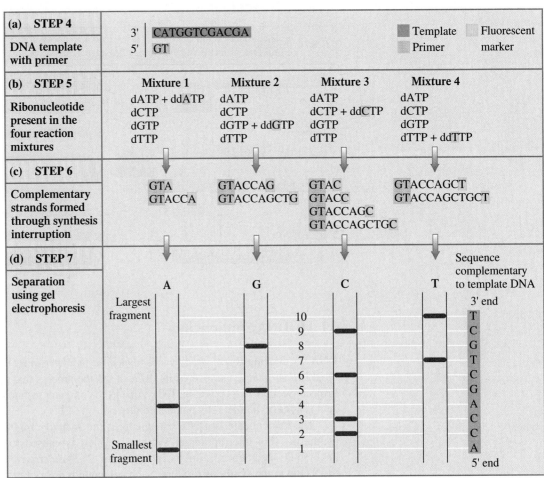

ddNTPs

The basic steps involved in the DNA sequencing process are as follows:

Step 1: **Cleavage using restriction enzymes.** Restriction enzymes (Section 11.14) are used to cleave a DNA molecule, which is too large to be sequenced as a whole, into smaller fragments (100–200 base pairs). These smaller fragments are the DNA actually sequenced. By later identifying the points of overlap among the fragments sequenced, it is possible to determine the base sequence of the entire original DNA molecule.

Step 2: **Separation into individual components.** The mixture of small DNA fragments generated by the restriction enzymes is separated into individual components. Each component type is then sequenced independently. Separation of the fragment mixture is accomplished via gel electrophoresis techniques (Section 9.4).

FIGURE 11.30 A schematic diagram of selected steps in the DNA sequencing procedure for the 10-base DNA segment 5′ AGCAGCTGGT 3′.

Step 3: **Separation into single strands.** Using chemical methods, a given DNA fragment is separated into its two strands, and one strand is then used as a template to create complementary strands of varying lengths via the "interruption of synthesis" process.

Step 4: **Addition of primer to the single strands.** The single-stranded DNA to be sequenced (the template) is mixed with short polynucleotides that serve as a primer for the complementary strands (see Figure 11.30a).

Step 5: **Separation of the reaction mixture into four parts.** The mixture of template DNA and primer is divided into four portions, and four parallel synthesis reactions are carried out. Each reaction mixture contains all four deoxyribonucleotide triphosphates: dATP, dCTP, dGTP, and dTTP (Section 11.2). Each test tube also contains a unique ingredient—one of the ddNTPs that has been labeled with a fluorescent material that can be detected using instrumentation (see Figure 11.30b).

Step 6: **Polynucleotide synthesis with interruption.** As DNA complementary strand synthesis proceeds, nucleotides from the solution are added to the growing polynucleotide chain. Elongation of the growing chain takes place without complication until a ddNTP is incorporated into the chain. Synthesis stops at this point because a ddNTP lacks a hydroxyl group at carbon 3 and hence cannot participate in a 3′-to-5′ phosphodiester linkage, a necessary requirement for chain elongation (Section 11.3). Thus the portion of the reaction mixture that contains ddATP will be a mixture of all possible lengths of DNA complementary strands that terminate in ddA. Similarly, all of the complementary strands in the portion that contains ddGTP will terminate in ddG, and so on (see Figure 11.30c).

Step 7: **Identification of the reaction mixture components.** The newly synthesized complementary DNA strands of the four portions of the reaction mixture are then subjected to gel electrophoresis. Smaller DNA fragments move more rapidly through the gel than do larger ones, which is the basis for the separation. Fluorescence from the four differently marked ddNTPs present in the complementary strands is the basis for identification; the labeling pattern observed indicates the sequence of bases. Figure 11.30d shows the gel separation of the complementary nucleotide strands for the 10-base DNA segment shown in Figure 11.30a.

CONCEPTS TO REMEMBER

Nucleic acids. Nucleic acids are polymeric molecules in which the repeating units are nucleotides. Cells contain two kinds of nucleic acids—deoxyribonucleic acids (DNA) and ribonucleic acids (RNA). The major biochemical functions of DNA and RNA are, respectively, transfer of genetic information and synthesis of proteins (Section 11.1).

Nucleotides. Nucleotides, the monomers of nucleic acid polymers, are molecules composed of a pentose sugar bonded to both a phosphate group and a nitrogen-containing heterocyclic base. The pentose sugar must be either ribose or deoxyribose. Five nitrogen-containing bases are found in nucleotides: adenine (A), guanine (G), cytosine (C), thymine (T), and uracil (U) (Section 11.2).

Primary nucleic acid structure. The "backbone" of a nucleic acid molecule is a constant alternating sequence of sugar and phosphate groups. Each sugar unit has a nitrogen-containing base attached to it (Section 11.3).

Complementary bases. Complementary bases are specific pairs of bases in nucleic acid structures that hydrogen-bond to each other (Section 11.4).

Secondary DNA structure. A DNA molecule exists as two polynucleotide chains coiled around each other in a double-helix arrangement. The double helix is held together by hydrogen bonding between complementary pairs of bases. Only two base-pairing combinations occur: A with T, and C with G (Section 11.4).

DNA replication. DNA replication occurs when the two strands of a parent DNA double helix separate and act as templates for the synthesis of new chains using the principle of complementary base pairing (Section 11.5).

Chromosome. A chromosome is a cell structure that consists of an individual DNA molecule bound to a group of proteins (Section 11.5).

RNA molecules. Five important types of RNA molecules, distinguished by their function, are ribosomal RNA (rRNA), messenger RNA (mRNA), heterogeneous nuclear RNA (hnRNA), transfer RNA (tRNA), and small nuclear RNA (snRNA) (Section 11.7).

Transcription. Transcription is the process in which the genetic information encoded in the base sequence of DNA is copied into RNA molecules (Section 11.8).

Gene. A gene is a portion of a DNA molecule that contains the base sequences needed for the production of a specific hnRNA/mRNA molecule. Genes are segmented, with portions called exons that contain genetic information and portions called introns that do not convey genetic information (Section 11.8).

Codon. A codon is a three-nucleotide sequence in mRNA that codes for a specific amino acid needed during the process of protein synthesis (Section 11.9).

Genetic code. The genetic code consists of all the mRNA codons that specify either a particular amino acid or the termination of protein synthesis (Section 11.9).

Anticodon. An anticodon is a three-nucleotide sequence in tRNA that binds to a complementary sequence (a codon) in mRNA (Section 11.10).

Translation. Translation is the stage of protein synthesis in which the codons in mRNA are translated into amino acid sequences of new proteins. Translation involves interactions between the codons of mRNA and the anticodons of tRNA (Section 11.11).

Mutations. Mutations are changes in the base sequence in DNA molecules (Section 11.12).

Recombinant DNA. Recombinant DNA molecules are synthesized by splicing a segment of DNA, usually a gene, from one organism into the DNA of another organism (Section 11.14).

Polymerase chain reaction. The polymerase chain reaction is a method for rapidly producing many copies of a DNA sequence (Section 11.15).

DNA sequencing. DNA sequencing is a multistep process for determining the sequence of bases in a DNA segment (Section 11.16).

KEY REACTIONS AND EQUATIONS

1. Formation of a nucleotide (Section 11.2)

    Pentose sugar (ribose or deoxyribose) + phosphate group

    + nitrogen-containing heterocyclic base $\longrightarrow$

$$\text{Phosphate} - \text{Sugar} \underset{|}{\overset{}{\text{Base}}} + 2H_2O$$

2. Formation of a nucleic acid (Section 11.3)

    Many deoxyribose-containing nucleotides $\longrightarrow$ DNA

    Many ribose-containing nucleotides $\longrightarrow$ RNA

3. Protein synthesis (Section 11.6)

    DNA $\xrightarrow{\text{Transcription}}$ RNA $\xrightarrow{\text{Translation}}$ protein

EXERCISES AND PROBLEMS

The members of each pair of problems in this section test similar material.

■ **Nucleotides (Section 11.2)**

11.1 What is the structural difference between the pentose sugars ribose and 2-deoxyribose?

11.2 What are the names of the pentose sugars present, respectively, in DNA and RNA molecules?

11.3 Characterize each of the following nitrogen-containing bases as a purine derivative or a pyrimidine derivative.
a. Thymine   b. Cytosine
c. Adenine   d. Guanine

11.4 Characterize each of the following nitrogen-containing bases as a component of (1) both DNA and RNA, (2) DNA but not RNA, or (3) RNA but not DNA.
a. Adenine   b. Thymine
c. Uracil   d. Cytosine

11.5 How many different choices are there for each of the following subunits in the specified type of nucleotide?
a. Pentose sugar subunit in DNA nucleotides
b. Nitrogen-containing base subunit in RNA nucleotides
c. Phosphate subunit in DNA nucleotides

11.6 How many different choices are there for each of the following subunits in the specified type of nucleotide?
a. Pentose sugar subunit in RNA nucleotides
b. Nitrogen-containing base subunit in DNA nucleotides
c. Phosphate subunit in RNA nucleotides

11.7 Which nitrogen-containing base is present in each of the following nucleotides?
a. AMP   b. dGMP   c. dTMP   d. UMP

11.8 Which nitrogen-containing base is present in each of the following nucleotides?
a. GMP   b. dAMP
c. CMP   d. dCMP

11.9 Which pentose sugar is present in each of the nucleotides in Problem 11.7?

11.10 Which pentose sugar is present in each of the nucleotides in Problem 11.8?

11.11 Characterize as true or false each of the following statements about the given nucleotide.

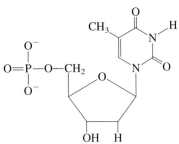

a. The nitrogen-containing base is a purine derivative.
b. The phosphate group is attached to the sugar unit at carbon 3'.
c. The sugar unit is ribose.
d. The nucleotide could be a component of both DNA and RNA.

11.12 Characterize as true or false each of the following statements about the given nucleotide.

a. The sugar unit is 2-deoxyribose.
b. The sugar unit is attached to the nitrogen-containing base at nitrogen 3.
c. The nitrogen-containing base is a pyrimidine derivative.
d. The nucleotide could be a component of both DNA and RNA.

11.13 Draw the structures of the three products produced when the nucleotide in Problem 11.11 undergoes hydrolysis.

11.14 Draw the structures of the three products produced when the nucleotide in Problem 11.12 undergoes hydrolysis.

■ **Primary Nucleic Acid Structure (Section 11.3)**

11.15 What are the two repeating subunits present in the *backbone* portion of a nucleic acid?

11.16 To which type of subunit in a nucleic acid *backbone* are the nitrogen-containing bases attached?

11.17 What distinguishes various DNA molecules from each other?

11.18 What distinguishes various RNA molecules from each other?

11.19 What is the difference between a nucleic acid's 3′ end and its 5′ end?

11.20 In the lengthening of a polynucleotide chain, which type of nucleotide subunit would bond to the 3′ end of the polynucleotide chain?

11.21 What are the nucleotide subunits that participate in a nucleic acid 3′,5′-phosphodiester linkage?

11.22 How many 3′,5′-phosphodiester linkages are present in a tetranucleotide segment of a nucleic acid?

11.23 Draw the structure of the dinucleotide product obtained by combining the nucleotides of Problems 11.11 and 11.12 such that the Problem 11.11 nucleotide is the 5′ end of the dinucleotide.

11.24 Draw the structure of the dinucleotide product obtained by combining the nucleotides of Problems 11.11 and 11.12 such that the Problem 11.11 nucleotide is the 3′ end of the dinucleotide.

■ **The DNA Double Helix (Section 11.4)**

11.25 Describe the DNA double helix in terms of
a. general shape.
b. what is on the outside of the helix and what is within the interior of the helix.

11.26 Describe the DNA double helix in terms of
a. the directionality of the two polynucleotide chains present.
b. a comparison of the total number of nitrogen-containing bases present in each of the two polynucleotide chains.

11.27 The base content of a particular DNA molecule is 36% thymine. What is the percentage of each of the following bases in the molecule?
a. Adenine   b. Guanine   c. Cytosine

11.28 The base content of a particular DNA molecule is 24% guanine. What is the percentage of each of the following bases in the molecule?
a. Adenine   b. Cytosine   c. Thymine

11.29 In terms of hydrogen bonding, a G–C base pair is more stable than an A–T base pair. Explain why this is so.

11.30 What structural consideration prevents the following bases from forming complementary base pairs?
a. A and G   b. T and C

11.31 What is the relationship between the total number of purine bases (A and G) and the total number of pyrimidine bases (C and T) present in a DNA double helix?

11.32 The base composition for one of the strands of a DNA double helix is 19% A, 34% C, 28% G, and 19% T. What is the percent base composition for the other strand of the DNA double helix?

11.33 Identify the 3′ and 5′ ends of the DNA base sequence TAGCC.

11.34 The two-base DNA sequences TA and AT represent different dinucleotides. Explain why this is so.

11.35 Using the concept of complementary base pairing, write the complementary DNA strands, with their 5′ and 3′ ends labeled, for each of the following DNA base sequences.
a. 5′ ACGTAT 3′   b. 5′ TTACCG 3′
c. 3′ GCATAA 5′   d. AACTGG

11.36 Using the concept of complementary base pairing, write the complementary DNA strands, with their 5′ and 3′ ends labeled, for each of the following DNA base sequences.
a. 5′ CCGGTA 3′   b. 5′ CACAGA 3′
c. 3′ TTTAGA 5′   d. CATTAC

11.37 How many total hydrogen bonds would exist between the DNA strand 5′ AGTCCTCA 3′ and its complementary strand?

11.38 How many total hydrogen bonds would exist between the DNA strand 5′ CCTAGGAT 3′ and its complementary strand?

■ **Replication of DNA Molecules (Section 11.5)**

11.39 What is the function of the enzyme *DNA helicase* in the DNA replication process?

11.40 What are two functions of the enzyme *DNA polymerase* in the DNA replication process?

11.41 In the replication of a DNA molecule, two daughter molecules, Q and R, are formed. The following base sequence is part of the newly formed strand in daughter molecule Q.

5′ ACTTAG 3′

Indicate the corresponding base sequence in
a. the newly formed strand in daughter molecule R.
b. the "parent" strand in daughter molecule Q.
c. the "parent" strand in daughter molecule R.

11.42 In the replication of a DNA molecule, two daughter molecules, S and T, are formed. The following base sequence is part of the "parent" strand in daughter molecule S.

5′ TTCAGAG 3′

Indicate the corresponding base sequence in
a. the newly formed strand in daughter molecule T.
b. the newly formed strand in daughter molecule S.
c. the "parent" strand in daughter molecule T.

11.43 During DNA replication, one of the newly formed strands grows continuously, whereas the other grows in segments that are later connected together. Explain why this is so.

11.44 DNA replication is most often a bidirectional process. Explain why this is so.

11.45 What is a chromosome?

11.46 Chromosomes are nucleoproteins. Explain.

■ RNA Molecules (Section 11.7)

11.47 What are the four major differences between RNA molecules and DNA molecules?

11.48 What are the names and abbreviations for the five major types of RNA molecules?

11.49 State whether each of the following phrases applies to hnRNA, mRNA, tRNA, rRNA, or snRNA.
a. Material from which messenger RNA is made
b. Delivers amino acids to protein synthesis sites
c. Smallest of the RNAs in terms of nucleotide units present
d. Also goes by the designation ptRNA

11.50 State whether each of the following phrases applies to hnRNA, mRNA, tRNA, rRNA, or snRNA.
a. Associated with a series of proteins in a complex structure
b. Contains genetic information needed for protein synthesis
c. Most abundant type of RNA in a cell
d. Involved in the editing of hnRNA molecules

11.51 For each of the following types of RNA, indicate whether the predominant cellular location for the RNA is the nuclear region, the extranuclear region, or both the nuclear and the extranuclear regions.
a. hnRNA   b. tRNA   c. rRNA   d. mRNA

11.52 Indicate whether each of the following processes occurs in the nuclear or the extranuclear region of a cell.
a. DNA transcription
b. Processing of hnRNA to mRNA
c. mRNA translation (protein synthesis)
d. DNA replication

■ Transcription: RNA Synthesis (Section 11.8)

11.53 What serves as a template in the process of *transcription*?

11.54 What is the initial product of the *transcription* process?

11.55 What are two functions of the enzyme *RNA polymerase* in the transcription process?

11.56 What is a *gene*?

11.57 What are the complementary base pairs in DNA–RNA interactions?

11.58 In DNA–DNA interactions there are two complementary base pairs, and in DNA–RNA interactions there are three complementary base pairs. Explain.

11.59 Write the base sequence of the hnRNA formed by transcription of the following DNA base sequence.

5′ ATGCTTA 3′

11.60 Write the base sequence of the hnRNA formed by transcription of the following DNA base sequence.

5′ TAGTGAT 3′

11.61 From what DNA base sequence was the following hnRNA sequence transcribed?

5′ UUCGCAG 3′

11.62 From what DNA base sequence was the following hnRNA sequence transcribed?

5′ GCUUAUC 3′

11.63 What is the relationship between an exon and a gene?

11.64 What is the relationship between an intron and a gene?

11.65 What mRNA base sequence would be obtained from the following portion of a gene?

| exon | intron | exon |
5′ TCAG–TAGC–TTCA 3′

11.66 What mRNA base sequence would be obtained from the following portion of a gene?

| intron | exon | intron |
5′ TTAC–AACG–GCAT 3′

11.67 In the process of splicing, which type of RNA
a. undergoes the splicing?
b. is present in the spliceosomes?

11.68 What is the difference between snRNA and snRNPs?

11.69 What is *alternative splicing*?

11.70 How many different mRNAs can be produced from an hnRNA that contains three exons, one of which is an "alternative" exon?

■ The Genetic Code (Section 11.9)

11.71 What is a codon?

11.72 On what type of RNA molecule are codons found?

11.73 Using the information in Table 11.2, determine what amino acid is coded for by each of the following codons.
a. CUU   b. AAU   c. AGU   d. GGG

11.74 Using the information in Table 11.2, determine what amino acid is coded for by each of the following codons.
a. GUA   b. CCC   c. CAC   d. CCA

11.75 Using the information in Table 11.2, determine the synonyms, if any, of each of the codons in Problem 11.73.

11.76 Using the information in Table 11.2, determine the synonyms, if any, of each of the codons in Problem 11.74.

11.77 Explain why the base sequence ATC could not be a codon.

11.78 Explain why the base sequence AGAC could not be a codon.

11.79 Predict the sequence of amino acids coded by the mRNA sequence

5′ AUG–AAA–GAA–GAC–CUA 3′

11.80 Predict the sequence of amino acids coded by the mRNA sequence

5′ GGA–GGC–ACA–UGG–GAA 3′

■ Anticodons and tRNA Molecules (Section 11.10)

11.81 Describe the general structure of a tRNA molecule.

11.82 Where is the anticodon site on a tRNA molecule?

11.83 By what type of bond is an amino acid attached to a tRNA molecule?

11.84 What principle governs the codon–anticodon interaction that leads to proper placement of amino acids in proteins?

11.85 What is the anticodon that would interact with each of the following codons?
a. AGA   b. CGU   c. UUU   d. CAA

11.86 What is the anticodon that would interact with each of the following codons?
a. CCU   b. GUA   c. AUC   d. GCA

11.87 Identify the amino acid associated with each of the following anticodons:
a. UGG   b. GAC   c. GGA   d. AGA

11.88 Identify the amino acid associated with each of the following anticodons.
a. UGU   b. ACG   c. AGU   d. CAC

Translation: Protein Synthesis (Section 11.11)

11.89 What are the five major steps in translation (protein synthesis)?

11.90 What is a ribosome, and what role do ribosomes play in protein synthesis?

11.91 In the growth step of protein synthesis, at which site in the ribosome does new peptide bond formation actually take place?

11.92 What two changes occur at a ribosome during protein synthesis immediately after peptide bond formation?

11.93 Write a possible mRNA base sequence that would lead to the production of this pentapeptide. (There is more than one correct answer.)

Gly–Ala–Cys–Val–Tyr

11.94 Write a possible mRNA base sequence that would lead to the production of this pentapeptide. (There is more than one correct answer.)

Lys–Met–Thr–His–Phe

Mutations (Section 11.12)

11.95 For the codon sequence

5′ GGC–UAU–AGU–AGC–CCC 3′

write the amino acid sequence produced in each of the following ways.
a. Translation proceeds in a normal manner.
b. A mutation changes CCC to CCU.
c. A mutation changes CCC to ACC.

11.96 For the codon sequence

5′ GGA–AUA–UGG–UUC–CUA 3′

write the amino acid sequence produced in each of the following ways.
a. Translation proceeds in a normal manner.
b. A mutation changes GGA to GGG.
c. A mutation changes GGA to CGA.

Viruses and Vaccines (Section 11.13)

11.97 Describe the general structure of a virus.

11.98 What is the only function of a virus?

11.99 What is the most common method by which viruses invade cells?

11.100 Why must a virus infect another organism in order to reproduce?

Recombinant DNA and Genetic Engineering (Section 11.14)

11.101 How does recombinant DNA differ from normal DNA?

11.102 Give two reasons why bacterial cells are used for recombinant DNA procedures.

11.103 What role do plasmids play in recombinant DNA procedures?

11.104 Describe what occurs when a particular restriction enzyme operates on a segment of double-stranded DNA.

11.105 Describe what happens during transformation.

11.106 How are plasmids obtained from *E. coli* bacteria?

11.107 A particular restriction enzyme will cleave DNA between A and A in the sequence AAGCTT in the 5′-to-3′ direction. Draw a diagram showing the structural details of the "sticky ends" that result from cleavage of the following DNA segment.

$$
\begin{array}{c}
5' \\
\text{C C A A G C T T G} \\
\text{G G T T C G A A C} \\
3' \\
\end{array}
\begin{array}{c}
3' \\
\\
\\
5' \\
\end{array}
$$

11.108 A particular restriction enzyme will cleave DNA between A and A in the sequence AAGCTT in the 5′-to-3′ direction. Draw a diagram showing the structural details of the "sticky ends" that result from cleavage of the following DNA segment.

$$
\begin{array}{c}
5' \\
\text{G G A A G C T T A} \\
\text{C C T T C G A A T} \\
3' \\
\end{array}
\begin{array}{c}
3' \\
\\
\\
5' \\
\end{array}
$$

Polymerase Chain Reaction (Section 11.15)

11.109 What is the function of the polymerase chain reaction?

11.110 What is the function of the enzyme *DNA polymerase* in the PCR process?

11.111 What is a *primer* and what is its function in the PCR process?

11.112 What are the four types of substances needed to carry out the PCR process?

DNA Sequencing (Section 11.16)

11.113 How do the notations dATP and ddATP differ in meaning?

11.114 What role do dideoxynucleotides play in the DNA sequencing process?

11.115 Assume that the red lines in the first and second columns of Figure 11.30d are interchanged, while the other labels stay the same.
a. Given this change, what would be the sequence of bases in the DNA fragment under study?
b. Given this change, what would be the sequence of bases in the original DNA fragment that was to be sequenced?

11.116 Assume that the "dark lines" in the second and fourth columns of Figure 11.30d are interchanged, while the other labels stay the same.
a. Given this change, what would be the sequence of bases in the DNA fragment under study?
b. Given this change, what would be the sequence of bases in the original DNA fragment that was to be sequenced?

ADDITIONAL PROBLEMS

11.117 With the help of the structures given in Section 11.2, describe the structural differences between the following pairs of nucleotide bases.
a. Thymine and uracil   b. Adenine and guanine

11.118 The following is a sequence of bases for an exon portion of a strand of a gene.

5′ CATACAGCCTGGAAGCTA 3′

a. What is the sequence of bases on the strand of DNA complementary to this segment?
b. What is the sequence of bases on the mRNA molecule synthesized from this strand?

c. What codons are present on the mRNA molecule from part b?

d. What anticodons will be found on the tRNA molecules that interact with the codons from part c?

e. What is the sequence of amino acids in the peptide formed using these protein synthesis instructions?

11.119 Which of these RNA types, (1) mRNA, (2) hnRNA, (3) rRNA, or (4) tRNA, is most closely associated with each of the following terms?

    a. Codon   b. Anticodon

    c. Intron   d. Amino acid carrier

11.120 Which of these processes, (1) translation phase of protein synthesis, (2) transcription phase of protein synthesis, (3) replication of DNA, or (4) formation of recombinant DNA, is associated with each of the following events?

    a. Complete unwinding of a DNA molecule occurs.

    b. Partial unwinding of a DNA molecule occurs.

    c. An mRNA–ribosome complex is formed.

    d. Okazaki fragments are formed.

11.121 Which of these base-pairing situations, (1) between two DNA segments, (2) between two RNA segments, (3) between a DNA segment and an RNA segment, or (4) between a codon and an anticodon, fits each of the following base-pairing sequences? More than one response may apply to a given base-pairing situation.

a. A G T   b. A C T
    ⋮ ⋮ ⋮   ⋮ ⋮ ⋮
    U C A   T G A

c. A G U   d. C C G
    ⋮ ⋮ ⋮   ⋮ ⋮ ⋮
    U C A   G G C

11.122 Which of these characterizations, (1) found in DNA but not RNA, (b) found in RNA but not DNA, (3) found in both DNA and RNA, or (4) not found in DNA or RNA, fits each of the following mono-, di- or trinucleotides?

    a. 5′ dAMP–dAMP 3′

    b. 5′ AMP–AMP–CMP 3′

    c. 5′ dAMP–CMP 3′

    d. 5′ GGA 3′

11.123 Suppose that 28% of the nucleotides of a DNA molecule are deoxythymidine 5′-monophosphate, and during replication the relative amounts of available nucleotide bases are 22% A, 28% T, 22% C, and 28% G. What base would be depleted first in the replication process?

11.124 On the basis of the most recent results of the Human Genome Project concerning the DNA present in a human cell

    a. How many base-pairs are present in the DNA?

    b. How many genes are present in the DNA?

    c. What percentage of the base-pairs are accounted for by the genes?

MULTIPLE-CHOICE PRACTICE TEST

11.125 Which of the following is not a structural subunit of a nucleotide?

    a. A nitrogen-containing heterocyclic base

    b. A pentose sugar

    c. An amino acid

    d. A phosphate

11.126 The number of kinds of RNA nucleotides is which of the following?

    a. The same as the number of kinds of DNA nucleotides

    b. Double the number of kinds of DNA nucleotides

    c. Less than the number of kinds of DNA nucleotides

    d. Greater than the number of kinds of DNA nucleotides

11.127 In which of the following pairs of nucleotide bases are both members of the pair "single-ring" bases?

    a. A and C   b. G and T

    c. T and U   d. A and G

11.128 Which of the following elements is not present in the "backbone" of a nucleic acid molecule?

    a. Nitrogen   b. Carbon

    c. Oxygen   d. Hydrogen

11.129 In a DNA double helix, the base pairs are which of the following?

    a. Part of the backbone structure

    b. Located inside the helix

    c. Located outside the helix

    d. Covalently bonded to each other

11.130 Which of the following types of RNA has a "cloverleaf shape" with three hairpin loops?

    a. mRNA   b. rRNA

    c. hnRNA   d. tRNA

11.131 Which of the following types of RNA contains introns and exons?

    a. hnRNA   b. mRNA

    c. tRNA   d. rRNA

11.132 Which of the following events occurs during the *translation* phase of protein synthesis?

    a. mRNA is converted to hnRNA

    b. tRNAs carry amino acids to the site for protein synthesis

    c. A partial unwinding of a DNA double helix occurs

    d. rRNAs interact with ribosomes

11.133 The genetic code is a listing that gives the relationships between which of the following?

    a. Codons and anticodons

    b. Codons and amino acids

    c. Anticodons and amino acids

    d. Codons and genes

11.134 What is the complementary hnRNA sequence to the DNA sequence CTA–TAC?

    a. TCG–CGT   b. GAU–AUG

    c. UGC–GCU   d. TCG–CGT

12 Biochemical Energy Production

The energy consumed by these scarlet ibises in flight is generated by numerous sequences of biochemical reactions that occur within their bodies.

This chapter is the first of four dealing with the chemical reactions that occur in a living organism. In this first chapter, we consider those molecules that are repeatedly encountered in biological reactions, as well as those reactions that are common to the processing of carbohydrates, lipids, and proteins. The three following chapters consider the reactions associated uniquely with carbohydrate, lipid, and protein processing, respectively.

12.1 Metabolism

Metabolism *is the sum total of all the biochemical reactions that take place in a living organism.* Human metabolism is quite remarkable. An average human adult whose weight remains the same for 40 years processes about 6 *tons* of solid food and 10,000 gallons of water, during which time the composition of the body is essentially constant. Just as we must put gasoline in a car to make it go or plug in a kitchen appliance to make it run, we also need a source of energy to think, breathe, exercise, or work. As we have seen in previous chapters, even the simplest living cell is continually carrying on energy-demanding processes such as protein synthesis, DNA replication, RNA transcription, and membrane transport.

Metabolic reactions fall into one of two subtypes: catabolism and anabolism. **Catabolism** *is all metabolic reactions in which large biochemical molecules are broken down to smaller ones.* Catabolic reactions usually release energy. The reactions involved in the oxidation of glucose are catabolic. **Anabolism** *is all metabolic reactions in which*

Catabolism is pronounced ca-TAB-o-lism, and *anabolism* is pronounced an-ABB-o-lism. *Catabolic* is pronounced CAT-a-bol-ic, and *anabolic* is pronounced AN-a-bol-ic.

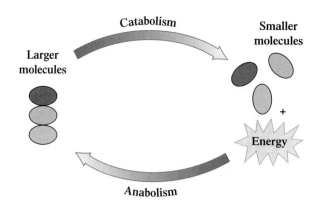

small biochemical molecules are joined together to form larger ones. Anabolic reactions usually require energy in order to proceed. The synthesis of proteins from amino acids is an anabolic process. Figure 12.1 contrasts catabolic and anabolic processes.

The metabolic reactions that occur in a cell are usually organized into sequences called *metabolic pathways.* A **metabolic pathway** *is a series of consecutive biochemical reactions used to convert a starting material into an end product.* Such pathways may be *linear,* in which a series of reactions generates a final product, or *cyclic,* in which a series of reactions regenerates the first reactant.

Linear metabolic pathway: $\quad A \longrightarrow B \longrightarrow C \longrightarrow D$

Cyclic metabolic pathway:

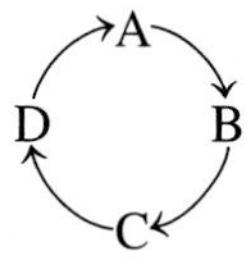

$$
\begin{array}{ccc}
 & A & \\
D & & B \\
 & C & \\
\end{array}
$$

The major metabolic pathways for all life forms are similar. This enables scientists to study metabolic reactions in simpler life forms and use the results to help understand the corresponding metabolic reactions in more complex organisms, including humans.

12.2 Metabolism and Cell Structure

Knowledge of the major structural features of a cell is a prerequisite to understanding *where* metabolic reactions take place.

Cells are of two types: prokaryotic and eukaryotic. *Prokaryotic cells* have no nucleus and are found only in bacteria. The DNA that governs the reproduction of prokaryotic cells is usually a single circular molecule found near the center of the cell in a region called the *nucleoid.* A **eukaryotic cell** *is a cell in which the DNA is found in a membrane-enclosed nucleus.* Cells of this type, which are found in all higher organisms, are about 1000 times larger than bacterial cells. Our focus in the remainder of this section will be on eukaryotic cells, the type present in humans. Figure 12.2 shows the general internal structure of a eukaryotic cell. Note the key components shown: the outer membrane, nucleus, cytosol, ribosomes, lysosomes, and mitochondria.

The **cytoplasm** *is the water-based material of a eukaryotic cell that lies between the nucleus and the outer membrane of the cell.* Within the cytoplasm are several kinds of small structures called *organelles.* An **organelle** *is a minute structure within the cytoplasm of a cell that carries out a specific cellular function.* The organelles are surrounded by the *cytosol.* The **cytosol** *is the water-based fluid part of the cytoplasm of a cell.*

Three important types of organelles are ribosomes, lysosomes, and mitochondria. We have considered ribosomes before; they are the sites where protein synthesis occurs (Section 11.11) A **lysosome** *is an organelle that contains hydrolytic enzymes needed for cellular rebuilding, repair, and degradation.* Some lysosome enzymes hydrolyze proteins

The term *eukaryotic,* pronounced you-KAHR-ee-ah-tic, is from the Greek *eu,* meaning "true," and *karyon,* meaning "nucleus." The term *prokaryotic,* which contains the Greek *pro,* meaning "before," literally means "before the nucleus."

Eukaryotic and prokaryotic cells differ in that the former contain a well-defined nucleus, set off from the rest of the cell by a membrane.

FIGURE 12.2 A schematic representation of a eukaryotic cell with selected internal components identified.

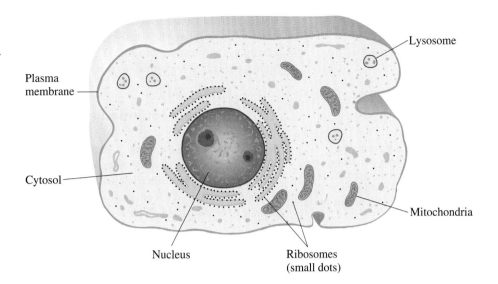

A protective mechanism exists to prevent lysosome enzymes from destroying the cell in which they are found if they should be accidently released (via membrane rupture or leakage). The optimum pH (Section 10.6) for lysosome enzyme activity is 4.8. The cytoplasmic pH of 7.0–7.3 renders them inactive.

Mitochondria, pronounced my-toe-KON-dree-ah, is plural. The singular form of the term is *mitochondrion.* The threadlike shape of the inner membrane of the mitochondria is responsible for this organelle's name; *mitos* is Greek for "thread," and *chondrion* is Greek for "granule."

to amino acids; others hydrolyze polysaccharides to monosaccharides. Bacteria and viruses "trapped" by the body's immune system (Section 9.16) are degraded and destroyed by enzymes from lysosomes.

A **mitochondrion** *is an organelle that is responsible for the generation of most of the energy for a cell.* Much of the discussion of this chapter deals with the energy-producing chemical reactions that occur within mitochondria. Further details of mitochondrion structure will help us understand more about how these reactions occur.

Mitochondria are sausage-shaped organelles containing both an *outer membrane* and a *multifolded inner membrane* (see Figure 12.3). The outer membrane, which is about 50% lipid and 50% protein, is freely permeable to small molecules. The inner membrane, which is about 20% lipid and 80% protein, is highly impermeable to most substances. The nonpermeable nature of the inner membrane divides a mitochondrion into two separate compartments—an interior region called the *matrix* and the region between the inner and outer membranes, called the *intermembrane space.* The folds of the inner membrane that protrude into the matrix are called *cristae.*

The invention of high-resolution electron microscopes allowed researchers to see the interior structure of the mitochondrion more clearly and led to the discovery, in 1962, of

FIGURE 12.3 (a) A schematic representation of a mitochondrion, showing key features of its internal structure. (b) An electron micrograph of a single mitochondrial crista, showing the ATP synthase knobs extending into the matrix.

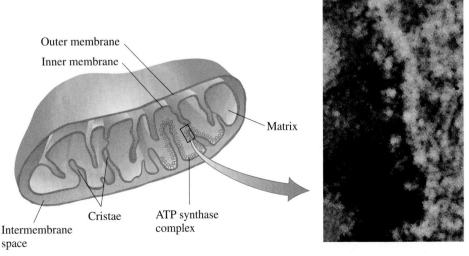

(a)

(b)

small spherical knobs attached to the cristae called *ATP synthase complexes*. As their name implies, these relatively small knobs, which are located on the matrix side of the inner membrane, are responsible for ATP synthesis, and their association with the inner membrane is critically important for this task. More will be said about ATP in the next section.

12.3 Important Intermediate Compounds in Metabolic Pathways

Nucleotides, besides being the monomer units from which nucleic acids are made, are also present in several *nonpolymeric* molecules that are important in energy production in living things.

As a prelude to an overview presentation (Section 12.5) of the metabolic processes by which our food is converted to energy, we now consider several compounds that repeatedly function as key intermediates in these metabolic pathways. Knowing about these compounds will make it easier to understand the details of metabolic pathways. The compounds to be discussed all have *nucleotides* (Section 11.2) as part of their structures.

■ Adenosine Phosphates (ATP, ADP, and AMP)

Several adenosine phosphates exist. Of importance in metabolism are adenosine *mono*phosphate (AMP), adenosine *di*phosphate (ADP), and adenosine *tri*phosphate (ATP). AMP is not a new molecule to us; it is one of the nucleotides present in RNA molecules (Section 11.2). ADP and ATP differ structurally from AMP only in the number of phosphate groups present. Block structural diagrams for these three adenosine phosphates follow.

			Phosphate	Ribose	Adenine	AMP
		Phosphate	Phosphate	Ribose	Adenine	ADP
Phosphate	Phosphate	Phosphate	Ribose	Adenine	ATP

Figure 12.4 shows actual structural formulas for these three adenosine phosphates.

ATP and ADP molecules readily undergo hydrolysis reactions in which phosphate groups (P_i, inorganic phosphate) are released

$$\text{ATP} + H_2O \longrightarrow \text{ADP} + P_i$$
$$\text{ADP} + H_2O \longrightarrow \text{AMP} + P_i$$
$$\text{ATP} + 2H_2O \longrightarrow \text{AMP} + 2P_i$$

In metabolic pathways in which they are involved, the adenosine phosphates continually change back and forth among the various forms:

$$\text{ATP} \rightleftharpoons \text{ADP} \rightleftharpoons \text{AMP}$$

These hydrolyses are energy-producing reactions that are used to drive cellular processes that require energy input. The phosphate–phosphate bonds in ATP and ADP are *very reactive* bonds that require less energy than normal to break. The presence of such reactive bonds, which are often called *strained bonds* (see Section 12.4), is the basis for the net energy

FIGURE 12.4 Structures of the various phosphate forms of adenosine.

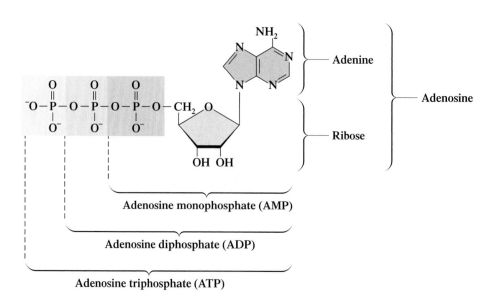

production that accompanies hydrolysis. Greater-than-normal electron–electron repulsive forces at specific locations within a molecule are the cause for bond strain; in ATP and ADP, it is the highly electronegative oxygen atoms in the additional phosphate groups that cause the increased repulsive strain.

ATP is not the only nucleotide triphosphate present in cells, although it is the most prevalent. The other nitrogen-containing bases associated with nucleotides (Section 11.2) are also present in triphosphate form. Uridine triphosphate (UTP) is involved in carbohydrate metabolism, guanosine triphosphate (GTP) participates in protein and carbohydrate metabolism, and cytidine triphosphate (CTP) is involved in lipid metabolism.

■ Flavin Adenine Dinucleotide (FAD, FADH$_2$)

Flavin adenine dinucleotide (FAD) is a coenzyme (Section 10.3) required in numerous metabolic redox reactions. Structurally, FAD can be visualized as containing either three subunits or six subunits. A block diagram of FAD from the three-subunit viewpoint is

| Flavin | Ribitol | ADP |

Flavin and ribitol, the two components attached to the ADP unit, together constitute the B vitamin riboflavin (Section 10.13). The block diagram for FAD from the six-subunit viewpoint is

| Flavin | Ribitol | Phosphate |
| Adenine | Ribose | Phosphate |

This block diagram shows the basis for the name *f*lavin *a*denine *d*inucleotide. Ribitol is a reduced form of ribose; a —CH$_2$OH group is present in place of the —CHO group (Section 7.12).

$$
\begin{array}{cc}
\text{CHO} & \text{CH}_2\text{OH} \\
\text{H}{-}\!\!-\text{OH} & \text{H}{-}\!\!-\text{OH} \\
\text{H}{-}\!\!-\text{OH} & \text{H}{-}\!\!-\text{OH} \\
\text{H}{-}\!\!-\text{OH} & \text{H}{-}\!\!-\text{OH} \\
\text{CH}_2\text{OH} & \text{CH}_2\text{OH} \\
\text{D-Ribose} & \text{D-Ribitol}
\end{array}
$$

The complete structural formula of FAD is given in Figure 12.5a.

FIGURE 12.5 Structural formulas of the molecules flavin adenine dinucleotide, FAD (a) and nicotinamide adenine dinucleotide, NAD$^+$ (b).

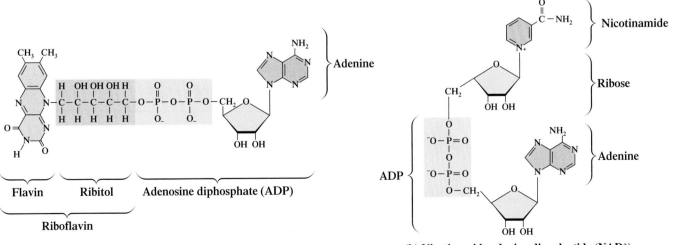

(a) Flavin adenine dinucleotide (FAD)

(b) Nicotinamide adenine dinucleotide (NAD$^+$)

The active portion of FAD in metabolic redox reactions is the flavin subunit of the molecule. The flavin is reduced, converting the FAD to $FADH_2$, a molecule with two additional hydrogen atoms. Thus FAD is the *oxidized* form of the molecule, and $FADH_2$ is the *reduced* form.

$$+ 2H^+ + 2e^- \rightleftharpoons$$

FAD
(oxidized form)

FADH$_2$
(reduced form)

$$R = \text{—} \boxed{\text{Ribitol}} \text{—} \boxed{\text{ADP}}$$

A typical cellular reaction in which FAD serves as the oxidizing agent involves a $-CH_2-CH_2-$ portion of a substrate being oxidized to produce a carbon–carbon double bond.

$$\underset{\substack{\text{Saturated}\\\text{(reduced)}}}{R\text{—}\overset{\text{H}}{\underset{\text{H}}{\text{C}}}\text{—}\overset{\text{H}}{\underset{\text{H}}{\text{C}}}\text{—}R} + FAD \longrightarrow \underset{\substack{\text{Unsaturated}\\\text{(oxidized)}}}{R\text{—}CH\text{=}CH\text{—}R} + FADH_2$$

For an enzyme-catalyzed redox reaction involving removal of two hydrogen atoms, such as this, each removed hydrogen atom is equivalent to a hydrogen *ion*, H^+, plus an electron, e^-.

$$2 \text{ H atoms (removed)} \rightarrow 2H^+ + 2e^-$$

On the basis of this equivalency, the summary equation relating the oxidized and reduced forms of flavin adenine dinucleotide is usually written as

$$\underbrace{2H^+ + 2e^-}_{\text{2 H atoms}} + FAD \rightleftharpoons FADH_2$$

> In metabolic pathways in which it is involved, flavin adenine dinucleotide continually changes back and forth between its oxidized form and its reduced form.
>
> $$2H^+ + 2e^- + \boxed{FAD} \rightleftharpoons \boxed{FADH_2}$$

■ Nicotinamide Adenine Dinucleotide (NAD$^+$, NADH)

Several parallels exist between the characteristics of nicotinamide adenine dinucleotide (NAD$^+$) and those of FAD. Both have coenzyme functions in metabolic redox pathways, both have a B vitamin as a structural component, and both can be represented structurally by using a three-subunit or a six-subunit formulation. In the case of NAD$^+$, the B vitamin present is nicotinamide (Section 10.13).

The three-subunit block diagram for the structure of NAD$^+$ is

$$\boxed{\text{Nicotinamide}} \text{—} \boxed{\text{Ribose}} \text{—} \boxed{\text{ADP}}$$

The six-subunit block diagram, which emphasizes the dinucleotide nature of the coenzyme, as well as the origin of its name, is

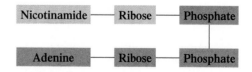

Examination of the detailed structure of NAD$^+$ (Figure 12.5b) reveals the basis for the positive electrical charge. The + sign refers to the positive charge on the nitrogen

atom in the nicotinamide component of the structure; this nitrogen atom has four bonds instead of the usual three (Section 6.6).

The active portion of NAD^+ in metabolic redox reactions is the nicotinamide subunit of the molecule. The nicotinamide is reduced, converting the NAD^+ to NADH, a molecule with one additional hydrogen atom and two additional electrons. Thus NAD^+ is the *oxidized* form of the molecule, and NADH is the *reduced* form.

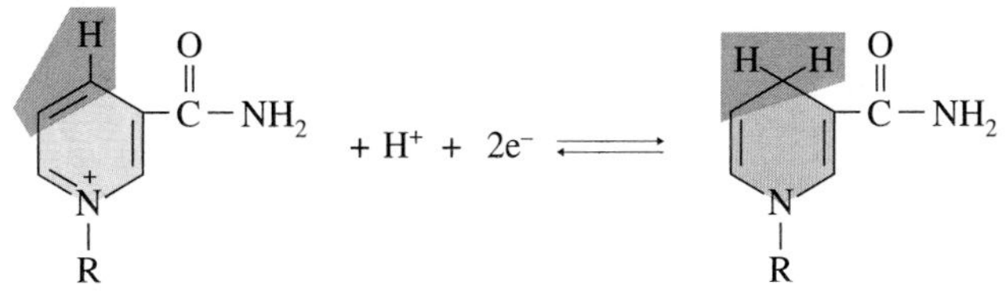

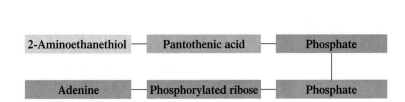

NAD$^+$
(oxidized form)

NADH
(reduced form)

$$R = \text{—} \boxed{\text{Ribose}} \text{—} \boxed{\text{ADP}}$$

A typical cellular reaction in which NAD^+ serves as the oxidizing agent is the oxidation of a secondary alcohol to give a ketone.

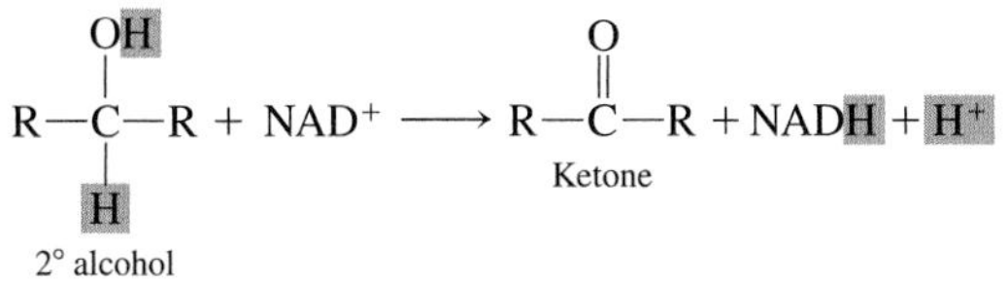

In metabolic pathways, nicotinamide adenine dinucleotide continually changes back and forth between its oxidized form and its reduced form.

$$2H^+ + 2e^- + NAD^+ \rightleftharpoons NADH + H^+$$

In this reaction, one hydrogen atom of the alcohol substrate is directly transferred to NAD^+, whereas the other appears in solution as H^+ ion. Both electrons lost by the alcohol go to the nicotinamide ring in NADH. (Two electrons are required, rather than one, because of the original positive charge on NAD^+.) Thus the summary equation relating the oxidized and reduced forms of nicotinamide adenine dinucleotide is written as

$$\underbrace{2H^+ + 2e^-}_{\text{2 H atoms}} + NAD^+ \rightleftharpoons NADH + H^+$$

■ Coenzyme A (CoA–SH)

Another important coenzyme in metabolic pathways is coenzyme A, a derivative of the B vitamin pantothenic acid (Section 10.13). The three-subunit and six-subunit block diagrams for coenzyme A are

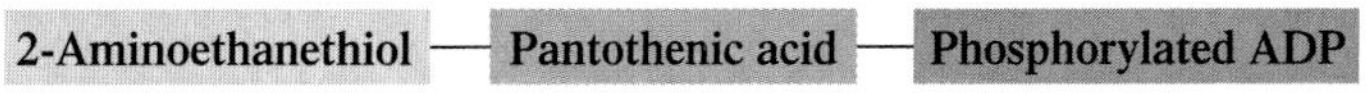

$$\boxed{\text{2-Aminoethanethiol}} \text{—} \boxed{\text{Pantothenic acid}} \text{—} \boxed{\text{Phosphorylated ADP}}$$

and

$$\boxed{\text{2-Aminoethanethiol}} \text{—} \boxed{\text{Pantothenic acid}} \text{—} \boxed{\text{Phosphate}}$$
$$\boxed{\text{Adenine}} \text{—} \boxed{\text{Phosphorylated ribose}} \text{—} \boxed{\text{Phosphate}}$$

Note, in the three-subunit block diagram, that the ADP subunit present is phosphorylated. As shown in Figure 12.6, the complete structural formula for coenzyme A, the phosphorylated version of ADP carries an extra phosphate group attached to carbon 3′ of its ribose.

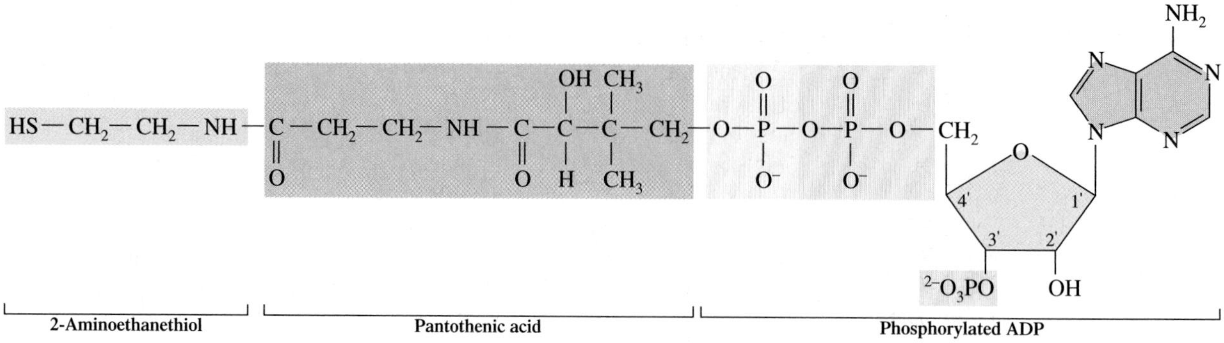

FIGURE 12.6 Structural formula for coenzyme A (CoA–SH).

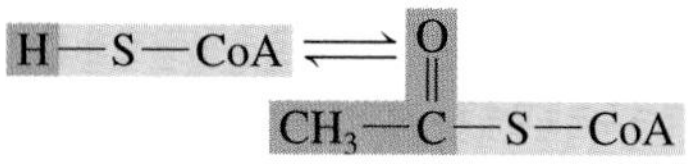

The active portion of coenzyme A is the sulfhydryl group (—SH group; Section 3.20) in the ethanethiol subunit of the coenzyme. For this reason, the abbreviation CoA–SH is used for coenzyme A.

Think of the letter A in the name *coenzyme A* as reflecting a general metabolic function of this substance; it is the transfer of *acetyl* groups in metabolic pathways. An **acetyl group** *is the portion of an acetic acid molecule* (CH_3–COOH) *that remains after the* —*OH group is removed from the carbonyl carbon atom.* An acetyl group bonds to CoA–SH through a thioester bond (Section 5.17) to give acetyl CoA.

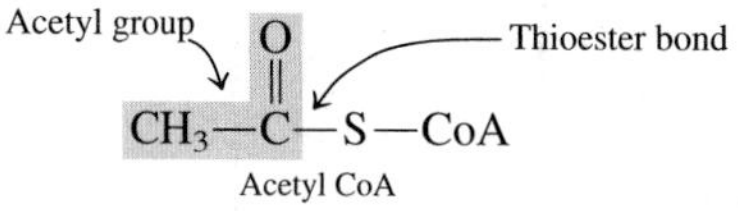

An acetyl group, which can be considered to be derived from acetic acid, has the structure

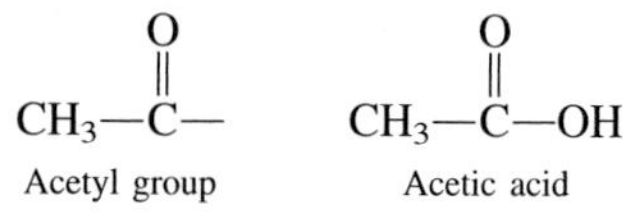

In metabolic pathways, coenzyme A is continually changing back and forth between its CoA form and its acetyl CoA form.

■ Classification of Metabolic Intermediate Compounds

The metabolic intermediate compounds considered in this section can be classified into three groups based on function. The classifications are:

1. Intermediates for the storage of energy and transfer of phosphate groups
2. Intermediates for the transfer of electrons in metabolic redox reactions
3. Intermediates for the transfer of acetyl groups

Figure 12.7 shows the category assignment for the intermediates previously considered.

12.4 High-Energy Phosphate Compounds

In the previous section, we noted that knowing about several key intermediate compounds in metabolic reactions makes it easier to understand the yet-to-come details of metabolic processes. In like manner, knowing about a particular type of bond present in certain

FIGURE 12.7 Classification of metabolic intermediate compounds in terms of function.

Intermediates for the storage of energy and transfer of phosphate groups	ATP ⇌ ADP ⇌ AMP
Intermediates for the transfer of electrons in metabolic redox reactions	FAD ⇌ $FADH_2$ NAD^+ ⇌ NADH
Intermediates for the transfer of acetyl groups	CoA ⇌ acetyl CoA

phosphate-containing metabolic intermediates makes the details of metabolic processes easier to understand.

Several phosphate-containing compounds found in metabolic pathways are known as high-energy compounds. A **high-energy compound** *is a compound that has a greater free energy of hydrolysis than that of a typical compound.* High-energy compounds differ from other compounds in that they contain one or more *very reactive* bonds, often called *strained bonds.* The energy required to break these strained bonds during hydrolysis is less than that generally required to break a chemical bond. Consequently, the balance between the energy needed to break bonds in the reactants and that released by bond formation in the products is such that more than the typical amount of free energy is released during the hydrolysis reaction.

Greater-than-normal electron–electron repulsive forces at specific locations within a molecule are the cause of bond strain. Highly electronegative atoms and/or highly charged atoms occurring together in a molecule cause increased repulsive forces and thus increase bond strain.

Let us specifically consider bond strain as it is related to phosphate-containing organic molecules involved in metabolic pathways. The parent molecule for phosphate groups is phosphoric acid, H_3PO_4, a weak triprotic inorganic acid. This acid exists in aqueous solution in several forms, the dominant form at cellular pH being HPO_4^{2-} ion.

$$\begin{array}{c} O \\ \parallel \\ HO-P-O^- \\ \mid \\ O^- \end{array}$$

Diphosphate and triphosphate ions can also exist in cellular fluids.

$$\begin{array}{cc} O \quad\quad O \\ \parallel \quad\quad \parallel \\ HO-P-O-P-O^- \\ \mid \quad\quad \mid \\ O^- \quad\quad O^- \end{array} \quad\quad \begin{array}{ccc} O \quad\quad O \quad\quad O \\ \parallel \quad\quad \parallel \quad\quad \parallel \\ HO-P-O-P-O-P-O^- \\ \mid \quad\quad \mid \quad\quad \mid \\ O^- \quad\quad O^- \quad\quad O^- \end{array}$$

Note the presence in these three phosphate structures of highly electronegative oxygen atoms, many of which bear negative charges. The factors that can produce bond strain are present when phosphates (mono-, di-, and tri-) are bonded to certain organic molecules.

Table 12.1 gives the structures of commonly encountered phosphate-containing compounds, as well as a numerical parameter—the free energy of hydrolysis—that can be considered a measure of the extent of bond strain in the molecules. The more negative the free energy of hydrolysis, the greater the bond strain. A free-energy release greater than 6.0 kcal/mole is generally considered indicative of bond strain. In Table 12.1, strained bonds within the molecules are noted with a squiggle (~), a notation often employed to denote strained bonds.

12.5 An Overview of Biochemical Energy Production

The energy needed to run the human body is obtained from ingested food through a multistep process that involves several different catabolic pathways. There are four general stages in the biochemical energy production process, and numerous reactions are associated with each stage.

Stage 1: The first stage, *digestion,* begins in the mouth (saliva contains starch-digesting enzymes), continues in the stomach (gastric juices), and is completed in the small intestine (the majority of digestive enzymes and bile salts). The end products of digestion—glucose and other monosaccharides from carbohydrates, amino acids from proteins, and fatty acids and glycerol from fats and oils—are small enough to pass across intestinal membranes and into the blood, where they are transported to the body's cells.

In the definition for a high-energy compound, the term *free energy* rather than simply *energy* was used. Free energy is the amount of energy released by a chemical reaction that is actually available for further use at a given temperature and pressure. In reality, the energy released in a chemical reaction is divisible into two parts. One part, lost as heat, is not available for further use. The other part, the free energy, is available for further use; in cells, it can be used to "drive" reactions that require energy.

In a chemical reaction, the energy balance between bond breaking among reactants (energy input) and new bond formation among products (energy release) determines whether there is a net loss or a net gain of energy.

The designation *high-energy compound* does not mean that a compound is different from other compounds in terms of bonding. High-energy compounds obey the normal rules for chemical bonding. The only difference between such compounds and other compounds is the presence of one or more *strained bonds.* The breaking of such bonds requires lower-than-normal amounts of energy.

The first stage of biochemical energy production, digestion, is not considered part of metabolism because it is extracellular. Metabolic processes are intracellular.

TABLE 12.1
Free Energies of Hydrolysis of Common Phosphate-Containing Metabolic Compounds

Type	Example	Free Energy of Hydrolysis (kcal/mole)
enol phosphates	phosphoenolpyruvate	-14.8
acyl phosphates	1,3-bisphosphoglycerate acetyl phosphate	-11.8 -11.3
guanidine phosphates	creatine phosphate arginine phosphate	-10.3 -9.1
triphosphates	ATP $\longrightarrow$ AMP + PP$_i$^{a} ATP $\longrightarrow$ ADP + P$_i$^{a}	-7.7 -7.5
diphosphates	PP$_i$ $\longrightarrow$ 2P$_i$ ADP $\longrightarrow$ AMP + P$_i$	-7.8 -7.5
sugar phosphates	glucose 1-phosphate fructose 6-phosphate AMP $\longrightarrow$ adenosine + P$_i$ glucose 6-phosphate glycerol 3-phosphate	-5.0 $+->8$ $+->4$ $+->3$ -2.2

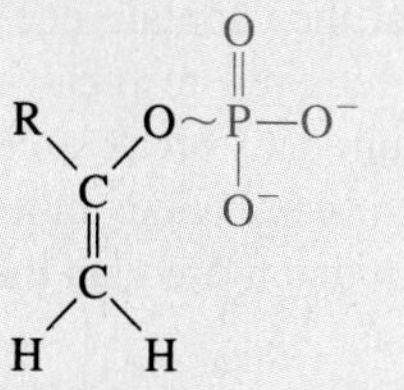

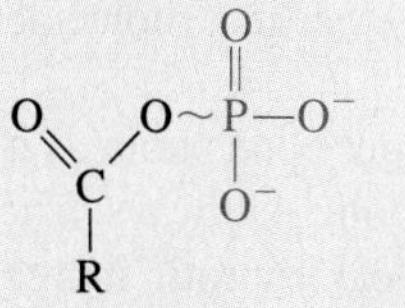

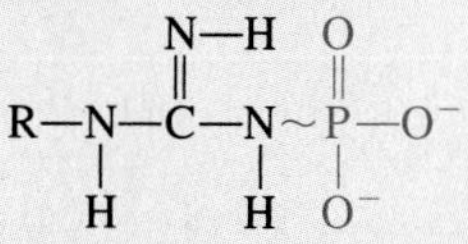

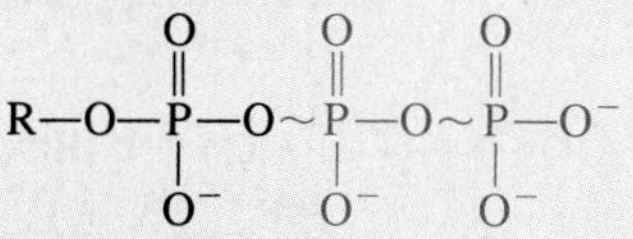

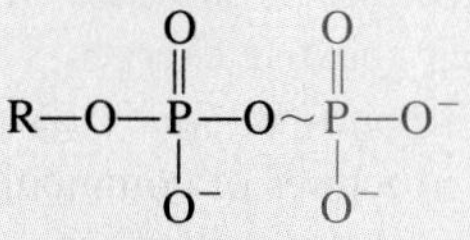

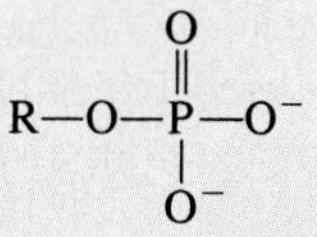

The —PO$_3$$^{2-}$ group as part of a larger organic phosphate molecule is referred to as a *phosphoryl group.*

^{a}The notation P$_i$ is used as a general designation for any free monophosphate species present in cellular fluid. Free diphosphate ions are designated as PP$_i$ ("i" stands for *inorganic*).

Stage 2: The second stage, *acetyl group formation,* involves numerous reactions, some of which occur in the cytosol of cells and some in cellular mitochondria. The small molecules from digestion are further oxidized during this stage. Primary products include two-carbon acetyl units (which become attached to coenzyme A to give acetyl CoA) and the reduced coenzyme NADH.

Stage 3: The third stage, the *citric acid cycle,* occurs inside mitochondria. Here acetyl groups are oxidized to produce CO_2 and energy. Some of the energy released by these reactions is lost as heat, and some is carried by the reduced coenzymes NADH and $FADH_2$ to the fourth stage. The CO_2 that we exhale as part of the breathing process comes primarily from this stage.

Stage 4: The fourth stage, the *electron transport chain and oxidative phosphorylation*, also occurs inside mitochondria. NADH and FADH$_2$ supply the "fuel" (hydrogen ions and electrons) needed for the production of ATP molecules, the primary energy carriers in metabolic pathways. Molecular O$_2$, inhaled via breathing, is converted to H$_2$O in this stage.

The reactions in stages 3 and 4 are the same for all types of foods (carbohydrates, fats, proteins). These reactions constitute the common metabolic pathway. The **common metabolic pathway** *is the sum total of the biochemical reactions of the citric acid cycle, the electron transport chain, and oxidative phosphorylation.* The remainder of this chapter deals with the common metabolic pathway. The reactions of stages 1 and 2 of biochemical energy production differ for different types of foodstuffs. They are discussed in Chapters 13–15, which cover the metabolism of carbohydrates, fats (lipids), and proteins, respectively.

The Chemistry at a Glance feature on page 430 summarizes the four general stages in the process of production of biochemical energy from ingested food. This diagram is a *very simplified* version of the "energy generation" process that occurs in the human body, as will become clear from the discussions presented in later sections of this chapter, which give further details of the process.

12.6 The Citric Acid Cycle

The **citric acid cycle** *is the series of biochemical reactions in which the acetyl portion of acetyl CoA is oxidized to carbon dioxide and the reduced coenzymes FADH$_2$ and NADH are produced.* This cycle, stage 3 of biochemical energy production, gets its name from the first intermediate product in the cycle, citric acid. It is also known as the *Krebs cycle,* after its discoverer Hans Adolf Krebs (see Figure 12.8), and as the *tricarboxylic acid cycle,* in reference to the three carboxylate groups present in citric acid. Figure 12.9 lists the compounds produced in all eight steps of the citric acid cycle.

We shall now consider the individual steps of the cycle in detail. As we go through these steps, we will observe two important types of reactions: (1) oxidation, which produces NADH and FADH$_2$, and (2) decarboxylation, wherein a carbon chain is shortened by the removal of a carbon atom as CO$_2$.

■ Reactions of the Citric Acid Cycle

Step 1: *Formation of Citrate.* Acetyl CoA, the two-carbon degradation product of carbohydrates, fats, and proteins (Section 12.5), enters the cycle by combining with the four-carbon keto dicarboxylate species oxaloacetate. This results in the transfer of the acetyl group from coenzyme A to oxaloacetate, producing the C$_6$ citrate species and free coenzyme A.

FIGURE 12.8 Hans Adolf Krebs (1900–1981), a German-born British biochemist, received the 1953 Nobel Prize in medicine for establishing the relationships among the different compounds in the cycle that carries his name, the Krebs cycle.

The formation of citryl CoA involves addition of acetyl CoA to the carbon–oxygen double bond. A hydrogen atom of the acetyl —CH$_3$ group adds to the oxygen atom of the double bond, and the remainder of the acetyl CoA adds to the carbon atom of the double bond. Citryl CoA formation is a condensation reaction (Section 3.9) because a new carbon–carbon bond is formed.

A *synthase* is an enzyme that makes a new covalent bond during a reaction without the direct involvement of an ATP molecule.

There are two parts to the reaction: (1) the condensation of acetyl CoA and oxaloacetate to form citryl CoA, a process catalyzed by the enzyme *citrate synthase,* and (2) hydrolysis of the thioester bond in citryl CoA to produce CoA—SH and citrate, also catalyzed by the enzyme *citrate synthase.*

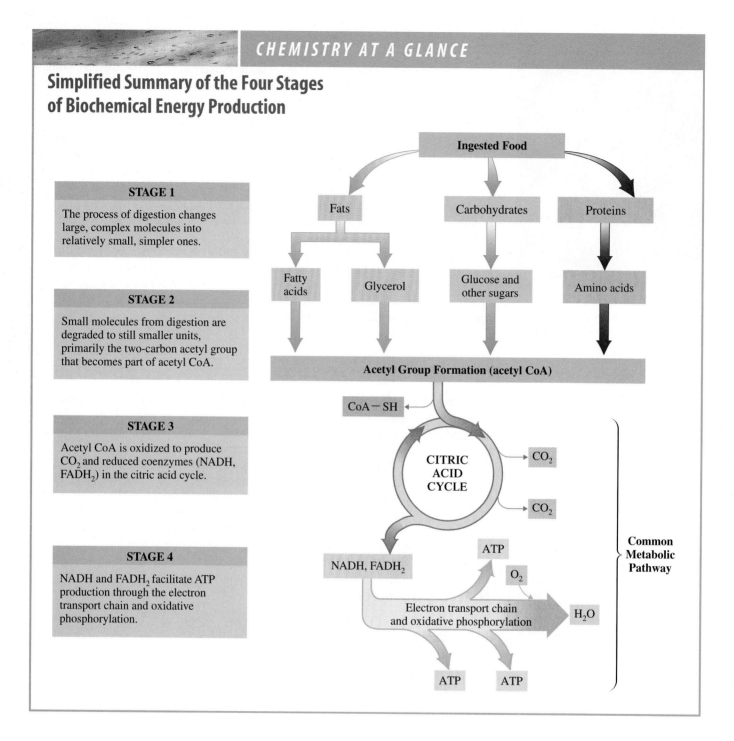

CHEMISTRY AT A GLANCE

Simplified Summary of the Four Stages of Biochemical Energy Production

STAGE 1

The process of digestion changes large, complex molecules into relatively small, simpler ones.

STAGE 2

Small molecules from digestion are degraded to still smaller units, primarily the two-carbon acetyl group that becomes part of acetyl CoA.

STAGE 3

Acetyl CoA is oxidized to produce CO_2 and reduced coenzymes (NADH, $FADH_2$) in the citric acid cycle.

STAGE 4

NADH and $FADH_2$ facilitate ATP production through the electron transport chain and oxidative phosphorylation.

Step 2: *Formation of Isocitrate.* Citrate is converted to its less symmetrical isomer isocitrate in an isomerization process that involves a dehydration followed by a hydration, both catalyzed by the enzyme *aconitase*. The net result of these reactions is that the —OH group from citrate is moved to a different carbon atom.

FIGURE 12.9 The citric acid cycle. Details of the numbered steps are given in the text.

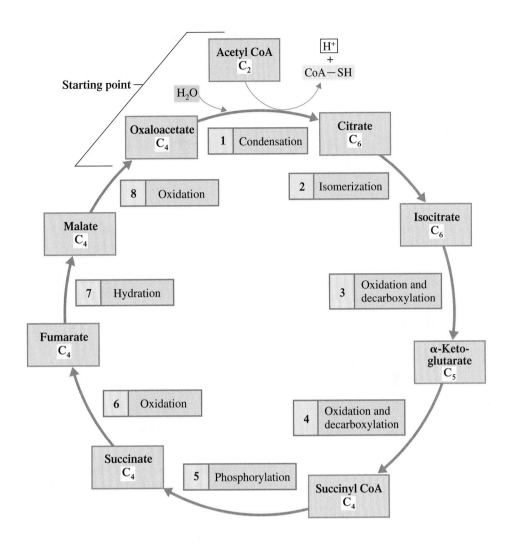

Citrate is a tertiary alcohol and isocitrate a secondary alcohol. Tertiary alcohols are not readily oxidized; secondary alcohols are easier to oxidize (Section 3.9). The next step in the cycle involves oxidation.

All acids found in the citric acid cycle exist as carboxylate ions (Section 5.8) at cellular pH.

Step 3: *Oxidation of Isocitrate and Formation of CO_2.* This step involves oxidation–reduction (the first of four redox reactions in the citric acid cycle) and decarboxylation. The reactants are a NAD^+ molecule and isocitrate. The reaction, catalyzed by *isocitrate dehydrogenase*, is complex: (1) Isocitrate is oxidized to a ketone (oxalosuccinate) by NAD^+, releasing two hydrogens. (2) One hydrogen and two electrons are transferred to NAD^+ to form NADH; the remaining hydrogen ion (H^+) is released. (3) The oxalosuccinate remains bound to the enzyme and undergoes decarboxylation (loses CO_2), which produces the C_5 α-ketoglutarate (a keto dicarboxylate species).

This step yields the first molecules of CO_2 and NADH in the cycle.

Step 4: *Oxidation of α-Ketoglutarate and Formation of CO_2.* This second redox reaction of the cycle involves one molecule each of NAD^+, CoA—SH, and α-ketoglutarate.

The CO_2 molecules produced in Steps 3 and 4 of the citric acid cycle are the CO_2 molecules we exhale in the process of respiration.

The catalyst is an aggregate of three enzymes called the *α-ketoglutarate dehydrogenase* complex. As in Step 3, both oxidation and decarboxylation occur. There are three products: CO_2, NADH, and the C_4 species succinyl CoA.

$$\begin{array}{c} COO^- \\ | \\ CH_2 \\ | \\ CH_2 \\ | \\ C{=}O \\ | \\ COO^- \end{array} + NAD^+ + H^+ + CoA{-}SH \xrightarrow[\text{complex}]{\text{α-Ketoglutarate}\atop\text{dehydrogenase}} \begin{array}{c} COO^- \\ | \\ CH_2 \\ | \\ CH_2 \\ | \\ C{=}O \\ | \\ S{-}CoA \end{array} + NADH + CO_2 + H^+$$

α-Ketoglutarate · Succinyl CoA

The thioester bond in succinyl CoA is a strained bond. Its hydrolysis releases energy, which is trapped by GTP formation. The function of the GTP produced is similar to that of ATP: to store energy in the form of a high-energy phosphate bond (Section 12.4).

Step 5: *Thioester bond cleavage in Succinyl CoA and Phosphorylation of GDP.* Two molecules react with succinyl CoA—a molecule of GDP (similar to ADP; Section 12.3) and a free phosphate group (P_i). The enzyme *succinyl CoA synthase* removes coenzyme A by thioester bond cleavage. The energy released is used to combine GDP and P_i to form GTP. Succinyl CoA has been converted to succinate.

$$\begin{array}{c} COO^- \\ | \\ CH_2 \\ | \\ CH_2 \\ | \\ C{=}O \\ | \\ S{-}CoA \end{array} + GDP + P_i \xrightarrow[\text{CoA synthase}]{\text{Succinyl}} \begin{array}{c} COO^- \\ | \\ CH_2 \\ | \\ CH_2 \\ | \\ COO^- \end{array} + GTP + CoA{-}SH$$

Succinyl CoA · Succinate

Steps 6 through 8 of the citric acid cycle involve a sequence of functional group changes that we have encountered several times in the organic sections of the text. The reaction sequence is

$$\text{Alkane} \xrightarrow[\text{(dehydrogenation)}]{\overset{①}{\text{Oxidation}}} \text{alkene} \xrightarrow{\overset{②}{\text{Hydration}}} \begin{array}{c}\text{secondary}\\\text{alcohol}\end{array} \xrightarrow[\text{(dehydrogenation)}]{\overset{③}{\text{Oxidation}}} \text{ketone}$$

Step 6: *Oxidation of Succinate.* This is the third redox reaction of the cycle. The enzyme involved is *succinate dehydrogenase,* and the oxidizing agent is FAD rather than NAD^+. Two hydrogen atoms are removed from the succinate to produce fumarate, a C_4 species with a *trans* double bond. FAD is reduced to $FADH_2$ in the process.

Fumarate, with its *trans* double bond, is an essential metabolic intermediate in both plants and animals. Its isomer, with a *cis* double bond, is called maleate, and it is toxic and irritating to tissues. *Succinate dehydrogenase* produces only the *trans* isomer of this unsaturated diacid.

$$\begin{array}{c} COO^- \\ | \\ CH_2 \\ | \\ CH_2 \\ | \\ COO^- \end{array} + FAD \xrightarrow[\text{dehydrogenase}]{\text{Succinate}} \begin{array}{c} H \quad COO^- \\ \diagdown \; / \\ C \\ \| \\ C \\ / \; \diagdown \\ ^-OOC \quad H \end{array} + FADH_2$$

Succinate · Fumarate

Step 7: *Hydration of Fumarate.* The enzyme *fumarase* catalyzes the addition of water to the double bond of fumarate. The enzyme is stereospecific, so only the L isomer of the product malate is produced.

$$\begin{array}{c} COO^- \\ | \\ C{-}H \\ \| \\ H{-}C \\ | \\ COO^- \end{array} + H_2O \xrightarrow{\text{Fumarase}} \begin{array}{c} COO^- \\ | \\ HO{-}C{-}H \\ | \\ H{-}C{-}H \\ | \\ COO^- \end{array}$$

Fumarate · L-Malate

Step 8: *Oxidation of L-Malate to Regenerate Oxaloacetate.* In the fourth oxidation–reduction reaction of the cycle, a molecule of NAD^+ reacts with malate, picking up two hydrogen atoms with their associated energy to form $NADH + H^+$. The product of this reaction is oxaloacetate, so we are back where we started. The oxaloacetate formed in this step can combine with another molecule of acetyl CoA (Step 1), and the cycle can begin again.

$$
\underset{\text{L-Malate}}{
\begin{array}{c}
COO^- \\
| \\
HO-C-H \\
| \\
CH_2 \\
| \\
COO^-
\end{array}
} + NAD^+
\xrightarrow[\text{dehydrogenase}]{\text{Malate}}
\underset{\text{Oxaloacetate}}{
\begin{array}{c}
COO^- \\
| \\
C=O \\
| \\
CH_2 \\
| \\
COO^-
\end{array}
} + NADH + H^+
$$

Summary of the Citric Acid Cycle

An overall summary equation for the citric acid cycle is obtained by adding together the individual reactions of the cycle:

$$\text{Acetyl CoA} + 3NAD^+ + FAD + GDP + P_i + 2H_2O \longrightarrow$$
$$2CO_2 + CoA{-}SH + 3NADH + 2H^+ + FADH_2 + GTP$$

Important features of the cycle include the following:

1. The reactions of the cycle take place in the mitochondrial matrix, except the succinate dehydrogenase reaction that involves FAD. The enzyme that catalyzes this reaction is an integral part of the inner mitochondrial membrane.
2. The "fuel" for the cycle is acetyl CoA, obtained from the breakdown of carbohydrates, fats, and proteins.
3. Four of the cycle reactions involve oxidation and reduction. The oxidizing agent is either NAD^+ (three times) or FAD (once). The operation of the cycle depends on the availability of these oxidizing agents.
4. In redox reactions, NAD^+ is the oxidizing agent when a carbon–oxygen double bond is formed; FAD is the oxidizing agent when a carbon–carbon double bond is formed.
5. The three NADH and one $FADH_2$ that are formed during the cycle carry electrons and H^+ to the electron transport chain (Section 14.7) through which ATP is synthesized.
6. Two carbon atoms enter the cycle as the acetyl unit of acetyl CoA, and two carbon atoms leave the cycle as two molecules of CO_2. The carbon atoms that enter and leave are not the same ones. The carbon atoms that leave during one turn of the cycle are carbon atoms that entered during the previous turn of the cycle.
7. Four B vitamins are necessary for the proper functioning of the cycle: riboflavin (in both FAD and the α-ketoglutarate dehydrogenase complex), nicotinamide (in NAD^+), pantothenic acid (in CoA—SH), and thiamin (in the α-ketoglutarate dehydrogenase complex).
8. One high-energy GTP molecule is produced by phosphorylation.

The Chemistry at a Glance feature on page 434 gives a detailed diagrammatic summary of the reactions that occur in the citric acid cycle.

The eight B vitamins and their structures were discussed in Section 10.13.

Regulation of the Citric Acid Cycle

The rate at which the citric acid cycle operates is controlled by the body's need for energy (ATP). When the body's ATP supply is high, the ATP present inhibits the activity of citrate synthase, the enzyme in Step 1 of the cycle. When energy is being used at a high rate, a state of low ATP and high ADP concentrations, the ADP activates citrate synthase and the cycle speeds up. A similar control mechanism exists at Step 3, which involves isocitrate dehydrogenase; here NADH acts as an inhibitor and ADP as an activator.

CHEMISTRY AT A GLANCE

Summary of the Reactions of the Citric Acid Cycle

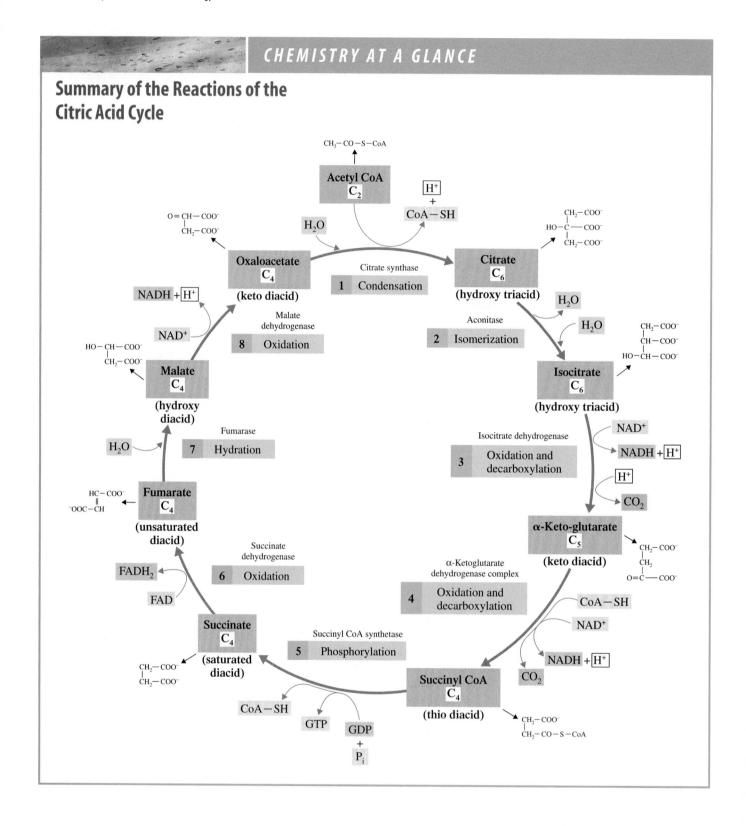

12.7 The Electron Transport Chain

The *electron transport chain* is also frequently called the *respiratory chain*.

The NADH and $FADH_2$ produced in the citric acid cycle pass to the electron transport chain. The **electron transport chain** *is a series of biochemical reactions in which electrons and hydrogen ions from NADH and $FADH_2$ are passed to intermediate carriers and* *then ultimately react with molecular oxygen to produce water.* NADH and $FADH_2$ are oxidized in this process.

$$NADH + H^+ \longrightarrow NAD^+ + 2H^+ + 2e^-$$
$$FADH_2 \longrightarrow FAD + 2H^+ + 2e^-$$

The oxygen involved in the water formation associated with the electron transport chain is the oxygen we breathe.

Water is formed when the electrons and hydrogen ions that originate from these reactions react with molecular oxygen.

$$O_2 + 4e^- + 4H^+ \longrightarrow 2H_2O$$

The electrons that pass through the various steps of the electron transport chain (ETC) lose some energy with each transfer along the chain. Some of this "lost" energy is used to make ATP from ADP (oxidative phosphorylation), as we will see in Section 12.8.

The enzymes and electron carriers needed for the ETC are located along the inner mitochondrial membrane. Within this membrane are four distinct protein complexes, each containing some of the molecules needed for the ETC process to occur. These four protein complexes, which are tightly bound to the membrane, are

Complex I: NADH–coenzyme Q reductase
Complex II: Succinate–coenzyme Q reductase
Complex III: Coenzyme Q–cytochrome *c* reductase
Complex IV: Cytochrome *c* oxidase

Two electron carriers, coenzyme Q and cytochrome *c,* which are not firmly attached to the membrane, serve as mobile electron carriers that shuttle electrons between the various complexes.

Our discussion of the individual reactions that occur in the ETC is divided into four parts, each part dealing with the reactions associated with one of the four protein complexes.

■ Complex I: NADH–Coenzyme Q Reductase

NADH, from the citric acid cycle, is the source for the electrons that are processed through complex I, the largest of the four protein complexes. Complex I contains over 40 subunits, including flavin mononucleotide (FMN) and several iron–sulfur proteins (FeSP). The net result of electron movement through complex I is the transfer of electrons from NADH to coenzyme Q (CoQ), a result implied by the name of complex I: *NADH–coenzyme Q reductase.* The actual electron transfer process is not, however, a single-step direct transfer of electrons from NADH to CoQ; several intermediate carriers are involved.

The first electron transfer step that occurs in complex I involves the interaction of NADH with flavin mononucleotide (FMN). The NADH is oxidized to NAD^+ (which can again participate in the citric acid cycle) as it passes two hydrogen ions and two electrons to FMN, which is reduced to $FMNH_2$.

$$NADH + H^+ \xrightarrow{\text{Oxidation}} NAD^+ + 2H^+ + 2e^-$$

$$2H^+ + 2e^- + FMN \xrightarrow{\text{Reduction}} FMNH_2$$

The $FMN/FMNH_2$ pair is the third biochemical situation we have encountered in which a flavin molecule is present. The other two are the $FAD/FADH_2$ pair and the B vitamin riboflavin. FMN differs from FAD in not having an adenine nucleotide. Both FMN and FAD are synthesized within the body from riboflavin.

NADH supplies both electrons and one of the H^+ ions that are transferred; the other H^+ ion comes from the cellular solution. The actual changes that occur within the structure of FMN as it accepts the two electrons and two H^+ ions are shown in Figure 12.10a.

The next steps involve transfer of electrons from the reduced $FMNH_2$ through a series of iron/sulfur proteins (FeSPs). The iron present in these FeSPs is Fe^{3+}, which is reduced to Fe^{2+}. The two H atoms of $FMNH_2$ are released to solution as two H^+ ions.

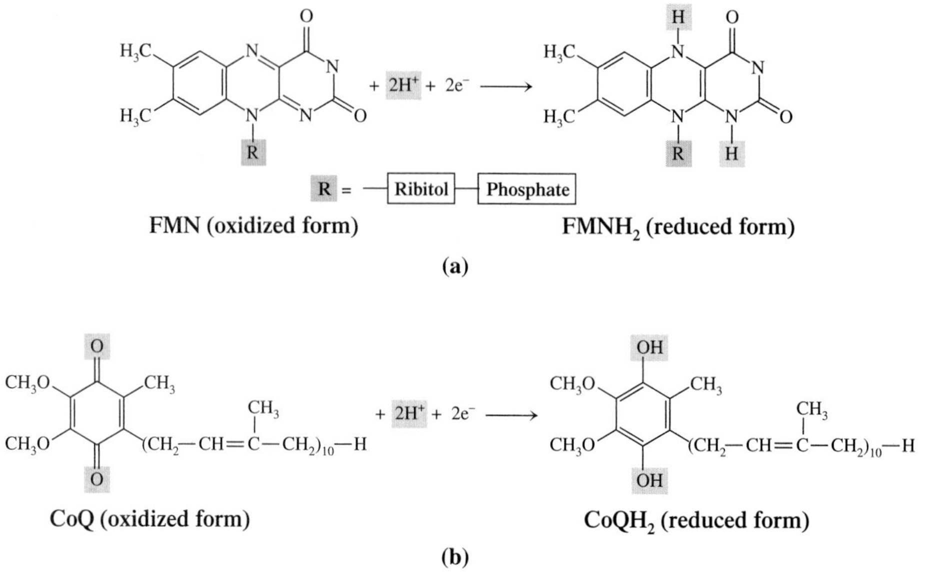

FIGURE 12.10 Structural characteristics of the electron carriers flavin mononucleotide and coenzyme Q. (a) The oxidized form (FMN) and reduced form ($FMNH_2$) of the electron carrier flavin mononucleotide. (b) The oxidized form (CoQ) and reduced form ($CoQH_2$) of the electron carrier coenzyme Q.

FMN (oxidized form) $+ 2H^+ + 2e^- \longrightarrow$ $FMNH_2$ (reduced form)

$R =$ | Ribitol | Phosphate |

(a)

CoQ (oxidized form) $+ 2H^+ + 2e^- \longrightarrow$ $CoQH_2$ (reduced form)

(b)

Two FeSP molecules are needed to accommodate the two electrons released by $FMNH_2$ because an Fe^{3+}/Fe^{2+} reduction involves only one electron.

$$FMNH_2 \xrightarrow{\text{Oxidation}} FMN + 2H^+ + 2e^-$$

$$2e^- + 2Fe(III)SP \xrightarrow{\text{Reduction}} 2Fe(II)SP$$

In the final complex I reaction, Fe(II)SP is reconverted into Fe(III)SP as each of two Fe(II)SP units passes an electron to CoQ, changing it from its oxidized form (CoQ) to its reduced form ($CoQH_2$).

$$2Fe(II)SP \xrightarrow{\text{Oxidation}} 2Fe(III)SP + 2e^-$$

$$2e^- + 2H^+ + CoQ \xrightarrow{\text{Reduction}} CoQH_2$$

Coenzyme Q, in both its oxidized and reduced forms, is lipid soluble and can move laterally within the mitochondrial membrane. Its function is to shuttle its newly acquired electrons to complex III, where it becomes the initial substrate for reactions at this complex.

The Q in the designation coenzyme Q comes from the molecule quinone. Structurally, coenzyme Q is a quinone derivative. In its most common form, coenzyme Q has a long carbon chain containing 10 isoprene units (Section 2.6) attached to its quinone unit. The actual changes that occur within the structure of CoQ as it accepts the two electrons and the two H^+ ions involve the quinone part of its structure, as is shown in Figure 12.10b. The two H^+ ions that CoQ picks up in forming $CoQH_2$ come from solution.

The molecule quinone, a cyclic ketone (Section 4.2), has the structure

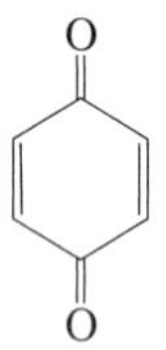

■ Complex II: Succinate–Coenzyme Q Reductase

Complex II, which is much smaller than complex I, contains only 4 subunits, including two FeSPs. This complex is used to process the $FADH_2$ that is generated in the citric acid cycle when succinate is converted to fumarate. (Thus the use of the term *succinate* in the name of complex II.)

CoQ is associated with the operations in complex II in a manner similar to its actions in complex I. It is the final recipient of the electrons from $FADH_2$, with iron–sulfur proteins serving as intermediaries.

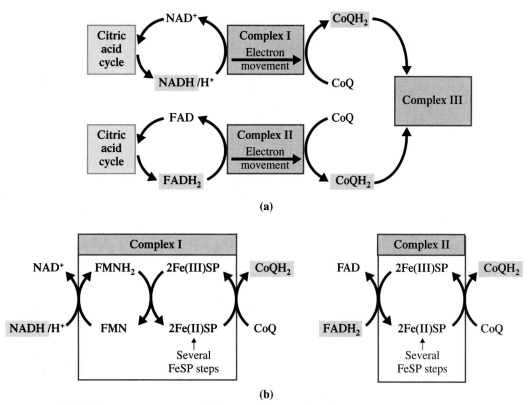

FIGURE 12.11 An overview of electron movement through complexes I and II of the electron transport chain. (a) $CoQH_2$ carries electrons from both complexes I and II to complex III. (b) NADH is the substrate for complex I and $FADH_2$ is the substrate for complex II. $CoQH_2$ is the common product from both electron transfer processes.

Thus complexes I and II produce a common product, the reduced form of coenzyme Q ($CoQH_2$) As was the case with complex I, the reduced $CoQH_2$ shuttles electrons to complex III.

$$FADH_2 \xrightarrow{\text{Oxidation}} FAD + 2H^+ + 2e^-$$

$$2e^- + 2Fe(III)SP \xrightarrow{\text{Reduction}} 2Fe(II)SP$$

$$2Fe(II)SP \xrightarrow{\text{Oxidation}} 2Fe(III)SP + 2e^-$$

$$2e^- + 2H^+ + COQ \xrightarrow{\text{Reduction}} CoQH_2$$

Figure 12.11 summarizes the electron transport chain reactions associated with complexes I and II. In Figure 12.11a the net process is shown with only starting and end products shown. In Figure 12.11b individual reaction detail is shown. Note the general pattern that is developing for the electron carriers. They are oxidized in one step (accept electrons) and then regenerated (reduced; lose electrons) in the next step so that they can again participate in electron transport chain reactions.

All H^+ ions required for the reactions of NADH, CoQ, and O_2 in the ETC come from the matrix side of the inner mitochondrial membrane.

■ Complex III: Coenzyme Q–Cytochrome *c* Reductase

Complex III contains 11 different subunits. Electron carriers present include several iron–sulfur proteins as well as several cytochromes. A **cytochrome** *is a heme-containing*

FIGURE 12.12 Electron movement through complex III is initiated by the electron carrier CoQH₂. In several steps the electrons are passed to cyt c.

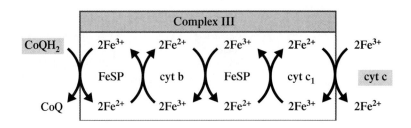

FIGURE 12.12 Electron movement through complex III is initiated by the electron carrier $CoQH_2$. In several steps the electrons are passed to cyt c.

protein in which reversible oxidation and reduction of an iron atom occur. Heme, a compound also present in hemoglobin and myoglobin (Section 9.12), has the structure

> In cytochromes the iron of the heme is involved in redox reactions in which the iron changes back and forth between the +2 and +3 oxidation states.

> In cytochromes the heme present is bound to protein in such a way as to prevent the heme from combining with oxygen as it does when it is present in hemoglobin.

> Iron/sulfur protein (FeSP) is a *non-heme iron protein.* Most proteins of this type contain sulfur, as is the case with FeSP. Often the iron is bound to the sulfur atom in the amino acid cysteine.

> A feature that all steps in the ETC share is that as each electron carrier passes electrons along the chain, it becomes reoxidized and thus able to accept more electrons.

Heme-containing proteins function similarly to FeSP; iron changes back and forth between the +3 and +2 oxidation states.

Various cytochromes, abbreviated cyt a, cyt b, cyt c, and so on, differ from each other in (1) their protein constituents, (2) the manner in which the heme is bound to the protein, and (3) attachments to the heme ring. Again, because the Fe^{3+}/Fe^{2+} system involves only a one-electron change, two cytochrome molecules are needed to move two electrons along the chain.

The initial substrate for complex III is $CoQH_2$ molecules carrying the electrons that have been processed through complex I (from NADH) and also those processed through complex II (from $FADH_2$). The electron transfer process proceeds from $CoQH_2$ to an FeSP, then to cyt b, then to another FeSP, then to cyt c_1, and finally to cyt c. Cyt c, like $CoQH_2$, can move laterally in the intermembrane space; it delivers its electrons to complex IV. Cyt c is the only one of the cytochromes that is water soluble. All of the electron transfer steps in complex III involve just electrons; no H^+ ions are involved. Figure 12.12 shows diagrammatically the electron transfer steps associated with complex III.

■ Complex IV: Cytochrome c Oxidase

Complex IV contains 13 subunits, including two cytochromes. The electron movement flows from cyt c (carrying electrons from complex III) to cyt a to cyt a_3. In the final step

FIGURE 12.13 The electron transfer pathway through complex IV (cytochrome c oxidase). Electrons pass through both copper and iron centers and in the last step interact with molecular O_2. Reduction of one O_2 molecule requires the passage of four electrons through complex IV, one at a time.

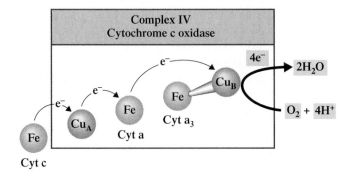

CHEMISTRY AT A GLANCE

Summary of the Flow of Electrons Through the Four Complexes of the Electron Transport Chain

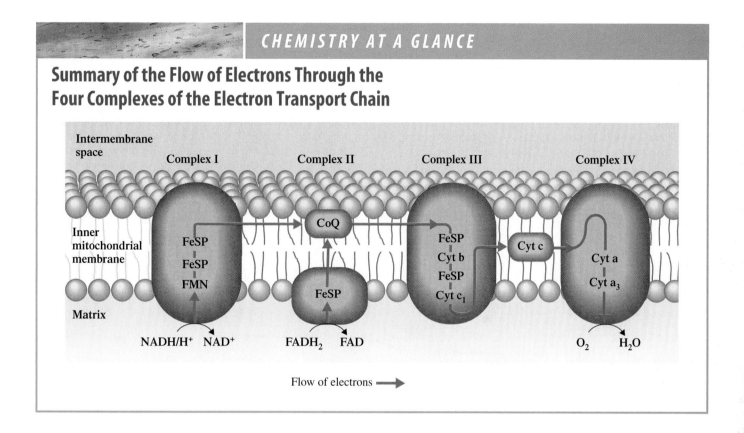

Flow of electrons ⟶

of electron transfer, the electrons from cyt a_3 and hydrogen ions combine with oxygen (O_2) to form water.

$$O_2 + 4H^+ + 4e^- \longrightarrow 2H_2O$$

It is estimated that 95% of the oxygen used by cells serves as the final electron acceptor for the ETC.

The two cytochromes present in cytochrome c oxidase (a and a_3) differ from previously encountered cytochromes in that each has a copper atom associated with it in addition to its iron center. The copper atom sites participate in the electron transfer process as do the iron atom sites, with the copper atoms going back and forth between the reduced Cu^+ state and the oxidized Cu^{2+} state. Figure 12.13 shows the electron transfer sequence through these copper and iron sites.

The Chemistry at a Glance feature is a schematic diagram summarizing the flow of electrons through the four complexes of the electron transport chain.

12.8 Oxidative Phosphorylation

Oxidative phosphorylation *is the biochemical process by which ATP is synthesized from ADP as a result of the transfer of electrons and hydrogen ions from NADH or FADH$_2$ to O$_2$ through the electron carriers involved in the electron transport chain.* Oxidative phosphorylation is conceptually simple but mechanistically complex. Learning the "details" of oxidative phosphorylation has been—and still is—one of the most challenging research areas in biochemistry.

One concept central to the oxidative phosphorylation process is that of *coupled* reactions. **Coupled reactions** *are pairs of biochemical reactions that occur concurrently in which energy released by one reaction is used in the other reaction.* Oxidative phosphorylation and the oxidation reactions of the electron transport chain are coupled systems.

CHEMICAL CONNECTIONS Cyanide Poisoning

Inhalation of hydrogen cyanide gas (HCN) or ingestion of solid potassium cyanide (KCN) rapidly inhibits the electron transport chain in all tissues, making cyanide one of the most potent and rapidly acting poisons known. The attack point for the cyanide ion (CN^-) is cytochrome c oxidase, the last of the four protein complexes in the electron transport chain. Cyanide inactivates this complex by bonding itself to the Fe^{3+} in the complex's heme portions. As a result, Fe^{3+} is unable to transfer electrons to oxygen, blocking the cell's use of oxygen. Death results from tissue asphyxiation, particularly of the central nervous system. Cyanide also binds to the heme group in hemoglobin, blocking oxygen transport in the bloodstream.

One treatment for cyanide poisoning is to administer various nitrites NO_2^-, which oxidize the iron atoms of hemoglobin to Fe^{3+}. This form of hemoglobin helps draw CN^- back into the bloodstream, where it can be converted to thiocyanate (SCN^-) by thiosulfate ($S_2O_3^{2+}$), which is administered along with the nitrite (see the accompanying figure).

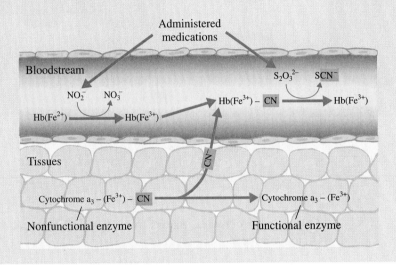

Oxidative phosphorylation is not the only process by which ATP is produced in cells. A second process, *substrate phosphorylation* (Section 13.2), can also be an ATP source. However, the amount of ATP produced by this second process is much less than that produced by oxidative phosphorylation.

The interdependence (coupling) of ATP synthesis with the reactions of the ETC is related to the movement of protons (H^+ ions) across the inner mitochondrial membrane. Three of the four protein complexes involved in the ETC chain (I, III, and IV) have a second function besides electron transfer down the chain. They also serve as "proton pumps," transferring protons from the matrix side of the inner mitochondrial membrane to the intermembrane space (Figure 12.14).

Some of the H^+ ions crossing the inner mitochondrial membrane come from the reduced electron carriers, and some come from the matrix; the details of how the H^+ ions cross the inner mitochondrial membrane are not fully understood.

For every two electrons passed through the ETC, four protons cross the inner mitochondrial membrane through complex I, four through complex III, and two more through

FIGURE 12.14 A second function for protein complexes I, III, and IV involved in the electron transport chain is that of proton pump. For every two electrons passed through the ETC, 10 H^+ ions are transferred from the mitochondrial matrix to the intermembrane space through these complexes.

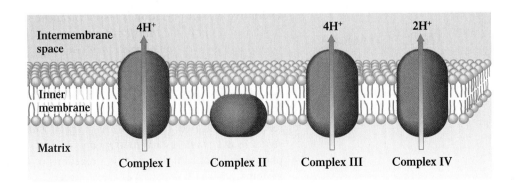

FIGURE 12.15 Formation of ATP accompanies the flow of protons from the intermembrane space back into the mitochondrial matrix. The proton flow results from an electrochemical gradient across the inner mitochondrial membrane.

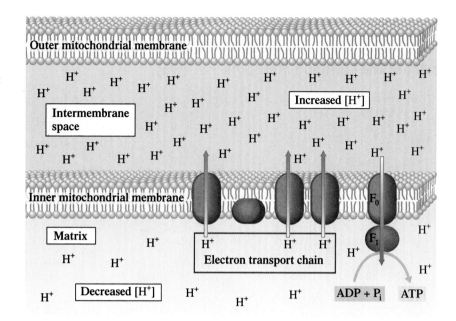

complex IV. This proton flow causes a buildup of H^+ ions (protons) in the intermembrane space; this high concentration of protons becomes the basis for ATP synthesis (Figure 12.15).

The "proton flow" explanation for ATP–ETC coupling is formally called chemiosmotic coupling. **Chemiosmotic coupling** *is an explanation for the coupling of ATP synthesis with electron transport chain reactions that requires a proton gradient across the inner mitochondrial membrane.* The main concepts in this explanation for coupling follow.

The difference in H^+ ion concentration between the two sides of the inner mitochondrial membrane causes a pH difference of about 1.4 units. A pH difference of 1.4 units means that the intermembrane space, the more acidic region, has 25 times more protons than the matrix.

1. The result of the pumping of protons from the mitochondrial matrix across the inner mitochondrial membrane is a higher concentration of protons in the intermembrane space than in the matrix. This concentration difference constitutes an *electrochemical (proton) gradient.* A chemical gradient exists whenever a substance has a higher concentration in one region than in another. Because the proton has an electrical charge (H^+ ion), an electrical gradient also exists. Potential energy is always associated with an electrochemical gradient.

2. A spontaneous flow of protons from the region of high concentration to the region of low concentration occurs because of the electrochemical gradient. This proton flow is not through the membrane itself (it is not permeable to H^+ ions) but rather through enzyme complexes called *ATP synthases* located on the inner mitochondrial membrane (Section 12.2). This proton flow through the ATP synthases "powers" the synthesis of ATP. ATP synthases are thus the *coupling factors* that link the processes of oxidative phosphorylation and the electron transport chain.

Some of the energy released at each of the protein complexes I, III, and IV is consumed in the movement of H^+ ions across the inner membrane from the matrix into the intermembrane space. Movement of ions from a region of lower concentration (the matrix) to one of higher concentration (the intermembrane space) requires the expenditure of energy because it opposes the natural tendency, as exhibited in the process of osmosis, to equalize concentrations.

3. ATP synthase has two subunits, the F_0 and F_1 subunits (Figure 12.15). The F_0 part of the synthase is the channel for proton flow, whereas the formation of ATP takes place in the F_1 subunit. As protons return to the mitochondrial matrix through the F_0 subunit, the potential energy associated with the electrochemical gradient is released and used in the F_1 subunit for the synthesis of ATP.

$$ADP + P_i \xrightarrow{\text{ATP synthase}} ATP + H_2O$$

The Chemistry at a Glance feature on page 442 brings together into one diagram the three processes that constitute the common metabolic pathway: the citric acid cycle, the electron transport chain, and oxidative phosphorylation. These three processes operate together. Discussing them separately, as we have done, is a matter of convenience only.

CHEMISTRY AT A GLANCE

Summary of the Common Metabolic Pathway

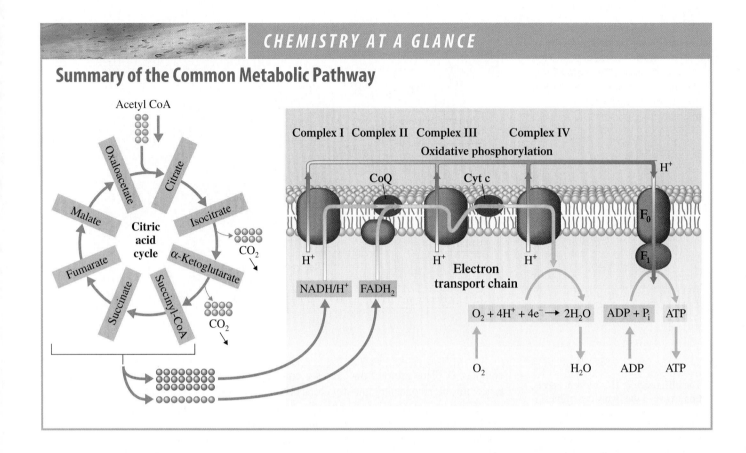

12.9 ATP Production for the Common Metabolic Pathway

Without oxygen, the biochemical systems of the human body quickly shut down and death occurs. Why? Without oxygen as the final electron acceptor in the ETC, the ETC chain shuts down and ATP production stops. Without ATP to power life's processes (Chapters 13–15) these processes stop.

For each mole of NADH oxidized in the ETC, 2.5 moles of ATP are formed. $FADH_2$, which does not enter the ETC at its start, produces only 1.5 moles of ATP per mole of $FADH_2$ oxidized. $FADH_2$'s entrance point into the chain, complex II, is beyond the first "proton-pumping" site, complex I. Hence fewer ATP molecules are produced from $FADH_2$ than from NADH.

The energy yield, in terms of ATP production, can now be totaled for the common metabolic pathway (Section 12.5). Every acetyl CoA entering the citric acid cycle (CAC) produces three NADH, one $FADH_2$, and one GTP (which is equivalent in energy to ATP; Section 12.6). Thus 10 molecules of ATP are produced for each acetyl CoA catabolized.

Biochemistry textbooks published before the mid-1990s make the following statements:

1 NADH produces 3 ATP in the ETC.
1 $FADH_2$ produces 2 ATP in the ETC.

As more has been learned about the electron transport chain and oxidative phosphorylation, these numbers have had to be reduced. The overall conversion process is more complex than was originally thought, and not as much ATP is produced.

$$
\begin{aligned}
3\ \text{NADH} &\longrightarrow 7.5\ \text{ATP} \\
1\ \text{FADH}_2 &\longrightarrow 1.5\ \text{ATP} \\
1\ \text{GTP} &\longrightarrow \underline{1\ \text{ATP}} \\
& \ \ 10\ \text{ATP}
\end{aligned}
$$

12.10 The Importance of ATP

The cycling of ATP and ADP in metabolic processes is the principal medium for energy exchange in biochemical processes. The conversion

$$\text{ATP} \longrightarrow \text{ADP} + P_i$$

powers life processes (the biosynthesis of essential compounds, muscle contraction, nutrient transport, and so on). The conversion

$$P_i + \text{ADP} \longrightarrow \text{ATP}$$

FIGURE 12.16 The interconversion of ATP and ADP is the principal medium for energy exchange in biochemical processes.

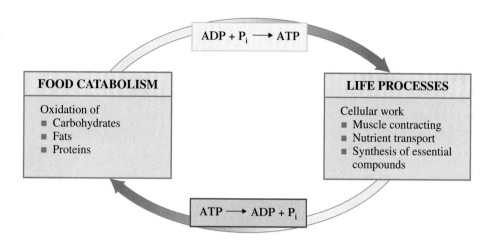

ATP molecules in cells have a high turnover rate. Normally, a given ATP molecule in a cell does not last more than a minute before it is converted to ADP. The concentration of ATP in a cell varies from 0.5 to 2.5 milligram per milliliter of cellular fluid.

which occurs in food catabolism cycles, regenerates the ATP expended in cell operation. Figure 12.16 summarizes the ATP–ADP cycling process.

ATP is a high-energy phosphate compound (Section 12.4). Its hydrolysis to ADP produces an *intermediate* amount of free energy (-7.5 kcal/mole; Table 12.1) compared with hydrolysis energies for other organophosphate compounds (Table 12.1). Of major importance, the energy derived from ATP hydrolysis is a *biochemically useful* amount of energy. It is larger than the amount of energy needed by compounds to which ATP donates energy, and yet it is smaller than that available in compounds used to form ATP. If the ATP hydrolysis energy were *unusually high,* the body would not be able to convert ADP back to ATP because ATP synthesis requires an energy input equal to or greater than the hydrolysis energy, and such an unusually high amount of energy would not be available.

12.11 Non-ETC Oxygen-Consuming Reactions

The electron transport chain/oxidative phosphorylation phase of metabolism consumes more than 90% of the oxygen taken into the human body via respiration. What happens to the remainder of the inspired O_2?

As a normal part of metabolic chemistry, significant amounts of this remaining O_2 are converted into several highly reactive oxygen species (ROS). Among these ROSs are hydrogen peroxide (H_2O_2), superoxide ion (O_2^-), and hydroxyl radical (OH). The latter two of these substances are free radicals, substances that contain an unpaired electron. Reactive oxygen species have beneficial functions within the body, but they can also cause problems if they are not eliminated when they are no longer needed.

White blood cells have a significant concentration of superoxide free radicals. Here, these free radicals aid in the destruction of invading bacteria and viruses. Their formation reaction is

$$2O_2 + NADPH \longrightarrow 2O_2^- + NADP^+ + H^+$$

(NADP is a phosphorylated version of the coenzyme NADH; see Section 13.8.)

Superoxide ion that is not needed is eliminated from cells in a two-step process governed by the enzymes *superoxide dismutase* and *catalase,* two of the most rapidly working enzymes known (see Table 10.2). In the first step, superoxide ion is converted to hydrogen peroxide, which is then, in the second step, converted to H_2O.

$$2O_2^- + 2H^+ \xrightarrow{\text{superoxide dismutase}} H_2O_2 + O_2$$
$$2H_2O_2 \xrightarrow{\text{catalase}} 2H_2O + O_2$$

CHEMICAL CONNECTIONS — Brown Fat, Newborn Babies, and Hibernating Animals

Ordinarily, metabolic processes generate enough heat to maintain normal body temperature. In certain cases, however, including newborn infants and hibernating animals, normal metabolism is not sufficient to meet the body's heat requirements. In these cases, a supplemental method of heat generation, which involves *brown fat tissue,* occurs.

Brown fat tissue, as the name implies, is darker in color than ordinary fat tissue, which is white. Brown fat is specialized for heat production. It contains many more blood vessels and mitochondria than white fat. (The increased number of mitochondria gives brown fat its color.)

Another difference between the two types of fat is that the mitochondria in brown fat cells contain a protein called *thermogenin,* which functions as an *uncoupling agent.* This protein "uncouples" the ATP production associated with the electron transport chain. The ETC reactions still take place, but the energy that would ordinarily be used for ATP synthesis is simply released as heat.

Brown fat tissue is of major importance for newborn infants. Newborns are immediately faced with a temperature regulation problem. They leave an environment of constant 37°C temperature and enter a much colder environment (25°C). A supply of *active* brown fat, present at birth, helps the baby adapt to the cooler environment.

Very limited amounts of brown fat are present in most adults. However, stores of brown fat increase in adults who are regularly exposed to cold environs. Thus the production of

Hibernating bears rely on brown fat tissue to help meet their bodies' heat requirements.

brown fat is one of the body's mechanisms for adaptation to cold.

Thermogenin, the uncoupling agent in brown fat, is a protein bound to the inner mitochondrial membrane. When activated, it functions as a proton channel through the inner membrane. The proton gradient produced by the electron transport chain is dissipated through this "new" proton channel, and less ATP synthesis occurs because the normal proton channel, ATP synthase, has been bypassed. The energy of the proton gradient, no longer useful for ATP synthesis, is released as heat.

Immediate destruction of the hydrogen peroxide produced in the first of these two steps is critical, because if it persisted, then unwanted production of hydroxyl radical would occur via hydrogen peroxide's reaction with superoxide ion.

$$H_2O_2 + O_2^- + H^+ \longrightarrow H_2O + O_2 + OH$$

Hydroxyl radicals quickly react with other substances by taking an electron from them. Such action usually causes bond breaking. Lipids in cell membranes are particularly vulnerable to such attack by hydroxyl radicals.

It is estimated that 5% of the ROSs escape destruction through normal channels (superoxide dimutase and catalase). Operating within a cell is a backup system—a network of antioxidants—to deal with this problem. Participating in this antioxidant network are glutathione (Section 9.7), vitamin C (Section 10.13), and beta-carotene and vitamin E (Section 10.14), as well as other compounds obtained from plants through dietary intake. Particularly important in this latter category are compounds called *flavonoids* (see the Chemical Connections feature on page 445). The vitamin antioxidants as well as the other antioxidants present prevent oxidative damage by reacting with the harmful oxidizing agents before they can react with other biologically important substances.

Reactive oxygen species can also be formed in the body as the result of external influences such as polluted air, cigarette smoke, and radiation exposure (including solar radiation). Vitamin C is particularly active against such free-radical damage.

> Antioxidant molecules provide electrons to convert free radicals and other ROSs into less-reactive substances.

CHEMICAL CONNECTIONS Flavonoids: An Important Class of Dietary Antioxidants

Numerous studies indicate that diets high in fruits and vegetables are associated with a "healthy lifestyle." One reason for this is that fruits and vegetables contain compounds called *phytochemicals.* Phytochemicals are compounds found in plants that have biochemical activity in the human body even though they have no nutritional value. The functions that phytochemicals perform in the human body include antioxidant activity, cancer inhibition, cholesterol regulation, and anti-inflammatory activity.

Each fruit and vegetable is a unique package of phytochemicals, so consuming a wide variety of fruits and vegetables provides the body with the broadest spectrum of benefits. In such a situation, many phytochemicals are consumed in *small* amounts. This approach is much safer than taking supplemental doses of particular phytochemicals; in larger doses, some phytochemicals are toxic.

A major group of phytochemicals are the *flavonoids,* of which over 4000 individual compounds are known. All flavonoids are antioxidants (Section 3.12), but some are stronger antioxidants than others, depending on their molecular structure. About 50 flavonoids are present in foods and in beverages derived from plants (tea leaves, grapes, oranges, and so on).

The core *flavonoid* structure is

Both aromatic and cyclic ether ring systems are present. Of particular importance as antioxidants in foods are flavonoids known as *flavones* and *flavonols,* flavonoids whose core structures are enhanced by the presence of ketone and/or hydroxyl groups and a double bond in the oxygen-containing ring system.

Flavones   Flavonols

The formation of flavone and flavonol compounds depends normally on the action of light, so in general the highest concentration of these compounds occurs in leaves or in the skins of fruits, whereas only traces are found in parts of plants that grow below the ground. The common onion is, however, a well-known exception to this generalization.

The most widespread flavonoid in food is the flavonol *quercetin.*

It is predominant in fruits, vegetables, and the leaves of various vegetables. In fruits, apples contain the highest amounts of quercetin, the majority of it being found in the outer tissues (skin, peel). A small peeled apple contains about 5.7 mg of the antioxidant vitamin C. But the same amount of apple *with the skin* contains flavonoids and other phytochemicals that have the effect of 1500 mg of vitamin C. Onions are also major dietary sources of quercetin.

In addition to their antioxidant benefits, flavonoids may also help fight bacterial infections. Recent studies indicate that flavonoids can stop the growth of some strains of drug-resistant bacteria.

 CONCEPTS TO REMEMBER

Metabolism. Metabolism is the sum total of all the biochemical reactions that take place in a living organism. Metabolism consists of catabolism and anabolism. Catabolic biochemical reactions involve the breakdown of large molecules into smaller fragments. Anabolic biochemical reactions synthesize large molecules from smaller ones (Section 12.1).

Mitochondria. Mitochondria are membrane-enclosed subcellular structures that are the site of energy production in the form of ATP molecules. Enzymes for both the citric acid cycle and the electron transport chain are housed in the mitochondria (Section 12.2).

Important coenzymes. Three very important coenzymes involved in catabolism are NAD^+, FAD, and CoA. NAD^+ and FAD are oxidizing agents that participate in the oxidation reactions of the citric

acid cycle. They transport hydrogen atoms and electrons from the citric acid cycle to the electron transport chain. CoA interacts with acetyl groups produced from food degradation to form acetyl CoA. Acetyl CoA is the "fuel" for the citric acid cycle (Section 12.3).

High-energy compounds. A high-energy compound liberates a larger-than-normal amount of free energy upon hydrolysis because structural features in the molecule contribute to repulsive strain in one or more bonds. Most high-energy biochemical molecules contain phosphate groups (Section 12.4).

Common metabolic pathway. The common metabolic pathway includes the reactions of the citric acid cycle and those of the electron transport chain and oxidative phosphorylation. The degradation products from all types of foods (carbohydrates, fats, and proteins) participate in the reactions of the common metabolic pathway (Section 12.5).

Citric acid cycle. The citric acid cycle is a cyclic series of eight reactions that oxidize the acetyl portion of acetyl CoA, resulting in the production of two molecules of CO_2. The complete oxidation of one acetyl group produces three molecules of NADH, one of $FADH_2$, and one of GTP besides the CO_2 (Section 12.6).

Electron transport chain. The electron transport chain is a series of reactions that passes electrons from NADH and $FADH_2$ to molecular oxygen. Each electron carrier that participates in the chain has an increasing affinity for electrons. Upon accepting the electrons and hydrogen ions, the O_2 is reduced to H_2O (Section 12.7).

Oxidative phosphorylation. Oxidative phosphorylation is the biochemical process by which ATP is synthesized from ADP as the result of a proton gradient across the inner mitochondrial membrane. Oxidative phosphorylation is coupled to the reactions of the electron transport chain (Section 12.8).

Chemiosmotic coupling. Chemiosmotic coupling explains how the energy needed for ATP synthesis is obtained. Synthesis takes place because of a flow of protons across the inner mitochondrial membrane (Section 12.8).

Importance of ATP. ATP is the link between energy production and energy use in cells. The conversion of ATP to ADP powers life processes, and the conversion of ADP back to ATP regenerates the energy expended in cell operation (Section 12.10).

KEY REACTIONS AND EQUATIONS

1. Oxidation by FAD (Section 12.3)
$$FAD + 2H^+ + 2e^- \longrightarrow FADH_2$$

2. Oxidation by NAD^+ (Section 12.3)
$$NAD^+ + 2H^+ + 2e^- \longrightarrow NADH + H^+$$

3. The citric acid cycle (Section 12.6)
$$Acetyl\ CoA + 3NAD^+ + FAD + GDP + P_i + 2H_2O \longrightarrow$$
$$2CO_2 + CoA{-}SH + 3NADH + 2H^+ + FADH_2 + GTP$$

4. The electron transport chain (Section 12.7)
$$NADH + H^+ \longrightarrow NAD^+ + 2H^+ + 2e^-$$
$$FADH_2 \longrightarrow FAD + 2H^+ + 2e^-$$
$$O_2 + 4H^+ + 4e^- \longrightarrow 2H_2O$$

5. Oxidative phosphorylation (Section 12.8)
$$ADP + P_i \xrightarrow[\text{from ETC}]{\text{Energy}} ATP + H_2O$$

EXERCISES AND PROBLEMS

The members of each pair of problems in this section test similar material.

■ Metabolism (Section 12.1)

12.1 Classify anabolism and catabolism as synthetic or degradative processes.

12.2 Classify anabolism and catabolism as energy-producing or energy-consuming processes.

12.3 What is a metabolic pathway?

12.4 What is the difference between a linear and a cyclic metabolic pathway?

12.5 What general characteristics are associated with a catabolic pathway?

12.6 What general characteristics are associated with an anabolic pathway?

■ Cell Structure (Section 12.2)

12.7 List several differences between prokaryotic cells and eukaryotic cells.

12.8 What kinds of organisms have prokaryotic cells and what kinds have eukaryotic cells?

12.9 What is an organelle?

12.10 What is the general function of each of the following types of organelles?
a. Ribosome   b. Lysosome   c. Mitochondrion

12.11 In a mitochondrion, what separates the matrix from the intermembrane space?

12.12 In what major way do the inner and outer mitochondrial membranes differ?

12.13 What is the intermembrane space of a mitochondrion?

12.14 Where are ATP synthase complexes located in a mitochondrion?

■ Intermediate Compounds in Metabolic Pathways (Section 12.3)

12.15 What does each letter in ATP stand for?

12.16 What does each letter in ADP stand for?

12.17 Draw a block diagram structure for ATP.

12.18 Draw a block diagram structure for ADP.

12.19 What is the structural difference between ATP and AMP?

12.20 What is the structural difference between ADP and AMP?

12.21 What is the structural difference between ATP and GTP?

12.22 What is the structural difference between ATP and CTP?

12.23 In terms of hydrolysis, what is the relationship between ATP and ADP?

12.24 In terms of hydrolysis, what is the relationship between ADP and AMP?

12.25 What does each letter in FAD stand for?

12.26 What does each letter in NAD^+ stand for?

12.27 Draw a block diagram structure for FAD based on the presence of an ADP core (three-block diagram).

12.28   Draw a block diagram structure for FAD based on the presence of two nucleotides (six-block diagram).

12.29   Draw a block diagram structure for NAD^+ based on the presence of two nucleotides (six-block diagram).

12.30   Draw a block diagram structure for NAD^+ based on an ADP core (three-block diagram).

12.31   Which part of an NAD^+ molecule is the active participant in redox reactions?

12.32   Which part of an FAD molecule is the active participant in redox reactions?

12.33   Give the letter designation for
a. the reduced form of FAD.
b. the oxidized form of NADH.

12.34   Give the letter designation for
a. the oxidized form of $FADH_2$.
b. the reduced form of NAD^+.

12.35   Name the vitamin B molecule that is part of the structure of
a. NAD^+ b. FAD

12.36   Indicate whether or not the vitamin B portion of the following molecules is the "active" portion of the molecule in redox processes.
a. NAD^+ b. FAD

12.37   Draw the three-block diagram structure for coenzyme A.

12.38   Which part of a coenzyme A molecule is the active participant in a redox reaction?

■ **High-Energy Phosphate Compounds (Section 12.4)**

12.39   What is a high-energy compound?

12.40   What factors contribute to a strained bond in high-energy phosphate compounds?

12.41   What does the designation P_i denote?

12.42   What does the designation PP_i denote?

12.43   With the help of Table 12.1, determine which compound in each of the following pairs of phosphate-containing compounds releases more free energy upon hydrolysis.
a. ATP and phosphoenolpyruvate
b. Creatine phosphate and ADP
c. Glucose 1-phosphate and 1,3-diphosphoglycerate
d. AMP and glycerol 3-phosphate

12.44   With the help of Table 12.1, determine which compound in each of the following pairs of phosphate-containing compounds releases more free energy upon hydrolysis.
a. ATP and creatine phosphate
b. Glucose 1-phosphate and glucose 6-phosphate
c. ADP and AMP
d. Phosphoenolpyruvate and PP_i

■ **Biochemical Energy Production (Section 12.5)**

12.45   Describe the four general stages of the process by which biochemical energy is obtained from food.

12.46   Of the four general stages of biochemical energy production from food, which are part of the common metabolic pathway?

■ **The Citric Acid Cycle (Section 12.6)**

12.47   What are two other names for the citric acid cycle?

12.48   What is the basis for the name *citric acid cycle*?

12.49   What is the "fuel" for the citric acid cycle?

12.50   What are the products of the citric acid cycle?

12.51   Consider the reactions that occur during *one turn* of the citric acid cycle in answering each of the following questions.
a. How many CO_2 molecules are formed?
b. How many molecules of $FADH_2$ are formed?
c. How many times is a secondary alcohol oxidized?
d. How many times does water add to a carbon–carbon double bond?

12.52   Consider the reactions that occur during *one turn* of the citric acid cycle in answering each of the following questions.
a. How many molecules of NADH are formed?
b. How many GTP molecules are formed?
c. How many decarboxylation reactions occur?
d. How many oxidation–reduction reactions occur?

12.53   There are eight steps in the citric acid cycle. List those steps that involve
a. oxidation
b. isomerization
c. hydration

12.54   There are eight steps in the citric acid cycle. List those steps that involve
a. oxidation and decarboxylation.
b. phosphorylation.
c. condensation.

12.55   There are four C_4 dicarboxylic acid species in the citric acid cycle. What are their names and structures?

12.56   There are two keto carboxylic acid species in the citric acid cycle. What are their names and structures?

12.57   What type of reaction occurs in the citric acid cycle whereby a C_6 compound is converted to a C_5 compound?

12.58   What type of reaction occurs in the citric acid cycle whereby a C_5 compound is converted to a C_4 compound?

12.59   Identify the oxidized coenzyme (NAD^+ or FAD) that participates in each of the following citric acid cycle reactions.
a. Isocitrate $\longrightarrow$ α-ketoglutarate
b. Succinate $\longrightarrow$ fumarate

12.60   Identify the oxidized coenzyme (NAD^+ or FAD) that participates in each of the following citric acid cycle reactions.
a. Malate $\longrightarrow$ oxaloacetate
b. α-Ketoglutarate $\longrightarrow$ succinyl CoA

12.61   List the two citric acid cycle intermediates involved in the reaction governed by each of the following enzymes. List the reactant first.
a. Isocitrate dehydrogenase
b. Fumarase
c. Malate dehydrogenase
d. Aconitase

12.62   List the two citric acid cycle intermediates involved in the reaction governed by each of the following enzymes. List the reactant first.
a. α-Ketoglutarate dehydrogenase
b. Succinate dehydrogenase
c. Citrate synthase
d. Succinyl CoA synthase

■ **The Electron Transport Chain (Section 12.7)**

12.63   By what other name is the electron transport chain known?

12.64   Give a one-sentence summary of what occurs during the reactions known as the electron transport chain.

12.65   What is the final electron acceptor of the electron transport chain?

12.66 Which substances generated in the citric acid cycle participate in the electron transport chain?

12.67 Give the abbreviation for each of the following electron carriers.
a. The oxidized form of flavin mononucleotide
b. The reduced form of coenzyme Q

12.68 Give the abbreviation for each of the following electron carriers.
a. The reduced form of flavin mononucleotide
b. The oxidized form of coenzyme Q

12.69 Indicate whether each of the following electron carriers is in its oxidized form or its reduced form.
a. Fe(III)SP   b. Cyt b (Fe^{3+})
c. NADH   d. FAD

12.70 Indicate whether each of the following electron carriers is in its oxidized form or its reduced form.
a. $FMNH_2$ b. Fe(II)SP
c. Cyt c_1 (Fe^{2+})   d. NAD^+

12.71 Indicate whether each of the following changes represents oxidation or reduction.
a. $CoQH_2 \longrightarrow CoQ$
b. $NAD^+ \longrightarrow NADH$
c. Cyt c (Fe^{2+}) $\longrightarrow$ cyt c (Fe^{3+})
d. Cyt b (Fe^{3+}) $\longrightarrow$ cyt b (Fe^{2+})

12.72 Indicate whether each of the following changes represents oxidation or reduction.
a. $FADH_2 \longrightarrow FAD$
b. $FMN \longrightarrow FMNH_2$
c. Fe(III)SP $\longrightarrow$ Fe(II)SP
d. Cyt c_1 (Fe^{3+}) $\longrightarrow$ cyt c_1 (Fe^{2+})

12.73 With which of the protein complexes (I, II, III, and IV) of the ETC is each of the following electron carriers associated? More than one answer may apply in a given situation.
a. NADH   b. CoQ   c. Cyt b   d. Cyt a

12.74 With which of the protein complexes (I, II, III, and IV) of the ETC is each of the following electron carriers associated? More than one answer may apply in a given situation
a. $FADH_2$ b. FeSP   c. Cyt c   d. Cyt c_1

12.75 Which electron carrier shuttles electrons between protein complexes I and III?

12.76 Which electron carrier shuttles electrons between protein complexes II and III?

12.77 How many electrons does the electron carrier between complexes II and III carry per "trip"?

12.78 How many electrons does the electron carrier between complexes III and IV carry per "trip"?

12.79 Fill in the missing substances in the following electron transport chain reaction sequences.

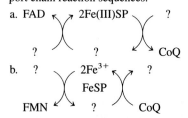

12.80 Fill in the missing substances in the following electron transport chain reaction sequences.

a.   ?   ?   ?
NADH FMN 2Fe(II)SP
b. $CoQH_2$?   ?
FeSP Cyt b
? $2Fe^{2+}$ $2Fe^{3+}$

Oxidative Phosphorylation (Section 12.8)

12.81 What is oxidative phosphorylation?

12.82 What are coupled reactions?

12.83 The coupling of ATP synthesis with the reactions of the ETC is related to the movement of what chemical species across the inner mitochrondial membrane?

12.84 At what protein complex location(s) in the electron transport chain does proton pumping occur?

12.85 At what mitochondrial location does H^+ ion buildup occur as the result of proton pumping?

12.86 How many protons cross the inner mitochondrial membrane for every two electrons that are passed through the electron transport chain?

12.87 What is the name of the enzyme that catalyzes ATP production during oxidative phosphorylation?

12.88 What is the location of the enzyme that uses stored energy in a proton gradient to drive the reaction that produces ATP?

12.89 How is the proton gradient associated with chemiosmotic coupling dissipated during ATP synthesis?

12.90 What are the "starting materials" from which ATP is synthesized as the proton gradient associated with chemiosmotic coupling is dissipated?

ATP Production (Section 12.9)

12.91 How many ATP molecules are formed for each NADH molecule that enters the electron transport chain?

12.92 How many ATP molecules are formed for each $FADH_2$ molecule that enters the electron transport chain?

12.93 NADH and $FADH_2$ molecules do not yield the same number of ATP molecules. Explain why.

12.94 What is the energy yield, in terms of ATP molecules, from one turn of the citric acid cycle, assuming that the products of the cycle enter the electron transport chain?

Non-ETC Oxygen-Consuming Reactions (Section 12.11)

12.95 What does the designation ROS stand for?

12.96 Give the chemical formula for each of the following.
a. Superoxide ion   b. Hydroxyl radical

12.97 Give the chemical equation for the reaction by which
a. superoxide ion is generated within cells.
b. superoxide ion is converted to hydrogen peroxide within cells.

12.98 Give the chemical equation for the reaction by which
a. hydrogen peroxide is converted to desirable products within cells.
b. hydrogen peroxide is converted to an undesirable product within cells.

12.99 Classify each of the following substances as (1) a reactant in the citric acid cycle, (2) a reactant in the electron transport chain, or (3) a reactant in both the CAC and the ETC.
a. NAD^+ b. NADH   c. O_2
d. H_2O e. Fumarate   f. Cytochrome a

12.100 Classify each of the following substances as (1) a product in the citric acid cycle, (2) a product in the electron transport chain, or (3) a product in both the CAC and the ETC.
a. $FADH_2$ b. FAD   c. CO_2
d. H_2O e. Malate   f. Flavin mononucleotide

12.101 Which of these substances, (1) ATP, (2) CoA, (3) FAD, and (4) NAD^+, contain the following subunits of structure? More than one choice may apply in a given situation.
a. Contains two ribose subunits
b. Contains two phosphate subunits
c. Contains one adenine subunit
d. Contains one ribitol subunit

12.102 Characterize, in terms of number of carbon atoms present, each of the following citric acid cycle changes as (a) a C_6 to C_6 change, (b) a C_6 to C_5 change, (c) a C_5 to C_4 change, or (d) a C_4 to C_4 change.

a. Citrate to isocitrate   b. Succinate to fumarate
c. Malate to oxaloacetate   d. Isocitrate to α-ketoglutarate

12.103 In what way are the processes of the citric acid cycle and the electron transport chain interrelated?

12.104 Where within a cell does each of the following take place?
a. Citric acid cycle
b. Electron transport chain and oxidative phosphorylation

12.105 One of the oxidation steps that occurs when lipids are metabolized is

$$R-CH_2-CH_2-\overset{\overset{\displaystyle O}{\|}}{C}-S-CoA \longrightarrow R-CH=CH-\overset{\overset{\displaystyle O}{\|}}{C}-S-CoA$$

Would you expect this reaction to require FAD or NAD^+ as the oxidizing agent?

12.106 In oxidative phosphorylation, what is oxidized and what is phosphorylated?

12.107 Which of the following is true for a mitochondrion?
a. The inner membrane separates the matrix from the intermembrane space.
b. The inner membrane is more permeable than the outer membrane.
c. The outer membrane has ATP-synthase complexes attached to it.
d. The outer membrane has a highly folded structure.

12.108 Which of the following molecules has two unsubstituted ribose molecules as structural subunits?
a. FAD   b. NAD^+ c. CoA   d. ATP

12.109 Which of the following are products of the citric acid cycle?
a. Acetyl CoA and NADH   b. Acetyl CoA and CO_2
c. CO_2 and H_2O d. CO_2 and $FADH_2$

12.110 Which of the following citric acid cycle intermediates is not a C_4 species?
a. Fumarate   b. Citrate
c. Malate   d. Oxaloacetate

12.111 Which are the first two intermediates, respectively, in the citric acid cycle?
a. Isocitrate and α-ketoglutarate
b. Citrate and α-ketoglutarate
c. Citrate and isocitrate
d. Isocitrate and succinate

12.112 Which of the following is an electron carrier that shuttles electrons between various protein complexes in the electron transport chain?
a. FMN   b. NADH
c. Cyt c   d. Cyt a_3

12.113 What is the substrate that interacts with protein complex III in the electron transport chain?
a. CoQ   b. Cyt c_1
c. FMN   d. FeSP

12.114 Which of the following is both a reactant and a product in the operation of the electron transport chain?
a. O_2 b. H_2O
c. $FADH_2$ d. Cyt b

12.115 At how many protein complex sites in the electron transport chain does proton pumping occur?
a. One   b. Two
c. Three   d. Four

12.116 How many moles of ATP result from the entry of one mole of NADH into the electron transport chain?
a. 1 mole ATP   b. 1.5 moles ATP
c. 2 moles ATP   d. 2.5 moles ATP

13 Carbohydrate Metabolism

Carbohydrates are the major energy source for human beings.

In this chapter we explore the relationship between carbohydrate metabolism and energy production in cells. The molecule glucose is the focal point of carbohydrate metabolism. Commonly called blood sugar, glucose is supplied to the body via the circulatory system and, after being absorbed by a cell, can be either oxidized to yield energy or stored as glycogen for future use. When sufficient oxygen is present, glucose is totally oxidized to CO_2 and H_2O. However, in the absence of oxygen, glucose is only partially oxidized to lactic acid. Besides supplying energy needs, glucose and other six-carbon sugars can be converted into a variety of different sugars (C_3, C_4, C_5, and C_7) needed for biosynthesis. Some of the oxidative steps in carbohydrate metabolism also produce NADH and NADPH, sources of reductive power in cells.

13.1 Digestion and Absorption of Carbohydrates

Digestion *is the biochemical process by which food molecules, through hydrolysis, are broken down into simpler chemical units that can be used by cells for their metabolic needs.* Digestion is the first stage in the processing of food products.

The digestion of carbohydrates begins in the mouth, where the enzyme *salivary α-amylase* catalyzes the hydrolysis of α-glycosidic linkages (Section 7.13) in starch from plants and glycogen from meats to produce smaller polysaccharides and the disaccharide maltose.

Only a small amount of carbohydrate digestion occurs in the mouth because food is swallowed so quickly. Although the food mass remains longer in the stomach, very little further carbohydrate digestion occurs there either, because *salivary α-amylase* is inactivated

Salivary α-amylase is a constituent of saliva, the fluid secreted by the salivary glands. Saliva is 99% water plus small amounts of several inorganic ions and organic molecules. Saliva secretion can be triggered by the taste, smell, sight, and even thought of food. Average saliva output is about 1.5 L per day.

FIGURE 13.1 A section of the small intestine, showing its folds and the villi that cover the inner surface of the folds. Villi greatly increase the inner intestinal surface area.

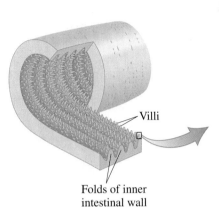

Villi

Folds of inner intestinal wall

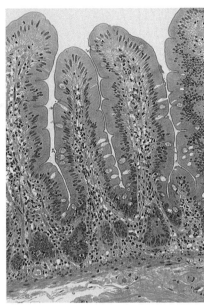

by the acidic environment of the stomach, and the stomach's own secretions do not contain any carbohydrate-digesting enzymes.

The primary site for carbohydrate digestion is within the small intestine, where α-amylase, this time secreted by the pancreas, again begins to function. The *pancreatic α-amylase* breaks down polysaccharide chains into shorter and shorter segments until the disaccharide maltose (two glucose units; Section 7.13) and glucose itself are the dominant species.

The final step in carbohydrate digestion occurs on the outer membranes of intestinal mucosal cells, where the enzymes that convert disaccharides to monosaccharides are located. The important disaccharidase enzymes are *maltase, sucrase,* and *lactase.* These enzymes convert, respectively, maltose to two glucose units, sucrose to one glucose and one fructose unit, and lactose to one glucose and one galactose unit (Section 7.13). (The disaccharides sucrose and lactose present in food are not digested until they reach this point.)

The three major breakdown products from carbohydrate digestion are thus glucose, galactose, and fructose. These monosaccharides are absorbed into the bloodstream through the intestinal wall. The folds of the intestinal wall are lined with fingerlike projections called *villi,* which are rich in blood capillaries (Figure 13.1). Absorption is by *active transport* (Section 8.10), which, unlike passive transport, is an energy-requiring process. In this case, ATP is needed. Protein carriers mediate the passage of the monosaccharides through cell membranes. Figure 13.2 summarizes the different phases in the digestive process for carbohydrates.

After their absorption into the bloodstream, monosaccharides are transported to the liver, where fructose and galactose are rapidly converted into compounds that are metabolized by the same pathway as glucose. Thus the central focus of carbohydrate metabolism is the pathway by which glucose is further processed, a pathway called *glycolysis* (Section 13.2)—a series of ten reactions, each of which involves a different enzyme.

The term *glycolysis,* pronounced "gligh-KOLL-ih-sis," comes from the Greek *glyco,* meaning "sweet," and *lysis,* meaning "breakdown."

Pyruvate, pronounced "PIE-roo-vate," is the carboxylate ion (Section 5.8) produced when pyruvic acid (a three-carbon keto acid) loses its acidic hydrogen atom.

$$\underset{\text{Pyruvic acid}}{\begin{matrix} CH_3 \\ | \\ C=O \\ | \\ COOH \end{matrix}} \longrightarrow \underset{\text{Pyruvate ion}}{\begin{matrix} CH_3 \\ | \\ C=O \\ | \\ COO^- \end{matrix}} + H^+$$

13.2 Glycolysis

Glycolysis *is the metabolic pathway by which glucose (a C_6 molecule) is converted into two molecules of pyruvate (a C_3 molecule), chemical energy in the form of ATP is produced, and NADH-reduced coenzymes are produced.* This metabolic pathway functions in almost all cells.

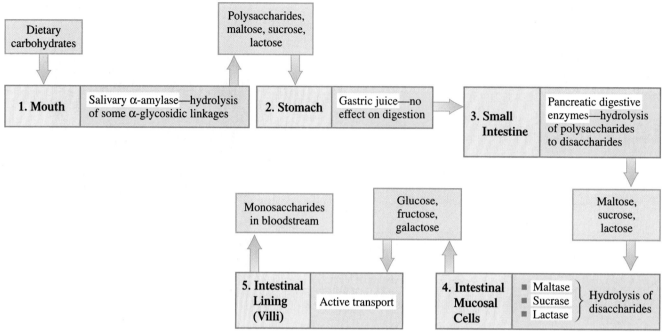

FIGURE 13.2 Summary of carbohydrate digestion in the human body.

Anaerobic is pronounced "AN-air-ROE-bic." *Aerobic* is pronounced "air-ROE-bic."

Glycolysis is also called the *Embden–Meyerhof pathway* after the German chemists Gustav Embden (1874–1933) and Otto Meyerhof (1884–1951), who discovered many of the details of the pathway in the early 1930s.

A *kinase* is an enzyme that catalyzes the transfer of a phosphoryl group (PO_3^{2-}) from ATP (or some other high-energy phosphate compound) to a substrate (Section 10.2).

The conversion of glucose to pyruvate is an oxidation process in which no molecular oxygen is utilized. The oxidizing agent is the coenzyme NAD^+. Metabolic pathways in which molecular oxygen is not a participant are called *anaerobic* pathways. Pathways that require molecular oxygen are called *aerobic* pathways. Glycolysis is an anaerobic pathway.

Glycolysis is a ten-step process (compared to the eight steps of the citric acid cycle; Section 12.6) in which every step is enzyme-catalyzed. Figure 13.3 gives an overview of glycolysis. There are two stages in the overall process, a *six-carbon stage* (Steps 1–3) and a *three-carbon stage* (Steps 4–10). All of the enzymes needed for glycolysis are present in the cell cytosol (Section 12.2), which is where glycolysis takes place. Details of the individual steps within the glycolysis pathway are now considered.

Six-Carbon Stage of Glycolysis (Steps 1–3)

The intermediates of the six-carbon stage of glycolysis are all either *glucose* or *fructose* derivatives in which phosphate groups are present.

Step 1: *Formation of Glucose 6-phosphate.* Glycolysis begins with the phosphorylation of glucose to yield glucose 6-phosphate, a glucose molecule with a phosphate group attached to the hydroxyl oxygen on carbon 6 (the carbon atom outside the ring). The phosphate group is from an ATP molecule. *Hexokinase,* an enzyme that requires Mg^{2+} ion for its activity, catalyzes the reaction.

The symbol Ⓟ is a shorthand notation for a PO_3^{2-} unit.

Glucose → Glucose 6-phosphate
ATP ADP
Hexokinase

This reaction requires energy, which is provided by the breakdown of an ATP molecule. This energy expenditure will be recouped later in the cycle. Phosphorylation of glucose provides a way of "trapping" glucose within a cell. Glucose can cross cell membranes, but glucose 6-phosphate cannot.

FIGURE 13.3 An overview of glycolysis.

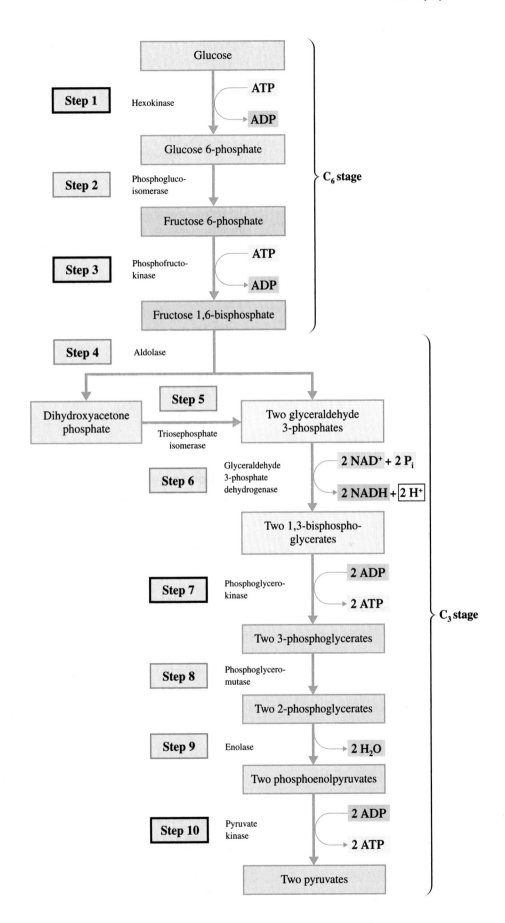

Step 2: *Formation of Fructose 6-phosphate.* Glucose 6-phosphate is isomerized to fructose 6-phosphate by *phosphoglucoisomerase.*

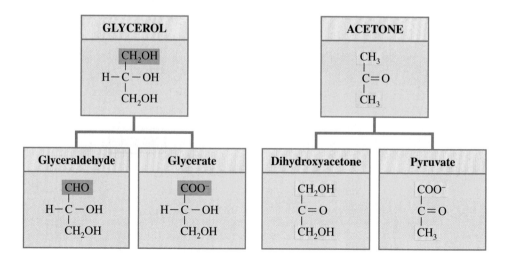

The net result of this change is that carbon 1 of glucose is no longer part of the ring structure. [Glucose, an aldose, forms a six-membered ring, and fructose, a ketose, forms a five-membered ring (Section 7.10); both sugars, however, contain six carbon atoms.]

> Step 3 of glycolysis commits the original glucose molecule to the glycolysis pathway. Glucose 6-phosphate (Step 1) and fructose 6-phosphate (Step 2) can enter other metabolic pathways, but fructose 1,6-bisphosphate can enter only glycolysis.

Step 3: *Formation of Fructose 1,6-bisphosphate.* This step, like Step 1, is a phosphorylation reaction and therefore requires the expenditure of energy. ATP is the source of the phosphate and the energy. The enzyme involved, *phosphofructokinase,* is another enzyme that requires Mg^{2+} ion for its activity. The fructose molecule now contains two phosphate groups.

> The term *bisphosphate* is used instead of *diphosphate* to indicate the two phosphates are on different carbon atoms in fructose and not connected to each other.

■ Three-Carbon Stage of Glycolysis (Steps 4–10)

All intermediates in the three-carbon stage of glycolysis are phosphorylated derivatives of *dihydroxyacetone, glyceraldehyde, glycerate,* or *pyruvate,* which in turn are derivatives of either glycerol or acetone. Figure 13.4 shows the structural relationships among these molecules.

Step 4: *Formation of Triose Phosphates.* In this step, the reacting C_6 species is split into two C_3 (triose) species. Because fructose 1,6-bisphosphate, the molecule being split, is unsymmetrical, the two trioses produced are not identical. One product is dihydroxyacetone phosphate, and the other is glyceraldehyde 3-phosphate. *Aldolase* is the enzyme that catalyzes this reaction. A better understanding of the structural relationships between reactant and products is obtained if the fructose 1,6-bisphosphate is written in its open-chain form (Section 7.10) rather than in its cyclic form.

FIGURE 13.4 Structural relationships among glycerol and acetone and the C_3 intermediates in the process of glycolysis.

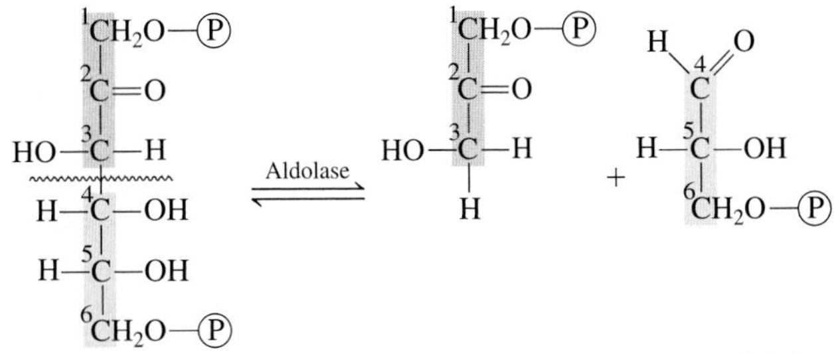

Fructose 1,6-bisphosphate (open-chain form)   Dihydroxyacetone phosphate   Glyceraldehyde 3-phosphate

Step 5: *Isomerization of Triose Phosphates.* Only one of the two trioses produced in Step 4, glyceraldehyde 3-phosphate, is a glycolysis intermediate. Dihydroxyacetone phosphate, the other triose, can, however, be readily converted into glyceraldehyde 3-phosphate. Dihydroxyacetone phosphate (a ketose) and glyceraldehyde 3-phosphate (an aldose) are isomers, and the isomerization process from ketose to aldose is catalyzed by the enzyme *triosephosphate isomerase.*

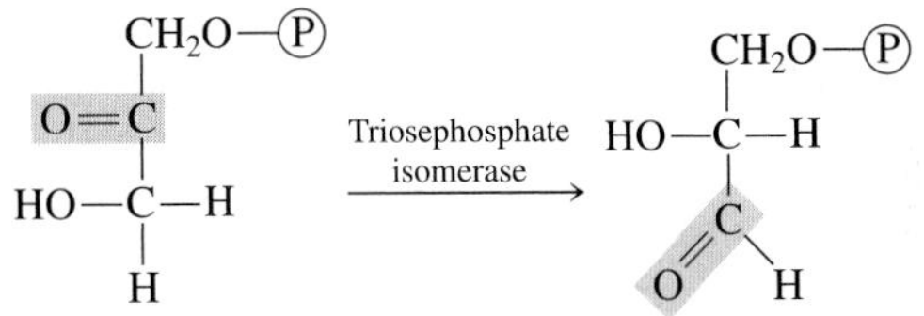

Dihydroxyacetone phosphate   Glyceraldehyde 3-phosphate

Step 6: *Formation of 1,3-Bisphosphoglycerate.* In a reaction catalyzed by *glyceraldehyde 3-phosphate dehydrogenase,* a phosphate group is added to glyceraldehyde 3-phosphate to produce 1,3-bisphosphoglycerate. The hydrogen of the aldehyde group becomes part of NADH.

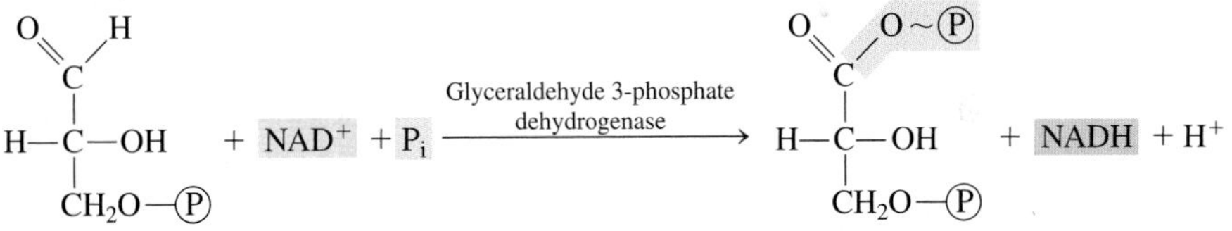

Glyceraldehyde 3-phosphate   1,3-Bisphosphoglycerate

Keep in mind that from Step 6 onward, two molecules of each of the C_3 compounds take part in every reaction for each original C_6 glucose molecule.

The newly added phosphate group in 1,3-bisphosphoglycerate is a high-energy phosphate group (Section 12.4). A high-energy phosphate group is produced when a phosphate group is attached to a carbon atom that is also participating in a carbon–carbon or carbon–oxygen double bond.

Note that a molecule of the reduced coenzyme NADH is a product of this reaction and also that the source of the added phosphate is inorganic phosphate (P_i).

Step 7: *Formation of 3-Phosphoglycerate.* In this step, the diphosphate species just formed is converted back to a monophosphate species. This is an ATP-producing step in which the C-1 phosphate group of 1,3-bisphosphoglycerate (the high-energy phosphate) is transferred to an ADP molecule to form the ATP. The enzyme involved is *phosphoglycerokinase.*

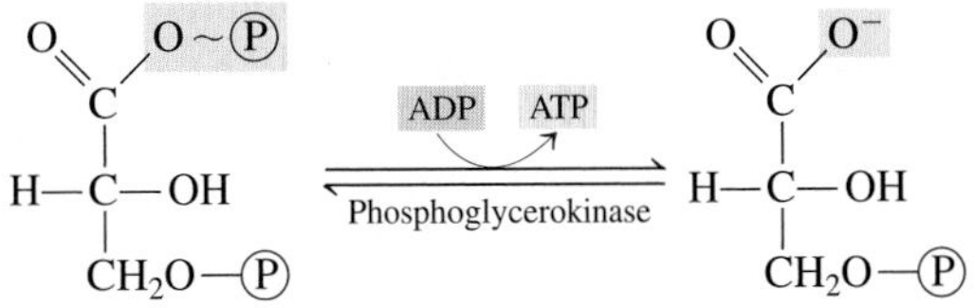

1,3-Bisphosphoglycerate   3-Phosphoglycerate

Remember that two ATP molecules are produced for each original glucose molecule because both C_3 molecules produced from the glucose react.

ATP production in this step involves substrate-level phosphorylation. **Substrate-level phosphorylation** *is the biochemical process by which a high-energy phosphate group from an intermediate compound (substrate) is directly transferred to ADP to produce ATP.* Substrate-level phosphorylation differs from oxidative phosphorylation (Section 12.8) in that the latter process involves the transfer of free phosphate ions in solution (P_i) to ADP molecules to form ATP.

A *mutase* is an enzyme that effects the shift of a phosphoryl group (PO_3^{2-}) from one oxygen atom to another within a molecule (Section 10.2).

Step 8: *Formation of 2-Phosphoglycerate.* In this isomerization step, the phosphate group of 3-phosphoglycerate is moved from carbon 3 to carbon 2. The enzyme *phosphoglyceromutase* catalyzes the exchange of the phosphate group between the two carbons.

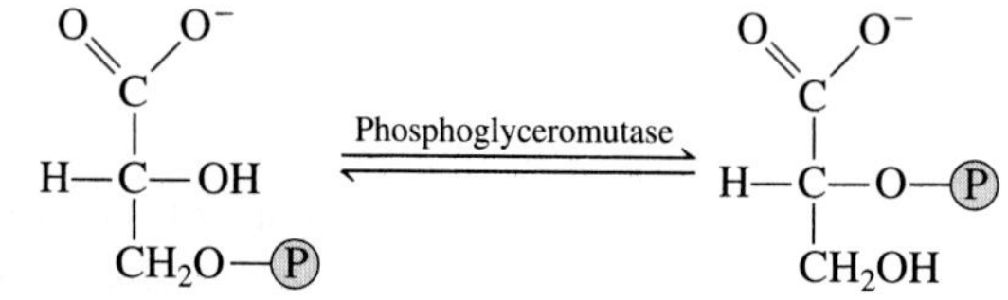

An *enol* (from *ene* + *ol*), as in phospho*enol*pyruvate, is a compound in which an —OH group is attached to a carbon atom involved in a carbon–carbon double bond. Note that in phosphoenolpyruvate, the —OH group has been phosphorylated.

Step 9: *Formation of Phosphoenolpyruvate.* This is an alcohol dehydration reaction that proceeds with the enzyme *enolase,* another Mg^{2+}-requiring enzyme. The result is another compound containing a high-energy phosphate group; the phosphate group is attached to a carbon atom that is involved in a carbon–carbon double bond.

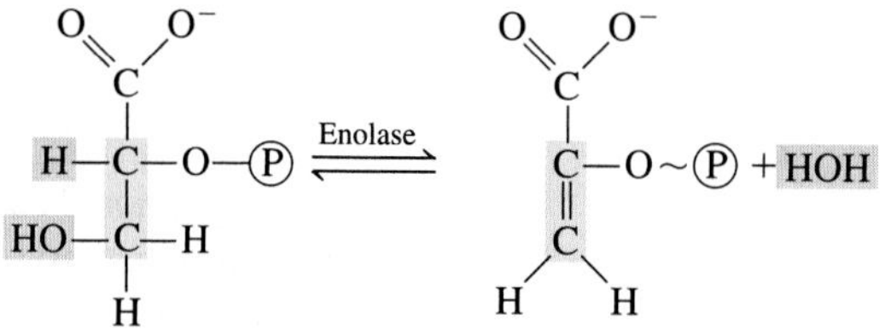

Step 10: *Formation of Pyruvate.* In this step, substrate-level phosphorylation again occurs. Phosphoenolpyruvate transfers its high-energy phosphate group to an ADP molecule to produce ATP and pyruvate.

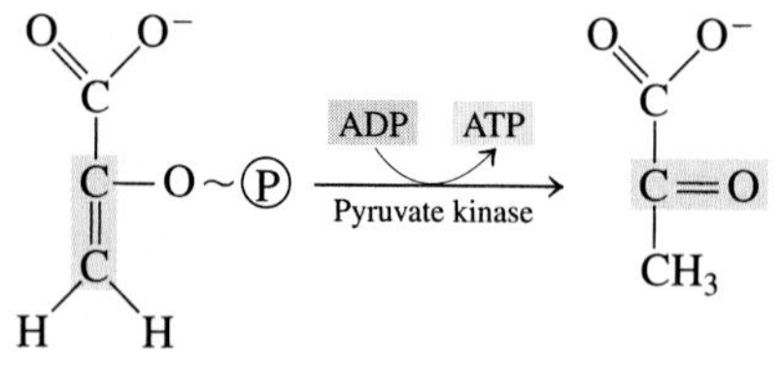

The enzyme involved, *pyruvate kinase,* requires both Mg^{2+} and K^+ ions for its activity. Again, because two C_3 molecules are reacting, two ATP molecules are produced.

ATP molecules are involved in Steps 1, 3, 7, and 10 of glycolysis. Considering these steps collectively shows that there is a net gain of two ATP molecules for every glucose molecule converted into two pyruvates (Table 13.1). Though useful, this is a small amount of ATP compared to that generated in oxidative phosphorylation (Section 12.8).

The net overall equation for the process of glycolysis is

Glucose + 2NAD$^+$ + 2ADP + 2P$_i$ $\longrightarrow$

2 pyruvate + 2NADH + 2ATP + 2H$^+$ + 2H$_2$O

TABLE 13.1
ATP Production and Consumption During Glycolysis

Step	Reaction	ATP Change per Glucose
1	Glucose $\rightarrow$ glucose 6-phosphate	−1
3	Fructose 6-phosphate $\rightarrow$ fructose 1,6-bisphosphate	−1
7	2(1,3-Bisphosphoglycerate $\rightarrow$ 3-phosphoglycerate)	+2
10	2(Phosphoenolpyruvate $\rightarrow$ pyruvate)	+2
		Net +2

■ Entry of Galactose and Fructose into Glycolysis

The breakdown products from carbohydrate digestion are glucose, fructose, and galactose (Section 13.1). Both fructose and galactose are converted, in the liver, to intermediates that enter into the glycolysis pathway.

The entry of fructose into the glycolytic pathway involves phosphorylation by ATP to produce fructose 1-phosphate, which is then split into two trioses—glyceraldehyde and dihydroxyacetone phosphate. Dihydroxyacetone phosphate enters glycolysis directly; glyceraldehyde must be phosphorylated by ATP to glyceraldehyde 3-phosphate before it enters the pathway (see Figure 13.5).

The entry of galactose into the glycolytic pathway begins with its conversion to glucose 1-phosphate (a four-step sequence), which is then converted to glucose 6-phosphate, a glycolysis intermediate (see Figure 13.5).

■ Regulation of Glycolysis

Glycolysis, like all metabolic pathways, must have control mechanisms associated with it. In glycolysis, the control points are Steps 1, 3, and 10 (see Figure 13.3).

Step 1, the conversion of glucose to glucose 6-phosphate, involves the enzyme *hexokinase*. This particular enzyme is inhibited by glucose 6-phosphate, the substance produced by its action (feedback inhibition; Section 10.8).

At Step 3, where fructose 6-phosphate is converted to fructose 1,6-bisphosphate by the enzyme *phosphofructokinase,* high concentrations of ATP and citrate inhibit enzyme activity. A high ATP concentration, which is characteristic of a state of low energy consumption, thus stops glycolysis at the fructose 6-phosphate stage. This stoppage also causes increases in glucose 6-phosphate stores because glucose 6-phosphate is in equilibrium with fructose 6-phosphate.

The third control point involves the last step of glycolysis, the conversion of phosphoenolpyruvate to pyruvate. *Pyruvate kinase,* the enzyme needed at this point, is inhibited by high ATP concentrations. Both pyruvate kinase (Step 10) and phosphofructokinase (Step 3) are allosteric enzymes (Section 10.8).

13.3 Fates of Pyruvate

The production of pyruvate from glucose (glycolysis) occurs in a similar manner in most cells. In contrast, the fate of the pyruvate so produced varies with cellular conditions and the nature of the organism. Three common fates for pyruvate are of prime importance: conversion into acetyl CoA, into lactate, and into ethanol (see Figure 13.6).

A key concept in considering these fates of pyruvate is the need for a continuous supply of NAD^+ for glycolysis. As glucose is oxidized to pyruvate in glycolysis, NAD^+ is reduced to NADH.

$$\text{Glucose} + 2NAD^+ \longrightarrow 2 \text{ pyruvate} + 2NADH + 2H^+$$
$$2ADP + 2P_i \qquad 2ATP$$

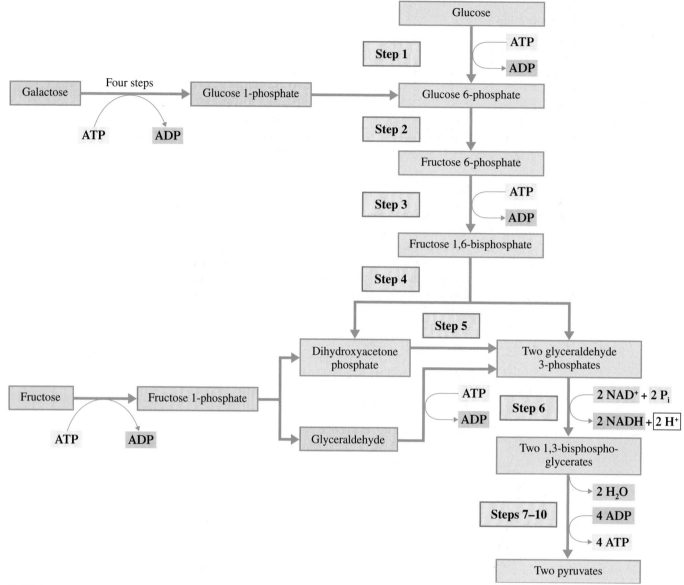

FIGURE 13.5 Entry points for fructose and galactose into the glycolysis pathway.

It is significant that each pathway of pyruvate metabolism includes provisions for regeneration of NAD^+ from NADH so that glycolysis can continue.

■ Oxidation to Acetyl CoA

Under *aerobic* (oxygen-rich) conditions, pyruvate is oxidized to acetyl CoA. Pyruvate formed in the cytosol through glycolysis crosses the two mitochondrial membranes and enters the mitochondrial matrix, where the oxidation takes place. The overall reaction, in simplified terms, is

$$CH_3-\overset{\overset{\text{O}}{\|}}{C}-COO^- + CoA-SH + NAD^+ \xrightarrow[\text{complex}]{\text{Pyruvate dehydrogenase}} CH_3-\overset{\overset{\text{O}}{\|}}{C}-S-CoA + NADH + CO_2$$

Pyruvate · · · Acetyl CoA

This reaction, which involves both oxidation and decarboxylation (CO_2 is produced), is far more complex than the simple stoichiometry of the equation suggests. The enzyme complex involved contains three different enzymes, each with numerous subunits. The

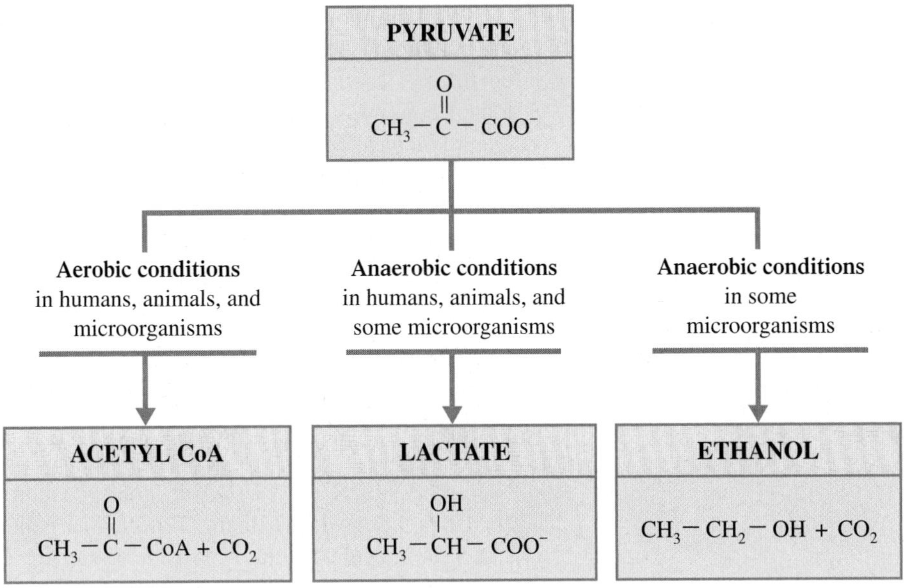

FIGURE 13.6 The three common fates of pyruvate generated by glycolysis.

An additional fate for pyruvate is conversion to oxaloacetate. This fate for pyruvate, which occurs during the process called *gluconeogenesis*, is discussed in Section 13.6.

Not all acetyl CoA produced from pyruvate enters the citric acid cycle. Particularly when high levels of acetyl CoA are produced (from excess ingestion of dietary carbohydrates), some acetyl CoA is used as the starting material for the production of the fatty acids needed for fat (triacylglycerol) formation (Section 14.7).

overall reaction process involves four separate steps and requires NAD^+, CoA—SH, FAD, and two other coenzymes (lipoic acid and thiamin pyrophosphate, the latter derived from the B vitamin thiamin).

Most acetyl CoA molecules produced from pyruvate enter the citric acid cycle. Citric acid cycle operations change more NAD^+ to its reduced form, NADH. The NADH from glycolysis, from the conversion of pyruvate to acetyl CoA, and from the citric acid cycle enters the electron transport chain directly or indirectly (Section 12.7). In the ETC, electrons from NADH are transferred to O_2, and the NADH is changed back to NAD^+. The NAD^+ needed for glycolysis, pyruvate–acetyl CoA conversion, and the citric acid cycle is regenerated.

The net overall reaction for processing one glucose molecule to two molecules of acetyl CoA is

$$\text{Glucose} + 2\text{ADP} + 2P_i + 4\text{NAD}^+ + 2\text{CoA—SH} \longrightarrow$$
$$2 \text{ acetyl CoA} + 2CO_2 + 2\text{ATP} + 4\text{NADH} + 4H^+ + 2H_2O$$

Fermentation Processes

When the body becomes oxygen deficient (anaerobic conditions), such as during strenuous exercise, the electron transport chain process slows down because its last step is dependent on oxygen. The result of this "slowing down" is a buildup in NADH concentration (it is not being consumed so fast) and a decreased amount of available NAD^+ (it is not being produced so fast). Decreased NAD^+ concentration then negatively affects the rate of glycolysis. An alternative method for conversion of NADH to NAD^+—a method that does not require oxygen—is needed if glycolysis is to continue, it being the only available source of *new* ATP under these conditions.

Fermentation processes solve this problem. **Fermentation** *is a biochemical process by which NADH is oxidized to NAD^+ without the need for oxygen.* We consider here two fermentation processes: lactate fermentation and ethanol fermentation.

Lactate Fermentation

Lactate fermentation *is the enzymatic anaerobic reduction of pyruvate to lactate.* The sole purpose of this process is the conversion of NADH to NAD^+. The lactate so formed is converted back to pyruvate when aerobic conditions are again established in a cell (Section 13.6).

Working muscles often produce lactate. If strenuous exercise is continued too long, the buildup of lactate in the muscles reaches a point beyond which fermentation cannot continue. This slows glycolysis and new ATP production, and the muscle action can no longer continue (fatigue and exhaustion; see the accompanying Chemical Connections feature on page 461) until oxygen supplies are re-established.

The equation for lactate formation from pyruvate is

$$CH_3-\underset{\text{Pyruvate}}{\overset{\overset{\displaystyle O}{\|}}{C}}-COO^- + NADH + H^+ \xrightarrow[\text{dehydrogenase}]{\text{Lactate}} CH_3-\underset{\text{Lactate}}{\overset{\overset{\displaystyle O\,H}{|}}{C}H}-COO^- + NAD^+$$

> Red blood cells have no mitochondria and therefore always form lactate as the end product of glycolysis.

When the reaction for conversion of pyruvate to lactate is added to the net glycolysis reaction (Section 13.2), an overall reaction for the conversion of glucose to lactate is obtained.

$$\text{Glucose} + 2ADP + 2P_i \longrightarrow 2\text{ lactate} + 2ATP + 2H_2O$$

Note that NADH and NAD$^+$ do not appear in this equation, even though the process cannot proceed without them. The NADH generated during glycolysis (Step 6) is consumed in the conversion of pyruvate to lactate. Thus there is no net oxidation–reduction in the conversion of glucose to lactate.

■ Ethanol Fermentation

Under anaerobic conditions, several simple organisms, including yeast, possess the ability to regenerate NAD$^+$ through ethanol, rather than lactate, production. Such a process is called ethanol fermentation. **Ethanol fermentation** *is the enzymatic anaerobic conversion of pyruvate to ethanol and carbon dioxide.* Ethanol fermentation involving yeast causes bread and related products to rise as a result of CO$_2$ bubbles being released during baking. Beer, wine, and other alcoholic drinks are produced by ethanol fermentation of the sugars in grain and fruit products.

> With bread and other related products obtained using yeast, the ethanol produced by fermentation evaporates during baking.

The first step in conversion of pyruvate to ethanol is a decarboxylation reaction to produce acetaldehyde.

$$CH_3-\underset{\text{Pyruvate}}{\overset{\overset{\displaystyle O}{\|}}{C}}-COO^- + H^+ \xrightarrow[\text{decarboxylase}]{\text{Pyruvate}} CH_3-\underset{\text{Acetaldehyde}}{\overset{\overset{\displaystyle O}{\|}}{C}}-H + CO_2$$

The second step involves acetaldehyde reduction to produce ethanol.

$$CH_3-\underset{\text{Acetaldehyde}}{\overset{\overset{\displaystyle O}{\|}}{C}}-H + NADH + H^+ \xrightarrow[\text{dehydrogenase}]{\text{Alcohol}} CH_3-\underset{\underset{\displaystyle H}{|}}{\overset{\overset{\displaystyle O\,H}{|}}{C}}-H + NAD^+$$
$$\text{Ethanol}$$

The overall equation for the conversion of pyruvate to ethanol (the sum of the two steps) is

$$CH_3-\underset{\text{Pyruvate}}{\overset{\overset{\displaystyle O}{\|}}{C}}-COO^- + 2H^+ + NADH \xrightarrow{\text{Two steps}} CH_3-CH_2-\underset{\text{Ethanol}}{OH} + NAD^+ + CO_2$$

An overall reaction for the production of ethanol from glucose is obtained by combining the reaction for the conversion of pyruvate with the net reaction for glycolysis (Section 13.2).

$$\text{Glucose} + 2ADP + 2P_i \longrightarrow 2\text{ ethanol} + 2CO_2 + 2ATP + 2H_2O$$

Again note that NADH and NAD$^+$ do not appear in the final equation; they are both generated and consumed.

Lactate Accumulation

During strenuous exercise, conditions in muscle cells can change from aerobic to anaerobic as the oxygen supply becomes inadequate to meet demand. Such conditions cause pyruvate to be converted to lactate rather than acetyl CoA. (Lactate production can also be high at the start of strenuous exercise before the delivery of oxygen is stepped up via an increased respiration rate.)

The resulting lactate begins to accumulate in the cytosol of cells where it is produced. Some lactate diffuses out of the cells into the blood, where it contributes to a slight decrease in blood pH. This lower pH triggers fast breathing, which helps supply more oxygen to the cells.

Lactate accumulation is the cause of muscle pain and cramping during prolonged, strenuous exercise. As a result of such cramping, muscles may be stiff and sore the next day. Regular, hard exercise increases the efficiency with which oxygen is delivered to the body. Thus athletes can function longer than nonathletes under aerobic conditions without lactate production.

Lactate accumulation can also occur in heart muscle if it experiences decreased oxygen supply (from artery blockage). The heart muscle experiences cramps and stops beating (cardiac arrest). Massage of heart muscle often reduces such cramps, just as it does for skeletal muscle, and it is sometimes possible to start the heart beating again by using such a technique.

Premature infants born with underdeveloped lungs are often given increased amounts of oxygen to minimize lactate accumulation. They are also given bicarbonate (HCO_3^-) solution to counteract the acidity change in blood that accompanies lactate buildup.

Strenuous muscular activity can result in lactate accumulation.

Figure 13.7 summarizes the relationship between the fates of pyruvate and the regeneration of NAD^+ from NADH.

13.4 ATP Production for the Complete Oxidation of Glucose

We now assemble energy production figures for glycolysis, oxidation of pyruvate to acetyl CoA, the citric acid cycle, and the electron transport chain. The result, with one added piece of information, gives the ATP yield for the *complete* oxidation of one molecule of glucose.

FIGURE 13.7 All three of the common fates of pyruvate from glycolysis provide for the regeneration of NAD^+ from NADH.

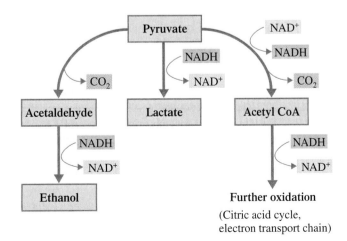

FIGURE 13.8 The dihydroxyacetone phosphate–glycerol 3-phosphate shuttle.

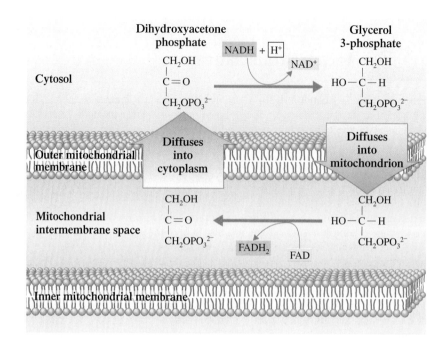

The new piece of information involves the NADH produced during Step 6 of glycolysis. This NADH, produced in the cytosol, cannot *directly* participate in the electron transport chain because mitochondria are impermeable to NADH (and NAD^+). A transport system shuttles the electrons from NADH, but not NADH itself, across the membrane. This shuttle involves dihydroxyacetone phosphate (a glycolysis intermediate) and glycerol 3-phosphate.

The first step in the shuttle is the cytosolic reduction of dihydroxyacetone phosphate by NADH to produce glycerol 3-phosphate and NAD^+ (see Figure 13.8). Glycerol 3-phosphate then crosses the outer mitochondrial membrane, where it is reoxidized to dihydroxyacetone phosphate. The oxidizing agent is FAD rather than NAD^+. The regenerated dihydroxyacetone phosphate diffuses out of the mitochondrion and returns to the cytosol for participation in another "turn" of the shuttle. The $FADH_2$ coproduced in the mitochondrial reaction can participate in the electron transport chain reactions. The net reaction of this shuttle process is

$$\underset{\text{(cytosolic)}}{NADH} + H^+ + \underset{\text{(mitochondrial)}}{FAD} \longrightarrow \underset{\text{(cytosolic)}}{NAD^+} + \underset{\text{(mitochondrial)}}{FADH_2}$$

The consequence of this reaction is that only 1.5 rather than 2.5 molecules of ATP are formed for each cytosolic NADH, because $FADH_2$ yields one less ATP than does NADH in the electron transport chain.

Table 13.2 shows ATP production for the complete oxidation of a molecule of glucose. The final number is 30 ATP, 26 of which come from the oxidative phosphorylation associated with the electron transport chain. This total of 30 ATP for complete oxidation contrasts markedly with a total of 2 ATP for oxidation of glucose to lactate and 2 ATP for oxidation of glucose to ethanol. Neither of these latter processes involves the citric acid cycle or the electron transport chain. Thus the aerobic oxidation of glucose is 15 times more efficient in the production of ATP than the anaerobic lactate and ethanol processes.

The production of 30 ATP molecules per glucose (Table 13.2) is for those cells where the dihydroxyacetone phosphate–glycerol 3-phosphate shuttle operates (skeletal muscle and nerve cells). In certain other cells, particularly heart and liver cells, a more complex shuttle system called the malate–aspartate shuttle functions. In this shuttle, 2.5 ATP molecules result from 1 cytosolic NADH, which changes the total ATP production to 32 molecules per glucose.

The net overall reaction for the *complete* metabolism (oxidation) of a glucose molecule is the simple equation

$$\text{Glucose} + 6O_2 + 30ADP + 30P_i \longrightarrow 6CO_2 + 6H_2O + 30ATP$$

TABLE 13.2
Production of ATP from the Complete Oxidation of One Glucose Molecule in a Skeletal Muscle Cell

Reaction	Comments	Yield of ATP
Glycolysis		
glucose $\rightarrow$ glucose 6-phosphate	consumes 1 ATP	-1
glucose 6-phosphate $\rightarrow$ fructose 1,6-bisphosphate	consumes 1 ATP	-1
2(glyceraldehyde 3-phosphate $\rightarrow$ 1,3-bisphosphoglycerate)	each produces 1 cytosolic NADH	—
2(1,3-bisphosphoglycerate $\rightarrow$ 3-phosphoglycerate)	each produces 1 ATP	$+2$
2(phosphoenolpyruvate $\rightarrow$ pyruvate)	each produces 1 ATP	$+2$
Oxidation of Pyruvate		
2(pyruvate $\rightarrow$ acetyl CoA + CO_2)	each produces 1 NADH	—
Citric Acid Cycle		
2(isocitrate $\rightarrow$ α-ketoglutarate + CO_2)	each produces 1 NADH	—
2(α-ketoglutarate $\rightarrow$ succinyl CoA + CO_2)	each produces 1 NADH	—
2(succinyl CoA $\rightarrow$ succinate)	each produces 1 GTP	$+2$
2(succinate $\rightarrow$ fumarate)	each produces 1 $FADH_2$	—
2(malate $\rightarrow$ oxaloacetate)	each produces 1 NADH	—
Electron Transport Chain and Oxidative Phosphorylation		
2 cytosolic NADH formed in glycolysis	each produces 1.5 ATP	$+3$
2 NADH formed in the oxidation of pyruvate	each produces 2.5 ATP	$+5$
2 $FADH_2$ formed in the citric acid cycle	each produces 1.5 ATP	$+3$
6 NADH formed in the citric acid cycle	each produces 2.5 ATP	$+15$
	Net production of ATP	$+30$

Note that substances such as NADH, NAD^+, and $FADH_2$ are not part of this equation. Why? They cancel out—that is, they are consumed in one step (reactant) and regenerated in another step (product). Note also what the net equation does not acknowledge: the many dozens of reactions that are needed to generate the 30 molecules of ATP.

13.5 Glycogen Synthesis and Degradation

Glycogen, a branched polymeric form of glucose (Section 7.15), is the storage form of carbohydrates in humans and animals. It is found primarily in muscle and liver tissue. In muscles it is the source of glucose needed for glycolysis. In the liver, it is the source of glucose needed to maintain normal glucose levels in the blood.

■ Glycogenesis

Glycogenesis *is the metabolic pathway by which glycogen is synthesized from glucose.* Glycogenesis involves three reactions (steps).

Step 1: *Formation of Glucose 1-phosphate.* The starting material for this step is not glucose itself but rather glucose 6-phosphate (available from the first step of glycolysis). The enzyme *phosphoglucomutase* effects the change from a 6-phosphate to a 1-phosphate.

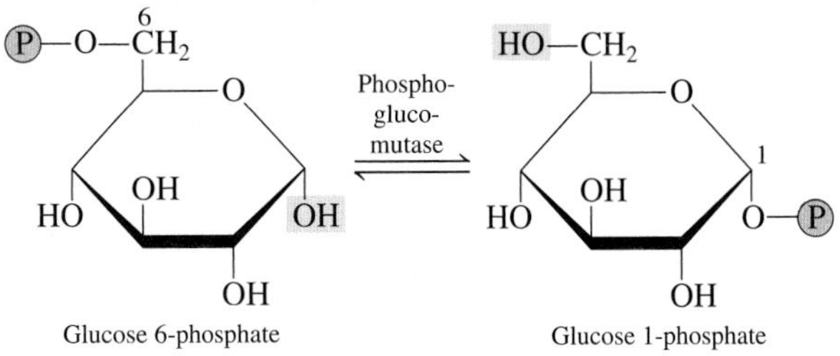

Glucose 6-phosphate   Glucose 1-phosphate

Step 2: *Formation of UDP-glucose.* Glucose 1-phosphate from Step 1 must be activated before it can be added to a growing glycogen chain. The activator is the high-energy compound UTP (uridine triphosphate). The UTP is hydrolyzed to UMP and PP$_i$, and then the PP$_i$ is further hydrolyzed to 2P$_i$. The UMP that is formed bonds to the glucose 1-phosphate to form UDP-glucose.

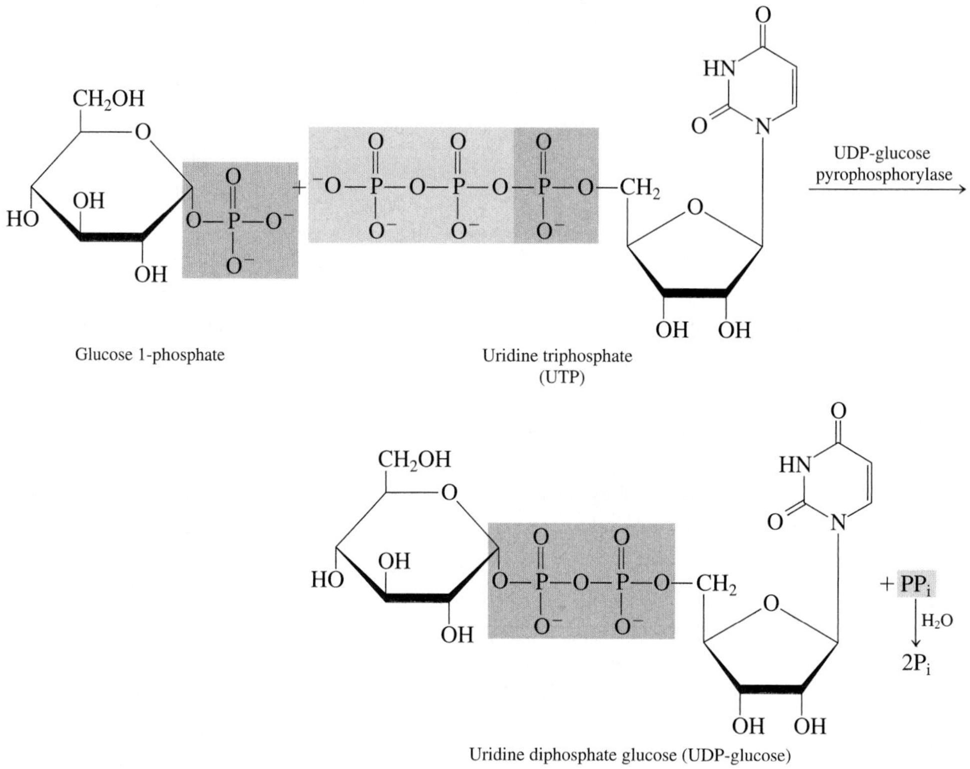

Glucose 1-phosphate

Uridine triphosphate (UTP)

UDP-glucose pyrophosphorylase

Uridine diphosphate glucose (UDP-glucose)

+ PP$_i$

H$_2$O

2P$_i$

Step 3: *Glucose Transfer to a Glycogen Chain.* The glucose unit of UDP-glucose is then attached to the end of a glycogen chain.

$$\text{UDP-glucose} + (\text{glucose})_n \longrightarrow (\text{glucose})_{n+1} + \text{UDP}$$

Glucogen chain

Glycogen with an additional glucose unit

In a subsequent reaction, the UDP produced in Step 3 is converted back to UTP, which can then react with another glucose 1-phosphate (Step 2). The conversion reaction requires ATP.

$$\text{UDP} + \text{ATP} \longrightarrow \text{UTP} + \text{ADP}$$

Adding a single glucose unit to a growing glycogen chain requires the investment of two ATP molecules: one in the formation of glucose 6-phosphate and one in the regeneration of UTP.

■ Glycogenolysis

Glycogenolysis *is the metabolic pathway by which free glucose units are obtained from glycogen.* This process is not simply the reverse of glycogen synthesis (glycogenesis), because it does not require UTP or UDP molecules. Glycogenolysis, like glycogenesis, is a three-step process.

Step 1: *Phosphorylation of a Glucose Residue.* The enzyme *glycogen phosphorylase* effects the removal of an end glucose unit from a glycogen molecule as glucose 1-phosphate.

A *phosphorylase* is an enzyme that catalyzes the cleavage of a bond by P$_i$ (in contrast to hydrolysis, which refers to bond cleavage by water), such as removal of a glucose unit from glucogen to give glucose 1-phosphate.

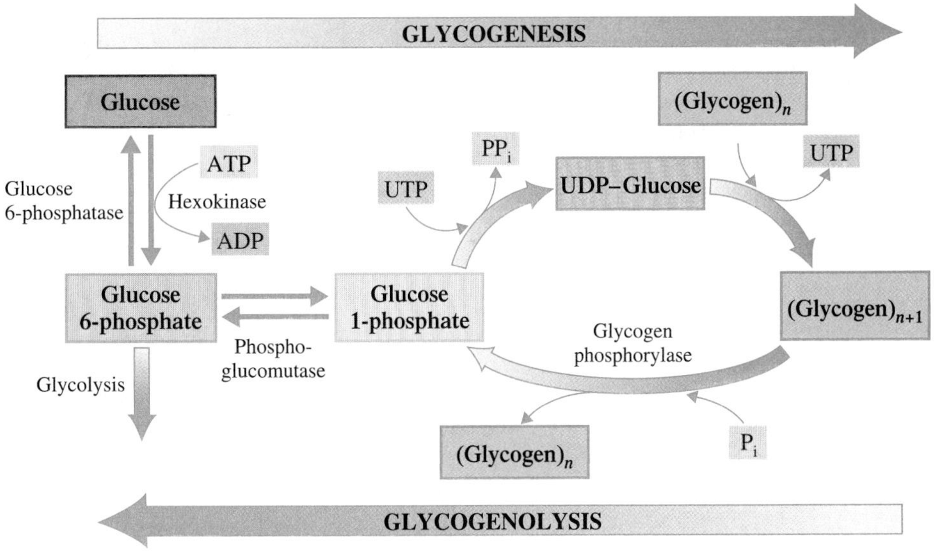

FIGURE 13.9 The processes of glycogenesis and glycogenolysis contrasted. The intermediate glucose–UDP is part of glycogenesis but not of glycogenolysis.

$$(Glucose)_n + P_i \longrightarrow (glucose)_{n-1} + glucose\ 1\text{-phosphate}$$

Glycogen   Glycogen with one
                               fewer glucose unit

Step 2: *Glucose 1-phosphate Isomerization.* The enzyme *phosphoglucomutase* catalyzes the isomerization process whereby the phosphate group of glucose 1-phosphate is moved to the carbon 6 position.

$$Glucose\ 1\text{-phosphate} \rightleftharpoons glucose\ 6\text{-phosphate}$$

This process is the reverse of the first step of glycogenesis.

Step 3: *Hydrolysis of Glucose 6-phosphate to Glucose.* The reaction for this step is

$$Glucose\ 6\text{-phosphate} + H_2O \longrightarrow glucose + P_i$$

The enzyme needed for this reaction, *glucose 6-phosphatase,* is found only in the liver, kidneys, and intestine. Thus *complete* glycogenolysis occurs only in these tissues.

A *phosphatase* is an enzyme that effects the removal of a phosphate group (P_i) from a molecule, such as converting glucose 6-phosphate to glucose, with H_2O as the attacking species.

Muscle and brain cells, which lack glucose 6-phosphatase, cannot form free glucose from glucose 6-phosphate. The first two steps of glycogenolysis do, however, occur in these tissues. The glucose 6-phosphate so produced can contribute to energy production through glycolysis and the common metabolic pathway because glucose 6-phosphate is the first intermediate in the glycolytic pathway (see Figure 13.3). Thus brain and muscle cells can use glycogen for energy production only. The liver, however, has the capacity to supply additional glucose to the blood.

The fact that glycogen synthesis (glycogenesis) and glycogen degradation (glycogenolysis) are not totally reverse processes has significance. In fact, it is almost always the case in biochemistry that "opposite" biosynthetic and degradative pathways differ in some steps. This allows for separate control of the pathways.

Figure 13.9 contrasts the "opposite" processes of glycogenesis and glycogenolysis. Both processes involve the intermediate glucose 6-phosphate. Glucose–UDP is unique to glycogenesis.

13.6 Gluconeogenesis and the Cori Cycle

Gluconeogenesis *is the metabolic pathway by which glucose is synthesized from noncarbohydrate materials.* Glycogen stores in muscle and liver tissue are depleted within 12–18 hours from fasting or in even less time from heavy work or strenuous exercise. Without gluconeogenesis, the brain, which is dependent on glucose as a fuel, would have problems functioning if food intake were restricted for even one day.

The noncarbohydrate starting materials for gluconeogenesis are lactate (from hardworking muscles and from red blood cells), glycerol (from triacylglycerol hydrolysis),

FIGURE 13.10 The "opposite" processes of gluconeogenesis (pyruvate to glucose) and glycolysis (glucose to pyruvate) are not exact opposites. The reversal of the last step of glycolysis requires two steps in gluconeogenesis. Therefore, gluconeogenesis has 11 steps, whereas glycolysis has only 10 steps.

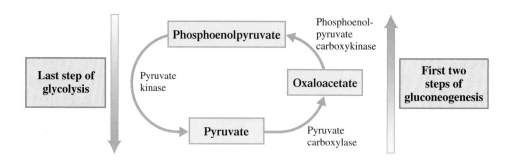

and certain amino acids (from dietary protein hydrolysis or from muscle protein during starvation). About 90% of gluconeogenesis takes place in the liver. Hence gluconeogenesis helps to maintain normal blood-glucose levels in times of inadequate dietary carbohydrate intake (such as between meals).

The processes of gluconeogenesis (pyruvate to glucose) and glycolysis (glucose to pyruvate) are not exact opposites. The most obvious difference between these two processes is that 11 compounds are involved in gluconeogenesis and only 10 in glycolysis. Why the difference? The last step of glycolysis is the conversion of the high-energy compound phosphoenolpyruvate to pyruvate. The reverse of this process, which is the beginning of gluconeogenesis, cannot be accomplished in a single step because of the large energy difference between the two compounds and the slow rate of the reaction. Instead, a two-step process by way of oxaloacetate is required to effect the change, and this adds an extra compound to the gluconeogenesis pathway (see Figure 13.10). Both an ATP molecule and a GTP molecule are needed to drive this two-step process.

$$
\begin{array}{c}
\text{COO}^- \\
| \\
\text{C}{=}\text{O} \\
| \\
\text{CH}_3 \\
\text{Pyruvate}
\end{array}
+ \text{CO}_2 + \text{ATP} + \text{H}_2\text{O}
\underset{\text{Biotin}}{\overset{\text{Pyruvate carboxylase}}{\rightleftharpoons}}
\begin{array}{c}
\text{COO}^- \\
| \\
\text{C}{=}\text{O} \\
| \\
\text{CH}_2 \\
| \\
\text{COO}^- \\
\text{Oxaloacetate}
\end{array}
+ \text{ADP} + \text{P}_i
$$

$$
\begin{array}{c}
\text{COO}^- \\
| \\
\text{C}{=}\text{O} \\
| \\
\text{CH}_2 \\
| \\
\text{COO}^- \\
\text{Oxaloacetate}
\end{array}
+ \text{GTP}
\overset{\text{Phosphoenolpyruvate carboxykinase}}{\rightleftharpoons}
\begin{array}{c}
\text{Phosphoenolpyruvate}
\end{array}
+ \text{CO}_2 + \text{GDP}
$$

The oxaloacetate intermediate in this two-step process provides a connection to the citric acid cycle. In the first step of this cycle, oxaloacetate combines with acetyl CoA. If energy rather than glucose is needed, then oxaloacetate can go directly into the citric acid cycle.

As is shown in Figure 13.11, there are two other locations where gluconeogenesis and glycolysis differ. In Steps 9 and 11 of gluconeogenesis (Steps 1 and 3 of glycolysis), the reactant–product combinations match between pathways. However, different enzymes are required for the forward and reverse processes. The new enzymes for gluconeogenesis are *fructose 1,6-bisphosphatase* and *glucose 6-phosphatase*.

The overall net reaction for gluconeogenesis is

$$2 \text{ Pyruvate} + 4\text{ATP} + 2\text{GTP} + 2\text{NADH} + 2\text{H}_2\text{O} \longrightarrow$$
$$\text{glucose} + 4\text{ADP} + 2\text{GDP} + 6\text{P}_i + 2\text{NAD}^+$$

Thus to reconvert pyruvate to glucose requires the expenditure of 4 ATP and 2 GTP. Whenever gluconeogenesis occurs, it is at the expense of other ATP-producing metabolic processes.

Glycolysis has a net production of 2 ATP (Section 13.2). Gluconeogenesis has a net expenditure of 4 ATP and 2 GTP, which is equivalent to the expenditure of 6 ATP.

FIGURE 13.11 The pathway for gluconeogenesis is similar, but not identical, to the pathway for glycolysis.

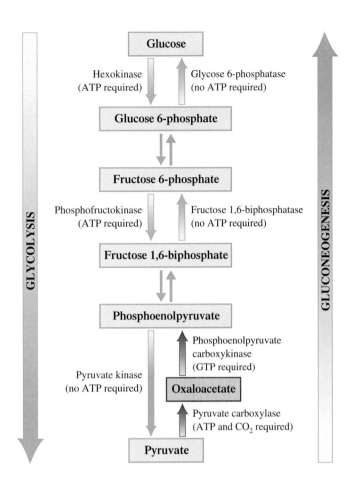

The Cori Cycle

Gluconeogenesis using lactate as a source of pyruvate is particularly important because of lactate formation during strenuous exercise. The lactate so produced (Section 13.3) diffuses from muscle cells into the blood, where it is transported to the liver. Here the enzyme lactate dehydrogenase (the same enzyme that catalyzes lactate formation in muscle) converts lactate back to pyruvate.

$$\underset{\text{Lactate}}{\underset{\text{CH}_3}{\overset{\text{COO}^-}{\text{H}-\text{C}-\text{OH}}}} + \text{NAD}^+ \xrightarrow[\text{dehydrogenase}]{\text{Lactate}} \underset{\text{Pyruvate}}{\underset{\text{CH}_3}{\overset{\text{COO}^-}{\text{C}=\text{O}}}} + \text{NAD H} + \text{H}^+$$

The newly formed pyruvate is then converted via gluconeogenesis to glucose, which enters the bloodstream and goes to the muscles. This cyclic process, which is called the Cori cycle, is diagrammed in Figure 13.12. The **Cori cycle** *is a cyclic biochemical process in which glucose is converted to lactate in muscle tissue, the lactate is reconverted to glucose in the liver, and the glucose is returned to the muscle tissue.*

The Cori cycle is named in honor of Gerty Radnitz Cori (1896–1957) and Carl Cori (1896–1984), the husband-and-wife team who discovered it. They were awarded a Nobel Prize in 1947, the third husband-and-wife team to be so recognized. Marie and Pierre Curie were the first, Irene and Frederic Joliot-Curie the second.

13.7 Terminology for Glucose Metabolic Pathways

In the preceding three sections we considered the processes of glycolysis, glycogenesis, glycogenolysis, and gluconeogenesis. Because of their like-sounding names, keeping the terminology for these four processes "straight" is often a problem. Figure 13.13 shows

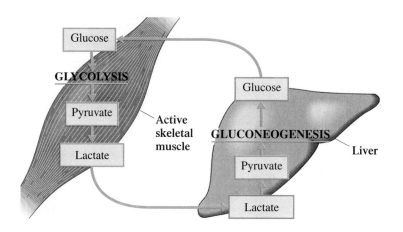

FIGURE 13.12 The Cori cycle. Lactate, formed from glucose under anaerobic conditions in muscle cells, is transferred to the liver, where it is reconverted to glucose, which is then transferred back to the muscle cells.

the relationships among these processes. Note that the glycogen degradation pathways (left side of Figure 13.13) have names ending in *-lysis,* which means "breakdown." The pathways associated with glycogen synthesis (right side of Figure 13.13) have names ending in *-genesis,* which means "making."

13.8 The Pentose Phosphate Pathway

Glycolysis is not the only pathway by which glucose may be degraded. Depending on the type of cell, various amounts of glucose are degraded by the pentose phosphate pathway, a pathway whose main focus is *not* subsequent ATP production as is the case for glycolysis. Major functions of this alternative pathway are (1) synthesis of the coenzyme NADPH needed in lipid biosynthesis (Section 14.7), and (2) production of ribose 5-phosphate, a pentose derivative needed for the synthesis of nucleic acids and many coenzymes. The **pentose phosphate pathway** *is the metabolic pathway by which glucose is used to produce NADPH, ribose 5-phosphate (a pentose phosphate), and numerous other sugar phosphates.* The operation of the pentose phosphate pathway is significant in cells that produce lipids: fatty tissue, the liver, mammary glands, and the adrenal cortex (an active producer of steroid lipids).

NADPH, the coenzyme produced in the pentose phosphate pathway, is the reduced form of $NADP^+$ (nicotinamide adenine dinucleotide phosphate). Structurally, $NADP^+$/ NADPH is a phosphorylated version of NAD^+/NADH (see Figure 13.14).

FIGURE 13.13 The relationships among four common metabolic pathways that involve glucose.

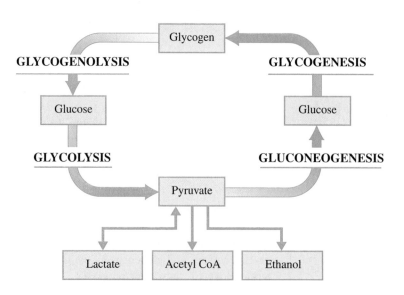

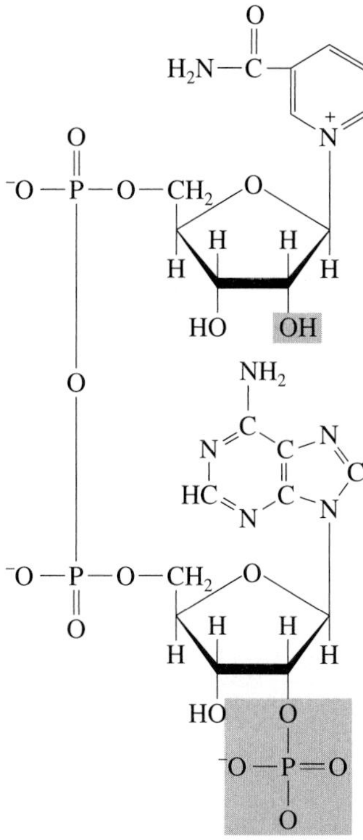

FIGURE 13.14 The Structure of NADPH. The phosphate group shown in color is the structural feature that distinguishes NADPH from NADH.

The nonphosphorylated and phosphorylated versions of this coenzyme have significantly different functions. The nonphosphorylated version is involved, mainly in its oxidized form (NAD^+/NADH), in the reactions of the common metabolic pathway (Section 12.5). The phosphorylated version is involved, mainly in its reduced form (NADPH/$NADP^+$), in biosynthetic reactions of lipids and nucleic acids.

There are two stages within the pentose phosphate pathway—an oxidative stage and a nonoxidative stage. The oxidative stage, which occurs first, involves three steps through which glucose 6-phosphate is converted to ribulose 5-phosphate and CO_2.

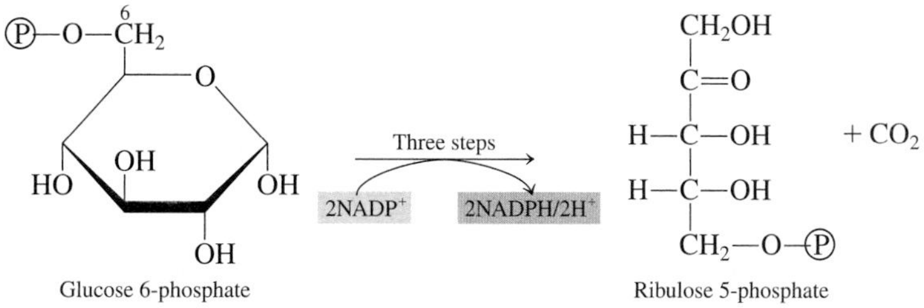

The net equation for the oxidative stage of the pentose phosphate pathway is

$$\text{Glucose 6-Phosphate} + 2NADP^+ + H_2O \longrightarrow$$
$$\text{ribulose 5-phosphate} + CO_2 + 2NADPH + 2H^+$$

Note the production of two NADPH molecules per glucose 6-phosphate processed during this stage.

In the first step of the nonoxidative stage of the pentose phosphate pathway, ribulose 5-phosphate (a ketose) is isomerized to ribose 5-phosphate (an aldose).

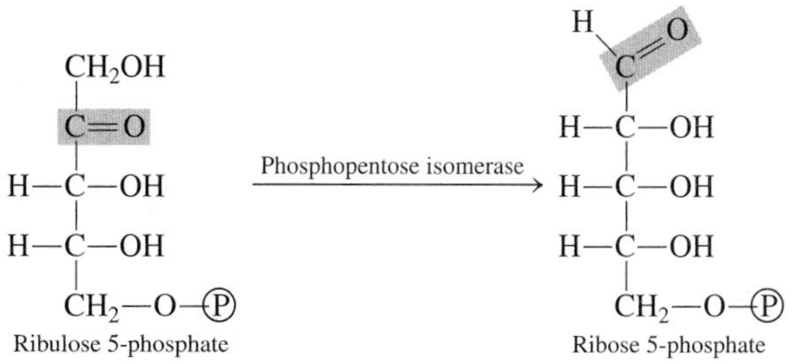

The pentose ribose is a component of ATP, GTP, UTP, CoA, NAD^+/NADH, FAD/$FADH_2$, and RNA. Further steps in the nonoxidative stage contain provision for the conversion of ribose 5-phosphate to numerous other sugar phosphates. Ultimately, glyceraldehyde 3-phosphate and fructose 6-phosphate (both glycolysis intermediates) are formed. The overall net reaction for the pentose phosphate pathway is

$$3\text{Glucose 6-phosphate} + 6NADP^+ + 3H_2O \longrightarrow$$
$$2\text{fructose 6-phosphate} + 3CO_2 + \text{glyceraldehyde 3-phosphate} + 6NADPH + 6H^+$$

The pentose phosphate pathway, with its many intermediates, helps meet cellular needs in numerous ways.

1. When ATP demand is high, the pathway continues to its end products, which enter glycolysis.
2. When NADPH demand is high, intermediates are recycled to glucose 6-phosphate (the start of the pathway), and further NADPH is produced.
3. When ribose 5-phosphate demand is high, for nucleic acid and coenzyme production, most of the nonoxidative stage is nonfunctional, leaving ribose 5-phosphate as a major product.

The Chemistry at a Glance feature on page 470 shows how the pentose phosphate pathway is related to the other major pathways of glucose metabolism that we have considered.

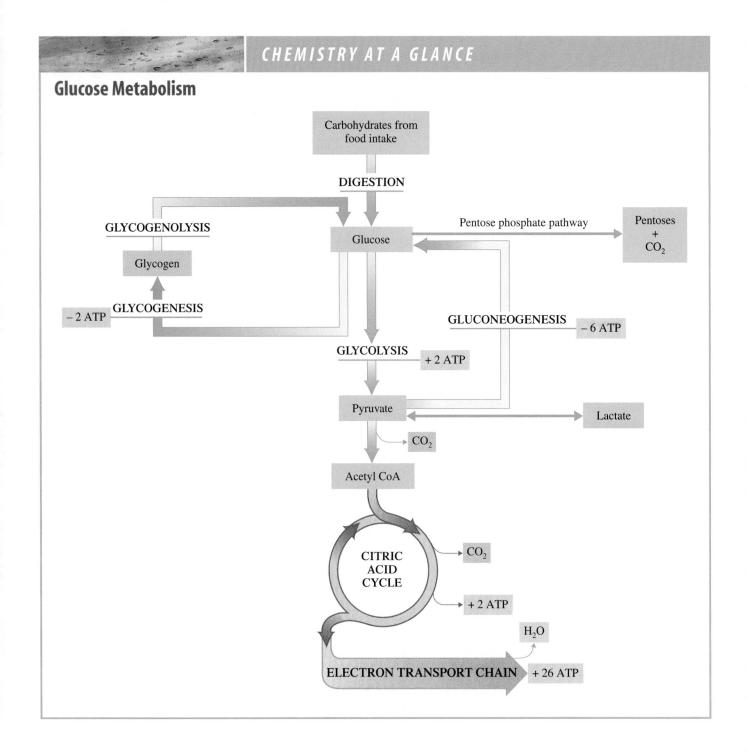

CHEMISTRY AT A GLANCE

Glucose Metabolism

13.9 Hormonal Control of Carbohydrate Metabolism

A second major method for controlling carbohydrate metabolism, besides enzyme inhibition by metabolites (Section 13.2), is hormonal control. Among others, three hormones—insulin, glucagon, and epinephrine—affect carbohydrate metabolism.

■ Insulin

Insulin, a 51-amino-acid protein whose structure we considered in Section 9.11, is a hormone produced by the beta cells of the pancreas. Insulin promotes the uptake and utilization

CHEMICAL CONNECTIONS Diabetes Mellitus

Diabetes mellitus is the best-known and most prevalent metabolic disease in humans, affecting approximately 4% of the population. There are two major forms of this disease: insulin-dependent (type I) and non–insulin-dependent (type II) diabetes.

Type I diabetes, which often appears in children, is the result of inadequate insulin production by the beta cells of the pancreas. Control of this condition involves insulin injections and special dietary programs. A risk associated with the insulin injections is that too much insulin can produce severe hypoglycemia (insulin shock); blackout or a coma can result. Treatment involves a quick infusion of glucose. Diabetics often carry candy bars (quick glucose sources) for use if they feel any of the symptoms that signal the onset of insulin shock.

In Type II diabetes, which usually occurs in overweight individuals more than 40 years old, body insulin production is normal, but the cells do not respond to it normally. Some of the insulin receptors on the cell membranes are not functioning properly and fail to recognize the insulin. Treatment involves drugs that increase body insulin levels and a carefully regulated diet (to

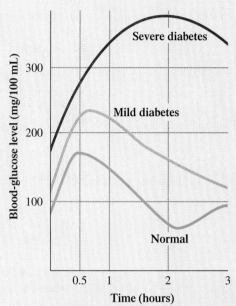

Typical Tolerance Curves for Glucose

reduce obesity). More efficient use of undamaged receptors occurs at increased insulin levels.

The effects of both types of diabetes are the same—inadequate glucose uptake by cells. The result is blood-glucose levels much higher than normal (hyperglycemia). With an inadequate glucose intake, cells must resort to other procedures for energy production, procedures that involve the breakdown of fats and protein.

A frequently used diagnostic test for diabetes is the glucose tolerance test. A patient who has fasted for 10–16 hours is given a single dose of glucose, typically in a fruit-flavored drink. Blood-glucose levels are then monitored at regular intervals over several hours. As the accompanying diagram shows, glucose levels drop to a fasting level in a nondiabetic in about 2 hours but remain high in a diabetic.

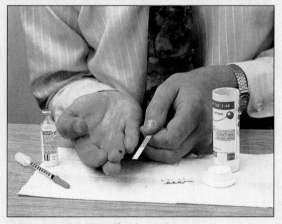

A diabetic giving himself a blood glucose test.

FIGURE 13.15 The series of events by which the hormone epinephrine stimulates glucose production.

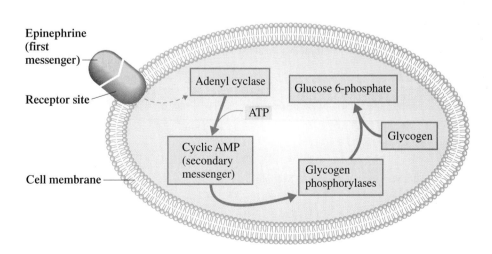

of glucose by cells. Thus its function is to lower blood glucose levels. It is also involved in lipid metabolism.

The release of insulin is triggered by *high* blood-glucose levels. The mechanism for insulin action involves insulin binding to protein receptors on the outer surfaces of cells, which facilitates entry of glucose into the cells. Insulin also produces an increase in the rate of glycogen synthesis.

■ Glucagon

Glucagon is a polypeptide hormone (29 amino acids) produced in the pancreas by alpha cells. It is released when blood-glucose levels are *low*. Its principal function is to increase blood-glucose concentrations by speeding up the conversion of glycogen to glucose (glycogenolysis) in the liver. Thus glucagon's effects are opposite to those of insulin.

■ Epinephrine

Epinephrine (Section 6.10), also called adrenaline, is released by the adrenal glands in response to anger, fear, or excitement. Its function is similar to that of glucagon—stimulation of glycogenolysis, the release of glucose from glycogen. Its primary target is muscle cells, where energy is needed for quick action. It also functions in lipid metabolism.

Epinephrine acts by binding to a receptor site on the outside of the cell membrane, stimulating the enzyme *adenyl cyclase* to begin production of a *secondary messenger,* cyclic AMP (cAMP) from ATP. The cAMP is released in the cell interior, where, in a series of reactions, it activates *glycogen phosphorylase,* the enzyme that initiates glycogenolysis. The glucose 6-phosphate that is produced from the glycogen breakdown provides a source of quick energy. Figure 13.15 shows the series of events initiated by the release of the hormone epinephrine. Cyclic AMP also inhibits glycogenesis, thus preventing glycogen production at the same time.

 CONCEPTS TO REMEMBER

Glycolysis. Glycolysis, a series of ten reactions that occur in the cytosol, is a process in which one glucose molecule is converted into two molecules of pyruvate. A net gain of two molecules of ATP and two molecules of NADH results from the metabolizing of glucose to pyruvate (Section 13.2).

Fates of pyruvate. With respect to energy-yielding metabolism, the pyruvate produced by glycolysis can be converted to acetyl CoA under aerobic conditions or to lactate under anaerobic conditions. Some microorganisms convert pyruvate to ethanol, an anaerobic process (Section 13.3).

Glycogenesis. Glycogenesis is the process whereby excess glucose is converted into glycogen. The glycogen is stored in the liver and in muscle tissue (Section 13.5).

Glycogenolysis. Glycogenolysis is the breakdown of glycogen into glucose. This process occurs when muscles need energy and when the liver is restoring a low blood-sugar level to normal (Section 13.5).

Gluconeogenesis. Gluconeogenesis is the formation of glucose from lactate and certain other substances. This process takes place in the liver when glycogen supplies are being depleted and when carbohydrate intake is low (Section 13.6).

Cori cycle. The Cori cycle is the cyclic process involving the transport of lactate from muscle tissue to the liver, the resynthesis of glucose by gluconeogenesis, and the return of glucose to muscle tissue (Section 13.6).

Pentose phosphate pathway. The pentose phosphate pathway metabolizes glucose to produce ribose (a pentose), NADPH, and other sugars needed for biosynthesis (Section 13.8).

Carbohydrate metabolism and hormones. Insulin decreases blood-glucose levels by promoting the uptake of glucose by cells. Glucagon increases blood-glucose levels by promoting the conversion of glycogen to glucose. Epinephrine stimulates the release of glucose from glycogen in muscle cells (Section 13.9).

 KEY REACTIONS AND EQUATIONS

1. Glycolysis (Section 13.2)

$$\text{Glucose} + 2P_i + \boxed{2ADP} + \boxed{2NAD^+} \longrightarrow$$
$$2 \text{ pyruvate} + 2ATP + \boxed{2NADH} + 2H^+ + 2H_2O$$

2. Oxidation of pyruvate to acetyl CoA (Section 13.3)

$$CH_3-\overset{\overset{\displaystyle O}{\|}}{C}-COO^- + \boxed{CoA-SH} + \boxed{NAD^+} \xrightarrow[\text{steps}]{\text{Four}}$$
$$CH_3-\overset{\overset{\displaystyle O}{\|}}{C}-S-CoA + \boxed{NADH} + CO_2$$

3. Reduction of pyruvate to lactate (Section 13.3)

$$CH_3-\overset{\overset{\displaystyle O}{\|}}{C}-COO^- + \boxed{NADH} + H^+ \longrightarrow$$
$$CH_3-\overset{\overset{\displaystyle OH}{|}}{CH}-COO^- + \boxed{NAD^+}$$

4. Reduction of pyruvate to ethanol (Section 13.3)

$$CH_3-\overset{O}{\overset{||}{C}}-COO^- + 2H^+ + \boxed{NADH} \xrightarrow{\text{Two} \atop \text{steps}}$$
$$CH_3-CH_2-OH + \boxed{NAD^+} + CO_2$$

5. Complete oxidation of glucose (Section 13.4)

Glucose $+ 6O_2 + 30ADP + 30P_i \longrightarrow 6CO_2 + 6H_2O + 30ATP$

6. Glycogenesis (Section 13.5)

$$\text{Glucose} \xrightarrow{\text{Three} \atop \text{steps}} \text{glycogen}$$

7. Glycogenolysis (Section 13.5)

$$\text{Glycogen} \xrightarrow{\text{Three} \atop \text{steps}} \text{glucose}$$

8. Gluconeogenesis (Section 13.6)

$$\left.\begin{array}{l}\text{Lactate, certain}\\ \text{amino acids,}\\ \text{citric acid cycle}\\ \text{intermediates}\end{array}\right\} \longrightarrow \text{pyruvate} \xrightarrow{\text{Eleven} \atop \text{steps}} \text{glucose}$$

EXERCISES AND PROBLEMS

The members of each pair of problems in this section test similar material.

■ Carbohydrate Digestion (Section 13.1)

13.1 Where does starch digestion begin in the body, and what is the name of the enzyme involved in this initial digestive process?

13.2 Very little digestion of starch occurs in the stomach. Why?

13.3 What is the primary site for carbohydrate digestion, and what organ produces the enzymes that are active at this location?

13.4 Where does the final step in carbohydrate digestion take place, and in what form are carbohydrates as they enter this final step?

13.5 Where does the digestion of sucrose begin, and what is the reaction that occurs?

13.6 Where does the digestion of lactose begin, and what is the reaction that occurs?

13.7 Identify the three major monosaccharides produced by digestion of carbohydrates.

13.8 The various stages of carbohydrate digestion all involve the same general type of reaction. What is this reaction type?

■ Glycolysis (Section 13.2)

13.9 What is the starting material for glycolysis?

13.10 What is the end product from glycolysis?

13.11 What coenzyme functions as the oxidizing agent in glycolysis?

13.12 What is meant by the statement that glycolysis is an anaerobic pathway?

13.13 What is the first step of glycolysis, and why is it important in retaining glucose inside the cell?

13.14 Step 3 of glycolysis is the commitment step. Explain.

13.15 What two C_3 fragments are formed by the splitting of a fructose 1,6-bisphosphate molecule?

13.16 In one step of the glycolysis pathway, a C_6 chain is broken into two C_3 fragments, only one of which can be further degraded. What happens to the other C_3 fragment?

13.17 How many pyruvate molecules are produced per glucose molecule during glycolysis?

13.18 How many molecules of ATP and NADH are produced per glucose molecule during glycolysis?

13.19 How many steps in the glycolysis pathway produce ATP?

13.20 How many steps in the glycolysis pathway consume ATP?

13.21 Of the 10 steps of glycolysis, which ones involve phosphorylation?

13.22 Of the 10 steps of glycolysis, which ones involve oxidation?

13.23 Where in a cell does glycolysis occur?

13.24 Do the reactions of glycolysis and the citric acid cycle occur at the same location in a cell? Explain.

13.25 Replace the question mark in each of the following word equations with the name of a substance.

a. Glucose $+$ ATP $\xrightarrow{\text{Hexokinase}}$? $+$ ADP $+ H^+$

b. ? $\xrightarrow{\text{Enolase}}$ phosphoenolpyruvate $+$ water

c. 3-Phosphoglycerate $\xrightarrow{?}$ 2-phosphoglycerate

d. 1,3-Bisphosphoglycerate $+$? $\xrightarrow{\text{Phosphoglycero-} \atop \text{kinase}}$
3-phosphoglycerate $+$ ATP

13.26 Replace the question mark in each of the following word equations with the name of a substance.

a. Glucose 6-phosphate $\xrightarrow{\text{Phosphogluco-} \atop \text{isomerase}}$?

b. ? $\xrightarrow{\text{Aldolase}}$ dihydroxyacetone phosphate $+$
glyceraldehyde 3-phosphate

c. Phosphoenolpyruvate $+$? $+ H^+ \xrightarrow{\text{Pyruvate} \atop \text{kinase}}$
pyruvate $+$ ATP

d. Dihydroxyacetone phosphate $\xrightarrow{?}$
glyceraldehyde 3-phosphate

13.27 In which step of glycolysis does each of the following occur?
a. Second substrate-level phosphorylation reaction
b. First ATP-consuming reaction
c. Third isomerization reaction
d. Use of NAD$^+$ as an oxidizing agent

13.28 In which step of glycolysis does each of the following occur?
a. First energy-producing reaction
b. First ATP-producing reaction
c. A dehydration reaction
d. First isomerization reaction

13.29 What is the net ATP production when each of the following molecules is processed through the glycolysis pathway?
   a. One glucose molecule
   b. One sucrose molecule

13.30 What is the net ATP production when each of the following molecules is processed through the glycolysis pathway?
   a. One lactose molecule
   b. One maltose molecule

13.31 Draw structural formulas for each of the following pairs of molecules.
   a. Pyruvic acid and pyruvate
   b. Dihydroxyacetone and dihydroxyacetone phosphate
   c. Fructose 6-phosphate and fructose 1,6-bisphosphate
   d. Glyceric acid and glyceraldehyde

13.32 Draw structural formulas for each of the following pairs of molecules.
   a. Glyceric acid and glycerate
   b. Glycerate and pyruvate
   c. Glucose 6-phosphate and fructose 6-phosphate
   d. Dihydroxyacetone and glyceric acid

13.33 Number the carbon atoms of fructose 1,6-bisphosphate 1 through 6, and show the location of each carbon in the two trioses produced during Step 4 of glycolysis.

13.34 Number the carbon atoms of glucose 1 through 6, and show the location of each carbon in the two molecules of pyruvate produced by glycolysis.

Fates of Pyruvate (Section 13.3)

13.35 What are the three common possible fates for pyruvate produced from glycolysis?

13.36 Compare the fates of pyruvate in the body under aerobic and anaerobic conditions.

13.37 What is the overall reaction equation for the conversion of pyruvate to acetyl CoA?

13.38 What is the overall reaction equation for the conversion of pyruvate to lactate?

13.39 Explain how lactate fermentation allows glycolysis to continue under anaerobic conditions.

13.40 How is the ethanol fermentation in yeast similar to lactate fermentation in skeletal muscle?

13.41 In ethanol fermentation, a C_3 pyruvate molecule is changed to a C_2 ethanol molecule. What is the fate of the third pyruvate carbon?

13.42 What are the structural differences between pyruvate and lactate ions?

13.43 What is the net reaction for the conversion of one glucose molecule to two lactate molecules?

13.44 What is the net reaction for the conversion of one glucose molecule to two ethanol molecules?

Complete Oxidation of Glucose (Section 13.4)

13.45 How does the fact that cytosolic $NADH/H^+$ cannot cross the mitochondrial membranes affect ATP production from cytosolic $NADH/H^+$?

13.46 What is the net reaction for the shuttle mechanism involving glycerol 3-phosphate by which NADH electrons are shuttled across the mitochondrial membrane?

13.47 Contrast, in terms of ATP production, the oxidation of glucose to CO_2 and H_2O with the oxidation of glucose to pyruvate.

13.48 Contrast, in terms of ATP production, the oxidation of glucose to CO_2 and H_2O with the oxidation of glucose to ethanol.

13.49 How many of the 30 ATP molecules produced from the complete oxidation of 1 glucose molecule are produced during glycolysis?

13.50 How many of the 30 ATP molecules produced from the complete oxidation of 1 glucose molecule are the result of the oxidation of pyruvate to acetyl CoA?

Glycogen Metabolism (Section 13.5)

13.51 Compare the meanings of the terms *glycogenesis* and *glycogenolysis.*

13.52 Where is most of the body's glycogen stored?

13.53 Glucose 1-phosphate is the product of the first step of glycogenesis. What is the reactant?

13.54 Glucose 1-phosphate is the product of the first step of glycogenolysis. What are the reactants?

13.55 What is the source of the PP_i produced during the second step of glycogenesis?

13.56 What is the function of the PP_i produced during the second step of glycogenesis?

13.57 How is ATP involved in glycogenesis?

13.58 How many ATP molecules are needed to attach a single glucose molecule to a growing glycogen chain?

13.59 Which step of glycogenolysis is the reverse of Step 1 of glycogenesis?

13.60 What reaction determines whether glucose formed by glycogenolysis can leave a cell?

13.61 What is the difference between glycogenolysis in liver cells and in muscle cells?

13.62 The liver, but not the brain or muscle cells, has the capacity to supply free glucose to the blood. Explain.

13.63 In what form does glycogen enter the glycolysis pathway?

13.64 Explain why one more ATP is produced when glucose is obtained from glycogen than when it is obtained directly from the blood.

Gluconeogenesis (Section 13.6)

13.65 What organ is primarily responsible for gluconeogenesis?

13.66 What is the physiological function of gluconeogenesis?

13.67 How does gluconeogenesis get around the three irreversible steps of glycolysis?

13.68 Although gluconeogenesis and glycolysis are "reverse" processes, there are 11 steps in gluconeogenesis and only 10 steps in glycolysis. Explain.

13.69 What intermediate in gluconeogenesis is also an intermediate in the citric acid cycle?

13.70 What are the sources of high-energy bonds in gluconeogenesis?

13.71 What is the fate of lactate formed by muscular activity?

13.72 What is the physiological function of the Cori cycle?

■ **The Pentose Phosphate Pathway (Section 13.8)**

13.73 What is the starting material for the pentose phosphate pathway?

13.74 What are two major functions of the pentose phosphate pathway?

13.75 How do the biochemical functions of NADH and NADPH differ?

13.76 How do the structures of NADH and NADPH differ?

13.77 Write a general equation for the oxidative stage of the pentose phosphate pathway.

13.78 Write a general equation for the entire pentose phosphate pathway.

13.79 What compound contains the carbon atom lost from glucose (a hexose) in its conversion to ribose (a pentose)?

13.80 How many molecules of NADPH are produced per glucose 6-phosphate in the pentose phosphate pathway?

■ **Control of Carbohydrate Metabolism (Section 13.9)**

13.81 What effect does insulin have on glycogen metabolism?

13.82 What effect does insulin have on blood-glucose levels?

13.83 What effect does glucagon have on blood-glucose levels?

13.84 What effect does glucagon have on glycogen metabolism?

13.85 What organ is the source of insulin?

13.86 What organ is the source of glucagon?

13.87 The hormone epinephrine generates a "second messenger." Explain.

13.88 What is the relationship between cAMP and the hormone epinephrine?

13.89 Compare the target tissues for glucagon and epinephrine.

13.90 Compare the biological functions of glucagon and epinephrine.

ADDITIONAL PROBLEMS

13.91 Indicate in which of the four processes *glycolysis, glycogenesis, glycogenolysis,* and *gluconeogenesis* each of the following compounds is encountered. There may be more than one correct answer for a given compound.
   a. Glucose 6-phosphate
   b. Glucose 1-phosphate
   c. Dihydroxyacetone phosphate
   d. Oxaloacetate

13.92 Indicate in which of the four processes *glycolysis, glycogenesis, glycogenolysis,* and *gluconeogenesis* each of the following situations is encountered. There may be more than one correct answer for a given situation.
   a. NAD^+ is consumed.   b. ATP is produced.
   c. ATP is consumed.   d. UDP is involved.

13.93 Indicate in which of the four processes *glycolysis, glycogenesis, glycogenolysis,* and *glyconeogenesis* each of the following characterizations applies.
   a. Glucose is converted to two pyruvates.
   b. Glycogen is synthesized from glucose.
   c. Glycogen is broken down into free glucose units.
   d. Glucose is synthesized from pyruvate.

13.94 What is the ATP yield *per glucose molecule* in each of the following processes?
   a. Glycolysis
   b. Glycolysis, acetyl CoA formation, and the common metabolic pathway

   c. Glycolysis plus oxidation of pyruvate to acetyl CoA
   d. Glycolysis plus reduction of pyruvate to lactate

13.95 Which one of these characterizations, (1) Cori cycle, (2) an anaerobic process, (3) oxidative stage of pentose phosphate pathway, or (4) nonoxidative stage of pentose phosphate pathway, applies to each of the following chemical changes?
   a. Pyruvate to lactate
   b. Pyruvate to ethanol
   c. Glucose 6-phosphate to ribulose 5-phosphate
   d. Ribulose 5-phosphate to ribose 5-phosphate

13.96 What condition or conditions determine that pyruvate is involved in each of the following?
   a. Gluconeogenesis   b. Converted to lactate
   c. Citric acid cycle   d. Converted to ethanol

13.97 In the complete metabolism of 1 mole of sucrose, how many moles of each of the following are produced?
   a. CO_2 b. Pyruvate
   c. Acetyl CoA   d. ATP

13.98 Under what conditions does glucose 6-phosphate enter each of the following pathways?
   a. Glycogenesis
   b. Glycolysis
   c. Pentose phosphate pathway
   d. Hydrolysis to free glucose

MULTIPLE-CHOICE PRACTICE TEST

13.99 Which of the following statements concerning glycolysis is *correct*?
   a. It is an oxidation process in which molecular oxygen is used.
   b. All reactions take place in the cytosol of a cell.
   c. There are two stages, each of which involves a series of five reactions.
   d. The overall process converts a C_6 molecule into three C_2 molecules.

13.100 What are the two steps in glycolysis in which ATP is converted to ADP?
   a. 1 and 2   b. 1 and 3
   c. 2 and 3   d. 7 and 10

13.101 Intermediates in the glycolysis pathway include two derivatives of which of the following?
   a. Glucose   b. Fructose
   c. Pyruvate   d. Glyceraldehyde

13.102 What are the total number of steps in the C_6 stage and C_3 stage of glycolysis, respectively?
   a. 10 and 10
   b. 5 and 5
   c. 4 and 6
   d. 3 and 7

13.103 During the overall process of glycolysis, which of the following occurs for each glucose molecule processed?
   a. Net loss of two ATP molecules
   b. Net loss of four ATP molecules
   c. Net gain of two ATP molecules
   d. Net gain of four ATP molecules

13.104 Lactate fermentation can occur in which of the following?
   a. Humans, animals, and microorganisms
   b. Humans and animals but not in microorganisms
   c. Microorganisms but not in humans and animals
   d. Microorganisms and animals but not in humans

13.105 What is the name of the process in which glycogen is converted to glucose?
   a. Glycolysis
   b. Glycogenolysis
   c. Glycogenesis
   d. Glyconeogenesis

13.106 The compound oxaloacetate is an intermediate in which of the following conversions?
   a. Glycogen to glucose
   b. Glucose to glycogen
   c. Pyruvate to glucose
   d. Pyruvate to acetyl CoA

13.107 As part of the Cori cycle, which of the following occurs in liver cells?
   a. Glucose is converted to pyruvate.
   b. Glucose is converted to lactate.
   c. Pyruvate is converted to lactate.
   d. Lactate is converted to pyruvate.

13.108 Which of the following are products of the first stage of the pentose phosphate pathway?
   a. Ribose 5-phosphate and ribulose 5-phosphate
   b. Ribose 5-phosphate and carbon dioxide
   c. Ribulose 5-phosphate and carbon dioxide
   d. Ribose and carbon dioxide

14 Lipid Metabolism

Migrating birds, when flying long distances without stopping, use lipids (stored fat) as their major source of energy.

Certain classes of lipids play an extremely important role in cellular metabolism, because they represent an energy-rich "fuel" that can be stored in large amounts in adipose (fat) tissue. Between one-third and one-half of the calories present in the diet of the average U.S. resident are supplied by lipids. Furthermore, excess energy derived from carbohydrates and proteins beyond normal daily needs is stored in lipid molecules (in adipose tissue), later to be mobilized and used when needed.

14.1 Digestion and Absorption of Lipids

Because 98% of total *dietary* lipids are triacylglycerols (fats and oils; Section 8.4), this chapter focuses on triacylglycerol metabolism. Like all lipids, triacylglycerols (TAGs) are insoluble in water. Hence, water-based salivary enzymes in the mouth have little effect on them. The *major* change that TAGs undergo in the stomach is physical rather than chemical. The churning action of the stomach breaks up triacylglycerol materials into small globules, or droplets, which float as a layer above the other components of swallowed food. The resulting material is called *chyme*.

High-fat foods remain in the stomach longer than low-fat foods. The conversion of high-fat materials into chyme takes longer than the breakup of low-fat materials. This is why a high-fat meal causes a person to feel "full" for a longer period of time.

Lipid digestion also begins in the stomach. Under the action of *gastric lipase* enzymes, hydrolysis of TAGs occurs. Normally, about 10% of TAGs undergo hydrolysis

The saliva of infants contains a lipase that can hydrolyze TAGs, so digestion begins in the mouth for nursing infants. Because mother's milk is already a lipid-in-water emulsion, emulsification by stomach churning is a much less important factor in an infant's processing of fat. Mother's milk also contains a lipase that supplements the action of the salivary lipases the infant itself produces. After weaning, infants cease to produce salivary lipases.

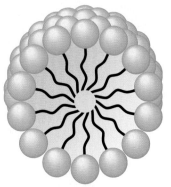

in the stomach, but regular consumption of a high-fat diet can induce the production of higher levels of gastric lipases.

The arrival of chyme from the stomach triggers in the small intestine, through the action of the hormone *cholecystokinin,* the release of bile stored in the gallbladder. The bile (Section 8.11), which contains no enzymes, acts as an emulsifier (Section 8.11). Colloid particle formation through bile emulsification "solubilizes" the triacylglycerol globules, and digestion of the TAGs resumes. The major enzymes involved at this point are the *pancreatic lipases,* which hydrolyze ester linkages between the glycerol and fatty acid units of the TAGs. *Complete* hydrolysis does not usually occur; only two of the three fatty acid units are liberated, producing a monoacylglycerol and two free fatty acids. Occasionally, enzymes remove all three fatty acid units, leaving a free glycerol molecule.

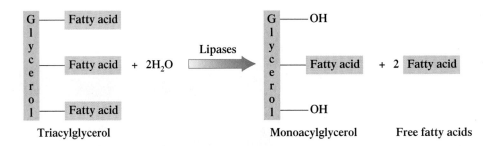

With the help of bile, the free fatty acids and monoacylglycerols produced from hydrolysis are combined into tiny spherical droplets called micelles (Section 8.6). A **fatty acid micelle** *is a micelle in which fatty acids and/or monoacylglycerols and some bile are present.* Fatty acid micelles are very small compared to the original triacylglycerol globules, which contain thousands of triacylglycerol molecules. Figure 14.1 shows a cross section of the three-dimensional structure of a fatty acid micelle.

Micelles, containing free fatty acid and monoacylglycerol components, are small enough to be readily absorbed through the membranes of intestinal cells. Within the intestinal cells, a "repackaging" occurs in which the free fatty acids and monoacylglycerols are reassembled into triacylglycerols. The newly formed triacylglycerols are then combined with membrane lipids (phospholipids and cholesterol) and water-soluble proteins to produce a type of lipoprotein (Section 9.17) called a *chylomicron* (see Figure 14.2). A **chylomicron** *is a lipoprotein that transports triacylglycerols from intestinal cells, via the lymphatic system, to the bloodstream.* Triacylglycerols constitute 95% of the core lipids present in a chylomicron.

Chylomicrons are too large to pass through capillary walls directly into the bloodstream. Consequently, delivery of the chylomicrons to the bloodstream is accomplished through the body's lymphatic system. Chylomicrons enter the lymphatic system through tiny lymphatic vessels in the intestinal lining. They enter the bloodstream through the thoracic duct (a large lymphatic vessel just below the collarbone), where the fluid of the lymphatic system flows into a vein, joining the bloodstream.

Once the chylomicrons reach the bloodstream, the TAGs they carry are again hydrolyzed to produce glycerol and free fatty acids. TAG release from chylomicrons and their ensuing hydrolysis is mediated by *lipoprotein lipases.* These enzymes are located on the lining of blood vessels in muscle and other tissues that use fatty acids for fuel and in fat synthesis. The fatty acid and glycerol hydrolysis products from TAG hydrolysis are absorbed by the cells of the body and are either broken down to acetyl CoA for energy or stored as lipids (they are again repackaged as TAGs). Figure 14.3 summarizes the events that must occur before triacylglycerols can reach the bloodstream through the digestive process.

Soon after a meal heavily laden with TAGs is ingested, the chylomicron content of both blood and lymph increases dramatically. Chylomicron concentrations usually begin

When freed of the triacylglycerol molecules they "transport" during digestion, bile acids are mostly recycled. Small amounts are excreted.

Chylomicron is pronounced "kye-lo-MY-cron."

FIGURE 14.2 A three-dimensional model of a chylomicron, a type of lipoprotein. Chylomicrons are the form in which TAGs are delivered to the bloodstream via the lymphatic system.

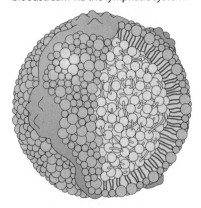

⌒ **Triacylglycerols (TAGs)**

■ **Protein**

Membrane lipids

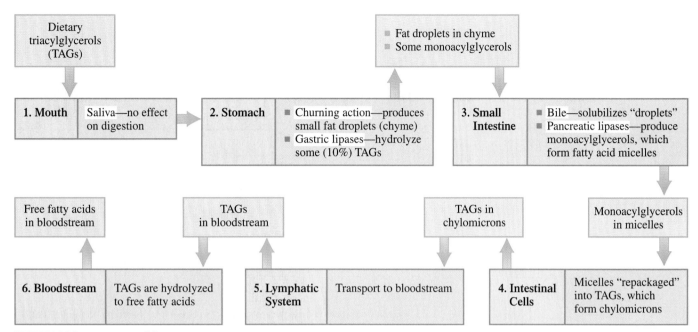

FIGURE 14.3 A summary of the events that must occur before triacylglycerols (TAGs) can reach the bloodstream through the digestive process.

Dietary TAGs deposited in adipose tissue have undergone hydrolysis two times (to form free fatty acids and/or monoacylglycerols) and are repackaged twice (to re-form TAGs) in reaching that state. They undergo hydrolysis for a third time when *triacylglycerol mobilization* occurs.

Adipose tissue is the only tissue in which *free* TAGs occur in appreciable amounts. In other types of cells and in the bloodstream, TAGs are part of lipoprotein particles.

FIGURE 14.4 Structural characteristics of an adipose cell.

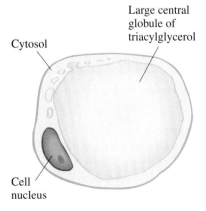

to rise within 2 hours after a meal, reach a peak in 4–6 hours, and then drop rather rapidly to a normal level as they move into adipose cells (Section 14.2) or into the liver.

14.2 Triacylglycerol Storage and Mobilization

Most cells in the body have limited capability for storage of TAGs. However, this activity is the major function of specialized cells called adipocytes, found in adipose tissue. *An* **adipocyte** *is a triacylglycerol-storing cell.* **Adipose tissue** *is tissue that contains large numbers of adipocyte cells.*

Adipose tissue is located primarily directly beneath the skin (subcutaneous), particularly in the abdominal region, and in areas around vital organs. Besides its function as a storage location for the chemical energy inherent in TAGs, subcutaneous adipose tissue also serves as an insulator against excessive heat loss to the environment and provides organs with protection against physical shock.

Adipose cells are among the largest cells in the body. They differ from other cells in that most of the cytoplasm has been replaced with a large triacylglycerol droplet (Figure 14.4). This droplet accounts for nearly the entire volume of the cell. As newly formed TAGs are imported into an adipose cell, they form small droplets at the periphery of the cell that later merge with the large central droplet.

Use of the TAGs stored in adipose tissue for energy production is triggered by several hormones, including epinephrine and glucagon. Hormonal interaction with adipose cell membrane receptors stimulates production of cAMP from ATP inside the adipose cell. In a series of enzymatic reactions, the cAMP activates *hormone-sensitive lipase (HSL)* through phosphorylation. HSL is the lipase needed for triacylglycerol hydrolysis, a prerequisite for fatty acids to enter the bloodstream from an adipose cell. This cAMP activation process is illustrated in Figure 14.5.

The overall process of tapping the body's triacylglycerol energy reserves (adipose tissue) for energy is called triacylglycerol mobilization. **Triacylglycerol mobilization** *is the hydrolysis of triacylglycerols stored in adipose tissue, followed by release into the bloodstream of the fatty acids and glycerol so produced.* Triacylglycerol mobilization is an ongoing process. On the average, about 10% of the TAGs in adipose tissue are replaced daily by new triacylglycerol molecules.

The use of cAMP in the activation of hormone-sensitive lipase in adipose cells is similar to cAMP's role in the activation of the glycogenolysis process (Section 13.9).

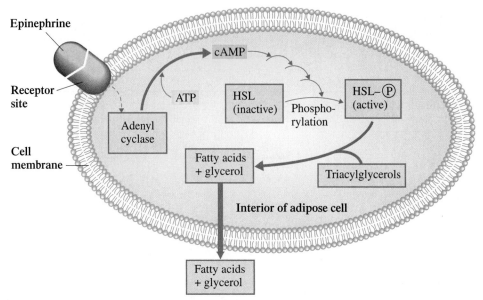

FIGURE 14.5 Hydrolysis of stored triacylglycerols in adipose tissue is triggered by hormones that stimulate cAMP production within adipose cells.

14.3 Glycerol Metabolism

During triacylglycerol mobilization, one molecule of glycerol is produced for each triacylglycerol completely hydrolyzed. Glycerol metabolism primarily involves processes considered in the previous chapter. After entering the bloodstream, glycerol travels to the liver or kidneys, where it is converted, in a two-step process, to dihydroxyacetone phosphate.

$$
\begin{array}{c}
\text{H}_2\text{C—OH} \\
\text{HC—OH} \\
\text{H}_2\text{C—OH}
\end{array}
\quad
\underset{\text{ATP} \quad \text{ADP}}{\overset{\text{Glycerol kinase}}{\longrightarrow}}
\quad
\begin{array}{c}
\text{H}_2\text{C—OH} \\
\text{HC—OH} \\
\text{H}_2\text{C—O—}\textcircled{P}
\end{array}
\quad
\underset{\text{NAD}^+ \quad \text{NADH/H}^+}{\overset{\text{Glycerol 3-phosphate dehydrogenase}}{\longrightarrow}}
\quad
\begin{array}{c}
\text{H}_2\text{C—OH} \\
\text{C=O} \\
\text{H}_2\text{C—O—}\textcircled{P}
\end{array}
$$

Glycerol — Glycerol 3-phosphate — Dihydroxyacetone phosphate

The first step involves phosphorylation of a primary hydroxyl group of the glycerol. In the second step, glycerol's secondary alcohol group (C-2) is oxidized to a ketone.

Dihydroxyacetone phosphate is an intermediate in both glycolysis (Section 13.2) and gluconeogenesis (Section 13.6). It can be converted to pyruvate, then acetyl CoA, and finally carbon dioxide, or it can be used to form glucose. Dihydroxyacetone phosphate formation from glycerol represents the first of several situations we will consider wherein carbohydrate and lipid metabolism are connected.

14.4 Oxidation of Fatty Acids

The stored TAGs in adipose tissue supply approximately 60% of the body's energy needs when the body is in a resting state.

There are three parts to the process by which fatty acids are broken down to obtain energy.

1. The fatty acid must be *activated* by bonding to coenzyme A.
2. The fatty acid must be *transported* into the mitochondrial matrix by a shuttle mechanism.
3. The fatty acid must be repeatedly *oxidized,* cycling through a series of four reactions, to produce acetyl CoA, FADH$_2$, and NADH.

Fatty Acid Activation

The outer mitochondrial membrane is the site of fatty acid *activation,* the first stage of fatty acid oxidation. Here the fatty acid is converted to a high-energy derivative of coenzyme A. Reactants are the fatty acid, coenzyme A, and a molecule of ATP.

$$R-\overset{\overset{\displaystyle O}{\|}}{C}-O^- + HS-CoA \xrightarrow[\text{ATP}\quad\text{AMP}\,+\,2P_i]{\text{Acyl CoA synthetase}} R-\overset{\overset{\displaystyle O}{\|}}{C}-S-CoA$$

Free fatty acid   Acyl CoA

This reaction requires the expenditure of two high-energy phosphate bonds from a single ATP molecule; the ATP is converted to AMP rather than ADP, and the resulting pyrophosphate (PP_i) is hydrolyzed to $2P_i$.

The activated fatty acid–CoA molecule is called *acyl* CoA. The difference between the designations *acyl* CoA and *acetyl* CoA is that *acyl* refers to a random-length fatty acid carbon chain that is covalently bonded to coenzyme A, whereas *acetyl* refers to a two-carbon chain covalently bonded to coenzyme A.

$$R-\overset{\overset{\displaystyle O}{\|}}{C}-S-CoA \qquad\qquad CH_3-\overset{\overset{\displaystyle O}{\|}}{C}-S-CoA$$

Acyl CoA   Acetyl CoA
R = carbon chain of any length   R = CH_3 group

Fatty Acid Transport

Acyl CoA is too large to pass through the inner mitochondrial membrane to the mitochondrial matrix, where the enzymes needed for fatty acid oxidation are located. A shuttle mechanism involving the molecule carnitine effects the entry of acyl CoA into the matrix (see Figure 14.6). The acyl group is transferred to a carnitine molecule, which carries it through the membrane. The acyl group is then transferred from the carnitine back to a CoA molecule.

The Fatty Acid Spiral

In the mitochondrial matrix, a sequence of four reactions *repeatedly* cleaves two-carbon units from the carboxyl end of a saturated fatty acid. This process is called the *fatty acid spiral* because of its repetitive nature, or *β oxidation spiral,* because the second, or beta,

---

Triacylglycerol reserves would enable the average person to survive starvation for about 30 days, given sufficient water. Glycogen reserves (stored glucose) would be depleted within 1 day.

Acyl is a generic term for

$$R-\overset{\overset{\displaystyle O}{\|}}{C}-$$

which is the species formed when the carboxyl —OH is removed from a carboxylic acid (Section 8.4). The R group can involve a carbon chain of any length.

FIGURE 14.6 Fatty acids are transported across the inner mitochondrial membrane in the form of acyl carnitine.

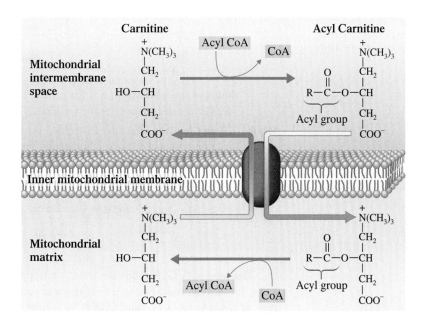

carbon from the carboxyl end of the chain is oxidized. The **fatty acid spiral** *is the metabolic pathway that degrades fatty acids, by removing two carbon atoms at a time, to acetyl CoA, with FADH$_2$ and NADH also being produced.*

For a *saturated* fatty acid, the fatty acid spiral involves the following functional group changes at the β carbon and the following reaction types.

We have encountered an identical set of functional group changes before, in the back side of the citric acid cycle (Section 12.6), Steps 6–8 of this cycle.

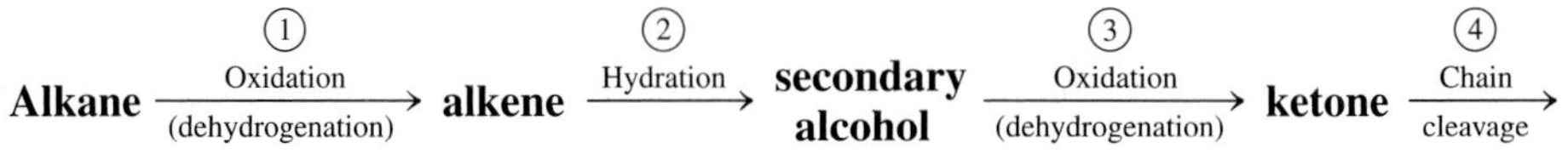

$$\textbf{Alkane} \xrightarrow[\text{(dehydrogenation)}]{\text{Oxidation} \textcircled{1}} \textbf{alkene} \xrightarrow{\text{Hydration} \textcircled{2}} \textbf{secondary alcohol} \xrightarrow[\text{(dehydrogenation)}]{\text{Oxidation} \textcircled{3}} \textbf{ketone} \xrightarrow{\text{Chain cleavage} \textcircled{4}}$$

Details about Steps 1–4 of the fatty acid spiral follow.

Step 1: *Oxidation (dehydrogenation).* Hydrogen atoms are removed from the α and β carbons, creating a double bond between these two carbon atoms. FAD is the oxidizing agent, and a FADH$_2$ molecule is a product.

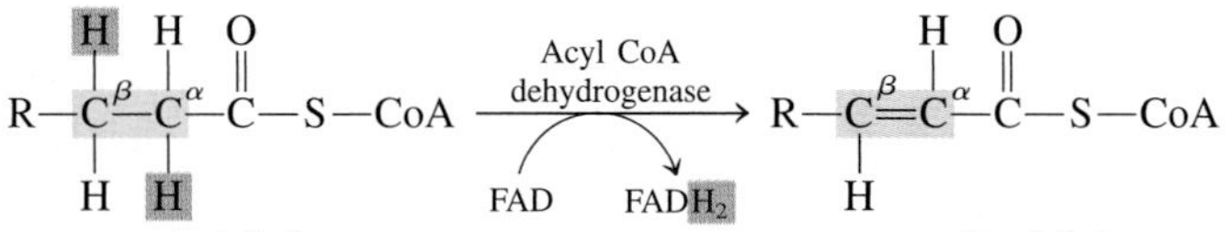

The enzyme involved is stereospecific in that only *trans* double bonds are produced.

Step 2: *Hydration.* A molecule of water is added across the *trans* double bond, producing a secondary alcohol at the β-carbon position. Again, the enzyme involved is stereospecific in that only the L-hydroxy isomer is produced from the *trans* double bond.

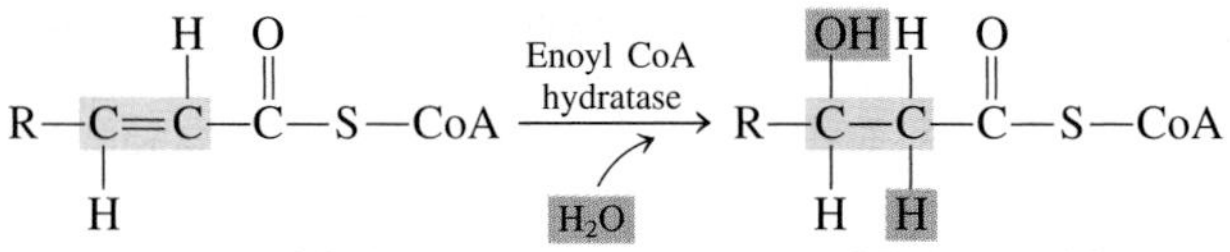

The enzyme involved in this hydration will also hydrate a *cis* double bond, but the product then is the D isomer. We shall return to this point later in considering how unsaturated fatty acids are oxidized.

The reaction sequence dehydrogenation–hydration–dehydrogenation in the fatty acid spiral has a parallel in Steps 6–8 of the citric acid cycle (Section 12.6), where succinate is dehydrogenated to fumarate, which is hydrated to malate, which is dehydrogenated to oxaloacetate.

Step 3: *Oxidation (dehydrogenation).* The β-hydroxy group is oxidized to a ketone functional group with NAD$^+$ serving as the oxidizing agent. The required enzyme exhibits absolute stereospecificity for the L isomer.

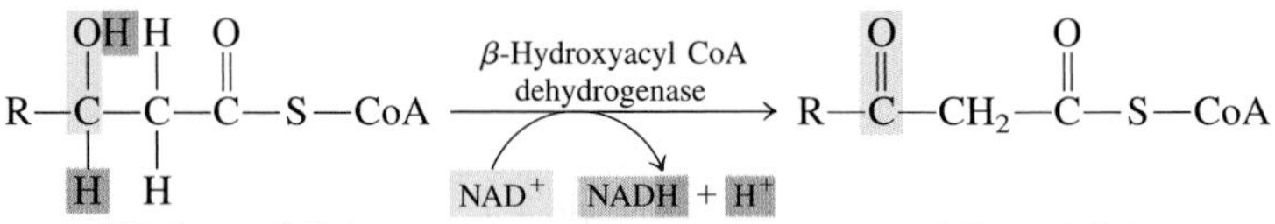

It is now apparent why one of the names for this series of reactions is β oxidation spiral. The β-carbon atom has been oxidized from a —CH$_2$— group to a ketone group.

Step 4: *Chain Cleavage.* The fatty acid chain is broken between the α and β carbons by reaction with a coenzyme A molecule. The result is an acetyl CoA molecule and a new acyl CoA molecule that is shorter by two carbon atoms than its predecessor.

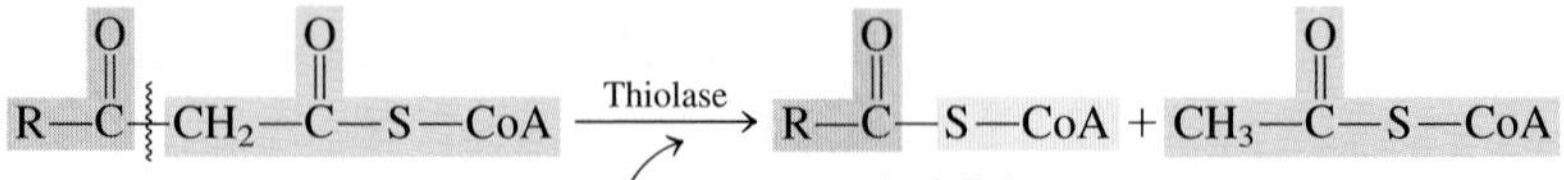

The new acyl CoA molecule (now shorter by two carbons) is *recycled* through the same set of four reactions again. This yields another acetyl CoA, a two-carbon-shorter new acyl CoA, FADH$_2$, and NADH. Recycling occurs again and again, until the entire

FIGURE 14.7 Reactions of the fatty acid spiral for an 18:0 fatty acid (stearic acid).

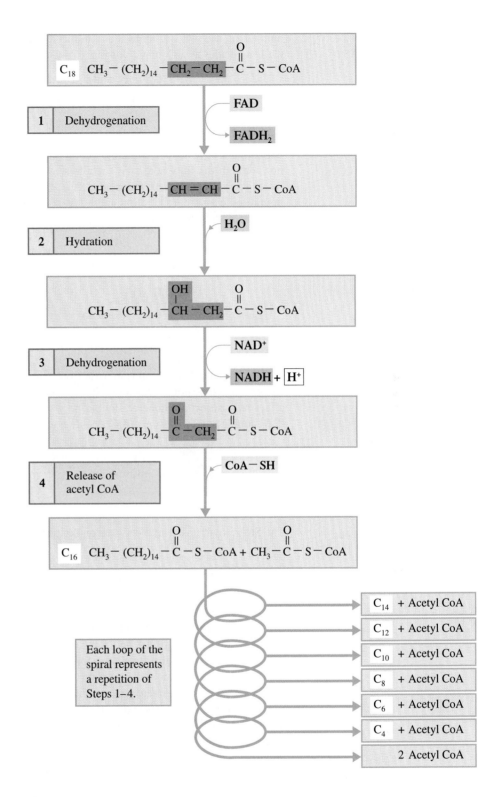

This sequence of reactions is called the fatty acid *spiral* rather than the fatty acid *cycle* because a different product results from each turn.

fatty acid is converted to acetyl CoA. Thus the fatty acid carbon chain is sequentially degraded, two carbons at a time.

Figure 14.7 summarizes the reactions of the fatty acid spiral for stearic acid (18:0) as the starting fatty acid.

The fatty acids normally found in dietary triacylglycerols contain an *even* number of carbon atoms. Thus the number of acetyl CoA molecules produced in the fatty acid spiral is equal to half the number of carbon atoms in the fatty acid. The number of *turns* of the fatty acid spiral that are needed to produce the acetyl CoA is always one less than the number of acetyl CoA molecules produced because the last turn produces two acetyl CoA molecules as a C_4 unit splits into two C_2 units.

$$C_{18} \text{ fatty acid} \longrightarrow 9 \text{ acetyl CoA (8 cycles)}$$
$$C_{14} \text{ fatty acid} \longrightarrow 7 \text{ acetyl CoA (6 cycles)}$$

■ Unsaturated Fatty Acids

Unsaturated fatty acids are common components of dietary triacylglycerols. Their oxidation through the fatty acid spiral requires two additional enzymes besides those needed for oxidation of saturated fatty acids. These two—an epimerase that can change a D configuration to an L configuration and a *cis–trans* isomerase—are needed for two reasons. First, the double bonds in naturally occurring unsaturated fatty acids are nearly always *cis* double bonds, which yield on hydration a D-hydroxy product rather than the L-hydroxy product needed for Step 3 of the spiral. The epimerase enzyme effects a configuration change from the D form to the L form.

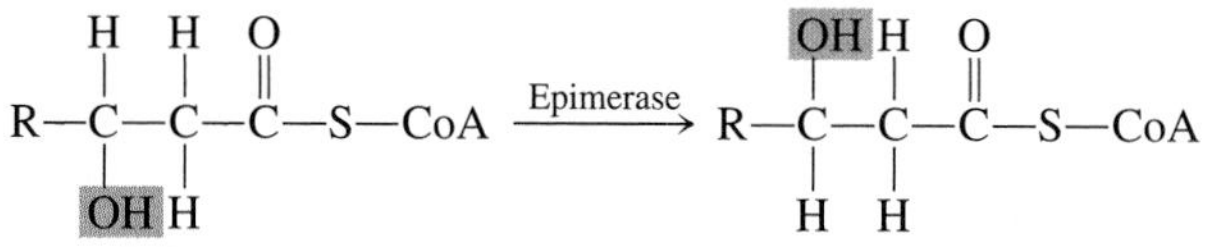

D-β-Hydroxyacyl CoA　　　　　　　　　L-β-Hydroxyacyl CoA

Second, the double bonds in naturally occurring unsaturated fatty acids often occupy odd-numbered positions (Section 8.2). The hydratase in Step 2 of the fatty acid spiral can affect only an even-numbered double bond. The *cis–trans* isomerase produces a *trans*-(2,3) double bond from a *cis*-(3,4) double bond.

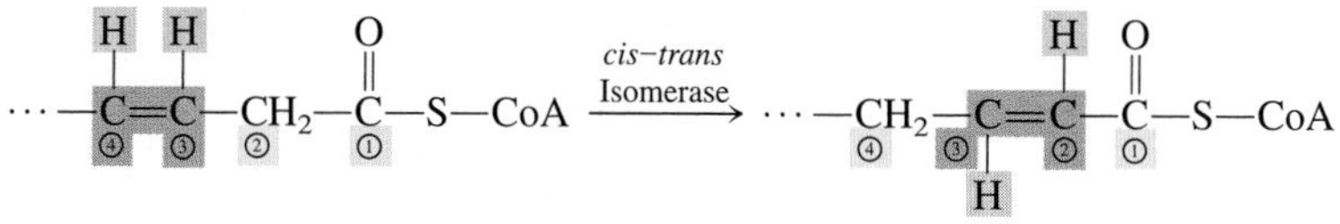

cis-(3,4)　　　　　　　　　　　　　　*trans*-(2,3)

The Step 2 hydratase can then work on the *trans*-(2,3) double bond in the normal fashion.

14.5 ATP Production from Fatty Acid Oxidation

How does the total energy output from fatty acid oxidation compare to that of glucose oxidation? Let us calculate ATP production for the oxidation of a specific fatty acid molecule, stearic acid (18:0), and compare it with that from glucose.

Figure 14.7 shows that for all turns of the fatty acid spiral except the last turn, one $FADH_2$ molecule, one NADH molecule, and one acetyl CoA molecule are produced. In the final turn, two acetyl CoA molecules are produced in addition to the $FADH_2$ and NADH molecules.

Eight turns of the fatty acid spiral are required for the oxidation of stearic acid, an 18-carbon acid. These eight turns of the spiral produce 9 acetyl CoA molecules, 8 $FADH_2$ molecules, and 8 NADH molecules. Further processing of these products through the common metabolic pathway (citric acid cycle, electron transport chain, and oxidative phosphorylation) leads to ATP production as follows:

$$9 \text{ acetyl CoA} \times \frac{10 \text{ ATP}}{1 \text{ acetyl CoA}} = 90 \text{ ATP}$$

$$8 \text{ FADH}_2 \times \frac{1.5 \text{ ATP}}{1 \text{ FADH}_2} = 12 \text{ ATP}$$

$$8 \text{ NADH} \times \frac{2.5 \text{ ATP}}{1 \text{ NADH}} = \underline{20 \text{ ATP}}$$

$$122 \text{ ATP}$$

The conversion factors used in this calculation were first presented in Section 12.9.

This *gross* production of 122 ATP must be decreased by the ATP needed to activate the fatty acid before it enters the fatty acid spiral. The activation consumes two high-energy phosphate bonds of an ATP molecule. For accounting purposes, this is equivalent to hydrolyzing 2 ATP molecules to ADP. Thus the *net* ATP production from oxidation of stearic acid is 120 ATP (122 minus 2).

The comparison between complete fatty acid oxidation and complete glucose oxidation (Section 13.4) shows that a stearic acid molecule produces four times as much ATP as a glucose molecule.

$$1 \text{ glucose} \longrightarrow \boxed{30 \text{ ATP}}$$
$$1 \text{ stearic acid} \longrightarrow \boxed{120 \text{ ATP}}$$

Taking into account the fact that glucose has only 6 carbon atoms and stearic acid has 18 carbon atoms still shows more ATP production from the fatty acid.

$$3 \text{ glucose (18 C)} \longrightarrow \boxed{90 \text{ ATP}}$$
$$1 \text{ stearic acid (18 C)} \longrightarrow \boxed{120 \text{ ATP}}$$

Thus, on the basis of equal numbers of carbon atoms, lipids are 33% more efficient than carbohydrates as energy-storage systems.

On an equal-mass basis, fatty acids produce 2.5 times as much energy per gram as carbohydrates (glucose); this is shown by the following calculation involving 1.00 gram of stearic acid and 1.00 gram of glucose.

$$1.00 \text{ g stearic acid} \times \left(\frac{1 \text{ mole stearic acid}}{284 \text{ g stearic acid}} \right) \times \left(\frac{120 \text{ moles ATP}}{1 \text{ mole stearic acid}} \right) = 0.423 \text{ mole ATP}$$

$$1.00 \text{ g glucose} \times \left(\frac{1 \text{ mole glucose}}{180 \text{ g glucose}} \right) \times \left(\frac{30 \text{ moles ATP}}{1 \text{ mole glucose}} \right) = 0.167 \text{ mole ATP}$$

The fact that fatty acids (stearic acid) yield 2.5 times as much energy per gram as carbohydrates (glucose) means that the former "do 2.5 times as much damage" to a person on a diet.

In dietary considerations, nutritionists say that 1 gram of carbohydrate equals 4 kcal and that 1 gram of fat equals 9 kcal. We now know the basis for these numbers. The value of 9 kcal for fat takes into account the fact that not all fatty acids present in fat contain 18 carbon atoms (the basis for our preceding calculations) and also the fact that fats contain glycerol, which produces ATP when degraded.

Is the preferred fuel for "running" the human body fatty acids, which yield 2.5 times as much energy per gram as glucose, or is it glucose? In a normally functioning human body, certain organs use both fuels, others prefer glucose, and still others prefer fatty acids. Here are some generalizations about "fuel" use:

1. Skeletal muscle uses glucose (from glycogen) when in an active state. In a resting state, it uses fatty acids.
2. Cardiac muscle depends first on fatty acids and secondarily on ketone bodies (Section 14.6), glucose, and lactate.
3. The liver uses fatty acids as the preferred fuel.
4. Brain function is maintained by glucose and ketone bodies (Section 14.6). Fatty acids cannot cross the blood–brain barrier and thus are unavailable.

14.6 Ketone Bodies

Ordinarily, when there is adequate balance between lipid and carbohydrate metabolism, most of the acetyl CoA produced from the fatty acid spiral is further processed through the citric acid cycle. The first step of the citric acid cycle (Section 12.6) involves the

CHEMICAL CONNECTIONS   High-Intensity Versus Low-Intensity Workouts

In a resting state, the human body burns more fat than carbohydrate. The fuel consumed is about one-third carbohydrate and two-thirds fat.

Information about fuel consumption ratios is obtainable from respiratory gas measurements, specifically from the respiratory exchange ratio (RER). The RER is the ratio of carbon dioxide to oxygen inhaled divided by the ratio of carbon dioxide to oxygen exhaled. For 100% fat burning, the RER would be 0.7; for 100% carbohydrate burning, the RER would be 1.0.

When a person at rest begins exercising, his or her body suddenly needs energy at a greater rate—more fuel and more oxygen are needed. It takes 0.7 L of oxygen to burn 1 gram of carbohydrate and 1.0 L of oxygen to burn 1 gram of fat. At the onset of exercise, the body is immediately short of oxygen. Also, there is a time delay in triacylglycerol mobilization. Triacylglycerols have to be broken down to fatty acids, which have

The initial stages of exercise are fueled primarily by glucose; in later stages, triacylglycerols become the primary fuel.

to be attached to protein carriers before they can be carried in the bloodstream to working muscles. At their destination, they must be released from the carriers and then undergo energy-producing reactions. By contrast, glycogen is already present in muscle cells, and it can release glucose 6-phosphate as an instant fuel.

Consequently, the initial stages of exercise are fueled primarily by glucose—it requires less oxygen and can even be burned anaerobically (to lactate). During the first few minutes of exercise, up to 80% of the fuel used comes from glycogen.

With time, increased breathing rates increase oxygen supplies to muscles, and triacylglycerol use increases. Continued activity for three-quarters of an hour achieves a 50–50 balance of triacylglycerol and glucose use. Beyond an hour, triacylglycerol use may be as high as 80%.

Suppose a person is exercising at a moderate rate and decides to speed up. Immediately, body fuel and oxygen needs are increased. The response is increased use of glycogen supplies.

The accompanying table compares exercise on a stationary cycle at 45% and 70% of maximum oxygen uptake sufficient to burn 300 calories.

	Low-Intensity Exercise	High-Intensity Exercise
percent of maximum oxygen uptake	45%	70%
time required to burn 300 calories	48 min	30 min
calories obtained from fat	133 cal	65 cal
percent of calories from fat	44%	22%
rate of fat burning per minute	2.8 cal/min	2.1 cal/min

reaction between oxaloacetate and acetyl CoA. Sufficient oxaloacetate must be present for the acetyl CoA to react with. Oxaloacetate concentration depends on pyruvate produced from glycolysis (Section 13.2); pyruvate can be converted to oxaloacetate by *pyruvate carboxylase* (Section 13.6).

Certain body conditions upset the lipid–carbohydrate balance required for acetyl CoA generated by fatty acids to be processed by the citric acid cycle. These conditions include (1) dietary intake high in fat and low in carbohydrates, (2) diabetic conditions where the body cannot adequately process glucose even though it is present, and (3) *prolonged* fasting conditions, including starvation, where glycogen supplies are exhausted. Under these conditions, the problem of inadequate oxaloacetate arises, which is compounded by the body's using oxaloacetate that is present to produce glucose through gluconeogenesis (Section 13.6).

What happens when oxaloacetate supplies are too low for all acetyl CoA present to be processed through the citric acid cycle? The excess acetyl CoA is diverted to the formation of ketone bodies. A **ketone body** *is one of three substances (acetoacetate,* *β-hydroxybutyrate, and acetone) produced from acetyl CoA when an excess of acetyl CoA from fatty acid degradation accumulates because of triacylglycerol–carbohydrate metabolic imbalances.* The structures for the three ketone bodies are

$$
\underset{\text{Acetoacetate}}{CH_3-\overset{\overset{\displaystyle O}{\|}}{C}-CH_2-\overset{\overset{\displaystyle O}{\|}}{C}-O^-}
\qquad
\underset{\beta\text{-Hydroxybutyrate}}{CH_3-\overset{\overset{\displaystyle OH}{|}}{C}H-CH_2-\overset{\overset{\displaystyle O}{\|}}{C}-O^-}
\qquad
\underset{\text{Acetone}}{CH_3-\overset{\overset{\displaystyle O}{\|}}{C}-CH_3}
$$

The structure of β-hydroxybutyrate does not actually include a ketone group, but it is still classified as a ketone body.

For a number of years, ketone bodies were thought of as degradation products that had little physiological significance. It is now known that ketone bodies can serve as sources of energy for various tissues and are very important energy sources in heart muscle and the renal cortex. Even the brain, which requires glucose, can adapt to obtain a portion of its energy from ketone bodies in dieting situations that involve a properly constructed low-carbohydrate diet.

◼ Ketogenesis

Ketogenesis *is the metabolic pathway by which ketone bodies are synthesized from acetyl CoA.* The primary site for ketogenesis is liver mitochondria. After they are produced, the ketone bodies diffuse from these structures into the bloodstream where they are transported to peripheral tissues. The reactions that constitute ketogenesis are shown in Figure 14.8.

Step 1: *Condensation.* Ketogenesis begins as two acetyl CoA molecules combine to produce acetoacetyl CoA, a reversal of the last step of the fatty acid spiral (Section 14.4).

$$
\underset{\text{Acetyl CoA}}{CH_3-\overset{\overset{\displaystyle O}{\|}}{C}-S-CoA} + \underset{\text{Acetyl CoA}}{CH_3-\overset{\overset{\displaystyle O}{\|}}{C}-S-CoA} \xrightarrow{\text{Thiolase}}
$$

$$
\underset{\text{Acetoacetyl CoA}}{CH_3-\overset{\overset{\displaystyle O}{\|}}{C}-CH_2-\overset{\overset{\displaystyle O}{\|}}{C}-S-CoA} + \; CoA-SH
$$

> Ketone bodies are produced when the amount of acetyl CoA is excessive compared with the amount of oxaloacetate available to react with it (Step 1 of the citric acid cycle).

> Even when ketogenic conditions are not present in the human body, the liver produces a *small amount* of ketone bodies.

FIGURE 14.8 Ketogenesis involves the production of ketone bodies from acetyl CoA.

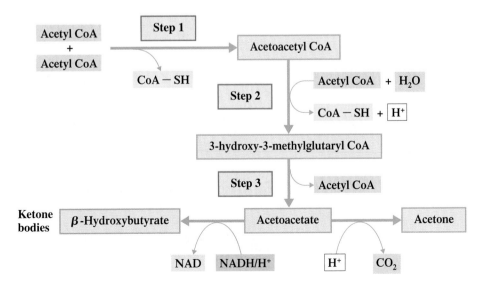

Step 2: *Condensation.* Acetoacetyl CoA then reacts with a third acetyl CoA and water to produce 3-hydroxy-3-methylglutaryl CoA (HMG-CoA) and CoA—SH.

$$CH_3-\overset{\overset{O}{\|}}{C}-CH_2-\overset{\overset{O}{\|}}{C}-S-CoA + CH_3-\overset{\overset{O}{\|}}{C}-S-CoA + H_2O \xrightarrow[\text{synthase}]{\text{HMG-CoA}}$$

Acetoacetyl CoA   Acetyl CoA

$$^-OOC-CH_2-\overset{\overset{OH}{|}}{\underset{\underset{CH_3}{|}}{C}}-CH_2-\overset{\overset{O}{\|}}{C}-S-CoA + CoA-SH + H^+$$

HMG-CoA

Step 3: *Chain cleavage.* HMG-CoA is then cleaved to acetyl CoA and acetoacetate.

$$^-OOC-CH_2-\overset{\overset{OH}{|}}{\underset{\underset{CH_3}{|}}{C}}-CH_2-\overset{\overset{O}{\|}}{C}-S-CoA \xrightarrow[\text{lyase}]{\text{HMG-CoA}}$$

HMG-CoA

$$^-OOC-CH_2-\overset{\overset{O}{\|}}{C}-CH_3 + CH_3-\overset{\overset{O}{\|}}{C}-S-CoA$$

Acetoacetate   Acetyl CoA

Summing these three reactions to obtain the net reaction for ketogenesis yields

$$2 \text{ Acetyl CoA} + H_2O \longrightarrow \text{acetoacetate} + 2 \text{ CoA} + H^+$$

The ketone body acetoacetate is the "parent" compound for the other two ketone bodies. Acetone arises from acetoacetate by the loss of the carboxyl group (as CO_2). Reduction of the keto group of acetoacetate to a hydroxyl group by NADH produces β-hydroxybutyrate. The amount of acetone present is usually small compared with the other two species.

For acetoacetate to be used as a fuel—in heart muscle, for example—it must first be activated. Acetoacetate is activated by transfer of a CoA group from succinyl CoA (a citric acid cycle intermediate). The resulting acetoacetyl CoA is then cleaved to give two acetyl CoA molecules that can enter the citric acid cycle (see Figure 14.9). In effect, acetoacetate is a water-soluble, transportable form of acetyl units.

> Heart muscle and the renal cortex use acetoacetate in preference to glucose. The brain adapts to the utilization of acetoacetate with starvation or diabetes. 75% of the fuel needs of the brain are obtained from acetoacetate during prolonged starvation.

FIGURE 14.9 The pathway for utilization of acetoacetate as a fuel. The required succinyl CoA comes from the citric acid cycle.

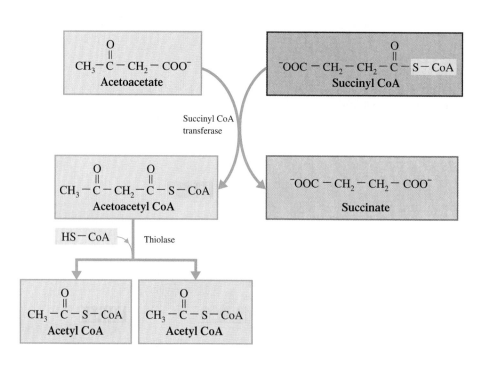

■ **Ketosis**

Under normal metabolic conditions (an appropriate glucose–fatty acid balance), the concentration of ketone bodies in the blood is very low—about 1 mg/100 mL. Abnormal metabolic conditions, such as those mentioned at the start of this section, produce elevated blood ketone levels, levels 50–100 times greater than normal. Excess accumulation of ketone bodies in blood (20 mg/100 mL) is called *ketonemia*. At a level of 70 mg/100 mL, the renal threshold is exceeded, and ketone bodies are excreted in the urine, a condition called *ketonuria*. The overall accumulation of ketone bodies in the blood and urine is called *ketosis*. Ketosis is often detectable by the smell of acetone on a person's breath; acetone is very volatile and is excreted through the lungs.

For the vast majority of persons following a low-carbohydrate diet, the effects of ketosis appear to be harmless or nearly so. The symptoms of the *mild* ketosis that occurs as the result of such dieting include headache, dry mouth, and sometimes foul-smelling breath.

Two of the three ketone bodies—acetoacetate and β-hydroxybutyrate—are acids. Their presence in blood causes a slight but significant decrease in blood pH. This can result in acidosis in *severe* ketosis situations. Symptoms include heavy breathing (because acidic blood can carry less oxygen) and increased urine output that can lead to dehydration. Ultimately, the condition can cause coma and death.

Acidosis from elevated ketone body levels is often called *keto* acidosis or *metabolic* acidosis to distinguish it from *respiratory* acidosis, which is not linked to ketone bodies.

14.7 Biosynthesis of Fatty Acids: Lipogenesis

Lipogenesis *is the metabolic pathway by which fatty acids are synthesized from acetyl CoA.* As was the case for the opposing processes of glycolysis and gluconeogenesis, lipogenesis is not simply a reversal of the steps for degradation of fatty acids (the fatty acid spiral). Before we look at the details of fatty acid synthesis, we will consider some differences between the synthesis and degradation of fatty acids.

1. Lipogenesis occurs in the cell cytosol, whereas degradation of fatty acids occurs in the mitochondrial matrix. Because they have different reaction sites, these two opposing processes can occur at the same time when necessary.
2. Different enzymes are involved in the two processes. Lipogenesis enzymes are collected into a multienzyme complex called *fatty acid synthase*. This enzyme complex ties the reaction steps of lipogenesis closely together. The enzymes involved in fatty acid degradation are not physically associated, so the reaction steps are independent.
3. Intermediates of the two processes are covalently bonded to different carriers. The carrier for fatty acid spiral intermediates is CoA. Lipogenesis intermediates are bonded to ACP (acyl carrier protein).
4. Fatty acid synthesis is dependent on the reducing agent NADPH. Fatty acid degradation is dependent on the oxidizing agents FAD and NAD^+.
5. Fatty acids are built up two carbons at a time during synthesis and are broken down two carbons at a time during degradation. The source of the two carbon units differs between the two processes. In lipogenesis, acetyl CoA is used to form malonyl ACP, which becomes the carrier of the two carbon units. CoA derivatives are involved in all steps of the fatty acid spiral.

In general, fatty acid biosynthesis (lipogenesis) occurs any time dietary intake provides more nutrients than are needed for energy requirements. The primary lipogenesis sites are the liver, adipose tissue, and mammary glands. The mammary glands show increased synthetic activity during periods of lactation.

FIGURE 14.10 The citrate–malate–pyruvate shuttle system for transferring acetyl CoA from a mitochondrion to the cytosol.

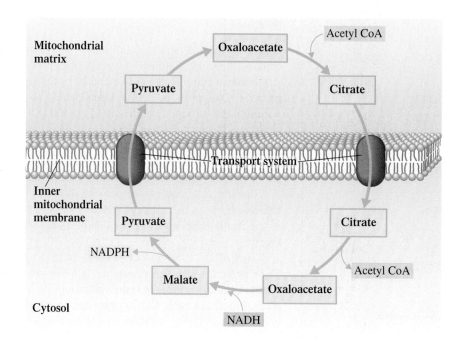

Formation of Malonyl CoA

Acetyl CoA is the starting material for lipogenesis. Because acetyl CoA is generated in mitochondria and lipogenesis occurs in the cytosol, the acetyl CoA must first be transported to the cytosol. It exits the mitochondria through a transport system that involves citrate ion.

Mitochondrial acetyl CoA reacts with oxaloacetate (the first step of the citric acid cycle) to produce citrate, which is then transported through the inner mitochondrial membrane.

$$\text{Acetyl CoA} + \text{oxaloacetate} + H_2O \longrightarrow \text{citrate} + \text{coenzyme A}$$

Once in the cytosol, the citrate undergoes the reverse reaction to regenerate acetyl CoA; an NADH molecule is involved in the process.

Figure 14.10 shows "details" concerning this shuttle system. Note that a number of different molecules are involved in regeneration of the citrate for use again and that the citrate is regenerated on the side of the membrane from which it originates. Compared to the carnitine shuttle system for long-chain fatty acid groups (acyl groups; Figure 14.6), the citrate shuttle system is much more complex. However, all of the intermediates in the shuttle have been encountered before, in glycolysis and the citric acid cycle.

Cytosolic acetyl CoA is then converted to malonyl CoA in a carboxylation reaction that involves carbon dioxide (CO_2) and ATP.

> The parent compound for the malonyl group is malonic acid, the C_3 dicarboxylic acid.
>
> $$\underset{O}{\overset{O}{\underset{\|}{\overset{\|}{HO-C-CH_2-C-OH}}}}$$

This reaction occurs only when cellular ATP levels are high. It is catalyzed by *acetyl CoA carboxylase complex,* which requires both Mn^{2+} ion and the B vitamin biotin for its activity.

ACP Complex Formation

Studies show that all intermediates in fatty acid synthesis are linked to acyl carrier proteins (ACP—SH) rather than to CoA—SH. Even the small C_2 acetyl and C_3 malonyl groups are bound to such carriers.

$$\text{Acetyl CoA} + \text{ACP—SH} \longrightarrow \text{acetyl ACP} + \text{CoA—SH}$$
$$\text{Malonyl CoA} + \text{ACP—SH} \longrightarrow \text{malonyl ACP} + \text{CoA—SH}$$

FIGURE 14.11 In the first cycle of the fatty acid biosynthetic pathway, acetyl ACP is converted to butyryl ACP. In the next cycle (not shown), the butyryl ACP reacts with another malonyl ACP to produce a 6-carbon acid. Continued cycles produce acids with 8, 10, 12, 14, and 16 carbon atoms.

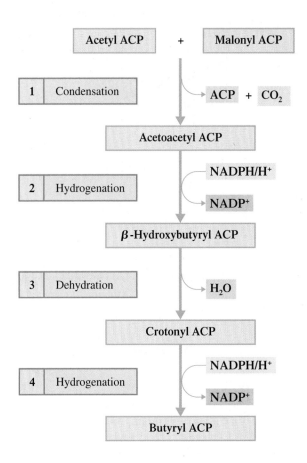

ACP—SH can be regarded as a "giant CoA—SH molecule." Involved in its structure are the 2-ethanethiol and pantothenic acid components of CoA—SH (Section 12.3) attached to a polypeptide chain containing 77 amino acid residues.

■ Chain Elongation

Four reactions that occur in a cyclic pattern within the multienzyme *fatty acid synthase complex* constitute the chain elongation process used for fatty acid synthesis. The reactions of the *first* turn of the cycle, in general terms, are shown in Figure 14.11. Specific details about this series of reactions follow.

Step 1: *Condensation.* Acetyl ACP and malonyl ACP condense together to form acetoacetyl ACP.

$$CH_3-\overset{\overset{O}{\|}}{C}-S-ACP \ + \ \overline{O}-\overset{\overset{O}{\|}}{C}-CH_2-\overset{\overset{O}{\|}}{C}-S-ACP \longrightarrow CH_3-\overset{\overset{O}{\|}}{C}-CH_2-\overset{\overset{O}{\|}}{C}-S-ACP \ + \ CO_2 \ + \ ACP-SH$$

Acetyl ACP   Malonyl ACP   Acetoacetyl ACP

Note that a C_2 species (acetyl) and a C_3 species (malonyl) react to produce a C_4 species (acetoacetyl) rather than a C_5 species. One carbon atom leaves the reaction in the form of a CO_2 molecule.

Steps 2 through 4 involve a sequence of functional group changes that we have encountered twice before—in the fatty acid spiral (Section 14.4) and in the citric acid cycle (Section 12.6). This time, however, the changes occur in the reverse sequence to that previously encountered. The functional group changes are

$$\textbf{Ketone} \xrightarrow[\text{(hydrogenation)}]{\overset{②}{\text{Reduction}}} \textbf{secondary alcohol} \xrightarrow[]{\overset{③}{\text{Dehydration}}} \textbf{alkene} \xrightarrow[\text{(hydrogenation)}]{\overset{④}{\text{Reduction}}} \textbf{alkane}$$

Step 2: *Hydrogenation.* The keto group of the acetoacetyl complex, which involves the β-carbon atom, is reduced to the corresponding alcohol by NADPH.

$$CH_3 - \overset{\displaystyle O}{\overset{\|}{C}} - CH_2 - \overset{\displaystyle O}{\overset{\|}{C}} - S - ACP \longrightarrow CH_3 - \overset{\displaystyle OH}{\overset{|}{CH}} - CH_2 - \overset{\displaystyle O}{\overset{\|}{C}} - S - ACP$$

Acetoacetyl ACP   NADPH/H⁺   NADP⁺   β-Hydroxybutyryl ACP

Step 3: *Dehydration.* The alcohol produced in Step 2 is dehydrated to introduce a double bond into the molecule (between the α and β carbons).

$$CH_3 - \overset{\displaystyle OH}{\overset{|}{CH}} - CH_2 - \overset{\displaystyle O}{\overset{\|}{C}} - S - ACP \longrightarrow CH_3 - \overset{trans}{CH} = CH - \overset{\displaystyle O}{\overset{\|}{C}} - S - ACP$$

β-Hydroxybutyryl ACP H_2O Crotonyl ACP

Step 4: *Hydrogenation.* The double bond introduced in Step 3 is converted to a single bond through hydrogenation. As in Step 2, NADPH is the reducing agent.

$$CH_3 - \overset{trans}{CH} = CH - \overset{\displaystyle O}{\overset{\|}{C}} - S - ACP \longrightarrow CH_3 - CH_2 - CH_2 - \overset{\displaystyle O}{\overset{\|}{C}} - S - ACP$$

Crotonyl ACP   NADPH/H⁺   NADP⁺   Butyryl ACP

Steps 2, 3, and 4 of fatty acid biosynthesis accomplish the reverse of Steps 3, 2, and 1 of the fatty acid spiral.

Further cycles of the preceding four-step process convert the four-carbon acyl group to a six-carbon acyl group, then to an eight-carbon acyl group, and so on (see Figure 14.12). Elongation of the acyl group chain through this procedure, which is tied to the fatty acid synthase complex, stops upon formation of the C_{16} acyl group (palmitic acid). Different enzyme systems and different cellular locations are required for elongation of the chain beyond C_{16} and for introduction of double bonds into the acyl group (unsaturated fatty acids).

FIGURE 14.12 The sequence of cycles needed to produce a C_{16} fatty acid from acetyl ACP. Each loop represents one cycle.

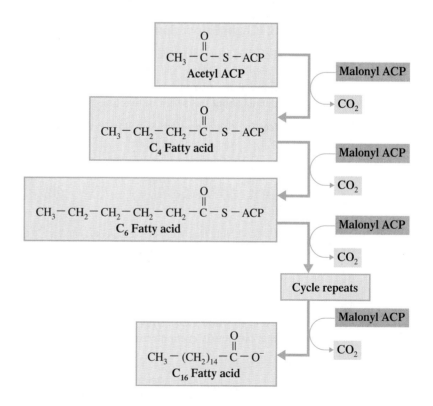

Reactants and Products in the Biosynthesis of One Molecule of Palmitic Acid, the 16:0 Fatty Acid

Reactants	Products
8 acetyl CoA	1 palmitate
7 ATP	8 CoA
14 NADPH	7 ADP
6 H$^+$	7 P$_i$
	14 NADP$^+$
	6 H$_2$O

A relatively large input of energy is needed to biosynthesize a fatty acid molecule, as can be seen from the data in Table 14.1, which gives a net summary of the reactants and products involved in the synthesis of one molecule of palmitic acid, the 16:0 fatty acid.

Production of unsaturated fatty acids (insertion of double bonds) requires molecular oxygen (O_2). In an oxidation step, hydrogen is removed and combined with the O_2 to form water.

$$R-\underset{\underset{H}{|}}{\overset{\overset{H}{|}}{C}}-\underset{\underset{H}{|}}{\overset{\overset{H}{|}}{C}}-(CH_2)_n-\overset{\overset{O}{\|}}{C}-O^- + O_2 \xrightarrow{\quad\text{NADPH/H}^+\quad\text{NADP}^+\quad}$$

$$R-\underset{\underset{H}{|}}{C}=\underset{\underset{H}{|}}{C}-(CH_2)_n-\overset{\overset{O}{\|}}{C}-O^- + 2H_2O$$

In humans and animals, enzymes can introduce double bonds only between C-4 and C-5 and between C-9 and C-10. Thus the important unsaturated fatty acids linoleic (C$_{18}$ with C-9 and C-12 double bonds) and linolenic (C$_{18}$ with C-9, C-12, and C-15 double bonds) cannot be biosynthesized. They must be obtained from the diet. (Plants have the enzymes necessary to synthesize these acids.) Acids such as linoleic and linolenic (Section 8.2), which cannot be synthesized by the body but are necessary for its proper functioning, are called *essential fatty acids.*

Lipogenesis can be used to convert glucose to fatty acids via acetyl CoA. The reverse process, conversion of fatty acids to glucose, is not possible within the human body. Fatty acids can be broken down to acetyl CoA, but there is no enzyme present for the conversion of acetyl CoA to pyruvate or oxaloacetate, starting materials for gluconeogenesis (Section 13.6). Plants and some bacteria do possess the needed enzymes and thus can convert fatty acids to carbohydrates.

14.8 Biosynthesis of Cholesterol

So far in this chapter, our discussion of lipid metabolism has focused on fats and oils (triacylglycerols) and their hydrolysis products, fatty acids and glycerol. We now consider another very important lipid—cholesterol.

Every membrane of every cell in the body has cholesterol as a necessary component. This substance is also the precursor for bile salts, sex hormones, and adrenal hormones (Sections 8.11 and 8.12).

In today's health-conscious world, dietary intake of cholesterol is of great interest because of correlations between high serum cholesterol levels and coronary heart disease. Average daily dietary intake of cholesterol is approximately 0.3 gram. This amount, though important, is small compared to the 1.5–2.0 grams of cholesterol that the body synthesizes every day from acetyl CoA units.

The biosynthesis of cholesterol, a C$_{27}$ molecule, occurs primarily in the liver. Its production consumes 15 molecules of acetyl CoA and involves at least 27 separate enzymatic steps. An overview of cholesterol synthesis is given in Figure 14.13.

In the first phase of cholesterol synthesis, three molecules of acetyl CoA are condensed into a C$_6$ mevalonate ion.

The "parent" compound for mevalonate ion is mevalonic acid (3,5-dihydroxy-3-methylpentanoic acid).

$$\underset{\underset{OH}{|}}{CH_2}-\underset{\underset{OH}{|}}{CH_2}-\overset{\overset{CH_3}{|}}{C}-CH_2-\overset{\overset{O}{\|}}{C}-OH$$

$$3\ CH_3-\overset{\overset{O}{\|}}{C}-S-CoA \xrightarrow{\text{Several steps}} \underset{\underset{\underset{\underset{CH_2OH}{|}}{CH_2}}{\underset{|}{HO-C-CH_3}}}{\overset{\overset{\overset{\overset{COO^-}{|}}{CH_2}}{|}}{}}$$

Acetyl CoA

Mevalonate

CHEMICAL CONNECTIONS
Statins: Drugs That Lower Plasma Levels of Cholesterol

Over half of all deaths in the United States are directly or indirectly related to heart disease, in particular to atherosclerosis. Atherosclerosis results from the buildup of plaque (fatty acid deposits) on the inner walls of arteries. Cholesterol, obtained from low-density-lipoproteins (LDL) that circulate in blood plasma, is also a major component of plaque.

Because most of the cholesterol in the human body is synthesized in the liver, from acetyl CoA, much research has focused on finding ways to inhibit its biosynthesis. The rate-determining step in cholesterol biosynthesis involves the conversion of 3-hydroxy-3-methylglutaryl CoA (HMG-CoA) to mevalonate, a process catalyzed by the enzyme HMG-CoA reductase.

3-Hydroxy-3-methylglutaryl-CoA
(HMG-CoA)

In 1976, as the result of screening more than 8000 strains of microorganisms, a compound now called *mevastatin*—a potent inhibitor of HMG-CoA reductase—was isolated from culture broths of a fungus. Soon thereafter, a second, more active compound called *lovastatin* was isolated.

R₁ = R₂ = H, mevastatin
R₁ = H, R₂ = CH₃, lovastatin (Mevacor)
R₁ = R₂ = CH₃, simvastatin (Zocor)

These "statins" are very effective in lowering plasma concentrations of LDL by functioning as competitive inhibitors of HMG-CoA reductase.

After years of testing, the statins are now available as prescription drugs for lowering blood cholesterol levels. Clinical studies indicate that use of these drugs lowers the incidence of heart disease in individuals with mildly elevated blood cholesterol levels. A later-generation statin with a ring structure distinctly different from that of earlier statins—atorvastatin (Lipitor)—became the most prescribed medication in the United States in the year 2000. Note the structural resemblance between part of the structure of Lipitor and that of mevalonate.

Mevalonate

Atorvastatin (Lipitor)

Recent research studies have unexpectedly shown that the cholesterol-lowering statins have two added benefits.

Laboratory studies with animals indicate that statins prompt growth cells to build new bone, replacing bone that has been leached away by osteoporosis ("brittle-bone disease"). A retrospective study of osteoporosis patients who also took statins shows evidence that their bones became more dense than did bones of osteoporosis patients who did not take the drugs.

Statins have also been shown to function as antiinflammatory agents that counteract the effects of a common virus, cytomegalovirus, which is now believed to contribute to the development of coronary heart disease. Researchers believe that by age 65, more than 70% of all people have been exposed to this virus. The virus, along with other infecting agents in blood, may actually trigger the inflammation mechanism for heart disease.

FIGURE 14.13 An overview of the biosynthetic pathway for cholesterol synthesis.

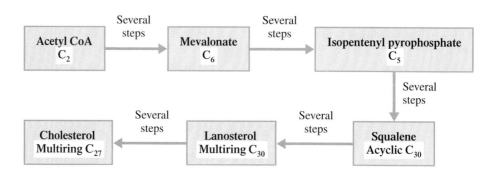

The C_6 mevalonate undergoes a decarboxylation to yield a C_5 isoprene derivative called isopentenyl pyrophosphate and CO_2. Three ATP molecules are needed in accomplishing this process.

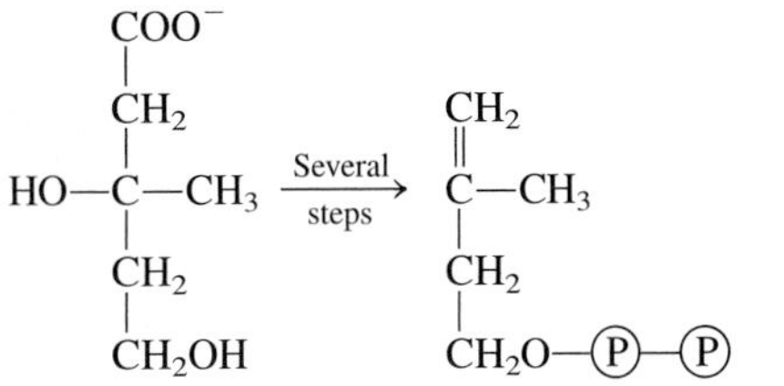

Mevalonate Isopentenyl pyrophosphate

Compounds whose structures are based on the five-carbon isoprene unit are called *terpenes* (Section 2.6).

The isoprene structural unit (Section 2.6), present in isoprene derivatives in a modified form, is a commonly used five-carbon building block in biosynthetic processes.

The next stage of cholesterol biosynthesis involves the condensation of six isoprene units to give the C_{30} squalene molecule.

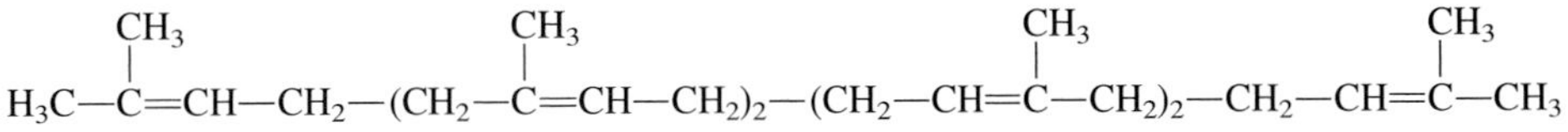

Squalene

A "redrawing" of the squalene structure, with numerous twists and bends in it, is helpful in visualizing the next stage of cholesterol biosynthesis, the formation of the four-ring steroid nucleus (Section 8.12) associated with lanosterol (and cholesterol).

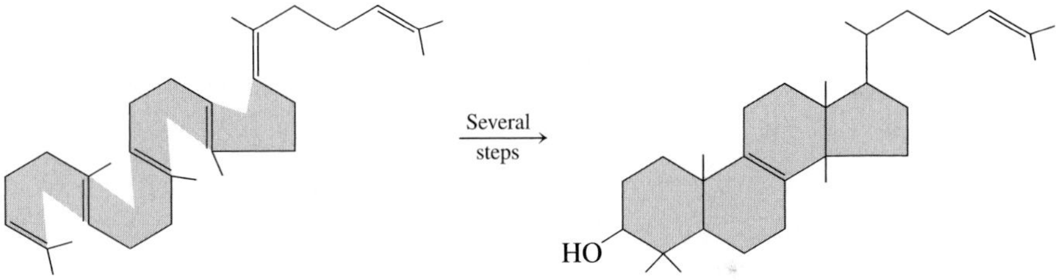

Squalene Lanosterol

The multistep squalene-to-lanosterol transition involves the formation of four ring systems, a decrease in double bonds from six to two, the migration of two methyl groups to new locations, and the addition of an —OH group to the C_{30} system. Addition of the —OH group requires the use of molecular oxygen; the O of the —OH group comes from the molecular O_2.

The transition from lanosterol to cholesterol involves removal of three methyl groups (C_{30} to C_{27}), reduction of the double bond in the side chain, and migration of the other double bond to a new location.

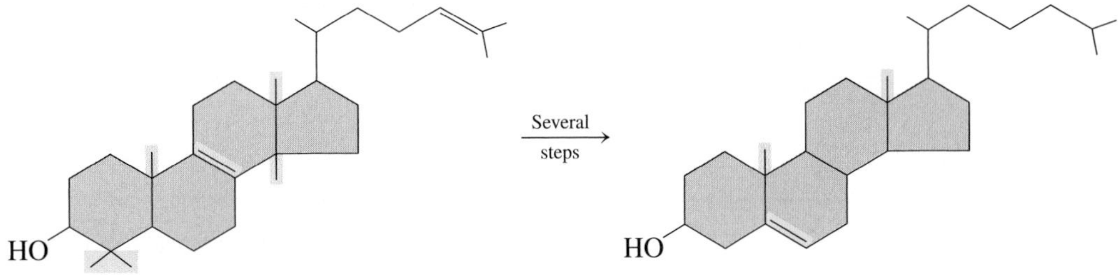

Lanosterol (C_{30}) Cholesterol (C_{27})

Once cholesterol has been formed, biosynthetic pathways are available to convert it to each of the five major classes of steroid hormones: progestins, androgens, estrogens, glucocorticoids, and mineralocorticoids (see Figure 14.14), as well as to bile acids and vitamin D (Section 10.14).

CHEMISTRY AT A GLANCE

Interrelationships Between Carbohydrate and Lipid Metabolism

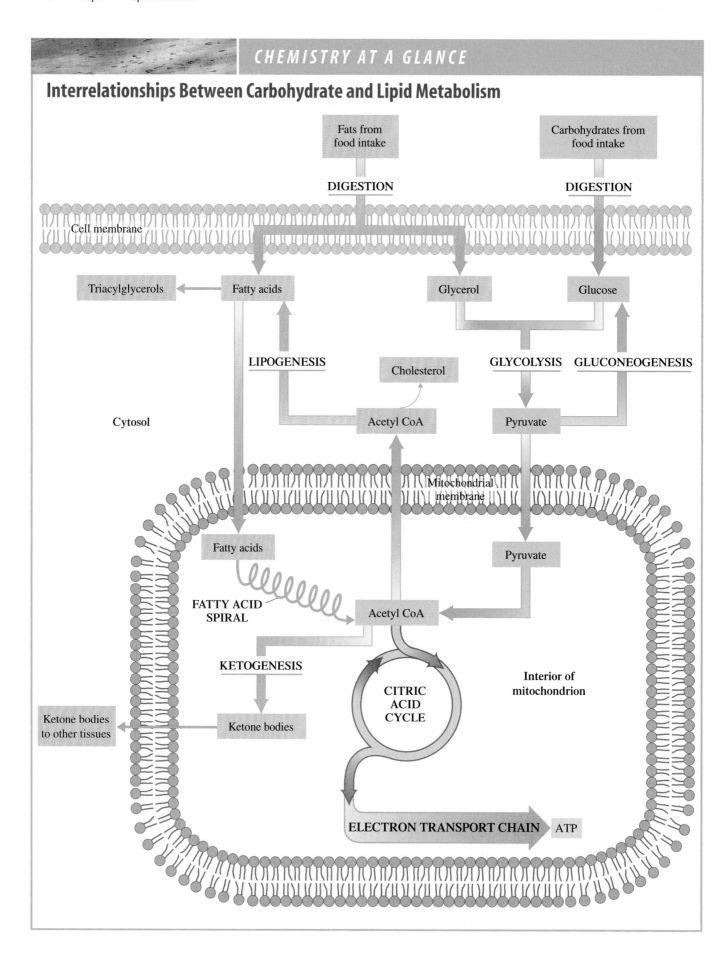

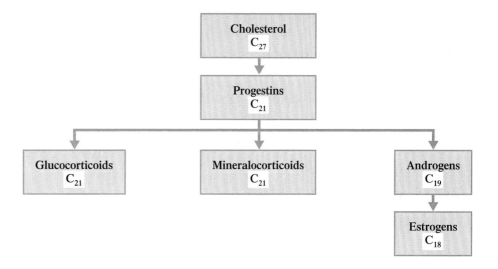

FIGURE 14.14 Biosynthetic relationships among steroid hormones.

14.9 Relationships Between Lipid and Carbohydrate Metabolism

Acetyl CoA is the primary link between lipid and carbohydrate metabolism. As shown in the Chemistry at a Glance feature on page 496, acetyl CoA is the degradation product for glucose, glycerol, and fatty acids, and it is also the starting material for the biosynthesis of fatty acids, cholesterol, and ketone bodies.

Note the four possible fates of acetyl CoA produced from fatty acid, glycerol, and glucose degradation processes.

1. *Oxidation in the citric acid cycle.* Both lipids (fatty acids and glycerol) and carbohydrates (glucose) supply acetyl CoA for the operation of this cycle.
2. *Ketone body formation.* This process is of major importance when there is imbalance between lipid and carbohydrate metabolic processes. The imbalance is caused by inadequate glucose metabolism during times of adequate lipid metabolism.
3. *Fatty acid biosynthesis.* The buildup of excess acetyl CoA when dietary intake exceeds energy needs leads to accelerated fatty acid biosynthesis.
4. *Cholesterol biosynthesis.* As with fatty acid biosynthesis, cholesterol biosynthesis occurs primarily when the body is in an acetyl CoA–rich state.

CONCEPTS TO REMEMBER

Triacylglycerol digestion and absorption. Triacylglycerols are digested (hydrolyzed) in the intestine and then reassembled after passage into the intestinal wall. Chylomicrons transport the reassembled triacyl-glycerols from intestinal cells to the bloodstream (Section 14.1).

Triacylglycerol storage and mobilization. Triacylglycerols are stored as fat droplets in adipose tissue. When they are needed for energy, enzyme-controlled hydrolysis reactions liberate the fatty acids, which then enter the bloodstream and travel to tissues where they are utilized (Section 14.2).

Glycerol metabolism. Glycerol is first phosphorylated and then oxidized to dihydroxyacetone phosphate, a glycolysis pathway intermediate. Through glycolysis and the common metabolic pathway, the glycerol can be converted to CO_2 and H_2O (Section 14.3).

Fatty acid degradation. Fatty acid degradation is accomplished through the fatty acid (β oxidation) spiral. The degradation process involves removal of carbon atoms, two at a time, from the carboxyl end of the fatty acid. There are four repeating reactions that accompany

the removal of each two-carbon unit. A turn of the cycle also produces one molecule each of acetyl CoA, NADH, and $FADH_2$ (Section 14.4).

Ketone bodies. Acetoacetate, β-hydroxybutyrate, and acetone are known as ketone bodies. They are synthesized in the liver from acetyl CoA as a result of excessive fatty acid degradation. During starvation and in unchecked diabetes, the level of ketone bodies in the blood becomes very high (Section 14.6).

Fatty acid biosynthesis. Fatty acid biosynthesis, lipogenesis, occurs through the addition of two-carbon units to a growing acyl chain. The added two-carbon units come from malonyl CoA. A multienzyme complex, an acyl carrier protein (ACP), and NADPH are important parts of the biosynthetic process (Section 14.7).

Biosynthesis of cholesterol. Cholesterol is biosynthesized from acetyl CoA in a complex series of reactions in which isoprene units are key intermediates. Cholesterol is the precursor for the various classes of steroid hormones (Section 14.8).

KEY REACTIONS AND EQUATIONS

1. Digestion of triacylglycerols (Section 14.1)

$$\text{Triacylglycerol} + H_2O \xrightarrow{\text{Lipase}}$$

$$\text{fatty acids} + \text{glycerol} + \text{monoacylglycerols}$$

2. Mobilization of triacylglycerols (Section 14.2)

$$\text{Triacylglycerol} + 3H_2O \xrightarrow{\text{Lipase}} 3 \text{ fatty acids} + \text{glycerol}$$

3. Glycerol metabolism (Section 14.3)

$$\text{Glycerol} + \boxed{\text{ATP}} + \boxed{\text{NAD}^+} \xrightarrow[\text{steps}]{\text{Two}}$$

$$\text{dihydroxyacetone} + \text{ADP} + \text{NADH} + H^+$$
$$\text{phosphate}$$

4. One cycle of the fatty acid spiral (Section 14.4)

$$R-CH_2-CH_2-\overset{\overset{\textstyle O}{\|}}{C}-S-CoA + \boxed{\text{NAD}^+} + \boxed{\text{FAD}} + CoA-SH \longrightarrow$$

$$R-\overset{\overset{\textstyle O}{\|}}{C}-S-CoA + CH_3-\overset{\overset{\textstyle O}{\|}}{C}-S-CoA + \boxed{\text{NADH}} + \boxed{\text{FADH}_2}$$

5. Ketone body formation (Section 14.6)

$$2 \text{ Acetyl CoA} + H_2O \xrightarrow[\text{steps}]{\text{Three}} \text{acetoacetate} + 2CoA-SH$$

6. First turn of lipogenesis (Section 14.7)

$$\text{Acetyl ACP} + \text{malonyl ACP} \xrightarrow{\overset{\text{2NADPH/H}^+ \quad \text{2NADP}^+}{\curvearrowright}}$$

$$\text{butyryl ACP} + \text{ACP} + CO_2 + H_2O$$

 EXERCISES AND PROBLEMS

The members of each pair of problems in this section test similar material.

■ **Digestion and Absorption of Lipids (Section 14.1)**

14.1 What percent of dietary lipids are triacylglycerols?

14.2 What are the solubility characteristics of triacylglycerols?

14.3 What effect do salivary enzymes have on triacylglycerols?

14.4 What effect do stomach fluids have on triacylglycerols?

14.5 Why does ingestion of lipids make one feel "full" for a long time?

14.6 The process of lipid digestion occurs primarily at two sites within the human body.
a. What are the identities of these two sites?
b. What is the relative amount of TAG digestion that occurs at each site?
c. What type of digestive enzyme functions at each site?

14.7 What function does bile serve in lipid digestion?

14.8 What are the major products of triacylglycerol digestion?

14.9 *Complete* hydrolysis of triacylglycerols during digestion is unusual. Explain.

14.10 What is a fatty acid micelle?

14.11 What happens to the products of triacylglycerol digestion after they pass through the intestinal wall?

14.12 What is a chylomicron, and what is its function?

■ **Triacylglycerol Storage and Mobilization (Section 14.2)**

14.13 What is the distinctive structural feature of *adipocytes*?

14.14 What is the major metabolic function of adipose tissue?

14.15 What is triacylglycerol mobilization?

14.16 What situation signals the need for mobilization of triacylglycerols from adipose tissue?

14.17 What role does cAMP play in triacylglycerol mobilization?

14.18 Triacylglycerols in adipose tissue do not enter the bloodstream as triacylglycerols. Explain.

■ **Glycerol Metabolism (Section 14.3)**

14.19 In what order are the compounds glycerol 3-phosphate and dihydroxyacetone phosphate encountered in the degradation of glycerol?

14.20 How many reactions are needed to convert glycerol into a glycolysis intermediate?

14.21 How many ATP molecules are expended in the conversion of glycerol to a glycolysis intermediate?

14.22 What are the two fates of glycerol after it has been converted to a glycolysis intermediate?

■ **Oxidation of Fatty Acids (Section 14.4)**

14.23 Where in a cell does fatty acid activation take place?

14.24 What is the chemical form for an activated fatty acid?

14.25 Only one molecule of ATP is used to activate fatty acids before oxidation occurs, yet we count this expenditure as *two* ATP molecules for "accounting" purposes. Explain.

14.26 What is the difference between an acetyl CoA molecule and an acyl CoA molecule?

14.27 What is the function of carnitine in the fatty acid degradation process?

14.28 The locations in a cell for fatty acid activation and fatty acid oxidation differ. Explain.

14.29 Explain what functional group change occurs, during one turn of the fatty acid spiral, in
a. Step 1   b. Step 2   c. Step 3

14.30 For one turn of the fatty acid spiral, arrange the following β-carbon functional groups in the order in which they are encountered: secondary alcohol, ketone, alkane, and alkene.

14.31 What is the configuration of the unsaturated enoyl CoA formed by dehydrogenation during a turn of the fatty acid spiral?

14.32 What is the configuration of the β-hydroxyacyl CoA formed by hydration during a turn of the fatty acid spiral?

14.33 In which step (of Steps 1 through 4) and in which turn (first or second) of the fatty acid spiral is each of the following compounds encountered as a reactant if the fatty acid to be degraded is hexanoic acid?

a.
$$CH_3-CH_2-CH_2-\overset{\overset{\displaystyle OH}{|}}{CH}-CH_2-\overset{\overset{\displaystyle O}{||}}{C}-S-CoA$$

b.
$$CH_3-CH=CH-\overset{\overset{\displaystyle O}{||}}{C}-S-CoA$$

c.
$$CH_3-\overset{\overset{\displaystyle O}{||}}{C}-CH_2-\overset{\overset{\displaystyle O}{||}}{C}-S-CoA$$

d.
$$CH_3-CH_2-CH_2-CH_2-CH_2-\overset{\overset{\displaystyle O}{||}}{C}-S-CoA$$

14.34 In which step (of Steps 1 through 4) and in which turn (first or second) of the fatty acid spiral is each of the following compounds encountered as a reactant if the fatty acid to be degraded is hexanoic acid?

a.
$$CH_3-CH_2-CH_2-\overset{\overset{\displaystyle O}{||}}{C}-CH_2-\overset{\overset{\displaystyle O}{||}}{C}-S-CoA$$

b.
$$CH_3-CH_2-CH_2-\overset{\overset{\displaystyle O}{||}}{C}-S-CoA$$

c.
$$CH_3-CH_2-CH_2-CH=CH-\overset{\overset{\displaystyle O}{||}}{C}-S-CoA$$

d.
$$CH_3-\overset{\overset{\displaystyle OH}{|}}{CH}-CH_2-\overset{\overset{\displaystyle O}{||}}{C}-S-CoA$$

14.35 Which compound(s) in Problem 14.33 undergo(es) a dehydrogenation reaction during a turn of the fatty acid spiral?

14.36 Which compound(s) in Problem 14.34 undergo(es) a chain-cleavage reaction during a turn of the fatty acid spiral?

14.37 How many turns of the fatty acid spiral would be needed to degrade each of the following fatty acids to acetyl CoA?
a. 16:0 fatty acid   b. 12:0 fatty acid

14.38 How many turns of the fatty acid spiral would be needed to degrade each of the following fatty acids to acetyl CoA?
a. 20:0 fatty acid   b. 10:0 fatty acid

14.39 The degradation of *cis*-3-hexenoic acid, a 6:1 acid, requires one more step than the degradation of hexanoic acid, a 6:0 acid. Describe the nature of this extra step.

14.40 The degradation of *cis*-4-hexenoic acid, a 6:1 acid, requires one more step than the degradation of hexanoic acid, a 6:0 acid. Describe the nature of this extra step.

■ **ATP Production from Fatty Acid Oxidation (Section 14.5)**

14.41 Identify the major fuel for skeletal muscle in
a. an active state   b. a resting state

14.42 Explain why fatty acids cannot serve as fuel for the brain.

14.43 Consider the conversion of a C_{10} saturated acid entirely to acetyl CoA.
a. How many turns of the fatty acid spiral are required?
b. What is the yield of acetyl CoA?
c. What is the yield of NADH?
d. What is the yield of $FADH_2$?
e. How many high-energy ATP bonds are consumed?

14.44 Consider the conversion of a C_{14} saturated acid entirely to acetyl CoA.
a. How many turns of the fatty acid spiral are required?
b. What is the yield of acetyl CoA?
c. What is the yield of NADH?
d. What is the yield of $FADH_2$?
e. How many high-energy ATP bonds are consumed?

14.45 What is the net ATP production for the complete oxidation to CO_2 and H_2O of the fatty acid in Problem 14.43?

14.46 What is the net ATP production for the complete oxidation to CO_2 and H_2O of the fatty acid in Problem 14.44?

14.47 Which yield more $FADH_2$, saturated or unsaturated fatty acids? Explain.

14.48 Which yield more NADH, saturated or unsaturated fatty acids? Explain.

14.49 Compare the energy released when 1 g of carbohydrate and 1 g of lipid are completely degraded in the body.

14.50 Compare the net ATP produced from 1 molecule of glucose and 1 molecule of hexanoic acid when they are completely degraded in the body.

■ **Ketone Bodies (Section 14.6)**

14.51 What three body conditions are conducive to ketone body formation?

14.52 Why does a deficiency of carbohydrates in the diet lead to ketone body formation?

14.53 What is the relationship between oxaloacetate concentration and ketone body formation?

14.54 What is the relationship between pyruvate concentration and ketone body formation?

14.55 Draw the structures of the three compounds classified as ketone bodies.

14.56 Two of the three ketone bodies can be synthesized from the third one. Write equations for the formation of these two compounds.

14.57 What is the primary site for ketone body formation?

14.58 What is the first reaction step in the process of ketogenesis?

14.59 What reaction step is necessary to activate the ketone body acetoacetate before it can be used as a fuel?

14.60 In what order are the compounds acetoacetyl CoA and 3-hydroxy-3-methylglutaryl CoA encountered in the process of using ketone bodies as fuel. Explain.

14.61 What is ketosis?

14.62 Severe ketosis situations produce acidosis. Explain.

■ **Biosynthesis of Fatty Acids (Section 14.7)**

14.63 Compare the locations of the enzymes for fatty acid biosynthesis and fatty acid degradation.

14.64 How does the structure of fatty acid synthase differ from that of the enzymes that degrade fatty acids?

14.65 Coenzyme A plays an important role in fatty acid degradation. What is its counterpart in fatty acid biosynthesis, and how does its structure differ from that of coenzyme A?

14.66 What does the designation ACP stand for?

14.67 What are the primary locations within the human body where lipogenesis occurs?

14.68 What is the starting material for lipogenesis?

14.69 What is the role of each of the following compounds in the citrate shuttle system associated with fatty acid biosynthesis?
   a. Oxaloacetate   b. Citrate

14.70 In the citric acid shuttle system associated with fatty acid biosynthesis, what molecule crosses mitochondrial membranes in the direction from
   a. cytosol to mitochondrial matrix?
   b. mitochondrial matrix to cytosol?

14.71 What is the role of malonyl ACP in fatty acid biosynthesis?

14.72 Write an equation for the reaction by which malonyl ACP is formed from acetyl ACP.

14.73 What type of reaction occurs in each of the four steps in the elongation of a fatty acid chain?

14.74 Why do almost all fatty acids in the human body contain an even number of carbon atoms?

14.75 In which step (of Steps 1 through 4) and in which cycle (first or second turn) of fatty acid biosynthesis is each of the following compounds encountered as a product?
   a.
$$CH_3-\overset{\overset{\displaystyle O}{\|}}{C}-CH_2-\overset{\overset{\displaystyle O}{\|}}{C}-S-ACP$$
   b.
$$CH_3-CH_2-CH_2-\overset{\overset{\displaystyle OH}{|}}{CH}-CH_2-\overset{\overset{\displaystyle O}{\|}}{C}-S-ACP$$
   c.
$$CH_3-CH_2-CH_2-CH=CH-\overset{\overset{\displaystyle O}{\|}}{C}-S-ACP$$
   d.
$$CH_3-CH_2-CH_2-\overset{\overset{\displaystyle O}{\|}}{C}-S-ACP$$

14.76 In which step (of Steps 1 through 4) and in which cycle (first or second turn) of fatty acid biosynthesis is each of the following compounds encountered as a product?
   a.
$$CH_3-CH_2-CH_2-CH_2-CH_2-\overset{\overset{\displaystyle O}{\|}}{C}-S-ACP$$
   b.
$$CH_3-CH=CH-\overset{\overset{\displaystyle O}{\|}}{C}-S-ACP$$
   c.
$$CH_3-\overset{\overset{\displaystyle OH}{|}}{CH}-CH_2-\overset{\overset{\displaystyle O}{\|}}{C}-S-ACP$$
   d.
$$CH_3-CH_2-CH_2-\overset{\overset{\displaystyle O}{\|}}{C}-CH_2-\overset{\overset{\displaystyle O}{\|}}{C}-S-ACP$$

14.77 Which of the compounds in Problem 14.75 is produced by a hydrogenation reaction?

14.78 Which of the compounds in Problem 14.76 is produced by a dehydration reaction?

14.79 What is the longest fatty acid that can be produced by the fatty acid synthase complex?

14.80 What central role does palmitic acid play in fatty acid biosynthesis?

14.81 What role does molecular oxygen, O_2, play in fatty acid biosynthesis?

14.82 What is the characteristic structural feature of an essential fatty acid?

14.83 Consider the biosynthesis of a C_{14} saturated fatty acid from acetyl CoA molecules.
   a. How many turns of the fatty acid biosynthetic pathway are needed?
   b. How many molecules of malonyl ACP must be formed?
   c. How many high-energy ATP bonds are consumed?
   d. How many NADPH molecules are needed?

14.84 Consider the biosynthesis of a C_{16} saturated fatty acid from acetyl CoA molecules.
   a. How many turns of the fatty acid biosynthetic pathway are needed?
   b. How many molecules of malonyl ACP must be formed?
   c. How many high-energy ATP bonds are consumed?
   d. How many NADPH molecules are needed?

■ **Biosynthesis of Cholesterol (Section 14.8)**

14.85 Approximately what percent of the total amount of cholesterol in your body is derived from the following?
   a. Your diet
   b. Biosynthesis

14.86 What is the starting material for the biosynthesis of cholesterol?

14.87 In each of the following pairs of intermediates in the biosynthetic pathway for cholesterol, specify which one is encountered first in the pathway.
   a. Mevalonate and squalene
   b. Isopentenyl pyrophosphate and lanosterol
   c. Lanosterol and squalene

14.88 In each of the following pairs of intermediates in the biosynthetic pathway for cholesterol, specify which one is encountered first in the pathway.
   a. Mevalonate and lanosterol
   b. Isopentenyl pyrophosphate and squalene
   c. Mevalonate and isopentenyl pyrophosphate

14.89 For each pair of compounds in Problem 14.87, tell whether the number of carbon atoms in the first compound is less than, the same as, or greater than the number of carbon atoms in the second compound.

14.90 For each pair of compounds in Problem 14.88, tell whether the number of carbon atoms in the first compound is less than, the same as, or greater than the number of carbon atoms in the second compound.

ADDITIONAL PROBLEMS

14.91 With which of these processes, (1) glycerol catabolism, (2) fatty acid spiral, (3) lipogenesis, or (4) ketogenesis, is each of the following molecules associated?
  a. Acyl CoA
  b. Enoyl CoA
  c. Malonyl ACP
  d. Dihydroxyacetone phosphate
  e. β-Hydroxybutyrate
  f. Acetoacetyl CoA

14.92 With which of these processes, (1) fatty acid catabolism, (2) lipogenesis, (3) ketogenesis, or (4) consumption of molecular O_2, is each of the following situations associated?
  a. Carnitine shuttle system
  b. Citrate shuttle system
  c. Fatty acid synthase complex
  d. Conversion of acetoacetyl CoA to HMG-CoA
  e. Conversion of squalene to cholesterol
  f. Conversion of a saturated fatty acid to an unsaturated fatty acid

14.93 Identify the step (among Steps 1 through 4) of the fatty acid chain elongation process in lipogenesis to which each of the following characterizations applies.
  a. Malonyl ACP is a reactant.
  b. CO_2 is a product.
  c. A dehydration reaction occurs.
  d. A carbon–carbon double bond is converted to a carbon–carbon single bond.

14.94 Indicate in what order the following events occur in the digestion of triacylglycerols (TAGs).
  (1) Bile emulsifies TAG "droplets."
  (2) TAGs incorporated into chylomicrons enter the lymph system.

  (3) TAGs are hydrolyzed to monoacylglycerols.
  (4) Free fatty acids are "repackaged" into TAGs.

14.95 Indicate whether each of the following statements is true or false.
  a. Chylomicrons are lipoproteins.
  b. Acetoacetate is an intermediate in the conversion of glycerol to dihydroxyacetone phosphate.
  c. The molecule carnitine is involved in fatty acid activation.
  d. One turn of the fatty acid spiral produces two molecules of ATP.

14.96 Indicate whether each of the following pairings of terms is correct or incorrect for reactions in the fatty acid spiral.
  a. Alkene functional group; dehydrogenation
  b. Ketone functional group; chain cleavage
  c. Alkane functional group; hydration
  d. Secondary alcohol functional group; oxidation

14.97 Indicate whether each of the following pairings of terms is correct or incorrect for reactions in the chain elongation phase of lipogenesis.
  a. Alkene functional group; hydrogenation
  b. Secondary alcohol group; dehydration
  c. Ketone group; reduction
  d. Ketone group; hydrogenation

14.98 Arrange the four molecules (1) glucose, (2) sucrose, (3) C_8 unsaturated fatty acid, and (4) C_{14} unsaturated fatty acid in order of increasing biochemical energy content (ATP production) per mole.

MULTIPLE-CHOICE PRACTICE TEST

14.99 Which of the following statements concerning digestion of dietary triacylglycerols in adults is *correct*?
  a. It begins in the mouth.
  b. It occurs to a small extent (10%) in the stomach.
  c. It occurs to a large extent (90%) in the stomach.
  d. It occurs only in the small intestine.

14.100 Monoacylglycerols are the predominant constituent in which of the following?
  a. Fatty acid micelles
  b. Chylomicrons
  c. Adipocytes
  d. Bile

14.101 The first stage of glycerol metabolism is the two-step conversion of glycerol to dihydroxyacetone phosphate. What is the intermediate in this process?
  a. Dihydroxyacetone
  b. Monohydroxyacetone phosphate
  c. Glycerol 3-phosphate
  d. 3-phosphoglycerate

14.102 In the oxidation of fatty acids, what is the molecule that shuttles the activated fatty acid across the inner mitochondrial membrane?
  a. CoA
  b. Acetyl CoA
  c. Carnitine
  d. Citrate

14.103 What is the first functional group change that occurs in the fatty acid spiral?
  a. Alkane to alkene
  b. Alkene to 2° alcohol

  c. Alkane to 2° alcohol
  d. 2° alcohol to ketone

14.104 Which of the following pairings of terms is *correct* for reactions in the fatty acid spiral?
  a. Alkene functional group; dehydrogenation
  b. Ketone functional group; chain cleavage
  c. Alkane functional group; hydration
  d. 2° alcohol functional group; hydrogenation

14.105 How many turns of the fatty acid spiral are needed to "process" a C_{16} fatty acid molecule?
  a. Seven
  b. Eight
  c. Fourteen
  d. Sixteen

14.106 Which of the following compounds is a *ketone body?*
  a. Carnitine
  b. Oxaloacetate
  c. Acetoacetate
  d. Acetyl CoA

14.107 What are the starting materials for the processes of ketogenesis and lipogenesis, respectively?
  a. Acetyl CoA and a fatty acid
  b. A fatty acid and acetyl CoA
  c. Acetyl CoA and acetyl CoA
  d. A fatty acid and a fatty acid

14.108 Which of the following is an intermediate in the process of lipogenesis?
  a. Isopentyl pyrophosphate
  b. Malonyl ACP
  c. Oxaloacetate
  d. Acetoacetate

15 Protein Metabolism

Fish, such as the Atlantic salmon, and other aquatic species process (eliminate) the nitrogen from protein in a manner different from that which occurs in human beings.

From an energy production standpoint, proteins supply only a small portion of the body's needs. With a normal diet, carbohydrates and fats supply 90% of the body's energy, and only 10% comes from proteins. However, despite its minor role in energy production, protein metabolism plays an important role in maintaining good health. The amino acids obtained from proteins are needed for both protein synthesis and synthesis of other nitrogen-containing compounds in the cell. In this chapter, we examine protein digestion, the oxidative degradation of amino acids, and amino acid biosynthesis.

15.1 Protein Digestion and Absorption

Protein digestion begins in the stomach rather than in the mouth because saliva contains no enzymes that affect proteins. Both protein denaturation (Section 9.15) and protein hydrolysis (Section 9.14) occur in the stomach. The partially digested protein (large polypeptides) passes from the stomach into the small intestine, where digestion is completed (Figure 15.1).

Proteins are denatured in the stomach by the hydrochloric acid present in gastric juice. The acid gives gastric juice a pH of between 1.5 and 2.0. The enzyme pepsin effects the hydrolysis of about 10% of peptide bonds in proteins, producing a variety of polypeptides. In the small intestine, *trypsin, chymotrypsin,* and *carboxypeptidase* in pancreatic juice attack peptide bonds. The pH of pancreatic juice is between 7.0 and 8.0, and it neutralizes the acidity of the material from the stomach. *Aminopeptidase,* secreted by

A very small number of people are unable to synthesize enough stomach acid, and these individuals must ingest capsules of dilute hydrochloric acid with every meal.

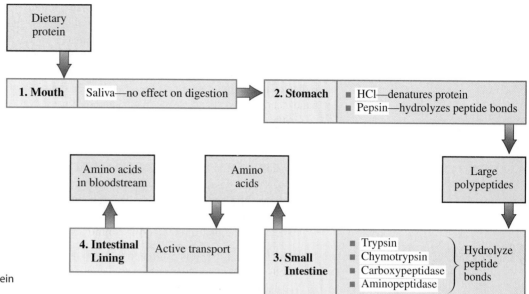

FIGURE 15.1 Summary of protein digestion in the human body.

The passage of polypeptide chains and small proteins across the intestinal wall is uncommon in adults. In infants, however, such transport allows the passage of antibodies (proteins) in colostral milk from a mother to a nursing infant to build up immunologic protection in the infant.

intestinal mucosal cells, also attacks peptide bonds. Pepsin, trypsin, chymotrypsin, carboxypeptidase, and aminopeptidase are all examples of *proteolytic* enzymes (Section 10.9). Enzymes of this type are produced in inactive forms called *zymogens* that are activated at their site of action (Section 10.9).

The net result of protein digestion is the release of the protein's constituent amino acids. Absorption of these "free" amino acids through the intestinal wall requires active transport with the expenditure of energy (Section 8.10). Different transport systems exist for the various kinds of amino acids. After passing through the intestinal wall, the free amino acids enter the bloodstream, which distributes them throughout the body.

15.2 Amino Acid Utilization

Amino acids produced from the digestion of proteins enter the amino acid pool of the body. The **amino acid pool** is *the total supply of free amino acids available for use in the human body.* Dietary protein is one of three sources that contributes amino acids to the amino acid pool. The other two sources are *protein turnover* and *biosynthesis* of amino acids in the liver.

Within the human body, proteins are continually being degraded (hydrolyzed) to amino acids and resynthesized. Disease, injury, and "wear and tear" are all causes of degradation. The degradation–resynthesis process is called protein turnover. **Protein turnover** is *the repetitive process in which proteins are degraded and resynthesized within the human body.*

The rate of protein turnover varies from a few minutes to several hours. Proteins with short turnover rates include many enzymes and regulatory hormones. In a healthy adult, about 2% of the body's protein is broken down and resynthesized every day.

Biosynthesis of amino acids by the liver also supplies the amino acid pool with amino acids. However, only the *nonessential* amino acids (Sections 9.2 and 15.6) can be produced in this manner.

In a healthy adult, the amount of nitrogen taken into the body each day (dietary proteins) equals the amount of nitrogen excreted from the body. Such a person is said to be in a state of nitrogen balance. **Nitrogen balance** is *the state that results when the amount of nitrogen taken into the human body as protein equals the amount of nitrogen excreted from the body in waste materials.*

Two types of nitrogen imbalance can occur. When protein degradation exceeds protein synthesis, the amount of nitrogen in the urine exceeds the amount of nitrogen ingested (dietary protein). This condition of *negative nitrogen balance* accompanies a state of "tissue wasting," because more tissue proteins are being catabolized than are being replaced by protein synthesis. Protein-poor diets, starvation, and wasting illnesses, for example, produce a negative nitrogen balance.

A *positive nitrogen balance* (nitrogen intake exceeds nitrogen output) indicates that the rate of protein anabolism (synthesis) exceeds that of protein catabolism. This state indicates that large amounts of tissue are being synthesized, such as during growth, pregnancy, and convalescence from an emaciating illness.

Although the overall nitrogen balance in the body often varies, the relative concentrations of amino acids within the amino acid pool remain essentially constant. No specialized storage forms for amino acids exist in the body, as is the case for glucose (glycogen) and fatty acids (triacylglycerols). Therefore, the body needs a relatively constant source of amino acids to maintain normal metabolism. During negative nitrogen balance, the body must resort to degradation of proteins that were synthesized for other functions.

The amino acids from the body's amino acid pool are used in four different ways.

Higher plants and certain microorganisms are capable of synthesizing all the protein amino acids from carbon dioxide, water, and inorganic salts.

There are approximately 100 grams of free amino acids present in the amino acid pool. Two amino acids, glutamic acid and glutamine, account for half of the amino acids present in the pool. The essential amino acids constitute approximately 10 grams of the pool.

1. *Protein synthesis.* It is estimated that about 75% of the free amino acids in a healthy, well-nourished adult go into protein synthesis. Proteins are continually needed to replace old tissue (protein turnover) and also to build new tissue (growth). The subject of protein synthesis was considered in Section 11.11.
2. *Synthesis of nonprotein nitrogen-containing compounds.* Amino acids are regularly withdrawn from the amino acid pool for the synthesis of nonprotein nitrogen-containing compounds. Such molecules include the purines and pyrimidines of nucleic acids, the heme of hemoglobin, neurotransmitters such as acetylcholine and serotonin, the choline and ethanolamine of phosphoglycerides, and hormones such as epinephrine.
3. *Synthesis of nonessential amino acids.* When required, the body draws on the amino acid pool for raw materials for the production of *nonessential* amino acids that are in short supply. The "roadblock" preventing the synthesis of the *essential* amino acids is not lack of nitrogen but lack of a correct carbon skeleton upon which enzymes can work. In general, the essential amino acids contain carbon chains or aromatic rings not present in other amino acids or the intermediates of carbohydrate or lipid metabolism. Table 15.1 lists the essential amino acids and the nonessential amino acids with the precursors needed to form the latter.
4. *Production of energy.* Because excess amino acids cannot be stored for later use, the body's response is to degrade them. The degradation process is complex because each of the 20 standard amino acids has a different degradation pathway.

In all the degradation pathways, the amino nitrogen atom is removed and converted to ammonium ion, which ultimately is excreted from the body as urea. The remaining

TABLE 15.1
Essential and Nonessential Amino Acids

Nutritionally Essential Amino Acids	Nutritionally Nonessential Amino Acids	
	Amino Acid	Precursor
histidine	alanine	pyruvate
isoleucine	arginine	glutamate
leucine	asparagine	aspartate
lysine	aspartic acid	oxaloacetate
methionine	cysteine	serine
phenylalanine	glutamic acid	α-ketoglutarate
threonine	glutamine	glutamate
tryptophan	glycine	serine
valine	proline	glutamate
	serine	3-phosphoglycerate
	tyrosine	phenylalanine

FIGURE 15.2 Possible fates for amino acid degradation products.

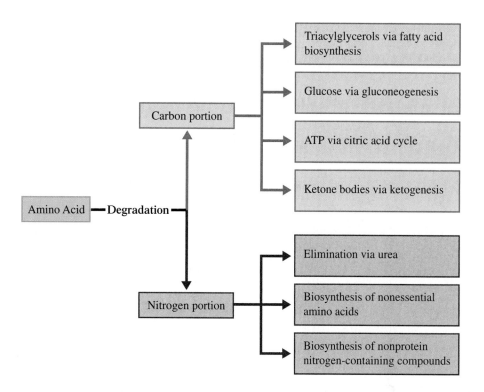

carbon skeleton is then converted to pyruvate, acetyl CoA, or a citric acid cycle intermediate, depending on its makeup, with the resulting energy production or energy storage. Figure 15.2 shows the various pathways available for the products of amino acid catabolism. Subsequent sections of this chapter give further details about these processes.

15.3 Transamination and Oxidative Deamination

Degradation of an amino acid has two stages: (1) the removal of the α-amino group and (2) the degradation of the remaining carbon skeleton. In this section and the next, we consider what happens to the amino group; in Section 15.5, the fate of the carbon skeleton is considered.

The release of an amino group from most amino acids requires a two-step process involving *transamination* followed by *oxidative deamination*. The following two procedures will make these processes easier to visualize.

1. Draw amino acid structures in the general format

$$\overset{\overset{+}{N}H_3}{\underset{|}{R-CH-COO^-}}$$

Remember that the ordering of the four groups attached to the carbon in an amino acid is not critical except in stereochemical considerations (Fischer projections; Section 9.3).

2. Review the structural relationships among six molecules—three pairs of keto/amino acids. In Section 5.5, we noted that the derivatives of three carboxylic acids—propionic (a three-carbon monoacid), succinic (a four-carbon diacid) and glutaric (a five-carbon diacid)—are particularly important in metabolic reactions. It is the α-keto and α-amino derivatives of these three acids that are the "key players"

FIGURE 15.3 Key compounds in the transamination/oxidative deamination process include three keto acid/amino acid pairs.

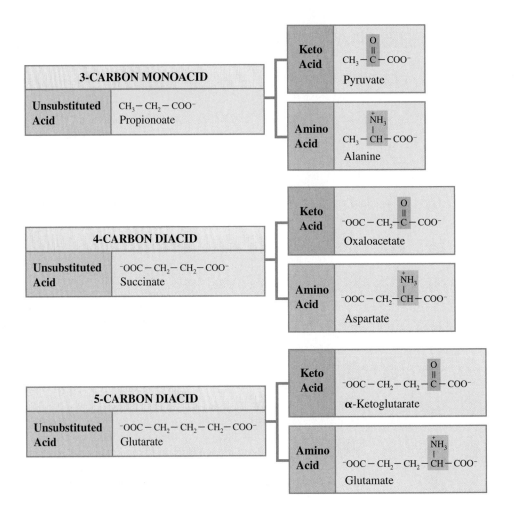

in the transamination/oxidative deamination process. Figure 15.3 gives the structural relationships among these compounds.

■ Transamination

A **transamination reaction** *is a biochemical reaction that involves the interchange of the amino group of an α-amino acid with the keto group of an α-keto acid.* The general equation for a transamination reaction is

$$\underset{\alpha\text{-Amino acid}}{R-\overset{\overset{+}{N}H_3}{\underset{|}{C}}H-COO^-} + \underset{\alpha\text{-Keto acid}}{R'-\overset{O}{\overset{||}{C}}-COO^-} \longrightarrow \underset{\text{New }\alpha\text{-keto acid}}{R-\overset{O}{\overset{||}{C}}-COO^-} + \underset{\text{New }\alpha\text{-amino acid}}{R'-\overset{\overset{+}{N}H_3}{\underset{|}{C}}H-COO^-}$$

There are at least 50 transaminase enzymes associated with transamination reactions. Most have a specificity for α-ketoglutarate as the amino group acceptor. Glutamate is the amino acid product from the action of these α-ketoglutarate specific enzymes.

The purpose of transamination is to remove amino groups from the various α-amino acids and collect them in a single amino acid, glutamate. Glutamate then acts as the source of amino groups for continued nitrogen metabolism (excretion or biosynthesis).

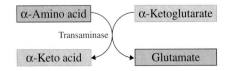

A specific example of this type of reaction is

$$CH_3\!-\!\overset{\overset{+}{N}H_3}{\underset{|}{CH}}\!-\!COO^- + {}^-OOC\!-\!CH_2\!-\!CH_2\!-\!\overset{\overset{O}{\parallel}}{C}\!-\!COO^- \xrightarrow{\text{Trans-}\atop\text{aminase}}$$

Alanine   α-Ketoglutarate

$$CH_3\!-\!\overset{\overset{O}{\parallel}}{C}\!-\!COO^- + {}^-OOC\!-\!CH_2\!-\!CH_2\!-\!\overset{\overset{+}{N}H_3}{\underset{|}{CH}}\!-\!COO^-$$

Pyruvate   Glutamate

A few transaminases are specific for the ketoacids pyruvate and oxaloacetate. They produce the amino acids alanine and aspartate, respectively. Ultimately, the alanine and aspartate so produced react with α-ketoglutarate, via transamination, to give glutamate. Such a "double" transamination sequence involving oxaloacetate would be diagrammed as follows:

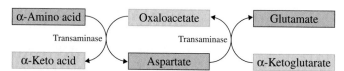

The net effect of transamination is to collect the amino groups from a variety of amino acids into a single compound—the amino acid glutamate—and to regenerate pyruvate and oxaloacetate for use in further transamination reactions.

Although the transamination reaction appears to involve the simple transfer of a $-\overset{+}{N}H_3$ group between two molecules, the reaction involves several steps and requires the presence of pyridoxal phosphate, a coenzyme produced from pyridoxine (vitamin B_6).

The concentration of transaminases in blood is used to diagnose liver and heart disorders. Liver damage releases the enzyme alanine aminotransferase (ALT) into the blood. Aspartate aminotransferase (AST) is abundant in heart muscle, and increased blood levels of this enzyme indicate heart damage (myocardial infarction).

Pyridoxine
(vitamin B_6)

Pyridoxal phosphate
(coenzyme)

This coenzyme is an integral part of the transamination process. The amino group of the amino acid is transferred first to the pyridoxal phosphate and then from the pyridoxal phosphate to the α-keto acid. Figure 15.4 shows the role of this coenzyme in the transamination process, where alanine is the amino acid and α-ketoglutarate is the α-keto acid.

Transamination reactions are reversible and can go easily in either direction, depending on the reactant concentrations. This reversibility is the basis for regulation of amino acid concentrations in the body.

■ Oxidative Deamination

In the second step of amino acid degradation, ammonium ion (NH_4^+) is liberated from the glutamate formed by transamination. This step involves oxidative deamination. An **oxidative deamination reaction** *is a biochemical reaction in which an α-amino acid is converted into an α-keto acid with release of an ammonium ion.* Oxidative deamination occurs primarily in liver and kidney mitochondria.

Oxidative deamination of glutamate requires the enzyme *glutamate dehydrogenase.* This enzyme is unusual in that it can function with either $NADP^+$ or NAD^+ as a coenzyme. With NAD^+ as the coenzyme, the reaction is

$${}^-OOC\!-\!CH_2\!-\!CH_2\!-\!\overset{\overset{+}{N}H_3}{\underset{|}{CH}}\!-\!COO^- + NAD^+ + H_2O \xrightarrow{\text{Glutamate}\atop\text{dehydrogenase}}$$

Glutamate

$$NH_4^+ + {}^-OOC\!-\!CH_2\!-\!CH_2\!-\!\overset{\overset{O}{\parallel}}{C}\!-\!COO^- + NADH + H^+$$

α-Ketoglutarate

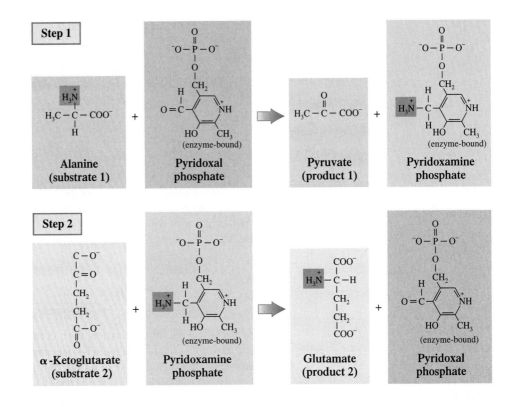

Note that α-ketoglutarate is a product of this process. It can be reused in the transamination process (first step). The NADH$^+$/H$^+$ formed can participate in the electron transport chain and oxidative phosphorylation to produce ATP molecules (Sections 12.7 and 12.8).

The sum of the transamination and deamination steps of the degradation of amino acids is

$$\alpha\text{-Amino acid} + NAD^+ + H_2O \longrightarrow \alpha\text{-keto acid} + NH_4^+ + NADH^+ + H^+$$

The NH$_4^+$ so produced, a toxic substance if left to accumulate in the body, is then converted to urea in the urea cycle (Section 15.4).

Two amino acids, serine and threonine, exhibit different behavior from the other amino acids. They undergo *direct deamination* by a dehydration–hydration process rather than *oxidative deamination*. This different behavior results from the presence of a side-chain β-hydroxyl group, a feature unique to these two acids. The direct deamination reaction for serine is

Threonine goes through a similar series of steps.

15.4 The Urea Cycle

From a nitrogen standpoint, the net effect of amino acid degradation is the production of ammonium ion. The accumulation of this ion in the body has potential toxic effects. Consequently, the ammonium ions are converted to urea, a less toxic nitrogen-containing compound, in the liver by a series of metabolic reactions called the urea cycle. The **urea cycle** *is the series of biochemical reactions in which urea is produced from ammonium ions and carbon dioxide.* The urea so produced is then transported in the blood from the liver to the kidneys and eliminated from the body in urine.

In the pure state, urea is a white solid with a melting point of 133°C. Its structure is

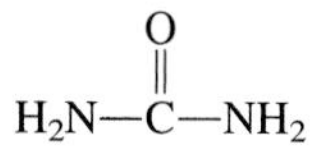

$$H_2N-\overset{\overset{\displaystyle O}{\|}}{C}-NH_2$$

Urea is very soluble in water (1 g per 1 mL), is odorless and colorless, and has a salty taste. (Urea does not contribute to the odor or color of urine.) With normal metabolism, an adult excretes about 30 g of urea daily in urine, although the exact amount varies with the protein content of the diet.

Three amino acids are involved as intermediates in the conversion of ammonium ions to urea through the urea cycle. These acids are arginine, ornithine, and citrulline, the latter two of which are nonstandard amino acids—that is, amino acids not found in protein. Structurally, all three of these amino acids have the same carbon chain.

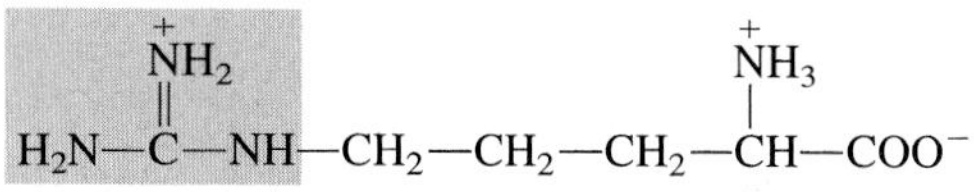

Arginine
(standard amino acid)

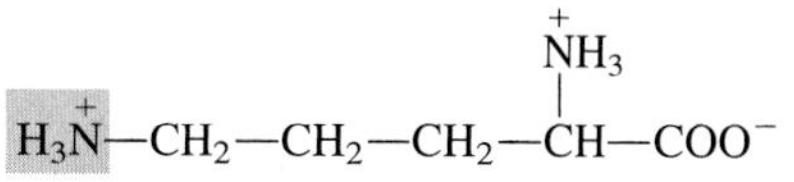

Ornithine
(nonstandard amino acid)

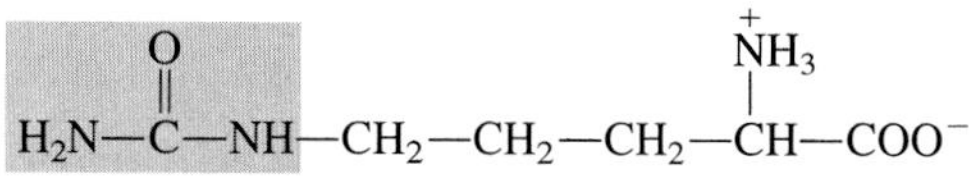

Citrulline
(nonstandard amino acid)

Carbamoyl Phosphate

The "fuel" for the urea cycle is the compound *carbamoyl phosphate.* This fuel is formed from ammonium ion (from oxidative deamination; Section 15.3), carbon dioxide (from the citric acid cycle), water, and two ATP molecules. The formation equation for carbamoyl phosphate is

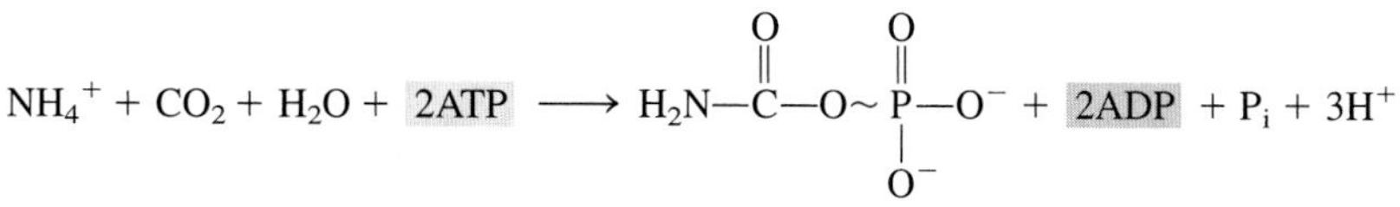

$$NH_4^+ + CO_2 + H_2O + \boxed{2ATP} \longrightarrow H_2N-\overset{\overset{\displaystyle O}{\|}}{C}-O\sim\overset{\overset{\displaystyle O}{\|}}{\underset{\underset{\displaystyle O^-}{|}}{P}}-O^- + \boxed{2ADP} + P_i + 3H^+$$

Carbamoyl phosphate

Note that two ATP molecules are expended in the formation of one carbamoyl phosphate molecule and that carbamoyl phosphate contains a high-energy phosphate bond. The carbamoyl phosphate formation reaction, like the reactions of the citric acid cycle, takes place in the mitochondrial matrix.

Steps of the Urea Cycle

Figure 15.5 shows the four-step urea cycle in outline form. Note that the urea cycle occurs partially in the mitochondria and partially in the cytosol and that ornithine and

---

The toxicity of ammonium ion is related to the oxidative deamination reaction by which it is formed, the conversion of glutamate to α-ketoglutarate. This reaction, which is an equilibrium situation, is shifted to the glutamate side by increased ammonium ion levels. This shift decreases α-ketoglutarate levels significantly, which affects the citric acid cycle of which α-ketoglutarate is an intermediate. Cellular ATP production drops, and the lack of ATP causes central nervous system problems.

Arginine is the most nitrogen-rich of the standard amino acids. It contains four nitrogen atoms.

The functional group attached to the phosphate in carbamoyl phosphate is the simple amide functional group

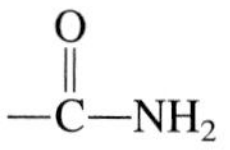

$$-\overset{\overset{\displaystyle O}{\|}}{C}-NH_2$$

The term *carbamoyl* is the *prefix* that denotes an amide group. Most often, amide groups are named by using the *suffix* system, in which case the suffix is *amide.*

FIGURE 15.5 The four-step urea cycle in which carbamoyl phosphate is converted to urea.

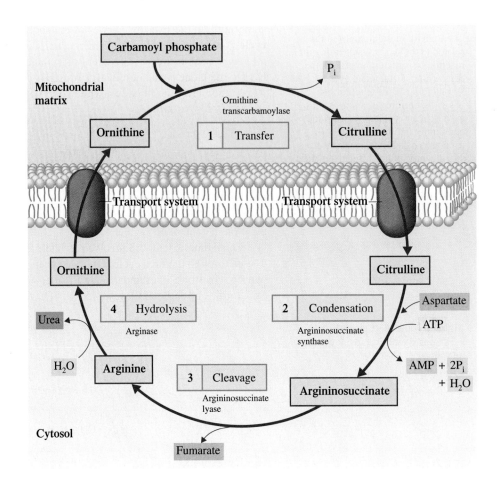

citrulline must be transported across the mitochondrial membrane. We will now consider in detail the individual steps of the urea cycle.

Step 1: *Carbamoyl group transfer.* The carbamoyl group of carbamoyl phosphate is transferred to ornithine to form citrulline in a reaction catalyzed by *ornithine transcarbamoylase.*

The standard amino acid lysine and the nonstandard amino acid ornithine, both basic amino acids, have closely related structures.

$$H_3\overset{+}{N}-(CH_2)_4-\overset{\overset{\displaystyle \overset{+}{N}H_3}{|}}{C}H-COO^-$$
Lysine

$$H_3\overset{+}{N}-(CH_2)_3-\overset{\overset{\displaystyle \overset{+}{N}H_3}{|}}{C}H-COO^-$$
Ornithine

Lysine has one more CH_2 group than does ornithine.

The breaking of the high-energy phosphate bond in carbamoyl phosphate drives the transfer process. With the carbamoyl transfer, the first of the two nitrogen atoms and the carbon atom needed for the formation of urea have been introduced into the cycle.

Step 2: *Citrulline–aspartate condensation.* Citrulline is transported into the cytosol where a condensation reaction between citrulline and aspartate (a standard amino acid) produces argininosuccinate. This condensation, catalyzed by *argininosuccinate synthase,* is driven by the expenditure of ATP.

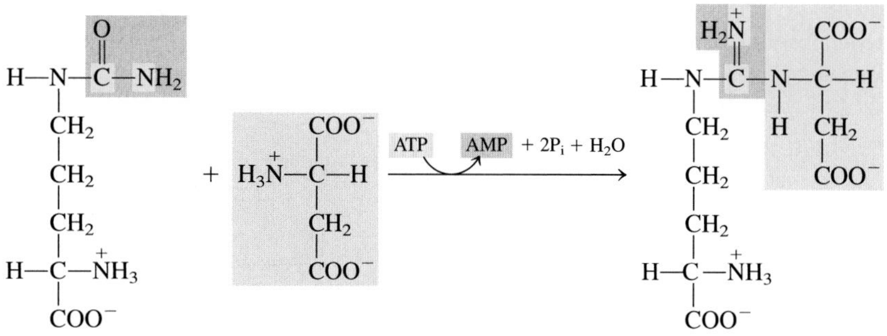

Citrulline   Aspartate   Argininosuccinate

With this reaction, the second of the two nitrogen atoms that will be part of the end-product urea has been introduced into the cycle. One nitrogen atom comes from carbamoyl phosphate, the other from aspartate.

However, the original source of both nitrogens is glutamate, the collecting agent for amino acid nitrogen atoms (Section 15.3). The flow of the nitrogen can be shown by these reactions:

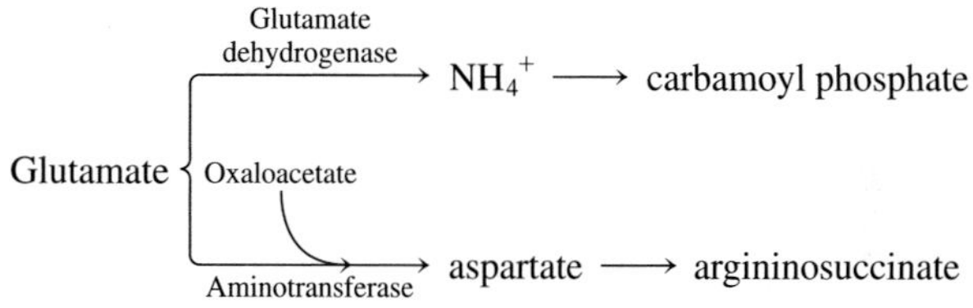

Step 3: *Argininosuccinate cleavage.* The enzyme *argininosuccinate lyase* catalyzes the cleavage of argininosuccinate into arginine, a standard amino acid, and fumarate, a citric acid cycle intermediate. The significance of this will be considered shortly.

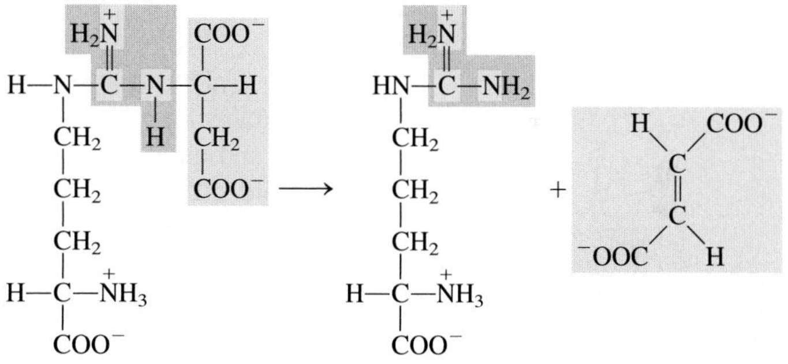

Argininosuccinate   Arginine   Fumarate

Step 4: *Urea from arginine hydrolysis.* Hydrolysis of arginine produces urea and regenerates ornithine, one of the cycle's starting materials. The enzyme involved is *arginase.*

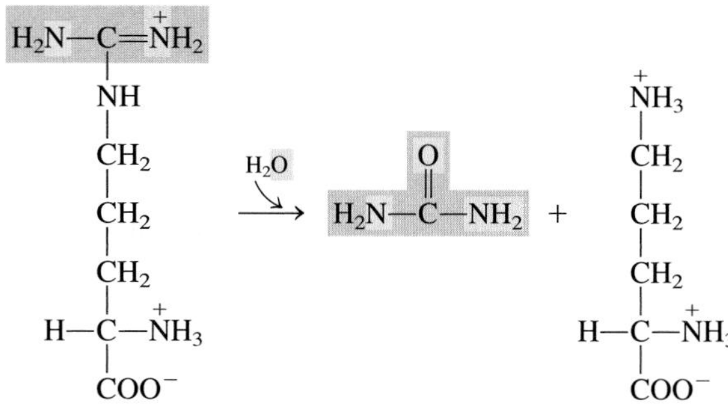

Arginine   Urea   Ornithine

The oxygen atom present in the urea comes from the water involved in the hydrolysis. The ornithine is transported back into the mitrochondria, where it becomes available to participate in the urea cycle again.

FIGURE 15.6 The nitrogen content of the various compounds that participate in the urea cycle.

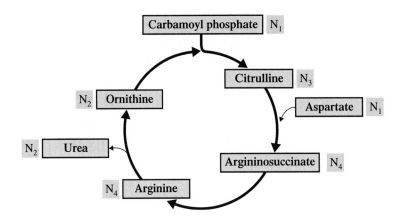

FIGURE 15.6 The nitrogen content of the various compounds that participate in the urea cycle.

Humans and most terrestrial animals excrete excess nitrogen as urea. Urea is not, however, the only biochemical means for disposing of excess nitrogen. Aquatic species (bacteria and fish) release ammonia directly into the surrounding water. Birds, terrestrial reptiles, and many insects secrete nitrogen as uric acid; it is the familar white solid in bird droppings. The structure of uric acid, a compound with a purine ring system (Section 6.9), is

Uric acid

Sulfur-containing amino acids (cysteine and methionine) contain both sulfur and nitrogen. The nitrogen-containing group is lost through transamination and processed to urea. The sulfur-containing group is processed to sulfur dioxide (SO_2), which is then oxidized to sulfate (SO_4^{2-}). The SO_4^{2-} ion, the negative ion from sulfuric acid, is eliminated in urine.

Figure 15.6 analyzes the urea cycle in terms of the nitrogen content of the various compounds that participate in it. The fuel, the N_1 carbamoyl phosphate, condenses with the N_2 ornithine to produce the N_3 citrulline. Next come two N_4 compounds, argininosuccinate and arginine. The N_4 arginine undergoes hydrolysis to produce the N_2 urea and regenerate the N_2 ornithine.

Urea Cycle Net Reaction

The net reaction for urea formation, in which all of the urea cycle intermediates cancel out of the equation, is

$$2NH_4^+ + CO_2 + 3ATP + 2H_2O + \text{aspartate} \longrightarrow$$
$$\text{urea} + 2ADP + AMP + 4P_i + \text{fumarate}$$

The equivalent of a total of four ATP molecules is expended in the production of one urea molecule. Two ATP molecules are consumed in the production of carbamoyl phosphate, and the equivalent of two ATP molecules is consumed in Step 2 of the urea cycle, where an ATP is hydrolyzed to AMP and PP_i and the PP_i is then further hydrolyzed to two P_i.

Linkage Between the Urea and Citric Acid Cycles

The net equation for urea formation shows fumarate, a citric acid cycle intermediate, as a product. This fumarate enters the citric acid cycle, where it is converted to malate and then to oxaloacetate, which can then be converted to aspartate through transamination. The aspartate then re-enters the urea cycle at Step 2 (see Figure 15.7).

Besides undergoing transamination, the oxaloacetate produced from fumarate of the urea cycle can be (1) converted to glucose via gluconeogenesis, (2) condensed with acetyl CoA to form citrate, or (3) converted to pyruvate.

15.5 Amino Acid Carbon Skeletons

The removal of the amino group of an amino acid by transamination or oxidative deamination (Section 15.3) produces an α-keto acid that contains the carbon skeleton from the amino acid. Each of the 20 amino acid carbon skeletons undergoes a different degradation process. For alanine and serine, the degradation requires a single step. For most carbon arrangements, however, multistep reaction sequences are required. We will not consider the details of these various degradation procedures in this text. It is important, however, to

CHEMICAL CONNECTIONS   The Chemical Composition of Urine

Urine is a dilute aqueous solution containing many solutes whose concentrations are dependent on the diet and state of health of the individual. On average, about 4 g of solutes are present in a 100-g urine sample; thus urine is an approximately 4%-by-mass aqueous solution of materials eliminated from the body.

The solutes present in urine are of two general types: organic compounds and inorganic ions. Generally, the organic compounds are more abundant because of the dominance of urea, as shown in the following composition data.

Major Constituents of Urine (for a 1400-mL specimen obtained over a 24-hour period)

Organic constituents		Inorganic constituents	
urea	25.0 g	chloride (Cl^-)	6.3 g
creatinine	1.5 g	sodium (Na^+)	3.0 g
amino acids	0.8 g	potassium (K^+)	1.7 g
uric acid	0.7 g	sulfate (SO_4^{2-})	1.4 g
		dihydrogen phosphate ($H_2PO_4^-$)	1.2 g
		ammonium (NH_4^+)	0.8 g
		calcium (Ca^{2+})	0.2 g
		magnesium (Mg^{2+})	0.2 g

Urea, the solute present in the greatest quantity in urine, is odorless and colorless in solution (Section 15.4). (The pale yellow color of urine is due to small amounts of urobilin and related compounds, as discussed in Section 15.7.) Urea is the principal nitrogen-containing end product of protein metabolism.

Creatinine, the second most abundant organic product in urine, is produced from the amino acids arginine, methionine, and glycine. Uric acid is a product of the metabolism of purines from nucleic acids.

The most abundant inorganic constituent of urine is chloride ion. Its primary source is dietary table salt (NaCl). Correspondingly, the second most abundant ion present is sodium ion, the positive ion in table salt. The sulfate ion present in urine comes primarily from the metabolism of sulfur-containing amino acids. Ammonium ions come primarily from the hydrolysis of urea.

Urine is normally slightly acidic, having an average pH value of 6.6. However, the pH range is wide—from 4.5 to 8.0. Fruits and vegetables in the diet tend to raise urine pH, and high-protein foods tend to lower urine pH.

A normal adult excretes 1000–1500 mL of urine daily. Actual urine volume depends on liquid intake and weather. During hot weather, urine volume decreases as a result of increased water loss through perspiration.

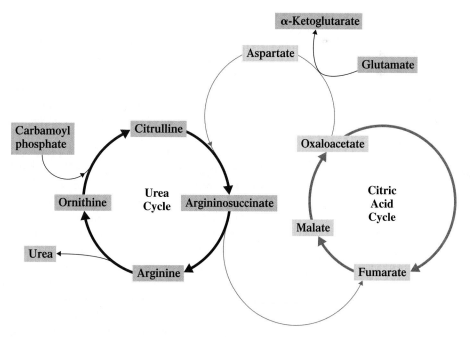

FIGURE 15.7 Fumarate from the urea cycle enters the citric acid cycle, and aspartate produced from oxaloacetate of the citric acid cycle enters the urea cycle.

The 20 standard amino acids are degraded by 20 different pathways that converge to produce just 7 metabolic intermediates.

consider the products of these degradation sequences. There are only seven, and each is a compound that we have previously encountered in our discussions of metabolism. The seven degradation products are pyruvate, acetyl CoA, acetoacetyl CoA, α-ketoglutarate, succinyl CoA, fumarate, and oxaloacetate. The last four products are intermediates in the citric acid cycle. Figure 15.8 relates these seven degradation products to the amino

CHEMICAL CONNECTIONS   Arginine, Citrulline, and the Chemical Messenger Nitric Oxide

A somewhat startling biochemical discovery made during the early 1990s was the existence within the human body of a *gaseous* chemical messenger, the simple diatomic molecule nitric oxide (NO). Its production involves two of the amino acid intermediates of the urea cycle—arginine and citrulline. Arginine reacts with oxygen to produce citrulline and NO. The reaction requires NADPH and the enzyme *nitric oxide synthase* (NOS).

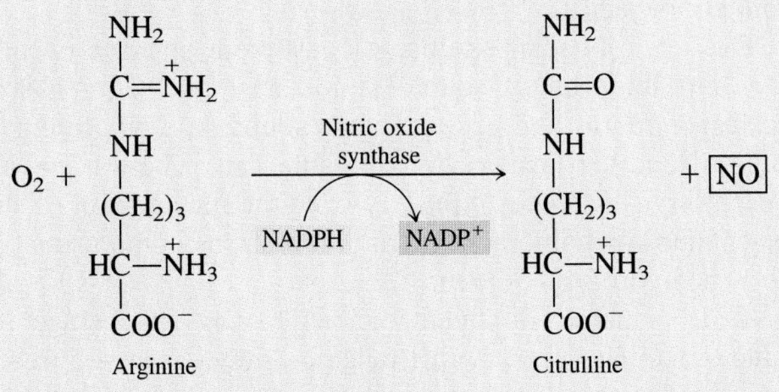

1. NO helps maintain blood pressure by dilating blood vessels.
2. NO is a chemical messenger in the central nervous system.
3. NO is involved in the immune system's response to invasion by foreign organisms or materials.
4. NO is found in the brain and may be a major biochemical component of long-term memory.

In humans, nitric oxide is the first known biochemical messenger compound that is a gas. It can easily pass through cell membranes by diffusion. No specific receptor or transport system is needed. Because of its extreme reactivity, NO exists for less than 10 seconds before undergoing reaction. This high reactivity prevents it from getting more than 1 millimeter from its site of synthesis.

The action of nitroglycerin, when it is used as a heart medication (for angina pectoris), is now known to be related to NO. Nitric oxide is the active metabolite from nitroglycerin.

Before the discovery of nitric oxide's role as a biochemical messenger, this gas was thought of mainly as a noxious atmospheric gas found in cigarette smoke and smog, as a destroyer of ozone, and as a precursor of acid rain. The contrast between nitric oxide's role in environmental pollution and its function in the human body as a chemical messenger is indeed startling.

Even though this reaction involves urea cycle intermediates, it is completely independent of the urea cycle.

Nitric oxide affects many kinds of cells and has particularly striking effects in the following areas:

acids from which they are obtained. Some amino acids appear in more than one box in Figure 15.8. This means either that there is more than one pathway for degradation or that some of the carbon atoms of the skeleton emerge as one product and others as another product.

Amino acids that are degraded to citric acid cycle intermediates can serve as glucose precursors and are called glucogenic. A **glucogenic amino acid** *is an amino acid that has a carbon-containing degradation product that can be used to produce glucose via gluconeogenesis.*

Amino acids that are degraded to acetyl CoA or acetoacetyl CoA can contribute to the formation of fatty acids or ketone bodies and are called ketogenic. A **ketogenic amino acid** *is an amino acid that has a carbon-containing degradation product that can be used to produce ketone bodies.* Even though acetyl CoA can enter the citric acid cycle, there can be no *net* production of glucose from it. Acetyl groups are C_2 species, and such species only maintain the carbon count in the cycle because two CO_2 molecules exit the cycle (Section 12.6). Thus amino acids that are degraded to acetyl CoA (or acetoacetyl CoA) are not glucogenic.

Amino acids that are degraded to pyruvate can be either glucogenic or ketogenic. Pyruvate can be metabolized to either oxaloacetate (glucogenic) or acetyl CoA (ketogenic).

Only two amino acids are purely ketogenic: leucine and lysine. Nine amino acids are both glucogenic and ketogenic: those degraded to pyruvate (see Figure 15.8), as well as tyrosine, phenylalanine, and isoleucine (which have two degradation products). The remaining nine amino acids are purely glucogenic.

Our discussion of glucogenicity and ketogenicity for amino acids points out that ATP production (common metabolic pathway) is not the only fate for amino acid degradation products. They can also be converted to glucose, ketone bodies, or fatty acids (via acetyl CoA).

FIGURE 15.8 Fates of the carbon skeletons of amino acids. Glucogenic amino acids are shaded blue, and ketogenic amino acids are shaded green. Some amino acids (marked with an asterisk) have more than one degradation pathway, and thus are present more than once in the diagram.

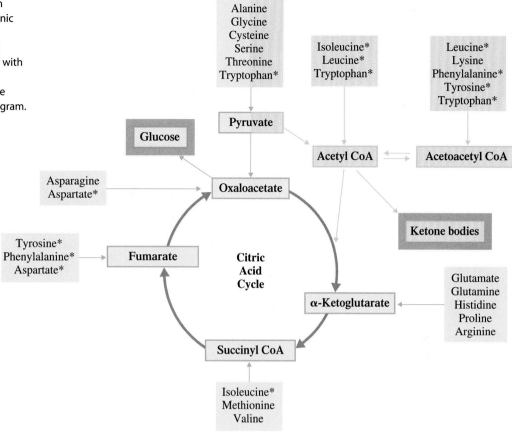

15.6 Amino Acid Biosynthesis

There is considerable variation in biosynthetic pathways for amino acids among different species. By contrast, the basic pathways of carbohydrate and lipid metabolism are almost universal.

The classification of amino acids as essential or nonessential for humans (Section 9.2) roughly parallels the number of steps in their biosynthetic pathways and the energy required for their synthesis. The nonessential amino acids can be made in 1–3 steps. The essential ones have biosynthetic pathways that require 7–10 steps, judging on the basis of observations of their synthesis in microorganisms. Most bacteria and plants can synthesize all the amino acids by pathways not present in humans. Plants, consumed as food, are the major source of the essential amino acids in humans and animals.

The starting materials for the biosynthesis of the 11 nonessential amino acids are the glycolysis intermediates 3-phosphoglycerate and pyruvate and the citric acid cycle intermediates oxaloacetate and α-ketoglutarate (see Figure 15.9).

Three of the nonessential amino acids—alanine, aspartate, and glutamate—are biosynthesized by transamination (Section 15.3) of the appropriate α-keto acid starting material.

FIGURE 15.9 A summary of the
starting materials for the biosynthesis
of the 11 nonessential amino acids.

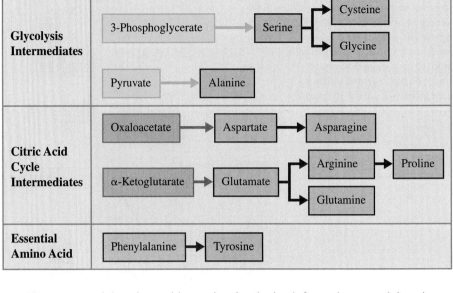

The nonessential amino acid tyrosine is obtained from the essential amino acid phenylalanine in a one-step oxidation that involves molecular O_2, NADPH, and the enzyme *phenylalanine hydroxylase*. Lack of this enzyme causes the metabolic disease phenylketonuria (PKU).

PKU is characterized by elevated blood levels of phenylalanine and phenylpyruvate. The physical consequence of PKU is damage to *developing* brain cells. In children up to six years old, PKU leads to retarded mental development. The major defense against PKU is mandatory screening of newborns to identify the one in every 20,000 who is afflicted and then restricting those children's dietary phenylalanine intake to that needed for protein synthesis until they are six years old. After that age, brain cells are not so susceptible to the toxic effect of phenylpyruvate.

15.7 Hemoglobin Catabolism

Red blood cells are highly specialized cells whose primary function is to deliver oxygen to, and remove carbon dioxide from, body tissues. Mature red blood cells have no nucleus or DNA. Instead, they are filled with the red pigment hemoglobin. Red blood cell formation occurs in the bone marrow, and approximately 200 billion new red blood cells are formed daily. The life span of a red blood cell is about 4 months.

The oxygen-carrying ability of red blood cells is due to the protein hemoglobin present in such cells (see Figure 15.10). Hemoglobin is a conjugated protein (Section 9.8); the protein portion is called *globin,* and the prosthetic group (nonprotein portion) is *heme.* Heme contains four pyrrole groups (Section 6.9) joined together with an iron atom in the center.

FIGURE 15.10 A molecular model of
the protein hemoglobin.

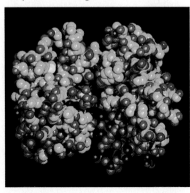

It is the iron atom in heme that interacts with O_2, forming a reversible complex with it. This complexation increases the amount of O_2 that the blood can carrry by a factor of 80 over that which simply "dissolves" in the blood.

The tetrapyrrole heme ring is the only component of hemoglobin that is not reused by the body.

Old red blood cells are broken down in the spleen (primary site) and liver (secondary site). Part of this process is degradation of hemoglobin. The globin protein is hydrolyzed to amino acids, which become part of the amino acid pool (Section 15.2). The iron atom of heme becomes part of *ferritin,* an iron-storage protein, which saves the iron for use in the biosynthesis of new hemoglobin molecules. The tetrapyrrole carbon arrangement of heme is degraded to *bile pigments* that are eliminated in feces and to a lesser extent in urine.

Degradation of heme begins with a ring-opening reaction in which a single carbon atom is removed. The product is called *biliverdin.*

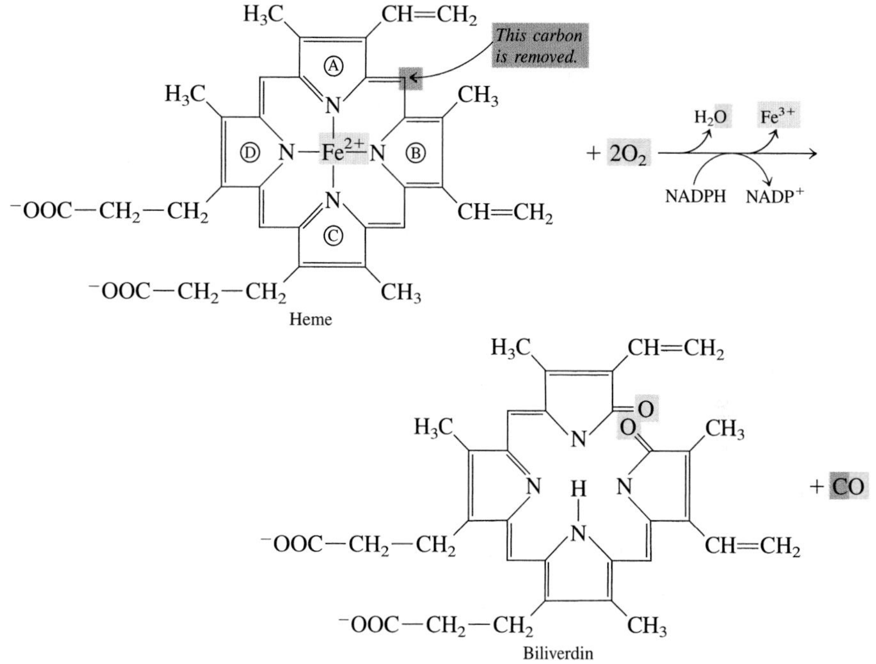

The level of carbon monoxide produced in the first step of hemoglobin degradation is sufficient to complex 1% of the oxygen-binding sites of the blood's hemoglobin.

This reaction has several important characteristics. (1) Molecular oxygen, O_2, is required as a reactant. (2) Ring opening releases the iron atom to be incorporated into ferritin. (3) The product containing the excised carbon atom is *carbon monoxide* (a substance toxic to the human body). The carbon monoxide so produced reacts with functioning hemoglobin, forming a CO–hemoglobin complex; this decreases the oxygen-carrying ability of the blood. CO–hemoglobin complexes are very stable; CO release to the lungs is a slow process.

An alternative rendering of the structure of biliverdin is

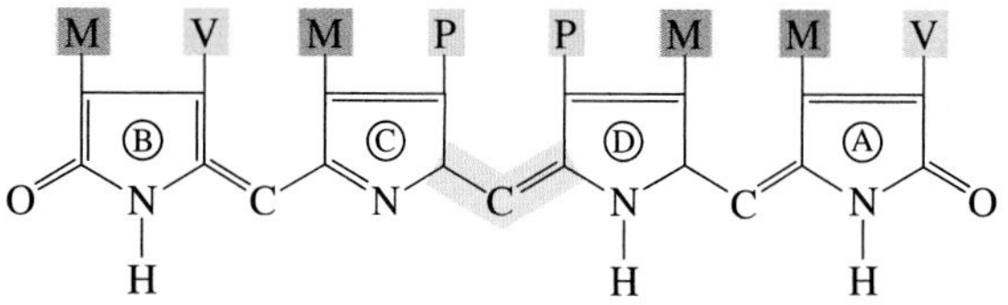

M = —CH₃ (methyl)
V = —CH=CH₂ (vinyl)
P = —CH₂—CH₂—COO⁻ (propionate)

This structure employs a notation, common in heme chemistry, in which letters are used to denote attachments to the pyrrole rings; such notation easily distinguishes the attachments. The structure's linear arrangement of pyrrole rings also saves space compared to the heme-like representation of the rings. However, the linear structure incorrectly implies that the arrangement of the pyrrole rings that results from the ring opening is linear (straight-line); rather, the pyrrole rings actually have a hemi-like arrangement.

In the second step of heme degradation, biliverdin is converted to bilirubin. This change involves reduction of the central methylene bridge of biliverdin.

In 2002 it was discovered that bilirubin has antioxidant properties. It protects against peroxyl radicals (Section 12.11) by being oxidized back to biliverdin. Its antioxidant properties are significantly better than those of glutathione (Section 9.7), the molecule believed for 80 years to be the most important cellular antioxidant.

Bilirubin is found only in low concentrations in cells but in higher concentrations in blood. This new research suggests that bilirubin is probably the major antioxidant protector for cell membranes, while glutathione protects components inside cells.

The first part of the names *biliverdin* and *bilirubin* and the last part of the names *stercobilin* and *urobilin* all come from the Latin *bilis,* which means "bile." As for the other parts of the names:

1. Latin *virdis* means "green"; biliverdin = "green bile."
2. Latin *rubin* means "red"; bilirubin = "red bile."
3. Latin *urina* means "urine"; urobilin = "urine bile."
4. Latin *sterco* means "dung"; stercobilin = "dung bile."

The change from heme to biliverdin to bilirubin usually occurs in the spleen. The bilirubin is then transported by serum albumin to the liver, where it is rendered more water-soluble by the attachment of sugar residues to its propionate side chains (P side chains). The solubilizing sugar is *glucuronide* (glucose with a —COO⁻ group on C-6 instead of a —CH₂OH group).

The solubilized bilirubin is excreted from the liver in bile, which flows into the small intestine. Here the bilirubin diglucuronide is changed, in a multistep process, to either stercobilin for excretion in feces or urobilin for excretion in urine. Both stercobilin and urobilin still have tetrapyrrole structures (Figure 15.11). Intestinal bacteria are primarily responsible for the changes that produce stercobilin and urobilin.

■ **Bile Pigments**

The tetrapyrrole degradation products obtained from heme are known as bile *pigments* because they are secreted with the bile (Section 14.1), and most of them are highly colored. A **bile pigment** *is a colored tetrapyrrole degradation product present in bile.*

FIGURE 15.11 Stercobilin and urobilin have structures closely resembling that of bilirubin. Changes include reduction of vinyl (V) groups to ethyl (E) groups and reduction of the —CH₂— bridge.

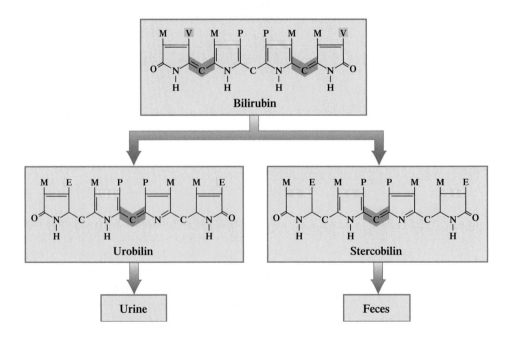

CHEMISTRY AT A GLANCE

Interrelationships Among Lipid, Carbohydrate, and Protein Metabolism

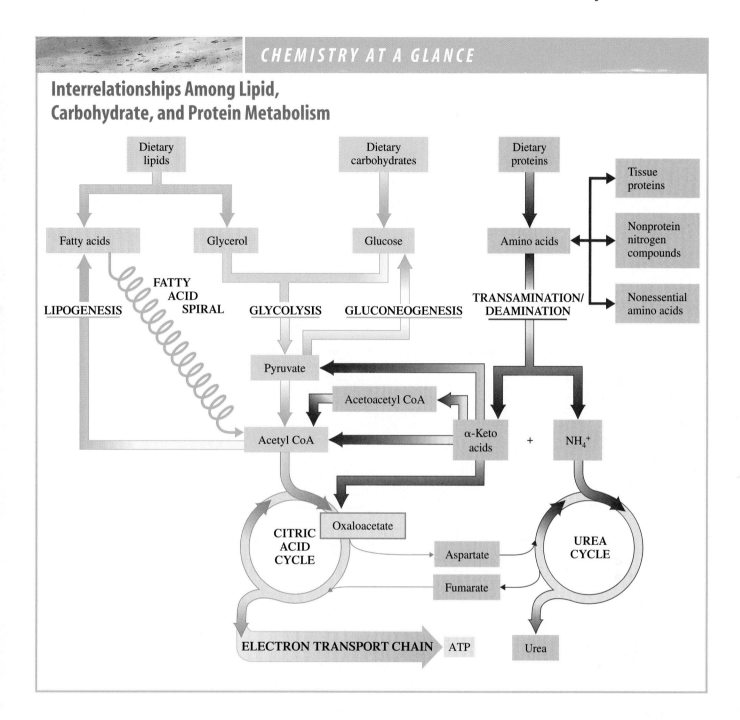

The word *jaundice* comes from the French *jaune,* which means "yellow."

A mild form of jaundice is common among premature infants because of underdeveloped liver function. Treatment involves the use of white or ultraviolet light, which breaks the bilirubin down to simpler compounds that are more easily excreted.

Biliverdin and bilirubin are, respectively, green and reddish orange in color. Stercobilin has a brownish hue and is the compound that gives feces their characteristic color. Urobilin is the pigment that gives urine its characteristic yellow color. Normally, the body excretes 1–2 mg of bile pigments in urine daily and 250–350 mg of bile pigments in feces daily.

When the body is functioning properly, the degradation of heme in the spleen to bilirubin and the removal of bilirubin from the blood by the liver balance each other.

Jaundice is the condition that occurs when this balance is upset such that bilirubin concentrations in the blood become higher than normal. The skin and the white of the eyes acquire a yellowish tint because of the excess bilirubin in the blood. Jaundice can occur as a result of liver diseases, such as infectious hepatitis and cirrhosis, that decrease the liver's ability to process bilirubin; from spleen malfunction, in which heme is degraded more rapidly than it can be adsorbed by the liver; and from gallbladder malfunction, usually from an obstruction of the bile duct.

The local coloration associated with a deep bruise is also related to the pigmentation associated with heme, biliverdin, and bilirubin. The changing color of the bruise as it heals reflects the dominant degradation product present at the time as the tissue repairs itself.

15.8 Interrelationships Among Metabolic Pathways

In this chapter and the previous two chapters, we have considered metabolic pathways of carbohydrates, lipids, and proteins. These pathways are not independent of each other but rather are integrally linked, as shown in the Chemistry at a Glance feature on page 519. The numerous connections among pathways mean that a change in one pathway can affect many other pathways.

A good illustration of the interrelationships among pathways emerges from comparing the processes of eating (feasting), not eating for a short period (fasting), and not eating for a prolonged period (starvation). Figure 15.12 shows how the body responds to each of these situations.

FIGURE 15.12 The human body's response to feasting, to fasting, and to starvation.

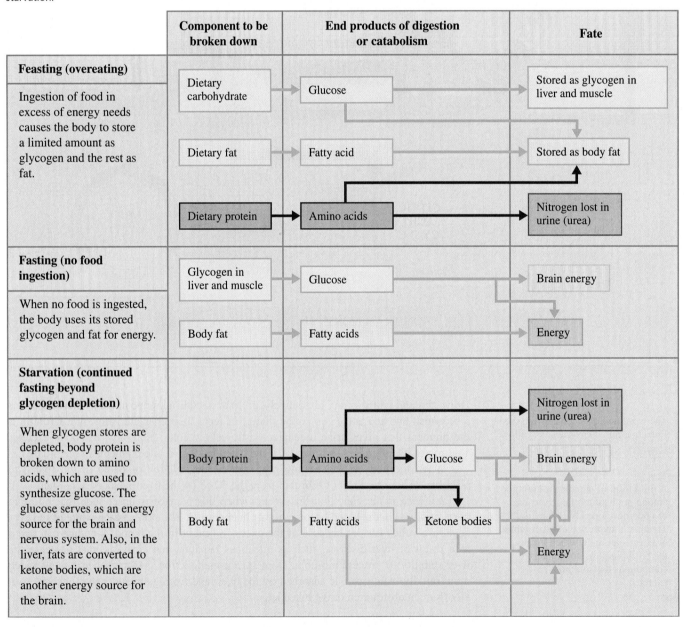

Protein digestion and absorption. Digestion of proteins involves the hydrolysis of the peptide bonds that link amino acids to each other. This process begins in the stomach and is completed in the small intestine. The amino acids released by digestion are absorbed through the intestinal wall into the bloodstream (Section 15.1).

Amino acid pool. The amino acid pool within cells consists of varying amounts of each of the 20 standard amino acids found in proteins (Section 15.2).

Amino acid utilization. Amino acids from the amino acid pool are used for protein synthesis, synthesis of nonprotein nitrogen compounds, synthesis of nonessential amino acids, and energy production (Section 15.2).

Transamination. A transamination reaction is an enzyme-catalyzed transfer of an amino group from an α-amino acid to an α-keto acid. Transamination is a step in obtaining energy from amino acids (Section 15.3).

Oxidative deamination. An oxidative deamination reaction is a reaction in which an α-amino acid is converted into an α-keto acid, accompanied by the release of a free ammonium ion. Oxidative deamination is a step in obtaining energy from amino acids (Section 15.3).

Urea cycle. The urea cycle is the metabolic pathway that converts ammonium ions into urea. This cycle processes the ammonium ions in the form of carbamoyl phosphate, a compound formed from CO_2, NH_4^+, ATP, and H_2O (Section 15.4).

Amino acid carbon skeletons. Amino acids are classified as glucogenic or ketogenic on the basis of their catabolic pathways. Glucogenic amino acids are degraded to intermediates of the citric acid cycle and can be used for glucose synthesis. Ketogenic amino acids are degraded into acetoacetyl CoA or acetyl CoA and can be used to make ketone bodies (Section 15.5).

Amino acid biosynthesis. Amino acid biosynthesis is the process in which the body synthesizes amino acids from intermediates of the glycolysis pathway and the citric acid cycle. Eleven amino acids can be synthesized by the body. The other nine amino acids, called essential amino acids, must be obtained from the diet (Section 15.6).

Hemoglobin catabolism. Hemoglobin from red blood cells undergoes a stepwise degradation to biliverdin, to bilirubin, and then to bile pigments that are excreted from the body (Section 15.7).

1. Digestion of protein (Section 15.1)

$$\text{Protein} \xrightarrow[\text{and HCl}]{\text{Pancreatic enzymes}} \text{amino acids}$$

2. Transamination (Section 15.3)

$$R_1\text{—}\overset{\overset{+}{NH_3}}{\underset{}{CH}}\text{—COO}^- + R_2\text{—}\overset{\overset{O}{\|}}{C}\text{—COO}^- \longrightarrow$$
$$R_1\text{—}\overset{\overset{O}{\|}}{C}\text{—COO}^- + R_2\text{—}\overset{\overset{+}{NH_3}}{\underset{}{CH}}\text{—COO}^-$$

3. Oxidative deamination (Section 15.3)

$$R\text{—}\overset{\overset{+}{NH_3}}{\underset{}{CH}}\text{—COO}^- + NAD^+ + H_2O \longrightarrow$$
$$R\text{—}\overset{\overset{O}{\|}}{C}\text{—COO}^- + NH_4^+ + NADH + H^+$$

4. Formation of urea (Section 15.4)

$$2\,NH_4^+ + CO_2 + 3ATP + 2H_2O + \text{aspartate} \longrightarrow$$
$$\text{urea} + \text{fumarate} + 2ADP + AMP + 4P_i$$

The members of each pair of problems in this section test similar material.

▓ Protein Digestion and Absorption (Section 15.1)

15.1 The first step in protein digestion is denaturation. Where does denaturation occur in the body, and what is the denaturant?

15.2 What is the first digestive enzyme that protein encounters, and where does this encounter take place?

15.3 What is the relationship between pepsinogen and pepsin?

15.4 What is the relationship between trypsinogen and trypsin?

15.5 Contrast gastric juice and pancreatic juice in terms of pH.

15.6 Contrast gastric juice and pancreatic juice in terms of enzymes present.

15.7 Absorption of amino acids through the intestinal wall requires a transport system. Explain.

15.8 The passage of small polypeptides through the intestinal wall is particularly important in infants. Explain.

15.9 What is the amino acid pool?

15.10 What are the three major sources of amino acids for the amino acid pool?

15.11 What is protein turnover?

15.12 The protein turnover rate is not the same for all proteins. Explain.

15.13 What is the difference between a positive nitrogen balance and a negative nitrogen balance?

15.14 What happens to the nitrogen balance during a period of fasting?

15.15 What happens to the nitrogen balance when the diet is lacking in one of the essential amino acids?

15.16 What happens to the nitrogen balance of a pregnant woman?

▓ Amino Acid Utilization (Section 15.2)

15.17 What four types of processes draw amino acids out of the amino acid pool?

15.18   What percent of amino acid utilization from the amino acid pool is for protein synthesis?

15.19   Classify each of the following amino acids as essential or nonessential.
   a. Lysine   b. Arginine
   c. Serine   d. Tryptophan

15.20   Classify each of the following amino acids as essential or nonessential.
   a. Proline   b. Asparagine
   c. Glutamic acid   d. Tyrosine

15.21   Which of the amino acids in Problem 15.19 can be biosynthesized in the human body?

15.22   Which of the amino acids in Problem 15.20 can be biosynthesized in the human body?

■ **Transamination and Oxidative Deamination (Section 15.3)**

15.23   In general terms, what are the two reactants in a transamination reaction?

15.24   In general terms, what are the two products in a transamination reaction?

15.25   Write structural equations for the transamination reactions that involve the following pairs of reactants.
   a. Threonine and pyruvate
   b. Alanine and oxaloacetate
   c. Glycine and α-ketoglutarate
   d. Threonine and α-ketoglutarate

15.26   Write structural equations for the transamination reactions that involve the following pairs of reactants.
   a. Threonine and oxaloacetate
   b. Glycine and pyruvate
   c. Alanine and oxaloacetate
   d. Isoleucine and α-ketoglutarate

15.27   What are the three α-keto acids that are usually reactants in transamination reactions?

15.28   The net effect of transamination is to collect the amino groups from a variety of α-amino acids into the compound glutamate. Explain.

15.29   What is the function of pyridoxal phosphate in transamination processes?

15.30   Which one of the B vitamins is important in the process of transamination?

15.31   Describe the process of oxidative deamination.

15.32   What coenzyme is required for an oxidative deamination reaction?

15.33   How does oxidative deamination differ from transamination?

15.34   What do the processes of oxidative deamination and transamination have in common?

15.35   Draw the structure of the α-keto acid produced from the oxidative deamination of each of the following amino acids.
   a. Glutamate   b. Cysteine
   c. Alanine   d. Phenylalanine

15.36   Draw the structure of the α-keto acid produced from the oxidative deamination of each of the following amino acids.
   a. Glycine   b. Leucine
   c. Aspartate   d. Tyrosine

15.37   The following α-keto acid can be used as a substitute for a particular *essential* amino acid in the diet. Explain how this is possible, and draw the structure of the essential amino acid.

$$CH_3\!-\!\underset{\underset{CH_3}{|}}{CH}\!-\!CH_2\!-\!\overset{\overset{O}{\|}}{C}\!-\!COO^-$$

15.38   The following α-keto acid can be used as a substitute for a particular *essential* amino acid in the diet. Explain how this is possible, and draw the structure of the essential amino acid.

$$CH_3\!-\!CH_2\!-\!\underset{\underset{CH_3}{|}}{CH}\!-\!\overset{\overset{O}{\|}}{C}\!-\!COO^-$$

15.39   Give the *name* of the compound produced from each reactant or the reactant needed to produce each product using transamination.
   a. Oxaloacetate $\longrightarrow$?   b. ? $\longrightarrow$ α-ketoglutarate
   c. Alanine $\longrightarrow$?   d. ? $\longrightarrow$ glutamate

15.40   Give the *name* of the compound produced from each reactant or the reactant needed to produce each product using transamination.
   a. Pyruvate $\longrightarrow$?   b. ? $\longrightarrow$ oxaloacetate
   c. Aspartate $\longrightarrow$?   d. ? $\longrightarrow$ alanine

■ **The Urea Cycle (Section 15.4)**

15.41   Draw the chemical structure of urea.

15.42   What are some of the physical characteristics of urea?

15.43   In what chemical form do ammonium ions enter the urea cycle?

15.44   What are the chemical reactants for the formation of carbamoyl phosphate?

15.45   What is a carbamoyl group?

15.46   Draw the structure of the molecule carbamoyl phosphate.

15.47   How do the structures of the three amino acids involved as intermediates in the urea cycle differ from each other?

15.48   Three amino acids are involved as intermediates in the urea cycle. Name them and classify them as standard or nonstandard amino acids.

15.49   Name the compound that enters the urea cycle by combining with
   a. ornithine.   b. citrulline.

15.50   Identify the first reaction of the urea cycle that occurs in the
   a. mitochondrial matrix.   b. cytosol.

15.51   What substance is the "fuel" for the urea cycle?

15.52   If the urea cycle were named in the same way as the citric acid cycle, what would the cycle's name be?

15.53   Characterize each of the following "urea cycle compounds" in terms of its nitrogen content (N_1, N_2, N_3, or N_4).
   a. Ornithine   b. Citrulline
   c. Aspartate   d. Argininosuccinate

15.54   Characterize each of the following "urea cycle compounds" in terms of its nitrogen content (N_1, N_2, N_3, or N_4).
   a. Carbamoyl phosphate   b. Ammonium ion
   c. Aspartate   d. Urea

15.55   In each of the following pairs of compounds associated with the urea cycle, specify which one is encountered first in the cycle.
   a. Citrulline and arginine
   b. Ornithine and aspartate
   c. Argininosuccinate and fumarate
   d. Carbamoyl phosphate and citrulline

15.56 In each of the following pairs of compounds associated with the urea cycle, specify which one is encountered first in the cycle.
 a. Carbamoyl phosphate and fumarate
 b. Argininosuccinate and arginine
 c. Ornithine and aspartate
 d. Citrulline and ATP

15.57 How much energy, in terms of ATP, is expended in the synthesis of a molecule of urea?

15.58 What are the sources of the carbon atom and the two nitrogen atoms in urea?

15.59 What is the fate of the fumarate formed in the urea cycle?

15.60 Explain how the urea cycle is linked to the citric acid cycle.

■ **Amino Acid Carbon Skeletons (Section 15.5)**

15.61 What are the four possible degradation products of the carbon skeletons of amino acids that are citric acid cycle intermediates?

15.62 What are the three possible degradation products of the carbon skeletons of amino acids that are not citric acid cycle intermediates?

15.63 With the help of Figure 15.7, write the name of the compound (or compounds) to which each of the following amino acid carbon skeletons is metabolized.
 a. Leucine   b. Isoleucine
 c. Aspartate   d. Arginine

15.64 With the help of Figure 15.7, write the name of the compound (or compounds) to which each of the following amino acid carbon skeletons is metabolized.
 a. Serine   b. Tyrosine
 c. Tryptophan   d. Histidine

15.65 What degradation characteristics do all purely glucogenic amino acids share?

15.66 What degradation characteristics do all purely ketogenic amino acids share?

■ **Amino Acid Biosynthesis (Section 15.6)**

15.67 What compound is a major source of amino groups in amino acid biosynthesis?

15.68 How does transamination play a role in both catabolism and anabolism of amino acids?

15.69 What are the five starting materials for the biosynthesis of the 11 nonessential amino acids?

15.70 What is a major difference between the biosynthesis pathways for the essential and the nonessential amino acids?

■ **Hemoglobin Catabolism (Section 15.7)**

15.71 What happens to the globin produced from the breakdown of hemoglobin?

15.72 What happens to the iron (Fe^{2+}) produced from the breakdown of hemoglobin?

15.73 What are the structural differences between heme and biliverdin?

15.74 What are the structural differences between biliverdin and bilirubin?

15.75 Arrange the following substances in the order in which they appear during the catabolism of heme: bilirubin, urobilin, biliverdin, and bilirubin diglucuronide.

15.76 Carbon monoxide is a by-product of the degradation of heme. At what point in the degradation process is it formed, and what happens to it once it is formed?

15.77 Which bile pigment is responsible for the yellow color of urine?

15.78 Which bile pigment is responsible for the brownish-red color of feces?

15.79 What chemical condition is responsible for jaundice?

15.80 What physical conditions cause jaundice?

■ **Interrelationships of Metabolic Pathways (Section 15.8)**

15.81 Briefly explain how the carbon atoms from amino acids can end up in ketone bodies.

15.82 Briefly explain how the carbon atoms from amino acids can end up in glucose.

15.83 How are the amino acids from protein "processed" when they are present in amounts that exceed the body's needs?

15.84 How are the amino acids from protein "processed" when an individual is in a state of starvation?

ADDITIONAL PROBLEMS

15.85 In which of the processes (1) urea cycle, (2) hemoglobin catabolism, and (3) transamination reactions would each of the following molecules be encountered?
 a. Citrulline   b. Carbon monoxide
 c. Pyruvate   d. Urobilin
 e. Arginine   f. Pyridoxal phosphate

15.86 Characterize each of the following molecules as a possible reactant, product, or enzyme of (1) transamination, (2) oxidative deamination, or (3) both transamination and oxidative deamination.
 a. Arginine   b. Glutamate
 c. α-Ketoglutarate   d. Ammonium ion
 e. Oxaloacetate   f. Glutamate dehydrogenase

15.87 Arrange the following events in the order in which they occur in the digestive process for proteins.
 (1) Peptide bonds are hydrolyzed with the help of pepsin.
 (2) Peptide bonds are hydrolyzed with the help of trypsin.

 (3) Large polypeptides pass from the stomach into the small intestine.
 (4) Amino acids pass through the intestinal wall into the bloodstream.

15.88 With the help of Figure 15.8 and the given conversion information, classify each of the following amino acids as (1) ketogenic but not glucogenic, (2) glucogenic but not ketogenic, (3) both ketogenic and glucogenic, or (4) neither ketogenic nor glucogenic.
 a. Alanine is converted to pyruvate.
 b. Aspartate is converted to either fumarate or oxaloacetate.
 c. Lysine is converted to acetoacetyl CoA.
 d. Isoleucine is converted to either succinyl CoA or acetyl CoA.

15.89 Indicate whether each of the following statements refers to *transamination* or *deamination*.
 a. Both an amino acid and a keto acid are reactants.
 b. An amino acid and water are reactants.

c. The ammonium ion is a product.

d. An amino acid is produced from a keto acid.

15.90 Which of the compounds (1) ornithine, (2) citrulline, (3) argininosuccinate, and (4) arginine is associated with each of the following urea cycle "occurrences"?

a. Reacts with carbamoyl phosphate.

b. Reacts with water to produce urea.

c. Reacts with aspartate.

d. Fumarate is a product of its "breakup."

15.91 Indicate whether each of the following statements is true or false.

a. Glutamate is the most abundant amino acid in the amino acid pool.

b. Pyruvate is a compound that participates in both the urea cycle and the citric acid cycle.

c. Citrulline, a participant in the urea cycle, is a nonstandard amino acid.

d. Glutamate is a reactant in oxidative deamination.

15.92 Which of the heme degradation products (1) bilirubin, (2) biliverdin, (3) stercobilin, and (4) urobilin is associated with each of the following heme degradation characterizations?

a. CO is produced at the same time as this substance.

b. The buildup of this substance in the blood produces jaundice.

c. Molecular O_2 is involved in the reaction that produces this substance.

d. This degradation product gives feces its characteristic color.

MULTIPLE-CHOICE PRACTICE TEST

15.93 Amino acid metabolism differs from that of carbohydrates and triacylglycerols in what way?

a. There is no storage form for amino acids in the body.

b. Amino acids cannot be used for energy production.

c. Amino acids cannot be converted to acetyl CoA.

d. All metabolic intermediates contain the element nitrogen.

15.94 Which of the following is not a use for the amino acids present in the body's amino acid pool?

a. Synthesis of proteins

b. Synthesis of nonprotein nitrogen-containing substances

c. Synthesis of nonessential amino acids

d. Synthesis of essential amino acids

15.95 Which of the following is always a product in a transamination reaction?

a. A keto acid   b. Glycerol

c. Ammonia   d. Ammonium ion

15.96 Which of the following is always a reactant in an oxidative deamination reaction?

a. Ammonium ion   b. Water

c. An amino acid   d. A keto acid

15.97 Which of the following statements concerning the compound urea is *incorrect*?

a. It is a white solid in the pure state.

b. It is very soluble in water.

c. It gives urine its odor and color.

d. Two —NH_2 groups are present in its structure.

15.98 Which of the following compounds is not a reactant in the formation of carbamoyl phosphate?

a. Carbon dioxide   b. Urea

c. Water   d. Ammonium ion

15.99 In which of the following pairs of amino acids are both members of the pair nonstandard amino acids?

a. Arginine and citrulline   b. Arginine and ornithine

c. Citrulline and ornithine   d. Aspartate and glutamate

15.100 Which of the following statements concerning amino acid "carbon skeleton" degradation is correct?

a. Each amino acid is degraded to a different product.

b. All amino acids are degraded to the same product.

c. Each amino acid is degraded by a different metabolic pathway.

d. All glucogenic amino acids are degraded to the same product.

15.101 Which of the following is produced in the first step of the degradation of the heme portion of hemoglobin?

a. Molecular O_2

b. Carbon monoxide

c. Individual pyrrole groups

d. Bilirubin

15.102 In which of the following is the compound citrulline encountered?

a. The urea cycle

b. Hemoglobin catabolism

c. Transamination reactions

d. Oxidative deamination reactions

Answers to Practice Exercises

Chapter 1 **1.1** (a) different conformations (b) different conformations (c) constitutional isomers **1.2** (a) 3,6-dimethyloctane (b) 3,4,4,5-tetramethyloctane

1.3

$$CH_3-CH_2-CH-\underset{\underset{\underset{CH_3CH_3}{|}}{\underset{CH_3CH_2}{|}}}{C}-\underset{\underset{CH_3}{|}}{C}-CH_2-CH_2-CH_3$$

1.4 (a) ③ ② ① ③ ③ $CH_3-CH-CH_2-CH_3$ $|$ CH_3

(b) ③ ① ② ① ② ③ ③ ③ $CH_3-CH-CH_2-CH-CH_2-CH_3$ $|$ CH_3 $|$ CH_3

1.5

(a) CH_2 CH_2 $CH-CH_2-CH_3$ CH_2-CH_2

(b) CH_3 CH_2 CH $CH-CH_2-CH_3$ CH_2 CH_2 CH CH_3

1.6 (a) methylcyclopropane (b) 1-ethyl-4-methylcyclohexane (c) 4-ethyl-1,2-dimethylcyclopentane **1.7** (a) not possible (b) not possible (c) (d) not possible

Cis isomer   *Trans* isomer

Chapter 2 **2.1** (a) 5-methyl-3-hexene (b) 3-ethyl-4-methylcyclohexene (c) 1,3-butadiene (d) 5-methyl-1,3-cyclopentadiene **2.2** (a) yes (b) no

2.3 (a) $CH_3-CH-CH_2-CH_3$; $|$ Cl (b) Br — CH_3

2.4 (a) $CH_3-CH_2-CH-CH_3$ $|$ Br (b)

(c) H_2 (d) $CH_3-\underset{\underset{CH_3}{|}}{\overset{\overset{Cl}{|}}{C}}-\overset{\overset{Cl}{|}}{C}H-CH_3$

2.5 (a) 1-bromo-3-propylbenzene (or *m*-bromopropylbenzene) (b) 1-chloro-4-propylbenzene (or *p*-chloropropylbenzene) (c) 3-phenylhexane (d) 4-bromo-1,2-dichlorobenzene

Chapter 3 **3.1** (a) 2,5-dimethyl-3-hexanol (b) 3-methyl-1-pentanol (c) 1,2-dimethylcyclopentanol (d) 3,4-dimethyl-1-heptano **3.2** (a) secondary (b) primary (c) secondary (d) secondary

3.3

(a) $CH_2-CH_2-CH_2-CH_3$ $|$ OH (b) $CH_3-\underset{\underset{CH_3CH_3}{|}}{\overset{\overset{OH}{|}}{C}}-CH-CH_3$

(c) $CH_3-CH_2-CH_2-OH$

3.4

(a) $CH_3-CH_2-\overset{\overset{O}{||}}{C}-H$, $CH_3-CH_2-\overset{\overset{O}{||}}{C}-OH$

(b) no reaction

(c) $CH_3-\overset{\overset{O}{||}}{C}-CH_2-CH_3$ (d)

3.5 (a) 1-propoxypropane (b) 1-methoxy-2-methylpropane (c) 1,3-dimethoxycyclohexane (d) methoxymethane **3.6** (a) methanethiol (b) 2-butanethiol (c) *tert*-butyl mercaptan (d) pentyl mercaptan

Chapter 4 **4.1** (a) 2-methylpropanal (b) 2-ethylpentanal (c) 2,3-dichlorobutanal (d) 2-methylbutanal **4.2** (a) 2-hexanone (b) 2,4-dimethyl-3-pentanone (c) cyclobutanone (d) 3-hydroxy-4-methylcyclohexanone

4.3

(a) $CH_3-\underset{\underset{CH_3}{|}}{C}H-\overset{\overset{O}{||}}{C}-H$ (b) $CH_3-CH_2-\overset{\overset{O}{||}}{C}-CH_3$

(c) $=O$ (d) $CH_3-\underset{\underset{CH_3}{|}}{\overset{\overset{CH_3}{|}}{C}}-\overset{\overset{O}{||}}{C}-H$

4.4 (a) yes (b) yes (c) no (d) yes

4.5 $CH_3-CH_2-CH_2-OH$ (alcohol), CH_3-CH_2-OH (alcohol),

$CH_3-\overset{\overset{O}{||}}{C}-CH_3$ (ketone)

Chapter 5 **5.1** (a) propanoic acid (b) 2,2-dimethylpropanoic acid (c) 2-ethylpentanoic acid

5.2

(a) $HO-\overset{\overset{O}{||}}{C}-CH_2-CH_2-CH_2-CH_2-\overset{\overset{O}{||}}{C}-OH$

(b) $CH_3-CH_2-\underset{\underset{Cl}{|}}{C}H-CH_2-\overset{\overset{O}{||}}{C}-OH$

(c) $HO-\overset{\overset{O}{||}}{C}-CH_2-\overset{\overset{O}{||}}{C}-OH$ (d) $CH_2-\overset{\overset{O}{||}}{C}-OH$

5.3 (a)

$H-\overset{\overset{O}{||}}{C}-OH + NaOH \longrightarrow H-\overset{\overset{O}{||}}{C}-O^-Na^+ + H_2O$

(b) $HO-\overset{\overset{O}{||}}{C}-CH_2-\overset{\overset{O}{||}}{C}-OH + 2KOH \longrightarrow$

$K^+{}^-O-\overset{\overset{O}{||}}{C}-CH_2-\overset{\overset{O}{||}}{C}-O^-K^+ + 2H_2O$

5.4 (a) ethyl ethanoate (IUPAC), ethyl acetate (common) (b) methyl pentanoate (IUPAC); methyl valerate (common) (c) propyl methanoate (IUPAC); propyl formate (common)

5.5

(a) $CH_3-CH_2-\overset{\overset{\displaystyle O}{\|}}{C}-O-CH_2-CH_2-CH_3 + H_2O \xrightarrow{H^+}$

$CH_3-CH_2-\overset{\overset{\displaystyle O}{\|}}{C}-OH + CH_3-CH_2-CH_2-OH$

(b) $CH_3-CH_2-\overset{\overset{\displaystyle O}{\|}}{C}-O-CH_2-CH_3 + KOH \xrightarrow{H_2O}$

$CH_3-CH_2-\overset{\overset{\displaystyle O}{\|}}{C}-O^-K^+ + CH_3-CH_2-OH$

(c) $CH_3-\overset{\overset{\displaystyle O}{\|}}{C}-OH + CH_3-CH_2-CH_2-OH \rightleftharpoons$

$CH_3-\overset{\overset{\displaystyle O}{\|}}{C}-O-CH_2-CH_2-CH_3 + H_2O$

Chapter 6   **6.1** (a) primary (b) secondary (c) primary (d) tertiary **6.2** (a) 3-hexanamine (b) N-propyl-1-propanamine (c) N,N-dimethyl-methanamine (d) N-methylanaline **6.3** (a) ethylmethylammonium ion (b) isopropylammonium ion (c) diethylmethylammonium ion (d) N-propylanilinium ion

6.4 (a) $CH_3-CH_2-\overset{\overset{\displaystyle +}{N}H}{\underset{\underset{\displaystyle CH_3}{|}}{}}-CH_3\ Cl^-$ (b) $CH_3-CH_2-\overset{+}{N}H_3\ HSO_4^-$

(c) $CH_3-CH_2-\underset{\underset{\displaystyle CH_3}{|}}{N}-CH_3 + NaBr + H_2O$

6.5 (a) α-bromobutyramide, 2-bromobutanamide (b) N-methylacetamide, N-methylethanamide (c) N,N-dimethylbenzamide (both common and IUPAC) (d) N-ethyl-β-methylbutyramide, N-ethyl-3-methylbutanamide

6.6

(a) $CH_3-\overset{\overset{\displaystyle O}{\|}}{C}-OH,\ CH_3-CH_2-NH-CH_3$

(b) $CH_3-CH_2-\overset{\overset{\displaystyle O}{\|}}{C}-OH,\ CH_3-NH_2$

(c) $\underset{\text{(benzene ring)}}{\bigcirc}-\overset{\overset{\displaystyle O}{\|}}{C}-OH,\ NH_3$

6.7

(a) $CH_3-\overset{\overset{\displaystyle O}{\|}}{C}-O^-Na^+,\ CH_3-NH_2$ (b) $CH_3-\overset{\overset{\displaystyle O}{\|}}{C}-OH,\ CH_3-\overset{+}{N}H_3\ Cl^-$

(c) $CH_3-\overset{\overset{\displaystyle O}{\|}}{C}-OH,\ CH_3-NH_2$ (d) $\underset{\text{(benzene ring)}}{\bigcirc}-\overset{\overset{\displaystyle O}{\|}}{C}-OH,\ NH_3$

Chapter 7   **7.1** (a) not a chiral center (b) not a chiral center (c) chiral center (d) not a chiral center

7.2 (a)

```
    CHO
HO──┼──H
 H──┼──OH
 H──┼──OH
    CH₂OH
```

(b)

```
    CH₂OH
    C═O
HO──┼──H
HO──┼──H
    CH₂OH
```

7.3 (a) D enantiomer (b) L entantiomer **7.4** (a) diastereomers (b) enantiomers (c) diastereomers **7.5** (a) ketohexose (b) aldohexose (c) aldotetrose (d) ketopentose

Chapter 8   **8.1** (a) MUFA (monounsaturated fatty acid) (b) 12:1 fatty acid (c) omega-3 fatty acid (ω-3) (d) delta-9 fatty acid (Δ^9) **8.2** (a) four products: glycerol and three fatty acids (b) four products: glycerol and three fatty acid salts (c) one product: a triacylglycerol in which all fatty acid residues are 18:0 residues

Chapter 9   **9.1** (a)

```
        H
        |
H₃N⁺──C──COO⁻
        |
        CH──CH₃
        |
        CH₃
```

(b)

```
        H
        |
H₂N──C──COO⁻
        |
        CH──CH₃
        |
        CH₃
```

(c)

```
        H
        |
H₃N⁺──C──COOH
        |
        CH──CH₃
        |
        CH₃
```

9.2 (a) toward positively charged electrode (b) toward negatively charged electron (c) isoelectric

9.3

```
        H   O   H   H   O   H   H
        |   ‖   |   |   ‖   |   |
H₃N⁺──C──C──N──C──C──N──C──COO⁻
        |           |           |
        CH──CH₃     CH₃         H
        |
        CH₃
```

Chapter 10   **10.1** (a) hydrolysis of maltose (b) removal of hydrogen from lactate ion (c) oxidation of fructose (d) rearrangement (isomerization) of maleate ion

Chapter 11   **11.1** 3′ T–T–A–C–G–T–C–G–A 5′ **11.2** 3′ A–U–U–G–G–A 5′ which becomes 5′ A–G–G–U–U–A 3′ **11.3** His–Pro–His–Thr–Val–Cys–Trp **11.4** (a) 3′ GCG–GCA–UCA–ACC–GGG–CCU–CCU 5′ (b) 3′ GCG–ACC–CCU–CCU 5′ which becomes 5′ UCC–UCC–CCA–GCG 3′ (c) Ser–Ser–His–Gly

Answers to Selected Exercises

Chapter 1 **1.1** (a) false (b) false (c) true (d) true **1.3** (a) meets (b) does not meet (c) does not meet (d) does not meet **1.5** Hydrocarbons contain C and H, and hydrocarbon derivatives contain at least one additional element besides C and H. **1.7** All bonds are single bonds in a saturated hydrocarbon, and at least one carbon–carbon multiple bond is present in an unsaturated hydrocarbon. **1.9** (a) saturated (b) unsaturated (c) unsaturated (d) unsaturated **1.11** (a) 18 (b) 4 (c) 13 (d) 22

1.13 (a) $CH_3-CH_2-CH_2-CH_3$

    (b) $CH_3-CH_2-\underset{\underset{CH_3}{|}}{CH}-CH_2-CH_3$

    (c) $CH_3-CH_2-\underset{\underset{CH_3}{|}}{CH}-CH_2-\underset{\underset{CH_3}{|}}{CH}-CH_3$

    (d) $CH_3-CH_2-\underset{\underset{\underset{\underset{CH_3}{|}}{CH_2}}{|}}{CH}-CH_2-CH_3$

1.15 (a) $CH_3-\underset{\underset{CH_3}{|}}{CH}-CH_2-CH_3$

    (b) $CH_3-\underset{\underset{CH_3}{|}}{CH}-\underset{\underset{CH_3}{|}}{CH}-\underset{\underset{CH_3}{|}}{CH}-CH_2-CH_3$

    (c) $CH_3-CH_2-CH_2-CH_2-CH_2-CH_3$

    (d) $CH_3-\underset{\underset{CH_3}{|}}{\overset{\overset{CH_3}{|}}{C}}-CH_2-CH_3$

1.17 (a)

    (b)

(c) $CH_3-(CH_2)_8-CH_3$ (d) C_6H_{14} **1.19** (a) different compounds that are not constitutional isomers (b) different compounds that are constitutional isomers (c) different conformations of the same molecule (d) different compounds that are constitutional isomers **1.21** (a) seven-carbon chain (b) eight-carbon chain (c) eight-carbon chain (d) seven-carbon chain **1.23** (a) 2-methylpentane (b) 2,4,5-trimethylheptane (c) 3-ethyl-2,3-dimethylpentane (d) 3-ethyl-2,4-dimethylhexane (e) decane (f) 4-propylheptane **1.25** horizontal chain, because it has more substituents (two)

1.27 (a) $CH_3-\underset{\underset{CH_3}{|}}{CH}-CH_2-CH_3$

    (b) $CH_3-CH_2-\underset{\underset{CH_3}{|}}{CH}-\underset{\underset{CH_3}{|}}{CH}-CH_2-CH_3$

(c) $CH_3-CH_2-\underset{\underset{\underset{\underset{CH_3}{|}}{CH_2}}{|}}{C}-CH_2-CH_3$

(d) $CH_3-\underset{\underset{CH_3}{|}}{CH}-\underset{\underset{CH_3}{|}}{CH}-\underset{\underset{CH_3}{|}}{CH}-\underset{\underset{CH_3}{|}}{CH}-CH_2-CH_3$

(e) $CH_3-CH_2-\underset{\underset{\underset{\underset{CH_3}{|}}{CH_2}}{|}}{CH}-CH_2-\underset{\underset{\underset{\underset{CH_3}{|}}{CH_2}}{|}}{CH}-CH_2-CH_2-CH_3$

(f) $CH_3-CH_2-CH_2-\underset{\underset{\underset{\underset{\underset{\underset{CH_3}{|}}{CH_2}}{|}}{CH_2}}{|}}{CH}-CH_2-CH_2-CH_2-CH_2-CH_3$

1.29 (a) 1, 1 (b) 2, 2 (c) 2, 2 (d) 4, 4 (e) 2, 2 (f) 1, 1 **1.31** (a) carbon chain numbered from wrong end; 2-methylpentane (b) not based on longest carbon chain; 2,2-dimethylbutane (c) carbon chain numbered from wrong end; 2,2,3-trimethylbutane (d) not based on longest carbon chain; 3,3-dimethylhexane (e) carbon chain numbered from wrong end and alkyl groups not listed alphabetically; 3-ethyl-4-methylhexane (f) like alkyl groups listed separately; 2,4-dimethylhexane

1.33 (a) $C-\underset{\underset{C}{|}}{C}-C-C-C-C-C$ (b) $C-\underset{\underset{C}{|}}{C}-\underset{\underset{C}{|}}{C}-C-C$

    (c) $C-C-\underset{\underset{C}{|}}{C}-C-C$ (d) $C-\underset{\underset{C}{|}}{C}-C-C-\underset{\underset{\underset{\underset{C}{|}}{C-C}}{|}}{C}-C-C$

1.35 (a) $CH_3-\underset{\underset{CH_3}{|}}{CH}-\underset{\underset{CH_3}{|}}{CH}-\underset{\underset{CH_3}{|}}{CH}-CH_3$

    (b) $CH_3-\underset{\underset{CH_3}{|}}{CH}-\underset{\underset{\underset{\underset{CH_3}{|}}{CH_2}}{|}}{CH}-CH_2-CH_3$

    (c) $CH_3-CH_2-\underset{\underset{CH_3}{|}}{CH}-\underset{\underset{\underset{\underset{CH_3}{|}}{CH_2}}{|}}{CH}-CH_2-CH_2-CH_3$

    (d) $CH_3-\underset{\underset{CH_3}{|}}{CH}-CH_2-CH_2-\underset{\underset{\underset{\underset{CH_3}{|}}{CH_2}}{|}}{CH}-CH_2-CH_2-CH_3$

1.37 (a) constitutional isomers (b) same compound

1.39 a.

    b.

c.

d.

e.

f.

1.41 (a) 2-methyloctane (b) 2,3-dimethylhexane (c) 3-methylpentane (d) 5-isopropyl-2-methyloctane   **1.43** (a) C_8H_{18} (b) C_9H_{20} (c) $C_{10}H_{22}$ (d) $C_{11}H_{24}$ **1.45** (a) 3, 2, 1, 0 (b) 5, 2, 3, 0 (c) 5, 2, 1, 1 (d) 5, 2, 3, 0 (e) 2, 8, 0, 0 (f) 3, 6, 1, 0   **1.47** (a) isopropyl (b) isobutyl (c) isopropyl (d) *sec*-butyl   **1.49**

(a) CH_3—CH_2—CH_2—CH_2—CH—CH_2—CH_2—CH_2—CH_2—CH_3
                                   |
                                  CH—CH_3
                                   |
                                  CH_2
                                   |
                                  CH_3

(b)
                                  CH_3
                                   |
                                  CH—CH_3
                                   |
CH_3—CH_2—CH_2—C—CH_2—CH_2—CH_2—CH_3
                                   |
                                  CH—CH_3
                                   |
                                  CH_3

(c) CH_3—CH—CH—CH_2—CH—CH_2—CH_2—CH_2—CH_3
             | | |
            CH_3 CH_3 CH_2
                                   |
                                  CH—CH_3
                                   |
                                  CH_3

(d) CH_3—CH_2—CH_2—CH—CH_2—CH_2—CH_2—CH_3
                          |
                         CH_3—C—CH_3
                                |
                               CH_3

1.51 (a) 16 (b) 6 (c) 5 (d) 15   **1.53** (a) C_6H_{12} (b) C_6H_{12} (c) C_4H_8 (d) C_7H_{14} **1.55** (a) cyclohexane (b) 1,2-dimethylcyclobutane (c) methylcyclopropane (d) 1,2-dimethylcyclopentane   **1.57** (a) must locate methyl groups with numbers (b) wrong numbering system for ring (c) no number needed (d) wrong numbering system for ring

1.59 (a) CH_2—CH_2—CH_3 (b) CH_3
                                              |
                                             CH—CH_3

(c) CH_2—CH_3
    CH_2—CH_3
    H
    H

(d) CH_2—CH_3
    H H
    H
    CH_2—CH_2—CH_3

1.61 (a) not possible

(b) CH_3—CH_2 CH_2—CH_3 CH_3—CH_2 H
            H   H   H CH_2—CH_3
            cis   *trans*

(c) not possible

(d) CH_3 CH_3
    CH_3 CH_3
        H   H
        H CH_3
    cis   *trans*

1.63 boiling point   **1.65** (a) octane (b) cyclopentane (c) pentane (d) cyclopentane   **1.67** (a) different states (b) same states (c) same states (d) same states   **1.69** (a) CO_2 and H_2O (b) CO_2 and H_2O (c) CO_2 and H_2O (d) CO_2 and H_2O **1.71** CH_3Br, CH_2Br_2, $CHBr_3$, CBr_4

1.73 (a) CH_3—CH_2
                        |
                       Cl

(b) CH_2—CH_2—CH_2—CH_3 CH_3—CH—CH_2—CH_3
      | |
     Cl   Cl

(c)   (d)   Cl
CH_2—CH—CH_3 CH_3—C—CH_3
  | | |
 Cl CH_3 CH_3

1.75 (a) iodomethane, methyl iodide (b) 1-chloropropane, propyl chloride (c) 2-fluorobutane, *sec*-butyl fluoride (d) chlorocyclobutane, cyclobutyl chloride

1.77 (a)   Cl   (b)   F   F
         H—C—Cl   F—C—C—F
              | | |
             Cl   Cl Cl

(c) CH_3—CH—Br   (d)   Br
              | H
             CH_3 H
                       Cl

1.79 (a) 16 (b) 6 (c) 5 (d) 22 (e) liquid (f) less dense (g) insoluble (h) flammable   **1.80** (a) no (b) yes (c) no (d) no

1.81 (à) F   H   (b) H CH_3
         H   F   Cl   H

(c) CH_3 (d) CH_3—CH—CH_2—I
    CH_3—C—Br |
         | CH_3
        CH_3

1.82 5-ethyl-2,2,6-trimethyl-4-(1,1-dimethylethyl)octane   **1.83** (a) $C_{18}H_{38}$ (b) C_7H_{14} (c) $C_7H_{14}F_2$ (d) $C_6H_{10}Br_2$

1.84 (a) CH_3 (b) CH_3—CH—CH_3
         CH_3—C—CH_2—CH_2—CH_2—CH_3 |
              | CH_3
             CH_3

(c)   Cl   (d)   Cl
    Cl—C—H   Cl—C—F
         | |
        H   F

1.85 (a) alkane (b) halogenated cycloalkane (c) halogenated alkane (d) cycloalkane

1.86 (a)

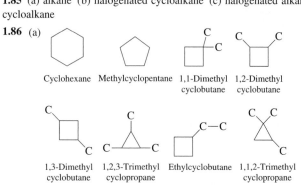

Cyclohexane   Methylcyclopentane   1,1-Dimethyl cyclobutane   1,2-Dimethyl cyclobutane

1,3-Dimethyl cyclobutane   1,2,3-Trimethyl cyclopropane   Ethylcyclobutane   1,1,2-Trimethyl cyclopropane

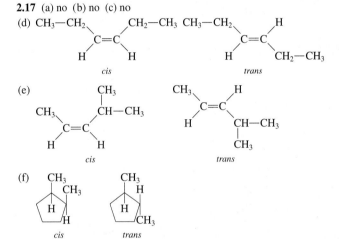

1-Ethyl-2-methyl cyclopropane   1-Ethyl-1-methyl cyclopropane   Propylcyclopropane

Isopropylcyclopropane

(b)

Hexane   2-Methylpentane

3-Methylpentane   2,3-Dimethyl butane   2,2-Dimethyl butane

(c)

1,1-Dibromo propane   2,2-Dibromo propane   1,2-Dibromo propane   1,3-Dibromo propane

1.87 (a) 1,2-diethylcyclohexane (b) 3-methylhexane (c) 2,3-dimethyl-4-propylnonane   (d) 1-isopropyl-3,5-dipropylcyclohexane   **1.88** 5-(1-ethylpropyl)nonane   **1.89** c   **1.90** a   **1.91** b   **1.92** b   **1.93** d   **1.94** c   **1.95** c   **1.96** c   **1.97** b   **1.98** c

Chapter 2   **2.1** (a) unsaturated, alkene with one double bond (b) saturated (c) unsaturated, alkene with one double bond (d) unsaturated, diene (e) unsaturated, triene (f) unsaturated, diene   **2.3** (a) C_4H_{10} (b) C_5H_{10} (c) C_5H_8 (d) C_7H_{10} **2.5** (a) C_nH_{2n-2} (b) C_nH_{2n-2} (c) C_nH_{2n-2} (d) C_nH_{2n-6} **2.7** (a) 2-butene (b) 2,4-dimethyl-2-pentene (c) cyclohexene (d) 1,3-cyclopentadiene (e) 2-ethyl-1-pentene (f) 2,4,6-octatriene   **2.9** (a) 2-pentene (b) pentane (c) 2,3,3-trimethyl-1-butene (d) 2-methyl-1,4-pentadiene (e) 1,3,5-hexatriene (f) 2,3-pentadiene

2.11 (a) $CH_2=CH-CH-CH_2-CH_3$
                                    $|$
                                   CH_3

(b)   (c) $CH_2=CH-CH=CH_2$

(d) $CH_2=CH-CH-CH=CH_2$
                    $|$
                  CH_2
                    $|$
                  CH_3

(e) $CH_3-CH=CH-CH-CH_2-CH_2-CH_3$
                              $|$
                            CH_2
                              $|$
                            CH_2
                              $|$
                            CH_3

(f)

2.13 (a) 3-methyl-3-hexene (b) 2,3-dimethyl-2-hexene (c) 1,3-cyclopentadiene (d) 4,5-dimethylcyclohexene

2.15 $C=C-C-C-C-C$ $C-C=C-C-C-C$
            1-Hexene   2-Hexene

$C-C-C=C-C-C$ $C=C-C-C-C$ $C=C-C-C-C$
     3-Hexene $|$ $|$
                                    C C
                            2-Methyl-1-pentene   3-Methyl-1-pentene

$C=C-C-C-C$ $C-C=C-C-C$ $C-C=C-C-C$
       $|$ $|$ $|$
      C C C
4-Methyl-1-pentene   2-Methyl-2-pentene   3-Methyl-2-pentene

$C-C=C-C-C$ $C=C-C-C$ $C=C-C-C$
       $|$ $|$ $|$ $|$
      C C C C
4-Methyl-2-pentene   2,3-Dimethyl-1-butene $|$
                                                     C
                                       3,3-Dimethyl-1-butene

$C-C=C-C$ $C=C-C-C$
   $|$ $|$ $|$
  C C C
2,3-Dimethyl-2-butene $|$
                              C
                   2-Ethyl-1-butene

2.17 (a) no (b) no (c) no
(d)

cis $trans$

(e)

cis $trans$

(f)

cis $trans$

2.19 (a) cis-2-pentene (b) $trans$-1-bromo-2-iodoethene (c) tetrafluoroethene (d) 2-methyl-2-butene

2.21 (a)   (b)

(c)

(d)

2.23 a compound used by insects (and some animals) to transmit messages to other members of the same species   **2.25** Isoprene, the building block for terpenes, contains 5 carbon atoms.   **2.27** (a) gas (b) liquid (c) liquid (d) liquid   **2.29** (a) yes (b) no (c) yes (d) no

2.31 (a) $CH_2=CH_2 + Cl_2 \longrightarrow CH_2-CH_2$
                                                           $|$ $|$
                                                         Cl Cl

(b) $CH_2{=}CH_2 + HCl \longrightarrow CH_3{-}CH_2$
                                           Cl

(c) $CH_2{=}CH_2 + H_2 \xrightarrow{Ni} CH_3{-}CH_3$

(d) $CH_2{=}CH_2 + HBr \longrightarrow CH_3{-}CH_2$
                                           Br

2.33 (a) $CH_2{=}CH{-}CH_3 + Cl_2 \longrightarrow CH_2{-}CH{-}CH_3$
                                                    Cl Cl

(b) $CH_2{=}CH{-}CH_3 + HCl \longrightarrow CH_3{-}CH{-}CH_3$
                                               Cl

(c) $CH_2{=}CH{-}CH_3 + H_2 \xrightarrow{Ni} CH_3{-}CH_2{-}CH_3$

(d) $CH_2{=}CH{-}CH_3 + HBr \longrightarrow CH_3{-}CH{-}CH_3$
                                                 Br

2.35

(a) $CH_3{-}CH{-}CH{-}CH_3$ (b)
                   Cl Cl

                                     Br
                                $CH_3{-}C{-}CH_3$
                                     CH_3

(c) $CH_3{-}CH_2{-}CH{-}CH_3$ (d)   (e)   (f) HO
                           Cl

2.37 (a) Br_2 (b) $H_2 + Ni$ catalyst (c) HCl (d) $H_2O + H_2SO_4$ catalyst
2.39 (a) 2 (b) 2 (c) 2 (d) 3   **2.41** (a) $CF_2{=}CF_2$
(b) $CH_2{=}C{-}CH{=}CH_2$ (c) $CH_2{=}CH$ (d) $CH_2{=}CH$
               Cl Cl

2.43 (a) $-CH_2{-}CH_2{-}CH_2{-}CH_2{-}CH_2{-}CH_2-$

(b) $-CH_2{-}CH{-}CH_2{-}CH{-}CH_2{-}CH-$
                   Cl Cl Cl

(c) $-CH{-}CH{-}CH{-}CH{-}CH{-}CH-$
        Cl Cl Cl Cl Cl Cl

(d) $-CH_2{-}CH{-}CH_2{-}CH{-}CH_2{-}CH-$
                   Cl Cl Cl

2.45 (a) 1-hexyne (b) 4-methyl-2-pentyne (c) 2,2-dimethyl-3-heptyne
(d) 1-butyne (e) 3-methyl-1,4-hexadiyne (f) 3,3-dimethyl-1-pentyne
2.47 $C{\equiv}C{-}C{-}C{-}C$ (1-pentyne)
       $C{-}C{\equiv}C{-}C{-}C$ (2-pentyne)
       $C{\equiv}C{-}C{-}C$ (3-methyl-1-butyne)
                C

2.49

(a) $CH_3{-}CH_3$ (b) Br Br (c) Br
                         $CH_3{-}C{-}CH$ $CH_3{-}C{-}CH_3$
                             Br Br Br

(d) $CH_2{=}CH$ (e) $CH_2{-}CH_3$ (f) $CH_3{-}CH_2{-}C{=}CH_2$
         Cl Br

2.51 (a) 1,3-dibromobenzene (b) 1-chloro-2-fluorobenzene (c) 1-chloro-4-fluorobenzene (d) 3-chlorotoluene (e) 1-bromo-2-ethylbenzene (f) 4-bromotoluene **2.53** (a) *m*-dibromobenzene (b) *o*-chlorofluorobenzene (c) *p*-chlorofluorobenzene (d) *m*-chlorotoluene (e) *o*-bromoethylbenzene (f)

p-bromotoluene **2.55** (a) 2,4-dibromo-1-chlorobenzene (b) 3-bromo-5-chlorotoluene (c) 1-bromo-3-chloro-2-fluorobenzene (d) 1,4-dibromo-2,5-dichlorobenzene **2.57** (a) 2-phenylbutane (b) 3-phenyl-1-butene (c) 3-methyl-1-phenylbutane (d) 2,4-diphenylpentane

2.59 (a) $CH_2{-}CH_3$ (b) CH_3 (c) CH_3
                                   CH_3
      $CH_2{-}CH_3$ $CH_2{-}CH_3$

(d)   (e) $-CH_2{-}CH_2-$

(f) CH_3
                   $CH_3{-}CH_2{-}C{-}CH_2{-}CH_3$

2.61 (a) substitution (b) addition (c) substitution (d) addition

2.63 (a) Br_2 (b) CH_3 (c) $CH_3{-}CH_2{-}Br$
                         $CH{-}CH_3$

2.65 (a) C_2H_4 (b) C_3H_4 (c) C_2H_2 (d) CH_4 **2.66** (a) more (b) more (c) more (d) the same number **2.67** (a) no (b) yes (c) no (d) yes **2.68** (a) All have six carbon atoms. (b) Cyclohexane has 12 H atoms, cyclohexene 10, and benzene 6. (c) Cyclohexane and benzene undergo substitution; cyclohexene undergoes addition. (d) All are liquids.

2.69 (a) $CH_3{-}C{\equiv}C{-}CH_2{-}CH{-}CH_3$
                                       CH_3

(b) $CH_2{-}CH{=}CH{-}CH_3$
    Cl

(c) $CH_3{-}C{\equiv}C{-}CH_2{-}CH{-}CH{-}CH_3$
                                CH_3 CH_3

(d) $CH_2{=}CH{-}CH{-}CH_2{-}CH_2{-}CH_3$
                     $CH{-}CH_3$
                     CH_3

(e) $CH_2{=}CH{-}CH_2{-}CH_2{-}CH_2{-}CH{=}CH_2$

(f) $CH_2{=}CH{-}CH{-}CH{=}CH_2$
                   CH_3

2.70 (a) two (b) one (c) two (d) one (e) two (f) four

2.71 (a)
     $CH_2{=}CH{-}$

(b) $CH_2{=}CH{-}CH_2{-}Cl$ (c) $CH_3{-}CH_2{-}CH_2{-}C{\equiv}CH$
(d) $CH_3{-}CH_2{-}CH_2{-}C{\equiv}C{-}CH_2{-}CH_2{-}CH_3$
(e) CH_3 (f) CH_3
                CH_3

2.72 It would require a carbon atom that formed five bonds.   **2.73** The substituted carbon atoms in 1,2-dichlorobenzene have only one substituent.

2.74 $CH_2=CH-CH_2-CH_2-CH_3$ $CH_3-CH=CH-CH_2-CH_3$
(*cis–trans* forms)

$CH_2=C-CH_2-CH_3$
$\quad\;\; |$
$\quad\; CH_3$

$CH_2=CH-CH-CH_3$
$\qquad\qquad\;\; |$
$\qquad\qquad CH_3$

$CH_3-C=CH-CH_3$
$\qquad |$
$\quad\; CH_3$

(*cis–trans* forms)

2.75 1,2,3-trimethylbenzene; 1,2,4-trimethylbenzene; 1,3,5-trimethylbenzene; 2-ethyltoluene; 3-ethyltoluene; 4-ethyltoluene; propylbenzene; isopropylbenzene   **2.76** (a) 3 (b) 3 (c) 11 (d) 3   **2.77** c   **2.78** d   **2.79** b   **2.80** a   **2.81** b   **2.82** a   **2.83** a   **2.84** c   **2.85** c   **2.86** a

Chapter 3   **3.1** (a) 2 (b) 1 (c) 4 (d) 1   **3.3** R—OH   **3.5** R—O—H versus H—O—H   **3.7** (a) 2-pentanol (b) ethanol (c) 3-methyl-2-butanol (d) 2-ethyl-1-pentanol (e) 2-butanol (f) 3,3-dimethyl-1-butanol   **3.9** (a) 1-hexanol (b) 3-hexanol (c) 5,6-dimethyl-2-heptanol (d) 2-methyl-3-pentanol

3.11 (a) $CH_3-CH_2-CH-CH_2-CH_3$
$\qquad\qquad\qquad\;\; |$
$\qquad\qquad\qquad OH$

(b) $CH_3-CH_2-\underset{\underset{\displaystyle CH_3}{\overset{\displaystyle |}{CH_2}}}{\overset{\overset{\displaystyle OH}{\displaystyle |}}{C}}-CH_2-CH_2-CH_3$

(c) $CH_2-CH-CH_3$
$\;\;\;\, |\quad\;\; |$
$\;\; OH\quad CH_3$

(d) $CH_3-CH-CH_2-CH-CH_3$
$\qquad\;\; |\qquad\qquad |$
$\qquad OH\qquad\;\; CH_3$

(e) $CH_3-\overset{\overset{\displaystyle OH}{\displaystyle |}}{\underset{\underset{\displaystyle C_6H_5}{}}{C}}-CH_3$

(f) (cyclobutanol with OH and CH₃)

3.13 (a) $CH_2-CH_2-CH_2-CH_2-CH_3$
$\quad\;\; |$
$\;\; OH$
1-Pentanol

(b) $CH_2-CH_2-CH_3$
$\;\;\, |$
$\; OH$
1-Propanol

(c) $CH_3-CH-CH_2-OH$
$\qquad\;\; |$
$\qquad CH_3$
2-Methyl-1-Propanol

(d) $CH_3-CH_2-CH-OH$
$\qquad\qquad\quad |$
$\qquad\qquad CH_3$
2-Butanol

3.15 (a) 1,2-propanediol (b) 1,4-pentanediol (c) 1,3-pentanediol (d) 3-methyl-1,2,4-butanetriol   **3.17** (a) cyclohexanol (b) *trans*-3-chlorocyclohexanol (c) *cis*-2-methylcyclohexanol (d) 1-methylcyclobutanol

3.19 (a) $CH_3-CH-CH_2-CH=CH_2$
$\qquad\;\; |$
$\qquad OH$

(b) $CH\equiv C-CH-CH_2-CH_3$
$\qquad\qquad |$
$\qquad\;\; OH$

(c) $CH_3-CH-C=CH_2$
$\qquad\;\; |\quad\;\; |$
$\qquad OH\; CH_3$

(d) $HO-CH_2\underset{}{\overset{\overset{\displaystyle CH_3}{}}{\diagdown}}C=C\overset{\diagup}{\underset{\underset{\displaystyle H}{}}{\diagup}}$
$\qquad\qquad H$

3.21 (a) $CH_2-CH-CH_3$
$\;\;\;\, |\quad\;\; |$
$\;\; OH\;\; CH_2$
$\qquad\qquad |$
$\qquad\;\; CH_3$
2-Methyl-1-butanol

(b) $CH_3-CH-CH_2-CH_2$
$\qquad\;\; |\qquad\qquad |$
$\qquad OH\qquad\;\; OH$
1,3-Butanediol

(c) $CH_3-CH-CH-CH_3$
$\qquad\;\; |\quad\;\; |$
$\qquad CH_3\; OH$
3-Methyl-2-butanol

(d) (cyclopentane with OH and HO) 1,3-Cyclopentanediol

3.23 (a) no (b) yes (c) yes (d) yes   **3.25** 1-heptanol, 2-heptanol, 3-heptanol, 4-heptanol   **3.27** $x = 1, 2, 3$ **3.29** (a) ethanol with all traces of H_2O removed (b) ethanol (c) 70% solution of isopropyl alcohol (d) ethanol   **3.31** (a) glycerol (b) ethanol (c) methanol (d) methanol   **3.33** Alcohol molecules can hydrogen-bond to each other; alkane molecules cannot.   **3.35** (a) 1-heptanol (b) 1-propanol (c) 1,2-ethanediol   **3.37** (a) 1-butanol (b) 1-pentanol (c) 1,2-butanediol   **3.39** (a) 3 (b) 3 (c) 3 (d) 3

3.41 (a) CH_2-CH_3
$\;\;\;\, |$
$\;\; OH$

(b) $CH_3-CH_2-CH_2$
$\qquad\qquad\quad |$
$\qquad\qquad OH$

(c) $CH_3-CH_2-\overset{\overset{\displaystyle OH}{\displaystyle |}}{\underset{\underset{\displaystyle CH_3}{\displaystyle |}}{C}}-CH_3$

(d) $CH_3-CH_2-CH-CH_2-CH_3$
$\qquad\qquad\qquad |$
$\qquad\qquad\qquad OH$

3.43 (a) 2° (b) 1° (c) 2° (d) 1° (e) 2° (f) 1°

3.45 (a) $CH_2=CH-CH_3$

(b) $CH_3-CH_2-\underset{\underset{\displaystyle CH_3}{\displaystyle |}}{C}=CH_2$

(c) $CH_3-CH=CH_2$

(d) $CH_3-CH_2-CH_2-O-CH_2-CH_2-CH_3$

3.47
(a) $CH_3-CH-CH-CH_3$
$\qquad\;\; |\quad\;\; |$
$\qquad OH\; CH_3$

(b) $CH_3-CH_2-CH_2$ or $CH_3-CH-CH_3$
$\qquad\qquad\qquad |\qquad\qquad\qquad\;\; |$
$\qquad\qquad\;\; OH\qquad\qquad\qquad OH$

(c) CH_3-CH_2-OH

(d) $CH_3-CH-CH_2-OH$
$\qquad\;\; |$
$\qquad CH_3$

3.49 (a) $CH_3-CH_2-CH-CH_3$
$\qquad\qquad\qquad |$
$\qquad\qquad\qquad OH$

(b) $CH_3-CH_2-CH_2$
$\qquad\qquad\quad\; |$
$\qquad\qquad\;\; OH$

(c) $CH_3-CH_2-CH_2$
$\qquad\qquad\quad\; |$
$\qquad\qquad\;\; OH$

(d) (cyclopentane)$-CH_2-OH$

3.51 (a) $CH_3-CH_2-CH_2-Cl$

(b) (cyclopentene with CH₃)

(c) $CH_3-\overset{\overset{\displaystyle O}{\displaystyle \|}}{C}-CH_2-CH_3$

(d) $CH_3-CH_2-CH-CH_2-CH_3$
$\qquad\qquad\qquad |$
$\qquad\qquad\qquad Cl$

(e) $CH_3-CH_2-O-CH_2-CH_3$

(f) CH_2-CH_2
$\;\;\; |\qquad |$
$\;\; Cl\qquad Cl$

3.53
$$\left(\begin{matrix} -CH-CH- \\ \;\;|\quad\;\; | \\ \;\;OH\;\;OH \end{matrix}\right)_n$$

3.55 Phenols require the —OH groups to be attached directly to the benzene ring. **3.57** (a) 3-ethylphenol (b) 2-chlorophenol (c) *o*-cresol (d) hydroquinone (e) 2-bromophenol (f) 2-bromo-3-ethylphenol

3.59

(a) OH phenyl ring with Cl para
(b) OH phenyl ring with CH₂—CH₃ ortho
(c) OH phenyl ring with Br ortho and Br para
(d) OH phenyl ring with CH₃ meta
(e) OH phenyl ring with OH meta
(f) OH phenyl ring with CH₃—CH₂— and —CH₂—CH₃ ortho, CH₃ para

3.61 An antiseptic kills microorganisms on living tissue; a disinfectant kills microorganisms on inanimate objects.

3.63

$$\text{(phenol)} + H_2O \rightleftharpoons \text{(phenoxide } O^-) + H_3O^+$$

3.65 (a) yes (b) no (c) yes (d) yes **3.67** (a) 1-methoxypropane (b) 1-ethoxypropane (c) 2-methoxypropane (d) methoxybenzene (e) cyclohexoxycyclohexane (f) ethoxycyclobutane **3.69** (a) methyl propyl ether (b) ethyl propyl ether (c) isopropyl methyl ether (d) methyl phenyl ether (e) dicyclohexyl ether (f) cyclobutyl ethyl ether **3.71** (a) 1-methoxypentane (b) 1-ethoxy-2-methylpropane (c) 2-ethoxybutane (d) 2-methoxybutane

3.73

(a) $CH_3-\underset{\underset{CH_3}{|}}{CH}-O-CH_2-CH_2-CH_3$ (b) CH_3-CH_2-O- (phenyl)

(c) (phenyl with CH₃ meta)—O—CH₃

(d) $CH_3-\underset{\underset{O-CH_2-CH_3}{|}}{CH}-CH_2-CH_2-CH_3$

(e) (cyclobutyl)—O—CH₂—CH₃ (f) $CH_3-O-CH_2-\underset{\underset{CH_3}{|}}{\overset{\overset{CH_3}{|}}{C}}-CH_3$

3.75 (a) no (b) no (c) yes (d) no **3.77** butyl methyl ether, *sec*-butyl methyl ether, isobutyl methyl ether, *tert*-butyl methyl ether, ethyl isopropyl ether

3.79 (a) $CH_3-O-CH_2-CH_2-CH_3$, $CH_3-O-\underset{\underset{CH_3}{|}}{CH}-CH_3$,
$CH_3-CH_2-O-CH_2-CH_3$
(b) $CH_3-CH_2-CH_2-CH_2-OH$, $CH_3-CH_2-\underset{\underset{CH_3}{|}}{CH}-OH$,
$CH_3-\underset{\underset{CH_3}{|}}{CH}-CH_2-OH$, $CH_3-\underset{\underset{CH_3}{|}}{\overset{\overset{CH_3}{|}}{C}}-OH$

3.81 $x = 1, 2,$ and 3 **3.83** Dimethyl ether molecules cannot hydrogen-bond to each other; ethanol molecules can. **3.85** flammability and peroxide formation **3.87** No oxygen–hydrogen bonds are present. **3.89**

(a) noncyclic ether (b) noncyclic ether (c) cyclic ether (d) cyclic ether (e) noncyclic ether (f) nonether **3.91** R—S—H versus R—O—H

3.93 (a) CH_3-SH (b) $CH_3-\underset{\underset{SH}{|}}{CH}-CH_3$ (c) $CH_3-CH_2-CH_2-\underset{\underset{SH}{|}}{CH_2}$

(d) $CH_3-CH_2-\underset{\underset{CH_3}{|}}{CH}-CH_2-\underset{\underset{SH}{|}}{CH_2}$ (e) (cyclopentyl)—SH (f) $\underset{\underset{SH}{|}}{CH_2}-\underset{\underset{SH}{|}}{CH_2}$

3.95 (a) methyl mercaptan (b) propyl mercaptan (c) *sec*-butyl mercaptan (d) isobutyl mercaptan **3.97** Alcohol oxidation produces aldehydes and ketones; thioalcohol oxidation produces disulfides. **3.99** (a) methylthioethane; ethyl methyl sulfide (b) 2-methylthiopropane; isopropyl methyl sulfide (c) methylthiocyclohexane; cyclohexyl methyl sulfide (d) cyclohexylthiocyclohexane; dicyclohexyl sulfide (e) 3-(methylthio)-1-propene; allyl methyl sulfide (f) 2-methylthiobutane; *sec*-butyl methyl sulfide **3.101** (a) 2-hexanol (b) 3-pentanol (c) 3-phenyoxy-1-propene (d) 2-methyl-1-propanol (e) 2-methyl-2-propanol (f) ethoxyethane

3.102

$CH_2-CH_2-CH_2-CH_2-CH_3$ with OH on C1

$\underset{\underset{OH}{|}}{CH_2}-\underset{\underset{CH_3}{|}}{CH}-CH_2-CH_3$

$CH_3-\underset{\underset{OH}{|}}{CH}-CH_2-CH_2-CH_3$

$CH_3-\underset{\underset{CH_3\;\;OH}{|\quad\;\;|}}{\overset{\overset{CH_3}{|}}{C}}-CH_2$

$CH_3-\underset{\underset{OH\;\;CH_3}{|\quad\;\;|}}{CH}-CH-CH_3$

$CH_3-\underset{\underset{OH}{|}}{\overset{\overset{CH_3}{|}}{C}}-CH_2-CH_3$

$CH_2-CH_2-\underset{\underset{CH_3}{|}}{CH}-CH$ with OH

$CH_3-CH_2-\underset{\underset{OH}{|}}{CH}-CH_2-CH_3$

3.103 1-pentanol **3.104** CH_3-O-CH_3, $CH_3-CH_2-CH_2-O-CH_2-CH_2-CH_3$, and $CH_3-O-CH_2-CH_2-CH_3$ **3.105** (a) disulfide (b) thiol, thioalcohol (c) alcohol (d) peroxide (e) alcohol, thiol, thioalcohol (f) ether, sulfide, thioether **3.106** (a) 1,2-ethanedithiol (b) 3-methoxy-1-propanol (c) 1-propanol (d) 1,2-dimethoxyethane (e) methylthioethane (f) 1-ethylthio-2-methoxyethane **3.107** d **3.108** c **3.109** b **3.110** c **3.111** b **3.112** a **3.113** a **3.114** d **3.115** a **3.116** d

Chapter 4 **4.1** (a) yes (b) no (c) yes (d) yes (e) no (f) no **4.3** similarity: both have bonds involving four shared electrons; difference: C=O is polar, C=C is not polar **4.5** (a) neither (b) aldehyde (c) ketone (d) neither (e) aldehyde (f) aldehyde

4.7 $\underset{O}{\overset{\|}{H-C}}-H$, $CH_3-\overset{\overset{O}{\|}}{C}-H$, $CH_3-\overset{\overset{O}{\|}}{C}-CH_3$, $CH_3-CH_2-\overset{\overset{O}{\|}}{C}-CH_3$

4.9 (a) neither (b) aldehyde (c) neither (d) ketone (e) ketone (f) aldehyde **4.11** (a) butanal (b) 2-methylbutanal (c) 4-methylheptanal (d) 3-phenylpropanal (e) propanal (f) 3,3-dimethylbutanal **4.13** (a) pentanal (b) 3-methylbutanal (c) 3-methylpentanal (d) 2-ethyl-3-methylpentanal

4.15 (a) $CH_3-CH_2-\underset{\underset{}{|}}{\overset{\overset{CH_3}{|}}{CH}}-CH_2-\overset{\overset{O}{\|}}{C}-H$

(b) $CH_3-CH_2-CH_2-CH_2-\underset{\underset{CH_2}{|}\;\;}{\underset{\underset{CH_3}{|}\;\;}{CH}}-\overset{\overset{O}{\|}}{C}-H$

(c)

$$CH_3-CH_2-CH_2-\underset{\underset{CH_3}{|}}{CH}-\underset{\underset{CH_3}{|}}{CH}-CH_2-\overset{\overset{O}{\|}}{C}-H$$

(d)

$$CH_3-\underset{\underset{Cl}{|}}{\overset{\overset{Cl}{|}}{C}}-\overset{\overset{O}{\|}}{C}-H$$

(e)

$$CH_3-CH_2-\underset{\underset{CH_3}{|}}{CH}-\underset{\underset{CH_3}{|}}{CH}-CH_2-\underset{\underset{CH_3}{|}}{CH}-\overset{\overset{O}{\|}}{C}-H$$

(f)

$$CH_3-CH_2-CH_2-CH_2-\underset{\underset{OH}{|}}{CH}-CH_2-\underset{\underset{CH_3}{|}}{CH}-\overset{\overset{O}{\|}}{C}-H$$

4.17 (a)

$$H-\overset{\overset{O}{\|}}{C}-H$$

(b)

$$CH_3-CH_2-\overset{\overset{O}{\|}}{C}-H$$

(c)

$$\underset{\underset{Cl}{|}}{CH_2}-\overset{\overset{O}{\|}}{C}-H$$

(d) benzaldehyde with Cl substituent and C—H, O

(e) benzaldehyde with CH₃ substituent and C—H, O

(f) benzaldehyde with two CH₃ substituents and C—H, O

4.19 (a) propionaldehyde (b) propionaldehyde (c) butyraldehyde (d) dichloroacetaldehyde (e) *o*-chlorobenzaldehyde (f) 3-chloro-4-hydroxy-benzaldehyde **4.21** (a) 2-butanone (b) 2,4,5-trimethyl-3-hexanone (c) 6-methyl-3-heptanone (d) 2-octanone (e) 1,5-dichloro-3-pentanone (f) 1,1-dichloro-2-butanone **4.23** (a) 2-hexanone (b) 5-methyl-3-hexanone (c) 2-pentanone (d) 4-ethyl-3-methyl-2-hexanone **4.25** (a) cyclohexanone (b) 3-methylcyclohexanone (c) 2-methylcyclohexanone (d) 3-chlorocyclopentanone

4.27 (a)

$$CH_3-\overset{\overset{O}{\|}}{C}-\underset{\underset{CH_3}{|}}{CH}-CH_2-CH_3$$

(b)

$$CH_3-CH_2-\overset{\overset{O}{\|}}{C}-CH_2-CH_2-CH_3$$

(c) cyclobutanone with =O

(d)

$$CH_3-\underset{\underset{CH_3}{|}}{CH}-\overset{\overset{O}{\|}}{C}-\underset{\underset{CH_3}{|}}{CH}-CH_3$$

(e)

$$\underset{\underset{Cl}{|}}{CH_2}-\overset{\overset{O}{\|}}{C}-CH_3$$

(f)

$$\underset{\underset{Cl}{|}}{CH_2}-\overset{\overset{O}{\|}}{C}-\underset{\underset{Cl}{|}}{CH_2}$$

4.29 (a)

$$CH_3-CH_2-\overset{\overset{O}{\|}}{C}-CH_2-CH_3$$

(b)

$$CH_3-\overset{\overset{O}{\|}}{C}-CH_3$$

(c)

$$CH_3-\underset{\underset{CH_3}{|}}{CH}-\overset{\overset{O}{\|}}{C}-CH_2-CH_2-CH_3$$

(d)

$$\underset{\underset{Cl}{|}}{CH_2}-\overset{\overset{O}{\|}}{C}-CH_3$$

(e)

$$CH_3-\overset{\overset{O}{\|}}{C}-\text{(phenyl)}$$

(f)

$$CH_3-\overset{\overset{O}{\|}}{C}-\text{(phenyl)}$$

4.31 (a) heptanal (b) 2-heptanone, 3-heptanone, 4-heptanone **4.33** (a) 1 aldehyde, no ketones (b) 1 aldehyde, 1 ketone **4.35** $x = 2, 4, 5$

4.37

$$C-C-C-C-\overset{\overset{O}{\|}}{C}-H \qquad C-C-C-\underset{\underset{C}{|}}{\overset{\overset{O}{\|}}{C}}-H \qquad C-C-\underset{\underset{C}{|}}{\overset{\overset{O}{\|}}{C}}-C-H$$

$$\underset{\underset{C}{|}}{C}-\overset{\overset{O}{\|}}{C}-C-H \qquad C-\overset{\overset{O}{\|}}{C}-C-C-C \qquad C-\underset{\underset{C}{|}}{\overset{\overset{O}{\|}}{C}}-C-C \qquad C-C-\overset{\overset{O}{\|}}{C}-C-C$$

4.39 Dipole–dipole attractions between molecules raise the boiling point.
4.41 2 **4.43** ethanal, because it has a shorter carbon chain
4.45 (a)

$$CH_3-CH_2-CH_2-CH_2-\overset{\overset{O}{\|}}{C}-H$$

(b)

$$CH_3-CH_2-\overset{\overset{O}{\|}}{C}-CH_3$$

(c)

$$CH_3-\underset{\underset{CH_3}{|}}{\overset{\overset{CH_3}{|}}{C}}-CH_2-\overset{\overset{O}{\|}}{C}-H$$

(d)

$$CH_3-CH_2-\overset{\overset{O}{\|}}{C}-CH_2-CH_3$$

(e) cyclopentanone with =O

(f) cyclohexanone with CH₃ substituent and =O

4.47 (a) CH_3-CH_2-OH (b)

$$CH_3-CH_2-\underset{\underset{OH}{|}}{CH}-CH_2-CH_3$$

(c)

$$\text{(phenyl)}-\underset{\underset{OH}{|}}{CH_2}-CH-CH_3$$

(d) CH_3-CH_2-OH

(e)

$$CH_3-\underset{\underset{OH}{|}}{CH}-CH_3$$

(f)

$$CH_3-CH_2-CH_2-CH_2-\underset{\underset{CH_3-CH_2}{|}}{CH}-\underset{\underset{OH}{|}}{CH_2}$$

4.49 (a)

$$CH_3-\overset{\overset{O}{\|}}{C}-OH$$

(b)

$$CH_3-CH_2-CH_2-CH_2-\overset{\overset{O}{\|}}{C}-OH$$

(c)

$$H-\overset{\overset{O}{\|}}{C}-OH$$

(d)

$$CH_3-CH_2-\underset{\underset{Cl}{|}}{CH}-\underset{\underset{Cl}{|}}{CH}-CH_2-\overset{\overset{O}{\|}}{C}-OH$$

4.51 appearance of a silver mirror **4.53** Cu^{2+} ion **4.55** (a) no (b) yes (c) yes (d) no
4.57
(a)

$$CH_3-CH_2-CH_2-\underset{\underset{OH}{|}}{CH_2}$$

(b)

$$CH_3-CH_2-\underset{\underset{OH}{|}}{CH}-CH_2-CH_3$$

(c)

$$CH_3-\underset{\underset{CH_3}{|}}{CH}-CH_2-\underset{\underset{OH}{|}}{CH_2}$$

(d)

$$CH_3-\underset{\underset{CH_3}{|}}{CH}-\underset{\underset{OH}{|}}{CH}-CH_2-CH_2-CH_3$$

4.59 R—O— and H— **4.61** (a) no (b) yes (c) no (d) yes (e) yes (f) no
4.63
(a)

$$CH_3-\underset{\underset{OH}{|}}{CH}-O-CH_2-CH_3$$

(b)

$$CH_3-\underset{\underset{O-CH_3}{|}}{\overset{\overset{OH}{|}}{C}}-CH_2-CH_2-CH_3$$

(c)

$$CH_3-CH_2-CH_2-\underset{\underset{O-CH_2-CH_3}{|}}{\overset{\overset{OH}{|}}{CH}}$$

(d)

$$CH_3-\underset{\underset{O-CH-CH_3}{|}}{\overset{\overset{OH}{|}}{C}}-CH_3$$

with CH_3 on the O—CH

4.65 (a)

$$CH_3-CH_2-CH_2-\underset{\underset{O-CH_2-CH_3}{|}}{\overset{\overset{OH}{|}}{CH}}$$

(b)

$$CH_3-CH_2-\overset{\overset{O}{\|}}{C}-H$$

(c)

$$CH_3-CH_2-\underset{\underset{O-CH_3}{|}}{\overset{\overset{OH}{|}}{C}}-CH_3$$

(d)

CH₂OH with pyranose ring, O, OH, OH

4.67 (a) yes (b) yes (c) no (d) yes

4.69 (a) CH_3-OH (b) $CH_3-\underset{\underset{OH}{|}}{CH}-O-CH_3$

(c)

$$CH_3-\underset{\underset{O-CH-CH_3}{|}}{CH}-O-CH_3$$ with $O-\underset{}{CH}-CH_3$ and CH_3

(d) $CH_3-\underset{\underset{OH}{|}}{CH}-O-CH_3$, CH_3-OH

4.71

(a) $CH_3-\overset{\overset{O}{||}}{C}-H$,   2 CH_3-OH **(b)** $CH_3-\overset{\overset{O}{||}}{C}-CH_3$,   2 CH_3-OH

(c) $CH_3-CH_2-\overset{\overset{O}{||}}{C}-CH_2-CH_3$, CH_3-OH, CH_3-CH_2-OH

(d) $CH_3-CH_2-CH_2-CH_2-\overset{\overset{O}{||}}{C}-H$, 2 CH_3-OH

4.73 (a) dimethyl acetal of ethanal (b) dimethyl acetal of propanone (c) ethyl methyl acetal of 3-pentanone (d) dimethyl acetal of pentanal **4.75** (a) By definition, the carbonyl carbon atom is numbered 1 in an aldehyde; therefore, the number does not have to be specified in the name. (b) There is only one possible location for the carbonyl group in propanone; therefore, its location does not have to be specified. **4.76** (a) A ketone carbonyl group cannot be on a terminal carbon atom. (b) It requires a carbon atom with five bonds. (c) It requires a carbon atom with five bonds. (d) It requires a carbon atom with five bonds. **4.77** (a) a carbon atom bonded to both a hydroxyl group and an alkoxy group (b) a carbon atom bonded to two alkoxy groups

4.78

(a) $CH_3-CH_2-\underset{\underset{OH}{|}}{CH}-O-CH_2-CH_3$ $CH_3-CH_2-\underset{\underset{O-CH_2-CH_3}{|}}{CH}-O-CH_2-CH_3$

(b) cyclohexane ring with OH and O—CH₃   cyclohexane ring with O—CH₃ and O—CH₃

4.79 ring: CH_2, O, CH with OH; CH_2-CH_2

4.80 (a) ketone, alkene (b) aldehyde, alcohol, ether (c) ketone, alkyne (d) aldehyde, ketone **4.81** (a) ketone (b) aldehyde (c) aldehyde (d) aldehyde **4.82** c **4.83** c **4.84** c **4.85** b **4.86** d **4.87** b **4.88** c **4.89** c **4.90** d **4.91** b

Chapter 5 **5.1** (a) yes (b) no (c) yes (d) yes (e) no (f) yes **5.3** (a) butanoic acid (b) heptanoic acid (c) 2,3-dimethylpentanoic acid (d) 4-bromopentanoic acid (e) 3-methylpentanoic acid (f) chloroethanoic acid **5.5** (a) hexanoic acid (b) 3-methylpentanoic acid (c) 2,3-dimethylbutanoic acid (d) 4,5-dimethylhexanoic acid

5.7 (a)

$$CH_3-CH_2-\underset{\underset{CH_3-CH_2}{|}}{CH}-\overset{\overset{O}{||}}{C}-OH$$

(b)

$$CH_3-\underset{\underset{CH_3}{|}}{CH}-CH_2-CH_2-\underset{\underset{CH_3}{|}}{CH}-\overset{\overset{O}{||}}{C}-OH$$

(c)

$$CH_3-\underset{\underset{CH_3}{|}}{CH}-\overset{\overset{O}{||}}{C}-OH$$

(d)

$$Cl-\underset{\underset{Cl}{|}}{CH}-\overset{\overset{O}{||}}{C}-OH$$

(e)

$$CH_3-CH_2-CH_2-\underset{\underset{Cl}{|}}{CH}-CH_2-\underset{\underset{Br}{|}}{CH}-CH_2-\overset{\overset{O}{||}}{C}-OH$$

(f)

$$CH_3-\underset{\underset{CH_3}{|}}{CH}-\underset{\underset{CH_3}{|}}{CH}-\overset{\overset{O}{||}}{C}-OH$$

5.9 (a) butanedioic acid (b) propanedioic acid (c) 3-methylpentanedioic acid (d) 2-chlorobenzoic acid (e) 2-bromo-4-chlorobenzoic acid (f) *m*-toluic acid

5.11

(a)

$$CH_3-CH_2-\underset{\underset{CH_3}{|}}{\overset{\overset{CH_3}{|}}{C}}-\overset{\overset{O}{||}}{C}-OH$$

(b)

$$HO-\overset{\overset{O}{||}}{C}-\underset{\underset{CH_3}{|}}{\overset{\overset{CH_3}{|}}{C}}-CH_2-\overset{\overset{O}{||}}{C}-OH$$

(c)

$$HO-\overset{\overset{O}{||}}{C}-\underset{\underset{CH_3}{|}}{\overset{\overset{CH_3}{|}}{C}}-CH_2-CH_2-\overset{\overset{O}{||}}{C}-OH$$

(d) benzene ring with COOH and Br (ortho)

(e) benzene ring with COOH, Cl, Cl

(f) benzene ring with COOH and CH₃ (para)

5.13 (a)

$$CH_3-CH_2-CH_2-CH_2-\overset{\overset{O}{||}}{C}-OH$$

(b)

$$CH_3-CH_2-\overset{\overset{O}{||}}{C}-OH$$

(c)

$$CH_3-\overset{\overset{O}{||}}{C}-OH$$

(d)

$$CH_3-CH_2-\underset{\underset{Cl}{|}}{CH}-\overset{\overset{O}{||}}{C}-OH$$

(e)

$$CH_3-CH_2-CH_2-\underset{\underset{Br}{|}}{CH}-CH_2-\overset{\overset{O}{||}}{C}-OH$$

(f)

$$CH_3-\underset{\underset{Cl}{|}}{CH}-CH_2-\underset{\underset{CH_3}{|}}{CH}-\overset{\overset{O}{||}}{C}-OH$$

5.15 (a)

$$HO-\overset{\overset{O}{||}}{C}-CH_2-\overset{\overset{O}{||}}{C}-OH$$

(b)

$$HO-\overset{\overset{O}{||}}{C}-CH_2-CH_2-\overset{\overset{O}{||}}{C}-OH$$

(c)

$$HO-\overset{\overset{O}{||}}{C}-CH_2-CH_2-CH_2-CH_2-\overset{\overset{O}{||}}{C}-OH$$

(d)

$$HO-\overset{\overset{O}{||}}{C}-CH_2-CH_2-\underset{\underset{Br}{|}}{CH}-CH_2-CH_2-\overset{\overset{O}{||}}{C}-OH$$

(e)

$$HO-\overset{\overset{O}{||}}{C}-\underset{\underset{CH_3}{|}}{CH}-CH_2-CH_2-\overset{\overset{O}{||}}{C}-OH$$

(f)

$$HO-\overset{\overset{O}{||}}{C}-\underset{\underset{Br}{|}}{\overset{\overset{Cl}{|}}{C}}-CH_2-\overset{\overset{O}{||}}{C}-OH$$

5.17 (a) 3 (b) 1 (c) 2 (d) 1   **5.19** (a) carbon–carbon double bond (b) hydroxyl group (c) carbon–carbon double bond (d) hydroxyl group **5.21** (a) propenoic acid (b) 2-hydroxy-propanoic acid (c) *cis*-butenedioic acid (d) 2-hydroxyethanoic acid

5.23

(a) $CH_3-CH_2-\underset{\underset{O}{\|}}{C}-CH_2-\underset{\underset{O}{\|}}{C}-OH$ (b) $CH_3-CH_2-\underset{\underset{OH}{|}}{CH}-\underset{\underset{O}{\|}}{C}-OH$

(c) CH_3
$CH=CH$
$CH_2-CH_2-\underset{\underset{O}{\|}}{C}-OH$

(d) $HO-\underset{\underset{O}{\|}}{C}-\underset{\underset{OH}{|}}{CH}-\underset{\underset{OH}{|}}{CH}-CH_2-\underset{\underset{O}{\|}}{C}-OH$

5.25 (a) propionic acid (b) propionic acid (c) succinic acid (d) glutaric acid   **5.27** (a) hydroxy, carboxy (b) hydroxy, carboxy (c) keto, carboxy (d) hydroxy, carboxy   **5.29** (a) 2 (b) 5   **5.31** (a) solid (b) solid (c) liquid (d) solid

5.33 (a) $CH_3-\underset{\underset{O}{\|}}{C}-OH$ (b) $CH_3-\underset{\underset{O}{\|}}{C}-OH$

(c) $CH_3-CH_2-\underset{\underset{CH_3}{|}}{CH}-CH_2-\underset{\underset{O}{\|}}{C}-OH$ (d) benzene ring $-\underset{\underset{O}{\|}}{C}-OH$

5.35 (a) 1 (b) 3 (c) 2 (d) 2   **5.37** (a) −1 (b) −3 (c) −2 (d) −2
5.39 (a) pentanoate ion (b) citrate ion (c) succinate ion (d) oxalate ion

5.41

(a) $CH_3-\underset{\underset{O}{\|}}{C}-OH + H_2O \rightarrow H_3O^+ + CH_3-\underset{\underset{O}{\|}}{C}-O^-$

(b) $HO-\underset{\underset{O}{\|}}{C}-CH_2-\underset{\underset{C-OH}{|}}{\underset{|}{\overset{OH}{|}{C}}}-CH_2-\underset{\underset{O}{\|}}{C}-OH + 3H_2O \rightarrow$

$3H_3O^+ + {}^-O-\underset{\underset{O}{\|}}{C}-CH_2-\underset{\underset{\underset{O}{\|}}{C-O^-}}{\overset{OH}{\underset{|}{C}}}-CH_2-\underset{\underset{O}{\|}}{C}-O^-$

(c) $CH_3-\underset{\underset{O}{\|}}{C}-OH + H_2O \rightarrow H_3O^+ + CH_3-\underset{\underset{O}{\|}}{C}-O^-$

(d) $CH_3-CH_2-\underset{\underset{CH_3}{|}}{CH}-\underset{\underset{O}{\|}}{C}-OH + H_2O \rightarrow$

$H_3O^+ + CH_3-CH_2-\underset{\underset{CH_3}{|}}{CH}-\underset{\underset{O}{\|}}{C}-O^-$

5.43 (a) potassium ethanoate (b) calcium propanoate (c) potassium butanedioate (d) sodium pentanoate
5.45

(a) $CH_3-\underset{\underset{O}{\|}}{C}-OH + KOH \rightarrow CH_3-\underset{\underset{O}{\|}}{C}-O^- K^+ + H_2O$

(b) $2\ CH_3-CH_2-\underset{\underset{O}{\|}}{C}-OH + Ca(OH)_2 \rightarrow$

$\left(CH_3-CH_2-\underset{\underset{O}{\|}}{C}-O^-\right)_2 Ca^{2+} + H_2O$

(c) $HO-\underset{\underset{O}{\|}}{C}-CH_2-CH_2-\underset{\underset{O}{\|}}{C}-OH + 2KOH \rightarrow$

$K^+\ {}^-O-\underset{\underset{O}{\|}}{C}-CH_2-CH_2-\underset{\underset{O}{\|}}{C}-O^-\ K^+ + 2H_2O$

(d) $CH_3-CH_2-CH_2-CH_2-\underset{\underset{O}{\|}}{C}-OH + NaOH \rightarrow$

$CH_3-CH_2-CH_2-CH_2-\underset{\underset{O}{\|}}{C}-O^-\ Na^+ + H_2O$

5.47

(a) $CH_3-CH_2-CH_2-\underset{\underset{O}{\|}}{C}-O^-\ Na^+ + HCl \rightarrow$

$CH_3-CH_2-CH_2-\underset{\underset{O}{\|}}{C}-OH + NaCl$

(b) $K^+\ {}^-O-\underset{\underset{O}{\|}}{C}-\underset{\underset{O}{\|}}{C}-O^-\ K^+ + 2HCl \rightarrow$

$HO-\underset{\underset{O}{\|}}{C}-\underset{\underset{O}{\|}}{C}-OH + 2KCl$

(c) $\left({}^-O-\underset{\underset{O}{\|}}{C}-CH_2-\underset{\underset{O}{\|}}{C}-O^-\right)_2 Ca^{2+} + 2HCl \rightarrow$

$HO-\underset{\underset{O}{\|}}{C}-CH_2-\underset{\underset{O}{\|}}{C}-OH + CaCl_2$

(d) benzene ring $-\underset{\underset{O}{\|}}{C}-O^-\ Na^+$
$+ HCl \rightarrow$ benzene ring $-\underset{\underset{O}{\|}}{C}-OH$ $+ NaCl$

5.49 (a) yes (b) yes (c) no (d) yes (e) yes (f) no

5.51 (a) $CH_3-CH_2-\underset{\underset{O}{\|}}{C}-O-CH_3$

(b) $CH_3-\underset{\underset{O}{\|}}{C}-O-CH_2-CH_2-CH_3$

(c) $CH_3-CH_2-\underset{\underset{CH_3}{|}}{CH}-\underset{\underset{O}{\|}}{C}-O-\underset{\underset{CH_3}{|}}{CH}-CH_3$

(d) $CH_3-CH_2-CH_2-CH_2-\underset{\underset{O}{\|}}{C}-O-\underset{\underset{CH_3}{|}}{CH}-CH_2-CH_3$

5.53

(a) $CH_3-CH_2-\underset{\underset{O}{\|}}{C}-OH;\ CH_3-CH_2-OH$

(b) $CH_3-CH_2-CH_2-\underset{\underset{O}{\|}}{C}-OH;\ CH_3-OH$

(c) $CH_3-CH_2-CH_2-\underset{\underset{O}{\|}}{C}-OH;\ CH_3-OH$

(d)
$$CH_3-\overset{\overset{\displaystyle O}{\|}}{C}-OH;$$ (phenol with OH)   (e)   (benzene ring)$-\overset{\overset{\displaystyle O}{\|}}{C}-OH;$ CH_3-OH

(f)
$$CH_3-\overset{\overset{\displaystyle Cl}{|}}{C}H-\overset{\overset{\displaystyle O}{\|}}{C}-OH;\ CH_3-CH_2-OH$$

5.55 (a) methyl propanoate (b) methyl methanoate (c) methyl ethanoate (d) propyl ethanoate (e) isopropyl propanoate (f) ethyl benzoate   **5.57** (a) methyl propionate (b) methyl formate (c) methyl acetate (d) propyl acetate (e) isopropyl propionate (f) ethyl benzoate   **5.59** (a) ethyl butanoate (b) propyl pentanoate (c) methyl 3-methylpropanoate (d) ethyl propanoate

5.61

(a)
$$H-\overset{\overset{\displaystyle O}{\|}}{C}-O-CH_3$$
(b)
$$CH_3-\overset{\overset{\displaystyle O}{\|}}{C}-O-CH_2-CH_2-CH_3$$

(c)
$$CH_3-(CH_2)_8-\overset{\overset{\displaystyle O}{\|}}{C}-O-(CH_2)_7-CH_3$$

(d)
(benzene ring)$-CH_2-\overset{\overset{\displaystyle O}{\|}}{C}-O-CH_2-CH_3$

(e)
$$CH_3-\overset{\overset{\displaystyle O}{\|}}{C}-O-\overset{\overset{\displaystyle CH_3}{|}}{C}H-CH_3$$
(f)
$$CH_3-\overset{\overset{\displaystyle O}{\|}}{C}-O-CH_2-\overset{\overset{\displaystyle Br}{|}}{C}H-CH_3$$

5.63 (a) ethyl ethanoate (b) methyl ethanoate (c) ethyl butanoate (d) propyl α-hydroxypropanoate (e) pentyl pentanoate (f) 1-methylpropyl hexanoate   **5.65** pentanoic acid, 2-methylbutanoic acid, 3-methylbutanoic acid, 2-2-dimethylpropanoic acid   **5.67** methyl pentanoate, methyl 2-methylbutanoate, methyl 3-methylbutanoate methyl 2,2-dimethylpropanoate   **5.69** nine (methyl butanoate, methyl 2-methylpropanoate, ethyl propanoate, propyl ethanoate, isopropyl ethanoate, butyl methanoate, *sec*-butyl methanoate, isobutyl methanoate, *tert*-butyl methanoate)

5.71

$$CH_3-CH_2-\overset{\overset{\displaystyle O}{\|}}{C}-OH,\ CH_3-\overset{\overset{\displaystyle O}{\|}}{C}-O-CH_3,\ H-\overset{\overset{\displaystyle O}{\|}}{C}-O-CH_2-CH_3$$

5.73 No oxygen–hydrogen bonds are present.   **5.75** There is no hydrogen bonding between ester molecules.

5.77 (a)
$$CH_3-CH_2-\overset{\overset{\displaystyle O}{\|}}{C}-OH;\ CH_3-CH_2-OH$$
(b)
$$CH_3-\overset{\overset{\displaystyle O}{\|}}{C}-OH;\ CH_3-CH_2-OH$$
(c)
$$CH_3-\overset{\overset{\displaystyle CH_3}{|}}{C}H-\overset{\overset{\displaystyle O}{\|}}{C}-OH;$$ (phenol with OH)

(d)
$$CH_3-CH_2-CH_2-\overset{\overset{\displaystyle O}{\|}}{C}-OH;\ CH_3-OH$$
(e)
$$H-\overset{\overset{\displaystyle O}{\|}}{C}-OH;\ CH_3-CH_2-OH$$
(f)
(benzene ring)$-\overset{\overset{\displaystyle O}{\|}}{C}-OH;\ CH_3-\overset{\overset{\displaystyle CH_3}{|}}{C}H-OH$

5.79 (a)
$$CH_3-CH_2-\overset{\overset{\displaystyle O}{\|}}{C}-O^-\ Na^+;\ CH_3-CH_2-OH$$
(b)
$$CH_3-\overset{\overset{\displaystyle O}{\|}}{C}-O^-\ Na^+;\ CH_3-CH_2-OH$$
(c)
$$CH_3-\overset{\overset{\displaystyle CH_3}{|}}{C}H-\overset{\overset{\displaystyle O}{\|}}{C}-O^-\ Na^+;$$ (phenol with OH)

(d)
$$CH_3-CH_2-CH_2-\overset{\overset{\displaystyle O}{\|}}{C}-O^-\ Na^+;\ CH_3-OH$$
(e)
$$H-\overset{\overset{\displaystyle O}{\|}}{C}-O^-\ Na^+;\ CH_3-CH_2-OH$$
(f)
(benzene ring)$-\overset{\overset{\displaystyle O}{\|}}{C}-O^-\ Na^+;\ CH_3-\overset{\overset{\displaystyle CH_3}{|}}{C}H-OH$

5.81 (a)
$$CH_3-\overset{\overset{\displaystyle CH_3}{|}}{C}H-\overset{\overset{\displaystyle O}{\|}}{C}-OH;\ CH_3-CH_2-OH$$
(b)
$$CH_3-\overset{\overset{\displaystyle CH_3}{|}}{C}H-\overset{\overset{\displaystyle O}{\|}}{C}-O^-\ Na^+;\ CH_3-CH_2-OH$$
(c)
$$H-\overset{\overset{\displaystyle O}{\|}}{C}-OH;\ CH_3-CH_2-CH_2-CH_2-OH$$
(d)
$$CH_3-\overset{\overset{\displaystyle O}{\|}}{C}-O^-\ Na^+;\ CH_3-\overset{\overset{\displaystyle CH_3}{|}}{C}H-\overset{\overset{\displaystyle CH_3}{|}}{C}H-CH_2-OH$$

5.83 (a)
$$CH_3-\overset{\overset{\displaystyle O}{\|}}{C}-S-CH_2-CH_3$$
(b)
$$CH_3-(CH_2)_8-\overset{\overset{\displaystyle O}{\|}}{C}-S-CH_3$$
(c)
(benzene ring)$-\overset{\overset{\displaystyle O}{\|}}{C}-S-\overset{\underset{\underset{\displaystyle CH_3}{|}}{}}{C}H-CH_3$
(d)
$$H-\overset{\overset{\displaystyle O}{\|}}{C}-S-CH_2-CH_2-CH_3$$

5.85
$$-\overset{\overset{\displaystyle O}{\|}}{C}-\overset{\overset{\displaystyle O}{\|}}{C}-O-(CH_2)_3-O-\overset{\overset{\displaystyle O}{\|}}{C}-\overset{\overset{\displaystyle O}{\|}}{C}-O-(CH_2)_3-O-$$

5.87
$$HO-\overset{\overset{\displaystyle O}{\|}}{C}-CH_2-CH_2-\overset{\overset{\displaystyle O}{\|}}{C}-OH;\ HO-CH_2-CH_2-CH_2-OH$$

5.89
(a)
$$CH_3-CH_2-\overset{\overset{\displaystyle O}{\|}}{C}-O-Cl$$
(b) $CH_3-\overset{\overset{\displaystyle }{}}{C}H-CH_2-\overset{\overset{\displaystyle O}{\|}}{C}-O-Cl$

(c)
$$CH_3-CH_2-CH_2-\overset{\overset{\displaystyle O}{\|}}{C}-O-\overset{\overset{\displaystyle O}{\|}}{C}-CH_2-CH_2-CH_3$$

(d) $CH_3-CH_2-CH_2-\overset{\overset{\displaystyle O}{\|}}{C}-O-\overset{\overset{\displaystyle O}{\|}}{C}-CH_3$

5.91 (a) ethanoic propanoic anhydride (b) pentanoyl chloride (c) 2,3-dimethylbutanoyl chloride (d) methanoic propanoic anhydride

5.93 (a)

$$CH_3—CH_2—CH_2—CH_2—\overset{\overset{O}{\|}}{C}—OH$$

(b) $CH_3—CH_2—CH_2—CH_2—\overset{\overset{O}{\|}}{C}—OH$

5.95 (a)

$$CH_3—\overset{\overset{O}{\|}}{C}—OH, CH_3—\overset{\overset{O}{\|}}{C}—O—CH_2—CH_3$$

(b)

$$CH_3—\overset{\overset{O}{\|}}{C}—OH, CH_3—\overset{\overset{O}{\|}}{C}—O—CH_2—CH_2—CH_2—CH_3$$

5.97

(a)
$$HO—\overset{\overset{O}{\|}}{\underset{\underset{OH}{|}}{P}}—O—CH_3$$

(b)
$$HO—\overset{\overset{O}{\|}}{\underset{\underset{O—CH_3}{|}}{P}}—O—CH_3$$

(c)
$$O—\overset{\overset{O}{\|}}{N}—O—CH_3$$

(d)
$$O—\overset{\overset{O}{\|}}{N}—O—CH_2—CH_2—O—\overset{\overset{O}{\|}}{N}—O$$

5.99 H_3PO_4 is a triprotic acid, and H_2SO_4 is a diprotic acid. **5.101** (a) 2,2 (b) 7,1 (c) 7,1 (d) 6,3 (e) 3,1 (f) 2,1

5.102

$$HO—\overset{\overset{O}{\|}}{C}—CH_2—\overset{\overset{O}{\|}}{C}—OH \qquad HO—\overset{\overset{O}{\|}}{C}—CH=CH—\overset{\overset{O}{\|}}{C}—OH$$
Malonic acid   Maleic acid (*cis* isomer)

$$HO—\overset{\overset{O}{\|}}{C}—\overset{\overset{OH}{|}}{CH}—CH_2—\overset{\overset{O}{\|}}{C}—OH$$
Malic acid

5.103 $C_nH_{2n-2}O_2$ **5.104** (a) ethyl 2-methylpropanoate (b) 2-methylbutanoic acid (c) ethyl thiobutanoate (d) sodium propanoate

5.105

$$CH_3—\overset{\overset{O}{\|}}{C}—O—CH_2—CH_3$$

5.106 (a)

$$CH_3—CH_2—\overset{\overset{O}{\|}}{C}—O^-\,Na^+; CH_3—OH$$

(b)
$$CH_3—CH_2—\overset{\overset{O}{\|}}{C}—S—CH_3$$

(c)
$$CH_3—\overset{\overset{O}{\|}}{C}—O^-\,Na^+$$

(d)
$$HO—\overset{\overset{O}{\|}}{C}—CH_2—CH_2—\overset{\overset{|}{CH_3}}{CH}—CH_2—\overset{\overset{OH}{|}}{CH_2}$$

5.107 b **5.108** a **5.109** b **5.110** a **5.111** b **5.112** a **5.113** d **5.114** c **5.115** d **5.116** d

Chapter 6 **6.1** (a) yes (b) yes (c) no (d) yes (e) no (f) yes **6.3** (a) 1° (b) 1° (c) 2° (d) 2° (e) 1° (f) 3° **6.5** (a) 2° (b) 3° (c) 3° (d) 1° (e) 2° (f) 2° **6.7** (a) ethylmethylamine (b) propylamine (c) diethylmethyl-amine (d) diphenylamine (e) isopropylmethylamine (f) diisopropylamine **6.9** (a) 3-pentanamine (b) 2-methyl-3-pentanamine (c) *N*-methyl-3-pentanamine (d) 1,5-pentanediamine (e) 2,3-butanediamine (f) *N,N*-dimethyl-1-butanamine **6.11** (a) 1-propanamine (b) *N*-ethyl-*N*-methylethanamine (c) *N*-methyl-1-propanamine (d) *N*-methyl-2-butanamine **6.13** (a) 2-bromoaniline (b) *N*-isopropylaniline (c) *N*-ethyl-*N*-methylaniline (d) *N*-methyl-*N*-phenylaniline (e) *N*-ethyl-*N*-methylaniline (f) *N*-(1-chloroethyl)aniline

6.15 (a) $CH_3—CH_2—NH_2$ (b)

$$CH_3—\overset{\overset{CH_3}{|}}{CH}—N—\overset{\overset{CH_3}{|}}{CH}—CH_3$$
$$\underset{CH_3—CH—CH_3}{}$$

(c) aniline with NH_2 and CH_3 substituent

(d) aniline with $NH—CH_3$ substituent

(e)
$$CH_3—\overset{\overset{NH_2}{|}}{\underset{\underset{CH_3}{|}}{C}}—CH_2—CH_3$$

(f) $H_2N—CH_2—CH_2—CH_2—CH_2—CH_2—CH_2—NH_2$

(g)
$$CH_3—\overset{\overset{NH_2}{|}}{CH}—\overset{\overset{O}{\|}}{C}—CH_2—CH_3$$

(h)
$$CH_3—\overset{\overset{|}{NH_2}}{CH}—\overset{\overset{O}{\|}}{C}—OH$$

6.17
$$\overset{}{\underset{\underset{NH_2}{|}}{CH_2}}—CH_2—CH_2—CH_2—CH_3, CH_3—\overset{\overset{|}{NH_2}}{CH}—CH_2—CH_2—CH_3,$$

$$CH_3—CH_2—\overset{\overset{|}{NH_2}}{CH}—CH_2—CH_3, \overset{}{\underset{\underset{NH_2}{|}\ \underset{CH_3}{|}}{CH_2}}—CH—CH_2—CH_3,$$

$$CH_3—\overset{\overset{CH_3}{|}}{\underset{\underset{NH_2}{|}}{C}}—CH_2—CH_3, CH_3—\overset{\overset{|}{CH_3}}{CH}—\overset{\overset{|}{NH_2}}{CH}—CH_3,$$

$$CH_3—\overset{\overset{|}{CH_3}}{CH}—CH_2—CH_2, CH_3—\overset{\overset{CH_3}{|}}{\underset{\underset{CH_3}{|}}{C}}—CH_2—NH_2$$
$$\underset{NH_2}{}$$

6.19 dimethylpropylamine, isopropyldimethylamine, diethylmethylamine **6.21** 1-propanamine, 2-propanamine, *N*-methylethanamine, *N,N*-dimethylmethanamine **6.23** (a) liquid (b) gas (c) gas (d) liquid **6.25** (a) 3 (b) 3 **6.27** Hydrogen bonding is possible for the amine. **6.29** (a) $CH_3—CH_2—NH_2$; it has fewer carbon atoms. (b) $H_2N—CH_2—CH_2—CH_2—NH_2$; it has two amino groups rather than one.

6.31 (a) $CH_3—CH_2—NH_3$ (b) OH^-
(c) $CH_3—\overset{\overset{|}{CH_3}}{CH}—NH—CH_3$
(d) $CH_3—CH_2—\overset{+}{N}H_2—CH_2—CH_3$; OH^-

6.33 (a) dimethylammonium ion (b) triethylammonium ion (c) *N,N*-diethylanilinium ion (d) dimethylpropylammonium ion (e) propylammonium ion (f) *N*-isopropylanilinium ion
6.35

(a) $CH_3—NH—CH_3$

(b) $CH_3—CH_2—\overset{\overset{|}{CH_2—CH_3}}{N}—CH_2—CH_3$

(c) $CH_3—CH_2—\overset{\overset{|}{\text{phenyl}}}{N}—CH_2—CH_3$ (d) $CH_3—CH_2—CH_2—\overset{\overset{|}{CH_3}}{N}—CH_3$

(e) $CH_3-CH_2-CH_2-NH_2$ (f)

$NH-CH-CH_3$ (attached to benzene ring, with CH_3 below)

6.37 (a) $CH_3-CH_2-\overset{+}{N}H_3\ Cl^-$

(b) (benzene ring)$-\overset{+}{N}H_3\ Br^-$

(c) $CH_3-\overset{\underset{\displaystyle CH_3}{|}}{\underset{\displaystyle CH_3}{\overset{\displaystyle CH_3}{C}}}-NH_2$ (d) HCl

6.39 (a) $CH_3-\underset{\displaystyle CH_3}{\overset{|}{CH}}-NH_2$ (b) $CH_3-\underset{\displaystyle CH_3}{\overset{|}{\overset{+}{N}H_2}}\ Cl^-$

(c) (benzene ring)$-\underset{\displaystyle CH_3}{\overset{|}{N}}-CH_3$ (d) $CH_3-NH-CH_3$

6.41 (a) propylammonium chloride (b) methylpropylammonium chloride (c) ethyldimethylammonium bromide (d) *N,N*-dimethylanilinium bromide **6.43** to increase water solubility **6.45** ethylmethylamine hydrochloride

6.47 (a) $CH_3-CH_2-CH_2-NH_2$, NaCl, H_2O

(b) $CH_3-\underset{\displaystyle CH_3}{\overset{|}{CH}}-\underset{\displaystyle CH_3}{\overset{|}{N}}-CH_3$, NaBr, H_2O

(c) $CH_3-CH_2-NH-CH_2-CH_3$, NaCl, H_2O

(d) $CH_3-\underset{\displaystyle CH_3}{\overset{\displaystyle CH_3}{\overset{|}{\underset{|}{C}}}}-NH_2$, NaBr, H_2O

6.49 ethylmethylamine and propyl chloride, ethylpropylamine and methyl chloride, methylpropylamine and ethyl chloride

6.51 (a) $CH_3-\underset{\displaystyle CH_3}{\overset{\displaystyle CH_3}{\overset{|}{\underset{|}{\overset{+}{N}}}}}-CH_2-CH_3\ Br^-$

(b) $CH_3-\underset{\displaystyle CH_3}{\overset{|}{CH}}-\underset{\displaystyle CH_3}{\overset{\displaystyle CH_3}{\overset{|}{\underset{|}{N}}}}-CH-CH_3$

(c) $CH_3-CH_2-\underset{\displaystyle CH_3}{\overset{\displaystyle CH_3}{\overset{|}{\underset{|}{\overset{+}{N}}}}}-CH_2-CH_2-CH_3\ Cl^-$

(d) $CH_3-CH_2-NH-CH_2-CH_3$

6.53 (a) amine salt (b) quaternary ammonium salt (c) amine salt (d) quaternary ammonium salt **6.55** (a) trimethylammonium bromide (b) tetramethylammonium chloride (c) ethylmethylammonium bromide (d) diethyldimethylammonium chloride **6.57** (a) purine (b) pyrrole (c) imidazole (d) indole **6.59** (a) true (b) false (c) true (d) false (e) false (f) false **6.61** (a) yes (b) yes (c) no (d) yes (e) no (f) yes **6.63** (a) monosubstituted (b) disubstituted (c) unsubstituted (d) monosubstituted **6.65** (a) secondary amide (b) tertiary amide (c) primary amide (d) secondary amide **6.67** (a) *N*-ethylethanamide (b) *N,N*-dimethylpropanamide (c) butanamide (d) *N*-methylmethanamide (e) 2-chloropropanamide (f) 2,*N*-dimethylpropanamide **6.69** (a) *N*-ethylacetamide

(b) *N,N*-dimethylpropionamide (c) butyramide (d) *N*-methylformamide (e) 2-chloropropionamide (f) 2,*N*-dimethylpropionamide **6.71** (a) propanamide (b) *N*-methylpropanamide (c) 3,5-dimethylhexanamide (d) *N,N*-dimethylbutanamide

6.73 (a) $CH_3-\overset{\displaystyle O}{\overset{||}{C}}-\underset{\displaystyle CH_3}{\overset{\displaystyle CH_3}{\overset{|}{\underset{|}{N}}}}-CH_3$

(b) $CH_3-CH_2-\underset{\displaystyle ... }{\overset{\displaystyle CH_3}{\overset{|}{CH}}}-\overset{\displaystyle O}{\overset{||}{C}}-NH_2$

(c) $CH_3-\underset{\displaystyle CH_3}{\overset{|}{CH}}-CH_2-\overset{\displaystyle O}{\overset{||}{C}}-NH-CH_3$

(d) $H-\overset{\displaystyle O}{\overset{||}{C}}-NH_2$

(e) (benzene ring)$-\overset{\displaystyle O}{\overset{||}{C}}-NH-$(benzene ring)

(f) $H-\overset{\displaystyle O}{\overset{||}{C}}-NH_2$

6.75 An electronegativity effect induced by the carbonyl oxygen atom makes the lone pair of electrons on the nitrogen atom unavailable.
6.77 (a) 5 (b) 5
6.79 (a) CH_3-NH_2 (b) (c) NH_3

(b) $CH_3-\underset{\displaystyle CH_3}{\overset{\displaystyle CH_3}{\overset{|}{\underset{|}{C}}}}-\overset{\displaystyle O}{\overset{||}{C}}-\underset{\displaystyle CH_3}{\overset{|}{N}}-CH_3$

(d) (benzene ring)$-\overset{\displaystyle O}{\overset{||}{C}}-OH$

6.81 (a) $CH_3-\overset{\displaystyle O}{\overset{||}{C}}-OH$, $CH_3-NH-\underset{\displaystyle ...}{\overset{\displaystyle CH_3}{\overset{|}{CH}}}-CH_3$

(b) $CH_3-CH_2-CH_2-CH_2-\overset{\displaystyle O}{\overset{||}{C}}-OH$, CH_3-NH_2

(c) $CH_3-\underset{\displaystyle ...}{\overset{\displaystyle CH_3}{\overset{|}{CH}}}-\overset{\displaystyle O}{\overset{||}{C}}-OH$, CH_3-NH_2

(d) $CH_3-\underset{\displaystyle CH_3}{\overset{|}{CH}}-\underset{\displaystyle CH_3}{\overset{|}{CH}}-\overset{\displaystyle O}{\overset{||}{C}}-OH$, CH_3-NH_2

6.83 (a) $CH_3-CH_2-CH_2-\overset{\displaystyle O}{\overset{||}{C}}-OH$, CH_3-NH_2

(b) $CH_3-CH_2-CH_2-\overset{\displaystyle O}{\overset{||}{C}}-OH$, $CH_3-\overset{+}{N}H_3\ Cl^-$

(c) $CH_3-CH_2-CH_2-\overset{\displaystyle O}{\overset{||}{C}}-O^-\ Na^+$, CH_3-NH_2

(d) (benzene ring)$-\overset{\displaystyle O}{\overset{||}{C}}-OH$, (benzene ring)$-NH-CH_3$

6.85 diacid and diamine
6.87

$$\left(-\overset{\displaystyle O}{\overset{||}{C}}-CH_2-CH_2-\overset{\displaystyle O}{\overset{||}{C}}-\overset{\displaystyle H}{\overset{|}{N}}-CH_2-CH_2-CH_2-CH_2-\overset{\displaystyle H}{\overset{|}{N}}-\right)_n$$

6.89 (a) $H-\overset{\displaystyle O}{\overset{||}{C}}-NH_2$

(b) $CH_3-\underset{\displaystyle NH_2}{\overset{|}{CH}}-CH_2-CH_2-CH_3$

(c) $CH_3-CH_2-CH_2-\underset{\displaystyle ...}{\overset{\displaystyle CH_3}{\overset{|}{CH}}}-\overset{\displaystyle O}{\overset{||}{C}}-NH_2$

(d) $CH_3-\overset{\displaystyle O}{\overset{||}{C}}-NH-\underset{\displaystyle CH_3}{\overset{\displaystyle CH_3}{\overset{|}{\underset{...}{CH}}}}-CH_3$

(e) $CH_3-CH_2-\overset{+}{N}H_2-CH_2-CH_3$ Cl^-

(f) $CH_3-\overset{+}{\underset{\underset{}{|}}{N}H}-CH_3$ with CH_3 and a phenyl group, Cl^-

6.90 (a) $CH_3-CH_2-NH-CH_3$

(b) $CH_3-CH_2-\underset{\underset{CH_3}{|}}{CH}-\overset{\overset{O}{||}}{C}-\underset{\underset{CH_3}{|}}{N}-CH_3$ (c) $CH_3-\overset{+}{\underset{\underset{CH_3}{|}}{\overset{\overset{CH_3}{|}}{N}}}-CH_3$ Cl^-

(d) $CH_3-CH_2-\underset{\underset{CH_3}{|}}{CH}-\overset{\overset{O}{||}}{C}-O^-$ Na^+

(e) $CH_3-NH-CH_3$ (f) $CH_3-CH_2-NH_3^+$

6.91

$CH_3-CH_2-\overset{\overset{O}{||}}{C}-NH_2$ Propanamide

$CH_3-\overset{\overset{O}{||}}{C}-NH-CH_3$ *N*-methylethanamide

$H-\overset{\overset{O}{||}}{C}-NH-CH_2-CH_3$ *N*-ethylmethanamide

$H-\overset{\overset{O}{||}}{C}-\underset{\underset{CH_3}{|}}{N}-CH_3$ (with CH₃)   *N,N*-dimethylmethanamide

6.92 $CH_3-\overset{+}{\underset{\underset{CH_3}{|}}{\overset{\overset{CH_3}{|}}{N}}}-CH_2-CH_3$ Cl^-

6.93 (a) unsubstituted (b) monosubstituted (c) monosubstituted (d) disubstituted (e) unsubstituted (f) unsubstituted **6.94** (a) amide (b) amine (c) amide (d) amine (e) amine (f) amide **6.95** (a) 1-butanamine (b) 2-methyl-1-pentanamine (c) 2,*N*-dimethylpentanamine (d) 3-methylpentanamide (e) 1,4-pentanediamine (f) 4-bromo-*N*-ethyl-*N*-methylpentanamide **6.96** c **6.97** c **6.98** c **6.99** b **6.100** c **6.101** b **6.102** c **6.103** a **6.104** b **6.105** b

Chapter 7 **7.1** Biochemistry is the study of the chemical substances found in living systems and the chemical interactions of these substances with each other. **7.3** proteins, lipids, carbohydrates, and nucleic acids

7.5 $CO_2 + H_2O + \text{solar energy} \xrightarrow[\text{Plant enzymes}]{\text{Chlorophyll}} \text{carbohydrates} + O_2$

7.7 serve as structural elements, provide energy reserves **7.9** Carbohydrates are polyhydroxy aldehydes, polyhydroxy ketones, or compounds that yield such substances upon hydrolysis. **7.11** (a) one unit versus a few units (b) two units versus four units **7.13** Superimposable objects have parts that coincide exactly at all points when the objects are laid upon each other. **7.15** (a) drill bit (b) hand, foot, ear (c) PEEP, POP **7.17** (a) no (b) no (c) yes (d) yes

7.19 (a) $CH_2\overset{*}{-}CH-Br$ with Cl, Cl   (b) $CH_2-\overset{*}{C}-\overset{*}{CH}$ with Cl, Cl, Br, Br, Br

(c) $CH_2-\overset{*}{CH}-\overset{*}{CH}-\overset{*}{CH}-\overset{\overset{O}{||}}{C}-H$ with OH, OH, OH, OH

(d) $CH_2-\overset{*}{CH}-\overset{*}{CH}-\overset{*}{CH}-\overset{*}{CH}-CH_2$ with OH, OH, OH, OH, OH, OH

7.21 (a) zero (b) two (c) zero (d) zero **7.23** Constitutional isomers have a different connectivity of atoms. Stereoisomers have the same connectivity of atoms with different arrangements of the atoms in space.

7.25 (a), (b), (c), (d) Fischer projection structures

7.27 (a), (b), (c), (d) Fischer projection structures

7.29 (a) D enantiomer (b) D enantiomer (c) L enantiomer (d) L enantiomer **7.31** (a) diastereomers (b) neither enantiomers nor diastereomers (c) enantiomers (d) diastereomers **7.33** (d) effect on plane-polarized light **7.35** (a) same (b) different (c) same (d) different **7.37** (a) aldose (b) ketose (c) ketose (d) ketose **7.39** (a) aldohexose (b) ketohexose (c) ketotriose (d) ketotetrose **7.41** (a) D-galactose (b) D-psicose (c) dihydroxyacetone (d) L-erythrulose **7.43** (a) carbon 4 (b) carbons 1 and 2 (c) carbons 1 and 2 (d) carbon 2 **7.45** (a) aldoses, hexoses, aldohexoses (b) hexoses (c) hexoses (d) aldoses

7.47 (a), (b), (c), (d) Fischer projection structures

7.49 (a) D-fructose (b) D-glucose (c) D-galactose **7.51** (a) carbons 1 and 5 (b) carbons 1 and 5 (c) carbons 2 and 5 (d) carbons 1 and 4 **7.53** the hydroxyl group orientation on carbon 1 **7.55** In fructose the cyclization involves carbons 2 and 5, and in ribose the cyclization involves carbons 1 and 4; both processes give five-membered rings. **7.57** The cyclic and noncyclic forms interconvert; an equilibrium exists between the forms. **7.59** (a) α-D-monosaccharide (b) α-D-monosaccharide (c) β-D-monosaccharide (d) α-D-monosaccharide **7.61** All four structures are hemiacetals.

7.63 (a), (b) Fischer projection structures

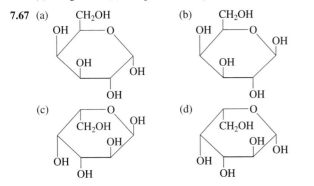

(c)

```
        CHO
   HO ─┼─ H
   HO ─┼─ H
    H ─┼─ OH
    H ─┼─ OH
       CH₂OH
```

(d)

```
       CH₂OH
       C═O
    H ─┼─ OH
   HO ─┼─ H
    H ─┼─ OH
       CH₂OH
```

7.65 (a) α-D-glucose (b) α-D-galactose (c) β-D-mannose (d) α-D-sorbose

7.67 (a) [structure] (b) [structure] (c) [structure] (d) [structure]

7.69 (a) reducing sugar (b) reducing sugar (c) reducing sugar (d) reducing sugar **7.71** The aldehyde group in glucose is oxidized to an acid group. The Ag⁺ in Tollens solution is reduced to Ag.

7.73

(a)
```
       COOH
    H ─┼─ OH
   HO ─┼─ H
   HO ─┼─ H
    H ─┼─ OH
       CH₂OH
```
(b)
```
       COOH
    H ─┼─ OH
   HO ─┼─ H
   HO ─┼─ H
    H ─┼─ OH
       COOH
```
(c)
```
       CHO
    H ─┼─ OH
   HO ─┼─ H
   HO ─┼─ H
    H ─┼─ OH
       COOH
```
(d)
```
       CH₂OH
    H ─┼─ OH
   HO ─┼─ H
   HO ─┼─ H
    H ─┼─ OH
       CH₂OH
```

7.75 (a) yes (b) yes (c) yes (d) yes **7.77** (a) alpha (b) beta (c) alpha (d) beta **7.79** (a) methyl alcohol (b) ethyl alcohol (c) ethyl alcohol (d) methyl alcohol **7.81** A glycoside is an acetal formed from a cyclic monosaccharide. A glucoside is a glycoside in which the monosaccharide is glucose.

7.83 (a) [structure with O—CH₂—CH₃] (b) [structure with O—CH₃]

7.85 (a) [phosphate structure] (b) [structure with NH, C═O, CH₃]

7.87 (a) glucose and fructose (b) glucose (c) glucose and galactose (d) glucose **7.89** The glucose part of the lactose structure has a hemiacetal carbon atom. **7.91** (a) negative (b) positive (c) positive (d) positive **7.93** (a) α(1 → 6) (b) β(1 → 4) (c) α(1 → 4) (d) α(1 → 4) **7.95** (a) alpha (b) beta (c) alpha (d) beta **7.97** (a) reducing sugar (b) reducing sugar (c) reducing sugar (d) reducing sugar **7.99** (a) glucose (b) galactose and glucose (c) glucose and altrose (d) glucose **7.101** (a) yes (b) yes (c) yes (d) no **7.103** (a) yes (b) yes (c) no (d) no **7.105** (a) yes (b) no (c) no (d) no **7.107** (a) Both are glucose polymers with α(1 → 4) and α(1 → 6) linkages. Glycogen is more highly branched than amylopectin. (b) Both are unbranched glucose polymers. Amylose has α(1 → 4) linkages, and cellulose has β(1 → 4) linkages. **7.109** (a) glycogen and amylopectin (b) amylopectin, amylose, glycogen, and cellulose (c) amylose, cellulose, and chitin (d) cellulose and chitin **7.111** The human body possesses enzymes for α(1 → 4) linkages (starch) but not for β(1 → 4) linkages (cellulose). **7.113** Simple carbohydrates are the mono- and disaccharides, and complex carbohydrates are the polysaccharides. **7.115** carbohydrates that provide energy but few other nutrients **7.117** (a) chiral (b) achiral (c) chiral (d) chiral **7.118** (a) no (b) no (c) no (d) yes **7.119** (a) no (b) no (c) yes (d) yes

7.120

```
   CH₂OH      CH₂OH      CH₂OH      CH₂OH
   C═O        C═O        C═O        C═O
 H ─┼─ OH   HO ─┼─ H    H ─┼─ OH   HO ─┼─ H
 H ─┼─ OH   HO ─┼─ H   HO ─┼─ H    H ─┼─ OH
   CH₂OH      CH₂OH      CH₂OH      CH₂OH
```

7.121 3-methylhexane (hydrogen, methyl, ethyl, and propyl groups attached to a carbon atom) **7.122** (a) glucose, fructose (b) glucose (c) glucose (d) glucose **7.123** (a) glucose-derivative (b) glucose (c) glucose-derivative (d) glucose **7.124** (a) strong oxidizing agent (b) ethyl alcohol, H⁺ ion (c) water (H⁺ ion or enzymes) (d) enzymes **7.125** b **7.126** d **7.127** b **7.128** c **7.129** c **7.130** a **7.131** c **7.132** c **7.133** d **7.134** c

Chapter 8 **8.1** All lipids are insoluble or only sparingly soluble in water. **8.3** (a) insoluble (b) soluble (c) insoluble (d) soluble **8.5** energy-storage lipids, membrane lipids, emulsification lipids, messenger lipids, and protective-coating lipids **8.7** (a) long-chain (b) short-chain (c) long-chain (d) medium-chain **8.9** (a) saturated (b) polyunsaturated (c) polyunsaturated (d) monounsaturated **8.11** In a SFA there are no double bonds in the carbon chain; in a MUFA there is one carbon–carbon double bond in the carbon chain. **8.13** (a) neither (b) omega-3 (c) omega-3 (d) neither

8.15 CH₃—(CH₂)₄—CH═CH—CH₂—CH═CH—(CH₂)₇—COOH

8.17 There are fewer attractions between fatty acid carbon chains because of bends in the chains caused by the presence of the double bonds. **8.19** (a) 18:1 acid (b) 18:3 acid (c) 14:0 acid (d) 18:1 acid **8.21** (a) tetradecanoic acid (b) cis-9-hexadecenoic acid **8.23** a glycerol molecule and three fatty acid molecules **8.25** one, ester

8.27
```
           O
           ‖
 H₂C─O─C─(CH₂)₁₄─CH₃
           O
           ‖
  HC─O─C─(CH₂)₁₄─CH₃
           O
           ‖
 H₂C─O─C─(CH₂)₁₄─CH₃
```

8.29
```
  ┌─S    ┌─L
  ├─L    ├─S
  └─L    └─L

  ┌─S    ┌─S
  ├─S    ├─L
  └─L    └─S
```

8.31 (a) palmitic, myristic, oleic (b) oleic, palmitic, palmitoleic **8.33** (top) 16 carbon atoms and 1 oxygen atom; (middle) 14 carbon atoms and 1 oxygen atom; (bottom) 18 carbon atoms and 1 oxygen atom **8.35** (a) no difference (b) A triacylglycerol may be a solid or a liquid; a fat is a triacylglycerol that is a solid. (c) A triacylglycerol can have fatty acid residues that are all the same, or two or more different kinds may be present. In a mixed triacylglycerol, two or more different fatty acid residues must be present. (d) A fat is a triacylglycerol that is a solid; an oil is a triacylglycerol that is a liquid. **8.37** (a) not correct (b) not correct **8.39** (a) correct (b) not correct **8.41** (a) nonessential fatty acid (b) essential fatty acid (c) nonessential fatty acid (d) nonessential fatty acid **8.43** (a) glycerol and three fatty acids (b) glycerol and three fatty acid salts

8.45 CH_2—CH—CH_2
　　　|　　|　　|
　　　OH　OH　OH

CH_3—$(CH_2)_{12}$—$COOH$

CH_3—$(CH_2)_{14}$—$COOH$

CH_3—$(CH_2)_7$—CH=CH—$(CH_2)_7$—$COOH$

8.47 glycerol, palmitic acid, myristic acid, oleic acid

8.49 CH_2—CH—CH_2
　　　|　　|　　|
　　　OH　OH　OH

CH_3—$(CH_2)_{12}$—COO^- Na^+

CH_3—$(CH_2)_{14}$—COO^- Na^+

CH_3—$(CH_2)_7$—CH=CH—$(CH_2)_7$—COO^- Na^+

8.51 glycerol, sodium palmitate, sodium myristate, sodium oleate **8.53** Carbon–carbon double bond(s) must be present. **8.55** six

8.57
(a)
18:0	18:0	18:1
18:0	18:1	18:0
16:1	16:0	16:0

(b)
18:0	18:0
18:1A	18:1B
16:0	16:0

| 18:0 |
| 18:0 |
| 16:1 |
There are two possibilities for converting the 18:2 acid to 18:1 acid depending on which double bond is hydrogenated (denoted as 18:1A and 18:1B).

8.59 Rancidity results from hydrolysis of ester linkages and oxidation of carbon–carbon double bonds. **8.61** glycerol and sphingosine

8.63
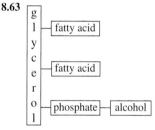

8.65 HO—CH_2—CH_2—$\overset{+}{N}(CH_3)_3$,　HO—CH_2—CH_2—$\overset{+}{N}H_3$,

　　and　HO—CH_2—CH—$\overset{+}{N}H_3$
　　　　　　　　　　|
　　　　　　　　　COO^-

8.67 The two tails are the carbon chain of sphingosine and the fatty acid carbon chain; the head is the phosphate–alcohol portion of the molecule. **8.69** the two tails **8.71** (a) four (b) two **8.73** They differ in the identity of the amino alcohol group; it is choline in a lecithin and serine in a phosphatidylserine.

8.75
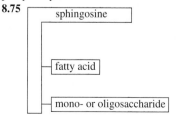

8.77 carbohydrate group versus a phosphate–alcohol group
8.79

$$\begin{array}{cccc} & 11 & \underset{13}{12} & 17 \\ 1 & 9 & 14 & 16 \\ 2 & 10 & 8 & 15 \\ 3 & 5 & 7 & \\ 4 & 6 & & \end{array}$$

8.81 —OH on carbon 3, —CH_3 on carbons 10 and 13, and a hydrocarbon chain on carbon 17 **8.83** "Good cholesterol" is that present in HDLs, and "bad cholesterol" is that present in LDLs. **8.85** phospholipids, sphingoglycolipids, and cholesterol **8.87** a two-layer structure of lipid molecules with nonpolar "tails" in the interior and polar "heads" on the exterior **8.89** creates "open" areas in the bilayer **8.91** Protein help is required in facilitated transport but not in passive transport. **8.93** (a) active transport (b) facilitated transport (c) active transport (d) passive transport and facilitated transport **8.95** tri- or dihydroxy versus monohydroxy; oxidized side chain amidified to an amino acid versus nonoxidized side chain **8.97** amino acid glycine versus amino acid taurine **8.99** bile fluid **8.101** gall bladder **8.103** sex hormones, adrenocortical hormones **8.105** estradiol has an —OH on carbon 3, while testosterone has a ketone group at this location; testosterone has an extra —CH_3 group at carbon 10 **8.107** Prostaglandins have a bond between carbons 8 and 12 that creates a cyclopentane ring structural feature. **8.109** inflammatory response, production of pain and fever, blood pressure regulation, induction of blood clotting, control of some reproductive functions, regulation of sleep/wake cycle

8.111
| long-chain alcohol | long-chain fatty acid |

8.113 mixture of esters involving a long-chain fatty acid and a long-chain alcohol versus a long-chain alkane mixture **8.115** (a) neither (b) glycerol-based (c) neither (d) sphingosine-based (e) neither (f) neither **8.116** (a) no (b) no (c) no (d) yes (e) yes (f) no **8.117** (a) sphingomyelins (b) triacylglycerols (c) steroids (d) leukotrienes (e) prostaglandins (f) cerebrosides **8.118** (a) energy-storage lipids (b) emulsification lipid (c) membrane lipid (d) messenger lipids (e) membrane lipids (f) messenger lipids **8.119** (a) glycerolipid (b) sphingolipid (c) glycerolipid and phospholipid (d) sphingolipid and phospholipid **8.120** (a) 3, 0, 0 (b) 4, 0, 0 (c) 2, 1, 0 (d) 1, 0, 0 (e) 0, 1, 1 (f) 4, 0, 0 **8.121** (a) no (b) yes (c) yes (d) yes (e) yes (f) no **8.122** b **8.123** a **8.124** c **8.125** a **8.126** c **8.127** d **8.128** c **8.129** a **8.130** c **8.131** a

Chapter 9 **9.1** (a) yes (b) no (c) no (d) yes **9.3** the identity of the R group (side chain) **9.5** (a) phenylalanine, tyrosine, tryptophan (b) methionine, cysteine (c) aspartic acid, glutamic acid (d) serine, threonine, tyrosine **9.7** An amino group is part of the side chain. **9.9** The side chain is part of a cyclic structure. **9.11** (a) alanine (b) leucine (c) methionine (d) tryptophan **9.13** asparagine, glutamine, isoleucine, tryptophan **9.15** (a) polar neutral (b) polar acidic (c) nonpolar (d) polar neutral **9.17** L family

9.19 (a)
$$\begin{array}{c} COOH \\ | \\ H_2N—\!\!\!\!\!\!-\!\!\!\!\!\!-H \\ | \\ CH_2 \\ | \\ OH \end{array}$$
(b)
$$\begin{array}{c} COOH \\ | \\ H—\!\!\!\!\!\!-\!\!\!\!\!\!-NH_2 \\ | \\ CH_2 \\ | \\ OH \end{array}$$

(c)
$$\begin{array}{c} COOH \\ | \\ H—\!\!\!\!\!\!-\!\!\!\!\!\!-NH_2 \\ | \\ CH_3 \end{array}$$
(d)
$$\begin{array}{c} COOH \\ | \\ H_2N—\!\!\!\!\!\!-\!\!\!\!\!\!-H \\ | \\ CH_2 \\ | \\ CH—CH_3 \\ | \\ CH_3 \end{array}$$

9.21 They exist as zwitterions.

9.23 (a)

$$H_3\overset{+}{N}-\overset{\overset{\displaystyle H}{|}}{\underset{\underset{\displaystyle CH-CH_3}{|}}{\underset{\underset{\displaystyle CH_3}{|}}{\underset{\displaystyle CH_2}{|}}}}-COO^-$$

(b)

$$H_3\overset{+}{N}-\overset{\overset{\displaystyle H}{|}}{\underset{\underset{\displaystyle CH_2}{|}}{\underset{\underset{\displaystyle CH_3}{|}}{\underset{\displaystyle CH-CH_3}{|}}}}-COO^-$$

(c)

$$H_3\overset{+}{N}-\overset{\overset{\displaystyle H}{|}}{\underset{\underset{\displaystyle CH_2}{|}}{\underset{\displaystyle SH}{|}}}-COO^-$$

(d)

$$H_3\overset{+}{N}-\overset{\overset{\displaystyle H}{|}}{\underset{\displaystyle H}{|}}-COO^-$$

9.25 (a)

$$H_3\overset{+}{N}-\overset{\overset{\displaystyle H}{|}}{\underset{\underset{\displaystyle CH_2}{|}}{\underset{\displaystyle OH}{|}}}-COO^-$$

(b)

$$H_3\overset{+}{N}-\overset{\overset{\displaystyle H}{|}}{\underset{\underset{\displaystyle CH_2}{|}}{\underset{\displaystyle OH}{|}}}-COOH$$

(c)

$$H_2N-\overset{\overset{\displaystyle H}{|}}{\underset{\underset{\displaystyle CH_2}{|}}{\underset{\displaystyle OH}{|}}}-COO^-$$

(d)

$$H_3\overset{+}{N}-\overset{\overset{\displaystyle H}{|}}{\underset{\underset{\displaystyle CH_2}{|}}{\underset{\displaystyle OH}{|}}}-COOH$$

9.27 the pH at which zwitterion concentration in a solution is maximized   **9.29** Two —COOH groups are present, which deprotonate at different times.   **9.31** (a) toward positive electrode (b) toward positive electrode (c) toward negative electrode (d) toward positive electrode   **9.33** Aspartic acid migrates toward the positive electrode, histidine migrates toward the negative electrode, and valine does not migrate.   **9.35** They react with each other to produce a covalent disulfide bond.   **9.37** —COOH and —NH₂

9.39

$$H_3\overset{+}{N}-\underset{\underset{\underset{\displaystyle CH_3}{|}}{\displaystyle CH-CH_3}}{CH}-\overset{\displaystyle O}{\overset{||}{C}}-\underset{\displaystyle H}{N}-\underset{\underset{\underset{\displaystyle \bigcirc}{|}}{\displaystyle CH_2}}{CH}-\overset{\displaystyle O}{\overset{||}{C}}-\underset{\displaystyle H}{N}-\underset{\underset{\underset{\displaystyle SH}{|}}{\displaystyle CH_2}}{CH}-COO^-$$

9.41 Ser is the N-terminal end of Ser–Cys and Cys is the N-terminal end of Cys–Ser.   **9.43** Ser–Val–Gly, Val–Ser–Gly, Gly–Ser–Val, Ser–Gly–Val, Val–Gly–Ser, Gly–Val–Ser   **9.45** (a) Ser–Ala–Cys (b) Asp–Thr–Asn   **9.47** two in each   **9.49** (a) serylcysteine (b) glycylalanylvaline (c) tyrosylaspartylglutamine (d) leucyllysyltryptophanylmethionine   **9.51** peptide bonds and α-carbon —CH groups   **9.53** (a) Both are nonapeptides with six of the residues held in the form of a loop by a disulfide bond. (b) They differ in the identity of the amino acid present at two positions in the nonapeptide.   **9.55** They bind at the same sites.   **9.57** Glu is bonded to Cys through the side-chain carboxyl group rather than through the α-carbon carboxyl group.   **9.59** Monomeric proteins contain a single peptide chain and multimeric proteins have two or more peptide chains.   **9.61** (a) true (b) false (c) true (d) true   **9.63** the sequence of amino acids in the protein chain   **9.65** α helix, β pleated sheet   **9.67** Intermolecular involves two separate chains and intramolecular involves a single chain bending back on itself.   **9.69** Yes, both α helix and β pleated sheet can occur at different regions in the same chain.   **9.71** Secondary-structure hydrogen bonding involves C=O···H—N interactions; tertiary-structure hydrogen bonding involves R group interactions.   **9.73** (a) hydrophobic (b) electrostatic (c) disulfide bond (d) hydrogen bonding   **9.75** (a) fibrous: generally water-insoluble; globular: generally water-soluble (b) fibrous: support and external protection; globular: involvement in metabolic reactions   **9.77** (a) fibrous (b) fibrous (c) globular (d) globular   **9.79** α-Keratin has a double-helix structure and collagen a triple-helix structure.   **9.81** Yes, both Ala and Val are products in each case.   **9.83** Drug hydrolysis would occur in the stomach.   **9.85** Ala–Gly–Met–His–Val–Arg   **9.87** five: Ala–Gly–Ser, Gly–Ser–Tyr, Ala–Gly, Gly–Ser, Ser–Tyr   **9.89** secondary, tertiary, and quaternary   **9.91** same primary structure   **9.93** 4-hydroxyproline and 5-hydroxylysine   **9.95** They are involved with cross-linking.   **9.97** An antigen is a substance foreign to the human body, and an antibody is a substance that defends against an invading antigen.   **9.99** four polypeptide chains that have constant and variable amino acid regions; two chains are longer than the other two; 1%–12% carbohydrates present; long and short chains are connected through disulfide linkages   **9.101** suspend and transport lipids in the bloodstream   **9.103** (a) tertiary (b) tertiary (c) secondary (d) primary   **9.104** (a) alanine (b) leucine (c) threonine (d) aspartic acid   **9.105** (a) +1 (b) +1 (c) +1 (d) +3   **9.106** (a) −1 (b) −1 (c) −4 (d) −1

9.107

$$\begin{array}{c|c|c|c}
COOH & COOH & COOH & COOH \\
H_2N-H & H-NH_2 & H_2N-H & H-NH_2 \\
H_3C-H & H-CH_3 & H-CH_3 & H_3C-H \\
CH-CH_3 & CH-CH_3 & CH-CH_3 & CH-CH_3 \\
CH_3 & CH_3 & CH_3 & CH_3
\end{array}$$

9.108 (a) 24 (b) 36 (c) 20

9.109

(a)

$$H_3\overset{+}{N}-\underset{\underset{\underset{\displaystyle OH}{|}}{\displaystyle CH_2}}{CH}-COOH; \quad H_3\overset{+}{N}-\underset{\displaystyle CH_3}{CH}-COOH; \quad H_3\overset{+}{N}-\underset{\underset{\underset{\displaystyle SH}{|}}{\displaystyle CH_2}}{CH}-COOH$$

(b)

$$H_2N-\underset{\underset{\underset{\displaystyle OH}{|}}{\displaystyle CH_2}}{CH}-COO^-; \quad H_2N-\underset{\displaystyle CH_3}{CH}-COO^-; \quad H_2N-\underset{\underset{\underset{\displaystyle SH}{|}}{\displaystyle CH_2}}{CH}-COO^-$$

9.110 (a) simple protein, fibrous protein (b) conjugated protein, globular protein (c) conjugated protein, globular protein (d) conjugated protein, fibrous protein, glycoprotein   **9.111** c   **9.112** b   **9.113** d   **9.114** b   **9.115** a   **9.116** d   **9.117** c   **9.118** a   **9.119** a   **9.120** a

Chapter 10   **10.1** catalyst   **10.3** more efficient, more specific   **10.5** (a) yes (b) no (c) yes (d) yes   **10.7** (a) add a carboxylate group to pyruvate (b) remove H₂ from an alcohol (c) reduce an L-amino acid (d) hydrolyze maltose   **10.9** (a) sucrase (or sucrose hydrolase) (b) pyruvate decarboxylase (c) glucose isomerase (d) lactate dehydrogenase   **10.11** (a) pyruvate (b) galactose (c) an alcohol (d) an L-amino acid   **10.13** (a) isomerase (b) lyase (c) ligase (d) transferase   **10.15** (a) isomerase (b) lyase (c) transferase (d) hydrolase   **10.17** (a) decarboxylase (b) lipase (c) phosphatase (d) dehydrogenase   **10.19** (a) conjugated (b) conjugated (c) simple (d) conjugated   **10.21** A coenzyme is a cofactor that is an organic substance. A cofactor can be an inorganic or an organic substance.   **10.23** to provide additional functional groups   **10.25** the portion of an enzyme actually involved in the catalysis process   **10.27** The substrate must have the same shape as the active site.   **10.29** interactions with amino acid R groups   **10.31** (a) accepts only one substrate (b) accepts substrate with a particular type of bond   **10.33** absolute specificity and stereochemical specificity   **10.35** (a) absolute (b) stereochemical   **10.37** Rate increases until enzyme denaturation occurs.   **10.39** Enzymes vary in the number of acidic and base amino acids present.

10.41

10.43 nothing; the rate remains constant   **10.45** no; only one molecule may occupy the active site at a given time   **10.47** (a) reversible competitive (b) reversible noncompetitive, irreversible (c) reversible noncompetitive, irreversible (d) reversible noncompetitive   **10.49** enzyme that has quaternary structure and more than one binding site   **10.51** The product of a subsequent reaction in a series of reactions inhibits a prior reaction. **10.53** A zymogen is an inactive precursor for a proteolytic enzyme. **10.55** so that they will not destroy the tissues that produce them   **10.57** competitive inhibition of the conversion of PABA to folic acid   **10.59** has absolute specificity for bacterial transpeptidase   **10.61** the threat of biological weapon use by terrorists   **10.63** dietary organic compound needed by the body in trace amounts   **10.65** (a) fat-soluble (b) water-soluble (c) water-soluble (d) water-soluble   **10.67** (a) likely (b) unlikely (c) unlikely (d) unlikely   **10.69** serves as a cosubstrate in the formation of collagen   **10.71** coenzymes   **10.73** (a) no (b) yes (c) yes (d) no   **10.75** alcohol, aldehyde, acid   **10.77** Cell differentiation is the process whereby immature cells change in structure and function to become specialized cells. Vitamin A binds to protein receptors in the process.   **10.79** They differ only in the identity of the side chain present. **10.81** to maintain normal blood levels of calcium and phosphorus ion so that bones can absorb these minerals   **10.83** α-tocopherol   **10.85** antioxidant effect   **10.87** in the length and degree of unsaturation of the side chain present   **10.89** Menaquinones are forms of vitamin K_1, and phylloquinones are forms of vitamin K_2.   **10.91** (a) An apoenzyme is the protein portion of a conjugated enzyme; a proenzyme is an inactive precursor of an enzyme. (b) A simple enzyme is pure protein; an allosteric enzyme has two or more protein chains and two binding sites. (c) A coenzyme is an organic cofactor, and an isoenzyme is one of several similar forms of an enzyme. (d) A conjugated enzyme has both a protein and a nonprotein portion; holoenzyme is just another name for a conjugated enzyme.   **10.92** (a) alcohol, ketone (b) double bond, alcohol (c) double bond, alcohol (d) double bond, ketone   **10.93** (a) no (b) no (c) no (d) yes (e) yes (f) yes   **10.94** (a) vitamin C (b) vitamin E (c) vitamin D (d) vitamin K (e) vitamin A (f) vitamin A (g) vitamin C (h) vitamin E **10.95** (a) oxidation–reduction reactions (b) addition of a group to, or removal of a group from, a double bond in a manner that does not involve hydrolysis or oxidation–reduction (c) conversion of a compound into another isomeric with it (d) bonding together of two molecules with the involvement of ATP (e) hydrolysis reactions (f) transfer of functional groups between two molecules   **10.96** (a) enzyme plus substrate produces an enzyme–substrate complex that breaks apart to regenerate the enzyme and a product molecule   **10.97** (a) ethanol (b) zinc ion (c) protein molecule (d) alcohol dehydrogenase   **10.98** (a) tissue plasminogen activator (b) lactate dehydrogenase (c) creatine phosphokinase (d) aspartate transaminase   **10.99** b   **10.100** b   **10.101** c   **10.102** d **10.103** d   **10.104** d   **10.105** c   **10.106** d   **10.107** c   **10.108** a

Chapter 11   **11.1** Ribose has both an —H group and an —OH group on carbon 2; deoxyribose has 2 —H atoms on carbon 2.   **11.3** (a) pyrimidine (b) pyrimidine (c) purine (d) purine   **11.5** (a) one (b) four (c) one   **11.7** (a) adenine (b) guanine (c) thymine (d) uracil   **11.9** (a) ribose (b) deoxyribose (c) deoxyribose (d) ribose   **11.11** (a) false (b) false (c) false (d) false

11.13

11.15 a pentose sugar and a phosphate   **11.17** base sequence   **11.19** 5′ end has a phosphate group attached to the 5′ carbon; 3′ end has a hydroxyl group attached to the 3′ carbon   **11.21** a phosphate group and two pentose sugars

11.23

11.25 (a) two polynucleotide chains coiled around each other in a helical fashion (b) The nucleic acid backbones are the outside, and the nitrogen-containing bases are on the inside.   **11.27** (a) 36% (b) 14% (c) 14%   **11.29** A G–C pairing involves 3 hydrogen bonds, and an A–T pairing involves 2 hydrogen bonds.   **11.31** They are the same. **11.33** 5′ TAGCC 3′   **11.35** (a) 3′ TGCATA 5′ (b) 3′ AATGGC 5′ (c) 5′ CGTATT 3′ (d) 3′ TTGACC 5′   **11.37** 20 hydrogen bonds **11.39** catalyzes the unwinding of the double helix structure   **11.41** (a) 3′ TGAATC 5′ (b) 3′ TGAATC 5′ (c) 5′ ACTTAG 3′   **11.43** The unwound strands are antiparallel (5′ → 3′ and 3′ → 5′). Only the 5′ → 3′ strand can grow continuously.   **11.45** a DNA molecule bound to a group of small proteins   **11.47** (1) RNA contains ribose instead of deoxyribose, (2) RNA contains the base U instead of T, (3) RNA is single-stranded rather than double-stranded, and (4) RNA has a lower molecular mass.   **11.49** (a) hnRNA (b) tRNA (c) tRNA (d) hnRNA **11.51** (a) nuclear region (b) extranuclear region (c) extranuclear region (d) both nuclear and extranuclear regions   **11.53** a strand of DNA   **11.55** causes a DNA helix to unwind; links aligned ribonucleotides together   **11.57** T–A, A–U, G–C, C–G   **11.59** 3′ UACGAAU 5′   **11.61** 3′ AAGCGTC 5′   **11.63** Exons convey genetic information whereas introns do not.   **11.65** 3′ AGUCAAGU 5′ **11.67** (a) hnRNA (b) snRNA   **11.69** a mechanism by which a number of proteins that are variations of a basic structural motif can be produced from a single gene   **11.71** A three-nucleotide sequence in mRNA that codes for a specific amino acid   **11.73** (a) Leu (b) Asn (c) Ser (d) Gly   **11.75** (a) CUC, CUA, CUG, UUA, UUG (b) AAC (c) AGC, UCU, UCC, UCA, UCG (d) GGU, GGC, GGA   **11.77** The base T cannot be present in a codon.   **11.79** Met–Lys– Glu–Asp–Leu   **11.81** A cloverleaf shape with three hairpin loops and one open side.   **11.83** covalent bond   **11.85** (a) UCU (b) GCA (c) AAA (d) GUU   **11.87** (a) Thr (b) Leu (c) Pro (d) Ser   **11.89** (1) activation of tRNA, (2) initiation, (3) elongation, (4) termination, and (5) post-translational processing   **11.91** A site   **11.93** Gly: GGU, GGC, GGA or GGG; Ala: GCU, GCC, GCA or GCG; Cys: UGU or UGC; Val: GUU, GUC, GUA or GUG; Tyr: UAU or UAC **11.95** (a) Gly–Tyr–Ser–Ser–Pro (b) Gly–Tyr–Ser–Ser–Pro (c) Gly–Tyr–Ser–Ser–Thr   **11.97** a DNA or a RNA molecule with a protein coating   **11.99** (1) attaches itself to cell membrane, (2) opens a hole in the membrane, and (3) injects itself into the cell   **11.101** contains a "foreign" gene   **11.103** host for a "foreign" gene   **11.105** Recombinant DNA is incorporated into a host cell.

11.107

11.109 to produce many copies of a specific DNA sequence in a relatively short time   **11.111** a short nucleotide chain bound to the template DNA strand to which new nucleotides can be attached   **11.113** dATP stands for an ATP in which deoxyribose is present; ddATP stands for an ATP in which dideoxyribose is present.   **11.115** (a) 5′ GCCGACTACT 3′ (b) 5′ AGTAGTCGGC 3′   **11.117** (a) Thymine has a methyl group on carbon-5 that uracil lacks. (b) Adenine is 6-aminopurine, and guanine is 2-amino-6-oxopurine.   **11.118** (a) 3′ GTATGTCGGACCTTCGAT 5′ (b) 3′ GUAUGUCGGACCUUCGAU 5′ (c) 3′ GUA–UGU–CGG–ACC–UUC–GAU 5′ (d) 5′ CAU–ACA–GCC–UGG–AAG–CUA 3′ (e) Val–Cys–Arg–Thr–Phe–Asp   **11.119** (a) 1 (b) 4 (c) 2 (d) 4   **11.120** (a) 3 (b) 2 (c) 1 (d) 3   **11.121** (a) 3 (b) 1 (c) 4 (d) 1, 2, 3, and 4   **11.122** (a) 1 (b) 2 (c) 4 (d) 3   **11.123** A   **11.124** (a) 2.9 billion base pairs (b) 20,000–25,000 genes (c) 2% of base pairs   **11.125** c   **11.126** a   **11.127** c   **11.128** a   **11.129** b   **11.130** d   **11.131** a   **11.132** c   **11.133** b   **11.134** b

Chapter 12   **12.1** anabolism—synthetic; catabolism—degradative   **12.3** a series of consecutive biochemical reactions   **12.5** Large molecules are broken down to smaller ones; energy is released.   **12.7** Prokaryotic cells have no nucleus, and the DNA is usually a single circular molecule. Eukaryotic cells have their DNA in a membrane-enclosed nucleus.   **12.9** An organelle is a small structure within the cell cytosol that carries out a specific cellular function.   **12.11** inner membrane   **12.13** region between inner and outer membranes   **12.15** adenosine triphosphate

12.17 phosphate–phosphate–phosphate–ribose–adenine

12.19 three phosphates versus one phosphate   **12.21** adenine versus guanine   **12.23** ADP is produced from the hydrolysis of ATP.   **12.25** flavin adenine dinucleotide   **12.27** flavin–ribitol–ADP

12.29 nicotinamide–ribose–phosphate
adenine–ribose–phosphate

12.31 nicotinamide subunit   **12.33** (a) $FADH_2$ (b) NAD^+ **12.35** (a) nicotinamide (b) riboflavin

12.37 2-aminoethanethiol–pantothenic acid–phosphorylated ADP

12.39 a compound with a greater free energy of hydrolysis than is typical for a compound   **12.41** free monophosphate species   **12.43** (a) phosphoenolpyruvate (b) creatine phosphate (c) 1,3-diphosphoglycerate (d) AMP   **12.45** (1) digestion, (2) acetyl group formation, (3) citric acid cycle, (4) electron transport chain and oxidative phosphorylation   **12.47** tricarboxylic acid cycle, Krebs cycle   **12.49** acetyl CoA   **12.51** (a) 2 (b) 1 (c) 2 (Steps 3, 8) (d) 2 (Steps 2, 7)   **12.53** (a) Steps 3, 4, 6, 8 (b) Step 2 (c) Step 7

12.55 succinate ($^-OOC–CH_2–CH_2–COO^-$);
fumarate ($^-OOC–CH=CH–COO^-$);
malate $^-OOC–CH_2–CH–COO^-$;
OH

oxaloacetate $^-OOC–CH_2–\overset{O}{\overset{\|}{C}}–COO^-$

12.57 oxidation and decarboxylation   **12.59** (a) NAD^+ (b) FAD   **12.61** (a) isocitrate, α-ketoglutarate (b) fumarate, malate (c) malate, oxaloacetate (d) citrate, isocitrate   **12.63** respiratory chain   **12.65** O_2 **12.67** (a) FMN (b) $CoQH_2$ **12.69** (a) oxidized (b) oxidized (c) reduced (d) oxidized   **12.71** (a) oxidation (b) reduction (c) oxidation (d) reduction   **12.73** (a) I (b) I, II, III (c) III (d) IV   **12.75** $CoQH_2$

12.77 two   **12.79** (a) $FADH_2$, 2 Fe(II)SP, $CoQH_2$ (b) $FMNH_2$, 2 Fe^{2+}, $CoQH_2$ **12.81** ATP synthesis from ADP using energy from the electron transport chain   **12.83** protons (H^+ ions)   **12.85** intermembrane space   **12.87** ATP synthase   **12.89** Protons flow through ATP synthase complex.   **12.91** 2.5 ATP molecules   **12.93** They enter the ETC at different stages.   **12.95** reactive oxygen species

12.97 (a) $2O_2 + NADPH \longrightarrow 2O_2^- + NADP^+ + H^+$
(b) $2O_2^- + 2H^+ \longrightarrow H_2O_2 + O_2$

12.99 (a) 1 (b) 2 (c) 2 (d) 1 (e) 1 (f) 2   **12.100** (a) 1 (b) 2 (c) 1 (d) 3 (e) 1 (f) 2   **12.101** (a) 4 (b) 2, 3, and 4 (c) 1, 2, 3, and 4 (d) 3   **12.102** (a) 1 (b) 4 (c) 4 (d) 2   **12.103** The products from the CAC, which are FAD and NAD^+, are the starting reactants for the ETC.   **12.104** (a) inside mitochondria (b) inside mitochondria   **12.105** FAD is the oxidizing agent for carbon–carbon double bond formation.   **12.106** Reduced coenzymes are oxidized, and ADP is phosphorylated.   **12.107** a   **12.108** b   **12.109** d   **12.110** b   **12.111** c   **12.112** c   **12.113** a   **12.114** d   **12.115** c   **12.116** d

Chapter 13   **13.1** mouth, salivary α-amylase   **13.3** small intestine, pancreas   **13.5** outer membranes of intestinal mucosal cells, sucrose hydrolysis   **13.7** glucose, galactose, fructose   **13.9** glucose   **13.11** NAD^+ **13.13** formation of glucose 6-phosphate, a species that cannot cross cell membranes   **13.15** dihydroxyacetone phosphate, glyceraldehyde 3-phosphate   **13.17** two   **13.19** two   **13.21** Steps 1, 3, and 6   **13.23** cytosol   **13.25** (a) glucose 6-phosphate (b) 2-phosphoglycerate (c) phosphoglyceromutase (d) ADP   **13.27** (a) Step 10 (b) Step 1 (c) Step 8 (d) Step 6   **13.29** (a) +2 (b) +4

13.31
(a) $CH_3–\overset{O}{\overset{\|}{C}}–COOH$ and $CH_3–\overset{O}{\overset{\|}{C}}–COO^-$

(b) $CH_2–OH$ $CH_2–O–\textcircled{P}$
     $C=O$ and $C=O$
     $CH_2–OH$ $CH_2–OH$

(c) $\textcircled{P}–O–CH_2$, O, CH_2OH, OH, OH, OH
and
$\textcircled{P}–O–CH_2$, O, $CH_2–O–\textcircled{P}$, OH, OH, OH

(d) COOH
    CH–OH   and
    $CH_2–OH$

CHO
CH–OH
$CH_2–OH$

13.33
$^1CH_2–O–\textcircled{P}$ ^{4}CHO
$^2C=O$ and ^{5}CH–OH
$^3CH_2–OH$ $^6CH_2–O–\textcircled{P}$

13.35 acetyl CoA, lactate, ethanol   **13.37** pyruvate + CoA + $NAD^+ \rightarrow$ acetyl CoA + NADH + CO_2 **13.39** NADH is oxidized to NAD^+, a substance needed for glycolysis.   **13.41** CO_2 **13.43** glucose + 2ADP + $2P_i \rightarrow$ 2lactate + 2ATP   **13.45** decreases ATP production by 2   **13.47** 30 ATP versus 2 ATP   **13.49** two   **13.51** Glycogenesis converts glucose to glycogen and glycogenolysis is the reverse process.   **13.53** glucose 6-phosphate   **13.55** UTP   **13.57** UDP + ATP → UTP + ADP   **13.59** Step 2   **13.61** In liver cells the product is glucose, and in muscle cells it is glucose 6-phosphate.   **13.63** as glucose 6-phosphate   **13.65** the liver   **13.67** two-step pathway for Step 10; different enzymes for Steps 1 and 3   **13.69** oxaloacetate   **13.71** goes to the liver, where it is converted to glucose   **13.73** glucose 6-phosphate   **13.75** NADPH is consumed in its reduced form; NADH is consumed in its oxidized form (NAD^+).   **13.77** Glucose 6-phosphate + $2NADP^+$ + H_2O → ribulose 5-phosphate +

CO_2 + 2NADPH + 2H$^+$ **13.79** CO_2 **13.81** increases rate of glycogen synthesis   **13.83** increases blood glucose levels   **13.85** pancreas   **13.87** Epinephrine attaches to cell membrane and stimulates the production of cAMP, which activates glycogen phosphorylase.   **13.89** glucagon (liver cells) and epinephrine (muscle cells)   **13.91** (a) all four (b) glycogenesis and glycogenolysis (c) glycolysis and gluconeogenesis (d) gluconeogenesis   **13.92** (a) glycolysis (b) glycolysis (c) glycolysis, gluconeogenesis, and glycogenesis (d) glycogenesis   **13.93** (a) glycolysis (b) glycogenesis (c) glycogenolysis (d) glyconeogenesis   **13.94** (a) 2 ATP (b) 30 ATP (c) 2 ATP (d) 2 ATP   **13.95** (a) 2 (b) 2 (c) 3 (d) 4   **13.96** (a) when the body requires free glucose (b) anaerobic conditions in muscle; red blood cells (c) when the body requires energy (d) anaerobic conditions in yeast   **13.97** (a) 12 moles (b) 4 moles (c) 4 moles (d) 60 moles   **13.98** (a) The glucose supply is adequate, and the body does not need energy. (b) The glucose supply is adequate, and the body needs energy. (c) Ribose 5-phosphate or NADPH is needed. (d) The free glucose supply is not adequate.   **13.99** b   **13.100** b   **13.101** b   **13.102** d   **13.103** c   **13.104** a   **13.105** b   **13.106** c   **13.107** d   **13.108** c

Chapter 14   **14.1** 98%   **14.3** no effect   **14.5** because lipids have a long residence time in the stomach   **14.7** acts as an emulsifier   **14.9** monoacylglycerols are the major product   **14.11** reassembled into triacylglycerols; converted to chylomicrons   **14.13** They have a large storage capacity for triacylglycerols.   **14.15** hydrolysis of triacylglycerols in adipose tissue; entry of hydrolysis products into bloodstream   **14.17** activates hormone-sensitive lipase   **14.19** glycerol 3-phosphate, dihydroxyacetone phosphate   **14.21** one   **14.23** outer mitochondrial membrane   **14.25** ATP is converted to AMP and 2P$_i$ **14.27** shuttles acyl groups across the inner mitochondrial membrane   **14.29** (a) alkane to alkene (b) alkene to 2° alcohol (c) 2° alcohol to ketone   **14.31** *trans* isomer   **14.33** (a) Step 3, turn 1 (b) Step 2, turn 2 (c) Step 4, turn 2 (d) Step 1, turn 1   **14.35** compounds a and d   **14.37** (a) 7 turns (b) 5 turns   **14.39** *Cis–trans* isomerase converts a *cis*-(3,4) double bond to a *trans*-(2,3) double bond.   **14.41** (a) glucose (b) fatty acids   **14.43** (a) 4 turns (b) 5 acetyl CoA (c) 4 NADH (d) 4 FADH$_2$ (e) 2 high-energy bonds   **14.45** 64 ATP   **14.47** yield the same amount   **14.49** 4 kcal versus 9 kcal   **14.51** (1) dietary intakes high in fat and low in carbohydrates, (2) inadequate processing of glucose present, and (3) prolonged fasting   **14.53** Ketone body formation occurs when oxaloacetate concentrations are low.

14.55

14.57 liver mitochondria   **14.59** formation of acetoacetyl CoA   **14.61** accumulation of ketone bodies in blood and urine   **14.63** cytosol versus mitochondrial matrix   **14.65** acyl carrier protein; polypeptide chain replaces phosphorylated ADP   **14.67** liver, adipose tissue, mammary glands   **14.69** (a) Oxaloacetate reacts with acetyl CoA to produce citrate and CoA. (b) Citrate crosses the mitrochondrial membranes, functioning as an acetyl group carrier.   **14.71** source of C$_2$ units for the growing fatty acid chain   **14.73** (1) condensation, (2) hydrogenation, (3) dehydration, and (4) hydrogenation   **14.75** (a) Step 1, cycle 1 (b) Step 2, cycle 2 (c) Step 3, cycle 2 (d) Step 4, cycle 1   **14.77** compounds b and d   **14.79** C$_{16}$ fatty acid   **14.81** needed to convert saturated fatty acids to unsaturated fatty acids   **14.83** (a) 6 rounds (b) 6 malonyl ACP (c) 6 ATP bonds (d) 12 NADPH   **14.85** (a) 13–17% (b) 83–87%   **14.87** (a) mevalonate (b) isopentenyl pyrophosphate (c) squalene   **14.89** (a) fewer than (b) fewer than (c) same as   **14.91** (a) fatty acid spiral (b) fatty acid spiral (c) lipogenesis (d) glycerol catabolism (e) ketogenesis (f) ketogenesis   **14.92** (a) fatty acid catabolism (b) lipogenesis (c) lipogenesis (d) ketogenesis (e) consumption of molecular O$_2$ (f) consumption of molecular O$_2$ **14.93** (a) Step 1 (b) Step 1 (c) Step 3 (d) Step 4   **14.94** (1),

(3), (4), and (2)   **14.95** (a) true (b) false (c) false (d) false   **14.96** (a) incorrect (b) correct (c) incorrect (d) correct   **14.97** (a) incorrect (b) correct (c) correct (d) correct   **14.98** glucose, C$_8$ fatty acid, sucrose, C$_{14}$ fatty acid   **14.99** b   **14.100** a   **14.101** c   **14.102** c   **14.103** a   **14.104** b   **14.105** a   **14.106** c   **14.107** c   **14.108** b

Chapter 15   **15.1** Denaturation occurs in the stomach with gastric juice as the denaturant.   **15.3** Pepsinogen is the inactive precursor of pepsin.   **15.5** Gastric juice is acidic (1.5–2.0 pH) and pancreatic juice is basic (7.0–8.0 pH).   **15.7** Membrane protein molecules facilitate the passage of amino acids through the intestinal wall.   **15.9** total supply of free amino acids available for use   **15.11** cyclic process of protein degradation and resynthesis   **15.13** A positive nitrogen balance has nitrogen intake exceeding nitrogen output; a negative nitrogen balance has nitrogen output exceeding nitrogen intake.   **15.15** negative balance; proteins are degraded to get the needed amino acid   **15.17** protein synthesis; synthesis of nonprotein nitrogen-containing compounds; nonessential amino acid synthesis; energy production   **15.19** (a) essential (b) nonessential (c) nonessential (d) essential   **15.21** b and c   **15.23** an amino acid and an α-keto acid

15.25
(a)

(b)

(c)

(d)

15.27 pyruvate, α-ketoglutarate, oxaloacetate   **15.29** coenzyme that participates in the amino group transfer   **15.31** conversion of an amino acid into a keto acid with the release of ammonium ion   **15.33** Oxidative deamination produces ammonium ion, and transamination produces an amino acid.

15.35 (a)

(b)

(c)

(d)

15.37 Transamination of the α-keto acid produces the amino acid.

15.39 (a) aspartate (b) glutamate (c) pyruvate (d) α-ketoglutarate

15.41

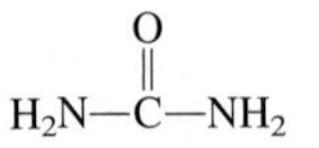

$$\underset{H_2N-C-NH_2}{\overset{O}{\|}}$$

15.43 carbamoyl phosphate   **15.45** an amide group,

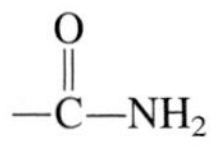

$$\overset{O}{\underset{-C-NH_2}{\|}}$$

15.47

$$\overset{+}{H_3N}-, \quad \underset{H_2N-C-NH-}{\overset{O}{\|}}, \quad \underset{H_2N-C-NH-}{\overset{\overset{+}{NH_2}}{\|}}$$

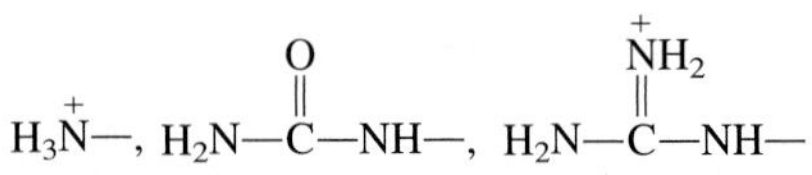

15.49 (a) carbamoyl phosphate (b) aspartate   **15.51** carbamoyl phosphate   **15.53** (a) N_2 (b) N_3 (c) N_1 (d) N_4 **15.55** (a) citrulline (b) ornithine (c) argininosuccinate (d) carbamoyl phosphate   **15.57** equivalent of four ATP molecules   **15.59** goes to the citric acid cycle where it is converted to oxaloacetate, which is then converted to aspartate   **15.61** α-ketoglutarate, succinyl CoA, fumarate, oxaloacetate   **15.63** (a) acetoacetyl CoA and acetyl CoA (b) succinyl CoA and acetyl CoA (c)

fumarate and oxaloacetate (d) α-ketoglutarate   **15.65** Degradation products can be used to make glucose.   **15.67** glutamate   **15.69** pyruvate, α-ketoglutarate, 3-phosphoglycerate, oxaloacetate, and phenylalanine   **15.71** hydrolyzed to amino acids   **15.73** In biliverdin the heme ring has been opened and one carbon atom has been lost (as CO).   **15.75** biliverdin, bilirubin, bilirubin diglucuronide, urobilin   **15.77** urobilin   **15.79** excess bilirubin   **15.81** Amino acid carbon skeletons are degraded to acetyl CoA or acetoacetyl CoA; ketogenesis converts these degradation products to ketone bodies.   **15.83** converted to body fat stores   **15.85** (a) 1 (b) 2 (c) 3 (d) 2 (e) 1 and 3 (f) 3   **15.86** (a) 1 (b) 3 (c) 3 (d) 2 (e) 1 (f) 2   **15.87** (1), (3), (2), and (4)   **15.88** (a) 3 (b) 2 (c) 1 (d) 3   **15.89** (a) transamination (b) deamination (c) deamination (d) transamination   **15.90** (a) 1 (b) 4 (c) 2 (d) 3   **15.91** (a) true (b) false (c) true (d) true   **15.92** (a) 2 (b) 1 (c) 2 (d) 3   **15.93** a   **15.94** d   **15.95** a   **15.96** b   **15.97** c   **15.98** b   **15.99** c   **15.100** c   **15.101** b   **15.102** a

Photo Credits

1: © Bill Ross/CORBIS. **5:** Doug Martin/Photo Researchers. **20:** © Richard Megna/Fundamental Photographs, NYC. **21:** © Daryl Solomon/Envision. **23:** Michael Newman/PhotoEdit. **24:** Phil Degginger/Color-Pic. **36:** Junebug Clark/Photo Researchers. **42:** Alan Detrick/Photo Researchers. **45:** Donald C. Booth/Color-Pic. **46:** © Martha Cooper/Peter Arnold, Inc. **48:** PhotoDisc. **50:** Edgar Fahs Smith Collection, University of Pennsylvania Library. **55:** © Bill Stanton/Rainbow. **72:** © Ed Reschke/Peter Arnold, Inc. **76:** Jonathan Ferrey/Getty Images. **79: (top)** © Hank Morgan/Rainbow. **79: (bottom)** James Cotier/Getty Images. **88:** Bob Daemmrich/Stock Boston/PictureQuest. **94: (top)** © Henryk T. Kaiser/Envision. **94: (bottom)** © Michael Viard/Peter Arnold Inc. **97:** © SIU/Peter Arnold Inc. **102:** Jeff Lepore/Photo Researchers. **104:** Getty Images. **113:** PhotoDisc. **119:** © Harvey Lloyd/Peter Arnold Inc. **120: (top)** © 2005 Norbert Wu/www.norbertwu.com. **120: (bottom)** © Steven Needham/Envision. **122:** Copyright S. McCutcheon/Visuals Unlimited. **132: (bottom)** © Coco McCoy/Rainbow. **142:** © Bilderberg/Peter Arnold Inc. **146:** AP/Wide World Photos. **147:** © George Mattel/Envision. **149:** F. Stewart Westmoreland/Photo Researchers. **161:** Fred Whitehead/Animals Animals. **171:** Custom Medical Stock Photo. **181:** © Royalty-Free/CORBIS. **193:** Copyright Inge Spence/Visuals Unlimited. **195: (top)** Nardin/Jacana/Photo Researchers. **195: (bottom)** Scott Camazine/Photo Researchers. **209: (top)** Peter Skinner/Photo Researchers. **209: (bottom)** © Dan McCoy/Rainbow. **220:** Copyright Norris Blake/Visuals Unlimited. **221:** © PictureNet/CORBIS. **227:** Edgar Fahs Smith Collection, University of Pennsylvania Library. **237:** Tom Raymond/Medichrome. **241:** © Hulton-Deutsch Collection/CORBIS. **244:** © Larry Mulvehill/Rainbow. **249:** © Erica Stone/Peter Arnold Inc. **251:** © Paul Skelcher/Rainbow. **255:** Dr. Dennis Kunkel/Phototake. **256: (top)** © Steven Needham/Envision. **256: (bottom)** © Don Kreuter/Rainbow. **257:** J. M. Barey/Vandystadt/Photo Researchers. **269:** Dan Guravich/Photo Researchers. **275:** © Manfred Kage/Peter Arnold Inc. **279:** © IFA/Peter Arnold Inc. **287:** BRI/Vision/Photo Researchers. **294:** Howard Socurek/Medichrome. **299:** C. James Webb/Phototake. **304:** © Kevin Schaefer/Peter Arnold Inc. **312:** Copyright F. H. Kolwicz/Visuals Unlimited. **325:** Hulton Archive/Getty Images. **334:** PhotoDisc. **336:** Prof. P. M. Motta & E. Vizza/Photo Researchers. **338: (top)** E. R. Degginger. **338: (bottom)** Rosenfield Images, Ltd/SPL/Photo Researchers. **349:** © Mark E. Gibson/CORBIS. **350:** © Steven Needham/Envision. **356:** Meckes/Ottawa/Photo Researchers. **357:** © Leonard Lessin/Peter Arnold Inc. **365:** Saturn Stills/SPL/Photo Researchers. **368:** © Jeff Greenberg/Peter Arnold Inc. **373:** Melissa Grimes-Guy/Photo Researchers. **381:** Dr. Nikos/James Burns/Phototake. **392:** © Erica Stone/Peter Arnold Inc. **407:** NIBSC/SPL/Photo Researchers. **419:** © Luiz Marigo/Peter Arnold Inc. **421:** © R. Bhatnagar/Visuals Unlimited. **429:** Hulton Archive/Getty Images. **444:** © Lynn Rogers/Peter Arnold Inc. **450:** © Jason Reed/Reuters/CORBIS. **451:** © Ed Reschke/Peter Arnold Inc. **461:** © Dan McCoy/Rainbow. **471:** Saturn Stills/SPL/Photo Researchers. **477:** © Ron Sanford/CORBIS. **486:** Adam G. Sylvester/Photo Researchers. **502:** Paul Nicklen/Getty Images. **516:** © Leonard Lessin/Peter Arnold Inc.

Index/Glossary